WILDLIFE MANAGEMENT AND CONSERVATION

WILDLIFE MANAGEMENT and CONSERVATION

Contemporary Principles and Practices

EDITED BY Paul R. Krausman and James W. Cain III

Published in Affiliation with The Wildlife Society

THE JOHNS HOPKINS UNIVERSITY PRESS | BALTIMORE

© 2013 The Johns Hopkins University Press
All rights reserved. Published 2013
Printed in the United States of America on acid-free paper
9 8 7 6 5 4 3 2 1

The Johns Hopkins University Press
2715 North Charles Street
Baltimore, Maryland 21218-4363
www.press.jhu.edu

Library of Congress Cataloging-in-Publication Data

Wildlife management and conservation : contemporary principles and
practices / edited by Paul R. Krausman and James W. Cain III.
       pages cm
    Includes bibliographical references and index.
    ISBN-13: 978-1-4214-0986-3 (hardcover : acid-free paper)
    ISBN-10: 1-4214-0986-0 (hardcover : acid-free paper)
    ISBN-13: 978-1-4214-0987-0 (electronic)
    ISBN-10: 1-4214-0987-9 (electronic)
    1. Wildlife management—North America. 2. Wildlife conservation—
North America. I. Krausman, Paul R., 1946– II. Cain, James W.
    SK361.W49 2013
    639.9097—dc23   2012045474

A catalog record for this book is available from the British Library.

The findings and conclusions in Chapter 18 are those of the authors and do
not necessarily represent the views of the U.S. Fish and Wildlife Service.

*Special discounts are available for bulk purchases of this book. For more informa-
tion, please contact Special Sales at 410-516-6936 or specialsales@press.jhu.edu.*

The Johns Hopkins University Press uses environmentally friendly book
materials, including recycled text paper that is composed of at least 30
percent post-consumer waste, whenever possible.

CONTENTS

CONTRIBUTORS

Bart M. Ballard
Caesar Kleberg Wildlife
Research Institute
Texas A&M University–Kingsville
Kingsville, Texas, USA

Warren B. Ballard (deceased)
Department of Natural
Resources Management
Texas Tech University
Lubbock, Texas, USA

John A. Bissonette
U.S. Geological Survey,
Utah Cooperative Fish and
Wildlife Research Unit
Utah State University
Logan, Utah, USA

Clint Boal
U.S. Geological Survey
Texas Cooperative Fish and
Wildlife Research Unit
Texas Tech University
Lubbock, Texas, USA

Leonard A. Brennan
Caesar Kleberg Wildlife
Research Institute
Texas A&M University–Kingsville
Kingsville, Texas, USA

Robert D. Brown
College of Natural Resources
North Carolina State University
Raleigh, North Carolina, USA

Fred C. Bryant,
Caesar Kleberg Wildlife
Research Institute,
Texas A&M University–Kingsville,
Kingsville, Texas, USA

James W. Cain III
U.S. Geological Survey
New Mexico Cooperative Fish and
Wildlife Research Unit
New Mexico State University
Las Cruces, New Mexico, USA

Tyler A. Campbell
U.S. Department of Agriculture
Wildlife Services
Gainesville, Florida, USA

Michael R. Conover
Department of Wildland Resources
Utah State University
Logan, Utah, USA

Daniel J. Decker
Department of Natural Resources
Cornell University
Ithaca, New York, USA

Charles A. DeYoung
Caesar Kleberg Wildlife
Research Institute
Texas A&M University–Kingsville
Kingsville, Texas, USA

Jonathan B. Dinkins
Department of Wildland Resources
Utah State University
Logan, Utah, USA

W. Sue Fairbanks
Department of Natural Resource
Ecology and Management
Iowa State University
Ames, Iowa, USA

James B. Grand
U.S. Geological Survey
Alabama Cooperative Fish and
Wildlife Research Unit
Auburn University
Auburn, Alabama, USA

Michael J. Haney
Department of Wildland Resources
Utah State University
Logan, Utah, USA

James R. Heffelfinger
Arizona Game and Fish Department
Tucson, Arizona, USA

Scott E. Henke
Caesar Kleberg Wildlife
Research Institute
Texas A&M University–Kingsville
Kingsville, Texas, USA

Fidel Hernandez
Caesar Kleberg Wildlife
Research Institute
Texas A&M University–Kingsville
Kingsville, Texas, USA

David G. Hewitt
Caesar Kleberg Wildlife
Research Institute
Texas A&M University–Kingsville
Kingsville, Texas, USA

Marta A. Jarzyna
Department of Fisheries and Wildlife
Michigan State University
East Lansing, Michigan, USA

David A. Jessup
Wildlife Disease Association
Royal Oaks, California, USA

Heather E. Johnson
Colorado Division of Parks
and Wildlife
Durango, Colorado, USA

John L. Koprowski
School of Natural Resources
and the Environment
University of Arizona
Tucson, Arizona, USA

Paul R. Krausman
Wildlife Biology Program
University of Montana
Missoula, Montana, USA

William P. Kuvlesky Jr.
Caesar Kleberg Wildlife
Research Institute
Texas A&M University–Kingsville
Kingsville, Texas, USA

Roel R. Lopez
Department of Wildlife and
Fisheries Sciences
Texas A&M University
College Station, Texas, USA

R. William Mannan
School of Natural Resources
and the Environment
University of Arizona
Tucson, Arizona, USA

L. Scott Mills
Fisheries, Wildlife, and Conservation
Biology Program
North Carolina State University
Raleigh, North Carolina, USA

Michael S. Mitchell
U.S. Geological Survey
Montana Cooperative Wildlife
Research Unit
University of Montana
Missoula, Montana, USA

Michael L. Morrison
Department of Wildlife and
Fisheries Sciences
Texas A&M University
College Station, Texas, USA

Anna M. Muñoz
U.S. Fish and Wildlife Service
Arlington, Virginia, USA

John F. Organ,
U.S. Fish and Wildlife Service
Hadley, Massachusetts, USA

Katherine L. Parker
Natural Resources and
Environmental Studies
University of Northern
British Columbia
Prince George, British Columbia,
Canada

William F. Porter
Department of Fisheries and Wildlife
Michigan State University,
East Lansing, Michigan, USA

Shawn J. Riley
Department of Fisheries and Wildlife,
Michigan State University
East Lansing, Michigan, USA

Steven S. Rosenstock
Arizona Game and Fish Department
Flagstaff, Arizona, USA

Michael C. Runge
U.S. Geological Survey
Patuxent Wildlife Research Center
Laurel, Maryland, USA

Susan P. Rupp
Enviroscapes Ecological Consulting
Brookings, South Dakota, USA

William F. Siemer
Department of Natural Resources
Cornell University
Ithaca, New York, USA

Robert J. Steidl
School of Natural Resources
and the Environment
University of Arizona
Tucson, Arizona, USA

Benjamin Zuckerberg
Department of Forest and
Wildlife Ecology
University of Wisconsin–Madison
Madison, Wisconsin, USA

PREFACE

Wildlife management and conservation are at a crossroads. Wildlife habitat is being altered at unprecedented rates because of human influences. Efficient and effective management is critical for the future of wildlife habitats, but it can only be accomplished by incorporating all stakeholders into the process. Our textbook incorporates the animal, its habitat, and how human management influences them both—the wildlife management triad. These three components are central to wildlife management and form the core of the profession. Unfortunately, the human side of wildlife management has not received as much attention as wildlife species or their habitats, but it must be recognized for effective management. Wildlife in North America belongs to the public, which must be considered in management decisions. Simply discussing animals and their habitats will not address the complexity of wildlife management in the 21st century, especially with increasing human demands for landscape use. Wildlife habitat is decreasing, and human use of those lands is facing serious challenges. We need leaders in wildlife management and conservation who are able to understand the biology of wildlife species and how landscapes can be managed to ensure their survival and long-term viability. This book considers the ways wildlife is managed and explores how management is only successful when the animal, habitat, and people are equally considered. It emphasizes the importance of structured decision making and planning in the process of wildlife management.

Wildlife Management and Conservation is for future wildlife leaders, written by current leaders in the field who are members of The Wildlife Society (TWS), the professional society for wildlife biologists. No other course is as important for undergraduate wildlife biologists as the basics of management and conservation. As such, this text is endorsed by TWS and written by TWS professionals across the nation.

Designed for use in various wildlife curricula, this book can serve as a stand-alone text for programs that do not have classes in all of the topics covered, or it can be used as an introductory text for students in programs that have a more complete suite of wildlife classes. It also contains a solid review of the profession for graduate students and practicing professionals.

This volume consists of 19 chapters, beginning with definitions of the wildlife profession and history of wildlife conservation in North America (Chapters 1–3). We then discuss the human dimensions of wildlife management (Chapter 4) and include that important dynamic throughout the text. Chapter 5 addresses the importance of structured decision making, a critical component in management. Following those chapters is an exploration of the biological component of management (Chapters 6–14), habitat and wildlife restoration (Chapters 15 and 16), and how climate change influences wildlife populations (Chapter 17). The last two chapters address conservation planning and include case studies of wildlife population management. Each chapter introduces readers to key personalities in the profession through short biographies.

Future leaders in the wildlife profession must use numerous tools to be effective, relevant, and current. One of those tools is a solid education related to wildlife biology, habitat, and the human dimensions that dictate what we do. *Wildlife Management and Conservation* has been written to assist in that effort.

ACKNOWLEDGMENTS

A work of this scope is possible only because of the combined efforts of numerous researchers and managers who have dedicated their professional lives to the management of wildlife. Our work environments at the University of Montana and the U.S. Geological Survey, New Mexico Cooperative Fish and Wildlife Research Unit, were also instrumental toward the completion of this book. We appreciate the support of the Boone and Crockett Program in Wildlife Conservation, the Wildlife Biology Program at the University of Montana, the U.S. Geological Survey, New Mexico Cooperative Fish and Wildlife Research Unit, and the Department of Fish, Wildlife and Conservation Ecology at New Mexico State University.

We also thank numerous colleagues who willingly agreed to review book chapters, including Bill Bartush, John A. Bissonette, Vernon C. Bleich, Bill Block, R. Terry Bowyer, Melanie Bucci, Jamal N. Butler, Colin M. Callahan, Casey J. Cardinal, Kevin Crooks, Ashley D'Antonio, Steven De Stefano, T. Donovan, Colin Gillin, Kevin Gutzwiler, Matthew Kauffman, Bryan M. Kluever, Amy J. Kuenzi, Cole J. Lamoreaux, Ryan L. Lokteff, Mark Madison, Monika E. Maier, Jason P. Marshal, James E. Miller, Michael L. Morrison, James D. Nichols, Ron Regan, Gary W. Roemer, Steve Running, Dana Sanchez, M. Schrage, James H. Shaw, William W. Shaw, Lisa A. Shipley, Tom Smith, Donald E. Spalinger, Jennifer Szymanski, Terry Walshe, Gary C. White, Tamara L. Wright, Kyle W. Young, numerous students in wildlife classes, and several anonymous referees. Jeanne Franz and Anna Derey-Wilson assisted with verifying sources, and Melanie Bucci prepared the index. Daniel Edge, Terry Johnson, Scott Bonar, Ray Lee, Kevin Hurley, and James Earl Kennamer provided information for various chapters and biographies.

We appreciate the cooperation of all organizations that supported the production of this work, including the Caesar Kleberg Wildlife Research Institute, the Boone and Crockett Program in Wildlife Conservation, and the Michigan Department of Natural Resources through the Partnership for Ecosystem Research and Management. Mary Bissonette is gratefully thanked for putting up with John A. Bissonette, the household curmudgeon, while he prepared his chapter.

We are also thankful for the excellent assistance we received from the Johns Hopkins University Press, especially from Vincent Burke, Jennifer Malat, Ashleigh McKown, and Courtney Bond. We also thank Michael Hutchins, former executive director of The Wildlife Society.

Anyone attempting a work of this scope knows that long hours are required outside the office, and we are grateful to our spouses and families—Carol Lee Krausman, Ellie Cain, Muriel Cain, and Logan Cain—for tolerating the evenings and weekends we spent working on this project. Their support was instrumental in seeing it to completion. We wish to thank all of the budding professionals who will use the text to advance to leadership in the profession. After all, we wrote this book for you.

Finally, we express our personal appreciation to Jim Cain, Jane Cain, Karina Haaseth, and all those who share our love of the wild, as well as those who inspire the same in others.

WILDLIFE MANAGEMENT AND CONSERVATION

1

DEFINING WILDLIFE AND WILDLIFE MANAGEMENT

PAUL R. KRAUSMAN

INTRODUCTION

Wildlife is such a part of society that most people have an idea of what wildlife is, but the term has many definitions. To be able to discuss wildlife management throughout this text in a consistent manner, it is important to have a common definition. To that end, this chapter defines wildlife, makes distinctions between active and inactive wildlife management, introduces the goals of management, and concludes with classifications of wildlife that were initiated by Leopold (1933) and added to by others.

WHAT IS WILDLIFE?

Initially, wildlife was considered game—animals that were hunted. Aldo Leopold defined game management as "the art of making land produce sustained annual crops of wild game for recreational use" (Leopold 1933:3). Contemporary use of the term "game" is declining, but state fish and game agencies commonly designate game based on the legislature of the state in which the animal resides. The word can have a narrow, precise, legislated meaning like the one that first appeared in British law in the Qualification Act of 1389 (McKelvie 1985). The Qualification Act stated that one had to be a landowner worth at least 40 shillings or a clergyman earning at least ten pounds per year to legally take game (Lueck 1989). In England, two classes of animals were designated based on common law of property: *domitae naturae* (domestic animals) and *ferae naturae* (wild animals). For *domitae naturae*, absolute property rights were in place (i.e., they belonged to the owners even when they strayed from their property), but *ferae naturae* did not have owners until they were captured (Lueck 1989). The use of "game" found its way to North America, and numerous anomalies have arisen. For example, mourning doves (*Zenaida macroura*) are classified as "songbirds" (i.e., nongame) in parts of the eastern United States but are hunted as game in many western states. In addition, many species are not classified as game, but some states require a game license to collect them (e.g., lizards, snakes). The law in the United States followed a different path than that in England. The fundamental legal doctrine regarding wildlife in the United States implies that states—and in some cases the federal government—have authority over wildlife, not landowners. "Wildlife" is not universally defined; it changes with the viewpoint of the user (Caughley and Sinclair 1994).

In general, the wildlife profession considers wildlife to be free-living, wild animals (excluding feral or exotic species) of major significance to humans. This definition includes the associated plants and lower animals (e.g., micro-organisms) because wildlife habitats that support wildlife have to be considered. Species and their habitats are interlocked and cannot be considered separately.

For several decades, the profession of wildlife management concentrated on game species and their habitats. However, by the 1970s and 80s, there was increased emphasis on considering the interests of all citizens when management decisions were made. Because wildlife is considered as belonging to the public in the United States and Canada, it makes sense that human opinions be considered. A holistic view considers wildlife a triad of the animal, its habitat, and people, and the interactions between them (Giles 1978; Fig. 1.1). The animal component considers all aspects of biology, ecology, behavior, genetics, physiology, and life history characteristics and other important factors related to the species. The habitat component considers vegetation, soils, weather, topography, and other relationships within the community. Human dimensions (i.e., "how people value wildlife, how they want wildlife to be managed, and how they affect or are affected by wildlife and wildlife management decisions"; Decker et al. 2001:3) considers all anthropogenic influences on populations and is often the most important aspect of the triad, because humans dictate how species will be managed, what intrusions into their habitat are acceptable, and how management will be funded. Each part of the triad involves different education and expertise, but all three parts are necessary to be efficient.

One definition of wildlife is given by the U.S. Congress: "The term *fish* or *wildlife* means any member of the animal kingdom, including without limitation any mammal, fish,

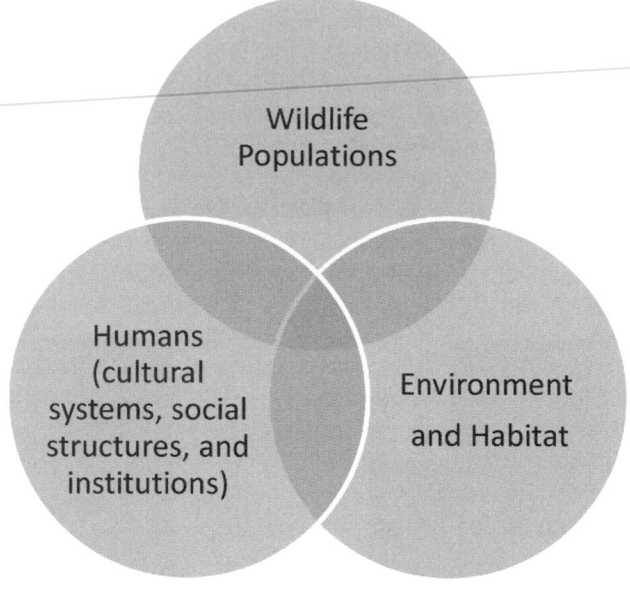

Figure 1.1. Wildlife management triad.

bird, (including any migratory, non-migratory, or endangered bird for which protection is also afforded by treaty or other international agreement), amphibian, reptile, mollusk, crustacean, arthropod or other invertebrate, and includes any part, products, egg, or off-spring thereof, or the dead body parts or parts thereof" (U.S. Government Printing Office 1975). This definition does not take into account important considerations of wildlife (e.g., whether it is free living, restricted by fences, or any aspects of human dimensions).

The Wildlife Society (the professional society for wildlifers) also defines "wildlife" as "free-ranging animals of major significance to man." Until the 1970s, wildlife was synonymous with animals that were hunted, but in the past four decades most free-living animals have become significant to humans, especially with the nation's emphasis on the conservation and management of biodiversity. In the wildlife profession, "wildlife" is often restricted to terrestrial and aquatic vertebrates other than fish because of a long political history.

THE POLITICAL DISCIPLINE OF WILDLIFE MANAGEMENT

Many wildlife departments in universities are part of, or associated with, departments of agriculture because they begin with an agricultural focus. Some of the earliest efforts to manage wildlife in the United States were established in the U.S. Department of Agriculture's Division of Entomology, which was funded by the American Ornithologists' Union. The Division of Entomology was established to determine the status of bird distributions and their migrations. This group was then transferred to the Division of Economic Ornithology and Mammalogy in 1885; their main function was to determine bird distributions and the damage they caused to agricultural crops. This agricultural base was expanded to address the relationship of all wildlife and agriculture and was placed

in the Bureau of Biological Survey in 1896. In 1940, a political decision by President Franklin Delano Roosevelt created the Bureau of Wildlife, which combined the Bureau of Biological Survey (which addressed birds and mammals) and the Bureau of Fisheries. However, fisheries biologists did not believe the newly formed bureau adequately represented them under this broad title, and the bureau name was changed to the Bureau of Sport Fish and Wildlife. It was later changed to the present name, the U.S. Fish and Wildlife Service. The name change implied that fish were to be treated differently than other wildlife, resulting in different disciplines and societies: one for fisheries and one for other vertebrates. Thus the term "wildlife" does not normally include fish.

Bureaucracies are rarely permanent, and in the late 1990s, Secretary of the Interior Bruce Babbitt attempted to create a freestanding organization that would combine all of the research conducted in the U.S. Department of the Interior into a single organization called the National Biological Survey. Babbitt's reason was that "the purpose of science is not to conquer the land, but to understand the mechanisms of ecosystems and to fit man into the resources he has available on the planet on which he has evolved." The evolution of the agencies that manage wildlife at the national level have expanded from understanding bird distributions and their influence on agricultural crops (i.e., Division of Entomology) to placing man in ecosystems (i.e., National Biological Survey) is a huge change over a short time period; however, the changes reflect how public attitudes change and how conservation organizations are formed and named. The National Biological Survey was eventually renamed the Biological Resources Division within the U.S. Geological Survey. As the political faces change at the federal level, so will the names of those "divisions" that are responsible for management. As wildlife populations are dynamic, so are the organizations that contain the people that manage them.

Regardless of the organizations that mange our wildlife, or the various definitions of wildlife, we consider wildlife as free-ranging, undomesticated animals in natural environments. Animals that are kept on property owned by private landowners with a barrier (most often a fence) are not free ranging and are not considered wildlife by most wildlifers. Leopold recognized this in 1933, when he described his theorems expressing the relationships between game and humans: "1. The denser the human population, the more intense the system of game management needed to supply the same proportion of people with hunting. 2. The recreational value of a head of game is inverse to the artificiality of its origin, and hence in a broad way to the intensiveness of the system of game management which produced it. 3. A proper game policy seeks a happy medium between the intensity of management necessary to maintain a game supply and that which would deteriorate its quality or recreational value" (Leopold 1933:394). Not everyone will agree, and there is certainly controversy about raising wildlife behind wire (Knox 2011). However, animals in captivity, regardless of the size of the enclosure, are often intensively managed. They are provided supplemental food, enclosed, often genetically manipulated (e.g., for trophy-size antlers), and are the subject of game ranching, which is developing

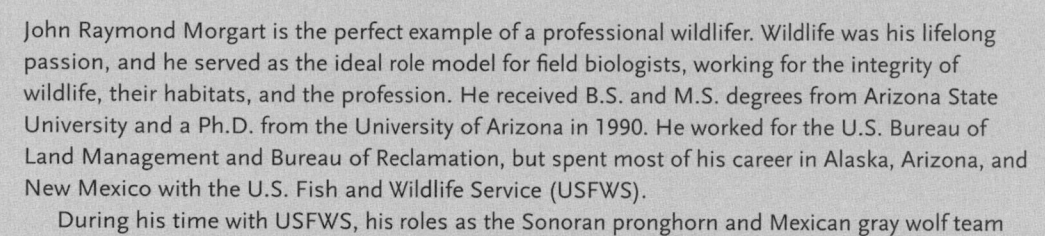

JOHN RAYMOND MORGART (1951–2009)

John Raymond Morgart is the perfect example of a professional wildlifer. Wildlife was his lifelong passion, and he served as the ideal role model for field biologists, working for the integrity of wildlife, their habitats, and the profession. He received B.S. and M.S. degrees from Arizona State University and a Ph.D. from the University of Arizona in 1990. He worked for the U.S. Bureau of Land Management and Bureau of Reclamation, but spent most of his career in Alaska, Arizona, and New Mexico with the U.S. Fish and Wildlife Service (USFWS).

During his time with USFWS, his roles as the Sonoran pronghorn and Mexican gray wolf team leader were instrumental in the recovery of both species. By involving multiple agencies and individuals, he was able to form coalitions that worked diligently to improve the status of Sonoran pronghorn, and their success today is in large part due to his leadership.

Morgart was active in local and national activities through The Wildlife Society (TWS), served as an associate editor of the *Journal of Wildlife Management*, and received the Jim McDonough Award from TWS.

Ethical, honest, compassionate, and friendly, Morgart loved to learn. He had little tolerance for ignorance or laziness. An avid hunter, outdoorsman, and naturalist, Morgart was the definition of a true wildlifer, one with drive, dedication, and devotion. His legacy is similar to that of other devoted professionals, whose achievements serve as a pathway to the successful conservation of wildlife.

into a separate discipline akin to animal husbandry. Those animals are not wildlife, and their management is not wildlife management. Knox (2011:45) states it well when he asks, "is shooting a privately-owned, half tame, semi-domesticated, supplementally-fed, genetically-engineered buck standing in a bait pile inside a pen the future of deer management?"

Caughley and Sinclair (1994) stated that defining wildlife management as the management of wildlife populations may be too restrictive because it could exclude the human aspect of the wildlife management triad. The authors of this text agree that the human aspect of wildlife management is very important, because management of wildlife incorporates human education, extension, law enforcement, and administration among many other related issues. Human dimensions are finally recognized as part of the core practice of manipulating or protecting wildlife populations to achieve a goal. Biologists and managers who understand animal ecology can make sound recommendations to those developing policy to achieve public goals. To manage populations effectively requires a combination of biological and sociological strategies. This book focuses on both, because a sound understanding of wildlife management is central to understanding the biology and habitat of the animal and the human dimensions associated with both. Leopold (1933:394) emphasized this philosophy by strongly encouraging management to consider the public in their activities (i.e., his fifth theorem, "only the landowner can practice game management cheaply").

For decades, wildlife management in the United States used the client model, which paid more attention to licensed hunters and anglers who paid for the services of management. It was not until Decker et al. (1996) called for a philosophical shift from the client model to the stakeholder model, which more closely followed Leopold's ideas for wildlife management to incorporate all wildlife and people. The stakeholder model of wildlife

management involved hunters and anglers but also included anyone who had a vested interest in a wildlife issue, program, action, or decision leading to an action (Decker et al. 1996). As the wildlife profession discusses the North American Model of Wildlife Conservation (Chapter 3), more of the public will need to be included in management of our natural resources, and the stakeholder model goes a long way toward that goal.

ACTIVE VERSUS INACTIVE MANAGEMENT

There are numerous ways wildlife is managed, but they all imply stewardship and can be classified into two broad categories: active management and inactive management. Active management does something to the population—such as increasing or decreasing its size—in a direct manner through strategies like translocations or hunting, respectively. Populations can also be actively managed by altering the habitat to the benefit or detriment of a population. If population numbers are too low for the goal of management agencies, other active management can be incorporated, such as predator control to minimize neonatal mortality or habitat improvement to provide required cover for neonates from predators. These efforts represent active approaches to management. Other populations may not be actively managed, like those in national parks. In such situations, management activities minimize external influences on populations and habitat, which often involves management of humans and not animals. Still, other populations may be so poorly understood that no action is taken because managers do not have enough information to make an informed decision. Thus the management is inactive. It is important to understand that all populations are managed either actively or inactively. For some populations, no active management occurs, but they are still managed because the decision to do nothing is a management decision; in these cases, management is often referred to as passive management or nonmanagement. There are yet other situations where populations (i.e., small vertebrates, insects) are essentially invisible to agencies because a decision to actively or passively manage is not even contemplated.

THE GOALS OF WILDLIFE MANAGEMENT

Most management goals for wildlife can be categorized into one of four options (Caughley 1977). When biologists are working with endangered species or declining populations, their goal is often to increase population numbers. For example, biologists have been working for years to increase the population of Sonoran pronghorn (*Antilocapra americana sonoriensis*) in southwest Arizona and northern Mexico (Krausman et al. 2005b) through active (Hervert et al. 2005) and inactive management (Krausman et al. 2005a). Indeed, most management plans for endangered populations aim to increase the population.

In other situations, the population may be too high and the management goal is to reduce the population. White-tailed deer (*Odocoileus virginianus*) in suburban and urban areas throughout the eastern United States have increased, and the public wants numbers reduced to minimize deer–vehicle collisions, disease, damage to vegetation, and threats to children (Krausman et al. 2011). Similar situations occur with elk (*Cervus canadensis*) in Arizona and Montana, American black bears (*Ursus americanus*) in national parks and housing areas (Merkle et al. 2011), and in other situations where wildlife poses a problem to the public, whether the concern is real or perceived. In some cases, even endangered species may be too abundant. A 2011 controversy with the endangered gray wolf (*Canis lupus*) has created nationwide concern, because some individuals in some states (i.e., Montana, Idaho, Wyoming) believe that they have too many wolves and need to reduce their numbers, while others claim more wolves are needed for the establishment of viable populations. In most of these situations the overall management goal is to reduce the population.

The third option is to manage the population for a sustained yield, as in the case of game animals throughout the United States. Hunting can be a tool for management, can provide game and recreation for hunters, and is a common management goal (Chapter 9).

The final option is to do nothing except, in some cases, to monitor the population as is done with some bats, reptiles, amphibians, small mammals, and other wildlife in national parks. At a minimum, management goals should include keeping track of the species under supervision. Unfortunately, monitoring is often time consuming and expensive and is not even considered until there is a problem that has to be addressed.

These four basic options (i.e., increase populations, decrease populations, harvest for sustained yield, monitor) are available to wildlife biologists and will vary depending on the goals of the organization. After goals are established in conjunction with the public, the manager can determine the appropriate action and how that action can be achieved in the best possible manner. The establishment of goals is a value judgment that is neither right nor wrong, good nor bad, but how the goals are achieved requires technical skills and decisions to ensure success.

THE WILDLIFE BEING MANAGED

All wildlife is managed actively or passively; even if there is no active management, the decision to do nothing is a management decision. Wildlife, however, has been classified numerous ways, and the way it is classified often dictates how it will be managed. There is a danger to classification schemes because, by setting up a classification criterion, some species will not be considered if they do not fall into the classification scheme. For example, funds set aside for threatened species cannot be used for other animals that are not threatened regardless of the need. However, classification of flora and fauna is an ongoing process. Estimates of the number of species on earth are as high as 30,000,000, yet fewer than 1,700,000 have been formally described since Linnaeus initiated the binomial system in 1753: there are approximately 750,000 insects; 440,000 plants, including algae and fungi; 47,000 vertebrates; and the rest is made up of assorted invertebrates and micro-organisms (Wilson 1985). Wildlifers concentrate their efforts on the ecology of vertebrates. Leopold (1933) established some of the more familiar classifications of types of wildlife, but other groupings are more recent (e.g., rare and endangered, urban, park) and are described below.

Farm Species
Farm species are wild species that do not migrate, can reside and reproduce on farms, and are a suitable by-product of farming. Typical farm species include quail, pheasants, squirrels, rabbits, raccoons (*Procyon lotor*), and Virginia opossums (*Didelphis virginiana*).

Forest and Range Species
Forest and range species are also sedentary but are compatible with forestry or livestock operations and are a suitable by-product of such. Typical examples are turkeys (*Meleagris gallopavo*), deer, elk, raccoons, foxes (*Vulpes* spp.), grouse, pronghorns, and collared peccaries (*Pecari tajacu*).

Wilderness Species
Wilderness species are those that are harmful to, or harmed by, economic land use and require special reserves of forests to be preserved as their habitat. Grizzly bears (*Ursus arctos horribilis*), mountain lions (*Puma concolor*), mountain goats (*Oreamnos americanus*), and bighorn sheep (*Ovis canadensis*) are true wilderness species. Others can include pronghorn, elk, deer, and moose (*Alces alces*).

Migratory Species
Migratory species are those that normally leave the land on which they were raised in the course of their seasonal movements. Ducks, geese, and swans are excellent examples, but

elk, deer, bats, and other mammals can also be placed into this category.

Furbearers

Furbearers are species that can be produced and marketed because of the commercial value of their pelts. Coyotes (*Canis latrans*), bobcats (*Lynx rufus*), muskrats (*Ondatra zibethicus*), otters (*Lontra canadensis*), and beavers (*Castor canadensis*) are just a few of those typically in this category.

Predators

Predators are animals that kill other species or are considered dangerous to livestock. Obvious examples are mountain lions, bobcats, coyotes, and feral cats and dogs.

Threatened Species

Threatened species are species of native fish and wildlife that are threatened with extinction. They are classified as such by the U.S. secretary of the interior whenever their existence is endangered; because their habitat faces destruction, drastic modification, or severe curtailment; or the species itself faces exploitation, disease, predation, or other factors. In such cases, their very survival requires assistance.

Urban Wildlife

More recently, emphasis has been placed on species that find habitat in towns, villages, and cities, and a relatively new discipline that studies habitat relationships with wildlife in cities is developing; these species are defined as urban wildlife. At one time, pigeons and rats were the species of cities, but deer, raccoons, falcons, and coyotes are other examples of wild species that are increasingly adapting to urban life.

Park Wildlife

Park wildlife is another classification for those species that exist in parks, which includes all those mentioned above plus others. The category has developed because of the confined management that occurs because of human-created boundaries and the philosophy of many parks (i.e., no harvest, limited management of wildlife, enhanced human management). Regardless of how humans classify wildlife, the basic goals for management are the same: influence the population to increase or decrease, manage for harvest, or simply monitor the population.

SUMMARY

Wildlife includes free-ranging, undomesticated animals in natural environments. Wildlife is not restricted to animals but includes their interrelationships with their habitats and with humans. Fish are usually considered separately from other animals because of political divisions and history. The goals of management are to increase populations, decrease populations, harvest the populations for sustained yield, or just monitor the population. Management can be active (i.e., influences populations directly) or inactive (i.e., no intentional management). There are numerous categories of wildlife, including farm, forest and range, wilderness, migratory, furbearers, predators, rare, endangered, urban, and park.

Literature Cited

Caughley, G. 1977. Analysis of vertebrate populations. John Wiley and Sons, New York, New York, USA.

Caughley, G., and A. R. E. Sinclair. 1994. Wildlife ecology and management. Blackwell Scientific, Boston, Massachusetts, USA.

Decker, D. J., T. L. Brown, and W. F. Siemer. 2001. Evolution of people–wildlife relations. Pages 3–22 *in* D. J. Decker, T. L. Brown, and W. F. Siemer, editors. Human dimensions of wildlife in North America. The Wildlife Society, Bethesda, Maryland, USA.

Decker, D. J., C. C. Krueger, R. A. Baer Jr., B. A. Knuth, and M. E. Richmond. 1996. From clients to stakeholders: a philosophical shift for fish and wildlife management. Human Dimensions of Wildlife 1:70–82.

Giles, R. H., Jr. 1978. Wildlife management. W. H. Freeman, San Francisco, California, USA.

Hervert, J. J., J. L. Bright, R. H. Henry, L. A. Piest, and M. T. Brown. 2005. Home-range and habitat-use patterns of Sonoran pronghorn in Arizona. Wildlife Society Bulletin 33:8–15.

Knox, W. M. 2011. The antler religion. Wildlife Society Bulletin 35:45–48.

Krausman, P. R., L. K. Harris, S. K. Haas, K. K. G. Koenen, P. Devers, D. Bunting, and M. Barb. 2005a. Wildlife Society Bulletin 33:16–23.

Krausman, P. R., J. R. Morgart, L. K. Harris, C. S. O'Brien, J. W. Cain III, and S. S. Rosenstock. 2005b. Introduction, management for the survival of Sonoran pronghorn in the United States. Wildlife Society Bulletin 33:5–7.

Krausman, P. R., S. M. Smith, J. Derbridge, and J. Merkle. 2011. The cumulative effects of suburban and exurban influences on wildlife. Pages 135–191 *in* P. R. Krausman, and L. K. Harris, editors. Cumulative effects in wildlife management: impact mitigation. CRC Press, Boca Raton, Florida, USA.

Leopold, A. 1933. Game management. Charles Scribner's Sons, New York, New York, USA.

Lueck, D. 1989. The economic nature of wildlife law. The Journal of Legal Studies 18:291–324.

McKelvie, C. L. 1985. A future for game? George Allen and Unwin, London, United Kingdom.

Merkle, J. A., P. R. Krausman, N. J. Decesare, and J. J. Jonkel. 2011. Predicting spatial distribution of human–black bear interactions in urban areas. Journal of Wildlife Management 75:1121–1127.

U.S. Government Printing Office. 1975. A compilation of federal laws. 052-070-02871-4. Washington, D.C., USA.

Wilson, E. O. 1985. The biological diversity crisis. BioScience 35:700–706.

THE HISTORY OF WILDLIFE CONSERVATION IN NORTH AMERICA

ROBERT D. BROWN

INTRODUCTION

The history of wildlife conservation in the United States, Canada, and Mexico must be discussed largely in the context of European migration. Some explorers, such as the Spaniards, came to the Americas looking for gold and silver. Others, like the French and English, came to escape an overcrowded Europe, looking for land and economic opportunity. Many came to escape religious and political persecution. Although initial settlements had very difficult times (about 90% of the Jamestown settlers perished), once established here, the emigrants saw unlimited land, timber, and wildlife with little or no government or laws to restrict their ambitions. As a result, the attitude of Americans, primarily in the United States, has been one of individual independence and the right to life, liberty, and the pursuit of happiness. These attitudes differed from those in Europe and eventually from other countries influenced by European traditions, including Canada and Mexico (Trefethen 1975, Colpitts 2002, Lopez-Hoffman et al. 2009). To this day, Americans bristle at attempts by local or federal government to restrict their personal freedom and independence. Nonetheless, the North American attitude toward wildlife gradually evolved from one of unrestricted abundance and harvest to one of regulated management and equitable access. We now call this the North American Model of Wildlife Conservation (Geist 1995, Geist et al. 2001; see Chapter 3), although its implementation has varied somewhat across Canada, the United States, and Mexico.

NATIVE AMERICANS AND WILDLIFE

Historians differ about the abundance of wildlife on our continent prior to the arrival of Europeans. Some writers speak of vast numbers of deer (*Odocoileus* spp.), elk (*Cervus canadensis*), moose (*Alces alces*), and beavers (*Castor canadensis*) in eastern forests; polar bears (*Ursus maritimus*), muskoxen (*Ovibos moschatus*), moose, and caribou (*Rangifer* spp.) in the Northwest; and millions of buffaloes (*Bison bison*) and pronghorns (*Antilocapra americana*) along with thousands of grizzly bears (*Ursus arctos horribilis*) and elk in the West. Others call this

the "pristine myth" (Denevan 2003:5) and note that English settlers had a difficult time finding game to hunt, and that early Spanish explorers do not even mention seeing bison. Many Native Americans, including the Sioux, Arapaho, Blackfoot, Cheyenne, Cree, Kiowa, Mandan, Pawnee, and Shawnee were hunter-gatherers, living off of meat from some large (but often small) mammals and the nuts, berries, and fruits they gathered. Some Native American societies, however, planted vast terraced and irrigated agricultural crops (Trefethen 1975, Krech 1999, Mann 2006). This was also true of the Huron in Canada and Aztec and Mayan cultures in Mexico (Simonian 1995).

Estimates of Native American populations prior to European exploration vary from 500,000 to five million in what is now the United States to as many as 100 million in both North and South America, with widely varying densities (Krech 1999, Mann 2006). Archeologists and anthropologists argue over the arrival of the first inhabitants to North America, variously dating it between 80,000 years ago and as recently as 10,000 years ago, with most settling on their appearance about 10,000 to 14,000 years ago. Most agree that the Paleoindians came across a continental ice bridge from Siberia and eventually worked their way eastward in North America and south to South America (Krech 1999), although some settlers of Canada may have come from Polynesia or across the North Atlantic from Europe (Taber and Payne 2003). Eventually, population densities were largest in Mexico and smallest in Canada (Simonian 1995, Druschka 2003).

Although the human population of North America increased prior to European arrival, some Native American populations disappeared, including the Anasazi in the 13th to 15th centuries in the American Southwest, the Cahokian in Missouri in the 12th to 14th centuries, and the Hohokam in the 15th century. Droughts, depletion of soil quality for growing crops or grazing animals, lack of firewood, warfare, pre-European diseases, or a combination of factors may have caused the disappearance or scattering of these populations (Krech 1999).

Archeologists examining animal and bird bones found in campfire sites believe that Native Americans may have had a significant impact on North American wildlife in the 10,000–15,000 years they were here before the arrival of Europeans. Over these many centuries, 35 genera of mammals and many bird species disappeared (Krech 1999). Surviving species were sometimes extirpated from localized areas. It is unclear whether predation by humans, increasing global temperatures, changing vegetation, or other factors caused the decline of these species (Fig. 2.1).

Native Americans hunted mammals for food and perhaps attempted to protect their crops from wildlife depredation. They diverted water for irrigation and cleared forests for crops, firewood, and construction materials. They terraced land and built dams. They built hundreds, if not thousands, of mounds for burials and temples. They commonly set fires, perhaps to clear underbrush for easier hunting, to improve wildlife habitat, to clear land for crops, and to herd wild animals. The extent of this burning and its impact on habitat is debated. Some scholars believe Native Americans may have decimated large game around their population centers, which were unevenly

Figure 2.1. Many Native American tribes depended on bison for their sustenance.

distributed. It is also believed that in some areas, including what is now Yellowstone National Park, occasional winter die-offs of game animals and localized overharvesting led to near starvation of the Native Americans (Kay 1998, Krech 1999, Mann 2006).

Scholars also debate how Native Americans valued and used wildlife. Because game was generally abundant, it was not always used wisely. Some Native Americans used the entire carcass of a bison, including the meat, hide, entrails, and bones. Others killed only female bison or took only their tongues or fetuses, which were considered delicacies. The practice of running herds of bison over cliffs was exceptionally wasteful, though some tribes stored carcasses under water for months to eventually consume them as "green bison soup." With such abundant wildlife, there may not have been a concept of waste. The practice of localized overexploitation may also have been supported by some Native American belief systems, such as those that believed bison were supernatural and emanated from underground, or that dead bison would be reincarnated (Krech 1999), or that their own ancestors were wild animals (Simonian 1995). In what is now the United States and Mexico, only dogs were domesticated; in Canada, however, caribou (reindeer) were domesticated over 3,000 years ago (Taber and Payne 2003).

Once European explorers arrived, their diseases quickly reduced the population of Native Americans. The English, Spanish, and French, having survived numerous plagues themselves, brought to the New World smallpox, typhus, bubonic plague, cholera, influenza, measles, mumps, yellow fever, common cold, diphtheria, chickenpox, scarlet fever, pneumonia, dysentery, and whooping cough. The explorers also brought hundreds of cattle, sheep, horses, and pigs, all with their own diseases such as anthrax and bovine tuberculosis. Diseases spread quickly, often ahead of the explorers. Native American runners and traders might have spread the word about the visitors to other villages, or captives might have escaped and returned home, taking the diseases with them. Some explorers came upon entire villages of the sick and dying. One estimate suggested that the North American population of about 3.8 million Native Americans was reduced by 74% to one million within 50 years of the European arrival. Later, diseases spread by individual trappers, explorers, and traders would increase that decline of Native Americans to over 90% of their pre-European population (Denevan 2003).

WILDLIFE REBOUND AND THEN DECLINE

As the Native American population dwindled, the wildlife population rebounded in those areas where it had been under harvest pressure. Logs from the Lewis and Clark expedition of 1805 stated that "the whole face of the country was covered with buffalo, elk and antelopes; deer were also abundant" (Krech 1999:74). As settlers arrived in the eastern United States and Canada, they cleared land for farming and cut forests for shipbuilding. Eastern tribes, such as the Sioux, migrated westward. At the same time, land clearing reduced habitat for wildlife. That habitat loss combined with hunting for subsistence, market hunting (harvesting wildlife for commercial meat, fur, and feather markets), and trapping for furs had a marked effect on eastern wildlife populations. In the Southwest, domestic livestock introduced by the Spaniards competed for forage with grazing wildlife. The Spaniards introduced the Native Americans to horses, which changed their lifestyle and improved their hunting success, but not to the extent that it had much impact on the wildlife populations.

To some extent, Europeans made modest efforts to conserve the New World's resources. From the 14th to 16th centuries, Spanish kings enacted laws for Mexico that prohibited the use of iron traps for bears, boars, or deer (1348), the use of poison for fishing (1435), killing mourning doves (*Zenaida macroura*; 1465), and hunting with hounds over snow (1516). Other laws limited timber cutting (1559), required reforestation (1763), and banned animal sacrifices. Even though Mexican kings put in place a few similar rules as early as the 13th century, the laws were rarely followed, and eventually the Spanish conquest cost Mexico 50–75% of its forests and much of its wildlife (Simonian 1995).

In Canada, after Inuits turned away Leif Ericksson from Newfoundland in the year 1000, explorers came from Portugal, Italy, France, Holland, England, and, on the west coast, Russia. Fisheries were the first resource to be exploited. As the British, Dutch, and French moved inland, they settled in the south of Canada, where winters were less severe. In 1728 the British appointed a Surveyor General of His Majesty's Woods and claimed Nova Scotia's forests for shipbuilding. In 1763, King George gave Canada's aboriginal people well-defined property rights for subsistence hunting. The extraction of natural resources in this part of the continent by the British, however, eventually helped spark the American Revolution (Taber and Payne 2003).

The growth of the European settler population from emigrants and their offspring was astounding, and human population growth had an increasing impact on wildlife and habitats. Emigration began in earnest after about 100 years of exploration in the 1400s and 1500s by the French, English, Spanish, and Dutch. By 1638 there were about 30,000 whites or Anglos in North America. That population grew to 1.3 million emigrants, their offspring, and slaves by 1700 and to 3.9 million by 1790. By 1850, the population of the United States was over 23 million Anglos with over four million additional slaves (Taylor 2003).

In addition to clearing land for farming and grazing, introducing domestic livestock, and hunting wildlife for food and markets, the new settlers discovered the value of exporting fur and feathers as a source of cash for the purchase of the other goods they needed. As early as 1607, Captain John Smith reported that the French were shipping 25,000 beaver pelts per year to Europe, and by 1650 most beavers were eliminated from the entire East Coast. The exploitation of furbearers in the Northeast and Canada was by the French and by England's Hudson's Bay Company. In the Pacific Northwest, the Russian-American Fur Company took seals (Phocidae) and sea

otters (*Enhydra lutris*) and by 1768 had extirpated the Steller's sea cow (*Hydrodamalis gigas*). Bird populations suffered from being taken for meat and for their plumage, which was used for ladies' hats in Europe. Deer and turkey (*Meleagris gallopavo*) populations also declined, again largely because of market hunting. In 1748 alone, South Carolina shipped 160,000 deer pelts to England. As wildlife populations declined, the settlers at first blamed it on predators. In 1630, the Massachusetts Bay Colony offered a bounty of one shilling for each wolf (*Canis lupus*) killed. When the deer did not rebound, the city of Portsmouth, Rhode Island, enacted the first closed season on deer hunting in 1646. This was only the beginning of American's game management efforts, as well as North America's never-ending conflict over our attitude toward predators (Trefethen 1975).

As the nation expanded, the new settlers moved westward, where land and wildlife seemed free and unlimited. An estimated 25 million buffaloes and ten million pronghorns resided in the West at their peak, although those numbers are obviously hard to confirm. Land purchases and other types of land acquisitions, such as the Louisiana Purchase, the Mexican–American War, the Oregon Compromise, the Gadsden Purchase of parts of Arizona and New Mexico, and the purchase of Alaska fulfilled the concept of "manifest destiny" and led to an America that spanned from coast to coast. Settlers followed, clearing more land and harvesting more wildlife. President Thomas Jefferson sent the famed Lewis and Clark Expedition from St. Louis, Missouri, to the Pacific Northwest in 1804. Although they encountered isolated stretches with little game, they found abundant herds of buffalo and deer, as well as grizzly bears and prairie dog (*Cynomys ludovicianus*) towns over 2.6 km^2 in size. By the early 1800s, trading posts had been established across the West, encouraging Native Americans to harvest game for their hides and to provide meat for settlers and trappers.

In Spain, parliament declared in 1813 that all lands in Mexico would be privatized, denying indigenous people access to public land, and guaranteeing land ownership for the wealthy. This policy helped to spark the Mexican revolution of 1810–1821 (Simonian 1995). Canada, on the other hand, was under the English policy, whereby most land was federal or owned by the Crown but managed by the provinces. The Crown managed land in the territories. In the United States in 1776, land not privately owned became property of the states. As states joined the Union, much state land was ceded to the federal government to pay for the revolution and in return for federal services (Taber and Payne 2003). Wildlife was believed to be superabundant and something to be used for food, fur, and income, or to be eradicated as pests (Dunlap 1988).

In 1830, Congress passed the Indian Removal Act, removing all title to lands from Indian tribes and moving many Native Americans to reservations in Oklahoma. By then, most of the beavers were gone from the continent, and silk had replaced beaver for the manufacture of men's hats. This collapse of the beaver hat market saved the beaver from total extirpation. Former beaver trappers became buffalo hunters.

In 1833 alone, the American Fur Company shipped 43,000 buffalo hides, most received in trade from Native Americans. Buffalo meat was also used for camp towns and for crews building railroads to the West. There was wanton wastefulness, with many buffalo again being killed solely for their hides or for their tongues. In 1845 the Hudson Bay Company shipped 4,300 buffalo tongues to England. By the mid-1840s a noticeable decline in buffalo numbers was already evident in the United States and in Canada, though some Canadians tried unsuccessfully to domesticate bison and to breed them with domestic cattle (Trefethen 1975, Krech 1999, Colpitts 2002).

The California Gold Rush of 1848 and later the Klondike Gold Rush of 1896 brought thousands of settlers westward, as did the Mormon migration to Salt Lake City. The population of the United States nearly doubled from 17 to 32 million between 1840 and 1860. The Civil War, from 1861 to 1865, slowed the western expansion, market hunting, and any thoughts of conservation. It also reduced the U.S. population by an estimated 600,000–750,000 more than all of the other U.S. wars combined. But the homestead laws of 1862 allowed anyone to mark out 259 ha of land for private ownership at no cost if they would live on it for five years. This of course led to the final Indian Wars, as settlers crowded Native Americans out of their traditional hunting grounds and onto reservations in Oklahoma and the Dakotas (Trefethen 1975, Mackie 2000).

MARKET HUNTING LEADS TO CONTROL EFFORTS

The late 1860s and 70s saw the final diminution of the bison herds. In 1871, Colonel R. I. Dodge reported a single herd in Colorado being 80.5 km wide and 32.1 km long, with an estimated four million head. However, the influx of hunters and railroads made shipment of hides, meat, and tongues very profitable. The famous "Buffalo Bill" Cody, a hunter for the railroad, once killed 69 in a single day and 4,240 in an 18-month period; other hunters were actually more successful. The annual kill in 1865 was one million; by 1871, it was five million. In 1864, Idaho imposed the first closed season on bison hunting, and Colorado and Kansas followed in 1875. In 1876, however, the annihilation of General George Armstrong Custer and 276 soldiers of the Seventh Cavalry at the infamous Battle of the Little Big Horn outraged the public and sealed the fate of both the Native Americans and the bison. Congress passed a bill that would have stopped the slaughter of the bison, but President Grant vetoed it. The U.S. Army knew that extermination of the bison would lead to control of the Native Americans. The last commercial shipment of buffalo hides was in 1884 (Krech 1999). It is estimated that between 1868 and 1881, over 31 million bison were killed (Taber and Payne 2003). A complete census in 1886 reported that there were only 540 bison left in the entire United States, mostly in the Yellowstone area of Montana (Trefethen 1975; Fig. 2.2).

The disappearance of the bison was the turning point for wildlife conservation in the United States and Canada. It became one of the rallying cries for those concerned about the future of wildlife in the United States. Another was the demise of the passenger pigeon (*Ectopistes migratorius*). Today it is hard to imagine the abundance of this bird. In 1806, Alexander Wilson, an ornithologist for whom the Wilson Society is named, recorded a flight 1.6 km wide and 64.4 km long, estimated to be over two billion birds. In 1813, James Audubon, famous artist and ornithologist, observed in one day a flight 88.5 km long which held an estimated one billion birds—and the flight continued two more days. These flocks often devoured agricultural crops, and market hunters—using cannon-like "punt guns," nets, and even clubs—decimated these flocks, as well as populations of ducks (Anatidae), swans (*Cygnus* spp.), and geese (*Anser* spp.) for the meat market in America and Europe. The last passenger pigeon died in the Cincinnati Zoo in 1914 (Trefethen 1975).

A CONSERVATION MOVEMENT BEGINS

The concept of conservation was clearly different across the three countries of North America. Mexicans had a utilitarian view of forests, and wildlife captured little economic interest. In the early 1800s, Mexico forbade foreigners from harvesting furbearers, passed a forestry act, protected sea otter pups, and licensed timber harvesting. Unfortunately, the laws were rarely heeded by the public or by government officials. The Treaty of Hidalgo, which ended the Mexican–American War in 1848, and the Gadsden Purchase of 1854 voided all previous Mexican laws in those territories. After the French intervention in Mexico (1862–1867), additional laws were passed to protect wildlife, except "ferocious and dangerous animals," but they, too, were ignored (Simonian 1995).

In Canada, as to some extent in the United States, the concept of wildlife was a cultural attitude shared with England. Subsistence hunting was accepted for settlers, but for the upper class, the hunting of game was associated with sport. Upper-class immigrants and Canadians and Americans who traveled to England were familiar with the pursuit of game on

Figure 2.2. Market hunting nearly extirpated bison from North America.

English estates. The evolution of hunters as conservationists ran parallel in the United States and Canada, though it developed more slowly in Canada because of its smaller human population and vast forested habitat unsuitable for farming (Taber and Payne 2003).

Most early settlers in the United States had little time for sport hunting, however, and wildlife was viewed as a source for sustenance and profit. Naturalist publications, such as the essay *Nature* (1836) by Ralph Waldo Emerson or the book *Walden* (1854) by Henry David Thoreau, had little impact until generations later. Nonetheless, the concept of wildlife as a "public trust," so different from European experience, became codified law. In 1842, the U.S. Supreme Court case *Martin v. Waddell* denied a landowner's claim to exclude others from taking oysters from some mudflats in New Jersey. The judge quoted the English Magna Carta of 1215 and codified the concept that in the United States, wildlife and fish belong to all the people, and stewardship of those fauna is entrusted to the individual states. This guaranteed the food supply at the time, although it continued to apply as wildlife became valued for sport, aesthetics, and for spiritual and cultural reasons. This decision, and others that followed, was the legal basis for the North American Model of Wildlife Conservation (Organ et al. 2010; see Chapter 3).

It was not until the nation became more prosperous that sport hunters would become the impetus for early conservation efforts. Wealthy, mostly eastern, landowners and businessmen who no longer had to hunt for subsistence formed clubs of like-minded friends to promote comradeship, a kinship with the pioneer spirit, and ethical hunting practices. The first sportsmen's club in the United States was the Carroll's Island Club, formed in 1832 near Baltimore, Maryland, largely for waterfowl hunting. In 1844 the New York Sportsmen's Club was formed, which drafted model game laws recommending closed hunting seasons on woodcock (*Scolopax minor*), quail (Odontophoridae), deer, and trout (Salmoninae) fishing. The Orange and Rockland counties of New York passed these laws in 1848. Many club members were attorneys, and they personally sued violators to encourage compliance with the law.

In 1859, British Columbia passed ordinances limiting mammal and bird hunting and forbidding the sales of some wild game meat to secure their food supply (Colpitts 2002). Eventually, hundreds of local sportsmen's clubs were formed across both countries, and similar game laws were passed. In addition to game limits and seasons, some states outlawed the use of hunting dogs and night hunting with lights, and others banned the use of traps, snares, and pitfalls, as well as poisoning wildlife to protect crops, all of which were common at the time. The laws were hard to enforce, however, and in 1852, Maine became the first state to employ a game warden (Trefethen 1975).

PRESERVING LANDS FOR PUBLIC USE

During the 1870s much of North America was still being explored and surveyed. In 1870, a group of explorers pleaded with the governor of Montana that the Yellowstone area was too beautiful to exploit for profit and that it should be held in trust for the entire public. The concept of a public, national park was first raised by a painter, George Catlin, who in 1832 proposed "a nation's park, containing man and beast, in all the wild and freshness of their natural beauty" (Adams 2004:77). In 1872, President Ulysses S. Grant established the first national park in the United States, Yellowstone, encompassing 8,671 km². (The first national park in North America was Chapultepec, established in what is now Mexico City, in the 13th century; Simonian 1995.) The Yellowstone law had little effect, however, and the U.S. Army was later dispatched to guard the park from squatters and poachers for more than 30 years. The protection of this land was particularly important, as it signified the concept of a national asset in the public interest, and it began a tradition of setting aside lands for public parks, forests, and wildlife refuges. Rather than develop a "Tragedy of the Commons" (Hardin 1968), where European public lands were overutilized to their detriment, Americans agreed to share their commons fairly and democratically: "In no other nation was nature seen as such an essential part of national identity" (Wellock 2007:17). Waterways were eventually included. In 1872 the U.S. Supreme Court case *Massachusetts v. Holyoke Water Power Company* decided that the company could not build a dam without a fish ladder. The court stated that the use of a river "may be regulated as public rights, subject to legislative control" (Wellock 2007:20). In 1871, Congress created the U.S. Commission on Fish and Fisheries, one of our first federal conservation agencies. The primary purpose of this agency was to establish fish hatcheries but, again, the move was significant in that it signaled the importance of federal management of wildlife and fisheries over that of states (Wellock 2007).

Canadians deplored the loss of the great auk (*Pinguinus impennis*), bison, Labrador ducks (*Camptorhynchus labradorius*), and the devastation of eastern sea bird colonies. In 1887, parliament established the first "Dominion park" in Canada, first called the Rocky Mountain Park and later known as Banff National Park. Parliament also established the first national wildlife refuge at Lost Mountain Lake in Saskatchewan. Numerous provincial parks were established, as were Glacier, Yoho, Jasper, and Waterton Lakes national parks. The Canadian Pacific Railway developed these protected areas with hotels and other amenities to entice tourism (Barnett 2003, Adams 2004). Eventually, the Dominion Forest Reserves and Parks Act standardized under a commission in 1911 management of Canada's parks (Adams 2004).

A young, affluent politician named Theodore Roosevelt aided the Yellowstone Park effort in the United States. Roosevelt was a 23-year-old Harvard graduate and member of the New York State legislature when in 1883 he went on his first buffalo hunting trip in North Dakota, where he would go on to become an avid and adventurous outdoorsman and hunter. The following year his wife and mother died a few hours apart, and he returned to North Dakota for three years to reflect on his life and then to purchase and establish the Elkhorn Ranch.

THEODORE ROOSEVELT (1858–1919)

Born into a wealthy New York family, weak and asthmatic as a child, Theodore Roosevelt rose to become America's most successful conservationist. Home schooled, he read voraciously about natural history, learned taxidermy, exercised regularly, and took hunting and camping trips to Maine and other states. After graduating from Harvard in 1880, he became New York's youngest state legislator at the age of 23. He wrote of his Dakota Territory hunting trips in his book *Hunting Trips of a Ranchman*. After the deaths of his mother from yellow fever and his wife from childbirth on the same day, he returned to the Dakotas for 16 months and purchased Elkhorn Ranch. He remarried in New York in 1886 and produced a plethora of books on hunting, ranching, and western lore. He was an active member of the Explorers Club, the Camp Fire Club, and he founded the Boone and Crockett Club and the New York Zoological Society, which established the Bronx Zoo.

In 1888, Roosevelt accepted a position on the federal Civil Service Commission. That was followed in 1895 at age 37 with his appointment as New York City police commissioner, and in 1897 he became assistant secretary of the Navy. He resigned in 1898 to form the Rough Riders and led the decisive battle of the Spanish–American War at San Juan Hill, Cuba, for which he was posthumously awarded the Medal of Honor by President Bill Clinton in 2001. He returned from the war to become governor of New York and then vice president of the United States under President William McKinley. When McKinley was assassinated in 1901, Roosevelt became the nation's youngest president at age 42. On a hunting trip to Mississippi in 1902, Roosevelt refused to shoot a captive bear, which would later be called "Teddy's Bear," and which became an icon for toy manufacturers. During his two terms in office, Roosevelt established 51 wildlife refuges, five national parks, and 16 national monuments, protecting 59,893,475 ha. He also enacted significant conservation legislation, building the Panama Canal and mediating the Russo-Japanese War, for which he was awarded the Nobel Prize. After his presidency, he took a three-month hunting trip to Africa, collecting dozens of animals for the Smithsonian collection.

In 1908 he again ran for president against Taft and Wilson as a candidate of the Progressive, or Bull Moose, Party. At one point he delivered an hour-long speech with a bullet in his lung from an assassination attempt. After losing the election to Wilson, he embarked on a three-month trip down the River of Doubt, a tributary of the Amazon, where he was seriously injured and contracted malaria. Roosevelt died in his sleep at his Sagamore Hill, New York, estate in 1919, at age 60. Roosevelt's legacy includes 26 books, over 1,000 magazine articles, and thousands of letters and speeches. He is remembered as one of our country's greatest historians, biographers, statesmen, hunters, naturalists, and orators.

Roosevelt was a cowboy and a hunter who loved trophies but who felt a near-spiritual kinship with nature. He detested the decline of the buffalo (Wilson 2009).

Roosevelt's enthusiasm for conservation was born in the Dakota Badlands during those years, and he became the most active president and leader in the history of North American conservation. In 1887 he gathered a group of influential American hunters in New York to form the Boone and Crockett Club, with a mission of preserving the big game of North America. One of the members, an outspoken editor named George Bird Grinnell, wrote numerous articles about the plight of Yellowstone in his magazine *Forest and Stream*. In 1894, President Grover Cleveland signed the Yellowstone Protection Act, making the park the first wildlife refuge in the United States, and provided it with guaranteed funding and administration. Another early conservationist, John Muir, advocated the establishment of Yosemite National Park and other parks in California. He later formed the Sierra Club in 1890. Muir was a preservationist who believed public lands should not be used for timber harvesting or hunting. Roosevelt, the conservationist, believed lands for public enjoyment should also be used for multiple purposes, including resource extraction. The philosophical argument over preservation—or strictly limiting access to public lands—versus conservation—to include recreation, grazing, logging, and hunting on public lands—continues to this day in the United States (Trefethen 1975).

THE IMPACT OF CLUBS AND THE MEDIA ON CONSERVATION

One cannot understate the influence of the formation of sportsmen's clubs, conservation organizations, scientific societies, and the print media on the North American conservation movement. The early apostle for hunting and conservation in America as a fair chase sport evoking manhood, character, and virtue was William Henry Herbert, who wrote under the pseudonym Frank Forester (Dunlap 1988). Magazines such as the *American Sportsman* (1871), *Forest and Stream* (1873), *Field and Stream* (1874), and the *American Angler* (1881) informed readers of the bounty and the plight of western wildlife. Public attitudes changed as they saw some species of wildlife disappearing because of market hunting. They came to realize that natural resources in America were not unlimited, and that conservation efforts should be used. In some circles, Charles Darwin's *On the Origin of Species* (1859) and *The Descent of Man* (1871) led people to realize that man had a genetic relationship with animals, and that animals might be able to reason and suffer (Dunlap 1988).

The American Association for the Advancement of Science (AAAS) in 1881 pressured the U.S. secretary of agriculture to form the Forestry Division, which later became the U.S. Forest Service. In 1891 the American Forestry Association and the AAAS convinced Congress to establish national forest reserves to assure the nation of a future timber supply. By then, the U.S. General Land Office and the U.S. Geological Survey had been formed to survey and keep track of federal lands (Trefethen 1975).

Gifford Pinchot, a friend of Roosevelt's, was named director of the U.S. Forestry Division in 1898. Pinchot was trained in Europe, as there were no forestry courses in American universities at that time, but he understood that forests could be used for timber, wildlife, and watershed, and conserved through "sustained yield" management. In fact, he and his staff coined the term "conservation," derived from the British "conservator." He led a survey of American forests with the AAAS, and in 1896 President Grover Cleveland added over 8.5 million ha to the forest reserves and established Grand Canyon and Mount Rainier as national parks. In 1904, Pinchot became the first director of the U.S. Forest Service and was widely known as "the father of American forestry." Although he, too, was part of the eastern elite, Pinchot strongly believed in equity in access to America's resources. He stated that "natural resources must be developed and preserved for the benefit of the many, not just the profit of the few" (Krech 1999:25). About the same time he formed the Society of American Foresters, an organization composed of forestry professionals. In 1905 they held the first American Forest Conference to bring together managers, educators, and scientists (Trefethen 1975).

During the 1870s and later, dozens of additional hunting, conservation, and scientific organizations were formed, including the League of American Sportsmen, the American Ornithologists' Union, the Camp Fire Club, the New York Zoological Society, the Audubon Society, and the American Bison Society. These groups, along with local sportsmen's clubs, lobbied for stricter laws to stop market hunting, to ban unethical sport hunting, and to begin game restoration efforts. They recognized that states had difficulty enforcing their game laws. In 1900, Congress passed the Lacey Game and Wild Birds Preservation and Disposition Act, the first national legislation for wildlife conservation. This law made it a federal offense to transport wild game across state borders if taken illegally. It also strictly controlled the importation of exotic species. This strengthened state game laws, and it helped stop the trade in plume and feathers, as well as the poaching and smuggling of wildlife meat products (Trefethen 1975).

In addition, sportsmen's clubs started wildlife restoration efforts. An interesting result was the importation in 1881 of 28 ring-neck pheasants (*Phasianus colchicus*) to Oregon from Shanghai, China. Game bird farms blossomed, raising birds in captivity and releasing them into the wild to be hunted. Pheasant hunting is now popular through much of the United States, and few Americans recognize this bird as an exotic species. Destructive logging techniques during that period actually led to better deer habitat, as more diverse vegetation replaced the mature forests. In 1878, a sportsmen's club in Vermont was the first to trap and restock deer. Other states followed (Trefethen 1975).

Pressure from these sportsmen's, conservation, and scientific organizations also impacted natural resource education. The Land Grant Act, signed by President Lincoln in the middle of the Civil War in 1862, established agricultural and technical colleges in all states, making higher education in agriculture and engineering affordable to "the sons and daughters of farmers and mechanics." Later legislation established the Agricultural Experiment Station System and the Cooperative Extension Service with joint funding from the state, federal, and county governments. In 1898 Cornell University began offering courses in forestry, and in 1900 Yale University established the School of Forestry, the first of its kind. Within three years, forestry schools opened at the universities of Maine, Michigan, Minnesota, and at Michigan State. Wildlife was not yet a scientific or management discipline of its own, as most studies of biota were still of a taxonomic sort found in botany and zoology programs (Trefethen 1975).

THE ROOSEVELT ERA

When Theodore Roosevelt became president in 1901, a new era of wildlife conservation began. He once stated, "there can be no greater issue than conservation in this country" (Wilson 2009:i). Roosevelt was advised by Gifford Pinchot, George Bird Grinnell, and to some extent John Muir. Roosevelt formed the Agriculture Department's Division of Economic Ornithology and Mammalogy, which soon became the Bureau of Biological Survey, and tasked it with surveying the nation's biota. It was initially a research organization but was given policing powers with the passage of the Lacey Act. Major John Wesley Powell, the well-known explorer of Grand Canyon, became director of the Geological Survey. He supported the Reclamation Act of 1902, which authorized over 30 federally funded reservoir and dam projects. Although this legislation was not directly associated with wildlife, it had a major impact on future wildlife populations. It recognized the authority of the U.S. government over the rights of states in national natural resources issues, though it led to the development of significant irrigated farming in the West (Trefethen 1975).

Although Roosevelt and John Muir were friends, their relationship was tested over the first major political battle over land preservation. After the tragic San Francisco earthquake and fire of 1906, it was clear that San Francisco needed a dependable water supply. A proposal was made to dam part of the Tuolumne

Figure 2.3. Pelican Island was the first national wildlife refuge. Photo courtesy of the U.S. Fish and Wildlife Service

River in California, which would flood part of Yosemite. Muir strenuously opposed the Hetch Hetchy Dam project, which Pinchot and Roosevelt supported. The dam was completed in 1914, and Muir died a year later. The battle over this use of a public resource split the conservationists from the preservationists. Eventually, however, the argument let to the formation of the U.S. National Park Service in 1916 under the auspices of the Department of the Interior (Cronon 2003; Fig. 2.3).

Theodore Roosevelt believed in a strong role for the federal government in protecting as much land as possible for public use. In 1903, Pelican Island, Florida, became the first unit of the National Wildlife Refuge System, which now encompasses over 60.7 million ha. Roosevelt used the National Antiquities Act of 1906 to declare scenic lands and wildlife habitat as parks and as forest reserves. In 1908, he added additional wildlife refuges in Alaska, Oregon, Florida, and Nebraska. That same year Roosevelt appointed the National Conservation Commission, which was tasked with inventorying the national forests, waters, and minerals and with recommending management strategies. In all, Roosevelt set aside 60 million ha during his presidency, over 20,000 ha for each day he was in office, including 16 national monuments, 51 wildlife refuges, and five national parks (Wellock 2007).

CONSERVATION EFFORTS CONTINUE

Congress was relatively slow to address the problem of market hunting waterfowl. The Weeks-McLean Migratory Bird Act, proposed in 1904 by Pennsylvania Congressman George Shiras III, asserted the federal government's power over states' rights and let the secretary of agriculture set hunting seasons and limits on migratory game birds. This was opposed for nine years as a states' rights issue, but it finally passed in 1913. Three years later the United States signed a treaty with Great Britain for the Protection of Migratory Birds in the United States and Canada. Mexico was invited to sign the treaty, but it was in revolution at the time. This landmark legislation was the first

between countries to protect wildlife. Similarly, after a decade of debate, Congress passed the Weeks Act of 1911, which authorized the federal government to purchase lands to protect stream flow and eventually led to the creation of the eastern national forests (Lewis 2011).

The Canadian wilderness lifestyle had evolved from that of hunter-trappers to a more agrarian type about the time of the confederation of the provinces in 1870. With it came a new ethic of wildlife conservation. The formation of the Historical Society of Manitoba (1879) was followed by the formation of the Winnipeg Game Preservation League (1882) and the Macleod Game Protective Association (1889). But as the railroads and developers enticed western emigration, they exaggerated the abundance of wild game available for sport hunting. Taxidermy exhibits went eastward to advertise the size and diversity of game in western Canada and Alaska. As the railroad, mining, and logging crews used more wildlife as a food source, conflicts with sport-hunting tourists became inevitable. In the 1890s, sportsmen successfully lobbied for controls on sport and meat hunting, including subsistence hunting by Native Americans. In 1909 the Canadian Commission on Conservation was formed to mirror Theodore Roosevelt's efforts in the U.S. conservation clubs formed in the United States, such as the Camp Fire Club, the Boone and Crockett Club, and the Audubon Society, and their individual members influenced Canadian legislatures. The 1911 Game Act prohibited the shooting and selling of 26 types of game species (Colpitts 2002).

Roosevelt, Pinchot, and Muir influenced the attitudes of Canadians, as well. In 1906, game protection and management acts were enacted in Saskatchewan, Alberta, Prince Edward Island, and New Brunswick; British Columbia followed in 1913. *Life Histories of North American Mammals*, by Seton and Roberts (1909), further enhanced public interest in protecting wildlife. U.S. and Canadian biologists and policy makers collaborated a great deal, and in 1909 parliament formed the 32-member Commission for the Conservation of Natural Resources, which later developed the National Parks Act, the Northwest Territories Game Act, and the Migratory Birds Convention Act. In 1911, Howard Douglas, superintendent of Rocky Mountain National Park, purchased 703 bison in Montana and released them in the park. In 1918, Hoyes Lloyd, supervisor of wildlife protection, authorized the Royal Canadian Mounted Police (RCMP) to enforce game laws (Burnett 2003). Two activist officials, Gordon Hewitt, dominion entomologist with the Canadian Department of Agriculture and author of *The Conservation of Wild Life in Canada* (1921), and James Harkin, commissioner of parks, were instrumental in keeping wildlife protection under the parks department rather than under the forestry department. They were also involved in lobbying for the Migratory Bird Treaty. Hewitt and Harkin were supported politically by Clifford Sifton, the minister of the interior, and by Prime Minister Wilfred Lauier (Taber and Payne 2003).

In Mexico, Miguel Angel de Quevedo, an engineer, architect, and founder of Mexico City's arboretum, was an advocate for conservation of Mexico's forests. He spoke at the North American Wildlife Conference in 1909, where he met Roosevelt and Pinchot. He promoted legislation that would preserve forests for watersheds, recreation, and scenic values, but the revolution undid all of his efforts. He later headed the effort to form the Mexican Forestry Society and headed the Commission for Protection of Wild Birds (Simonian 1995).

By 1910 every state in the United States had some sort of commission for the protection of wild game and fisheries. The National Association of Game Wardens and Commissioners became the International Association of Game, Fish and Conservation Commissioners (now known as the Association of Fish and Wildlife Agencies). But funding was still a problem. In 1913 Pennsylvania was the first state to issue a hunting license. When more than $300,000 came into the state's treasury the next year, many other states followed suit. These funds paid for wildlife restoration efforts, enforcement of game laws, and predator control. Predators interfered with livestock operations, and hunters also still believed that predators limited game abundance. Many states paid bounties for wolves, mountain lions (*Puma concolor*), foxes (*Vulpes* spp.), coyotes (*Canis latrans*), bobcats (*Lynx rufus*), hawks (Accipitridae), owls (Strigiformes), and even eagles (Accipitridae; Trefethen 1975).

Debates continued in the state houses and at the national level over the concepts of preservation versus conservation. Dr. William T. Hornaday, superintendent of the National Zoo in Washington, D.C., and head of the National Bison Society, published *Our Vanishing Wildlife* in 1913 and led his followers to oppose hunting. He was an elitist who once remarked, "all members of the lower classes of southern Europe are a dangerous menace to our wild life," and that protection was needed from "Italians, negroes and others who shoot songbirds for food" (Wellock 2007:87). He was also opposed to federal funding of waterfowl restoration through hunting fees. Hunters and conservationists created the organizations listed above plus new ones like the Izaak Walton League, Forests and Wild Life, and American Wild Fowlers (later to become Ducks Unlimited). Scientists formed the Ecological Society of America in 1914 and American Society of Mammalogists in 1919. Both societies were to then publish scientific journals. Wildlife science began to move beyond taxonomy. The scientific societies had a close relationship with the Biological Survey, and together they developed better methods of censusing wildlife and studying diets, cover requirements, and disease issues on the national parks and refuges. As ecology developed as a discipline, the concepts of plant succession, niche, community scales, trophic levels, and food chains were developed and debated in the United States and abroad (Trefethen 1975).

Preservation of public land continued, with the formation of national parks in the United States at Glacier (1910), Lassen (1916), Denali (1917), and Grand Canyon (1919). In 1917, Mexico's President Carranza declared Desierto de los Leones as the country's first modern national park (Simonian 1995). The U.S. National Park Service (NPS) was eager to sustain wildlife in their parks for visitors to see, a much different mis-

sion than that of the U.S. Forest Service, which was to preserve a sustainable timber supply. The NPS hired scientists to study the natural resources they managed. Efforts by the NPS were instrumental in preserving trumpeter swans (*Cygnus buccinator*), grizzly bears, bighorn sheep (*Ovis canadensis*), and wild burros (*Equus asinus*). Still, however, there was little funding at the national level to support wildlife research.

In the early 1900s, America's assault on predators continued on private and public lands, as it was believed they harmed domestic livestock and wildlife populations. In the Kaibab Plateau of Arizona, a herd of 3,000 deer on about 405,000 ha was thought to be in decline. The government hired hunters and trappers to kill 120 bobcats, 11 wolves, 674 mountain lions, and over 3,000 coyotes. By 1924 the deer herd had grown to 100,000. During that year's severe winter, 60,000 deer died of starvation. This episode was later to become a national example of the perils of predator control as a means of managing wildlife populations (Trefethen 1975).

In 1937 the scientists, with the help of the conservation groups, lobbied the U.S. Congress to pass the Pittman-Robertson Federal Aid in Wildlife Restoration Act (P-R Act). This act excised an 11% tax on all hunting weapons and ammunition. The act provided nearly $3 million in its first year. The funds are collected by the federal government and distributed to the states based on the number of hunting licenses they sell, their population, and their land area. The act provides 75% of the funding, which must be matched by 25% from the states. The funds are used for wildlife restoration projects, research, and education, and they cannot be used for any other purpose. This is still one of the major sources of funding for wildlife research in the United States (Trefethen 1975).

WILDLIFE RESEARCH BEGINS TO SUPPORT MANAGEMENT

The 1930s were the beginning of wildlife management research in the United States, led by Aldo Leopold. Although trained as a forester at Yale, he became "the father of wildlife management" in the United States. Working for the U.S. Forest Service in New Mexico, later for their Forest Products Lab in Wisconsin, and then as a consultant for the Sporting Arms and Ammunition Manufacturer's Institute, he conducted the first intensive analysis of wildlife populations in the Midwest. In 1933 he became the first professor of game management at the University of Wisconsin and published *Game Management*, the first book of its kind in America. In it he said, "we have the scientist, but not his science, employed as an instrument of game conservation" (Leopold 1933:20). Leopold was well known for his many essays on land ethics, and he developed a series of wildlife management principles: every species has a defined set of habitat requirements that sets it apart from all other species, all animal and plant biota are interconnected, and the habitat has a seasonal carrying capacity that should not be exceeded. He said the tools of wildlife management were "the ax, the plow, the cow, fire and the gun" (Leopold 1933:332).

The stock market crash of 1929 and the dust bowl era of the early 1930s were disastrous for American wildlife. Pressure from livestock growers and the public pushed the Biological Survey toward predator control rather than wildlife research as its primary mission. They formed the Division of Predator and Rodent Control (later to become Animal Damage Control and then Wildlife Services under the Department of Agriculture). Much of the federal funding for other wildlife-related programs was cut at the same time waterfowl and game bird habitat was disappearing in the Midwest. Wildlife conservation and restoration needed more funding, and in 1934 Congress passed the Migratory Bird Hunting Stamp Act (similarly adopted in Canada in 1966). Because waterfowl are migratory across state boundaries, the federal government limits their seasons and bag, and hunters must purchase a duck stamp to affix to their state hunting licenses. These funds have now protected over 1.8 million ha of waterfowl habitat (Trefethen 1975).

FRANKLIN DELANO ROOSEVELT AND A NEW DEAL FOR WILDLIFE

The presidents who followed Theodore Roosevelt were not nearly as supportive of conservation. President Taft (1909–1913) thought Roosevelt had gone too far, and he opposed habitat preservation and even forest fire control in Alaska. He fired Pinchot and Interior Secretary James Garfield (the former president's son) and replaced them with men more amenable to dealing favorably with developers and industrialists. Deals were cut to sell federal coalfields and to encourage development of railroads, copper mines, and steamship companies in Alaska. Presidents Wilson (1913–1921), Harding (1921–1923), Coolidge (1923–1929), and Hoover (1929–1933) showed little interest in conservation, though the latter two were fly fishers (Brinkley 2011).

Franklin D. Roosevelt, who was elected to office in 1933, commonly referred to himself as a tree farmer (Brinkley 2011) and made conservation a national jobs program. The Civilian Conservation Corps (CCC) employed over three million out-of-work factory workers and farmers. Between 1933 and 1942, the CCC built 800 new parks, 193,000 km of roads, and 1,147 fire lookout towers; reforested over 809,000 ha with two billion trees; and restored thousands of hectares of wetlands for waterfowl breeding. Not all of this construction was ecologically sound, but it popularized the concept of conservation to the American working classes (Wellock 2007). President Roosevelt hired Jay "Ding" Darling, a political cartoonist critical of the New Deal wildlife programs, to head the Bureau of Biological Survey. He was an enthusiastic leader, and he enhanced the morale of the bureau. He saw the need to develop wildlife research programs and to enhance educational programs to produce wildlife managers and biologists. In 1934 he urged Congress and private organizations to establish the Cooperative Wildlife Research Unit Program, the American Wildlife Institute, the North American Wildlife Federation, the North American Wildlife Institute, and the North American Wildlife and Natural Resources Conference (Trefethen 1975).

JAY NORWOOD "DING" DARLING (1876–1962)

Born in Norwood, Michigan, and raised in Sioux City, Iowa, Jay Darling intended to pursue a career in medicine. But at Beloit College in Wisconsin, he began newspaper writing, cartooning, and signing his name "Ding," a contraction of his last name. His professional newspaper career began in 1900, with posts at the *Sioux City Journal*, *Des Moines Register and Leader*, the *New York Globe*, and the *New York Herald Tribune*. Though he eventually moved back to Des Moines, the *Tribune* published his political cartoons from 1917 to 1949. His cartoons were wildly popular and syndicated in 130 papers across the country. They covered many topics, but as an avid hunter and fisher, Ding was especially concerned about pollution and loss of wildlife habitat. Ding received the Pulitzer Prize for editorial cartooning in 1924 and again in 1943, and was voted the best cartoonist in America by a group of leading editors in 1934.

Despite his inexperience, Ding was appointed head of the U.S. Biological Survey (later to become the U.S. Fish and Wildlife Service) in 1934 by President Franklin D. Roosevelt. During his one-year tenure at the bureau, Ding raised $17 million for wildlife habitat restoration. The idea of the Duck Stamp was his, and he drew its first design. The funds he raised led to the expansion of the National Wildlife Refuge System. He established the Migratory Bird Commission to bring together hunters and other conservationists, later founded the National Wildlife Federation, and the first North American Wildlife Conference was his idea. He often vacationed on Sanibel Island, Florida, and there he convinced his neighbors to purchase land that President Harry Truman named a national wildlife refuge in 1945. In 1967 the refuge was named the J. N. Ding Darling National Wildlife Refuge.

The Cooperative Wildlife Research Program established research units of two to four biologists at ten land grant universities, with funding supplied by the Bureau of Biological Survey, the universities, the American Wildlife Institute (later the Wildlife Management Institute), and state fish and game agencies. These units provided research on practical wildlife problems, taught university courses, and helped train thousands of biologists. There are now 40 such units in the country. The Wildlife Management Institute receives its funding from the sporting arms and ammunition industry, and it lobbies for conservation laws and conducts professional reviews of university wildlife programs and state wildlife agencies. It also annually convenes the North American Wildlife and Natural Resources Conservation Conference of scientists, state and federal agency personnel, and nongovernmental organizations (NGOs). The first meeting, held in 1935, produced more information in its proceedings than had ever before been accumulated in one volume on wildlife in North America. The North American Wildlife Federation later became the National Wildlife Federation, a private organization of over 2.5 million members, and the North American Wildlife Institute later became the North American Wildlife Foundation, a semiprivate granting agency of the federal government (Trefethen 1975).

Other accomplishments during the New Deal were the passage of the Fish and Wildlife Coordination Act to force federal agencies to communicate with each other; the establishment of the Soil Erosion Service, later to be called the Soil Conservation Service and now called the Natural Resources Conservation Service; and the Taylor Grazing Act, which increased management of public lands and charged fees for restoration of over 32 million ha (Wellock 2007). These improvements had a powerful, positive effect on wildlife in North America, restoring deer to the Northeast and waterfowl to the Midwest. Other initiatives, such as the Tennessee Valley Authority (TVA), had mixed impacts. The TVA built 16 dams and brought electricity to thousands, but it destroyed many streams and wildlife habitat in doing so. In 1930 a committee headed by Aldo Leopold developed the Model Game Law, recommending that states set up wildlife commissions of volunteers, appointed by governors, for staggered terms. That eliminated the problem of turnover of state wildlife agency personnel caused by new governors firing supervisors after each election. The P-R Act prohibited the governors from redirecting its funds to other uses (Trefethen 1975).

The Wildlife Society, an organization of professional wildlife biologists, was formed in 1936, and the first issue of the *Journal of Wildlife Management* was published the next year. The American Fisheries Society and the Society for Range Management also represented professionals and published scientific journals. Just before World War II, the Bureau of Fisheries in the Department of Commerce and the Biological Survey of the Department of Agriculture were merged into the U.S. Fish and Wildlife Service (USFWS) under the secretary of the Department of the Interior, a significant development, as governmental leaders recognized that fish and wildlife were more than crops to be grown and harvested (Trefethen 1975).

Things did not always go so smoothly in Canada. Subsistence hunting and fishing by indigenous people have been, and continue to be, an issue. In 1949, Newfoundland joined the Canadian confederation and challenged the Migratory Bird Act, hoping to reduce or eliminate hunting restrictions. The political debate continued into the 1950s, until new laws allowed needy residents to hunt for food but also outlawed the sale of game meat (Taber and Payne 2003). Unlike Canada, Mexico did not recognize indigenous communities or their rights to

land or hunting and fishing privileges. In 1936, Mexico signed the Protection of Migratory Birds and Animals Act, which established hunting regulations and wildlife refuges, and outlawed transportation of live or dead game, much like the Lacey Act in the United States (Lopez-Hoffman et al. 2009).

In 1935, Mexico's President Cardenas (1934–1940) appointed Angel de Quevedo to head the Department of Forestry, Fish and Game. Cardenas had made land reform a major platform of his election campaign. During his administration, Cardenas and Quevedo created 40 national parks, plus additional wildlife refuges, agricultural cooperatives, irrigation programs, and dams. This amounted to the largest land reform in Mexico's history. In 1935 the United States–Mexico International Parks Commission was formed. Unfortunately, in Mexico a strong hunter-conservationist lobby never formed, largely owing to lack of interest and the lack of money spent on sport hunting. Quevedo's initiatives were strongly opposed by agricultural interests. Game laws continued to be poorly enforced, and Mexico's grizzly bear and wolf populations were nearly extirpated. Finally, bowing to agricultural interests, President Cardenas abolished the Department of Forestry, Fish and Game in 1940 and accused Quevedo of being antirevolutionary. Quevedo died in 1946 (Simonian 1995).

WORLD WAR II AND THE 1950S

As with the Civil War and World War I, conservation efforts took a backseat during World War II. After the war, however, returning servicemen bought hunting licenses by the thousands. Sales in the United States increased from $7 million per year before the war to $12 million by 1947. States benefited from the influx of hunting license fees and the P-R Act funds they generated, as did federal agencies from the sale of duck stamps. Unused P-R Act funds had accumulated during the war, and were now available for larger projects. Large restocking efforts resulted for deer, pronghorns, elk, mountain goats (*Oreamnos americanus*) and sheep, bears, beavers, and turkeys. By then, most land grant universities had wildlife departments, and the GI Bill allowed returning military personnel to pursue wildlife biology as a profession. Aldo Leopold died in 1948, and his final essays were published in a book entitled *A Sand County Almanac* (Leopold 1949), which became required reading for Americans interested in conservation for generations to come (Trefethen 1975).

During the 1950s, support for conservation took a downturn in the United States as the federal government, concerned over military threats in Europe, redirected funds toward increasing the size and power of America's military. Funding for fish, wildlife, park, and forest programs diminished, while logging, grazing, and oil and gas exploration leases were granted on public lands. In 1950, concern over funding for restoration of our freshwater fisheries led to the Dingell-Johnson Federal Aid in Fisheries Restoration Act (D-J Act). The act was funded by an excise tax on fishing equipment and boats, and it functioned much as the P-R Act did for wildlife. The Magnuson Act was passed to separate commercial fisheries from the USFWS and to pass it to a new U.S. Fisheries Commission.

Americans during the 1950s flocked by the millions to our national parks, wildlife refuges, seashores, and forests, establishing recreation as an integral component of the natural resource value of our public lands. In response to the decrease in federal land acquisition, The Nature Conservancy was formed in 1951 to preserve lands the government could not afford. It is now the wealthiest conservation organization in the nation, protecting 6.1 million ha in the United States and over 41 million ha worldwide. Bird watching became a popular pastime with the continued publication and updates of Roger Tory Peterson's *A Field Guide to Birds* (1934). Disney Studios, with the feature film *Bambi* and television documentaries like *The Living Desert* in 1953 (Trefethen 1975), had a huge impact on the public's attitude toward nature.

But the new prosperity and growth of the 1950s brought more commercial development of land for housing and agriculture, loss of habitat, and concentrated farming and livestock operations. Farming efficiency was helped by mechanization as well as liberal use of pesticides and herbicides. One of the most notorious was dichlorodiphenyltrichloroethane (DDT). DDT was an effective pesticide, but it was released before being adequately tested. As early as 1946, it was found to be lethal to crustaceans like crabs (Brachyura) and crayfish (Astacoidea). As insects became impervious to DDT, other chemicals such as chlordane, dieldren, aldrin, and methoxychlor went on the market. Wildlife biologists first reported dying songbirds, followed by brown pelicans (*Pelecanus occidentalis*), ospreys (*Pandion haliaetus*), and bald eagles (*Haliaeetus leucocephalus*). Researchers at the USFWS Patuxent Wildlife Research Center in Laurel, Maryland, fed these chemicals to wildlife and proved that they were toxic (Gottlieb 2003).

WILDLIFE BECOME PART OF THE ENVIRONMENT

In 1962, Rachel Carson, a former USFWS editor, researcher, and internationally successful writer, published *Silent Spring*, which documented the impact of these chemicals on the environment, especially on wildlife. She spoke to the issue of science being ignored in public conservation policy and being corrupted by commercial interests. She stated that the increasing use of pesticides was indicative of "an era dominated by industry, in which the right to make money, at whatever costs to others, is seldom challenged" (Carson 1962:67). She predicted a future of spring seasons when no birds would be heard. This resonated with the public. No book before or after has had as great an effect on arousing the American public's awareness of environmental concerns. It set the stage for the future environmental movement for the next three decades. In Canada, the book and the public outcry that followed led to the formation of the Wildlife Toxicology Division of the Canadian Wildlife Service. Chemical companies tried to discredit Carson, but President Kennedy defended her and made the environment part of his political platform, though DDT

RACHEL LOUISE CARSON (1907–1964)

Rachel Carson was a nationally known scientist and author well before she published *Silent Spring*. Born in Springdale, Pennsylvania, Carson graduated from the Pennsylvania College for Women (now Chatham College) in 1925, studied at the Woods Hole Marine Biological Laboratory, and obtained an M.A. in zoology from the Johns Hopkins University in 1932. She began her professional career writing radio scripts for the U.S. Bureau of Fisheries in 1936 and eventually became editor-in-chief for all U.S. Fish and Wildlife Service publications. She wrote scientific articles and pamphlets, and published popular articles in *The Atlantic Monthly*, the *New Yorker*, and the *Baltimore Sun*. Her books *Under the Sea-wind* (1941), *The Sea around Us* (1952), and *The Edge of the Sea* (1955) brought her fame as a naturalist and author. She was awarded a National Book Award, the John Burroughs Medal, the Henry Grier Bryant Gold Medal, the New York Zoological Society Gold Medal, and a Simon Guggenheim Fellowship.

Carson resigned from federal service in 1952 to concentrate on writing, and later she changed her focus to the misuse of pesticides. In 1962 her book *Silent Spring* was published in three installments in the *New Yorker* and was a book-of-the-month club selection. She challenged agricultural practices and lack of government oversight as she linked the health of the natural world to that of humans. Though she called only for prudent use of DDT, rather than a total ban, she was vigorously attacked by the chemical industry. Her testimony before Congress in 1963 eventually led to the formation of the Environmental Defense Fund in 1967, the formation of the Environmental Protection Agency in 1970, and the passage of the federal Insecticide, Fungicide and Rodenticide Act in 1972, which banned the use of DDT in the United States. Prior to her death from cancer in 1964, she was awarded the National Audubon Society Medal and was inducted into the American Academy of Arts and Letters. She was posthumously awarded the Presidential Medal of Freedom by President Jimmy Carter.

was not actually banned until 1974. Kennedy also established the Land and Water Conservation Fund to acquire land for scenic, recreational, and public values, and he signed the Sikes Act of 1960, which required the Department of Defense to prepare natural resources management plans for all 12 million ha of military bases. In 1964, President Lyndon B. Johnson passed the Wilderness Act, incorporating areas of 2,023 or more roadless hectares untrammeled by man into wilderness areas. This legislation removed them from the National Forest Inventory, placing them into national parks and wildlife refuges (Cronon 2003). Next, Johnson signed the Wild and Scenic Rivers Act, setting aside seven major rivers for recreational and conservation purposes (Trefethen 1975).

NIXON: THE RELUCTANT ENVIRONMENTALIST

Concerns over pesticide toxins eventually led to public dissatisfaction over predator control techniques. Common methods for killing predators included the use of sodium floroacetate (compound 1080) and coyote getters, explosive shells filled with cyanide pellets, as well as the use of strychnine and thallium poisons. These devices and poisons killed all manner of small mammals, hawks, and eagles. Under the Nixon administration, an executive order prohibited the use of chemical poisons on all public lands. In response to public outcry over the potential loss of our national bird, the bald eagle, and other species, Congress passed the Endangered Species Conservation Act in 1969, but it had little in the way of enforcement power or funding. Congress followed with a tougher Endangered Species Act (ESA) in 1973. It defined and divided species and subspecies into threatened versus endangered. It eliminated all commercial traffic in live, dead, parts, or products from endangered species, and it funded research on why species were becoming threatened and how to recover them. The act defined "critical habitat," included harassing wildlife to be a "taking," set substantial fines, and required teams of scientists and managers to develop recovery plans for each endangered species. Early endangered species included wolves, whooping cranes (*Grus americana*), Key deer (*O.v. clavium*), Sonoran pronghorns (*Antilocapra americana sonoriensis*), peregrine falcons (*Falco peregrinus*), bald eagles, alligators (*Alligator mississippiensis*), Kirtland's warblers (*Dendroica kirtlandii*), and California condors (*Gymnogyps californianus*). Captive breeding programs were often begun as part of the restoration projects. Swift foxes (*Vulpes velox*), black-footed ferrets (*Mustela nigripes*), and California condors are examples of translocation and breeding and release projects. The USFWS and the National Marine Fisheries Service were given authority to enforce the act. In 1978, the U.S. Supreme Court case of *Tennessee Valley Authority v. Hill* delayed the federal Tellico Dam project to protect the snail darter (*Percina tanasi*), a tiny endangered fish. The court ruled that "the plain intent of the statute was to halt and reverse the trend towards extinction, whatever the cost" (Lueck 2008:154). The ESA is often touted as the most significant conservation legislation ever enacted in the United States. It continues to be the basis and the source of much

funding for wildlife research, though the funding is far below what is needed for this enterprise.

Surprisingly to many, the Nixon administration was one of the most environmentally progressive of all American presidencies. In addition to the ESA, Nixon signed the Marine Mammal Protection Act, the National Environmental Policy Act (NEPA), the Clean Air Act, and the Clean Water Act, and he established the Environmental Protection Agency (EPA). This legislation, taken together, had a monumental impact on guiding what could and could not be done with America's natural resources, and it clearly established the environment as an issue of national importance. The NEPA led to the Council on Environmental Quality, which requires environmental impact statements on all government projects. It also requires open hearings for public input, thus making environmental decision-making an open and democratic process. The EPA is now one of the major funding sources for wildlife research, in addition to being an environmental enforcement and research agency. In 1969, the United States signed the Convention on International Trade in Endangered Flora and Fauna Species Act (CITES), making it illegal to import or export items made from endangered species.

Nixon's impetus for signing all of this environmental and conservation legislation was strictly political. Public opposition to the war in Vietnam—in the form of street demonstrations and even riots—was growing during his administration. Simultaneously, the public had lost faith in the government to protect them from protracted war, environmental pollution, and nuclear proliferation. Nixon inherited Lyndon Johnson's unpopularity over the war, but Johnson had enjoyed the popular Secretary of the Interior Stuart Udall, and Lady Bird Johnson had made beautifying America by planting flowers a priority. Public concerns over the environment were heightened by television ads of Iron Eyes Cody, an actor who depicted an American Indian shedding a tear over highway trash. In addition, population growth became a concern in 1968 with the publication of Paul Ehrlich's book *The Population Bomb* (his coauthor and wife, Anne, was not credited). On 22 April 1970, the nation participated in the first Earth Day, a project suggested by Senator Gaylord Nelson, in which 2,000 universities, 10,000 schools, and over 20 million people voiced their frustrations and suggested solutions to the nation's environmental problems. During this time there were legal pressures on the federal government, as well. The Environmental Defense Fund was formed in 1967, to be followed by the Natural Resources Defense Council and the Sierra Club Legal Defense Fund, all of which commonly sued the government. Though Nixon did not initiate any of the environmental or conservation legislation he signed, he was so unpopular over the war (and especially his ordering of the invasion of Cambodia in 1970) that he saw environmentalists as perhaps the only friends he had (Flippen 2003).

During this period, numerous challenges to the Migratory Bird Act continued in Canada. Canadian Wildlife Service Director David Munro ordered the RCMP to allow indigenous people to hunt on Indian reservations and on unoccupied

Crown land, but legal guarantees of traditional hunting rights were not legislated until the 1980s (Burnett 2003). From the 1940s to the 1970s, Mexico's human population more than doubled, bringing massive pollution problems. The Green Revolution, led by future Nobel Prize winner Dr. Norman Borlaug and funded by the Rockefeller Foundation, increased Mexico's crop productivity by 22%, it but required heavy use of fertilizers, pesticides, and herbicides and led to clearing forests and draining wetlands. In the 1950s, Enrique Beltran, who held a Ph.D. in zoology from Columbia University and who served as Mexico's undersecretary of forestry and fauna, was director of the Institute de Recursos Naturales Renovables and sponsored wildlife and fisheries surveys, research, and conferences. Unfortunately, he opposed Leopold's land ethic, believing natural resources should be exploited primarily for their economic value. In fact, "between 1940 and 1970 . . . Mexican governments abandoned conservation altogether" (Simonian 1995:131).

In 1971, the United Nations Educational, Scientific, and Cultural Organization initiated the Man in the Biosphere Program for scientific study and the protection of biodiversity. Biosphere reserves consist of a core protected area surrounded by a managed use area encircled by zones of cooperation. All of these involve local people with attention to their cultural and economic needs to find a balance between nature and mankind. Mexico established their first two biosphere reserves in Durango in 1974, and eventually 85% of Mexico's protected areas fell under the biosphere reserve program. The World Wildlife Fund and Conservation International were instrumental in establishing the programs. Success varied, but the program contributed to the recovery of the Bolson tortoise (*Gopherus flavomarginatis*), Kemp's Ridley turtle (*Lepidochelys kempii*), and the Mexican crocodile (*Crocodylus moreletii*; Simonian 1995).

THE REAGAN ERA

Ronald Reagan, the former governor of California, became president in 1980, and he appointed James Watt as his secretary of interior. They sensed attitudes had changed in the previous decade. The conservation movement, interested in protecting scenic places and other resources, was largely outpaced by the environmental movement, interested in pollution, clean air, and clean water. The Earth First movement, some of which included environmental terrorism like burning logging equipment and spiking trees with nails, radicalized the environmental movement and led to a fracturing of mainstream and radical environmental groups. The Sagebrush Rebellion or Wise Use movement of western farmers and ranchers called for private control of public lands (Wellock 2007). The Reagan administration sought to dismantle the ESA via the Taskforce for Regulatory Reform. David Stockman, head of the federal Office of Management and Budget, recommended selling off national parks and forests to balance the federal budget. Morale in the federal agencies was low during this period, yet there was great support for established legislation from the

Figure 2.4. The bald eagle was one of the first animals on the endangered species list. Photo courtesy of the U.S. Fish and Wildlife Service

memberships of the Sierra Club, Audubon Society, and The Wilderness Society, all of which had doubled their memberships during the previous decade (Fig. 2.4).

Despite being re-elected in 1984, Reagan was not able to dismantle the federal agencies or sell off public lands, and he eventually retreated from these efforts. In 1980, the Alaska National Interest Lands Conservation Act expanded the National Wildlife Refuge System by 21.4 million ha. Twenty-nine new wildlife refuges were added to the system, 200 plants and animals were added to the Endangered and Threatened Species List, and the ESA and Clean Water Act were reauthorized during Reagan's eight years in office (Pope 2003) A Fish and Wildlife Conservation Act was passed, to do for nongame species what the P-R and D-J acts had done for game species. Unfortunately, no funding was provided. Funding for federal agencies and universities continued to lag for the next two decades. Nonetheless, conservation groups found a new source of revenue for preserving wildlife habitat and for restoration projects.

The federal Farm Bill establishes the authority and funding for the Department of Agriculture and its many crop subsidy and price support programs. It is renewed every five years. During the 1980s and 90s, conservation became more prominent each time the bill was renewed. Programs such as the Conservation Reserve Program, the Wetlands Reserve Program, the Wildlife Habitat Incentive Program, the Grasslands Reserve Program, and the Environmental Quality Incentive Program, all managed by the Natural Resources Conservation Service, provided private land owners with payments to take land out of production or to make conservation improvements. Most states developed similar programs (Wellock 2007).

THE 21ST CENTURY BECKONS

During the Clinton administration, Secretary of the Interior Bruce Babbitt felt that there was too much duplication among the scientific research projects by the many agencies of that department, so in 1996 he reassigned all Department of Interior scientists to a new organization called the National Biological Survey, and later renamed it the Biological Resource Division of the U.S. Geological Survey. Although that may have improved federal efficiency, the field resource managers of the NPS, the Bureau of Land Management, the USFWS, and others felt they had lost control over the research they needed. Subsequently, these agencies formed a coalition with university researchers to conduct their needed research, and additional educational and outreach programs. The organization, known as the Cooperative Ecosystem Studies Units, is composed of 270 universities organized into 17 regions, and allows funding from more than a dozen federal agencies to flow to the academic researchers (Brown 2010).

But as we entered the 21st century, funding for nongame species was still inadequate. A concerted effort to correct the shortfall was developed and supported by over 3,000 conservation organizations. Termed the Conservation and Reinvestment Act (CARA), it would have used a tax on outdoor recreational items like canoes, binoculars, and birdseed to provide $350 million a year to the states in much the same manner as the P-R and D-J acts. The national effort to support this legislation was called Teaming with Wildlife. A few equipment manufacturers supported the endeavor, but many others strenuously opposed it. A compromise was reached in 2001 with "CARA-lite," now called the State Wildlife Grants Program, which provided $50 million a year for the USFWS budget. The funds are awarded competitively and require a match of 25–50% from the state. Importantly, it required each state to develop a comprehensive wildlife conservation plan; plans were completed in 2005. This funding has increased each year, and some of the money is set aside specifically for Native American tribes (Brown 2010).

In Mexico, the end of the 20th century included the administrations of Miguel de la Madrid (1982–1988), Jose Lopez Portillo (1982–1988), and Carlos Salinas de Gortari (1988–1994). Public concerns over pollution and loss of forests and wildlife led to the formation of new NGOs, including Pronotura (1981), the Mexican Ecologists Movement (1981), and Biocenosis (1982). Under Portillo and Madrid, pollution continued unabated, largely due to the oil boom and poor control of PEMEX, the state-owned oil company. Illegal trade in jaguars (*Panthera onca*), pumas, ocelots (*Leopardus pardalis*), Mexican parrots (*Amazona viridigenalis*), and other birds continued. Mexico refused to sign the CITES agreement. President Salinas, however, passed the General Law on Ecological Balance and Environmental Protection. In 1989, when it was determined that Mexican slaughterhouses had processed 35,000 Kemp's Ridley turtles (*Lepidochelys kempii*), Salinas banned the process. In 1991 he signed CITES. The NGOs helped persuade the government to join Partners in Flight (1990), the United States–Canada Waterfowl Management Plan (1994), and to form the Sonoran and Rio Grande Joint Ventures. The North American Free Trade Agreement of 1994 required Mexico to establish the Commission on Environmental Protection, but its success has been hard to assess (Simonian 1995).

In 1993, Canada overhauled its Migratory Bird Conservation Act and passed the Wild Animal and Plant Protection and Regulation of International and Interprovincial Trade Act. This consolidated 40 years of wildlife protection policy. In 2002, parliament passed the Species at Risk Act, modeled after the ESA (Taber and Payne 2003).

CONTINUING PROBLEMS AND NEW OPPORTUNITIES

The U.S. government owns or controls 148 million ha, or about 16%, of the total landmass of the country; states own or control another 79 million ha (Lubowski et al. 2006). It is clear that the United States, Canada, and Mexico now have just about all the public land they will ever have, and that our wildlife resources are limited. Our population continues to grow, using more food, goods, and services. Forests and farmlands are being converted for commercial uses at alarming rates, removing wildlife habitat and fragmenting what is left. Water quality is a concern, and water quantity is restricting development and wildlife habitat in some areas. The infrastructure in our national parks and monuments is eroding, and decades of fire suppression have led to devastating fires in national parks and forests. Overpopulation of some species, game ranching, and international travel have led to outbreaks of wildlife–domestic and animal–human diseases in our wild mammal and bird populations, including the West Nile virus, avian influenza, chronic wasting disease, brucellosis, and tuberculosis. Today, Canada has lost over 16 million ha of forests and habitat to the mountain pine beetle (*Dendroctonus ponderosae*), Colorado and Wyoming have lost 1.5 million ha, and there is no end in sight.

Our economies thrive on energy, and there are increased political efforts to expand oil and gas exploration in our national parks and refuges, such as in the Arctic National Wildlife Refuge and off of our coasts. The process of fracking to release natural gas deposits, and extraction of oil from sand and shale, may be environmentally unviable. Hunting, although still generally supported by the nonhunting public, is declining as a recreational activity. Animal rights and welfare groups, such as the People for the Ethical Treatment of Animals, have organized public support to ban trapping in some states and hunting of carnivores such as bears and mountain lions. Human–wildlife conflicts have increased, including millions of deer killed on our highways and mountain lion and bear attacks on humans each year. There are still political threats to the future of the Endangered Species Act, and some hunting groups favor the private ownership of wildlife, even to the extent of allowing artificial insemination of wild deer and cloning of lucrative species. Arguments over listing polar bears or delisting wolves continue. Enrollment in natural resources academic programs has been declining for a decade, as state and federal agencies have not had the funds to hire new employees. Climate change threatens the world's ecosystems, and yet we are slow to respond to the threat of global warming. Funding for wildlife research in 2010 remained at about the

1980s level, as the wars in Iraq and Afghanistan continued to absorb more of the federal budget. The financial collapse of 2008 and subsequent economic recession continue to impact all aspects of funding for wildlife research and management, as our political representatives struggle to define priorities for use of public funds.

Mexican gray wolves (*Canis lupus baileyi*), lesser long-nosed bats (*Leptonycteris yerbabuenae*), black-tailed prairie dogs, and even Monarch butterflies (*Danaus plesippus*) share habitat across the border between the United States and Mexico, yet collaboration on recovery efforts is stymied by immigration and drug smuggling issues. Smugglers of people and drugs (and the law enforcement officers sent to interdict them) disrupt wildlife movement and damage habitat. The 2006 U.S. Secure Fence Act called for 1,127 km of 3.1-m fences separating Mexico and the United States, further restricting movement of wildlife.

On the other hand, the public feels the environment, natural resources, and wildlife are important to their quality of life. Recent books by Richard Louv, *Last Child in the Woods* (2006) and *The Nature Principle* (2011), have raised public awareness of the importance of both children and adults connecting with nature if we are to remain psychologically well in a complex and technological world. Funding for scientific research, such as through the National Science Foundation and the National Institutes of Health, continues to increase. New technologies like satellite radio tracking, geographic information system mapping, and DNA genetic analysis provide new tools for the study of conservation science. The development by the USFWS of the Safe Harbor concept, wherein landowners are not penalized if their managed lands attract endangered species, has provided flexibility to farmers and ranchers. Some species once at risk of disappearing—bison, timber and gray wolves, bald eagles, peregrine falcons, the American alligator, grizzly bears—have made spectacular comebacks. There are more deer and turkeys in the United States now than ever before. Canada reached its goal of 12% of its land in wilderness areas by 2000, and there are now numerous collaborative agreements across Canada, the United States, and Mexico to protect wildlife, from CITES to the North American Waterfowl Management Plan to the Black-tailed Prairie Dog Conservation Action Plan. Private landowners, looking toward hunting leases and ecotourism as means of income, see the value of conserving their wildlife and wild lands, and state and federal agencies acknowledge the importance and value of private lands for the conservation of game and nongame wildlife (Benson et al. 1999). The *Conservation Directory* (2010), published annually by the National Wildlife Federation, lists over 1,600 private conservation organizations in the United States and Canada. Hundreds of private land trusts and local governments raise money to purchase important habitat land for greenways, or to buy permanent conservation easements to restrict commercial development. Slowly, the concept of paying private landowners for "ecosystems services" is developing in the United States. A fledgling carbon-trading market has already developed, and some entities are beginning to

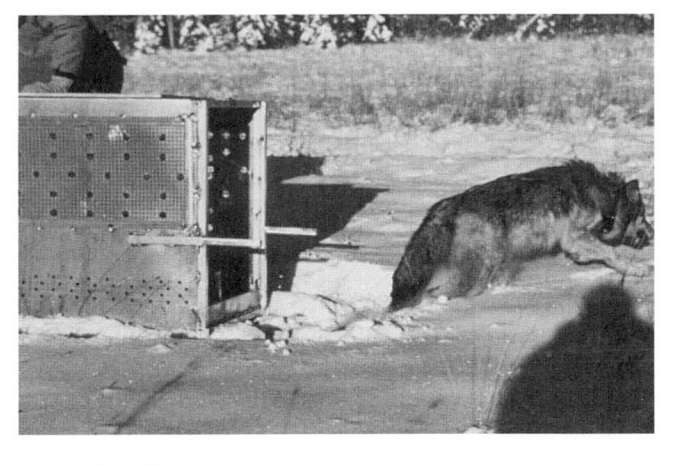

Figure 2.5. Wolf releases in the Yellowstone area have been remarkably successful, but not without controversy. Photo courtesy of the U.S. Fish and Wildlife Service

pay landowners for their watersheds and endangered species habitat (Fig. 2.5).

LESSONS LEARNED

The North American Model of Wildlife Conservation, based on the Public Trust Doctrine of public ownership of wildlife, has a checkered history, but in general it has served society well in the United States and in Canada. Clearly, wildlife in North America do not know international boundaries any more than they know state or provincial ones. Our three countries share thousands of kilometers of common borders, and if we are to conserve our shared wildlife, we need to understand each other and share at least some common goals.

The North American model holds that wildlife is a public and international resource managed by policies based on sound science. Under this model, the killing of wildlife should only occur for legitimate purposes, the allocation of wildlife is by law, markets for game are eliminated, and the democracy of hunting is standard (Organ et al. 2012). Though wildlife is considered an international resource, even in North America, there should be no commercial markets for wildlife, and wild birds and animals should only be killed for legitimate reasons. Hunting has been an essential component of this model, but democracy means that there is equitable access to game (The Wildlife Society 2010).

Unfortunately, few modern-day Americans, Canadians, or Mexicans know of or understand this history. Wildlife conservation is a political issue with multiple stakeholders. Conservation issues—such as drilling for oil in the Arctic National Wildlife Refuge, control of urban deer or feral cats, or shooting wolves—become highly emotional in our societies. As our population becomes more urbanized and children spend less time outdoors, the public, though supportive of conservation, becomes less knowledgeable and less able to make informed decisions. It is thus critical that we in the wildlife conservation profession provide public and private educational programs to inform our citizens about the choices before us. As Richard Stengle, editor of *Time* magazine, stated in 2006, "being an American is not based on a common ancestry, a common religion, even a common culture—it's based on accepting an uncommon set of ideas. And if we don't understand those ideas, we don't value them; and if we don't value them, we don't protect them" (Posewitz 2010:32).

SUMMARY

The history of wildlife conservation in Canada, the United States, and Mexico is a history of human migration, changing values, and associated behavior. Native Americans, having arrived about 10,000–14,000 years ago, exploited wildlife for their food, clothing, shelter, tools, weapons, and even fuel. Wildlife populations were sometimes diminished because of hunting, more so around areas of human concentration with large and elaborate agricultural developments. Upon the arrival of Europeans, Native American populations crashed, mostly because of diseases brought by immigrants and their livestock. Wildlife populations rebounded, only to be soon reduced significantly due to market hunting and trapping for meat, fur, and feathers. As early as the 13th century, Spanish kings enacted laws in Mexico to limit harvests and to protect animals, birds, and forests.

Those laws and similar ones enacted in Canada and the United States in later centuries had limited effect. It was not until the late 1800s, when more affluent North Americans noted the extirpation of some wildlife species and near loss of others, that effective laws were passed and enforced. Once game meat was replaced by domestic meat production and felt, cotton, and wool replaced fur and feathers, hunting became a sport rather than a commercial enterprise. Under the leadership of President Theodore Roosevelt, many influential individuals, sportsmen's groups, the media, and conservation groups rallied to protect and manage land and wildlife in the United States with laws and refuges. Canada followed with similar laws, but with a vast land area and relatively small human population, the stress of human impact on wildlife populations was only moderate. Mexico's wildlife populations suffered from greater human impact and weaker enforcement of wildlife protection laws. This may be partially due to Mexico's tumultuous political history, its lack of economic development as compared to its northern neighbors, and a continuing attitude among the public that natural resources need to be exploited for the survival of its people.

Overall, the history of wildlife conservation in North America is a timeline of the evolution of what we now call the North American Model of Wildlife Conservation. It is largely based on our European ancestry, with imbued values of public ownership of wildlife—management by policies based on science, and protected by laws enforced the state and federal government. As our human population continues to increase, and as our consumption of goods per capita continues to rise, only time will tell if these values can be sustained.

Literature Cited

Adams, W. M. 2004. Against extinction: the story of conservation. Earthscan, London, United Kingdom.

Benson, D. E., R. Shelton, and D. W. Steinbach. 1999. Wildlife stewardship and recreation on private lands. Texas A&M University Press, College Station, USA.

Brinkley, D. 2011. The quiet world: saving Alaska's wilderness kingdom, 1879–1960. Harper Collins, New York, New York, USA.

Brown, R. D. 2010. A conservation timeline: milestones of the model's evolution. The Wildlife Professional 4:28–31.

Burnett, J. A. 2003. A passion for wildlife: the history of the Canadian Wildlife Service. University of British Columbia Press, Vancouver, Canada.

Carson, R. 1962. Silent spring. Houghton Mifflin, Boston, Massachusetts, USA.

Colpitts, G. 2002. Game in the garden: a human history of wildlife in western Canada to 1940. University of British Columbia Press, Vancouver, Canada.

Cronon, W. 2003. The trouble with wilderness, or, getting back to the wrong nature. Pages 213–243 in L. S. Warren, editor. American environmental history. Blackwell, Malden, Massachusetts, USA.

Denevan, W. 2003. The pristine myth: the landscape of the Americas in 1492. Pages 5–42 in L. S. Warren, editor. American environmental history. Blackwell, Malden, Massachusetts, USA.

Druschka, K. 2003. Canada's forests: a history. Forest History Society Issues Series. McGill-Queen's University Press, Montreal, Quebec, Canada.

Dunlap, T. R. 1988. Saving America's wildlife. Princeton University Press, Princeton, New Jersey, USA.

Ehrlich, P. 1968. The population bomb. Ballantine Books, New York, New York, USA.

Emerson, R. W. 1836. Nature. James Monroe, Boston, Massachusetts, USA.

Flippen, J. B. 2003. Richard Nixon and the triumph of environmentalism. Pages 272–297 in L. S. Warren, editor. American environmental history. Blackwell, Malden, Massachusetts, USA.

Geist, V. 1995. North American policies of wildlife conservation. Pages 75–129 in V. Geist and I. McTaggert-Cowan, editors. Wildlife conservation policy. Detselig Enterprises, Calgary, Alberta, Canada.

Geist, V., S. Mahoney, and J. Organ. 2001. Why hunting has defined the North American model of wildlife conservation. Transactions of the North American Wildlife and Natural Resources Conference 66:175–185.

Gottlieb, R. 2003. Reconstructing environmentalism: complex movements, diverse roots. Pages 245–270 in L. S. Warren, editor. American environmental history. Blackwell, Malden, Massachusetts, USA.

Hardin, G. 1968. The tragedy of the commons. Science 162:1243–1248.

Hornaday, W. T. 1913. Our vanishing wildlife: its extermination and preservation. Kissinger Legacy Reprints, Whitefish, Montana, USA.

Kay, C. E. 1998. Are ecosystems structured from the top-down or bottom-up: a new look at an old debate. Wildlife Society Bulletin 23:484–498.

Krech, S., III. 1999. The ecological Indian: myth and history. W. W. Norton, New York, New York, USA.

Leopold, A. 1933. Game management. University of Wisconsin Press, Madison, USA.

Leopold, A. 1949. A Sand County almanac. Oxford University Press, New York, New York, USA.

Lewis, J. G. 2011. The Weeks Act at 100: the "organic act" of the eastern national forests. Forest Landowner 69:22–27.

Lopez-Hoffman, L., E. D. McGovern, R. G. Varady, and K. W. Flessa. 2009. Conservation of shared environments: learning from the U.S. and Mexico. University of Arizona Press, Tucson, USA.

Louv, R. 2006. Last child in the woods: saving our children from nature-deficit disorder. Algonquin Books, Chapel Hill, North Carolina, USA.

Louv, R. 2011. The nature principle. Algonquin Books, Chapel Hill, North Carolina, USA.

Lubowski, R. N., M. Vesterby, S. Bucholtz, A. Baez, and M. Roberts. 2006. Major uses of land in the United States, 2002. USDA Economic Information Bulletin May:1–54.

Lueck, D. 2008. Wildlife: sustainability and management. Pages 133–174 in R. A. Sedjo, editor. Perspectives on sustainable resources in America. Resources for the Future Press, Washington, D.C., USA.

Mackie, R. J. 2000. History of management of large mammals in North America. Pages 292–320 in S. Demarais and P. R. Krausman, editors. Ecology and management of large mammals in North America. Prentice-Hall, Upper Saddle River, New Jersey, USA.

Mann, C. C. 2006. 1491: new revelations of the Americas before Columbus. Vintage Books, New York, New York, USA.

National Wildlife Federation. 2010. Conservation directory. Island Press, Reston, Virginia, USA.

Organ, J. F., S. P. Mahoney, and V. Geist. 2010. Born in the hands of hunters: the North American model of wildlife conservation. The Wildlife Professional 4:22–27.

Organ, J. F., V. Geist, S. P. Mahoney, S. Williams, P. R. Krausman, et al. 2012. The North American Model of Wildlife Conservation. The Wildlife Society Technical Review 12-04. The Wildlife Society, Bethesda, Maryland, USA.

Peterson, R. T. 1934. A field guide to birds. Houghton Mifflin, Boston, Massachusetts, USA.

Pope, C. 2003. The politics of plunder. Pages 325–327 in L. S. Warren, editor. American environmental history. Blackwell, Malden, Massachusetts, USA. Reprinted from Boyer, P. 1990. Reagan as president: contemporary views of the man, his politics and his policies. Ivan R. Dee, Chicago, Illinois, USA.

Posewitz, J. 2010. The hunter's ethic: the past, the peril and the future. The Wildlife Professional 4:32–34.

Seton, E. T., and C. G. D. Roberts. 1909. Life histories of North American mammals: an account of the mammals of Manitoba. Volume 1. Grass eaters. Constable, London, United Kingdom.

Simonian, L. 1995. Defending the land of the jaguar: a history of conservation in Mexico. University of Texas Press, Austin, USA.

Taber, R. D., and N. F. Payne. 2003. Wildlife conservation and human welfare: a U.S. and Canadian perspective. Krieger, Malabar, Florida, USA.

Taylor, A. 2003. Wasty ways: stories of American settlement. Pages 102–124 in L. S. Warren, editor. American environmental history. Blackwell, Malden, Massachusetts, USA.

The Wildlife Society. 2010. The public trust doctrine: implications for wildlife management and conservation in the U.S. and Canada. Technical Review 10-01. Bethesda, Maryland, USA.

Thoreau, H. D. 1854. Walden. Tichnor and Fields, Boston, Massachusetts, USA.

Trefethen, J. B. 1975. An American crusade for wildlife. Boone and Crockett Club, Missoula, Montana, USA.

Wellock, T. R. 2007. Preserving the nation: the conservation and environmental movements 1870–2000. Harlan Davidson, Wheeling, Illinois, USA.

Wilson, R. L. 2009. Theodore Roosevelt, hunter-conservationist. Boone and Crockett Club, Missoula, Montana, USA.

3

THE WILDLIFE PROFESSIONAL

JOHN F. ORGAN

INTRODUCTION

Wildlife management is a young profession. Leopold (1933) identified precursors involving controls on hunting that date back to biblical times. Indeed, one might construe the first human attempts to domesticate wild animals circa 10,000 BP to be the original precursor of wildlife management (Diamond 1997). Wildlife management in its current form in North America arose out of a unique set of social and environmental circumstances. The wildlife conservation movement in North America began in earnest in the mid-19th century, when organized sport hunters paved the way for restrictive laws and regulations designed to eliminate commercial market hunting, a by-product of the Industrial Revolution, and to curtail "pot" hunting, the rural practice of shooting purely for food (Reiger 1975, Trefethen 1975).

Until around 1905 the dominant paradigm was to perpetuate wildlife through restrictive laws and regulations, predator control, protected areas, and restocking. Science as a tool in wildlife management was a novel concept at the onset of the 20th century. Early naturalists discovered, catalogued, and described wildlife. President Thomas Jefferson, in commissioning the Lewis and Clark Expedition, instructed them to describe "the animals of the country generally and especially those not known in the U.S.; the remains and accounts of any which may be deemed rare or extinct" (Lewis and Clark 2002:xxix). The Division of Economic Ornithology and Mammalogy was formed in the U.S. Department of Agriculture in 1885, and its first chief, C. Hart Merriam, dedicated its resources to the discovery and cataloging of birds and mammals of the western United States.

Many naturalists realized that species were not like planets and geologic strata; they were different because civilizations can destroy them. George Bird Grinnell, a Yale-educated naturalist, accompanied George Armstrong Custer on his first expedition into the Black Hills in present-day South Dakota in 1874. Nearly ten years later, Harvard-educated naturalist Theodore Roosevelt would embark on his own expedition to the Badlands of present-day North Dakota and publish a book about his experiences (Roosevelt 1885). Grinnell produced a review of the book (Grinnell 1885) that contained some criticisms, prompting a meeting between them. That meeting spawned a realization that exemplified the emerging paradigm shift among naturalists who recognized that civilization's conquest was leading to the extirpation of species.

Later, as president of the United States, Roosevelt articulated the need for scientifically based wildlife management in what Leopold termed the Roosevelt doctrine (Leopold 1933). In short, the Roosevelt doctrine states:

1. All outdoor resources are one integral whole.
2. Conservation through wise use is a public responsibility and private ownership is a public trust.
3. Science is the proper tool for discharging that responsibility.

Important milestones in the application of science to managing wildlife occurred in the 1920s with Herbert Stoddard's work on northern bobwhite (*Colinus virginianus*) in Georgia, and Aldo Leopold's game surveys funded by the Sporting Arms and Ammunition Manufacturers Institute. These science-based investigations led to another realization: restrictive laws, predator control, refuges, and stocking were not enough to stem the decline of wildlife. A program of active restoration was needed. Leopold chaired a committee of leading conservationists who published *The American Game Policy* (Leopold 1930), which called for advancements including the establishment of a wildlife management profession. At that time, there were no university programs teaching wildlife management or any organizations promoting professional standards. Within ten years, the first university programs would be established at the Universities of Wisconsin and Michigan, the first textbook on the topic published (Leopold 1933), and the first professional scientific society for wildlife biologists established (i.e., The Wildlife Society, or TWS, established in 1937).

At its inception as a profession, wildlife management was considered an art practiced by people with scientific training. During the early formative years of the wildlife management profession, three subdisciplines were recognized: game research, game administration, and game keeping (King 1938). This narrow scope was reflective of the dynamics surrounding the movement to refocus wildlife conservation on restoration programs. These subdisciplines were considered interrelated, yet many believed they could not be combined into a single undergraduate curriculum. Leopold (1939) described a wildlife professional as an individual with an intense conviction of the need for and usefulness of science as a tool for the accomplishment of conservation; the ability to diagnose the landscape to discern and predict trends in its biotic community and to modify them where necessary in the interest of conservation; knowledge of plants, animals, soil, and water; and familiarity with other professions and their influence and impact on the landscape. Wildlife professionals in these early years were primarily public employees, and the lack of private employment and private lands management was viewed as a weakness of the profession (Leopold 1940). Additionally, it was presumed that the wildlife professional was male, Caucasian, from a rural background, and a hunter.

During the 1950s, a second generation of wildlife professionals emerged, and the concept of the wildlife professional showed signs of maturation. McCabe (1954) redefined wildlife managers as wildlife ecologists and stated unequivocally that they are scientists and not artisans. He further stated that a wildlife professional must have an ethical code and sense of aesthetic values toward conservation as a whole in addition to knowledge and skills acquired from academic training. Murie (1954) expanded upon the notion of professional ethics and deemed it a responsibility of every member of the wildlife profession. Murie framed his argument in the context of ethical thinking that was ongoing in society at that time. The post–World War II era in America saw a new social consciousness brought about by recent genocide, nuclear proliferation, communist expansion, and a realization that the country was not as insulated from global conflict as once thought. Murie viewed this social consciousness as an effort by people to try to understand their proper place in nature, and he felt the highest calling of the wildlife professional was to contribute to that understanding. Leopold predicted that the fusion of wildlife biology and social sciences would be the outstanding accomplishment of the 20th century (Meine 1988); Murie viewed people to be on equal footing with animals, plants, soil, and water as the fundamental knowledge base and realm of responsibility for the wildlife professional.

During the 1960s, an increasing awareness of environmental issues arose in American society. A number of factors contributed to this, including the publication of Rachel Carson's *Silent Spring* (1962) and advances in media that brought issues of pollution and animal exploitation into people's homes (Organ et al. 1998). The number of nongovernmental wildlife organizations expanded from 56 in 1945 to more than 300 by the mid-1970s and to more than 400 by the 1980s (Dunlap 1988). In 1973 the Wildlife Management Institute published the North American Wildlife Policy in response to these social and environmental changes (Allen 1973). The policy called for greater federal oversight and broader environmental programs. Sweeping federal legislation enacted during the 1970s mandated clean air, clean water, protection of endangered species, and public input on federal actions that affect the environment. Many state fish and wildlife agencies became divisions or subunits of larger environmental agencies. The ranks of wildlife professionals were no longer confined to wildlife biologists, wildlife researchers, wildlife educators, wildlife administrators, wildlife managers, and wildlife law enforcement officers. The expanding profession now included within its ranks wildlife damage management specialists, wildlife toxicologists, wildlife pathologists, land-use planners, geographic information system analysts, statisticians, wetland scientists, community ecologists, wildlife veterinarians, and other practitioners. Human dimensions of wildlife management emerged as a formal discipline with sociologists, resource economists, political scientists, and cultural anthropologists contributing to wildlife management (Decker et al. 2001).

In the 21st century the term "wildlife professional" might be appropriately applied to professionals ranging from traditional wildlife biologists to filmmakers. Additionally, in contrast to the early years of the profession, the human diversity of the field has begun to broaden. Increased participation by a wider range of people from different backgrounds (e.g., gender, sexual orientation, race, ethnicity, culture, socioeconomic status, and others) promises to add valuable talent and insights as wildlife professionals are pressed to manage in evermore challenging landscapes and situations. Recent years have seen the number of women entering the workforce equal or exceed men, and with the majority of Americans living in urban areas, the proportion of wildlife professionals coming from cities has increased substantially. Ethnic and racial diversity within the wildlife profession increased also, although it remains far less than society in general. Data for members of TWS (Table 3.1) indicate females and minorities are underrepresented relative to the current labor force, although a substantial proportion of members did not identify their gender or ethnicity, and

Table 3.1. Percentage of the demographic composition of The Wildlife Society membership in 2011 in comparison with rough averaged data for the current labor force

Demographic	The Wildlife Society	Current labor force
Caucasian	56.7	38.0
Members who did not identify their ethnicity	38.7	
Other	1.3	
Hispanic	1.3	6.0
Asian	1.0	2.0
Native American/indigenous	0.5	0.3
African American	0.4	5.0
Male	45.0	53.0
Members who did not identify their sex	32.7	
Female	22.3	47.0

Source: Data courtesy of Ankit Mehta, The Wildlife Society; http://www.fws.gov/humancapital/pdf/MD_715_Servicewide_Tables_2010.pdf.

Table 3.2. Percentages of diversity trends within the Wildlife Biologist Series in the U.S. Fish and Wildlife Service, 1999–2011

Demographic	1999	2011	2011 labor force
Black	1.2	0.5	3.0
Hispanic	4.1	5.2	4.1
Asian	1.4	1.7	8.4
Native American	1.7	1.7	0.6
Women	27.7	35.5	44.1
Minority women	2.5	3.5	9.1
Total minority	8.5	9.9	17.7

Source: Courtesy of Charles Davis, statistician, U.S. Fish and Wildlife Service.

response bias is unknown. Gender and ethnicity within the Wildlife Biologist Series (GS-482) for U.S. Fish and Wildlife Service (USFWS) employees from 1999 to 2011 have not been representative of the American population from 1999 to 2011 (Table 3.2). These data represent less than 10% of the total USFWS workforce, and other series, such as Fish and Wildlife Biologist, Refuge Manager, and Biological Technician, are not represented. Declines in some categories could represent movement into other series. Nevertheless, these data suggest that women and minorities are underrepresented as wildlife biologists. Native Americans are the exception and are overrepresented relative to the current labor force (Table 3.2).

The wildlife profession in its current state, as indicated above, is broad and encompasses many disciplines. The focus hereafter will be on the core discipline (i.e., wildlife biology) and the educational requirements, ethical responsibilities, and employment opportunities associated with it. Part of being a professional involves maintaining currency with science, policy, and ethical standards. Membership in professional scientific societies is an indicator of one's professionalism.

THE WILDLIFE BIOLOGIST

Wildlife biology by its nature is an integrative discipline. It is a mixture of whole-animal biology, ecology, zoology, botany, policy, and social science with grounding in life, physical, and quantitative sciences. Background in forestry, range science, marine biology or other disciplines might be necessary depending on the species or landscape one focuses on. TWS has established educational requirements necessary to become a certified wildlife biologist. These requirements are fluid and subject to change; current information can be found at www.wildlife.org. Coursework requirements are nested within six core areas:

1. Biological sciences. Thirty-six semester credit hours in biological sciences are required and must include courses in the following subcategories:
   a. Wildlife management (six hours). Wildlife management courses focus on the principles and techniques of managing wildlife. Traditional wildlife biology curricula typically have a course dedicated to principles

of wildlife management and one dedicated to wildlife management techniques. A principles course covers population dynamics theory (e.g., population growth rates, *r* versus *K* species, compensatory versus additive mortality), habitat management principles (e.g., structural requirements, species/area dynamics, successional patterns, island biogeography), history, policies, laws, and other components that form the scientific basis for wildlife management. A techniques course will introduce students to the tools used to manage wildlife and how to use them. This can include capture devices, chemical immobilization, radio telemetry, geographic positioning systems, habitat mapping, survey and monitoring techniques, and other field and laboratory efforts. Courses in conservation biology can be considered wildlife management if they contain a specific focus on management and decision making.
   b. Wildlife biology (six hours). Wildlife biology courses focus on the biology and behavior of birds, mammals, reptiles, or amphibians. These courses should provide the student with an understanding of species biology and habitat requirements sufficient to provide a basic knowledge of management needs. As least one course in this category must deal solely with the science of mammalogy, ornithology, or herpetology.
   c. Ecology (three hours). A course in general animal or plant ecology will meet this requirement, which provides the wildlife biologist with an understanding of the interrelationships within and among ecological systems.
   d. Zoology (nine hours). Courses in taxonomy, biology, behavior, physiology, anatomy, and natural history of vertebrates and invertebrates will meet this requirement. A wildlife biologist should have a basic understanding of animal biology, systematics, and evolutionary mechanisms.
   e. Botany (nine hours). Courses in general botany, plant genetics, plant morphology, plant physiology, dendrology, or plant taxonomy will meet this requirement. A background in botany is essential for the wildlife biologist to understand and manage habitat.
2. Physical sciences. Nine credit hours in physical sciences are required and can include such courses as chemistry, physics, geology, or soil science, with at least two disciplines represented. A background in physical sciences is important for the wildlife biologist to understand ecosystem processes and physiological mechanisms.
3. Quantitative sciences. Nine semester hours in quantitative sciences are required. These must include:
   a. Basic statistics (three hours). A course in basic statistics is essential for a wildlife biologist. It would be rare to go through a career in wildlife biology without having to collect and analyze data, and even rarer to get an advanced degree in wildlife biology without doing so. Without this grounding, the wildlife biologist will be ill equipped to be a scientist.

b. Quantitative sciences (six hours). This can include courses in calculus, biometry, advanced algebra, systems analysis, mathematical modeling, sampling, and computer science. For the wildlife biologist who desires a career in research, expertise in quantitative sciences will pay dividends. White (2001) provides a compelling argument for the importance of quantitative science in the training of a wildlife biologist.

4. Humanities and social sciences. Nine credit hours in humanities and social sciences are required. This includes courses such as economics, sociology, psychology, political science, government, history, literature, or foreign language. Murie (1954:293) cautioned that "our training in universities should be such that we do not come out pretty good technicians but philosophical illiterates." Study in humanities is essential for development of critical thinking ability, a characteristic that will serve one well in any profession. Social science background has become a necessity in the wildlife profession (Organ et al. 2006). Many wildlife agencies have social scientists on their staff in recognition of this need. The wildlife profession, in striving to maintain relevancy within a changing society (Jacobson et al. 2010), increasingly applies to social science, with equal rigor in biological science, toward understanding wildlife stakeholder attitudes, values, normative behaviors, and preferences. Decision-making processes that integrate biological and social sciences are becoming more rigorous and transparent (Riley et al. 2003, Williams et al. 2009).

5. Communications. Twelve credit hours are required in courses designed to improve communication skills. This can include courses in English composition, technical writing, journalism, public speaking, and media. Effective communication is important in every aspect of a wildlife professional's work, but wildlife agencies are increasingly looking for individuals with skills who can assist them in communicating with a broadening group of stakeholders through diverse information streams.

6. Policy, administration, and law. Six credit hours are required, with courses focusing on natural resource policy, administration, wildlife or environmental law, or natural resource or land-use planning. The wildlife professional should have an understanding of the legal bedrock for wildlife conservation. In the United States, this is provided for in the Public Trust Doctrine (Geist and Organ 2004), the U.S. Constitution, and various state constitutions and laws (Batcheller et al. 2010). The wildlife professional should be aware of how law and policy differ, and how each is developed.

Graduate education has become a virtual necessity for the student desiring employment as a wildlife biologist. The diverse and specialized needs in the field of wildlife conservation make it very difficult for an individual with an undergraduate degree to compete, even in entry-level positions, with those who have advanced degrees and specialized training. During the course of undergraduate studies, the student should assess his or her interests and desires, and seek graduate education that will afford the skills and experience necessary for entering the workforce and embarking on a career directed toward those goals.

Conservation biology emerged as a discipline around 1978 and, by 1986 the Society for Conservation Biology had formed (Hunter 1996). Conservation biology is the applied science of maintaining the earth's biological diversity. As such, it is a multidisciplinary science that ideally integrates biology, ecology, physical sciences, and social sciences (economics, sociology, anthropology, political science) in efforts to sustain natural ecosystems and biological processes (Meffe and Carroll 1994). Conservation biology principles are an essential part of a wildlife professional's knowledge base.

PROFESSIONAL BEHAVIOR AND THE WILDLIFE PROFESSION'S CODE OF ETHICS

McCabe (1954) and Murie (1954) articulated the importance of professional ethics in the wildlife profession. TWS developed a code of ethics for its members (http://joomla.wildlife.org/documents/TWS_bylaws.pdf). The code demands that wildlife professionals:

1. Uphold the dignity and integrity of the wildlife profession. They shall endeavor to avoid even the suspicion of dishonesty, fraud, deceit, misrepresentation, or unprofessional demeanor.

2. Refrain from plagiarism in verbal or written communications and shall give credit to the works and ideas of others.

3. Refrain from fabrication, falsification, or suppression of results, and shall not deliberately misrepresent research findings, or otherwise commit scientific fraud.

4. Exercise high standards in the care and use of live vertebrate animals used for research, in accordance with accepted professional guidelines for the respective classes of animals under study.

5. Protect the rights and welfare of human subjects used in research and obtain the informed consent of those individuals, in accordance with approved professional guidelines for human subjects.

6. Be mindful of their responsibility to society, and seek to meet the needs of all people when seeking advice in wildlife-related matters. They shall studiously avoid discrimination in any form, or the abuse of professional authority for personal satisfaction.

7. Recognize and inform clients or employers of the wildlife professional's prime responsibility to the public interest, conservation of the wildlife resource, and the environment. They shall exercise professional judgment, and avoid actions or omissions that may compromise these broad responsibilities. They shall cooperate fully with other professionals in the best interest of the wildlife resource.

ROBERT A. MCCABE (1914–1995)

At a time when specialization became vogue in the wildlife profession, Robert McCabe was asked to describe his specialty. Without hesitation, he responded that he was a generalist, a matter-of-fact reply that typified him.

Bob McCabe was a product of impoverished South Milwaukee, Wisconsin. Of Germanic/Irish heritage, he strove to succeed at the few opportunities available during the Depression era. He proved to be a gifted athlete, but only an average student. With the same drive that earned him athletic achievements, McCabe sought to escape his hardscrabble neighborhood. His determination took him to higher education at Carroll College in Waukesha, Wisconsin, where he discovered three passions. First was Marie Stanfield, his wife for 54 years. Second was rabbit hunting. Third was an abiding curiosity in nature that fostered his academic enthusiasm.

McCabe went on to attend graduate school at the University of Wisconsin–Madison. Applying for admission in 1939 to zoological study in game conservation, he was directed to "see Leopold on the Ag campus." He did so reluctantly, but it was a fortuitous, momentous encounter. Aldo Leopold's graduate student, protégé, assistant, and occasional hunting companion, McCabe eventually succeeded him as chair of the department of wildlife management. More important was the deep friendship they shared—the inspiration for McCabe's career and a daily guidepost until his passing in 1995. In his copy of *A Sand County Almanac,* Leopold's wife poignantly inscribed: "A son by affection to Aldo."

During his 27 years as department chair, McCabe dedicated himself to enhancing its reputation of academic excellence by accepting only the most promising graduate students; hiring faculty of high intelligence, character, and outdoor experience; and by brooking no challenge, institutional or otherwise, to the Leopold legacy. Although an administrator, advisor, and teacher, McCabe relished being a researcher. The most enduring of his work concerned the alder flycatcher, subject of his book *The Little Green Bird* (1991), published several years after his cathartically crafted biography *Aldo Leopold: The Professor* (1987).

McCabe was an inveterate, indefatigable pursuer of upland game birds, particularly ruffed grouse and woodcock. He was a very good shot, nonpareil as a hunter. Time afield with him was a treat for his sons and his few other, favored hunting companions. A Fulbright scholar, advisor on international wildlife matters (Ireland, Africa, Canada, Russia), recipient of honorary degrees (Carroll College and University of Dublin), The Wildlife Society president, conferee of the Aldo Leopold Medal (1986) and other honors, McCabe most cherished his time with family, especially at Rusty Rock, their farm in Iowa County, Wisconsin.

To some, he was seen as a taciturn, even stern taskmaster. He insisted on quality effort in all endeavors and was unimpressed with those he deemed prima donnas and pretenders. To friends and family, Bob McCabe was sensitive, fun, and funny. And he was as unstintingly devoted to them as he was to his profession.

Photo courtesy of Richard E. McCabe

8. Provide maximum possible effort in the best interest of each client or employer, regardless of the degree of remuneration.

9. Accept employment to perform professional services only in areas of their own competence, and consistent with the code of ethics. They shall seek to refer clients or employers to other natural resource professionals when the expertise of such professionals shall best serve the interests of the public, wildlife, and the client or employer.

10. Maintain a confidential relationship between professionals and clients or employers except when specifically authorized by the client or employer or required by due process of law or the code of ethics to disclose pertinent information. They shall not use such confidences to their personal advantage or to the advantage of other parties, nor shall they permit personal interests or other client or employer relationships to interfere with their professional judgment.

11. Refrain from advertising in a self-laudatory manner—beyond statements intended to inform prospective clients or employers of one's qualifications—or in a manner detrimental to fellow professionals and the wildlife resource. They shall clearly distinguish among facts, hypotheses, and opinions. They shall provide professional advice and guidance only when qualified to do so by training and experience.

12. Refuse compensation or rewards of any kind intended to influence their professional judgment or advice or to secure preferential treatment. They shall not permit a person who recommends or employs them, directly or indirectly, to regulate or impair their professional judgment. They shall not accept compensation for the same

professional services from any source other than the client or employer without prior consent of all the clients or employers involved.

13. Avoid performing professional services for any client or employer when such service is judged to be contrary to the code of ethics or detrimental to the well-being of wildlife resources and their environments. If a wildlife professional believes that his or her employment activities conflict with the code of ethics, that person shall advise the client or employer of such conflict.

14. Advise against an action by a client or employer that violates any statute or regulation.

TWS has provisions for enforcing this code among its members, and violations could result in censure or suspension from membership in the society. An ethics board appointed by the society president will review allegations of ethical misconduct and will follow a process outlined in the society's bylaws in determining whether a violation occurred.

EMPLOYMENT OPPORTUNITIES FOR THE WILDLIFE BIOLOGIST

Employment within the wildlife profession is skewed toward the public sector, but private sector employment today represents a large percentage of available jobs. In recent decades, the growth of private lands, hunting-for-fee facilities, environmental consulting firms, wildlife organizations, and other private endeavors has broadened the employment field. In 2011, 22.1% of TWS members were employed in the private sector (i.e., consulting firm, nonprofit organization, corporation, self-employed; A. Mehta, TWS, personal communication). Another 20.7% of TWS members were employed by federal, state, or provincial authorities; 21.6% by universities (Table 3.3).

Students and others desiring careers as a wildlife biologists must be prepared for competition. Students should take advantage of opportunities to volunteer on research projects within their academic department. These opportunities will help develop applied skills that can enhance résumés and make an impression on potential future employers or references.

Table 3.3. Employment category data for members of The Wildlife Society, 2011

Employment category	Percentage of total
University	21.6
Federal	20.7
State or province	20.2
Consulting firm	9.1
Nonprofit organization	6.1
Retired	4.8
Corporation	3.5
Self-employed	3.4
Other	10.6

Source: Data courtesy of Ankit Mehta, The Wildlife Society.

Summer seasonal work opportunities are a valuable means to develop skills and to make contacts within agencies and organizations that could ultimately lead to full-time employment. Seasonal and full-time employment opportunities can be found on several Internet sites. Federal job opportunities (seasonal and full time) are listed at www.usajobs.gov. Texas A&M University's Department of Wildlife and Fisheries Sciences maintains an extensive employment website at wfscjobs .tamu.edu, and TWS has a job board at www.wildlife.org (career center). Outlined below is an overview of the major entities that employ wildlife biologists in the United States.

Federal Government

The legal public trust authority for most species of wildlife rests with the states; the federal government has primary legal authority for species captured within one of three clauses of the U.S. Constitution: the property clause (wildlife on federally owned land such as national wildlife refuges), the commerce clause (wildlife affected by interstate commerce such as trade in endangered species), and the supremacy clause, which makes federal treaties supreme over any other law of the land (e.g., the Migratory Bird Treaty of 1916).

U.S. Department of Agriculture

The USDA contains several bureaus with responsibilities affecting fish and wildlife management. The USDA has tremendous influence over management of private agricultural lands and public lands that collectively represent significant wildlife habitat.

U.S. Forest Service

The USFS manages a nationwide network of 78 million ha of forests and grasslands. Its mission is to sustain the health, diversity, and productivity of the nation's forests and grasslands to meet the needs of present and future generations. The USFS has a Watershed, Fish, Wildlife, Air, and Rare Plants Program that coordinates management activities and research for fish, wildlife, and endangered species nationwide. National forests are managed for multiple uses, and hunting and fishing are primary uses, with seasons and limits typically adhering to state regulations.

Animal and Plant Health Inspection Service: Wildlife Services

Wildlife Services provides leadership in managing human–wildlife conflicts. The agency has state-based field staff and major research facilities that develop tools and techniques for managing wildlife conflicts. Wildlife damage management has matured as a major subdiscipline within the wildlife profession, and Wildlife Services provides leadership in this dimension.

Animal and Plant Health Inspection Service: Veterinary Services

Veterinary Services works to protect and improve the health, quality, and marketability of our nation's animals, animal products, and veterinary biologics by preventing, controlling, and eliminating animal diseases, and monitoring and promoting animal health and productivity. Veterinary Services addresses wildlife zoonoses, such as avian influenza, that have potential to impact domestic agricultural animals.

Natural Resources Conservation Service
The NRCS works with landowners through conservation planning and assistance designed to benefit the soil, water, air, plants, and animals that result in productive lands and healthy ecosystems. Funding for much of NRCS activities comes through the federal Farm Bill.

Department of Commerce
The Department of Commerce, through the National Oceanic and Atmospheric Administration and its bureau the National Marine Fisheries Service (NMFS), has federal authority over certain species of marine organisms and regulates commerce in marine fisheries. Under the Marine Mammal Protection Act of 1972, NMFS has authority over the order Cetacea and all members, except walruses (*Odobenus rosmarus*), of the order Carnivora. The NMFS supports regional fisheries management councils and commissions that work collaboratively to regulate commerce in marine fisheries.

Department of Defense
The Department of Defense employs wildlife professionals on many of its installations, where they coordinate efforts to manage wildlife populations on their extensive land holdings, including hunting programs and endangered species recovery efforts. Wildlife professionals also provide guidance on the impacts of proposed field activities and exercises on wildlife resources. Information can be found at http://www.nmfwa.org/.

U.S. Army Corps of Engineers
The USACE has jurisdictional responsibility for portions of the federal Clean Water Act. It employs wildlife professionals to implement portions of the regulatory program overseeing impacts to navigable waterways, including wetlands, by land and water development activities.

U.S. Department of the Interior
The DOI has several bureaus with management responsibility for fish and wildlife. Similar to the USDA, the DOI has influence over a significant portion of public and private lands and some regulatory authority over land and water development activities.

U.S. Fish and Wildlife Service
The USFWS is the largest fish and wildlife conservation agency in the world. Its mission is to work with others to conserve, protect, and enhance fish, wildlife, plants, and their habitats for the continuing benefit of the American people. The USFWS manages a network of national wildlife refuges, fish hatcheries and fishery resource stations, law enforcement offices, and other field stations that provide a multitude of

Figure 3.1. Wildlife biologists Adam Vashon and Jennifer Vashon of the Maine Department of Inland Fisheries and Wildlife conducting research on Canada lynx (*Lynx canadensis*). Photo courtesy of John Organ

conservation services, including endangered species recovery activities, wetland protection, environmental contaminants, and private landowner conservation assistance (Fig. 3.1). The USFWS also has an international affairs office that assists conservation efforts abroad and oversees activities in the United States, and its Division of Wildlife and Sport Fish Restoration provides federal aid to state fish and wildlife agencies.

Bureau of Land Management
The mission of the BLM is to sustain the health, diversity, and productivity of public lands for the use and enjoyment of present and future generations. The BLM's Fish, Wildlife, and Plant Conservation Program oversees management activities on lands under its control.

National Park Service
The NPS manages a network of national parks across the United States. The NPS's Biological Resource Management Division provides policy, planning, and operational support to NPS personnel concerning the management of native vegetation and wildlife resources, the control of nonnative species, and the biological restoration of disturbed ecosystems. It formulates biological resource policy recommendations and conducts legislative, regulatory, and environmental reviews related to biological resource protection. It assists the field in vegetation and wildlife resource management activities.

U.S. Geological Survey
The USGS's Biological Resource Division provides science support to other DOI bureaus and partners. It has a network of science centers and cooperative research units that specialize in applied fish and wildlife research.

State Government
There are 56 states, territories, and insular areas (including the District of Columbia) within the United States, and each has an agency or bureau dedicated to the conservation and

management of fish and wildlife resources. Most species of wildlife, except for those reserved by (i.e., fall within federal oversight) the U.S. Constitution and its amendments (e.g., migratory birds, marine mammals, endangered species), are under the legal jurisdiction of the states. State fish and wildlife agencies are typically governed by a board or commission, or by a politically appointed director. Many state fish and wildlife bureaus are nested within larger environmental or natural resource agencies; a few, such as the Pennsylvania Game Commission and Idaho Fish and Game, are independent agencies. State fish and wildlife agencies have broad responsibilities ranging from research, management, and law enforcement to environmental review, education, policy, and administration.

Historically, these agencies focused on game species, but recent decades have seen an increase in staffing and effort for endangered species and wildlife diversity programs (Fig. 3.2). The development of state wildlife action plans in 2005 and funding from the State Wildlife Grants Program have enhanced the capacity of state fish and wildlife agencies to broaden programs (http://www.wildlifeactionplans.org/). Major funding for state fish and wildlife agencies comes from hunting license revenues and the Pittman-Robertson Wildlife Restoration Program, with funding derived from federal excise taxes on firearms, ammunition, handguns, and archery equipment (http://wsfrprograms.fws.gov/).

Wildlife biologists can find employment in other state government agencies, too, including state conservation and recreation agencies focused on parks and forests, and transportation and utility departments where biologists assist in planning and mitigation of adverse impacts of roads and other infrastructure on wildlife.

Tribal Nations

Federally recognized Indian tribes within the lower 48 states have jurisdiction over a reservation land base of more than 21 million ha (130,759 km^2). Alaskan Native lands comprise another 18 million ha. Some tribes control resources outside of reservations due to federal court decisions and voluntary cooperative agreements that allow a comanagement status between tribes and states. These lands are called ceded and usual and accustomed areas, and they equal over 15 million ha. In these areas, tribes maintain comanagement jurisdiction for fisheries and wildlife management and utilization. Tribal lands coupled with the ceded and usual and accustomed areas total a natural resource base of over 226,314 km^2, containing more than 295,420 ha of lakes and impoundments, and over 16,093 km of streams and rivers. This land combined would constitute the fifth-largest state in the United States. Many tribal nations employ wildlife professionals to manage their natural resources. Additional information can be found at http://www.nafws.org/.

Private Sector

Private sector employment for wildlife biologists is broad and diversified in contrast to the early years of the profession. Land management firms and private individuals hire wildlife biologists to manage wildlife for hunting, to restore rare species, and to ensure extractive uses comply with environmental regulations. Many environmental consulting firms provide services to private individuals and corporations in planning and environmental review and compliance (Fig. 3.3). In the southeastern United States, a number of private wildlife consulting firms specialize in developing land management programs for private land holdings to enhance hunting opportunities.

There are numerous nongovernmental wildlife organizations that focus on advocacy, policy, science, or some combination of the three. Depending upon the mission of the organization, a wildlife biologist employed in this realm could be immersed in legal court actions, working the halls of federal

Figure 3.2. A U.S. Fish and Wildlife Service wildlife biologist bands an endangered red-cockaded woodpecker (*Picoides borealis*). Photo courtesy of the U.S. Fish and Wildlife Service

Figure 3.3. Private consultant wildlife biologists work with individuals and corporations to ensure land management activities minimize adverse impacts to wildlife habitat. Photo courtesy of Jon Haufler

or state capitol buildings, or designing and implementing field research investigations. The larger national (and international) organizations include professional scientific societies such as TWS, professional organizations such as the Association of Fish and Wildlife Agencies and the Wildlife Management Institute, and member-based groups such as the Boone and Crockett Club, Defenders of Wildlife, Ducks Unlimited, National Wildlife Federation, National Wild Turkey Federation, National Audubon Society, Pheasants Forever, Quail Unlimited, Quality Deer Management Association, Rocky Mountain Elk Foundation, Safari Club International, The Nature Conservancy, Whitetails Unlimited, and the Wildlife Conservation Society. Many other groups operate at the regional and local levels.

In some parts of the United States, leasing lands for hunting has become big business. In many cases, wildlife biologists are employed to manage populations and habitats on these properties to optimize opportunity and quality of hunting for big game, upland game birds, and waterfowl. Within the wildlife profession, concern has been raised over the implication of this trend relative to privatizing wildlife (Organ and Batcheller 2009, Batcheller et al. 2010).

Wildlife veterinary work is a distinct subdiscipline within the wildlife profession (Fig. 3.4). Amendments to the Animal

Figure 3.4. Wildlife veterinarians provide training and expertise in capturing and handling wildlife for research and management purposes, in addition to other aspects of conservation medicine. Photo courtesy of Dave Jessup

Welfare Act promulgated in 1985 require veterinarians with expertise on wildlife for certain activities regulated under the act. Many federal and state conservation agencies, zoos, and private facilities employ wildlife veterinarians. Many zoos are active in captive breeding programs for endangered wildlife and employ wildlife veterinary professionals. The field of wildlife veterinary science has become established at many leading veterinary schools. The American Association of Wildlife Veterinarians (http://www.aawv.net/) provides a forum for the advocacy of veterinary medicine within the wildlife conservation field. Conservation medicine is gaining traction as an area of inquiry and management necessary for addressing contemporary conservation challenges. The emergence of white-nose syndrome in bats, chronic wasting disease in ungulates, avian influenza, and other diseases suggests there will be many opportunities for wildlife professionals interested in pursuing careers in wildlife veterinary science and conservation medicine.

Universities

Universities represent a major pillar within the wildlife profession (Gill 1996). The National Association of University Fisheries and Wildlife Programs (http://naufwp.org/index .html) lists 63 member colleges and universities in the United States, and this does not include many programs at smaller state colleges and private institutions. As a rule, faculty positions require a Ph.D. at a minimum. Many programs maintain a professional technical staff to assist in field and laboratory research.

The Cooperative Research Units Program (http://www .coopunits.org/Headquarters/) was established in 1935 to enhance graduate education in fisheries and wildlife sciences and to facilitate research between natural resource agencies and universities on topics of mutual concern. Today, there are 40 cooperative research units in 38 states. Each unit is a partnership among the U.S. Geological Survey, a state natural resource agency, a host university, and the Wildlife Management Institute. Staffed by federal personnel, cooperative research units conduct research on renewable natural resource questions, participate in the education of graduate students, provide technical assistance and consultation on natural resource issues, and provide continuing education for natural resource professionals.

SUMMARY

The wildlife professional of today is part of a diverse network of men and women who have expertise in a variety of disciplines and who work in virtually every employment sector. What was once a profession focused on restoration of game species by government biologists is now much too broad to characterize in simple terms. Individuals interested in a career as a wildlife professional would be well advised to focus their undergraduate education on meeting requirements for becoming a certified wildlife biologist, and then focusing their postgraduate education on a specialty field, whether it be wildlife population or habitat management, human dimensions, law, policy, medicine, or otherwise. An individual with solid educational grounding and a demonstrated commitment to the profession as evidenced by participation in professional scientific societies will be better able to compete for professional wildlife jobs.

Literature Cited

Allen, D. L. 1973. Report of the committee on North American wildlife policy. Wildlife Society Bulletin 1:73–92.

Batcheller, G. R., M. C. Bambery, L. Bies, T. Decker, S. Dyke, et al. 2010. The public trust doctrine: implications for wildlife management and conservation in the United States and Canada. Technical Review 10-1. The Wildlife Society, Bethesda, Maryland, USA.

Carson, R. 1962. Silent spring. Houghton Mifflin, Boston, Massachusetts, USA.

Decker, D. J., T. L. Brown, and W. E. Siemer. 2001. Human dimensions of wildlife management in North America. The Wildlife Society, Bethesda, Maryland, USA.

Diamond, J. M. 1997. Guns, germs, and steel: the fate of human societies. Norton, New York, New York, USA.

Dunlap, T. R. 1988. Saving America's wildlife. Princeton University Press, Princeton, New Jersey, USA.

Geist, V., and J. F. Organ. 2004. The public trust foundation of the North American model of wildlife conservation. Northeast Wildlife 58:49–56.

Gill, R. B. 1996. The wildlife professional subculture: the case of the crazy aunt. Human Dimensions of Wildlife 1:60–69.

Grinnell, G. B. 1885. New publications: hunting trips of a ranchman. Forest and Stream 24:450–451.

Hunter, M. L., Jr. 1996. Fundamentals of conservation biology Blackwell Science, Cambridge, Massachusetts, USA.

Jacobson, C. A., J. F. Organ, D. J. Decker, G. R. Batcheller, and L. Carpenter. 2010. A conservation institution for the 21st century: implications for state wildlife agencies. Journal of Wildlife Management 74:203–209.

King, R. T. 1938. What constitutes training in wildlife management. Transactions of the North American Wildlife and Natural Resources Conference 3:548–557.

Leopold, A. 1930. Report to the American game conference on an American game policy. Transactions of the American Game Conference 17:281–283.

Leopold, A. 1933. Game management. Charles Scribner's Sons, New York, New York, USA.

Leopold, A. 1939. Academic and professional training in wildlife work. Journal of Wildlife Management 3:156–161.

Leopold, A. 1940. The state of the profession. Journal of Wildlife Management 4:343–346.

Lewis, M., and W. Clark. 2002. The journals of Lewis and Clark. National Geographic Society, Washington, D.C., USA.

McCabe, R. A. 1954. Training for wildlife management. Journal of Wildlife Management 18:145–149.

Meffe, G. K., and C. R. Carroll. 1994. Principles of conservation biology. Sinauer, Sunderland, Massachusetts, USA.

Meine, C. 1988. Aldo Leopold: his life and work. University of Wisconsin Press, Madison, USA.

Murie, O. J. 1954. Ethics in wildlife management. Journal of Wildlife Management 18:289–293.

Organ, J. F., and G. R. Batcheller. 2009. Reviving the public trust doctrine as a foundation for wildlife management in North America. Pages 161–171 in M. J. Manfredo, J. J. Vaske, P. J. Brown, D. J. Decker, and E. A. Duke, editors. Wildlife and society: the science of human dimensions. Island Press, Washington, D.C., USA.

Organ, J. F., D. J. Decker, L. H. Carpenter, W. F. Siemer, and S. R. Riley. 2006. Thinking like a manager: reflections on wildlife management. Wildlife Management Institute, Washington, D.C., USA.

Organ, J. F., R. M. Muth, J. E. Dizard, S. J. Williamson, and T. A. Decker. 1998. Fair chase and humane treatment: balancing the ethics of hunting and trapping. Transactions of the North American Wildlife and Natural Resources Conference 63:528–543.

Reiger, J. F. 1975. American sportsmen and the origins of conservation. Winchester, New York, New York, USA.

Riley, S. J., W. F. Siemer, D. J. Decker, L. H. Carpenter, J. F. Organ, and L. Berchielli. 2003. Adaptive impact management: an integrative approach to wildlife management. Human Dimensions of Wildlife 8:81–95.

Roosevelt, T. 1885. Hunting trips of a ranchman. G. P. Putnam's Sons, New York, New York, USA.

Trefethen, J. B. 1975. An American crusade for wildlife. Winchester, New York, New York, USA.

White, G. C. 2001. Why take calculus? Rigor in wildlife management. Wildlife Society Bulletin 29:380–386.

Williams, B. K., R. C. Szaro, and C. D. Shapiro. 2009. Adaptive management: the U.S. Department of the Interior technical guide. Adaptive Management Working Group, U.S. Department of the Interior, Washington, D.C., USA.

HUMAN DIMENSIONS OF WILDLIFE MANAGEMENT

DANIEL J. DECKER, SHAWN J. RILEY, AND WILLIAM F. SIEMER

INTRODUCTION

Wildlife management systems are usually depicted as having three key elements: humans, wildlife, and wildlife habitats (Giles 1978; Fig. 1.1). Wildlife management addresses the interacting and interdependent components and processes that comprise such systems in their entirety, not just one part or another. Everything in a wildlife management system that is not about wildlife and habitat is about humans: individuals, groups, social structures, cultural systems, communities, and institutions. These components are the "human dimensions" of wildlife management. For purposes of this chapter, the concept of human dimensions does not include human resource management or personnel management functions in organizations, yet the topic of organizational and personnel alignment with appropriate tasks is an emerging area of study as agency transformation occurs in response to changes in society and funding for conservation. Human dimensions can be described as discovering and applying insight about how humans value wildlife, how humans want wildlife to be managed, and how humans affect or are affected by wildlife and wildlife management decisions.

Decker et al. (2012) address human dimensions considerations, from broad social ideals (e.g., about governance) to specific traits of individuals (e.g., motivations to hunt). In this chapter we introduce core principles and practices of human dimensions inquiry and describe applications of human dimensions insights in wildlife management. The idea of "impacts" is central because it reinforces the human purposes for wildlife management, and impacts are often topics of human dimensions research and engagement with stakeholders (i.e., any persons who significantly affect or are affected by wildlife or wildlife management directly or indirectly; Riley et al. 2002).

In most cases any approach taken to manage wildlife reflects the governance structure that the level of jurisdiction has adopted or has been mandated by law to follow. Because wildlife is a public trust resource in the United States, governments are responsible for wildlife management, a responsibility typically executed through state and federal wildlife agencies. Increasingly, however, partnerships among multiple levels of government (including local governments) and nongovernmental organizations (NGOs) are forged to achieve conservation through a form of participatory governance referred to as collaborative conservation (Lauber et al. 2011). To date, most efforts in human dimensions of wildlife management have focused on acquisition and application of insights from social sciences that enable and support governance of wildlife resources. Demand for human dimensions insight grows, and the sophistication of human dimensions research and application increases, as number and diversity of stakeholder interests in wildlife governance and management increases.

Humans and wildlife affect each other within complex, coupled sociocultural and ecological systems. Human experiences with wildlife can be direct or indirect, vary in intensity and type, and occur at many scales. Wildlife management enhances, regulates, or prohibits various experiences people might have with wildlife. Outcomes of these experiences are typically described as effects, the most important of which (i.e., those typically generating strong stakeholder reactions and prompting management attention) are impacts (Riley et al. 2002). Impacts take many forms and can be positive or negative (e.g., economic benefits or costs; threats to or enhancement of human health and safety; ecological services that wildlife provide; physical, mental, and social benefits produced by recreational enjoyment of wildlife). Impacts arise from many kinds of interactions between humans and wildlife, and interactions among humans because of wildlife. Understanding these interactions is one purpose of human dimensions research and citizen engagement practices. Influencing these interactions is the function of wildlife management. Informing management decisions and actions with relevant insight from human dimensions inquiry is the role of the wildlife manager. Taking this human values perspective, wildlife management can be defined as the guidance of decision-making processes and implementation of practices to influence interactions between people, wildlife, and wildlife habitats, and among people about wildlife to achieve impacts valued by stakeholders.

The emphasis of wildlife management is similar to any management activity: to turn complexity, information, and specialized activity into value-producing performance (Margretta 2002). Wildlife management is aimed at production of value defined by and for society, where value or benefits are the outcomes (i.e., positive impacts created or negative impacts reduced) experienced by stakeholders as the results of wildlife management (e.g., values associated with biodiversity, recreation, and economic activity). Fundamentally a process for providing a range of benefits to society, wildlife management is one of a set of processes that produces ecosystem goods and services (de Groot et al. 2002). Human dimensions of wildlife management emphasize the fundamentally human purpose of wildlife management (Decker et al. 2001), focusing on understanding and addressing the reasons for public and private interest in (or impacts desired from) the management of wildlife and wildlife habitat, including those arising from sustaining biodiversity.

The process of wildlife management is often misunderstood because the term itself implies protection or manipulation of wildlife populations, habitats, and use (e.g., for hunting, trapping, and wildlife viewing), which leaves out much of what wildlife managers do in practice. Simply describing wildlife management in terms of activities aimed at animals and habitats further contributes to an incomplete conception of the enterprise. Yet descriptions of wildlife management tend to emphasize specific, on-the-ground activities, such as manipulation of habitat through burning, influencing a wildlife population through hunting quotas, or introducing animals for purposes of establishing a new population. Wildlife management activities of these kinds are necessary to achieve many of the outcomes desired by society, but wildlife management includes a greater array of processes (e.g., informative communication, negotiation, development of strategic partnerships, decision making), most of which have human dimensions components. An impacts management approach promotes broader thinking and connections between the coupled human and natural systems.

Impacts Management

An impacts-oriented approach attempts to shape value created by human–wildlife interactions through managing coupled systems. The concept of coupled systems refers to sociocultural systems, ecological systems, and their interactions (Liu et al. 2007). Essentially, this integrative concept explicitly addresses how linked sociocultural and ecological systems interact to produce diverse outputs. Although the term "coupled systems" is relatively new in environmental management, it is not really a new idea in wildlife management. Coupled systems produce the human–wildlife interactions that create the need for wildlife management. Identifying the different effects (i.e., sociocultural, biological, ecological) resulting from those interactions and understanding how those effects are perceived by people as desirable or undesirable allows managers to identify impacts for management attention. Insights from natural and social sciences are required to understand

human–wildlife interactions and the positive and negative impacts experienced by stakeholders.

Stakeholders have mixed reactions to interactions with wildlife, reflecting perceptions of positive or negative impacts (Decker and Gavin 1987, Chase et al. 1999, Riley and Decker 2000a, Ericsson and Heberlein 2003, Decker et al. 2006, Campa et al. 2011). For example, research has described the influence of impacts on people's acceptance capacity for white-tailed deer (Odocoileus virginianus) in southern Michigan, where rural residents were conflicted in their desire to view and hunt deer (their stated reasons for living in a rural environment) and their concern about the risks of deer–vehicle collisions (i.e., nearly 33% of the respondents reported being involved in a recent deer–vehicle collision; Lischka et al. 2008). Attitudes about wildlife are complex, as evidenced in studies where stakeholders (sometimes the same individuals) report both positive and negative impacts from their experiences with wildlife such as elk (Cervus canadensis), black bears (Ursus americanus), coyotes (Canis latrans), beavers (Castor canadensis), and Canada geese (Branta canadensis). To be aware of the effects produced by the system, one has to understand the human and nonhuman components. This calls for public input and participation in deliberations about wildlife management (i.e., it requires stakeholder involvement in governance of wildlife resources).

Governance

Governments, alone or in partnership with NGOs, conduct most wildlife management. In the simplest sense, governance refers to mechanisms whereby governments and other organizations direct their activities, including the processes, laws, rules, and policies that collectively guide decisions. In wildlife management, governance has focused on various levels of government because they have played dominant roles in policy and decision making related to wildlife resources. Most Western democracies consider good governance to be participatory, transparent, and accountable. Government agencies and others involved in wildlife management rely on human dimensions insights to gauge citizens' expectations of institutions and processes of government. Thinking about wildlife management as a manifestation of good governance of wildlife resources is an alternative to a more controlling, top-down view often implicit in definitions of management. A desire to govern effectively in the public interest conveys a fundamental purpose for gaining insights about the human dimensions of wildlife management.

The legal basis for American wildlife management, referred to as the public trust doctrine, reflects attention to the public interest for current and future generations (The Wildlife Society 2010). This case law and state statutes define public (versus individual) ownership of wildlife and empowers the government to be the trustee of wildlife as a public trust resource. One of the obligations of a trustee (i.e., typically state legislatures in the United States for wildlife resources) is to engage the beneficiaries of the trust (i.e., citizens) when determining goals and objectives for the trust. In the United States, beneficiaries of the trust are the people of the states in which wildlife resides. The staff of state wildlife agencies serve as agents of the trustees (i.e., legislatures, wildlife commissions, and directors), mainly playing the critical role of technical analysts, who provide recommendations to decision makers and implementers of management consistent with policy and law (Smith 2011). This arrangement gives rise to a core function of human dimensions inquiry and stakeholder engagement: seeking to better understand the outcomes (i.e., impacts) of management that are desired by stakeholders (i.e., beneficiaries of wildlife trust management). Such insight can inform and help frame recommendations offered to decision makers.

Stakeholder Orientation

Stakeholders are central to why and how wildlife is managed. As defined earlier, a stakeholder is any person who significantly affects or is affected by wildlife or wildlife management decisions or actions (Decker et al. 1996). Stakeholders are people with various kinds of interests or stakes in wildlife, human–wildlife interactions, and management interventions. Stakeholders may be individuals who are well organized into formal interest groups; individuals joined in ad hoc, situation-specific grassroots groups; or simply a set of individuals who are unaffiliated and perhaps even unknown to one another yet who have a similar interest or stake in a management issue. People, however, need not be organized or even aware they have a stake to be stakeholders in wildlife management.

Stakes in wildlife management can be thought of as impacts of interest and typically take the form of recreational, cultural, psychological, social, economic, ecological, or health and safety impacts (Siemer and Decker 2006). Impacts of interest often are expressed as recreational benefits, esthetic benefits related to quality of life, economic costs and benefits, or a species' contribution to biological diversity. A wildlife issue might involve a range of stakeholder-identified impacts, and a variety of these and other factors can influence stakeholder expectations of management. In beaver management, for example, the range of stakeholders includes: forest owners with trees felled by beavers; farmers, homeowners, and highway superintendents who contend with the economic costs and nuisance associated with flooding caused by beavers; public health officials concerned with health threats posed by giardiasis; and fur trappers. Sometimes stakeholders themselves might not even recognize their stakes in wildlife management decisions because they are unaware of the impacts they will experience as a consequence of management, especially if impacts they never experienced previously arise from management actions (e.g., management resulting in more beavers and in turn greater cost for highway repairs).

Irrespective of the level of stakeholder organization around wildlife interests, management of wildlife usually occurs in response to stakeholders' expressed need for an agency or NGO to influence impacts associated with wildlife. Frequently, though, management activity in response to one stakeholder interest can produce a reaction from new stakeholders anticipating undesirable impacts from management actions. This

normally leads to resistance and disagreement. For example, calls for management of suburban wildlife that cause negative impacts (e.g., deer, Canada geese, coyotes) often are initiated by individuals perceiving direct negative impacts from the animals involved and seeking relief from those problems. Typically, the management of wildlife in such situations escalates to a public issue when proposed management objectives and actions are not acceptable to others in a community (Decker et al. 2004, Siemer et al. 2007). These secondary stakeholders are created by the management effort itself, a common scenario.

Diversity and complexity of stakeholder perspectives point to another area of human dimensions critical to success of wildlife management: cooperation, collaboration, and coalition building among stakeholders, and between stakeholders and wildlife managers. Collaboration is necessary in many situations to access required expertise, funding, and staff. It is also helpful to overcome jurisdictional impediments and to bridge values gaps between stakeholder groups (Yankelovich 1991a,b; Wondelleck and Yaffee 2000; Beierle and Cayford 2002). Collaborative ventures can be challenging for wildlife managers to create or work with, but knowledge of the values and motivations of collaborators, cooperators, and partners is necessary to forge successful working relationships (Decker et al. 2005). Strong, positive relationships among various stakeholders and between them and wildlife agencies are key ingredients of successful collaborative conservation efforts (Lauber et al. 2011).

Stakeholder Engagement: Evolution of Approaches to Input and Involvement

Wildlife managers are continuously improving their engagement of stakeholders in decision making (Decker and Chase 1997, Chase et al. 2000). Stakeholder perceptions of transparency and fairness in decision making, gained through participatory processes and coupled with accountability (i.e., good governance), increases satisfaction, trust, and support for management (Lauber and Knuth 1997).

Stakeholder engagement as practiced by wildlife agencies takes many forms. The trend over the last 25 years has been to seek stakeholder involvement in more aspects of management. We describe six general approaches or postures taken with respect to stakeholders: expert authority, passive-receptive, inquisitive, intermediary, transactional, and comanagerial (Fig. 4.1).

Expert Authority Approach

Expert authority is a top-down approach in which wildlife managers make decisions and take actions unilaterally. Such decisions were the norm when managers focused almost exclusively on hunters and trappers, a narrow set of stakeholders compared to what managers deal with today. Values differences among these stakeholder groups were less challenging than managers now typically encounter, because fewer people were actively concerned about wildlife, and they were less diverse in their expectations for wildlife conservation. In addi-

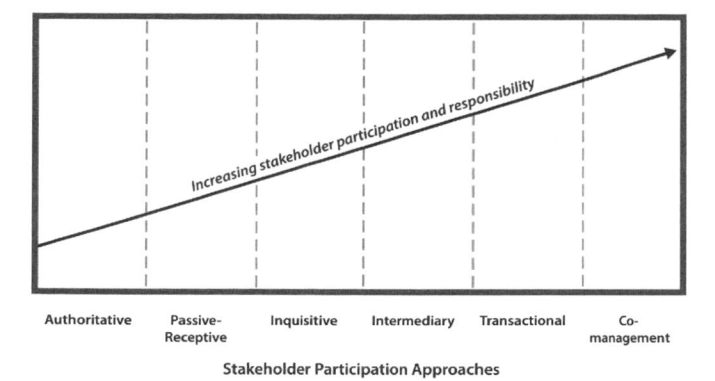

Figure 4.1. Approaches to public participation in wildlife resource governance and relative responsibility of partners and stakeholders versus wildlife management agencies. From Decker et al. (2012)

tion, many managers personally identified and shared values with hunters and trappers.

Today, the expert authority approach is still practiced, perhaps most noticeably under emergency circumstances, such as when a wildlife disease outbreak (e.g., chronic wasting disease) is discovered in a new area and what is referred to as an incident command structure is triggered to contain the disease.

Passive-receptive Approach

The passive-receptive approach is when wildlife managers welcome stakeholder input but do not seek it systematically. They consider concerns of those stakeholders who take the initiative to make their views known. When making decisions, the wildlife managers determine how much weight to give to various concerns voiced by stakeholders. Implicit in this approach is the assumption that if managers do not hear from stakeholders about their concerns, those stakeholders must not be interested in the outcome and therefore have little standing in decision making. Active stakeholders take advantage of this approach, because they organize themselves to voice their interests and concerns, sometimes to the exclusion of other legitimate stakeholders who have not figured out how to make their concerns known to managers or who lack the resources needed to organize and compete for the attention of decision makers. The passive-receptive approach is common, often manifested by brief statements at the end of news releases and websites offering opportunities for stakeholders to submit comments or questions.

Inquisitive Approach

An inquisitive approach by managers recognizes that unsolicited input alone can lead to bias because marginally important stakes can be magnified, and some stakes can be missed. Managers using the inquisitive approach actively seek information about stakeholders to inform an anticipated management decision (i.e., decisions about appropriate goals or means for achieving them). They also reach out to stakeholders to evaluate programs that are in place (i.e., to refine goals or management policies, regulations, or activities). This approach

involves social science research and systematic evaluation. For example, rather than assume what rural residents' preferences might be for deer management in southern Michigan, the Michigan Department of Natural Resources commissioned a study by researchers at Michigan State University to collect information about landowner attitudes and experiences across various regions of the state (Lischka et al. 2008). They found through inquiry that one of the primary reasons rural residents live in that environment is to interact with wildlife, of which white-tailed deer is the most common and largest species. On the other hand, rural residents also expressed a great concern about the risks associated with deer–vehicle collisions. This research defined the need to balance positive and negative impacts of deer to maintain broad stakeholder satisfaction. Based on insights from this study, the agency refined regional population objectives for deer and also became more involved in efforts to mitigate effects of deer–vehicle collisions.

Intermediary Approach

The intermediary approach emerged as managers recognized the value of more dialogue with stakeholders. It encourages two-way communication between individual stakeholder groups and the wildlife management agency but does not emphasize dialogue among stakeholder groups with different concerns. Instead, managers act as intermediaries in a form of shuttle diplomacy, where they assume much of the responsibility for deciphering similarities and differences in stakeholder positions and interests. Managers operating in this mode attempt to weigh and balance different, often competing, stakeholder concerns. Wildlife managers dealing with a controversial topic such as wolf (*Canis lupus*) conservation may choose to work separately with stakeholder groups who have different stakes (e.g., ranchers, hunters, environmentalists) in an attempt to find common ground, rather than or prior to convening representatives of the various interests. A difficulty with this approach, if carried to the extreme or used as a stand-alone approach, is that the responsibility for finding compromise falls largely on the shoulders of the manager, rather than being shared among the various stakeholder groups involved.

Transactional Approach

Managers take a transactional approach when a choice must be made about how to prioritize different stakes in management decisions, and they want to engage stakeholders in that process. When managers and diverse stakeholders need to find objectives and actions that are acceptable, they use processes where stakeholders engage one another directly to articulate their values and stakes. In the transactional approach, stakeholders describe their stakes to each other, rather than through the manager as intermediary, and they collaborate to rank or weigh these stakes. By learning about various perspectives on (or stakes in) the issue directly, conducting inquiry, discussing viewpoints, debating the issue, and compromising, stakeholder participants frequently reach consensus about appropriate objectives and courses of action (Nelson 1992).

A transactional approach was developed for deer population management decisions on a management unit basis in New York in the early 1990s (Nelson 1992). Ad hoc stakeholder committees called citizen task forces (CTFs) were established in each management unit for the purposes of identifying mutually acceptable population management objectives for white-tailed deer. In these CTFs, an independent facilitator (not an agency employee) was engaged to facilitate discussion among CTF members, who were chosen for the stake they represented, not an official affiliation with an interest group, thereby avoiding lingering effects of historical policy disputes between stakeholders and the agency or among stakeholder themselves (Stout et al. 1996).

Comanagerial Approach

In comanagement, wildlife conservation agencies engage other government agencies, NGOs, and local communities in decision making and share responsibility and authority for management. A fundamental distinction exists between the comanagerial approach and the approaches discussed above. In the expert authority, passive-receptive, inquisitive, intermediary, and transactional approaches, the role of stakeholders in decision making ranges from limited input to extensive input and involvement. Stakeholders may have greater or lesser degrees of influence on the decision, but it is the purview of the wildlife agency to decide how much that will be. In comanagement scenarios, however, the authority and resources necessary for effective management derive from partnerships. The specifics of partnerships are negotiated on a case-by-case basis, and the responsibility for wildlife conservation is shared. Because these partnerships are tailored to individual circumstances, they take many forms.

A comanagerial approach requires rethinking the role of state wildlife agencies and empowerment of other agencies, organizations, and local communities for greater responsibility in solving wildlife problems. Operational guidelines for partners (e.g., standards, criteria, and requirements for specific management efforts), oversight, accountability and evaluation processes, and assignment of responsibility must be negotiated. The role of the wildlife management agency varies widely in terms of context and might include providing biological and human dimensions expertise, managing processes, training representatives of communities or other organizations, approving management plans, certifying private consultants and community wildlife managers, and monitoring management activities. The approach also calls for educational programs on a level seldom seen in wildlife management. Nevertheless, the agency cannot abrogate or assign trustee authority and responsibility, as those powers reside with the legislature. Nor can the agency delegate all their responsibility to act as agents for the trust (i.e., wildlife resources). Comanagement is occurring with increasing frequency at the local level. Community-based wildlife management often requires cooperation and coordination in all stages of the management process, from planning through implementation and evaluation. For example, local communities

(either governments or individuals and groups) participate in a variety of ways to make a community's desires for management of white-tailed deer possible (Raik et al. 2005).

As the typology described above indicates, stakeholder engagement in contemporary wildlife management is a key part of governance of wildlife resources. Involvement may be as individuals, through NGOs representing member interests, or through local governments, rather than at state and federal levels of government. Increasingly, wildlife management is a shared responsibility where good wildlife management (i.e., governance) is not simply exercising authority over, steadfastly retaining control of, or even taking sole responsibility for wildlife resources. Good governance is wisely sharing responsibility for wildlife conservation with partners and other stakeholders. In all of the approaches identified, insight about stakeholders' beliefs and attitudes, patterns of behavior, and expectations of management with respect to wildlife are needed for success. In the more common approaches used today in complex issues, social science inquiry is required.

Wildlife Management as a Process

Wildlife management is not a tidy, linear process that unfolds predictably over time. Wildlife management is often described in the abstract as a cyclic or dynamic process, but it is even messier in practice. The overall process of wildlife management is multifaceted. Taking action, such as building water impounds and nest structures for waterfowl, is an important part of wildlife management. Yet wildlife management also is a more extensive process that includes development of broad goals and policies, setting specific objectives, selecting among actions, implementing those as management interventions, monitoring outcomes, evaluating the interventions, and then revisiting objectives, goals, and policies with new insights derived from evaluation. These facets of management require collecting information (research and monitoring); analysis; planning and decision making (including articulating fundamental and enabling objectives and selecting actions); implementing various kinds of actions directed at wildlife, habitat, and people; evaluation; and adjustment. Involving partners (e.g., other agencies, NGOs) and other stakeholders in the various facets of management adds to the complexity of the process in practice.

The management process takes place in a management environment composed of sociocultural, economic, political (e.g., governance structure), and ecological components. An information base, built upon research and experience, informs the management process. Wildlife management is seldom influenced solely by ecological components of the system. The human dimensions tend to be the prominent drivers, and they need to be well understood and integrated throughout all phases of the management process.

The diversity of expertise needed to address the complexity of issues in wildlife management is typically greater than what can be expected of any single person. Although some wildlife managers operate independently or in groups lacking needed expertise because of constraints imposed by budgets, geography, or other obstacles, much of the work in wildlife management gets done in teams. Ideally, teams of people from disciplines such as ecology, social psychology, sociology, economics, communication, law enforcement—as well as stakeholders—would be engaged in the multidisciplinary work of wildlife management.

Articulating Management Systems: Practice for Thinking Like a Manager

Articulating management systems from a manager's perspective, following some of the tenets of soft-systems methodology (Checkland 1981, Checkland and Scholes 1999) reveals the complex interdependency of the biological and human dimensions of wildlife management. Models of management systems developed by practitioners (i.e., managers' concept maps of the systems in which they work; Decker et al. 2006) help identify factors affecting management and the desired outcomes of management interventions (Decker et al. 2011). Such models typically describe objectives of management in terms of impacts desired by stakeholders, generally some combination of relief from negative outcomes of living with wildlife, direct benefits from human–wildlife interactions (including those associated with wildlife-related recreation), and indirect benefits from wildlife (e.g., biodiversity, ecosystem services). Taking time to articulate the system in which management occurs helps wildlife managers, partners, and stakeholders avoid jumping to favored, tried-and-true, or conventional actions without adequately considering long-term fundamental objectives and short-term enabling objectives (Riley et al. 2003).

Clear articulation of fundamental objectives for wildlife management, including those pertaining to desired human conditions (i.e., relating to individuals, communities, and institutions) and wildlife population and habitat conditions, makes possible evaluation of all other considerations in the management process. Fundamental objectives emerge when comparing what is known about actual condition (derived in part from analysis of the concerns raised by scientists, biologists, managers, and stakeholders) with the desired outcome to determine management need (i.e., the disparity or gap between actual and desired conditions). If analysis indicates need for management intervention, then more specific enabling objectives can be developed, and socially acceptable management actions selected.

In the process of seeking a socially acceptable and otherwise viable management action, collateral and subsequent effects of alternative management actions should be assessed. This is distinctly unlike an iterative approach that considers these effects when they arise rather than anticipating them and calculating their cost beforehand (in terms of impacts on stakeholders and cost to mitigate) such that decisions are better informed. Collateral effects are those that occur during implementation of management actions. Subsequent effects occur as a consequence of achieving enabling objectives. Wildlife management approached comprehensively includes

identification and mitigation of these two types of effects, revealed and addressed in management planning. Particularly difficult management issues arise when negative collateral or subsequent effects have the potential to offset the benefits of achieving fundamental or enabling objectives. An example of how collateral effects enter into consideration is the common management issue created by Canada geese. A first approximation of efficient management of overabundant suburban Canada geese can include some form of direct mortality (e.g., rounding up and euthanizing geese), but a study of the local human population may indicate that the collateral effects of taking that course of action would result in a huge controversy, and that alternative methods such as egg addling and egg oiling would be more acceptable, even if at greater cost to the community.

Identification of collateral and subsequent effects often reveals likely additional stakeholders for management—people not necessarily affected by, experiencing problems with, or deriving benefits from wildlife prior to a management intervention. These potential stakeholders are people likely to be affected by management actions as they are implemented or by cascading effects from management meeting its enabling objectives. Such stakeholders would be created by management interventions. In the Canada geese scenario, the initial thought of rounding up and killing geese was acceptable to city park managers and health department officials, yet that approach would have "created" a set of stakeholders concerned about the welfare of these birds.

A managers' model lays out these various considerations for review and analysis. Such a conceptual model is not meant to establish objectives or to select actions without stakeholder involvement; it is intended to help the manager or management team review the broad management situation comprehensively, particularly the many human dimensions that need to be considered. This preparatory work allows managers to engage with partners and stakeholders more effectively. During the external part of the management process that includes stakeholder engagement, structured decision making can aid the actual selection of actions.

Decisions

The most basic element of management, regardless of the type of management, is making decisions. Decisions are needed to set the course of wildlife management, to guide management actions, or to be purposeful in not taking any actions. Effective wildlife managers need to be capable of integrating biological and human dimensions insights into management decisions. Decision analyses have to integrate scientifically derived knowledge and experience-based insight about human and biological factors to determine which actions are likely to achieve fundamental objectives. If only one dimension or the other is considered, the chances that the management decision will be effective (i.e., successful and sustainable) are diminished. Combining adaptive resource management (Lancia et al. 1996) and adaptive impact management (Riley et al. 2003), where fundamental objectives are expressed in terms of

DANIEL J. WITTER (b. 1950)

Today the human dimensions of wildlife management are routinely considered in agency decisions and programs, but that was not always the case. As late as the 1970s, wildlife agencies remained wary that considering public opinion would supplant biological considerations and that wildlife management would devolve into a popular vote based on public opinion polling. It would take a generation of trained human dimensions experts years to allay those concerns and to demonstrate that integration of human dimensions considerations could be a turning point in wildlife management and conservation. Dan Witter was one of those pioneering experts.

Witter earned an associate's degree at Valley Forge Military College, a bachelor's degree in sociology from Millersville University, a master's degree in resource administration from Penn State, and a doctorate in watershed management from the University of Arizona. His Ph.D. research focused on birdwatchers, hunters, and wildlife professionals' beliefs about the importance and management of wildlife. His dissertation was part of the first wave of research to demonstrate how solid social science could help wildlife managers compare and contrast the social values underlying wildlife management issues.

In 1976, the state of Missouri passed a citizen-initiated referendum creating a sales tax to raise funds for the programs of the Missouri Department of Conservation (MDC). Years before the vote, the MDC unveiled Design for Conservation, a strategic plan explaining how it would use new revenues, beyond traditional angling and hunting income, to serve all Missourians through fisheries, forestry, and wildlife management programs. Witter was well prepared to accept a policy analyst position that the MDC created in 1978 with Design for Conservation funding. He brought social science expertise and a passion for public service that helped his agency understand its many stakeholders and learn how to serve them through the strategic plan. He implemented a comprehensive research program that yielded insights on the behavior and values of the state's diverse publics.

As his career progressed, he played a larger role in incorporating such information into agency planning, policies, and regulations. Those efforts, sustained over a 26-year career at MDC, helped the agency fulfill its public service mission and bolster political and financial support for progressive goals. Effective integration of human dimensions contributed directly to sustained growth in the MDC's programs and helped affirm the agency as a national conservation leader. Along the way, Witter has given back to the profession as an active member of The Wildlife Society and as a mentor to future professionals. Now a consultant with D. J. Case & Associates, he continues to provide guidance to wildlife management agencies striving to integrate human dimensions considerations into their decisions.

Photo courtesy of D. J. Case & Associates

impacts, is a method for integrating insights from biological and human dimensions of wildlife management. Structured decision making is discussed further in Chapter 5; adaptive management is described in Organ et al. (2012).

HUMAN DIMENSIONS INQUIRY AND APPLICATION

Since the 1970s, wildlife management has relied increasingly on social science to improve understanding of (1) how people value wildlife, (2) benefits stakeholders expect from wildlife management (i.e., desired impacts), (3) social acceptability of management practices, and (4) how various stakeholders affect or are affected by wildlife and wildlife management decisions. The following subsections describe primary social sciences used to create human dimensions insights, and three areas of wildlife management interest: wildlife-related activity participation; values, value orientations, beliefs, attitudes, norms, and motivations; and applications (i.e., integrating human dimensions insight into management).

Social Sciences Central to Human Dimensions of Wildlife Management

A variety of social sciences inform wildlife management decisions, but three disciplines most frequently provide insights about the human dimensions of wildlife management: social psychology, sociology, and economics.

Social Psychology

Improving the likelihood of achieving and sustaining outcomes or impacts desired by stakeholders can enhance wildlife managers' effectiveness. It requires knowledge of people's beliefs, attitudes, preferences, and expectations with respect to interactions with wildlife or management actions. The study of human beliefs, attitudes, and behaviors is the realm of social psychology, which offers wildlife managers insight about the basis for impacts perceived by stakeholders because impacts are typically expressed in terms of attitudes. Though not a perfect predictor, understanding aspects of beliefs, attitudes, and value orientations contributes to anticipating subsequent human behaviors.

One theoretical perspective in social psychology that provides a foundation for many human dimensions studies in wildlife management—the theory of reasoned action or planned behavior—posits that behavior is a function of three influences: people's attitude toward a behavior, assumptions about the likelihood of a behavior happening, and social norms (i.e., established behavioral standards or patterns typical of a social group). This theoretical perspective aids understanding variations in stakeholder support for management actions. For example, it helps explain why stakeholders support lethal control of beavers if the animals are perceived to carry disease, and why those same stakeholders do not support lethal control if the reason is to prevent beavers from dropping trees on golf courses.

Sociology

Values, value orientations, beliefs, and attitudes are influenced by the societies in which people live and interact. Sociology concerns how people behave as members of a group and how they interact with one another. Sociological inquiry reveals how individuals are influenced by society or social structure, and how individuals in turn continually shape the society of which they are part. Significant social differences occur between regions of the United States and even within states (e.g., urban versus rural communities). The complexity of wildlife management is caused in large part because the enterprise operates in a sociocultural context characterized by diverse value orientations with respect to wildlife management outcomes and methods. Fortunately, sociology can help managers predict why different value orientations emerge in the context of the society in which they exist and how value orientations come into play for wildlife management.

Wildlife managers are frequently concerned with how to disseminate knowledge (such as details about harvest regulations) or influence behavior (such as compliance with regulations based on understanding the basis for and goals of regulations). The rapidly emerging sociological technique of social network analysis (Lauber et al. 2008, Triezenberg 2010, Muter et al. 2011) is an example of how sociological considerations can increase wildlife managers' efficiency by directing their efforts toward key individuals or groups rather than toward less efficient broadcast approaches.

Economics

Economics is a core social science for an impacts approach to wildlife management because positive and negative impacts from human–wildlife interactions are essentially benefits and costs, respectively. Wildlife and management of wildlife have many direct, indirect, and induced economic effects on individuals, communities, states, regions, and even nations.

A partnership of state wildlife agencies and national conservation organizations conducts the National Survey of Fishing, Hunting, and Wildlife-Associated Recreation. This national survey quantifies expenditures for wildlife-based recreation every five years in the United States. The 2006 survey (U.S. Fish and Wildlife Service 2007) estimated that 87.5 million U.S. residents fished, hunted, and watched wildlife. In doing so, more than $122 billion was spent in wildlife-related activities. These estimates of direct expenditures are useful, but they incompletely describe the economic value of wildlife, which includes nonmarket values. Fortunately, even for attributes of wildlife that cannot be assigned a market value, economists have developed ways of nonmonetary and nonconsumptive valuation that are applicable to wildlife management.

Economic theory and methods can be invaluable when evaluating proposed management actions such as habitat restoration or educational outreach programs. For example, though limited by the assumption that all of the benefits and costs can be predicted and enumerated, benefit-cost analysis gauges efficiency within proposed projects by comparing estimated benefits of a project with the costs of conducting the project.

Wildlife-related Activity Participation

Studies of wildlife-related activity participation typically inventory and characterize wildlife users, monitor uses, and describe trends in wildlife recreation participation. Individual state wildlife agencies gather harvest data for various game species using hunter surveys to estimate population abundance and trends at state and substate levels. Following national trends in wildlife recreation in the United States has been facilitated by surveys of American households conducted approximately every five years since 1955 (U.S. Fish and Wildlife Service 2007). Data on hunter numbers, types of hunting, hunter days afield, and hunters' expenditures have been collected for more than a half century through this national survey. Information on nonconsumptive activities (e.g., wildlife observation, photography) has been collected in this survey since 1980. The methodology for these surveys changed at various times, limiting use of these survey data to chart precisely the long-term trends in wildlife recreation participation (Chu et al. 1992). The national surveys, however, indicate a long-term decline in hunting and trapping participation. The surveys also identified that a large portion of the American public enjoys wildlife viewing, as a focus of nature-based recreation or as a casual pastime.

National- and state-level surveys identified that numbers of hunters were static or declining beginning in the 1980s (Brown et al. 1987, Applegate 1989), but the reasons for this decline in participation initially were not fully elaborated. Subsequently, human dimensions researchers examined the dynamics of hunting participation in terms of recruitment, retention, continuous versus sporadic participation, and cessation. This line of inquiry indicated that hunting is strongly driven culturally from initiation through cessation (Decker et al. 1984). Broad social trends affecting hunting (e.g., urbanization) suggest a continuing decline in participation is likely over the long term (e.g., Decker et al. 1993, Heberlein and Thomson 1996, Brown et al. 2000). By describing the complexity of hunting as a social phenomenon, this research highlighted the difficulties agencies face when attempting to stimulate recruitment or retention at a level that would effectively curb the decline of hunters.

Nonconsumptive wildlife uses and users (i.e., people interested in wildlife-associated recreation for other than hunting and trapping purposes) became interests of the wildlife management and research communities beginning in the 1980s (Manfredo 2002). Wildlife observation near people's homes (i.e., viewing wildlife in one's yard or nearby green spaces) accounted for much of the time spent watching wildlife, though trips for the specific purpose of viewing wildlife were also common. The U.S. Fish and Wildlife Service survey identified a demand (i.e., number of participants) for wildlife viewing that far exceeded that for hunting and trapping. During this period, specific studies revealed new insight about potential new beneficiaries for wildlife programs, such as wildlife viewers (Brown et al. 1979, Witter and Shaw 1979) and donors to state and private wildlife program funds (Brown et al. 1979, Witter et al. 1981, Manfredo 1988).

Values, Beliefs, Attitudes, Norms, and Motivations

This area of inquiry seeks answers to questions such as: (1) how do people value and evaluate wildlife and human–wildlife interactions? (2) what makes management actions acceptable or not to various stakeholders in different contexts? and (3) why do people participate (or not participate) in various wildlife-related activities? These lines of inquiry reflect a theoretical framework referred to as the cognitive hierarchy: values, value orientations (basic belief patterns), attitudes, and norms (Fulton et al. 1996). Motivation theory also has been used to understand why people engage in wildlife-related activities.

Values, Beliefs, and Value Orientations

Despite some inconsistency in conceptual definition (Manfredo et al. 2004), wildlife values are typically regarded as stable, central modes of thought about wildlife. Early studies attempted to develop "values typologies" that described how segments of society relate to wildlife. Kellert (1980a,b), for example, developed an instrument to characterize groups of people based on a set of ten basic orientations toward animals: naturalistic, ecologistic, humanistic, moralistic, scientistic, aesthetic, utilitarian, dominionistic, negativistic, neutralistic. The Wildlife Attitudes and Values Scale (Purdy and Decker 1989) sought to describe people's orientations toward wildlife in human–wildlife conflict situations, emphasizing traditional conservation, social benefits, and problem acceptance perspectives. These early attempts to describe how people relate to wildlife at a general level were useful in reinforcing the idea that people were not homogeneous in their core perspectives about wildlife.

In the mid-1990s, Fulton et al. (1996) introduced the concept of wildlife value orientations. These orientations are the direction and pattern of basic beliefs about wildlife. Basic beliefs are part of a cognitive hierarchy hypothesized to align as follows: values → basic beliefs → attitudes and norms → behaviors. People with the same underlying values associated with wildlife could be oriented differently in the expression of those values. Research applying the values orientation idea identified a protection-use dimension and a wildlife appreciation dimension in people's overall thinking about wildlife. People with a more utilitarian view, for example, are more likely to engage in consumptive recreation and support more invasive forms of wildlife management than those with a protectionist orientation.

Attitudes

Values and value orientations (i.e., basic belief patterns) are important in shaping a person's attitudes, such as favorable or unfavorable evaluation of an object, behavior, or event (e.g., an encounter with a wild animal in one's yard, a class of wildlife such as predators, a specific management action). Attitudes are a combination of beliefs and negative or positive evaluations about an object (Heberlein 2012). Attitudes people hold about specific things (e.g., deer, gray wolf restoration, beaver trapping) are believed to influence their behavior (e.g., participation in an activity, reaction to an encounter with a wild animal, political support, opposition to a policy action).

Public attitudes about wildlife and wildlife management have been a major focus of human dimensions research because they provide insight into behaviors. Much of the research described as dealing with opinions, preferences, and perceptions are actually studies of attitudes. Some of these studies made important contributions to wildlife managers' understanding of longstanding issues, often yielding major surprises. The attitudes of rural landowners toward hunting, for example, have received considerable attention by managers over the decades because landowners play a key role in providing access for hunting in many states (Brown et al. 1984, 1987; Wright et al. 1989). Contrary to the assumptions of some state programs, these studies report most landowners who posted their land were not opposed to hunting per se. Instead, the access restrictions that landowners imposed reflected the poor image they had of hunter behavior and their concerns over liability risks. Influenced by this research, state wildlife agencies and hunting organizations embarked on efforts to address landowner concerns (e.g., improve hunter behaviors, assist landowners in controlling access, incentives for allowing access for hunting, laws to reduce landowner liability).

Managers' concerns about public attitudes toward potentially dangerous wildlife, especially among people living along the wildland–urban interface, have motivated human dimensions research (Wieczorek Hudenko et al. 2010). Examples include studies of public attitudes about mountain lions (*Puma concolor*) in Colorado (Manfredo et al. 1998) and Montana (Riley and Decker 2000a), black bears in Colorado (Loker and Decker 1995) and the Northeast (Organ and Ellingwood 2000, Siemer et al. 2009), wolves in Utah (Bruskotter et al. 2007) and Alaska (Decker et al. 2006), eastern massasauga rattlesnakes (*Sistrurus catenatus catenatus*) in Michigan (Christoffel 2007), and coyotes in suburban New York (Wieczorek Hudenko et al. 2008). Studies indicate that, despite human deaths, the public is split with respect to how to control large mammals, findings that can be partially explained by differences in risk perceptions (Riley and Decker 2000b). Public attitude research has also been used to examine the social feasibility of restoring large herbivores, such as elk and moose (*Alces alces*), that have potential for creating significant impacts for humans (Enck et al. 1998, McClafferty and Parkhurst 2001). This line of inquiry typically involves predicting respondents' voting intentions on restoration initiatives using attitudinal measures.

Norms

Normative beliefs, or social norms, are defined as shared beliefs about the acceptability of a situation, action, or outcome (Shelby et al. 1996). Norms are standards of acceptable behavior that inform people about what they should do or what most people in their social system are doing in a given context (Vaske and Whittaker 2004). The structural characteristics of norms (e.g., the intensity or strength of the norm and the level of agreement about the norm) have important management implications, so studying norms in a specific context has practical value to wildlife managers.

Normative theory has improved our understanding of the relationship among tolerance of wildlife problems, acceptance of management actions, and human values with respect to wildlife. The normative model has been used to understand interactions among wildlife recreationists (Heberlein et al. 1982, Whittaker 1997), and more recently a normative approach has been applied in studies about the acceptability of wildlife management actions (Wittmann et al. 1998, Zinn et al. 1998, Campbell and Mackay 2003, Dougherty et al. 2003, Jonker et al. 2009). Research on the latter has focused on "incident extremity" and "response extremity" relationships, where the emphasis is on predicting normative response to a range of specific human–wildlife interactions or incidents (e.g., seeing wildlife in residential areas, wildlife injuring a person; Zinn et al. 1998).

Tolerance for interactions with wildlife varies across stakeholder groups, situations, species of wildlife, and time. This makes it difficult to predict situation-specific stakeholder behavior and expectations of management without recent inquiry about the particular context or a very similar one. Managers report that stakeholder acceptance of management actions is as variable as tolerance of human–wildlife interactions. Traditional management methods (e.g., hunting, trapping) can be effective in removing individual problem-causing animals or reducing wildlife populations, but may not be socially acceptable to urban residents (Zinn et al. 1998). An understanding of how stakeholders perceive particular management actions can help wildlife agencies minimize controversy when choosing among management alternatives (Loker et al. 1999).

Differences in contexts where human–wildlife interactions occur influence norms of acceptable management actions. The wildlife acceptance capacity (WAC) concept, for example, proposes that there is some "maximum wildlife population level in an area that is acceptable to people" (Decker and Purdy 1988:53). Basically, WAC suggests that there is a threshold of acceptability for the impacts from human–wildlife interactions that are produced by a wildlife population of a certain density

and distribution in a particular context. Similar to the concept of "range of tolerable conditions" in the norm literature (Vaske and Whittaker 2004), the WAC concept suggests that a person's acceptance threshold is situation specific and dependent on the severity of impacts produced by human–wildlife interactions associated with the populations of humans and wildlife in question. These impacts, often focusing on negative outcomes (i.e., conflicts, problems), can be arranged along a continuum ranging from nuisance (e.g., bears raiding dumpsters), to economic or aesthetic impacts (e.g., elk eating ornamental plants), to health and safety threats (e.g., Lyme disease; Decker 1991). The more severe the problem, with severity being reflected by kind of impact (e.g., nuisance versus safety threat) or amount (e.g., infrequent versus nightly trash can dumping by raccoons), the more likely urban residents are to accept lethal methods for managing wildlife. Residents of suburban communities in New York, for example, were more willing to accept aesthetic or economic wildlife impacts (i.e., damage to ornamental plantings) than health risks like disease (Connelly et al. 1987). Similar findings have emerged in related studies (Decker and Gavin 1987, Stout et al. 1997, Jonker et al. 2009). Safety risks tend to be least acceptable (Wieczorek Hudenko et al. 2010).

The WAC concept was advanced with a more multifaceted view of tolerance, where WAC reflects the conditions necessary for a mix in types and level of human–wildlife interactions acceptable to stakeholders (Carpenter et al. 2000). This refinement recognizes that people perceive negative and positive impacts associated with wildlife, and that stakeholders weigh the positives and negatives differentially. Although prior approaches suggest fostering a mix of human–wildlife interactions that minimize negative impacts, this concept focuses on a level that yields the greatest net benefit while keeping negative impacts at socially acceptable levels (often with some type of mitigation involved).

Trying to understand public support or opposition for killing problem wildlife is a complex undertaking in even the most severe situations (Whittaker and Manfredo 1997, Zinn et al. 1998). Although the same management tactics can be applied in a variety of situations, the acceptability of the tactic can vary depending on the species of animal. Variations in the animal's image, perceived abundance, and impact potential may influence norms with respect to acceptability of killing the animal. Coyotes, for example, often are portrayed as scavengers or pests (McIvor and Conover 1994). As a result, destroying a coyote is more acceptable than killing another animal (Wittmann et al. 1998).

An animal's potential to create severe impacts also influences human norms about their management. Beavers, coyotes, and mountain lions can all influence human activities, but in different ways. Beavers alter suburban landscapes with dams and resultant flooding, coyotes kill pets, and mountain lions kill or injure livestock, pets, and occasionally humans. Each animal's behavior, in conjunction with its reputation and potential for human impact, influences the acceptability of killing the animal (Zinn et al. 1998).

The idea of situation-specific impact dependency as a determinant of management acceptability has been explored in the context of large predator and herbivore interactions in Alaska (Decker et al. 2006). This research demonstrated that acceptance of predator control depended on the effects of predation on caribou (*Rangifer tarandus*) and moose, and the perceived effects of this predation on people; that is, the impacts of fewer moose and caribou available for human consumption. Support for lethal control of grizzly bears (*Ursus arctos horribilis*) and wolves to reduce predation on moose and caribou was influenced by the perceived effect predation has on humans' access to moose and caribou. Alaskans were more likely to support the use of lethal methods to control predation in situations where the effect of predation on moose and caribou had the greatest impact on hunters' access to these resources. Conversely, lethal control of predators was less likely to be supported in situations where the impact of predators on moose and caribou was seen as limited to recreational interests in hunting (versus reliance on big game for food).

Overall, the structural norm approach (e.g., the WAC concept) has provided information useful for managers to define evaluative standards (i.e., acceptability) for specific management actions, to identify impacts about which people feel strongly, and to describe the amount of agreement about a policy (e.g., acceptability of killing an animal to avoid an interaction that causes a negative impact) or practice (e.g., shooting versus capture and euthanizing a "problem" animal) among the general public or various kinds of stakeholders.

Motivation and Satisfaction

A research thrust starting in the 1970s was directed toward understanding wildlife recreationists' (particularly hunters') motivations and satisfactions with respect to participation in an activity (particularly various kinds of hunting). Motivation theory suggests that people are driven (motivated) to take actions to achieve particular goals (i.e., they seek certain outcomes from their experiences). Two lasting approaches to conceptualizing and studying motivations emerged in the literature. One emphasized a multiple-satisfaction approach to big game management (Hendee 1974). This approach suggested that we can identify and manage for categories of hunters (and other wildlife recreationists) who differ based on the types of satisfaction they seek and receive. A second line of inquiry emphasized the importance of understanding the package of psychological outcomes recreationists desire and derive from participation (Driver et al. 1991). This work also emphasized the insight gained by examining the context (e.g., setting and activity) associated with those outcomes. Planning and managing for types of hunters were predicted to differ based on the mix of outcomes, activities, and settings sought by participants (Hautaluoma and Brown 1978, Manfredo and Larson 1993). A contributing line of research explored the concept of psychological investment and its importance in sustaining a person's participation in big game hunting (Barro and Manfredo 1996).

Understanding of hunters was deepened by motivation and satisfaction research, from which several key insights emerged. Research revealed that hunters seek multiple satisfactions from hunting (e.g., companionship, nature appreciation) and can be meaningfully segmented with respect to the satisfactions and benefits sought (Potter et al. 1973, Driver 1976). Hunters also were found to have a range of primary motivational orientations (i.e., affiliative, achievement, appreciative), often expressed differently by the same hunters in different contexts (Decker et al. 1987). The diversity of hunters' motivations is reflected in their many activity and experience preferences (Hautaluoma and Brown 1978). Knowing the primary motivational orientations of hunters informs development of practical applications to hunting regulations (e.g., liberalized deer permits to increase use by permit holders; Decker and Connelly 1989). These insights provide ways to improve hunter experiences and wildlife population management programs that depend on hunter participation.

While identifying the diversity of experiences desired by participants in wildlife-related activities, researchers uncovered the value of differentiating users into homogeneous and meaningful subgroups (i.e., segmentation). In one conceptualization of subgroups, Bryan (1977:29) defines recreation specialization as a "continuum of behavior from the general to the particular, reflected by equipment and skills used in the sport." Within the continuum, individuals range from the novice to the specialist. User classes differ with respect to motivations, the extent of prior experience with an activity, and commitment to an activity. As people become more specialized, they become more particular in their setting preferences, objectives for various experiences, and equipment. More specialized users are also more likely to have specific managerial requirements and more likely to communicate with managers. Research has applied the concept of specialization to angling (Bryan 1977), hunting (Miller and Graefe 2000), and wildlife viewing (McFarlane 1996). The work on motivational orientations revealed that an individual can have different motivations for participation in different kinds of hunting and in different social and environmental settings.

Application of Human Dimensions Insight

Insights arising from human dimensions inquiry and theory can improve governance of wildlife uses. This is accomplished by considering the stakeholders of wildlife management, and relevant social and institutional structures important to them and to management effectiveness. Much of the early human dimensions research focused on consumptive uses of wildlife. This body of work has influenced wildlife management in at least three broad ways:

1. Research has influenced wildlife policies. Revealing the extent of Americans' interest in wildlife, for example, served as a building block in the arguments for federal-level program initiatives that support nongame wildlife conservation. With respect to private land access, states have strengthened statutory protection against

liability for landowners who allow uncompensated wildlife recreation on their property. In addition, incentives encourage landowners to permit access for wildlife recreation (e.g., state-run hunting cooperatives on private lands, offering special signs indicating "access by permission only," increased law enforcement consideration for cooperating landowners). Finally, states have established policies to improve recruitment (e.g., lowered hunting age, enhanced access) and retention of hunters (e.g., mentoring programs, land access programs, reduced license fees for older hunters).

2. Human dimensions research has been used to develop and modify wildlife management practices. For example, evaluative research about public participation practices improved stakeholder involvement and improved decision making. Research based on understanding people's expectations for involvement guides wildlife managers in designing stakeholder engagement (i.e., citizen participation and public involvement) processes. The body of research in this area has identified key factors for community-based collaborative approaches and elements leading to participant satisfaction with the process (Raik et al. 2003, Schusler et al. 2003). Experience-based management (Manfredo 2002) and visitor impact management (Vaske et al. 1995) models have improved the framing of management decisions.

3. Human dimensions research has been used to design education efforts and to inform communication strategies. Landowner studies, for example, have led state hunter education programs to include hunting ethics and landowner–sportsperson relations. State agencies also have increased awareness among hunters of the importance of improving behavior and public image. Furthermore, studies of the general population and of population segments have led state wildlife agencies to adopt proactive strategies for improving public understanding of hunting and trapping in North America and programs designed specifically for youth and women.

ADDITIONAL HUMAN DIMENSIONS CONSIDERATIONS

Communication

Wildlife agencies often identify improvement of communication skills or "people" skills of wildlife professionals as a high priority. Interpersonal communication skills are always an asset, but what many wildlife agencies seek are employees who can develop and deliver informative communication that improves stakeholder knowledge of an issue and in turn leads to better informed input by stakeholders and better stakeholder understanding of management parameters and actions. Keeping in mind that important attributes of good governance are transparency and accountability, the legitimacy of decisions and decision-making bodies depends upon adequate opportunities for input (i.e., transparency) and efforts to ensure ac-

countability. Transparency and accountability are attained by more than just access to information. Those attributes, vital to earning and sustaining public trust in wildlife managers, are made possible by use of language and selection of channels of communication that stakeholders can understand and access.

Ethics

If you take nothing else away from your study of wildlife management, you must understand that all the sociological and ecological science in the world does not tell a wildlife manager what *should* be done in a given situation. At best, science informs managers of what *could* be done, what stakeholders desire, or what is likely to happen with and without particular interventions. Determining what should be done is in the realm of ethics. Ethics are best understood if analyzed rigorously using tools developed in an area of philosophy that examines human behavior, values (i.e., whether certain actions are right or wrong), and morals inherent in motives and actions. Human dimensions insights help wildlife professionals evaluate ethical components of wildlife management by clarifying pertinent values in a particular issue, but it is a different process to determine whose values (e.g., values of all voters, certain stakeholders, managers) should prevail in management, and who should ultimately judge the adequacy and acceptability of alternative management actions.

Ackoff (2001:345), a highly regarded scholar of management, articulated the following caution for managers attempting to do "the right thing":

> Paraphrasing Peter Drucker, there is a big difference between doing things right and doing the right thing. Efficiency is concerned with doing things right; effectiveness with doing the right thing.
>
> The righter one does the wrong thing, the wronger one becomes. If one corrects an error in the pursuit of the wrong thing one becomes wronger. Therefore, it is better to do the right thing wrong than the wrong thing right because it offers the possibility of improvement.

Much of a wildlife manager's training is geared toward how to do things right—how to measure a population, evaluate condition of a habitat, or reveal stakeholder attitudes accurately. These are very important skills, and they must be done correctly. Yet, as Ackoff notes, if those are not the things we should be concentrating on, if they do not address the most important decisions, trying harder to do a better job can have the effect of diverting attention away from what should be done. Arriving at a good decision about what should be done is not easy, but ethics provide a structured approach for defining and deliberating about the issues.

Unfortunately, ethical discourse does not guarantee a solution for decision making in messy wildlife management problems; that is, ethical discourse does not necessarily resolve value-laden problems for which there are multiple perspectives about what constitutes the right objectives and best outcomes of management. Stakeholders with competing value orientations often pull wildlife professionals in different directions. Aiding wildlife management decision makers in their quest to identify the "right" thing to do and how best to do it in the public interest gives a vital purpose to human dimensions of wildlife management.

Learning

Although much has been learned from the integration of social sciences into wildlife management, much more remains to be explored and understood. Human dimensions needs include developing generalizations that can be used as tenets of management, making existing knowledge more accessible to wildlife managers who do not have social science backgrounds, and developing management methods that integrate biological and human dimensions. The combination of more human dimensions research, more stakeholder engagement, and more effort to integrate human dimensions insights into management will lead to greater contributions to sound policy and management decision making.

One way to think about the application of knowledge in wildlife management is through the lens of learning—by managers, partners, and stakeholders participating in the management process. Several types of learning in many kinds of environmental decision-making processes have been identified (Sabatier 1988, Lauber and Brown 2006), but Glasbergen (1996) and Fiorino (2001) identify three basic types of learning relevant to wildlife management: conceptual learning, technical learning, and social learning. Human dimensions insight contributes to each of these essential kinds of learning for wildlife management.

Conceptual learning involves searching for new objectives and new ways of defining the problem that is being addressed. Conceptual learning occurs as fundamental objectives (i.e., ends) for wildlife management are proposed and publicly debated, and through that process, how people think about issues changes. In wildlife management, conceptual learning involves the selection and prioritization of particular fundamental objectives that enabling objectives and management actions (i.e., means) will be designed to accomplish. For example, conceptual learning can occur during stakeholder input and involvement processes, resulting in identifying the species, habitats, or ecosystems that need to be the focus of management.

Technical learning involves searching for new policies and actions to accomplish existing objectives. For example, in a wildlife management context, technical learning could contribute to the selection or refinement of actions that will help a population of a rare species become better established at particular sites. Both ecological and human dimensions insights can contribute to technical learning. Human dimensions knowledge helps managers identify which wildlife management needs should be given priority and which actions will be socially acceptable. Human dimensions research can also help the manager to understand why.

Social learning results from relationships and the quality of dialogue between partners, stakeholders, and wildlife managers. It entails learning about how to promote effective communication and interaction. In wildlife management, social learning involves identifying and engaging new stakeholders and collaborators or engaging existing stakeholders and collaborators in new ways. For example, social learning could contribute to the establishment of a task force of multiple local governments and the wildlife agency to guide coordinated deer management across a set of adjacent communities.

All three types of learning contribute to the management process, with the dominant types changing as conservation initiatives evolve (Glasbergen 1996, Fiorino 2001). Lauber and Brown (2006) demonstrated relationships between learning types and suggested that social learning provides the necessary foundation for technical and conceptual learning. They reported that developing effective conservation objectives and actions depends on fostering good working relationships between key organizations and individuals and effective forums for dialogue. The role of social learning in linking human dimensions knowledge to management action reflects the importance of collaboration in wildlife governance. Linking knowledge to action often requires many individuals and organizations—including stakeholders, scientists, and practitioners—to work together effectively.

SUMMARY

Wildlife management aims to produce value defined by and for society, where value or benefits are outcomes (i.e., positive impacts created or negative impacts reduced) experienced by stakeholders. Making decisions and taking actions to achieve management objectives require knowledge about wildlife organisms, their habitat, and people. Human dimensions of wildlife management focus on people and their institutions (Decker et al. 2012). Many kinds and scales of considerations, from broad social ideals such as governance and conservation, to specific traits of people such as beliefs and motivations to participate in wildlife-related activities, are the subject areas of human dimensions.

Public wildlife management occurs within a framework of governance that regards wildlife as public trust resources. Stakeholders, as beneficiaries of wildlife management, are a fundamental consideration in governance. A stakeholder is any person who is significantly affected by, or can significantly affect, wildlife or wildlife management processes (Decker et al. 1996). Stakeholders interpret the effects of events or interactions with wildlife through their values and beliefs as being either unimportant or important. Those effects considered important can be thought of as impacts (Riley et al. 2002). Understanding impacts and the reasons why some interactions with wildlife are viewed as creating impacts can help managers integrate biological and human dimensions of wildlife management.

Stakeholder expectations for managing the impacts from human–wildlife interactions prompt wildlife managers, NGOs, elected officials, and community leaders to seek acceptable management objectives and actions consistent with stakeholder value orientations. Insights from human dimensions studies and stakeholder engagement efforts play a major role in informing objectives for management and identifying kinds of management actions that are socially acceptable for different contexts. This alignment of management alternatives with stakeholder acceptance is certain to become even more daunting as new technologies (e.g., effective fertility control in wildlife) become available.

Social sciences such as social psychology, sociology, and economics provide ways of attaining reliable information on which to base management decisions. Knowledge of human values, beliefs, attitudes, and behaviors improves the likelihood of achieving desirable impacts—outcomes sought by stakeholders—through wildlife management. Sustainability of public support for wildlife management is tied to achieving desirable impacts and minimizing undesirable impacts. Some management actions are also the domain of human dimensions, such as communication, education, and stakeholder engagement in decision making. Wildlife management is a system of processes with feedback rather than a discrete, one-shot, linear event. The feedback comes from understanding the effects of management on humans, habitats, and wildlife populations, and ultimately humans as stakeholders evaluate those effects as impacts or not. Insights from rigorous application of the social sciences ensure those feedbacks in the management system are reliable.

Looking to the future, the role of community involvement and collaboration with wildlife management agencies will almost certainly become more important, particularly where overabundant wildlife become a major concern. In these and other situations, research is needed to better understand how local communities and agencies can work together for effective governance of wildlife, including the processes of joint (social) learning and capacity building for improved collaborative wildlife management. When the human and biological dimensions of wildlife management are well informed by research, vetted transparently with partners and stakeholders, and effectively integrated in decision making, learning in wildlife management improves its capacity to adapt to changing contexts. The only way for wildlife professionals to stay current and relevant in a changing sociocultural environment is through continual learning.

Literature Cited

Ackoff, R. L. 2001. OR: after the post mortem. Systems Dynamics Review 17:341–346.

Applegate, J. E. 1989. Patterns of early desertion among New Jersey hunters. Wildlife Society Bulletin 17:476–481.

Barro, S. C., and M. J. Manfredo. 1996. Constraints, psychological investment, and hunting participation: development and testing of a model. Human Dimensions of Wildlife 1:42–61.

Beierle, T. C., and J. Cayford. 2002. Democracy in practice: public participation in environmental decisions. Resources for the Future, Washington, D.C., USA.

Brown, T. L., C. P. Dawson, and R. L. Miller. 1979. Interests and attitudes of metropolitan New York residents about wildlife. Transactions of the North American Wildlife and Natural Resources Conference 44:289–297.

Brown, T. L., D. J. Decker, and J. W. Kelley. 1984. Access to private lands for hunting in New York: 1963–1980. Wildlife Society Bulletin 12:344–349.

Brown, T. L., D. J. Decker, K. G. Purdy, and G. F. Mattfeld. 1987. The future of hunting in New York. Transactions of the North American Wildlife and Natural Resources Conference 52:553–566.

Brown, T. L., D. J. Decker, W. F. Siemer, and J. W. Enck. 2000. Trends in hunting participation and implications for management of game species. Pages 145–154 in W. C. Gartner and D. W. Lime, editors. Trends in outdoor recreation, leisure, and tourism. CABI, New York, New York, USA.

Bruskotter, J. T., R. H. Schmidt, and T. Teel. 2007. Are attitudes toward wolves changing? A case study in Utah. Biological Conservation 139:211–218.

Bryan, H. 1977. Leisure value systems and recreational specialization: the case of trout fishermen. Journal of Leisure Research 9:174–187.

Campa, H., III, S. J. Riley, S. R. Winterstein, T. L. Hiller, S. A. Lischka, and J. P. Burroughs. 2011. Changing landscapes for white-tailed deer management in the 21st century: parcelization of land ownership and evolving stakeholder values in Michigan. Wildlife Society Bulletin 35:168–176.

Campbell, J. M., and K. J. Mackay. 2003. Attitudinal and normative influences on support for hunting as a wildlife management strategy. Human Dimensions of Wildlife 8:181–197.

Carpenter, L. H., D. J. Decker, and J. F. Lipscomb. 2000. Stakeholder acceptance capacity in wildlife management. Human Dimensions of Wildlife 5:5–19.

Chase, L. C., T. M. Schusler, and D. J. Decker. 2000. Innovations in stakeholder involvement: what's the next step? Wildlife Society Bulletin 28:208–217.

Chase, L. C., W. F. Siemer, and D. J. Decker. 1999. Suburban deer management: a case study in the village of Cayuga Heights, New York. Human Dimensions of Wildlife 4:59–60.

Checkland, P. B. 1981. Systems thinking, systems practice. John Wiley and Sons, New York, New York, USA.

Checkland, P. B., and J. Scholes. 1999. Soft systems methodology in action. John Wiley and Sons, New York, New York, USA.

Christoffel, R. A. 2007. Using human dimensions insights to improve conservation efforts for the eastern massasauga rattlesnake (Sistrurus catenatus catenatus) in Michigan and timber rattlesnake (Crotalus horridus horridus) in Minnesota. Dissertation, Michigan State University, East Lansing, USA.

Chu, A., D. Eisenhower, M. Hay, D. Morganstein, J. Neter, and J. Waksberg. 1992. Measuring the recall error in self-reported fishing and hunting activities. Journal of Official Statistics 8:19–39.

Connelly, N. A., D. J. Decker, and S. Wear. 1987. Public tolerance of deer in a suburban environment: implications for management and control. Eastern Wildlife Damage Control Conference 3:207–218.

Decker, D. J. 1991. Implications of the wildlife acceptance capacity concept for urban wildlife management. Pages 45–53 in E. A. Webb and S. Q. Foster, editors. Perspectives in urban ecology. Denver Museum of Natural History, Denver, Colorado, USA.

Decker, D. J., T. L. Brown, B. L. Driver, and P. J. Brown. 1987. Theoretical developments in assessing social values of wildlife: toward a comprehensive understanding of wildlife recreation involvement. Pages 76–95 in D. J. Decker and G. R. Goff, editors. Valuing wildlife: economic and social perspectives. Westview Press, Boulder, Colorado, USA.

Decker, D. J., T. L. Brown, and W. F. Siemer. 2001. Human dimensions of wildlife management in North America. The Wildlife Society, Bethesda, Maryland, USA.

Decker, D. J., and L. C. Chase. 1997. Human dimensions of living with wildlife—a management challenge for the 21st century. Wildlife Society Bulletin 25:788–795.

Decker, D. J., and N. A. Connelly. 1989. Motivations for deer hunting: implications for antlerless deer harvest as a management tool. Wildlife Society Bulletin 17:455–463.

Decker, D. J., J. W. Enck, and T. L. Brown. 1993. The future of hunting: will we pass on the heritage? Proceedings of the Annual Governor's Symposium on North American Hunting Heritage 2:22–46.

Decker, D. J., and T. Gavin. 1987. Public attitudes toward a suburban deer herd. Wildlife Society Bulletin 15:173–180.

Decker, D. J., C. A. Jacobson, and T. L. Brown. 2006. Situation-specific "impact dependency" as a determinant of management acceptability: insights from wolf and grizzly bear management in Alaska. Wildlife Society Bulletin 34:426–432.

Decker, D. J., C. C. Krueger, R. A. Baer Jr., B. A. Knuth, and M. E. Richmond. 1996. From clients to stakeholders: a philosophical shift for fish and wildlife management. Human Dimensions of Wildlife 1:70–82.

Decker, D. J., R. W. Provencher, and T. L. Brown. 1984. Antecedents to hunting participation: an exploratory study of the social-psychological determinants of initiation, continuation, and desertion in hunting. Outdoor Recreation Research Unit Series No. 84-6. Human Dimensions Research Unit, Cornell University, Ithaca, New York, USA.

Decker, D. J., and K. G. Purdy. 1988. Toward a concept of wildlife acceptance capacity in wildlife management. Wildlife Society Bulletin 1:53–57.

Decker, D. J., D. B. Raik, L. H. Carpenter, J. F. Organ, and T. M. Schusler. 2005. Collaborations for community-based wildlife management. Urban Ecosystems 8:227–236.

Decker, D. J., D. B. Raik, and W. F. Siemer. 2004. Community-based suburban deer management: a practitioner's guide. Northeast Wildlife Damage Management Research and Outreach Cooperative, Ithaca, New York, USA.

Decker, D. J., S. J. Riley, J. F. Organ, W. F. Siemer, and L. H. Carpenter. 2011. Applying impact management: a practitioner's guide. Human Dimensions Research Unit and Cornell Cooperative Extension, Department of Natural Resources, Cornell University, Ithaca, New York, USA.

Decker, D. J., S. J. Riley, and W. F. Siemer. 2012. Human dimensions of wildlife management. Johns Hopkins University Press. Baltimore, Maryland, USA.

Decker, D. J., M. A. Wild, S. J. Riley, W. F. Siemer, M. M. Miller, K. M. Leong, J. G. Powers, and J. C. Rhyan. 2006. Wildlife disease management: a manager's model. Human Dimensions of Wildlife 11:151–158.

de Groot, R. S., M. A. Wilson, and R. M. J. Boumans. 2002. A typology for the classification, description, and valuation of ecosystem functions, goods and services. Ecological Economics 41:393–408.

Dougherty, E. N., D. C. Fulton, and D. H. Anderson. 2003. The influence of gender on the relationship between wildlife value orientations, beliefs, and the acceptability of lethal deer control in Cuyahoga Valley National Park. Society and Natural Resources 16:603–623.

Driver, B. L. 1976. Toward a better understanding of the social benefits of outdoor recreation participation. Southeastern Forest Experiment Station, USDA Forest Service, North Carolina State University, Ashville, USA.

Driver, B. L., P. J. Brown, and G. L. Peterson, editors. 1991. Benefits of leisure. Venture, State College, Pennsylvania, USA.

Enck, J. W., D. J. Decker, S. J. Riley, J. F. Organ, L. H. Carpenter, and W. F. Siemer. 2006. Integrating ecological and human dimensions in adaptive management of wildlife-related impacts. Wildlife Society Bulletin 34:698–705.

Enck, J. W., W. F. Porter, K. A. Didier, and D. J. Decker. 1998. The feasibility of restoring elk to New York State. College of Agriculture and Life Sciences, Ithaca, New York, and New York State College of Environmental Science and Forestry, Syracuse, USA.

Ericsson, G., and T. Heberlein. 2003. Attitudes of hunters, locals and the general public in Sweden now that the wolves are back. Biological Conservation 111:149–159.

Fiorino, D. J. 2001. Environmental policy as learning: a new view of an old landscape. Public Administration Review 61:322–334.

Fulton, D. C, M. J. Manfredo, and J. Lipscomb. 1996. Wildlife value orientations: a conceptual and measurement approach. Human Dimensions of Wildlife 1:24–47.

Giles, R. H. 1978. Wildlife management. W. H. Freeman, San Francisco, California, USA.

Glasbergen, P. 1996. Learning to manage the environment. Pages 175–193 in W. M. Lafferty and J. Meadowcroft, editors. Democracy and the environment: problems and prospects. Edward Elgar, Brookfield, Vermont, USA.

Hautaluoma, J., and P. J. Brown. 1978. Attributes of the deer hunting experience: a cluster-analytic study. Journal of Leisure Research 10:271–287.

Heberlein, T. A. 2012. Navigating environmental attitudes. Oxford University Press, New York, New York, USA.

Heberlein, T. A., and E. Thomson. 1996. Changes in U.S. hunting participation, 1980–90. Human Dimensions of Wildlife 1:85–86.

Heberlein, T. A., J. N. Trent, and R. M. Baumgartner. 1982. The influence of hunter density on firearm deer hunters' satisfaction: a field experiment. Transactions of the North American Wildlife and Natural Resources Conference 47:665–676.

Hendee, J. C. 1974. A multiple satisfaction approach to game management. Wildlife Society Bulletin 1:24–47.

Jonker, S. A., J. F. Organ, R. M. Muth, R. R. Zwick, and W. F. Siemer. 2009. Stakeholder norms toward beaver management in Massachusetts. Journal of Wildlife Management 73:1158–1165.

Kellert, S. R. 1980a. Americans' attitudes and knowledge of animals. Transactions of the North American Wildlife and Natural Resources Conference 45:111–124.

Kellert, S. R. 1980b. Contemporary values of wildlife in American society. Pages 31–60 in W. W. Shaw and E. H. Zube, editors. Wildlife values. Center for Assessment of Non-commodity Natural Resource Values, Institute Series Report No. 1, University of Arizona, Tucson, USA.

Lancia, R. A., C. E. Braun, M. W. Callopy, R. D. Dueser, J. G. Kie, C. J. Martinka, J. D. Nichols, T. D. Nudds, W. R. Porath, and N. G. Tilghman. 1996. ARM! For the future: adaptive resource management in the wildlife profession. Wildlife Society Bulletin 24:436–442.

Lauber, T. B., and T. L. Brown. 2006. Learning by doing: policy learning in community-based deer management. Society and Natural Resources 19:411–428.

Lauber, T. B., D. J. Decker, and B. A. Knuth. 2008. Social networks and community-based natural resource management. Environmental Management 42:677–687.

Lauber, T. B., and B. A. Knuth. 1997. Fairness in moose management decision-making: the citizen's perspective. Wildlife Society Bulletin 25:776–787.

Lauber, T. B., R. C. Stedman, D. J. Decker, and B. A. Knuth. 2011. Linking knowledge to action in collaborative conservation. Conservation Biology 25:1186–1194. doi:10.1111/j.1523-1739.2011.01742.x.

Lischka, S. A., S. J. Riley, and B. A. Rudolph. 2008. Effects of impact perception on acceptance capacity for white-tailed deer. Journal of Wildlife Management 72:502–509.

Liu, J., T. Dietz, S. R. Carpenter, C. Folke, M. Alberti, et al. 2007. Coupled human and natural systems. Ambio 36:639–649.

Loker, C. A., and D. J. Decker. 1995. Colorado black bear hunting referendum: what was behind the vote? Wildlife Society Bulletin 23:370–376.

Loker, C. A., D. J. Decker, and S. J. Schwager. 1999. Social acceptability of wildlife management actions in suburban areas: 3 cases from New York. Wildlife Society Bulletin 27:152–159.

Manfredo, M. J. 1988. Second-year analysis of donors to Oregon's nongame tax checkoff. Wildlife Society Bulletin 16:221–224.

Manfredo, M. J. 2002. Planning and managing for wildlife viewing recreation: an introduction. Pages 1–8 in M. Manfredo, editor. Wildlife viewing in North America: a management planning handbook. Oregon State University Press, Corvallis, USA.

Manfredo, M. J., and R. A. Larson. 1993. Managing for wildlife viewing recreation experiences: an application in Colorado. Wildlife Society Bulletin 21:226–236.

Manfredo, M. J., T. Teel, and A. D. Bright. 2004. Application of the concepts of values and attitudes in human dimensions of natural resources research. Pages 271–282 in M. J. Manfredo, J. J. Vaske, D. R. Field, and P. Brown, editors. Society and natural resources: a summary of knowledge. Modern Litho, Jefferson City, Missouri, USA.

Manfredo, M. J., H. C. Zinn, L. Sikorowski, and J. Jones. 1998. Public acceptance of cougar management: a case study of Denver, Colorado, and nearby foothills areas. Wildlife Society Bulletin 26:964–970.

Margretta, J. 2002. What management is: how it works and why it's everyone's business. Free Press, New York, New York, USA.

McClafferty, J. A., and J. A. Parkhurst. 2001. Using public surveys and GIS to determine the feasibility of restoring elk to Virginia. Pages 83–98 in D. S. Maehr, R. F. Noss, and J. L. Larkin, editors. Large mammal restoration—ecological and social challenges in the 21st century. Island Press, Washington, D.C., USA.

McFarlane, B. L. 1996. Socialization influences of specialization among birdwatchers. Human Dimensions of Wildlife 1:35–50.

McIvor, D. E., and M. R. Conover. 1994. Perceptions of farmers and nonfarmers toward management of problem wildlife. Wildlife Society Bulletin 22:212–219.

Meffe, G. K., L. A. Nielsen, R. L. Knight, and D. A. Schenborn. 2002. Ecosystem management: adaptive, community-based conservation. Island Press, Washington, D.C., USA.

Miller, C. A., and A. R. Graefe. 2000. Degree and range of specialization across related hunting activities. Leisure Sciences 22:195–204.

Muter, B. A., M. A. Gore, and S. J. Riley. 2011. Toward exploring stakeholder and professional information sources about cormorant management in the Great Lakes. Human Dimensions of Wildlife 16:63–66.

Nelson, D. 1992. Citizen task forces on deer management: a case study. Northeast Wildlife 49:92–96.

Organ, J. F., and M. R. Ellingwood. 2000. Wildlife stakeholder acceptance capacity for black bears, beavers, and other beasts in the East. Human Dimensions of Wildlife 5:63–75.

Organ, J. F., D. J. Decker, S. J. Riley, J. E. McDonald Jr., and S. P. Mahoney. 2012. Adaptive management in conservation. Pages 43–54 in N. J. Silvy, editor. The wildlife techniques manual: management. Seventh edition. Volume 2. Johns Hopkins University Press, Baltimore, Maryland, USA.

Potter, D. R., J. C. Hendee, and R. N. Clark. 1973. Hunting satisfaction: game, guns, or nature? Transactions of the North American Wildlife and Natural Resources Conference 38:220–229.

Purdy, K. G., and D. J. Decker. 1989. Applying wildlife values information in management: the wildlife attitudes and values scale. Wildlife Society Bulletin 17:494–500.

Raik, D. B., D. J. Decker, and W. F. Siemer. 2003. Dimensions of capacity in community-based suburban deer management: the managers' perspective. Wildlife Society Bulletin 31:854–864.

Raik, D. B., W. F. Siemer, and D. J. Decker. 2005. Intervention and capacity considerations in community-based deer management: the stakeholders' perspective. Human Dimensions of Wildlife 10:259–272.

Riley, S. J., and D. J. Decker. 2000a. Wildlife stakeholder acceptance capacity for cougars in Montana. Wildlife Society Bulletin 28:931–939.

Riley, S. J., and D. J. Decker. 2000b. Risk perception as a factor in wildlife acceptance capacity for cougars in Montana. Human Dimensions of Wildlife 5:50–62.

Riley, S. J., D. J. Decker, L. H. Carpenter, J. F. Organ, W. F. Siemer, G. F. Mattfeld, and G. Parsons. 2002. The essence of wildlife management. Wildlife Society Bulletin 30:585–593.

Riley, S. J., W. F. Siemer, D. J. Decker, L. H. Carpenter, J. F. Organ, and L. T. Berchielli. 2003. Adaptive impact management: an integrative approach to wildlife management. Human Dimensions of Wildlife 8:81–95.

Sabatier, P. A. 1988. An advocacy coalition framework of policy change and the role of policy-oriented learning therein. Policy Science 21:129–168.

Schusler, T. M., D. J. Decker, and M. J. Pfeffer. 2003. Social learning for collaborative natural resource management. Society and Natural Resources 15:309–326.

Shelby, B., J. J. Vaske, and M. P. Donnelly. 1996. Norms, standards, and natural resources. Leisure Sciences 18:103–123.

Siemer, W. F., and D. J. Decker. 2006. An assessment of black bear impacts in New York. Human Dimensions Research Unit Series Publication 06-6. Department of Natural Resources, Cornell University, Ithaca, New York, USA.

Siemer, W. F., D. J. Decker, P. Otto, and M. L. Gore. 2007. Working through black bear management issues: a practitioners' guide. Northeast Wildlife Damage Management Research and Outreach Cooperative, Ithaca, New York, USA.

Siemer, W. F., P. S. Hart, D. J. Decker, and J. Shanahan. 2009. Factors that influence concern about human–black bear interactions in residential settings. Human Dimensions of Wildlife 14:185–197.

Smith, C. A. 2011. The role of state wildlife professionals under the Public Trust Doctrine. Journal of Wildlife Management 75:1539–1543.

Stout, R. J., D. J. Decker, B. A. Knuth, J. C. Proud, and D. H. Nelson. 1996. Comparison of three public-involvement approaches for stakeholder input into deer management decisions: a case study. Wildlife Society Bulletin 24:312–317.

Stout, R. J., B. A. Knuth, and P. D. Curtis. 1997. Preferences of suburban landowners for deer management techniques: a step towards better communication. Wildlife Society Bulletin 25:348–359.

The Wildlife Society. 2010. The public trust doctrine: implications for wildlife management and conservation in the United States and Canada. Technical Review 10-01. The Wildlife Society, Bethesda, Maryland, USA.

Triezenberg, H. A. 2010. Social networks and collective actions among wildlife management stakeholders: insights from furbearer trapping and waterfowl hunting conflicts in New York State. Dissertation, Cornell University, Ithaca, New York, USA.

U.S. Fish and Wildlife Service. 2007. 2006 national survey of fishing, hunting, and wildlife-associated recreation. U.S. Government Printing Office, Washington, D.C., USA.

Vaske, J. J., D. J. Decker, and M. J. Manfredo. 1995. Human dimensions of wildlife management: an integrated framework for coexistence. Pages 33–49 in R. L. Knight and K. J. Gutzwiller, editors. Wildlife and recreationists: coexistence through management and research. Island Press, Washington, D.C., USA.

Vaske, J. J., and D. Whittaker. 2004. Normative approaches to natural resources. Pages 283–294 in M. J. Manfredo, J. J. Vaske, D. R. Field, and P. Brown, editors. Society and natural resources: a summary of knowledge. Modern Litho, Jefferson City, Missouri, USA.

Whittaker, D. 1997. Capacity norms on bear viewing platforms. Human Dimensions of Wildlife 2:37–49.

Whittaker, D., and M. J. Manfredo. 1997. Living with wildlife in Anchorage: a survey of public attitudes. Human Dimensions in Natural Resources Unit Summary Report No. 35. Colorado State University, Fort Collins, USA.

Wieczorek Hudenko, H., D. J. Decker, and W. F. Siemer. 2008. Living with coyotes in suburban areas: insights from two New York State counties. Human Dimensions Research Unit Series Publication 08-8. Department of Natural Resources, Cornell University, Ithaca, New York, USA.

Wieczorek Hudenko, H., W. F. Siemer, and D. J. Decker. 2010. Urban carnivore conservation and management: the human dimension. Pages 21–33 in S. Gehrt, S. Riley, and B. Cypher, editors. Urban carnivores: ecology, conflict, and conservation. Johns Hopkins University Press, Baltimore, Maryland, USA.

Witter, D. J., and W. W. Shaw. 1979. Beliefs of birders, hunters, and wildlife professionals about wildlife management. Transactions of the North American Wildlife and Natural Resources Conference 44:298–305.

Witter, D. J., D. L. Tylka, and J. E. Werner. 1981. Values of urban wildlife in Missouri. Transactions of the North American Wildlife and Natural Resources Conference 46:424–431.

Wittmann, K., J. J. Vaske, M. J. Manfredo, and H. C. Zinn. 1998. Standards for lethal control of problem wildlife. Human Dimensions of Wildlife 3:29–48.

Wondolleck, J. M., and S. L. Yaffee. 2000. Making collaboration work: lessons from innovation in natural resources management. Island Press, Washington, D.C., USA.

Wright, B. A., T. L. Brown, H. K. Cordell, and A.L. Rowell. 1989. The national private land ownership study: establishing the benchmark. Pages 33–50 in A. H. Watson, editor. Outdoor recreation benchmark 1988: proceedings of the National Outdoor Recreation Forum. General Technical Report SE-52. Southeastern Forest Experiment Station, Asheville, North Carolina, USA.

Yankelovich, D. 1991a. Coming to public judgment. Syracuse University Press, Syracuse, New York, USA.

Yankelovich, D. 1991b. The magic of dialogue: transforming conflict into cooperation. Simon and Schuster, New York, New York, USA.

Zinn, H., M. J. Manfredo, J. J. Vaske, and K. Wittmann. 1998. Using normative beliefs to determine the acceptability of wildlife management actions. Society and Natural Resources 11:649–662.

5

STRUCTURED DECISION MAKING

MICHAEL C. RUNGE, JAMES B. GRAND, AND MICHAEL S. MITCHELL

INTRODUCTION

Wildlife management is an exercise in decision making. While wildlife science is the pursuit of knowledge about wildlife and its environment (including wildlife ecology, physiology, behavior, evolution, demography, genetics, disease, habitat, and population dynamics), wildlife management is the application of that knowledge in a human social context, application that typically requires a choice of management options. Decisions require the integration of science with values, because in the end any decision is an attempt to achieve some future condition that is desirable to the decision maker (Keeney 1996b). Wildlife management, particularly under the North American Model of Wildlife Conservation (Chapter 2), is often practiced by federal, state, or private agencies on behalf of the public and thus integrates science, law, and public values (Chapter 4). For example, the development of hunting regulations for white-tailed deer (*Odocoileus virginianus*) in Pennsylvania is a complicated choice among many possible permutations of regulations, a choice designed to balance many desires: hunting opportunity, the long-term conservation of deer, a sense of fair pursuit, fair public access, population levels commensurate with habitat capacity and predator density, wildlife viewing, and state and local economic benefits from hunting and tourism. Certainly, there are decades of wildlife science about deer and social science about deer hunters to support this decision, but they alone cannot identify the best regulations. The decision (i.e., the choice of hunting regulations) needs to integrate science- and values-based components (Wagner 1989).

Decision analysis is to wildlife management as the scientific method is to wildlife science, a framework and a theory to guide practice. The field of decision science is broad, with roots in economics stretching back to the 1940s, if not earlier (von Neumann and Morgenstern 1944), and the cross-disciplinary nature of the field became evident in the 1960s, with contributions from cybernetics (computer science), business administration, and mathematics (Raiffa and Schlaifer 1961, Howard 1968). Modern decision science has added expertise in many areas, including psychology, operations research, sociology, risk analysis, and statistics. Decision analysis has been applied in many contexts, including nuclear warfare planning (Dalkey and Helmer 1963), energy planning (Diakoulaki et al. 2005), adoption of health-care technologies (Claxton et al. 2002), and top-level political decisions in the Finnish parliament (Hämäläinen and Leikola 1996), to name a few. Formal decision analysis techniques are increasingly used in environmental fields (Kiker et al. 2005), particularly fisheries (Bain 1987, Gregory and Long 2009, Runge et al. 2011a), but also in wildlife management (Ralls and Starfield 1995, Johnson et al. 1997, Regan et al. 2005, Lyons et al. 2008, Runge et al. 2009, McDonald-Madden et al. 2010, Moore et al. 2011, Runge et al. 2011b). But it is perhaps surprising that, although wildlife management focuses on integrating values and science to make decisions, formal decision analysis is not applied more often, nor is it a core element in graduate education (van Heezik and Seddon 2005).

Is wildlife management an art or a science? There are wildlife managers who will vigorously argue the former, that the decisions they make are the result of years of experience, a deep sense of intuition, and scientific training. This is perhaps a traditional view; the language can be traced to the very beginning of our field. Leopold (1933:3) wrote, "game management is the art of making land produce sustained annual crops of wild game for recreational use." More recently, Bailey (1982:366) similarly described wildlife management: "As an art wildlife management is the application of knowledge to achieve goals . . . In selecting goals, [wildlife managers] compare and judge values." But note that the art that Leopold (1933) and Bailey (1982) describe is the integration of wildlife science with values-based judgments. Leopold's (1933) example embeds three main goals: providing recreational use of wild game, having that use be sustainable, and having that use be consistent (i.e., annual). A deeper question is whether the integration of science and values in making wildlife management decisions can be more than the informal and loosely structured judgment of a decision maker. Are wildlife management decisions transparent and replicable? Does the public know what values were balanced in choosing the decision, and what science was consulted? Would a different decision maker have weighed the evidence and the values in the same way, and would that person come to the same decision? Will the decision maker's successor be able to maintain continuity, or will knowledge be lost every time someone retires? Increasingly, the public is demanding more transparency of natural resource managers, and decision analysis provides the framework for this transparency. This is not to say that the intuitive decision making of experienced wildlife managers is without merit, only that modern demands of transparency, accountability, inclusiveness, and efficiency require structured approaches to wildlife management decisions.

A FRAMEWORK FOR DECISION MAKING

Making decisions is a hallmark of human existence, something we do every day. Decisions are not always difficult to make, but some (e.g., public sector decisions) are sufficiently complex and challenging that the common tools and rules of thumb used by humans in daily decision making are inadequate for achieving good decisions reliably. Decision analysis, or structured decision making (SDM), is "a formalization of common sense for decision problems which are too complex for informal use of common sense" (Keeney 1982:806). This section describes the elements of decision analysis in the context of wildlife management.

What is a decision? A decision is an "irrevocable allocation of resources . . . not a mental commitment to follow a course of action but rather the actual pursuit of the course of action" (Howard 1966:55). In the United States, the annual federal waterfowl hunting framework and the corresponding state waterfowl hunting seasons are decisions: they irrevocably set in motion harvest of waterfowl. State wildlife action plans (Fontaine 2011) are not themselves decisions, but they give rise to decisions when staff and fiscal resources are dedicated to carrying out actions in the plans. Likewise, recovery plans under the U.S. Endangered Species Act (ESA) are not decisions, but the actions taken under their auspices are.

The PrOACT Framework

There are two hallmarks of structured decision making: values-focused thinking and problem decomposition. Values-focused thinking emphasizes that all decisions are inherently statements about values, and so discussion of those values should precede other analysis (Keeney 1996a). Problem decomposition breaks a decision into its logical components, allowing identification of impediments to the decision, providing focus when and where needed, and creating an explicit, transparent, and replicable framework for decision making that improves performance and stands up to scrutiny. The logical components of decision analysis include defining the Problem, identifying Objectives, defining alternative Actions to be taken, evaluating Consequences of actions, and assessing Trade-offs among alternative actions (Fig. 5.1). These components constitute the PrOACT framework (Hammond et al. 1999). Problem framing is often an iterative process intended

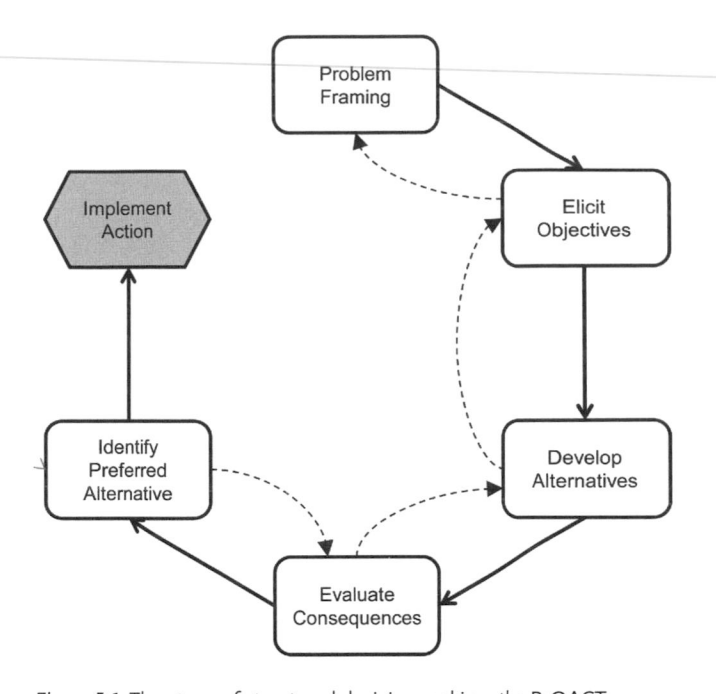

Figure 5.1. The steps of structured decision making: the PrOACT sequence.

to facilitate insights about a decision throughout development of the analysis. Each step benefits from re-evaluation at the completion of subsequent steps (Fig. 5.1).

Defining the problem is the critical first step of SDM that guides the process toward appropriate tools and information, determines appropriate levels of investment, and ensures that the right problem is being solved. Its importance cannot be overstated; time taken to craft a concise yet comprehensive and accurate problem definition pays off (Hammond et al. 1999). A good problem statement comprises the actions that need to be taken; legal considerations; who the decision maker is; the scope, frequency, and timing of the decision; goals that need to be met; and the role of uncertainty.

Objectives make explicit what the decision maker cares most about, defining what will constitute successful outcomes in the decision-making process. Along with the problem statement, well-defined objectives are critical to all subsequent steps in structured decision making, allowing the creation and assessment of alternative actions, identification of pertinent information for making the decision, and explanation of the decision-making process to others.

Actions represent choices available to a decision maker, or alternative approaches to achieving at least a subset of objectives. Good alternative actions address the future (not the past), are unique, encompass a broad range of possible actions, and can be implemented by the decision maker (i.e., are financially, legally, and politically reasonable).

Once alternative actions have been defined, the consequences of taking each action need to be predicted with respect to the objectives. All decisions involve prediction, whether implicit or explicit. One of the strengths of wildlife science is the wealth of tools (e.g., sampling protocols, data analysis methods, and modeling approaches) designed to help managers make predictions.

The final step in the PrOACT sequence is an analysis of trade-offs among alternatives based on their expected performance relative to the objectives, an analysis designed to identify an alternative that best achieves the set of objectives. This analysis can be anywhere from narrative to mathematical, depending on the complexity of the problem. The key role of a decision maker is to integrate the values- and science-based elements of the decision. Done well, this analysis should be transparent, should be comprehensive with respect to all fundamental objectives, should be explicit, should make use of best available information, and should address uncertainty directly.

The PrOACT sequence is simple but surprisingly powerful. In many decision settings, simply framing the problem helps to remove impediments to the decision. But the PrOACT framework also provides direction toward more advanced tools that may be needed in some circumstances.

When Is SDM Appropriate?

Structured decision making is a broad and flexible set of tools that can be applied in a variety of settings. The PrOACT model provides a useful framework for ordering and deploying these tools, but SDM is not appropriate in all settings. First, SDM assumes that there is a decision to be made, which is not always the case. Strategic planning processes, prioritization schemes, research design, species status assessment, and compiling of scientific findings are all activities in which a wildlife biologist might participate, and products a wildlife manager might want, but they are not always in service to a specific decision. In those cases, SDM might help guide thinking toward the decisions that might be downstream of those activities, but it might also be frustrating to apply. Second, SDM assumes either that there is a single decision maker, or a single decision-making body, or multiple decision makers who agree to a spirit of open-mindedness and discovery for the purposes of identifying a common path. In situations where multiple parties to a decision are in substantial conflict, the endeavor might be better served by other facilitation, mediation, joint fact finding, or conflict resolution techniques. In situations where there are multiple decision makers in competition with one another, who have no intention to openly reveal their objectives or search for common ground, another branch of the decision sciences—game theory—provides insights and methods for analysis.

There are a number of other processes meant to support decision making that wildlife managers will hear about, which have overlapping domains of application (Fig. 5.2). Structured decision making is useful when the objectives are known or can be developed, but conflict resolution methods are better when the objectives are deeply disputed. Structured decision making is broadly applicable whether the scientific aspects of the decision are well known or not; joint fact finding is sometimes used when the science is disputed, as a way to engage stakeholders and develop common ground (Karl et al. 2007). As discussed later in this chapter, adaptive management is a

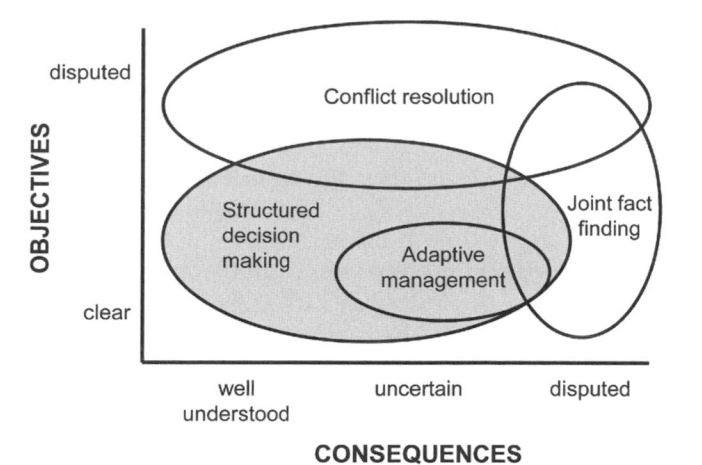

Figure 5.2. When is structured decision making appropriate?

special case of structured decision making, valuable for recurrent decisions that are impeded by uncertainty.

Classes of Decisions

One of the values of early attention to problem framing is the ability to recognize classes of decisions, which can in turn lead to identifying the best analytical tools to support the decision maker. Decisions can be classified on three axes: single-versus multiple-objective decisions; decisions in which uncertainty is, or is not, a major impediment; and stand-alone versus linked decisions (Table 5.1). The binary nature of these classes masks the complexity of true problems, so the reader should understand that there are some gray areas. Single-objective problems (or ones in which an objective carries significantly more weight than all others) that are not plagued by uncertainty (or for which the uncertainty in not consequential) are simple optimization problems for which a variety of tools (e.g., graphical, numerical, analytical) exist. Single-objective problems made in the face of uncertainty are the setting of classical decision analysis, and tools such as decision trees are valuable. Multiple-objective settings are supported by a broad array of multicriteria decision analysis (MCDA) techniques. Decisions that are linked to other decisions, either in a fixed sequence or in a recurrent pattern, require still more methods: dynamic optimization methods to address the linkages across time, and adaptive methods to account for resolution of uncertainty. Many of these methods are described in more detail later in this chapter, but it is helpful to have a context in which to place them.

THE VALUES-BASED ASPECTS OF DECISIONS

In the absence of a structured framework for coherently integrating value judgments and scientific judgments, decision makers tend to confound personal preferences and technical predictions (Failing et al. 2007). One of the key benefits of the problem decomposition embodied in PrOACT is the ability to separate the values- and science-based aspects of the decision,

Table 5.1. Eight classes of decisions and the common decision analytic tools associated with them

	Single objective	Multiple objective
Single stand-alone decision		
Not impeded by uncertainty	Optimization tools	MCDA[a]
Impeded by uncertainty	Decision trees	MCDA with sensitivity analysis
Linked decisions		
Not impeded by uncertainty	Dynamic optimization	Dynamic MCDA
Impeded by uncertainty	EVPI,[b] ARM[c]	Multiple-objective ARM

[a]MCDA, multicriteria decision analysis.

[b]EVPI, expected value of perfect information.

[c]ARM, adaptive resource management.

which allows those pieces to be analyzed by the right people with the appropriate tools. In the spirit of value-focused thinking (Keeney 1996a), we first discuss the values-based aspects before turning attention to the science-based aspects.

Defining the Problem

How a decision is framed affects how it should be analyzed, and this framing should reflect the values of the decision maker. Framing the decision can be surprisingly difficult and frustrating, but without a full definition of the problem and its context, considerable resources can be invested in solving the wrong problem. Further, a concise framing of the problem can aid clear communication with interested parties. For a simple, widely understood rubric to developing a problem statement, it is useful to refer to the five W's used in journalistic and technical writing. Many of the critical elements of the problem can be identified with explicit statements addressing the who, what, where, when, why, and how of a decision.

One way to begin is to ask, who needs to make a decision? Sometimes the decision maker is obvious (e.g., where mandated by law or regulations), but other times, identifying the decision maker can be challenging. First, it is useful to distinguish decision makers from those that implement a decision. The decision maker is the authority upon whom responsibility for the decision rests. Second, there may not be a single decision maker. In some collaborative settings, decision making is the joint responsibility of representatives from multiple agencies or interests; if that is the case, it is important for the decision analyst to understand the governance structure that supports that group. Third, in many public agency settings, the authority for the decision may be delegated. For example, in the United States, the secretary of interior has statutory responsibility under the ESA, but typically that authority is delegated to the director of the U.S. Fish and Wildlife Service (USFWS), who in turn may further delegate portions of that authority. This can create a challenge, because while the field office supervisor might be the decision maker with the motivation to analyze the deci-

sion, it is not clear at the outset how much consultation will be required up the delegated chain.

The question of who can be broadened considerably by asking, who is interested in the decision? Stakeholders include anyone with an interest in the outcome of the decision. These include individuals who could be directly or indirectly affected by the actions under consideration. In the case of the private landowner, it may be relatively simple to identify the stakeholders on the basis of familial and business relationships. However, many natural resource problems faced by public agencies affect a diverse group of stakeholders, including such consumers as hunters, anglers, hikers, and bird watchers, and groups that are seemingly detached from the natural resources in question but that are intensely interested in their status. For example, few individuals will ever visit the Alaskan arctic, but interest in the effects of such stressors as mineral extraction and climate change on arctic wildlife has evoked reactions from countless individuals across North America. The field of human dimensions offers methods to identify, understand, and involve stakeholders in decisions (Chapter 4).

The central question that a problem statement needs to address is, what is the decision to be made? To put it differently, what choice does the decision maker face? In wildlife management, decisions can be simple or exceedingly complex. For example, a wildlife manager might be faced with the relatively simple decision of whether to plant wildlife openings with native legumes or to allow old-field succession to take its course. The same manager may be tasked with developing a management plan that involves making decisions about dozens or hundreds of sites that will play out over many years.

Knowing explicitly where the affected resources are helps define the geographic and taxonomic scale of the problem. By asking when a decision is needed, we define two important aspects of the problem: timing and frequency. The first concerns the urgency for a decision; a short time scale may limit the complexity of the decision analysis. The second concerns whether the decision is made one time or recurrently. In many cases, the decision occurs once, such as the placement of infrastructure—roads, buildings, or dams. In other cases, decisions are recurrent, as in setting annual harvest regulations. In still others, a series of sequential decisions that hinge on the success of previous actions are considered.

The problem statement should address why the decision is important. To do so, the consequences of failing to make a decision can be examined. Will it result in strongly negative consequences, such as extinction, loss of hunting opportunities, loss of revenue, or litigation? In some cases, there may be a legal mandate related to agency mission, as in setting harvest regulations, listing species that are candidates for protection as threatened or endangered, or reviewing management alternatives (e.g., an Environmental Impact Statement under the National Environmental Policy Act, or NEPA). In other cases, decisions can be related to meeting an agency strategic objective such as providing public hunting or other recreational opportunities. In still other cases, a decision might relate to meeting tactical objectives of an agency, such as minimizing risk to natural resources, maximizing effectiveness of management, or meeting an agreed upon population objective.

The problem statement should also describe how to solve the problem. This description should be broad and conceptual; an explicit statement of alternatives and their relative value to solving the problem comes later in the process. A good way to think about this portion of the problem statement is a description of the natural resource management tools that could be implemented in reaching a solution. For example, manipulating harvest regulations at continental scales can maximize harvest of waterfowl. Meeting population objectives for nongame species can be achieved by enhancing habitat quality. These statements may put bounds on the alternatives that will be considered in the analysis, but they might also stimulate discussion and require revision during the development of the decision analysis.

Many insights about the nature of the decision arise out of the analysis, however, so problem definition often evolves. A well-constructed decision process allows the decision maker to revisit the elements of the decision framework repeatedly as the analysis proceeds.

Articulating the Objectives

In wildlife management the development of unambiguous, meaningful objectives of the decision makers and the stakeholders is a critical step in the decision-making process. Ambiguous, poorly formed, and hidden objectives often lead to poor decision making, as does the exclusion of objectives that are important to large or important segments of the community of stakeholders. Clear, concise objectives with measureable attributes are the key to making informed, smart decisions because they define the decision's purpose (Keeney 1996a). However, when forced to make decisions in natural resources management, few individuals actually take the time to fully describe the purpose of the actions under consideration. We find it useful to distinguish four steps in the development of objectives: eliciting objectives, classifying objectives, structuring objectives, and developing measurable attributes.

Eliciting Objectives

In developing objectives, it is often useful to start by eliciting the concerns of the decision makers and other stakeholders. Elicitation takes many forms, including workshops, public meetings, and one-on-one interviews. The important concept here is to be inclusive, empowering stakeholders and their representatives to articulate objectives that are important to making an informed decision. A variety of objectives is typical in wildlife management. Traditional concerns relate to the abundance and distribution of wildlife species, the health and quality of individual animals, the resources on which they depend, and their availability for consumptive or nonconsumptive uses. During the last several decades, new concerns related to maintaining or increasing biodiversity have made their way into wildlife conservation and management. And, increasingly, we recognize that wildlife management takes place in a sociopolitical context, and so a broader set of objectives is important, including economic, cultural, aesthetic, and spiritual concerns.

Objectives related to wildlife population abundance usually stem from worries about their viability (e.g., rare species), long-term persistence (e.g., many migratory songbirds), or harvestable surplus (e.g., most game species). Stakeholders often express these types of concerns in terms of declining populations or harvest levels. However, concerns over wildlife populations may also stem from overabundance, especially where there are large economic impacts—for example, cormorants (*Phalacrocorax* spp.), white-tailed deer, nutria (*Myocaster coypus*), muskrat (*Ondatra zibethicus*), and raccoons (*Procyon lotor*)—or environmental impacts—for example, western Canada geese (*Branta canadensis*) and lesser snow goose (*Chen caerulescens*).

Wildlife managers are often concerned about objectives above and beyond wildlife abundance, including distribution and quality of wildlife populations. For example, recovery criteria for listed species usually include a description of the number and distribution of distinct populations—like the red-cockaded woodpecker (RCW, *Picoides borealis*; USFWS 2003)—as an indication of viability and as a fundamental desire to see the species restored to its former range. The quality of the individuals in a population is also often a concern, both as an indication of the health of the population and also as a fundamental objective. For example, management of wildlife populations for trophy harvest will focus on elements such as age structure, size, and other indicators of individual health.

Concerns over biodiversity have increased as the field of wildlife management has been broadened beyond traditional game management. Large-scale programs such as gap analysis (Scott 1993) have increased awareness about the impacts of cumulative habitat loss by focusing on land management practices and areas of high biotic diversity. Federal aid programs like state wildlife grants have enabled many state agencies to identify concerns and to develop objectives related to the conservation of biodiversity and populations of concern.

The objectives related to wildlife management, however, transcend concerns about wildlife. Economic concerns, too, are deeply important to stakeholders. The development of the Northwest Forest Plan needed to consider old-growth habitat for spotted owls (*Strix occidentalis*), the viability of the forest products industry, and the livelihood of its employees (Thomas et al. 2006). Reintroduction of wolves (*Canis lupus*) into the northern Rocky Mountains needed to consider the viability of the wolves and the impact on hunting opportunity for big game, but also the economic concerns of cattle and sheep ranchers (Fritts et al. 1997). Social concerns related to the impacts of wildlife management go beyond economic considerations and include spiritual, aesthetic, cultural, and recreational objectives (Bengston 2000; Chapter 4). Wildlife and fish management in Grand Canyon needs to take into consideration the spiritual and cultural objectives of native tribes, the opportunity for wilderness recreation, and the provision of energy and water to the arid Southwest in addition to economic and strictly wildlife-related objectives (Runge et al. 2011a).

Classifying Objectives

Objectives can be classified into four broad categories: strategic, fundamental, means, and process objectives (Keeney 2007). Strategic objectives are the highest-level objectives and are often associated with the mission of the agency or individual. For example, the legal mandates of a state agency associated with the maintenance of imperiled species and productivity of game species would be considered strategic objectives. These objectives are frequently beyond the scope of the management decisions faced by wildlife managers, and as such they often do not help discern among management alternatives. But they do define the context of the fundamental objectives, which are perhaps the most important category. Fundamental objectives are the "ends" of the wildlife management problem and the highest-level objectives incorporated in a decision analysis. Means objectives are the methods by which we achieve the fundamental objectives, but they may not be necessary if there are multiple pathways to achieve the fundamental objectives. Finally, process objectives govern how the decision is made but do not affect discrimination among the alternatives. For example, a decision maker—for legal, strategic, or ethical reasons—may desire that public meetings and outreach are included in the decision-making process.

Fundamental objectives are the focus of decision analysis; they alone are used to distinguish among the alternatives. Good fundamental objectives have several key characteristics. First, they are measurable. Attributes can be developed for them that can be measured on an unambiguous scale. Second, good fundamental objectives are controllable; that is, they can be influenced by the management actions under consideration. Third, fundamental objectives are those the decision maker deems essential—there is no acceptable substitute.

It often requires careful thought to distinguish fundamental from means objectives. A useful way to make such distinctions is to ask why each objective is important, which frequently leads to the discovery of new, higher-level objectives that describe the most important, desired outcomes. For example, managers interested in wildlife populations in longleaf pine (*Pinus palustris*) habitats often identify concerns related to the absence or infrequent use of fire in those systems. A concise initial objective might be to increase the use of prescribed fire in longleaf pine. When asked why, managers often respond that it improves habitat quality; the restated objective may be to increase foraging habitat for RCW and northern bobwhite (NOBO; *Colinus virginianus*). Asking why again can reveal that there is concern over the productivity or abundance of those populations, suggesting an objective to increase populations of both species. Asking the question yet again may elicit concerns over the viability of the RCW population and the size of the harvest of NOBO. Asking why once more may reveal that the agency has a mandate to maintain populations of endangered species (e.g., RCW) and to increase harvestable populations of game species (e.g., NOBO). So, classifying objectives identifies two fundamental goals (i.e., to maintain a viable population of RCW and to maximize harvest potential of NOBO) from a nested set of means objectives.

Structuring Objectives

A fundamental objectives hierarchy illustrates the relationships among the most important objectives in a decision problem. A generic fundamental objectives hierarchy can be used to stimulate discussion and to identify problem-specific objectives. In many natural resource–related problems, useful generic fundamental objectives include: improving or maintaining wildlife populations, minimizing cost, and providing utilitarian and nonutilitarian benefits to stakeholders. A generic fundamental objectives hierarchy (Fig. 5.3) can be modified to develop specific objectives related to a specific problem. Depending on the problem at hand, objectives surrounding the status of wildlife populations may be more specifically defined as one or more of the following: abundance, distribution, health, genetic diversity, and species diversity. Cost is nearly always a consideration and, given a choice between two equally effective and likely solutions, the less expensive option is almost always more desirable. In other situations, where a budget is fixed or cost is viewed as a constraint, the solution that results in the best population status and stakeholder satisfaction for the same cost is the logical choice. A broader set of stakeholder concerns is often a crucial consideration for wildlife populations held in public trust. Notice that the elements of this fundamental objectives hierarchy do not overlap; they

1. Maximize ecological benefits
   a. Maximize persistence of native species (or communities)
      i. Maximize population size
      ii. Maximize distribution
      iii. Maximize individual quality
      iv. Maintain genetic and species diversity
   b. Minimize nonnative and invasive species (or communities)
   c. Maintain ecosystem function
2. Minimize costs
   a. Minimize capital (fixed) costs
   b. Minimize ongoing (variable) costs
3. Maximize public and private benefits (utilitarian benefits)
   a. Maximize consumptive recreational benefit
   b. Maximize nonconsumptive recreational benefit
   c. Maximize public services (e.g., energy generation, water delivery)
   d. Maximize public health and safety
   e. Maximize private economic opportunity
   f. Provide sustainable subsistence use, where appropriate
4. Facilitate cultural values and traditions (nonutilitarian benefits)
   a. Maximize aesthetic and spiritual values
   b. Minimize taking of life
   c. Treat animals in a humane manner

Figure 5.3. Hierarchy of generic fundamental objectives for wildlife management.

express independent elements of concern in the decision problem, so there is no double counting. A fundamental objectives hierarchy must be complete, including all of the concerns that bear on the decision.

Measurable Attributes

Attributes are the measurement scales for fundamental objectives. Identifying attributes not only allows measurement of achievement, it forces clarity in the definition of each objective. The purpose of decision analysis is to provide a transparent comparison of the alternatives, and the attributes provide the quantitative measure of the consequences of each alternative for each objective. The capacity to make informed trade-offs is severely compromised if attributes are not clearly described (Keeney 2002). Because fundamental objectives are the focus of decision analysis, measurable attributes should be developed for fundamental objectives. Attributes that might be used by a manager interested in wildlife populations in longleaf pine habitats vary (Table 5.2).

There are three types of measurable attributes: natural attributes, proxy attributes, and constructed attributes. Each of the examples (Table 5.2) is a natural attribute—the scales directly capture the objective of interest, they are easily interpreted by anyone familiar with wildlife management, and there are widely accepted techniques or guidelines for their empirical measurement or estimation. However, for many objectives, appropriate natural attributes do not exist or are impractical for assessing consequences (e.g., data may not be available). In some cases an attribute can be constructed based on a relative scale. For example, absent measurement of fitness of individuals in a habitat, no universal scale exists for measuring the degree to which an area provides habitat for a species, because habitat requirements vary among species, and for most species we can only measure what we perceive to be the important requisites for habitat. An attribute for measuring habitat quality must be constructed and scored on an ordinal scale. By their very nature, constructed scales are subjective; therefore clear definitions of the levels are required for repeatable, transparent scoring (Table 5.3). By contrast,

proxy attributes are usually natural attributes for quantities (sometimes associated with means objectives) that provide an indirect measure of the objective of interest. For example, if our true objective was to increase hunting opportunities on public lands, the number of hectares open to public hunting might be a useful proxy attribute. Although many other factors—weather, access, and habitat condition—influence hunting opportunity, we assume that the area available for public hunting is highly correlated with hunting opportunities on public lands. In general, natural attributes are preferable to proxies or constructed scales. But often this preference has to be relaxed to achieve a complete description of the decision problem (Keeney 2007).

Generating Alternative Actions

Generating alternatives is a values-based exercise and a scientific exercise. The values-based element recognizes that alternatives are the admissible ways of achieving the objectives. Alternative actions can vary from simple to complex. In some cases, the alternative actions are a small set of discrete options, such as whether to use prescribed fire, mowing, or herbicide to set back succession in a grassland. In other cases, the alternative actions come from a continuous set, such as possible sustained harvest rates for a waterfowl population, which could take any value between zero and the logistic growth rate for the population. But often in wildlife management, the alternatives have quite complex structures. Portfolios are alternative actions that are composed of permutations of like elements. For example, a management agency allocating resources to invasive species control could consider a large number of potential portfolios of invasive species, each portfolio a list of invasive species targeted by management control. The number of potential portfolios in this case would include all permutations of the set of invasive species in that ecosystem. Strategies (or strategy tables) are alternative actions composed of permutations of unlike elements. For example, the options considered in an analysis of potential responses to the emergence of white-nose syndrome in bats were strategies composed of such elements as the methods of addressing the fungal agent, methods of captive propagation, cave access restrictions, and management of disease spread (Szymanski et al. 2009).

Frequently, the need for structured decision making arises from the desire to compare alternatives that are developed

Table 5.2. Natural attributes for objectives in the longleaf pine example

Objective	Attribute
Increase use of prescribed fire (means)	Return interval or frequency of fires
Increase foraging habitat (means)	Ha of pine burned in the last four years
Increase NOBO[a] harvest (fundamental)	Number of birds shot by hunters annually
Increase viability of RCW[b] (fundamental)	Probability of persistence over 100 years

Measurable attributes are normally developed only for fundamental objectives, but the attributes for some means objectives are shown, too, for illustrative purposes.

[a]NOBO, northern bobwhite.

[b]RCW, red-cockaded woodpecker.

Table 5.3. Example of a constructed scale for habitat quality

Attribute level	Description of level
3	Very good: >80% canopy closure, >75% of canopy trees mast-producing oak, hickory, or beech
2	Good: 60–80% canopy closure, 26–75% of canopy trees mast-producing oak, hickory, or beech
1	Poor: <60% canopy closure, ≤25% of canopy trees mast-producing oak, hickory, or beech
0	No value: no mast-producing trees in forest canopy

before the problem is well defined, but a thorough analysis of any problem will attempt to consider a wide variety of alternatives. There are a number of pitfalls that limit our ability to develop creative, potentially valuable alternatives. One of the most common pitfalls is "anchoring." Anchoring is the tendency to conduct business as usual, choosing solutions to recently addressed problems, or grasping at the first suggested alternative (Keeney 1996a). Choices made by anchoring constrain creativity and thoughtful development of alternatives. There are many techniques that can be applied to avoid anchoring and to encourage development of good alternatives (Keeney 1996a). One method offers constructive insight: developing creative alternatives may result from broadening the decision context. This usually occurs when the decision maker or analyst determines that additional fundamental objectives exist. For instance, a game manager facing dissatisfied stakeholders (e.g., hunters) may assume that their objective is to harvest trophy animals and may perceive the trigger to be low harvest of trophy animals, which could result in a set of alternatives related to increasing the frequency of trophy characteristics in populations. But if the actual trigger is that hunters are seeing fewer deer, then broadening opportunities to view deer could lead to alternatives that do not result in increased harvest.

In summary, the intent is not to develop an exhaustive set of potential actions, but to develop a set of alternatives for impartial evaluation that represents the spectrum of potential solutions to the problem at hand. The set of alternatives must influence all of the fundamental objectives via means objectives, but it is not necessary to limit alternatives to just those that affect every fundamental objective. It is also possible to find that some important objectives are not controllable within the set of feasible alternatives and may require either consideration as sources of uncontrollable uncertainty, broadening the context of the problem, or elimination of those objectives from the analysis.

Evaluating the Trade-offs

The crux of any decision is the set of values placed on the objectives. In a single-objective problem, once the measurable attribute for the objective is established and the values-based aspects of the decision are expressed, the solution is the alternative that best achieves that objective. But a common wildlife management framework that might be cast as a single-objective problem—harvest management—reveals the complexity inherent in objectives. The solution of a maximum sustained yield problem is really a balance between two objectives: maximizing the short-term harvest and sustaining the population in perpetuity. The optimal harvest rate balances these two objectives to produce a maximum annual harvest that can be sustained indefinitely (Runge et al. 2009). But it is possible to ask whether these objectives might be balanced in some other way, or perhaps in deference to even more objectives; such has been the dialogue in the North American waterfowl management community in the 21st century (Runge et al. 2006).

Most wildlife management decisions involve trade-offs among multiple objectives, and meaningful evaluation of those trade-offs is grounded in values preferences among fundamental objectives. It is rare for all of the objectives to be achieved under a single alternative; typically the objectives compete, and the challenge for the decision maker is how to choose an alternative that best balances those objectives. The balancing of objectives is a values judgment that should reflect the preferences of the decision maker, preferences that often reflect societal priorities embodied in the organization the decision maker represents.

There are several tools from the field of decision analysis that are designed to elicit these value judgments from decision makers. A commonly used method is swing weighting (von Winterfeldt and Edwards 1986), which has the desirable property of encouraging decision makers to think about the range of consequences associated with alternatives together with their importance (Keeney 2002). In this method, the decision maker is asked to consider a series of hypothetical orthogonal scenarios in which the objectives are swung from their worst consequence to their best consequence one at a time; the decision maker ranks these scenarios and then assigns a score that represents how much any scenario is preferred over another. From these scores, weights are derived for each objective, and these weights are used in a multicriteria decision analysis (MCDA; see below). These weights on the individual objectives explicitly state how much one objective is valued over another, and can be used to balance the trade-offs in the analysis.

Another way to examine and value trade-offs is to look at the "efficiency frontier." The efficiency frontier, also called the Pareto frontier, is the set of possible actions for which no gain in one objective can be achieved without a loss in some other objective. For two-objective problems, the Pareto frontier is often depicted as a graph of performance on one objective against performance on the other objective. Such a graph makes the trade-off visually evident and can be used to engender discussion about which solution best balances the two objectives.

One important point to emphasize is that the judgment about how to balance competing objectives cannot be answered by science. At its heart, wildlife management is an expression of a rich array of societal objectives that speak to a complex set of economic, recreational, aesthetic, and spiritual values. How these values are expressed in decision analysis is one of the most important things a decision maker needs to be able to judge and communicate.

THE SCIENCE-BASED ASPECTS OF DECISIONS

Wildlife management is, of course, founded in wildlife science; our decisions about how to manage wildlife are, and should be, influenced by our understanding of how natural systems respond to management. The science-based aspects of decisions include three sets of activities: generating alternative actions, predicting the consequences of those actions, and coherently integrating value judgments and technical judgments through reasoned use of decision analysis tools.

Predicting the Consequences

One of the critical roles of science in a decision analysis is the evaluation of the alternatives against the objectives. Often, this involves predicting how the alternative actions will affect the resources in question, and how those effects will influence achievement of the fundamental objectives. These predictions are often made using empirical data, inferring future responses based on past observations, but increasingly we recognize the importance of expert elicitation for predicting consequences. In a full decision analysis, the consequences need to be predicted for all the fundamental objectives; while predictions about natural resources themselves are the mainstay of traditional wildlife management, predictions about the human responses to wildlife management are also critical.

A central theme in wildlife science is prediction of how individual animals, wildlife populations, and the ecosystems in which they reside respond to management actions (Chapters 7 and 19). The wildlife literature is rich with examples of predictive models based on empirical data, including age-structured population models (Caswell 2001), harvest and take models (Runge et al. 2009), population viability analyses (Beissinger and McCullough 2002), wildlife-habitat models (Morrison et al. 2006), resource selection functions (Boyce and McDonald 1999), and, increasingly, coupled climate-wildlife models (Hunter et al. 2010). There are two steps in the development of these predictive models: development of the model structure and estimation of the parameters. In an applied setting, the model structure is in part determined by the decision context; the alternative actions serve as the inputs to the model, and the measurable attributes of the fundamental objectives are the outputs. The innards of the model structure are an expression of the current understanding of the causal linkages between the actions and the outcomes. Methods for empirical estimation of parameters flourished in wildlife science since the 1990s (Williams et al. 2002) and require little comment here.

In a decision-making context, there are often other fundamental objectives besides wildlife resource objectives, and the consequences of the alternatives for these objectives need to be predicted, too. These objectives include economic, recreational, and spiritual objectives, and appropriate methods of prediction need to be found for each. Economic models related to wildlife management are being used more and more (Pickton and Sikorowski 2004). The nature of human satisfaction with recreational opportunities can be complex, but empirical models are increasingly available (Chapter 4). Models for predicting spiritual and aesthetic outcomes are not common, although some initial attempts have been made (Failing et al. 2007). One of the challenges the human dimensions field faces in incorporating its work into decision-making contexts is moving from descriptive to predictive models. Many of the current models describe patterns in economic, recreation, and aesthetic outcomes, but they are not yet able to predict those outcomes under alternative management actions.

For all types of outcomes that are important in wildlife management, there are often occasions when there is not enough empirical information to build predictive models, and not enough time to collect new data. In these settings, there is increasing use of methods of expert elicitation (Kuhnert et al. 2010). These methods typically rely on the accrued knowledge of a group of experts, rather than on empirical data, to structure a predictive model and to provide parameter estimates. There is a considerable literature on the reliability and fallibility of experts, and from this literature emerges some best practices in expert elicitation (Burgman 2005). Briefly, these methods seek to tap into the privileged knowledge of experts while avoiding common cognitive biases to which humans are prone. In the modified Delphi method (MacMillan and Marshall 2006), a group of experts makes individual judgments about a parameter, fact, or relationship; they share their initial responses (often anonymously) with the group; discussion ensues; and then the experts are asked to make a final, private judgment. The feedback step promotes clarity, eliminates linguistic uncertainty, and allows sharing of insights, and the private judgments allow individual insights to be retained, capture uncertainty as expressed by the range of experts, and avoid the effects of damaging group dynamics. Some other recent methods of elicitation guard against the overconfidence of experts by asking them to be explicit about their degree of uncertainty (Speirs-Bridge et al. 2010).

In most cases, there is uncertainty about the predictions from any model, whether empirically or expert based, and this uncertainty can affect the identification of a preferred alternative. Several taxonomies of uncertainty have been advanced (Morgan and Henrion 1990, Nichols et al. 1995, Regan et al. 2002); a combination of them is useful here. Broadly, uncertainty can be aleatory or epistemic. Aleatory uncertainty arises from stochastic processes that are outside of the manager's control. For example, environmental stochasticity (e.g., weather patterns), demographic stochasticity (i.e., the chance events that determine which animals survive), and partial controllability (i.e., our inability to completely control the implementation of our actions) give rise to aleatory uncertainty. Epistemic uncertainty arises from our lack of knowledge about the managed system. Structural uncertainty (i.e., uncertainty about how the system works), parametric uncertainty (i.e., imprecision in the model parameters), and partial observability (i.e., the inability to know exactly the condition of the resource) are examples of epistemic uncertainty. The distinction between aleatory and epistemic uncertainty is often important to a decision maker, because research or monitoring can theoretically reduce epistemic uncertainty. Incorporation and expression of uncertainty in the consequences are important aspects of prediction; they allow the decision maker to understand—and therefore manage—risk.

Generating Alternative Actions

In the section on values-based decisions, we discussed the generation of alternative actions, but there is also a technical side to this step of decision analysis. In some cases, one of the primary impediments to a decision is that none of the available actions can satisfactorily solve the problem, and the decision maker looks to scientists or engineers to craft a novel approach. For example, when white-nose syndrome emerged in cave-dwelling bats in eastern North America, no known method existed for controlling the fungus that causes the disease; one avenue of research was to identify a fungicide that might eradicate it (Chaturvedi et al. 2011).

The generation of novel alternatives through scientific investigation actually switches the order of analysis implied in PrOACT by putting the consequence analysis before the generation of alternatives. An engineering approach to decision analysis begins with the objectives, works backward through an understanding of how the system works, and then identifies an action that will achieve the objectives. A means-ends network is a useful graphical tool for this approach. The objectives (the ends) are identified, and then means to achieve those ends are drawn based on a current understanding of how the system works. Proceeding backward in this way to more proximate influences leads to actions that might be investigated as potential solutions.

Decision Analysis Tools

In addition to the array of tools it provides to help structure a problem, decision science also provides a diverse set of tools for analysis. These analytical tools offer insight into the nature of decisions and frequently motivate even deeper reflection by decision makers. The complete set of analytical tools is too large to be fully discussed here; what follows is a sampling of some of the most commonly applied techniques. Skilled decision analysts diagnose a decision problem and identify the most appropriate analytical tools to apply.

Multicriteria Decision Analysis

As the field of human dimensions has made evident, wildlife management decisions involve many objectives on the part of many stakeholders. Understanding these objectives, being able to measure these objectives, and predicting the consequences of alternative actions with regard to these objectives are critical steps in evaluation of a multiple-objective problem. Multicriteria decision analysis is a set of techniques to analyze and balance the trade-offs inherent in multiple objectives (Herath and Prato 2006). A consequence table, one that shows the consequences of each alternative action in units of the measurable attribute for each objective, embodies the central expression of the decision problem in MCDA. The analytical question is how to identify the single alternative that best achieves the array of objectives, recognizing that there are trade-offs.

The first step in MCDA is to simplify the problem by examining the structure of the consequence table. A dominated alternative is one that can be improved without sacrifice—one for which there is another alternative that is at least as good on all objectives. Given that there should be no reason to choose a dominated alternative, it can be removed from further consideration. An irrelevant objective is one that does not help distinguish the alternatives because they have similar scores on the corresponding measurable attribute. Irrelevant objectives may be important in the absolute sense (the deci-

sion maker may care very much about the performance of an action on an irrelevant objective), but they are not important in the relative sense (because they do not help the decision maker choose among the alternatives under consideration); thus they can also be dropped from further analysis. Often, an objective that is initially relevant may become irrelevant as dominated alternatives are identified and removed. When a consequence table no longer has any dominated alternative or irrelevant objectives, the remaining alternatives are said to be "pareto optimal"; for any alternative, no improvement can be made on one objective without sacrificing another.

To proceed further with analysis requires grappling with how to trade one objective with another. Some analysts will stop here, and simply ask the decision makers to make an intuitive judgment about the trade-offs and choose a preferred alternative. Quantitative analysis of the trade-offs requires expressing the objectives on a common scale. There are numerous ways to do this—including even swapping, pricing out, the analytical hierarchy process and outranking—but perhaps the most common is the weighted additive model embodied in the Simple Multi-attribute Rating Technique (SMART; Goodwin and Wright 2004). The consequences are first converted to a common scale by normalizing the scores on each objective to a range of zero (i.e., the worst-performing alternative) to one (i.e., the best-performing alternative). Second, the decision maker provides weights, which reflect value judgments about the relative importance of the different objectives, through a process such as swing weighting (von Winterfeldt and Edwards 1986). Third, the weighted sum of the normalized consequences is taken across objectives, using the swing weights, and used to rank the alternatives. Sensitivity analysis can be performed to evaluate the robustness of the preferred alternative to uncertainty in the weights or uncertainty in the consequence values.

Multicriteria decision analysis has been applied extensively in natural resource management (Kiker et al. 2005, Herath and Prato 2006). Specific applications in wildlife management are increasing (Redpath et al. 2004, Szymanski et al. 2009). Case study two (wolf hunting management in Montana) uses an MCDA approach.

Decision Trees

Some decisions have to be made in the face of uncertainty, without recourse to resolving the uncertainty first. This may be because the uncertainty is aleatory, or because the uncertainty is epistemic, but the decision has to be made before the uncertainty can be reduced. In either case, the decision maker must accept the possibility of regret associated with an undesired outcome. Decision trees make clear the risks (and regrets) associated with alternatives, effectively insulating against poor decisions. This setting was the genesis of decision theory in economics, but it is just as applicable in wildlife management. For example, imagine a manager of a 400-km^2 tract of arid land whose primary objective is to provide habitat for pronghorn (*Antilocapra americana*). Without prescribed fire, grasses increase and native forbs, which pronghorns thrive

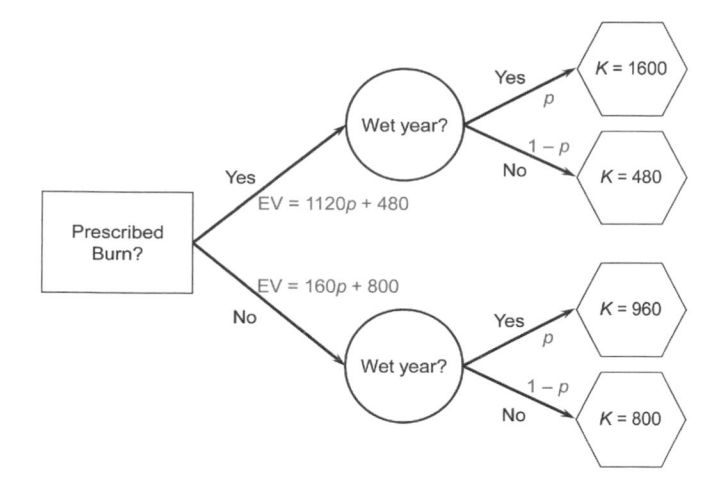

Figure 5.4. Decision tree for pronghorn habitat management. The outcomes are the expected carrying capacities of the refuge as a function of whether a prescribed fire was instituted and whether the year was wet. The expected value (EV) of the outcome depends on the probability of a wet year (*p*) and can be used to identify a preferred action. With the values shown here, if *p* > 33%, the expected capacity is higher with a prescribed burn than without.

on, decline. Prescribed fire returns nutrients to the soil and encourages growth of forbs, especially in a wet year, but in a dry year, prescribed fire can remove moisture from the system and substantially reduce the total biomass available for forage. The manager has a predictive model for the carrying capacity of the refuge, but whether a particular year will be wet is an uncontrollable uncertainty (Fig. 5.4). The manager can use the decision tree shown in Figure 5.4 to calculate the expected carrying capacity under either decision by taking a weighted average of the carrying capacities in each branch, where the weights describe the likelihood of a wet year. In this particular case, if the probability of a wet year is greater than 33%, the manager should institute the prescribed burn.

There are a number of more advanced methods to make a simple decision tree (Fig. 5.4) more realistic. The tree might have additional branches to represent the likelihood of wildfire occurring and restoring some habitat condition. The manager might assign nonlinear values to the outcomes, to reflect a nonneutral attitude toward risk. The tree could be extended to acknowledge that this decision can be made annually, and the manager might care more about the cumulative responses over many years. The value in all these methods is in helping the manager think about how to make decisions in the face of uncontrollable uncertainty.

Expected Value of Information

There may be recourse to resolve uncertainty before having to commit to a decision. As wildlife scientists, we are always interesting in reducing uncertainty, but a wildlife manager has a different perspective. The decision maker needs to ask whether the benefits that accrue from acquiring new information are worth the costs of obtaining that information.

Decision analysis offers a formal method for answering this question. The expected value of information is the amount by which the outcome can be improved by reducing uncertainty before making the decision (Runge et al. 2011b). A powerful and underutilized method in natural resource management, calculating the value of information can help decide what research is valuable, what monitoring should be instituted, and whether adaptive management is warranted.

ADAPTIVE MANAGEMENT

Although use of a broad set of formal decision analysis techniques in wildlife management is only beginning, for decades there has been a very widespread call for and use of a special class of decision analysis, namely adaptive management. Developed in the context of fisheries management in the 1970s (Holling 1978, Walters 1986), adaptive management is now a central tenet of all natural resource management, including wildlife management (Lancia et al. 1996, Callicott et al. 1999, Allen et al. 2011).

A Special Class of Decision

Adaptive management is a special case of structured decision making for recurrent decisions made under uncertainty (Fig. 5.2). Many wildlife management decisions have two key features: they are recurrent (a similar decision is made on a regular basis), and they are impeded by uncertainty (the consequences of the alternatives are not fully understood). To address the first feature, a wildlife manager needs to understand and anticipate the dynamics of the system — namely, the immediate costs or rewards from taking an action—and also the future opportunities, costs, and potential rewards attending subsequent actions that might be taken. System modeling and dynamic optimization are tools that can support recurrent decisions. To address the second feature, the wildlife manager needs to know how to make decisions in the face of uncertainty, by evaluating and balancing risks. Decision analytical techniques exist for making decisions in the face of uncertainty; in fact, they are the basis for the entire discipline. When these two key features occur together, when recurrent decisions need to be made in the face of uncertainty, there is an opportunity to learn from actions taken early on to reduce uncertainty so better decisions can be made in the future. The ability to adapt future decisions to information that arises during the course of management is the purpose and foundation of adaptive management.

The PrOACT sequence is central to adaptive management: objectives need to be expressed; alternative actions need to be developed; consequences need to be predicted; and a solution, through optimization or balancing trade-offs, needs to be found. To this sequence, adaptive management adds several details: developing dynamic predictive models, articulating and evaluating uncertainty, implementing monitoring to provide feedback, updating the predictive models based on new information, and adapting future decisions based on the updated understanding of how the resource responds to management.

CARL J. WALTERS (b. 1944)

Carl J. Walters is a fisheries biologist who pioneered the concept of adaptive management. He received his doctorate from Colorado State University in 1969, and has been a professor of zoology and fisheries at the University of British Columbia ever since. His 1986 book *Adaptive Management of Renewable Resources* offered a full decision-analytical treatment of natural resource management, and formally considered how to make optimal recurrent decisions in the face of epistemic uncertainty. His interest in adaptive environmental assessment led to the 2004 publication, with Steven Martell, of *Fisheries Ecology and Management*, a graduate-level textbook on the use of quantitative models in fisheries management. The influence of his work on the practice of wildlife and fisheries management cannot be overstated. Among other honors, Walters received The Wildlife Society's best paper award in 1976, the American Fisheries Society Award of Excellence in 2006, and the Volvo Environment Prize in 2006. He is a fellow of the Royal Society of Canada.

Photo courtesy of Sandra Buckingham

First, the predictive models that are constructed need to be dynamic; that is, they need to predict current rewards and future conditions of the system that could affect subsequent decision making. Predictive models need to incorporate the temporal linkage among decisions. Predictive models of habitat and population dynamics for wildlife have included such dynamics, even outside of formal decision analysis.

Second, uncertainty needs to be articulated and evaluated. What aspects of the predictions are not well known and might impede the decision? Nichols et al. (1995) describe four sources of uncertainty relevant to wildlife management: environmental variation, structural uncertainty, partial observability, and partial controllability. Two of these (i.e., environmental variation and partial controllability) are types of aleatory uncertainty (Helton and Burmaster 1996), uncertainty that cannot be reduced. The other two (i.e., structural uncertainty and partial observability) are types of epistemic uncertainty—uncertainty due to our lack of knowledge, which (at least theoretically) can be reduced through investment in monitoring. Formal approaches to decision analysis attempt to express these uncertainties quantitatively, so that the uncertainty in the predictions can be stated clearly. To evaluate the uncertainties, the decision maker wants to know whether reduction of

any uncertainty would improve the expected outcome of the decision. In the context of management, relevant uncertainty is uncertainty that affects the *decision*, not simply the *predictions*. The expected value of information measures how much a decision could improve if uncertainty could be reduced, and it is important for identifying the critical uncertainty to address in an adaptive program (Runge et al. 2011*b*). Key uncertainty is often expressed as a set of plausible models, each of which makes a different prediction about the effects of management actions on the outcomes that are relevant to the decision maker.

Third, an appropriate monitoring program that provides the necessary feedback to resolve critical uncertainty is central to meeting the promise of adaptive management. The needs of this monitoring program stem from the decision context and serve three fundamental purposes: evaluation of performance against the objectives, tracking of key variables that are tied to decision thresholds, and reduction of key uncertainty (Nichols and Williams 2006). This "targeted" monitoring is important to make efficient use of scarce resources, allocating funds and staff time only to monitoring that is expected to improve management outcomes in the long term. Lyons et al. (2008) provide examples of monitoring design for management on national wildlife refuges that reflect these principles.

Fourth, monitoring data are valuable only if they are analyzed. In an adaptive management setting, analysis consists of confronting the predictive models with the observed data (Hilborn and Mangel 1997). Each of the alternative models makes a prediction about the outcome associated with the action that was last implemented, and the monitoring system provides information about the actual outcome. The comparison of the observed response to the expected responses allows the predictive models to be updated, often through an application of Bayes' theorem. The degree of belief increases for those models whose predictions most closely matched the observed response, and decreases for models that performed poorly (Johnson et al. 2002).

Fifth, what makes adaptive management adaptive is the application of learning to subsequent decisions. This adaptation can be anticipated; that is, the decision maker can articulate in advance how future decisions will change as a result of monitoring outcomes. In "active adaptive management," this anticipation goes one step farther: in making a decision, the decision maker may choose an action that will accelerate learning, if the long-term gains from that learning are anticipated to offset the short-term costs (Walters and Hilborn 1978).

Single-, Double-, and Triple-loop Learning

One of the real challenges of decision analysis is correctly framing the decision. For recurrent decisions, each iteration provides the opportunity to learn and reflect about the framing of the decision, in addition to the predictions of the system models. There is another layer of learning, and hence another layer of adaptive management. This "double-loop" learning (Argyris and Shon 1978) focuses on emerging understanding

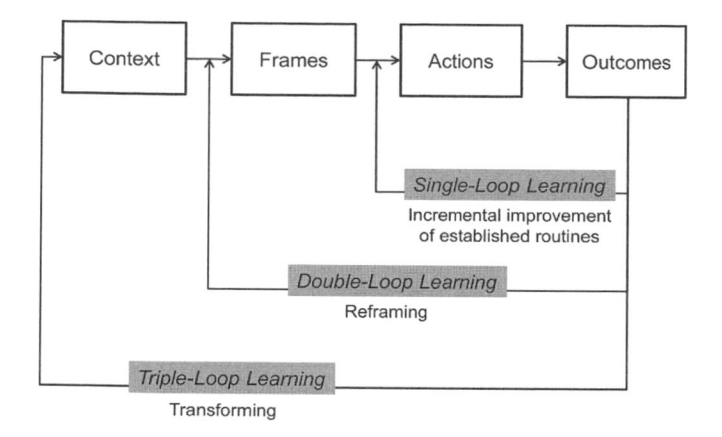

Figure 5.5. Adaptive learning cycles. From Pahl-Wostl (2009)

of the framing of the decision; in particular, the objectives, the set of potential actions, and the relevant uncertainties (Fig. 5.5). In the most challenging natural resource management problems, where the ecological and institutional dynamics are very complex, experience managing the system may give rise to insights about the context in which management is occurring. "Triple-loop" learning (Fig. 5.5; Pahl-Wostl 2009) can result in transformative adaptation (e.g., through changes to the institutional relationships, governance structures, regulatory frameworks, or even the social and organizational values that are associated with the managed resources).

Schools of Adaptive Management

Adaptive management has seized the imagination of natural resource managers since the phrase was coined, but in recent years a number of writers have decried its failure to live up to its promise, documenting the challenges and alleged failures of implementation (McLain and Lee 1996, Gregory et al. 2006, Allen and Gunderson 2011). One of the challenges is that there is not a single definition of adaptive management. There are many layers on which learning and adaptive management can occur (Fig. 5.5). So perhaps it is not surprising that there are very different schools of thought regarding adaptive management, each focused on a different layer of adaptation. McFadden et al. (2011) provide the beginnings of a long-needed taxonomy of adaptive management, identifying two primary schools of thought: the resilience-experimentalist school, exemplified by Gunderson et al. (1995), and the decision-theoretic school, exemplified by Williams et al. (2007).

The Resilience-experimentalist School

The resilience-experimentalist (RE) school has arisen from the management challenges in large-scale, complex socioecological systems, where framing the decisions and constructing effective institutional arrangements for collaborative management pose enormous challenges. The ecological dynamics are so complex that the notion of being able to articulate critical uncertainty seems ambitious. So the focus is on double- and triple-loop learning, with an emphasis on collaboration and adaptive governance, with reduction of uncertainty occur-

ring through experimental manipulation. The most-noted examples include management of the Columbia River (Lee and Lawrence 1985), the Everglades (Gunderson and Light 2006), and Grand Canyon (Hughes et al. 2007). All of these examples reveal the ecosystem focus of the RE school; this broader scope encompasses wildlife management.

The Decision-theoretic School

The decision-theoretic (DT) school is grounded in the seminal writings of Holling (1978) and Walters (1986), but it was perhaps most profoundly influenced by the adaptive harvest management of waterfowl in North America (Johnson et al. 1997, Nichols et al. 2007). The emphasis is on management of dynamic systems in the face of uncertainty, through explicit use of a decision-theoretic framing of the problem, with reduction of uncertainty not occurring through experimentation but through ongoing monitoring and management. The approach taken by the DT school emphasizes single-loop learning but has the flexibility to accommodate learning and adaptation at all three levels. The DT school is as useful for local decisions with a single decision maker as it is for broad-scale decisions with multiple management partners.

CASE STUDIES

Consider three case studies of the application of structured decision ranging in scale from local (Skyline Wildlife Management Area) to state (wolf harvest management in Montana) to continental (adaptive harvest management of waterfowl in North America). Each of the case studies exemplifies particular elements of the SDM process, but they share the underlying PrOACT structure.

Case Study One: Skyline Wildlife Management Area

The Alabama Department of Conservation and Natural Resources (ADCNR), like many state agencies, is obligated to implement management to improve the status of the species of greatest conservation need identified in the state wildlife action plan. This project sought to balance game and nongame wildlife population objectives for the J. D. Martin Skyline Wildlife Management Area (SWMA) as a test case for other state-owned lands. The SWMA (170 km²) is located in Jackson County, Alabama, in the Cumberland Plateau region. Most of SWMA was logged at the turn of the century, and only the most inaccessible slopes were spared. Agriculture grew in the region during the 1930s under the auspices of federal programs (Hammer 1967). The majority of the current forest vegetation is the result of natural regeneration, with some planted pine plantations on the plateaus and in the valleys. Even today, forested habitat exists only in narrow valleys, on steep hillsides, and on top of the Cumberland Plateau. Most of the lower, flatter areas in larger valleys have been converted to nonnative pasture or row crops.

Some portions of SWMA are owned and managed by the ADCNR Wildlife and Freshwater Fisheries Division (WFFD). Lands owned by WFFD were purchased with federal aid funds for the purposes of wildlife management and public hunting. Other portions are owned and managed by ADCNR Lands Division. Lands Division purchases were made with state funds for their potential contribution to parks, nature preserve, wildlife management, and recreation. Other portions of SWMA are under long-term lease from Alabama Power Company, but management decisions are delegated to WFFD in mitigation for the establishment of the R. L. Harris Dam, which flooded a substantial amount of forest habitat in central Alabama. Each of these entities has a different mandate and approach to wildlife management. The decision in this case was to identify and recommend alternatives for land use and forest management that would benefit greatest conservation need (GCN) species while providing adequate opportunity for hunters.

Objectives

1. Maintain or enhance populations of species of GCN identified in the Alabama Comprehensive Wildlife Conservation Strategy.
2. Provide hunting opportunities for large and small game, including white-tailed deer, eastern wild turkey (*Meleagris gallopavo*), eastern cottontail (*Sylvilagus floridanus*), northern bobwhite, and gray squirrel (*Sciurus carolinensis*).
3. Provide nonconsumptive recreational opportunities, including hiking, wildflower viewing, wildlife viewing, horseback riding, and primitive camping.

The measurable attribute for the first objective was the average occupancy of four representative nongame species—cerulean warbler (*Dendroica cerulean*), Kentucky warbler (*Oporornis formosus*), worm-eating warbler (*Helmitheros vermivorum*), and wood thrush (*Hylocichla mustelina*)—equally weighted. The measurable attribute for the second objective was the average occupancy of three representative game species—northern bobwhite, wild turkey, and mourning dove (*Zenaida macroura*)—equally weighted. A measurable attribute for the final objective was not developed for the initial analysis.

Alternatives

The management alternatives considered were combinations of landscape alternatives that increased the amount of nonforested areas, and treatments to forested and nonforested habitat (Fig. 5.6). The management alternatives considered in forested areas included four options.

1. Status quo: maintaining the current landscape and forest management practices.
2. Even-aged forest management with large (~60 ha) or small stands (~20 ha) by clear-cutting, seed tree, or shelter wood techniques.
3. Two-aged forest management system throughout the forest.
4. Uneven-aged management throughout the forest using either group or single-tree selection methods.

In nonforested areas the alternatives included:

1. Status quo: maintaining a mixture of plantings of green fields, row crops, and early successional habitat.
2. Increasing early successional habitats and native warm-season grass meadows.
3. Increasing early successional habitats and native warm-season grass meadows in some areas, and creating oak (*Quercus* spp.) savannah in others.

Modeling Consequences

As a prototype, areas were mapped that met an agreed-upon minimum area requirement for NOBO populations (404 ha) and cerulean warbler (6,000 ha). The consequences of the management actions were predicted and evaluated in terms of the expected population response by the game and nongame populations of interest. Uncertainties included the effect of management practices on the composition and structure of the vegetative cover and the response of the animal populations to the structure, availability, and distribution of suitable areas. Occupancy (i.e., probability of use by each species) was determined to be an acceptable population response for comparing alternatives. For many species of reptiles, amphibians, birds, and small mammals, recent research provided estimates of the relationship between occupancy and many forest characteristics including composition, structure, and context (Grand et al. 2008). For some game species where occupancy models were not available, expert judgments were elicited to predict wildlife responses. Experts ranked the alternative landscapes with respect to each of the objectives, but it was difficult to predict the effects of forest management on habitat structure and species responses. Therefore a system model employing a Bayesian belief network was developed as a second prototype (Fig. 5.7).

The Bayes net was used because it provided the means to start with a graphical representation (i.e., influence diagram) of the system, which could be parameterized and converted to a decision model using existing data or expert judgments. Each node in the network represents an important characteristic (i.e., state variable) of the system and the linkages among nodes represent relationships between variables. Uncertainty in the relationships between variables, and uncertainty in the estimates of the variables themselves, was incorporated. Decisions were modeled using the Bayes net by adding a decision node, used to manipulate the state of the system under candidate management actions, and a utility node, which was used to assign stakeholder values to the measurable attributes of each top-level objective (game and nongame species). The relative weights on these objectives were not formally elicited from the decision makers; rather, a range of values was explored to understand how the preferred alternative was affected by the weights.

Decision Recommendation and Implementation

Analysis using the second prototype suggested that managing forested areas using a two-aged system would achieve the greatest utility. The model was not sensitive to the size of for-

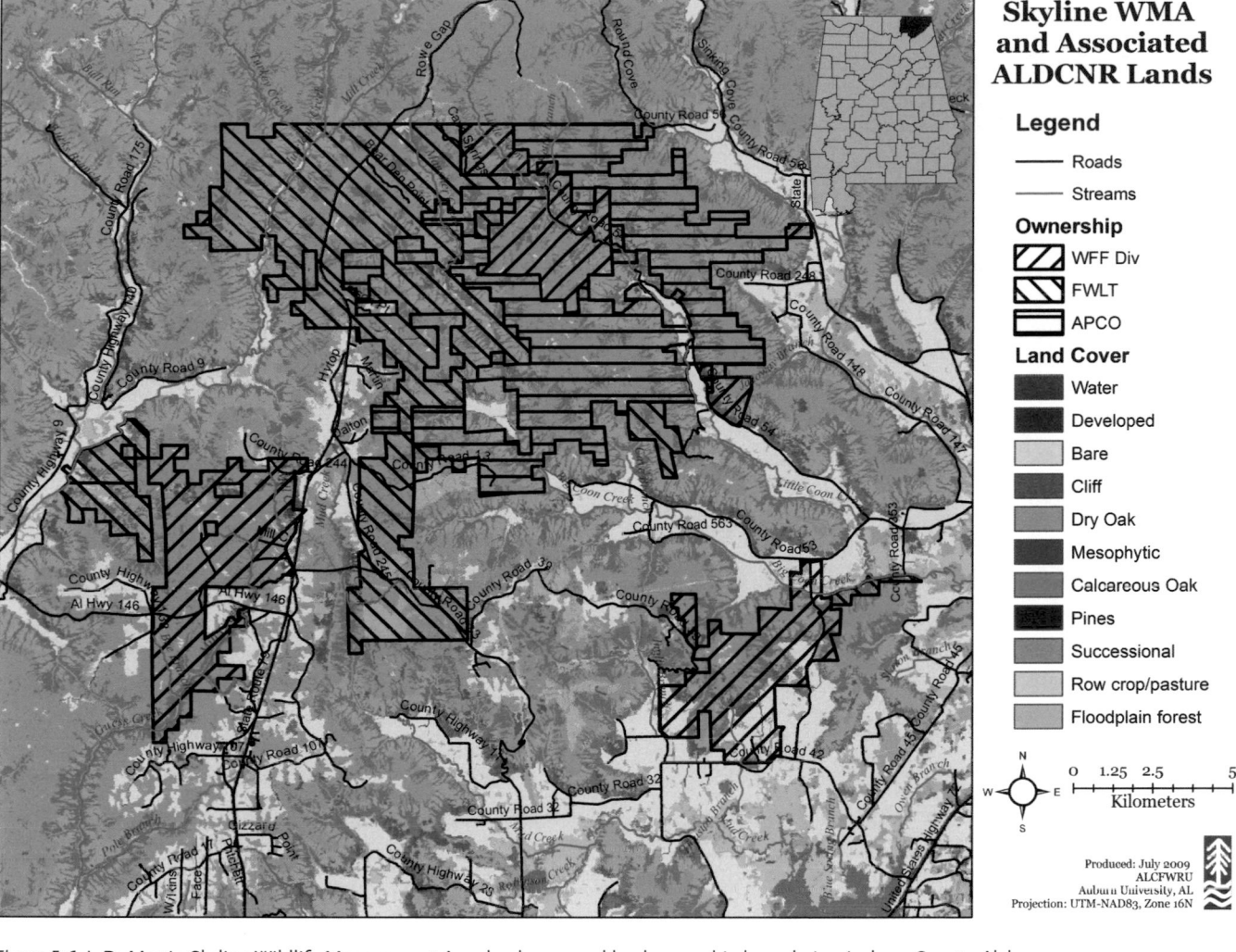

Figure 5.6. J. D. Martin Skyline Wildlife Management Area land cover and land ownership boundaries, Jackson County, Alabama.

est stands, nor did it indicate differences in utility between the alternatives for managing nonforested habitat. As of late 2011, the management alternatives had not been implemented, but the SDM process is being applied to develop and evaluate management alternatives for 12 additional wildlife management areas, parks, and nature preserves across the state.

Case Study Two: Wolf Hunting Management in Montana

Gray wolves in the U.S. northern Rocky Mountains (NRM) were first removed from the endangered species list in February 2008, at which point management authority for wolves passed from the USFWS to the states of Montana and Idaho. Wolf management in each state included setting harvest quotas and seasons. Lessons learned from the first wolf hunting season in Montana in 2009 suggested that Montana Fish, Wildlife and Parks (MFWP) needed to redefine its wolf management units (WMUs) to better allocate hunter opportunity and harvest and to manage wolf numbers. For the 2009 hunting season, MFWP defined three WMUs (Fig. 5.8). Be-

cause wolves are primarily located in the mountainous portions of western Montana, managers believed that smaller, redistributed WMUs in that portion of the state would be necessary to manage allocation of hunter opportunity and thus the distribution of harvest across the Montana wolf population. Statutory obligations for effective conservation of a game species and often-contentious public attitudes and expectations regarding wolf management in Montana combined to present a challenging context for deciding on new WMUs. The MFWP thus elected to use a structured process to ensure explicit consideration of all relevant factors affecting the designation of WMUs, and to provide transparency to the public.

Representatives from MFWP—including regional managers, biologists, and wolf specialists—developed the following problem statement:

MFWP must propose a 2010 wolf harvest strategy that maintains a recovered and connected wolf population, minimizes wolf–livestock conflicts, reduces wolf impacts on low or de-

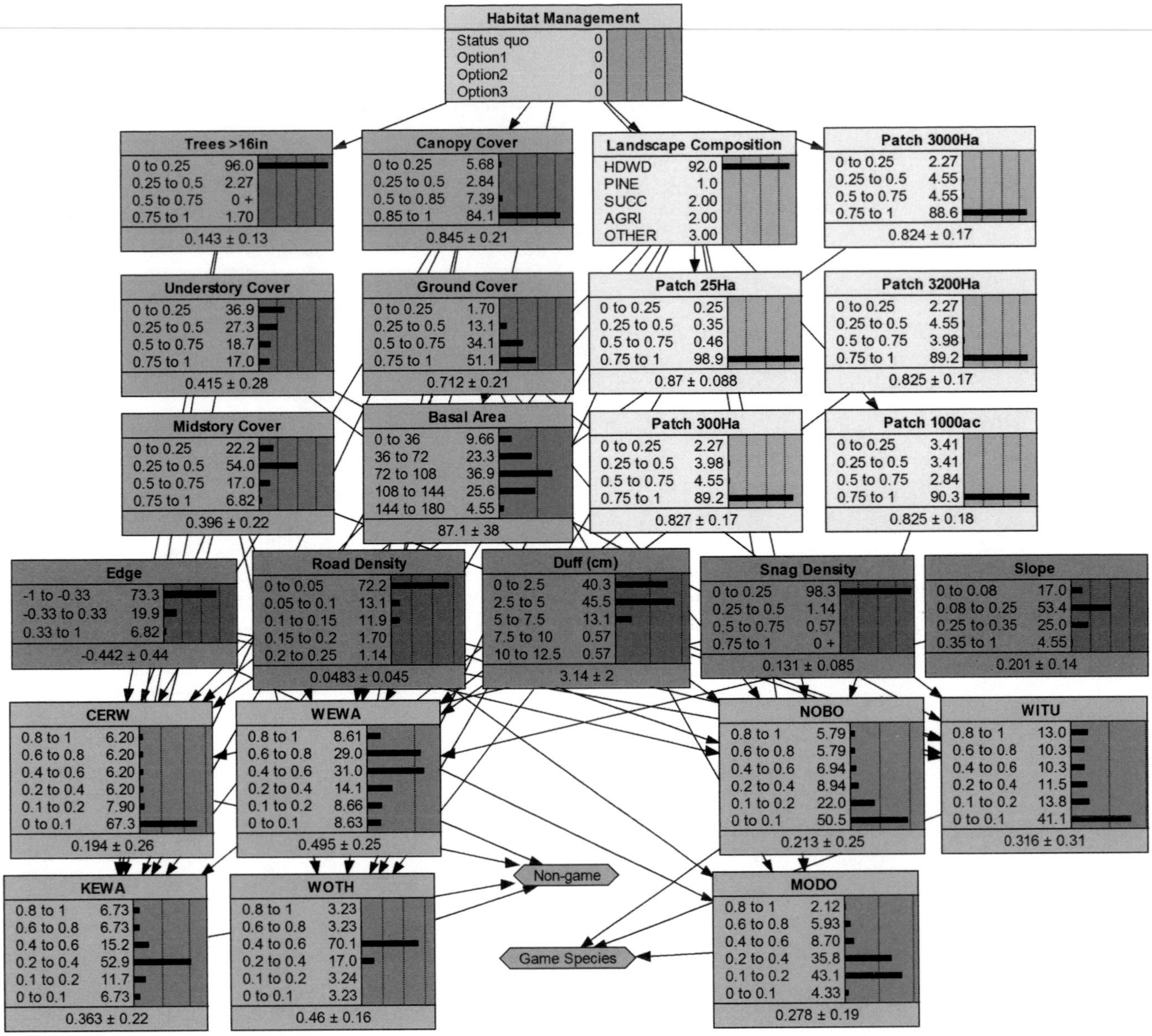

Figure 5.7. Bayes decision network for evaluating management alternatives for the J. D. Martin Skyline Wildlife Management Area, Jackson County, Alabama. The decision network (prototype 2) includes nodes that represent the decision (blue), habitat structure (green), land cover (yellow), physical characteristics (brown), species responses (gray), and utilities (red).

clining ungulate populations and ungulate hunting opportunities, and effectively communicates to all parties the relevance and credibility of the harvest while acknowledging the diversity of values among those parties.

The group developed a set of fundamental, process, and strategic objectives.

Fundamental objectives:
1. Maintain positive and effective working relationships with
   a) livestock producers,
   b) hunters, and
   c) other stakeholders.
2. Reduce wolf impacts on big game populations.

3. Reduce wolf impacts on livestock.
4. Maintain hunter opportunity for ungulates.
5. Maintain a viable and connected wolf population in Montana.
6. Maintain hunter opportunity for wolves.

Process objectives:
7. Enhance open and effective communication to better inform decisions.
8. Learn and improve as we go.

Strategic objectives:
9. Increase broad public acceptance of harvest and hunter opportunity as part of wolf conservation.
10. Gain and maintain authority for the state of Montana to manage wolves.

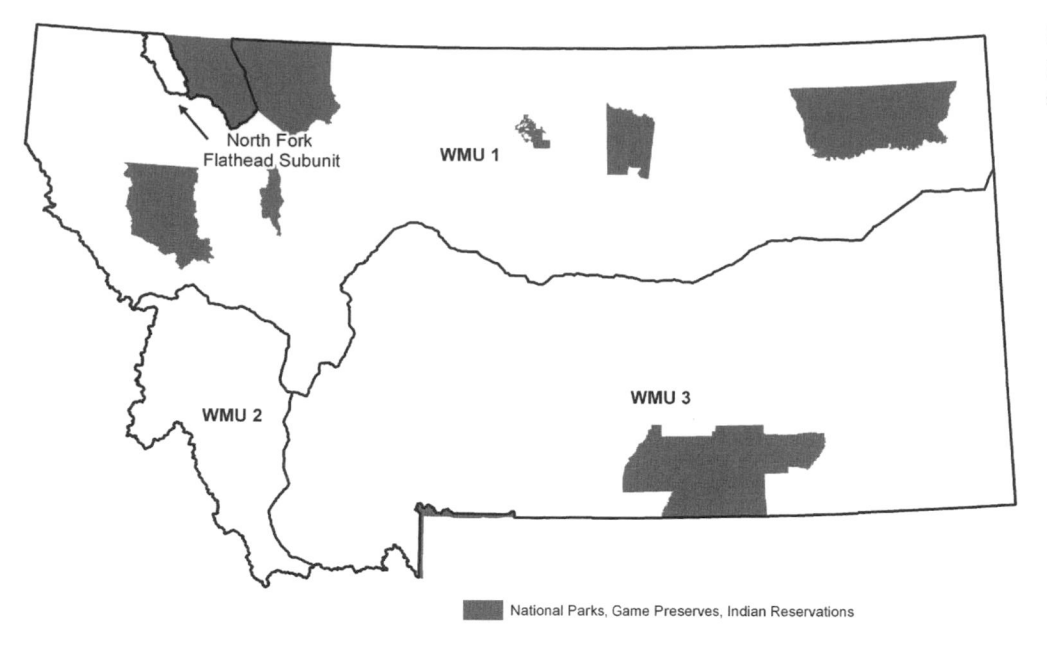

Figure 5.8. Wolf management units (WMUs) in Montana, hunting season 2009.

North Fork Flathead Subunit

WMU 1

WMU 2

WMU 3

■ National Parks, Game Preserves, Indian Reservations

The group developed five management alternatives to address the set of fundamental objectives. The number and distribution of WMUs affect how finely the state can control the distribution of wolf harvest, which in turn affects wolf density and distribution, and the various impacts associated with wolf density. The alternatives focused on the arrangement of WMUs. The first alternative represented the status quo, retaining the same three WMUs used during the 2009 hunting season. The remaining four options represented alternative ways of dividing Montana into WMUs.

Alternative 2. Fifteen WMUs, with eastern Montana incorporated into western units.
Alternative 3. Fourteen WMUs, with eastern Montana incorporated into western units.
Alternative 4. Thirteen WMUs, with eastern Montana incorporated into western units.
Alternative 5. Fifteen WMUs, with eastern Montana having its own management unit not incorporated into western units.

The measurable attributes for each fundamental objective were expressed on a constructed scale that ranged from zero (i.e., poor outcome) to one (i.e., ideal outcome). Two fundamental objectives (numbers 5 and 6 above) were not scored, because the group did not believe their consequences varied among the management alternatives and thus did not affect the decision. One of the strategic objectives (number 9 above) was viewed as critical enough to the decision that it was also scored. A panel of experts composed of wildlife managers, biologists, and wolf specialists from MFWP were asked to score individually each alternative against each objective. An average score for each response was taken across experts (Table 5.4).

The status quo (alternative 1) ranked high among alternatives for maintaining relationships with livestock producers, hunters, and other stakeholders, and for public acceptance,

but ranked relatively low for reducing impacts to big game and livestock while maintaining a sustainable ungulate harvest (Table 5.4). Alternative 2 scored relatively low for maintaining relationships but moderately well for reducing impacts of wolves, public acceptance, and maintaining sustainable ungulate harvest. Alternatives 3 and 4 scored comparably across all objectives. Alternative 5 was judged to have strong benefits for reducing impacts to big game and maintaining sustainable ungulate harvests, but would have the strongest negative impacts among alternatives on maintaining relationships with stakeholders and public opinion.

Inspection of the consequence table shows that alternative 3 dominates alternatives 2, 4, and 5, because it scores as well or better than those other alternatives on all objectives. Thus alternatives 2, 4, and 5 can be removed from further consideration, leaving only alternatives 1 and 3 as viable candidates.

At this point, formal multicriteria decision analysis could be used to place weights on the objectives to develop a composite score for each alternative. But the panel chose instead to proceed qualitatively on the basis that identification of dominated alternatives and redundant objectives provided a cognitively accessible trade-off. The group decided alternative 3 (Fig. 5.9) was most likely to satisfy the fundamental objectives for setting WMUs for the 2010 hunting season. This was because the relative benefits of maintaining relationships with hunters, reducing impacts, and maintaining sustainable ungulate populations in alternative 3 outweighed the slight advantages in maintaining relationships with livestock producers and stakeholders and public acceptance offered by alternative 1.

The SDM approach allowed decision makers to see the structure of the problem and the major trade-offs among the alternatives; those insights alone were enough to allow the decision to proceed. The decision was presented in July 2010 as a recommendation to the MFWP commission, which adopted it; the SDM process and product were considered clear assets

Table 5.4. Consequence table for case study 2, wolf hunting management in Montana

Fundamental objective	Measurable attribute	Preferred direction	Alternative 1	Alternative 2	Alternative 3	Alternative 4	Alternative 5
Maintain relationships							
Livestock producer	Perception 0 to 1	Maximize	**0.83**	*0.54*	0.66	0.66	0.63
Stakeholders	Perception 0 to 1	Maximize	**0.69**	0.60	0.66	0.66	*0.34*
Hunters	Perception 0 to 1	Maximize	0.80	*0.57*	**0.83**	0.77	0.60
Reduce impacts							
Big game	Ungulate populations at or near objectives Yes (1) / no (0)	Maximize	*0.60*	**1.00**	**1.00**	0.80	**1.00**
Livestock	Reduction in the number of livestock confirmed injured or killed by wolves 0 to 1	Maximize	*0.56*	0.72	**0.80**	**0.80**	0.76
Sustainable ungulate harvest	Quota in every WMU for foreseeable future Yes (1) / no (0)	Maximize	*0.60*	**1.00**	**1.00**	0.80	**1.00**
Public acceptance	Perception 0 to 1	Maximize	**0.80**	0.72	0.74	0.74	*0.37*

Consequences for each alternative were elicited individually, then averaged over a group of wildlife managers, biologists, and wolf specialists from Montana Fish, Wildlife and Parks. For each objective, the alternative that was predicted to perform best is indicated by boldface, moderately performing alternatives are in regular type, and the alternative predicted to perform worst is indicated by italics. Alternative 3 dominates all alternatives except alternative 1, which performs relatively well for maintaining relationships but poorest among the alternatives for reducing impacts and maintaining sustainable ungulate harvests.

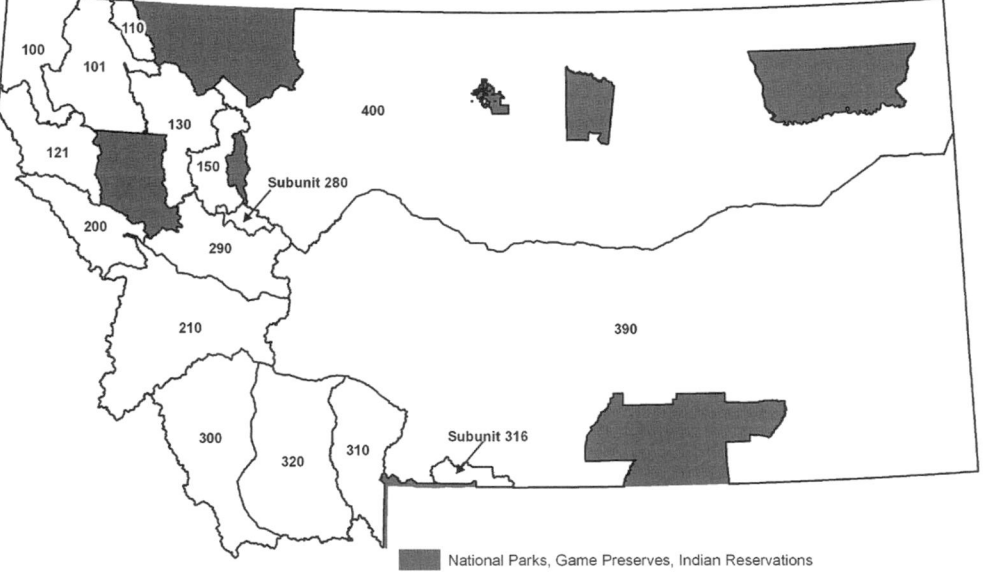

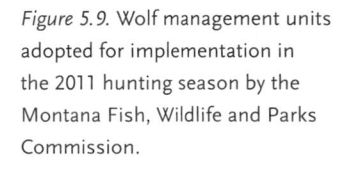

Figure 5.9. Wolf management units adopted for implementation in the 2011 hunting season by the Montana Fish, Wildlife and Parks Commission.

in the public presentation and review of the proposed season structure. The 2010 wolf hunting season was not implemented, however, because wolves in the NRM were returned to the endangered species list by court order in August 2010. With the legislated removal of wolves in the NRM from the endangered species list in May 2011, management of wolves under the 2010 WMUs was implemented in 2011 with minor adjustments.

Although there was a pressing need to make a recommendation for 2010, this decision can be revisited each year, creating an opportunity to improve the analysis and to reduce uncertainty over time. Future iterations of this process might address three topics: developing better measurable attributes

for the objectives, founded on natural scales that are tied to monitoring systems; reinstating the omitted fundamental objectives, which may be more relevant in subsequent years; and analyzing wolf monitoring data over time to evaluate the efficacy of the wolf hunting program.

Case Study Three: Adaptive Harvest Management of Waterfowl in North America

Each year, the USFWS sets harvest regulations for waterfowl based on population and habitat conditions. The USFWS has sole regulatory responsibility for this decision under the Migratory Bird Treaty Act (16 USC 703–712), and a number of

NEPA compliance documents govern the regulations-setting process (USFWS 1988). But the USFWS recognizes important management partnerships with the states and flyways, and has established a formal collaborative structure for garnering input from these partners. In 1995, a prescriptive decision-theoretic approach to setting harvest regulations for midcontinent mallards (*Anas platyrhynchos*) was established (Nichols et al. 1995, Johnson et al. 1997). Referred to as adaptive harvest management (AHM), this process recognizes the dynamic nature of the resource, the recurrent nature of the decisions, and the role that uncertainty plays in impeding decision making. Because mallards are the most abundant duck species in the midcontinent, AHM also serves as the framework around which regulations for hunting of other duck species is centered.

There are multiple objectives that AHM seeks to achieve: to maximize annual harvest of mallards, to maintain a sustainable level of harvest, to maintain the population size close to or higher than the North American Waterfowl Management Plan (NAWMP) goal, and to prevent closed seasons, except in extreme circumstances. These multiple—and competing—objectives have been combined into a single objective function. The objective of AHM is to maximize

$$\sum_{t=0}^{\infty} H_t \min\left(\frac{\hat{N}_{t+1}}{8.5}, 1\right)$$

where H_t is the annual harvest, $\hat{N}_{t+1}$ is the predicted breeding population size in the next year, and 8.5 is the NAWMP goal for midcontinent mallards (in millions). The minimization within the objective function devalues the harvest whenever the population size is predicted to be below the NAWMP goal. Summing the harvest over an infinite time horizon ensures sustainability; the only way to maximize a long-term cumulative harvest is to keep the population extant.

The alternatives are chosen from a small set of regulatory packages: closed, restrictive, moderate, and liberal seasons, which differ in the length of the season and the daily bag limit. The closed season is only permitted when the midcontinent mallard population size falls below 5.5 million. For each of the regulatory packages, an expected harvest rate has been estimated.

The consequences of the different packages are evaluated through predictive models of mallard population dynamics. These models take three input values: two state variables (mallard breeding population size and the number of ponds in prairie Canada) and one decision variable (the regulatory package). They predict two quantities: the expected harvest, H_t, and the breeding population size in the subsequent year, $\hat{N}_{t+1}$ (Runge et al. 2002). One of the motivations for an adaptive management approach was intense disagreement that arose out of uncertainty about the population dynamics. There is uncertainty about the degree of density dependence in recruitment (weak versus strong density dependence), and uncertainty about the effect of harvest mortality on annual mortality (additive versus compensatory harvest mortality); in combination, these uncertainties are captured in four alterna-

Table 5.5. Optimal regulatory strategy for midcontinent mallards for the 2010 hunting season

Bpop	Ponds									
	1.5	2.0	2.5	3.0	3.5	4.0	4.5	5.0	5.5	6.0
≤4.5	C	C	C	C	C	C	C	C	C	C
4.75–5.75	R	R	R	R	R	R	R	R	R	R
6	R	R	R	R	R	R	R	R	M	M
6.25	R	R	R	R	R	M	M	M	M	L
6.5	R	R	R	M	M	M	M	L	L	L
6.75	R	R	M	L	L	L	L	L	L	L
7	R	M	M	M	L	L	L	L	L	L
7.25	M	L	L	L	L	L	L	L	L	L
7.5	L	L	L	L	L	L	L	L	L	L
≥7.75	L	L	L	L	**L**	L	L	L	L	L

Source: USFWS (2010).

The two state variables are the breeding population size (Bpop, in millions) and the number of ponds in prairie Canada (ponds, in millions). The regulatory packages are closed (C), restrictive (R), moderate (M), and liberal (L). Boldface represents the regulatory prescription for 2010.

tive population models. This uncertainty matters; the four alternative models lead to very different harvest strategies, and the resolution of the uncertainty has a significant value of information (Johnson et al. 2002).

The optimal strategy is found each year through passive adaptive stochastic dynamic programming (Williams 1996), which produces a state-dependent harvest strategy that stipulates the optimal regulatory package for any combination of breeding population size and number of ponds (Table 5.5). It is a passive adaptive strategy, in that the optimization does not anticipate the effect of learning on future decisions.

The USFWS, Canadian Wildlife Service, U.S. states, and Canadian provinces collaboratively operate an extensive monitoring program for waterfowl, which includes aerial surveys to estimate abundance and habitat conditions, banding and band-recovery programs for survival and related estimates, and harvest surveys for harvest and reproductive estimates. From the standpoint of adaptation, the key annual monitoring data are the breeding population estimates, because these provide the feedback for evaluating the model uncertainty. The weights on the four models have evolved over time as a result of the observed responses to management (Fig. 5.10); the evidence for the weakly density-dependent model has increased significantly, and the evidence for the additive model has increased slightly. These changes in model weights have been accompanied by an evolution in the harvest strategy over time; thus the annual regulations have adapted to the new information.

The AHM program has undergone some technical adjustments and minor policy modifications over the years since its first implementation, but has largely remained intact. Currently, the waterfowl management community is engaged in a process of double-loop learning, examining the nature of the objectives, alternatives, and models that underlie the regulations setting process (Anderson et al. 2007).

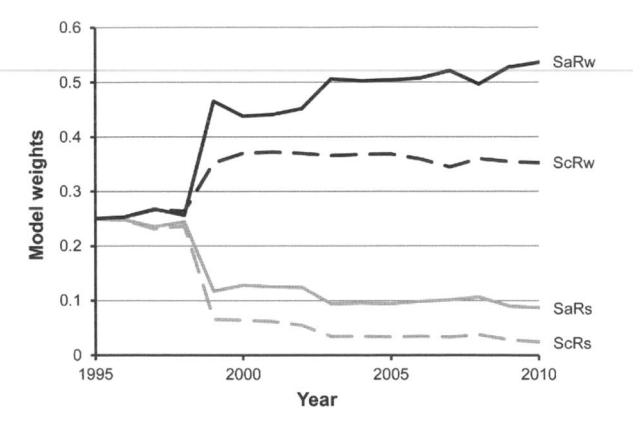

Figure 5.10. Weights on alternative predictive models for midcontinent mallard dynamics, 1995–2010 (USFWS 2010). The four models are distinguished by whether the survival model is compensatory (Sc) or additive (Sa), and whether the reproductive model is strongly (Rs) or weakly (Rw) density dependent.

SUMMARY

Wildlife management is a decision-focused discipline. It needs to integrate traditional wildlife science and social science to identify actions that are most likely to achieve the array of desires society has surrounding wildlife populations. Decision science, a vast field with roots in economics, operations research, and psychology, offers a rich set of tools to help wildlife managers frame, decompose, analyze, and synthesize their decisions. The nature of wildlife management as a decision science has been recognized since the inception of the field, but formal methods of decision analysis have been underused. There is tremendous potential for wildlife management to grow further through the use of formal decision analysis. First, the wildlife science and human dimensions of wildlife disciplines can be readily integrated. Second, decisions can become more efficient. Third, decisions makers can communicate more clearly with stakeholders and the public. Fourth, good, intuitive wildlife managers, by explicitly examining how they make decisions, can translate their art into a science that is readily used by the next generation.

Literature Cited

Allen, C. R., J. J. Fontaine, K. L. Pope, and A. S. Garmestani. 2011. Adaptive management for a turbulent future. Journal of Environmental Management 92:1339–1345.

Allen, C. R., and L. H. Gunderson. 2011. Pathology and failure in the design and implementation of adaptive management. Journal of Environmental Management 92:1379–1384.

Anderson, M. G., D. Caswell, J. M. Eadie, J. T. Herbert, M. Huang, et al. 2007. Report from the joint task group for clarifying North American Waterfowl Management Plan population objectives and their use in harvest management. U.S. Fish and Wildlife Service, Department of the Interior, U.S. Geological Survey, Washington, D.C., USA.

Argyris, C., and D. Shon. 1978. Organizational learning: a theory of action learning. Addison-Wesley, Reading, Massachusetts, USA.

Bailey, J. A. 1982. Implications of "muddling through" for wildlife management. Wildlife Society Bulletin 10:363–369.

Bain, M. B. 1987. Structured decision making in fisheries management. North American Journal of Fisheries Management 7:475–481.

Beissinger, S. R., and D. R. McCullough, editors. 2002. Population viability analysis. University of Chicago Press, Chicago, Illinois, USA.

Bengston, D. N. 2000. Environmental values related to fish and wildlife lands. Pages 126–132 *in* D. C. Fulton, K. C. Nelson, D. H. Anderson, and D. W. Lime, editors. Human dimensions of natural resource management: emerging issues and practical applications. Cooperative Park Studies Program, University of Minnesota, Department of Forest Resources, St. Paul, USA.

Boyce, M. S., and L. L. McDonald. 1999. Relating populations to habitats using resource selection functions. Trends in Ecology and Evolution 14:268–272.

Burgman, M. A. 2005. Risks and decisions for conservation and environmental management. Cambridge University Press, Cambridge, United Kingdom.

Callicott, J. B., L. B. Crowder, and K. Mumford. 1999. Current normative concepts in conservation. Conservation Biology 13:22–35.

Caswell, H. 2001. Matrix population models: construction, analysis, and interpretation. Sinauer Associates, Sunderland, Massachusetts, USA.

Chaturvedi, S., S. S. Rajkumar, X. Li, G. J. Hurteau, M. Shtutman, and V. Chaturvedi. 2011. Antifungal testing and high-throughput screening of compound library against *Geomyces destructans*, the etiologic agent of geomycosis (WNS) in bats. PLoS One 6:e17032.

Claxton, K., M. Sculpher, and M. Drummond. 2002. A rational framework for decision making by the National Institute for Clinical Excellence (NICE). Lancet 360:711–715.

Dalkey, N., and O. Helmer. 1963. An experimental application of the Delphi method to the use of experts. Management Science 9:458–467.

Diakoulaki, D., C. H. Antunes, and A. Gomes Martins. 2005. MCDA and energy planning. Pages 859–890 *in* J. Figueira, S. Greco, and M. Ehrogott, editors. Multiple criteria decision analysis: state of the art surveys. Springer, New York, New York, USA.

Failing, L., R. S. Gregory, and M. Harstone. 2007. Integrating science and local knowledge in environmental risk management: a decision-focused approach. Ecological Economics 64:47–60.

Fontaine, J. J. 2011. Improving our legacy: incorporation of adaptive management into state wildlife action plans. Journal of Environmental Management 92:1403–1408.

Fritts, S. H., E. E. Bangs, J. A. Fontaine, M. R. Johnson, M. K. Phillips, E. D. Koch, and J. R. Gunson. 1997. Planning and implementing a reintroduction of wolves to Yellowstone National Park and central Idaho. Restoration Ecology 5:7–27.

Goodwin, P., and G. Wright. 2004. Decision analysis for management judgment. 3rd edition. John Wiley and Sons, West Sussex, United Kingdom.

Grand, J. B., Y. Wang, and E. C. Soehren. 2008. Monitoring program for biodiversity of terrestrial vertebrates on conservation lands within the Cumberland Plateau Region of Alabama. U.S. Geological Survey Alabama Cooperative Fish and Wildlife Research Unit, Auburn University, Mobile, Alabama, USA.

Gregory, R., and G. Long. 2009. Using structured decision making to help implement a precautionary approach to endangered species management. Risk Analysis 29:518–532.

Gregory, R., D. Ohlson, and J. Arvai. 2006. Deconstructing adaptive management: criteria for applications to environmental management. Ecological Applications 16:2411–2425.

Gunderson, L., C. S. Holling, and S. S. Light, editors. 1995. Barriers and bridges to the renewal of ecosystems and institutions. Columbia University Press, New York, New York, USA.

Gunderson, L., and S. S. Light. 2006. Adaptive management and adaptive governance in the everglades ecosystem. Policy Sciences 39:323–334.

Hämäläinen, R., and O. Leikola. 1996. Spontaneous decision conferencing with top-level politicians. OR Insight 9:24–28.

Hammer, W. 1967. A pictorial walk thru ol' high Jackson: Scottsboro 1868–1968. Limited centennial edition. College Press, Collegedale, Tennessee, USA.

Hammond, J. S., R. L. Keeney, and H. Raiffa. 1999. Smart choices: a practical guide to making better life decisions. Broadway Books, New York, New York, USA.

Helton, J. C., and D. E. Burmaster. 1996. Guest editorial: treatment of aleatory and epistemic uncertainty in performance assessments for complex systems. Reliability Engineering and System Safety 54:91–94.

Herath, G., and T. Prato, editors. 2006. Using multi-criteria decision analysis in natural resource management. Ashgate, Hampshire, United Kingdom.

Hilborn, R., and M. Mangel. 1997. The ecological detective: confronting models with data. Volume 28. Princeton University Press, Princeton, New Jersey, USA.

Holling, C. S., editor. 1978. Adaptive environmental assessment and management. John Wiley and Sons, London, United Kingdom.

Howard, R. A. 1966. Decision analysis: applied decision theory. Pages 55–71 in D. B. Hertz and J. Melese, editors. Proceedings of the fourth international conference on operational research. John Wiley and Sons, New York, New York, USA.

Howard, R. A. 1968. The foundations of decision analysis. IEEE Transactions of Systems Science and Cybernetics 4:211–219.

Hughes, T. P., L. H. Gunderson, C. Folke, A. H. Baird, D. Bellwood, et al. 2007. Adaptive management of the great barrier reef and the Grand Canyon world heritage areas. AMBIO: A Journal of the Human Environment 36:586–592.

Hunter, C. M., H. Caswell, M. C. Runge, E. V. Regehr, S. C. Amstrup, and I. Stirling. 2010. Climate change threatens polar bear populations: a stochastic demographic analysis. Ecology 91:2883–2897.

Johnson, F. A., W. L. Kendall, and J. A. Dubovsky. 2002. Conditions and limitations on learning in the adaptive management of mallard harvests. Wildlife Society Bulletin 30:176–185.

Johnson, F. A., C. T. Moore, W. L. Kendall, J. A. Dubovsky, D. F. Caithamer, J. R. Kelley Jr., and B. K. Williams. 1997. Uncertainty and the management of mallard harvests. Journal of Wildlife Management 61:202–216.

Karl, H. A., L. E. Susskind, and K. H. Wallace. 2007. A dialogue, not a diatribe: effective integration of science and policy through joint fact finding. Environment 49:20–34.

Keeney, R. L. 1982. Decision analysis: an overview. Operations Research 30:803–838.

Keeney, R. L. 1996a. Value-focused thinking: identifying decision opportunities and creating alternatives. European Journal of Operational Research 92:537–549.

Keeney, R. L. 1996b. Value-focused thinking: a path to creative decision-making. Harvard University Press, Cambridge, Massachusetts, USA.

Keeney, R. L. 2002. Common mistakes in making value trade-offs. Operations Research 50:935–945.

Keeney, R. L. 2007. Developing objectives and attributes. Pages 104–128 in W. Edwards, R. F. J. Miles, and D. Von Winterfeldt, editors. Advances in decision analysis: from foundations to applications. Cambridge University Press, Cambridge, United Kingdom.

Kiker, G. A., T. S. Bridges, A. Varghese, T. P. Seager, and I. Linkov. 2005. Application of multicriteria decision analysis in environmental decision making. Integrated Environmental Assessment and Management 1:95–108.

Kuhnert, P. M., T. G. Martin, and S. P. Griffiths. 2010. A guide to eliciting and using expert knowledge in Bayesian ecological models. Ecology Letters 13:900–914.

Lancia, R. A., C. E. Braun, M. W. Collopy, R. D. Dueser, J. G. Kie, C. J. Martinka, J. D. Nichols, T. D. Nudds, W. R. Porath, and N. G. Tilghman. 1996. ARM! For the future: adaptive resource management in the wildlife profession. Wildlife Society Bulletin 24:436–442.

Lee, K. N., and J. Lawrence. 1985. Adaptive management: learning from the Columbia River basin fish and wildlife program. Environmental Law 16:431–460.

Leopold, A. 1933. Game management. Charles Scribner's Sons, New York, New York, USA.

Lyons, J. E., M. C. Runge, H. P. Laskowski, and W. L. Kendall. 2008. Monitoring in the context of structured decision-making and adaptive management. Journal of Wildlife Management 72:1683–1692.

MacMillan, D. C., and K. Marshall. 2006. The Delphi process—an expert-based approach to ecological modelling in data-poor environments. Animal Conservation 9:11–19.

McDonald-Madden, E., W. J. M. Probert, C. E. Hauser, M. C. Runge, H. P. Possingham, M. E. Jones, J. L. Moore, T. M. Rout, P. A. Vesk, and B. A. Wintle. 2010. Active adaptive conservation of threatened species in the face of uncertainty. Ecological Applications 20:1476–1489.

McFadden, J. E., T. L. Hiller, and A. J. Tyre. 2011. Evaluating the efficacy of adaptive management approaches: is there a formula for success? Journal of Environmental Management 92:1354–1359.

McLain, R. J., and R. G. Lee. 1996. Adaptive management: promises and pitfalls. Environmental Management 20:437–448.

Moore, C. T., E. V. Lonsdorf, M. G. Knutson, H. P. Laskowski, and S. K. Lor. 2011. Adaptive management in the U.S. National Wildlife Refuge System: science-management partnerships for conservation delivery. Journal of Environmental Management 92:1395–1402.

Morgan, M. G., and M. Henrion. 1990. Uncertainty: a guide to dealing with uncertainty in quantitative risk and policy analysis. Cambridge University Press, Cambridge, United Kingdom.

Morrison, M. L., B. G. Marcot, and R. W. Mannan. 2006. Wildlife–habitat relationships: concepts and applications. Island Press, Washington, D.C., USA.

Nichols, J. D., F. A. Johnson, and B. K. Williams. 1995. Managing North American waterfowl in the face of uncertainty. Annual Review of Ecology and Systematics 26:177–199.

Nichols, J. D., M. C. Runge, F. A. Johnson, and B. K. Williams. 2007. Adaptive harvest management of North American waterfowl populations: a brief history and future prospects. Journal of Ornithology 148:S343–S349.

Nichols, J. D., and B. K. Williams. 2006. Monitoring for conservation. Trends in Ecology and Evolution 21:668–673.

Pahl-Wostl, C. 2009. A conceptual framework for analysing adaptive capacity and multi-level learning processes in resource governance regimes. Global Environmental Change 19:354–365.

Pickton, T., and L. Sikorowski. 2004. The economic impacts of hunting, fishing and wildlife watching in Colorado. Final report prepared for the Colorado Division of Wildlife. BBC Research and Consulting, Denver, Colorado, USA.

Raiffa, H., and R. O. Schlaifer. 1961. Applied statistical decision theory. Graduate School of Business Administration, Harvard University, Cambridge, Massachusetts, USA.

Ralls, K., and A. M. Starfield. 1995. Choosing a management strategy: two structured decision-making methods for evaluating the predictions of stochastic simulation models. Conservation Biology 9:175–181.

Redpath, S. M., B. E. Arroyo, F. M. Leckie, P. Bacon, N. Bayfield, R. J. Gutiérrez, and S. J. Thirgood. 2004. Using decision modeling with stakeholders to reduce human–wildlife conflict: a raptor–grouse case study. Conservation Biology 18:350–359.

Regan, H. M., Y. Ben-Haim, B. Langford, W. G. Wilson, P. Lundberg, S. J. Andelman, and M. A. Burgman. 2005. Robust decision making under severe uncertainty for conservation management. Ecological Applications 15:1471–1477.

Regan, H. M., M. Colyvan, and M. A. Burgman. 2002. A taxonomy and treatment of uncertainty for ecology and conservation biology. Ecological Applications 12:618–628.

Runge, M. C., E. Bean, D. R. Smith, and S. Kokos. 2011a. Non-native fish control below Glen Canyon Dam—report from a structured decision making project. U.S. Geological Survey Open File Report 2011-1012:1–74.

Runge, M. C., S. J. Converse, and J. E. Lyons. 2011b. Which uncertainty? Using expert elicitation and expected value of information to design an adaptive program. Biological Conservation 144:1214–1223.

Runge, M. C., F. A. Johnson, M. G. Anderson, M. D. Koneff, E. T. Reed, and S. E. Mott. 2006. The need for coherence between waterfowl harvest and habitat management. Wildlife Society Bulletin 34:1231–1237.

Runge, M. C., F. A. Johnson, J. A. Dubovsky, W. L. Kendall, J. Lawrence, and J. Gammonley. 2002. A revised protocol for the adaptive harvest management of mid-continent mallards. Division of Migratory Bird Management, U.S. Fish and Wildlife Service, Laurel, Maryland, USA.

Runge, M. C., J. R. Sauer, M. L. Avery, B. F. Blackwell, and M. D. Koneff. 2009. Assessing allowable take of migratory birds. Journal of Wildlife Management 73:556–565.

Scott, J. M. 1993. Gap analysis: a geographic approach to protection of biological diversity. Wildlife Monographs 123:1–41.

Speirs-Bridge, A., F. Fidler, M. F. McBride, L. Flander, G. Cumming, and M. A. Burgman. 2010. Reducing overconfidence in the interval judgments of experts. Risk Analysis 30:512–523.

Szymanski, J. A., M. C. Runge, M. J. Parkin, and M. Armstrong. 2009. White-nose syndrome management: report from a structured decision making initiative. U.S. Fish and Wildlife Service, Department of the Interior, Fort Snelling, Minnesota, USA.

Thomas, J. W., J. F. Franklin, J. Gordon, and K. N. Johnson. 2006. The Northwest Forest Plan: origins, components, implementation experience, and suggestions for change. Conservation Biology 20:277–287.

USFWS. U.S. Fish and Wildlife Service. 1988. Final supplemental environmental impact statement: issuance of annual regulations permitting the sport hunting of migratory birds. Department of the Interior, Washington, D.C., USA.

USFWS. U.S. Fish and Wildlife Service. 2003. Recovery plan for the red-cockaded woodpecker (Picoides borealis). Second revision. Atlanta, Georgia, USA.

USFWS. U.S. Fish and Wildlife Service. 2010. Adaptive harvest management: 2010 hunting season. Department of the Interior, Washington, D.C., USA.

van Heezik, Y., and P. J. Seddon. 2005. Structure and content of graduate wildlife management and conservation biology programs: an international perspective. Conservation Biology 19:7–14.

von Neumann, J., and O. Morgenstern. 1944. Theory of games and economic behavior. Princeton University Press, Princeton, New Jersey, USA.

von Winterfeldt, D., and W. Edwards. 1986. Decision analysis and behavioral research. Cambridge University Press, Cambridge, United Kingdom.

Wagner, F. H. 1989. American wildlife management at the crossroads. Wildlife Society Bulletin 17:354–360.

Walters, C. J. 1986. Adaptive management of renewable resources. Macmillan, New York, New York, USA.

Walters, C. J., and R. Hilborn. 1978. Ecological optimization and adaptive management. Annual Review of Ecology and Systematics 9:157–188.

Williams, B. K. 1996. Adaptive optimization and the harvest of biological populations. Mathematical Biosciences 136:1–20.

Williams, B. K., J. D. Nichols, and M. J. Conroy. 2002. Analysis and management of animal populations: modeling, estimation, and decision making. Academic Press, San Diego, California, USA.

Williams, B. K., R. C. Szaro, and C. D. Shapiro. 2007. Adaptive management: the U.S. Department of the Interior technical guide. Adaptive Management Working Group, Department of the Interior, Washington, D.C., USA.

SCALE IN WILDLIFE MANAGEMENT
The Difficulty with Extrapolation

JOHN A. BISSONETTE

INTRODUCTION

State and federal land management agencies have legislative authority for natural resources over broad landscape extents. Given that conflicts over the use of resources are increasing worldwide (Bannon and Collier 2003, Humphreys 2005), decisions regarding land use must be based on valid data. Increasingly, agencies rely on research data to inform management (Holl et al. 2003, MacKenzie 2005). Attempting to understand the relationships between pattern, process, and wildlife response has been the *sine qua non* of wildlife ecology and management for almost a century. The problems with which managers are usually concerned are typically large scale, complex, and nonlinear. They may include system-wide changes like invasive species (complete.dochttp://www .invasivespeciesinfo.gov/), changes in species population vital rates caused by anthropogenic land cover fragmentation and change (Lindenmayer and Fischer 2006), and the spread of infectious disease (e.g., chronic wasting disease, or CWD; Conner et al. 2007). The setting of harvest limits for big game is most often based on hunt units, but with a statewide context; it is clear that the problems managers face are complex. At the same time, the data on which managers rely to understand the relevant dynamics are typically smaller scale, taken on site and over shorter time periods. Extrapolation from data collected at smaller resolution and extents to larger landscapes is often problematic. A fundamental mismatch of scale persists. Almost every student in any of the environmental sciences has heard the term "scale" and understands the importance of its consideration in wildlife studies. Perhaps fewer can explain its importance to wildlife ecology or understand its connection to system organization.

This chapter will: (1) define and briefly explain the different uses of the term "scale"; (2) examine what has been termed "the problem of scale" and what it means; (3) explain the concept of scale dependency; (4) discuss the idea of finding the "right" scale(s); (5) discuss the idea of space–time diagrams and domains of scale; (6) outline the idea of animal response and discuss the idea that how we measure and interpret our results is not a trivial exercise; (7) discuss how, in terrestrial ecology, the context in which the scale concept has developed is intimately tied to hierarchy theory—that is, system organization; and (8) suggest that incorporating landscape context may solve much of the problem involved with scale issues and animal response. Examples drawn from research studies will be used throughout to demonstrate these ideas.

WHAT EXACTLY IS SCALE?
DIFFERENT USES IN THE LITERATURE

To a wildlife ecologist, scale is a metric that refers to the spatial or temporal dimensions of an object, pattern, or process (Turner et al. 2001). Schneider (2001:546) described the term as arising from two different etymological roots: the Norse root *skal*, suggesting "measurement by means of pairwise comparisons of objects," and the Latin root *scala*, "measuring a length by counting steps or subdivisions." There is great variability in the interpretation and meaning of scale; its use in the literature can be confusing (Montello 2001) because it has several referents.

Withers and Meentemeyer (1999) wrote that scale can refer to nominal, ordinal, interval, or ratio measurement scales, each with permissible operations of addition, subtraction, division, multiplication, and appropriate statistical tests (Stevens 1946, Siegel 1956). This use of the term "scale" refers to the theory of measurement where the mathematical or statistical operations allowable on a given set of data are wholly dependent on the "level" of measurement. For example, range condition may be classified as poor, fair, good, or excellent (Dyksterhuis, 1949). This is an ordinal (ranking) level of measurement because the categories are not only different from each other (nominal), but they stand in some relationship (better or worse) to each other. Alternatively, one uses a ratio level of measure when measuring the mass of individual vertebrates: a ratio level of measurement has a definite zero point. Contrast these with interval measurements of temperature in Fahrenheit or Centigrade, which are characterized by a known interval between values, but the zero point and the unit of measurement are arbitrary. For example, the freezing and boiling points in Fahrenheit are 32° and 212°, respectively, and for Centigrade, 0° and 100°, respectively. Both systems measure temperature, but with different units.

The concept can also refer to isometric and allometric scaling (i.e., the relationship between two entities). A relationship where the slope of the line representing the relationship between two variables is equal to 1 (i.e., a 45° line) characterizes isometric scaling. An allometric relationship is disproportionate; one variable increases or decreases more rapidly than the other. For example, Kleiber (1947) described the allometric scaling relationship between body size and metabolic rate as $Y = Y_oM^b$, where Y is the metabolic rate, Y_o is a constant characteristic of the organism, M is body mass, and b is the scaling exponent. Kleiber reported that metabolic rates scaled as $M^{3/4}$ (West et al. 1997). Others have shown scaling relationships between home range area and body mass (McNab 1963) and between body mass and dispersal distance (Wolff 1999, Sutherland et al. 2000). Recently, using work by Bowman et al. (2002), who found an isometric relationship (exponent = 1) between home range size and dispersal distance, Bissonette and Adair (2008) determined the optimal spacing of wildlife crossings to restore habitat permeability to roaded landscapes for terrestrial species using scaling concepts. They identified six home range area scale domains; three quarters of the species clustered in the three smallest domains.

They used home range$^{0.5}$ (HR$^{0.5}$) to represent a daily movement distance metric; when individual species movements were plotted against road markers, greater than 71% of 72 species found in North America were included at distances of less than or equal to 1.6 km. They argued that if the spacing of wildlife crossings in road hotspots of wildlife–vehicle mortality was based on the HR$^{0.5}$ metric, then (along with appropriate auxiliary mitigation) landscape permeability would be restored, thereby facilitating wildlife movement across the roaded landscape and significantly improving road safety by reducing wildlife–vehicle collisions.

Scale can refer to map or cartographic scale. In geography, scale concerns are primarily spatial, but they can also be temporal or thematic (e.g., the grouping of entities such as weather variables; Montello 2001). The familiar components of scale are resolution (sometimes referred to as grain, but that term is less exact) and extent. Ecologists and geographers have used these terms in very different ways and mean opposing things when they refer to small and large scale. To a geographer, a cartographic scale is the degree of spatial reduction, represented as a ratio or representative fraction (Montello 2001) of map distance to Euclidean distance on the surface of the earth; large scale means fine resolution (e.g., 1:24,000). To an ecologist, large scale means coarse resolution (e.g., 1:62,500). The difficulty is in large part semantic. Turner et al. (2001) recommend using "fine" and "broad" scale over "small" and "large" scale. Fine scale would then refer to smaller areas, more detail, and greater resolution. Broad scale would then refer to larger areas, less detail, and lower resolution (Fig. 6.1).

Martinez and Dunne (1998:208) wrote that ecologists use scale in a spatiotemporal sense; the term is almost always explained in the context of space and time. They suggested that this is restrictive and argued that "scale may more broadly refer to any metric that can quantitatively measure difference between observations in order to create data." They stated that a variable such as temperature (an interval scale of measurement; see above) can have an extent (range of temperatures measured) and a resolution (preciseness of the measurement). Martinez and Dunne (1998) wrote that the metrics of resolution and extent can be applied to nonspatial scales, for example, species richness or primary productivity, in the same way. They argued that "non-spatio-temporal scales have already advanced food-web research through extent- and resolution-based sensitivity analyses of various food web properties," including species richness and primary productivity. They used the species richness—population stability arguments discussed by Hutchinson (1959) and May (1988) whereby species diversity was related (or not) to stability—because of the effects on food web structure. Simply put, Martinez and Dunne (1998) referred to the degree of aggregation of different taxa within a food web as its resolution, and the range of species richness values over which specific scaling laws applied as its extent. This is an unusual approach and has not been used as frequently in ecological research or management discussions as the more familiar spatial and temporal scales.

Perhaps the most common use of scale in the ecological literature and in wildlife science in general has been in a spatiotemporal framework. In this sense, scale has been used in at least five ways (Wu and Li 2006): (1) the resolution and extent that the physical problem or process under consideration exists in nature (characteristic, phenomenon, or intrinsic scale), implying that many, if not most, natural phenomena have their own distinctive spatial extent and event frequencies that characterize their behavior; (2) the scale of measurement and sampling (*observation scale*); (3) the scale that a problem is analyzed or modeled (*analysis* or *analytical scale*); (4) the scale of experiments (*experimental scale*); and (5) the scale of policy making (*policy scale*). In all of these uses, resolution and extent are ultimately selected or determined by the investigator or manager. In response to the many components of scale, Dungan et al. (2002) proposed a framework for considering scale terms and approaches (Fig. 6.2). They distinguished between the phenomena being studied, the sampling protocol units used to measure the phenomena, and the spatial analysis used to detect pattern. The idea of spatiotemporal scaling, although relatively new to ecology, has a long history. Schneider (2001) pointed out that, although the term "spatial scaling" first appeared in the journal *Ecology* in a paper on mouse tracking (Martens 1972), explicit treatment of the concept appeared much earlier, in work by Collins (1884) and Murphy (1914), who recorded the latitude and longitude for more than 500 birds he collected. Since the early 1970s, the use of scale, scaling, and spatial and temporal scales has increased exponentially in the ecology literature.

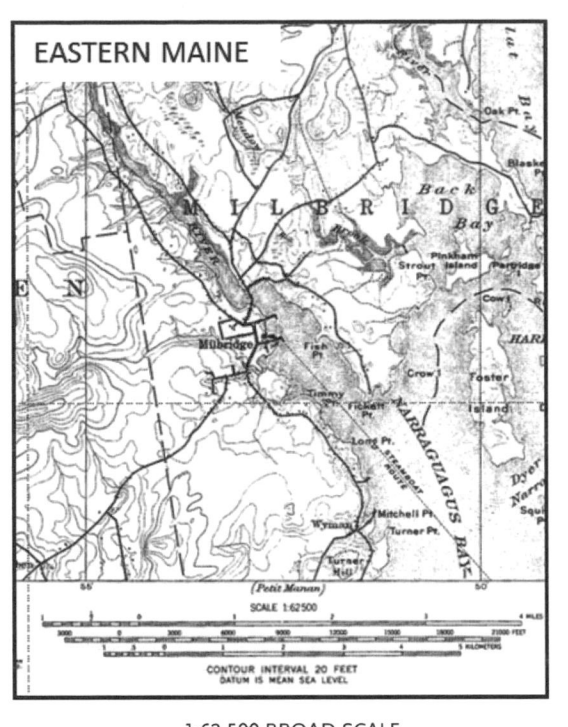

Figure 6.1. Fine and broad scales.

1:24,000 FINE SCALE
small spatial extent
(smaller area represented on map)
greater resolution
more detail

1:62,500 BROAD SCALE
large spatial extent
(larger area represented on map)
lower resolution
less detail

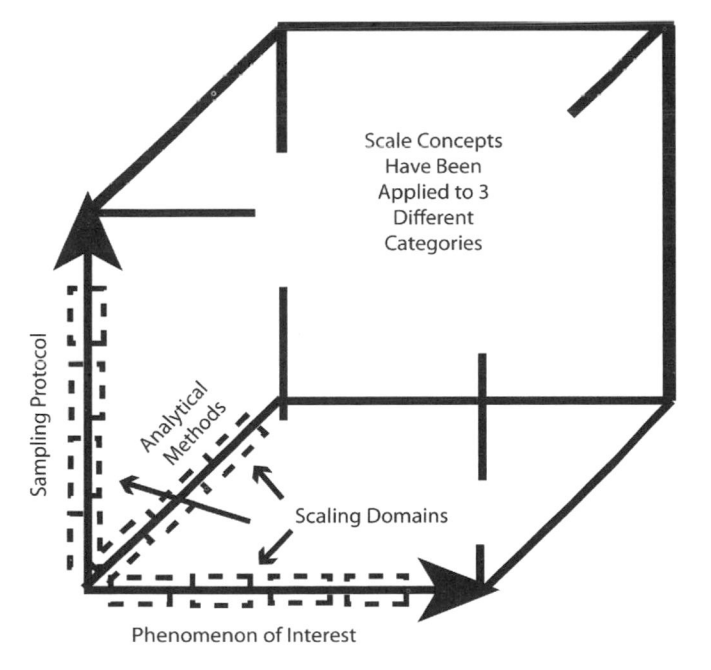

Figure 6.2. A framework in which to consider scale terms and the different components of an ecological question. Ideally, the sampling protocol and analytical methods should accurately reflect the intrinsic scales of the phenomenon of interest. If scale domains (see Fig. 6.4) are present, then matching sampling and analyses to the problem is theoretically simpler to achieve. Redrawn from Dungan et al. (2002)

THE PROBLEM OF SCALE

Why is scale a problem? The answer is threefold (Schneider 2001), playing on three system dynamics: complexity, the measurement of data, and scale determination. First, ecological problems are complex, exist at larger-scale extents, and are characterized by dynamics that have long time horizons. Examples are as diverse as the loss of biodiversity globally (Tilman 2000) and the decline of pronghorn (*Antilocapra americana*) in Grand Teton National Park (Berger 2003). Second, most variables can only be measured directly in small areas and on site (Schneider 2001). Even point data that are sampled regularly or irregularly in space or are spatially interpolated by kriging (a geostatistical technique that produces a trend surface that includes both sampled and nonsampled areas) are essentially finer-resolution data scaled to large extents (Gustafson 1998). Only a few variables have been measured continuously over large-scale extents. They include the normalized difference vegetation index, enhanced vegetation index, and fraction of absorbed photosynthetically active radiation. All are derived indices with often-questionable interpretations, in part because of the corrections required to interpret the raw data (see http://ivm.cr.usgs.gov/index.php). The relevant point here is to note that ecological problems have their characteristic scales of expression, while our methods for measuring variables are often limited to smaller spatial and temporal extents. Additionally, at least in the recent past, we are inclined to study familiar ecological problems at scales that accord with our perceptions (Wiens 1989). Indeed, the nature of research institutions with their characteristic time constraints for M.S. and Ph.D. research mitigates against longer-term studies (May 1994). Tilman (1989) reported that only 13 of 749 studies (1.7%) published in ecology lasted five years or more. Third, scale is a problem because it is very difficult to scale up measurements taken at finer-scale resolutions and extents to the larger scale of the real topic of investigation without transmutation or some qualitative change in the resulting pattern (O'Neill 1979, King 1991, King et al. 1991). Transmutation, also known as aggregation error, was originally defined in terms of hierarchy theory (see section on terminology below). It occurs most often when a process is changed as one moves from one hierarchical level to another (O'Neill 1979). For example, assume that the response of a body function in individual animals (e.g., respiration rate, metabolic function) is influenced by some critical temperature. Now suppose that the response is characterized by a threshold or a step function. If there is variation in the population (i.e., each individual responds to a slightly different critical temperature but in the same way), then by taking an average response across individuals and projecting it to the population, the shape of the response curve for the population is qualitatively different than for any of the individuals. O'Neill (1979) explained the idea in detail and described transmutations not only of threshold functions, but also of more complex discontinuous functions between trophic levels and at the ecosystem level.

I argue that transmutation can be used to describe results that are qualitatively different when data collected at smaller spatial extents are extrapolating to larger-scale extents. Miller et al. (2004:310) describe this mismatch as "one of the most formidable challenges confronting environmental scientists." The default assumption appears to be that the areas are either (1) homogeneous, (2) similarly heterogeneous, or (3) the dynamics are linear. For example, when biologists measure species loss at smaller-scale extents and then extrapolate to the species range, problems of data reliability often arise because species number often does not scale directly with area (Connor and McCoy 1979). The properties of the system change as scale extent increases (Rastetter et al. 1992). Additionally, where topography varies within a survey, the effects of area are often difficult to separate from habitat diversity effects (Johnson et al. 2003). The essence of the message is that caution is advisable when scaling essentially small-scale data to larger-scale extents.

One must consider these problems, because scaling is inevitable in research and practice whenever predictions need to be made at a scale extent that is different from the scale where data are acquired (Wu and Li 2006). Given the above-mentioned problems, Levin's (1992:1959) statement is clear:

> Two fundamental and interconnected themes in ecology are the development and maintenance of spatial and temporal pattern, and the consequences of that pattern for the dynamics of populations and ecosystems. Central to these questions is the issue of how the scale of observation influences the

description of pattern; each individual and each species experience the environment on a unique range of scales, and thus responds to variability individualistically. Thus, no descriptions of the variability and predictability of the environment make sense without reference to the particular range of scales that are relevant to the organisms or processes being examined.

For example, pronghorns spend summers in Grand Teton National Park (GTNP), and most migrate through a narrow corridor to southern Wyoming for the winter (Sawyer et al. 2005). They suffer mortality throughout the year. However, in GTNP, a trophic cascade and mesopredator release involving wolves (*Canis lupus*) and coyotes (*Canis latrans*) mediates fawn mortality. Areas with higher densities of wolves have lower pronghorn fawn mortalities, apparently because of a negative interaction between wolves and coyotes (Berger 2007, Berger et al. 2008). Scaling results of fawn mortality of this pronghorn population to the entire Wyoming pronghorn population would likely cause problems because of age-related difference in vulnerability and because wolves are either absent or very limited currently in southern Wyoming.

SCALE DEPENDENCY

Scale dependency and scale-dependent factors are common topics in the literature. Although the terms are easy to confuse, they refer to the idea that different patterns emerge when the scale resolution (sometimes referred to as the grain) or the scale extent (the total area under consideration) of the study changes. This is especially relevant when the *observation scale* (the scale of measurement and sampling) changes. There is pattern at every scale, so when the resolution or extent is changed, the patterns change. For example, Turner et al. (1989) found that rare cover types were lost or homogenized when the resolution of digital maps became coarser, but that the rate of loss was influenced by the landscape pattern. Indices of pattern (e.g., dominance and contagion) also were changed as grain and extent were varied. When measurements at finer resolutions and small extents are scaled up or extrapolated to broader landscape extents, transmutation or qualitatively different patterns can emerge (King et al. 1991). This is a manifestation of scale dependency. One might think of the broad- and fine-scale relationships between mule deer (*Odocoileus hemionus*) and white-tailed deer (*Odocoileus virginianus*). At the scale extent of species distributions, the pattern is sympatric; the species co-occur in many places. However, at the resolution of multiple home ranges, the pattern is allopatric, and the distribution of the species does not overlap. Krausman (1976) showed that in Big Bend National Park, Texas, white-tailed and mule deer distributions are separated by altitude: white-tailed locations occur at elevations above 1,633 m, mule deer locations below 1,633 m. There appeared to be little overlap in their distributions. However, Anthony and Smith (1977) did not find a similar pattern in southeastern Arizona. Whenever competition between two species results in competitive exclusion, one might expect to find a similar-scale relationship. For example, Sherry and Holmes (1988) reported that least flycatchers (*Empidonax minimus*) negatively influenced the distribution of American redstarts (*Setophaga ruticilla*) locally, but the distribution of the two species overlapped in the northeastern United States (Wiens 1989). Clearly, the extents over which the processes are viewed make a difference.

FINDING THE RIGHT SCALE

What is the "right" scale, and how can we find it? Perhaps the first lesson learned is that there is no single correct scale (Turner et al. 2001). As Wiens (1989:391) stated, "what is an 'appropriate' scale depends in part on the question one asks." Additionally, the specific processes of interest may have their own scales. Addicott et al. (1987) argued that there were no well-accepted general procedures or criteria for determining how organisms respond to landscape heterogeneity. They refined the concept of ecological neighborhoods developed earlier by Wright (1943, 1946), Southwood (1977), and Antonovics and Levin (1980). They characterized an ecological neighborhood as defined by three properties: a specific ecological process, a time scale appropriate to that process, and the organism's activity during that time period. They suggested that movement (scale extent) defined the neighborhood. For ex-

ample, foraging activities might typically cover a smaller area than dispersal or alert-alarm movements. The time periods for these activities would also be different. It is not difficult to understand that different processes may need to be measured at different spatial and temporal scales. Understanding foraging dynamics, breeding behavior, or dispersal would seem to require consideration about the resolution and extent associated not only with the spatial and temporal dimensions of each process but also with the variables or metrics selected for the sampling effort (Martinez and Dunne 1998). To deal with ecological complexity, a multiscale approach is often appropriate. Recent work on CWD by Conner et al. (2007) recognized that a multiscale approach was necessary to handle the types of data that were available (Fig. 6.3). First, the investigators developed and adapted methods to handle data appropriate to multijurisdictional or multistate modeling; they termed this scale and extent the regional scale. Then, methods and data appropriate for within-state areas (wildlife management units or metapopulations) were developed and called the landscape scale. Finally, data and methods adapted for population and individual-based models were developed, and they were termed the fine scale. Conner et al. (2007) recognized and reported that CWD data have been collected not only over different extents but also at different resolutions. Their approach was clearly and overtly multiscale.

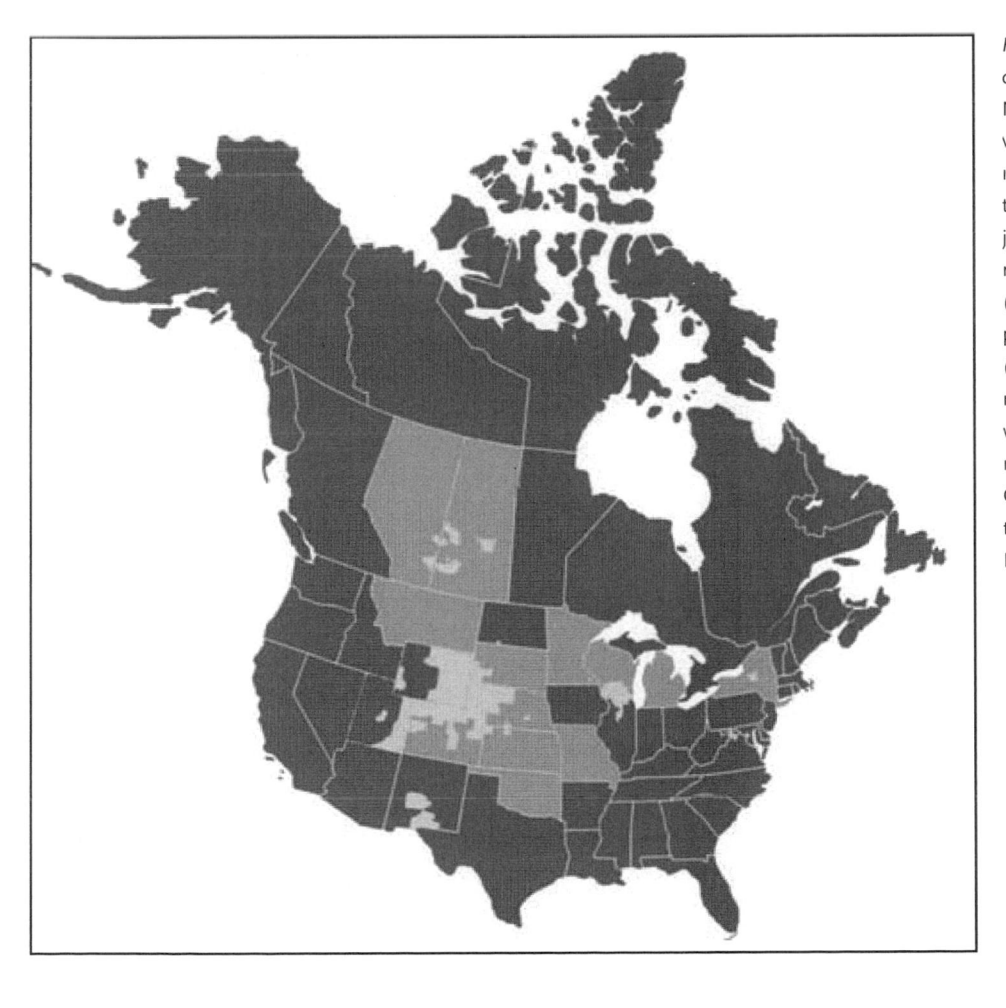

Figure 6.3. Presence of chronic wasting disease (CWD) in deer and elk across North America. Conner et al. (2007) developed and adapted methods to handle multiple-resolution data appropriate to three different scale extents: (1) multijurisdictional or multistate modeling (the regional scale); (2) within-state areas (i.e., wildlife management units or metapopulations, the landscape scale); and (3) population- and individual-based models (fine scale). Light gray, areas with CWD-infected cervid populations; medium gray, states/provinces where CWD has been found in captive populations. Redrawn from the Chronic Wasting Disease Alliance (www.cwd-info.org)

HENRY M. STOMMEL (1920–1992)

A leading theoretician on ocean currents, Henry M. Stommel graduated from Yale University in 1942 with a B.S. degree. After teaching mathematics and astronomy at Yale for two years, he became a research associate at the Woods Hole Oceanographic Institution, leaving in 1960 to become a professor of oceanography at Harvard University. In 1963 he joined the faculty of the Massachusetts Institute of Technology, remaining until 1978, when he returned to Woods Hole. Stommel was considered one of the most influential oceanographers of his time. His 1963 schematic diagram of the spectral distribution of sea level in space and time is considered to be one of the earliest depictions of a space–time relationship. Space–time diagrams clearly show the optimal matching of temporal and spatial scales needed to capture the relevant dynamics as closely as possible and underpin the conceptual theory of scaling.

Photo courtesy of the National Academy of Sciences

SPACE–TIME DIAGRAMS AND DOMAINS OF SCALE

Perhaps one way to think about appropriate scale is to review the history of its graphical expression: space–time diagrams. A space–time diagram is little more than a comparison of a pattern, phenomenon, or process on space (abscissa) and time (ordinate) axes. Stommel's (1963) schematic diagram of the spectral distribution of sea level in space and time is considered one of the earliest depictions of a space–time relationship. Familiar adaptations of space–time diagrams are given in Delcourt et al. (1983), Weins (1989), Holling (1992), and Bissonette (1997). A space–time diagram (Fig. 6.4) clearly shows the optimal matching of temporal and spatial scales needed to capture the relevant dynamics as closely as possible. In practice, however, it is almost always impossible to optimally match space and time scales, and space–time diagrams. Why? The time frames over which the dynamics operate are almost always longer than is practical to measure. When the time horizon of the pattern or process becomes very long (e.g., centuries or longer), then it is impossible to closely match sampling scale to phenomenon scale, and a space-for-time substitution is often employed (Pickett 1989). What this means is that, rather than studying a phenomenon (e.g., plant succession) over the time periods necessary to capture the relevant dynamics, different stages in the successional pattern are studied at different sites simultaneously. For example, the study of how forest-dwelling species are expected to change over time as forests grow and mature is done by examining multiple plots of varying age classes at one time (i.e., a snapshot

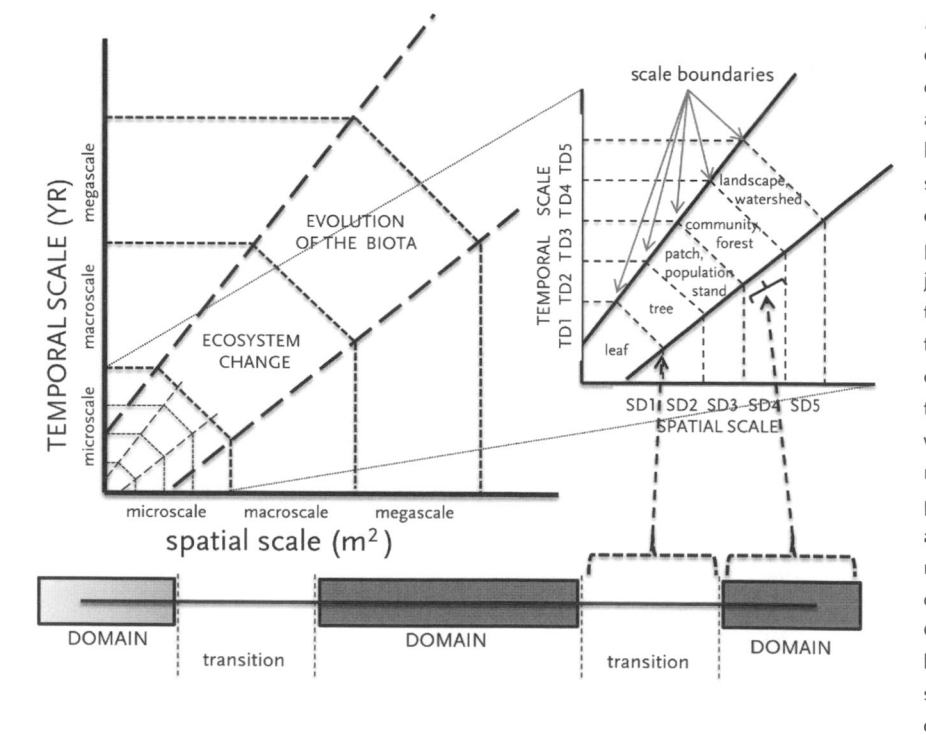

Figure 6.4. Space–time domains for ecological phenomena. The large graph shows scale domains ranging from small to very large spatial and temporal processes. The inset graph illustrates smaller spatial and shorter temporal scale domains for processes ranging from gas exchange through stomata on the underside of plant leaves to changes in landscape pattern. The junctions between scale domains—that is, the transitions from one scale to another—represent the locations on the scale gradient where rapid changes or thresholds are expected to occur. The theory suggests that observations and sampling within a scale domain are expected to result in reliable results for processes, phenomena, and patterns operating within that domain. A spatial and temporal domain that indicates the approximate matching of scales required to sample the dynamics characterizes each process or pattern. Given the very long time scales involved for larger-scale phenomena, a space-for-time substitution is often necessary. See the text for additional explanation. Adapted from Delcourt et al. (1983), Wiens (1989), and Bissonette (1997)

approach) and recording species patterns on each plot (e.g., Thompson et al. 1992). The inference of how the community will change over time is based on these short-term samples.

The Missouri Forest Ecosystem Project (http://mofep .mdc.mo.gov/overview/default_overview.htm) is an example of a longer-term study (data collection from 1990 to present) to understand the effects of changing forest dynamics on many components of forest ecology, including small mammals, reptiles and amphibians, forest invertebrates, and interior songbirds. However, longer-term studies are often not possible. What is one to do? As explained below, scaling domains appear to provide some leeway in selecting the appropriate scale.

Wiens (1989) pointed to a potentially crippling problem with scale dependency. If there is pattern at every scale, how is one to find generalizations? How can one extrapolate results if the spatial and temporal scale spectrum is a continuum? Understanding patterns would be difficult indeed. However, Wiens (1989) suggested that if the gradient of spatial and temporal scales is discontinuous with sections or domains over which patterns do not change or change very little or monotonically with changes in scale, then as ecologists we have a way to deal with extrapolation. As Wiens (1989:393) suggests, "domains of scale for particular pattern-process combinations define the boundaries of generalizations." As long as we are within a domain of scale, we can expect relative clean correspondence between the observational scales we use to measure system response and the patterns that emerge (Fig. 6.4). Scale dependence becomes somewhat muted.

UNDERSTANDING ANIMAL RESPONSE: A MAJOR PROBLEM IN WILDLIFE ECOLOGY

It is not uncommon to see phrases like these in the current literature: "at the landscape level"; "to the ecosystem scale"; "we studied habitat selection at the landscape, meso, and plot scales"; "work has been devoted to assessing the relative influence of landscape level"; "our goal was to identify brood rearing habitat at the landscape scale." These statements do not appear to distinguish between some very important concepts, and they are misleading in their strict interpretation. Two problems exist: one involves terminology; the other involves a disjunct between the scales that characterize animal movements and processes and how they are perceived and measured by the observer.

Terminology

In ecology, the context in which the scale concept has developed is intimately tied to hierarchy theory (i.e., system organization). This may account for some of the confusing use of terms. As suggested above, it appears that ecologists commonly confuse the terms "scale" and "level." For example, it is easy to find references to population scale and population level, and certainly to landscape scale and level, often in the same publication. Although the problem was recognized early (Allen and Hoekstra 1990) and often (King 1997, Allen 1998,

O'Neill and King 1998), the problem persists. It is a problem because the terms mean different things. Scale refers to some measureable dimension in space and time, while level is more appropriately applied in the sense of organizational level. "Landscape scale," when used in a journal article, implies some landscape extent that is large, but "large" is seldom defined, and the extent of the landscape is seldom given. There is no *one* landscape scale, so the term is imprecise and misleading at best. Because there are two relevant aspects of scale that are most often used (resolution and extent), for clarity it is preferable to use the term "landscape extent" to convey the meaning and give the areal extent of that landscape. One reason why landscape extent is seldom given may be related to the modifiable areal unit problem (Openshaw 1984). That is, the idea that choosing the appropriate areal unit (landscape extent) is often arbitrary and modifiable, meaning that the extent chosen is more often than not based on the decision of the investigator.

The confused use of terms, however, is more than just semantics and may stem from a lack of appreciation of the distinction between scale concepts and system organization. As O'Neill et al. (1986:39) stated, "the task of choosing an appropriate system for investigating a particular phenomenon is inseparable from consideration of underlying organization and complexity." Allen and Starr (1982) argued that hierarchy theory provides the organization framework with which to consider the concept of scale. A level defines a specific organization in a hierarchy definition—for example, an organism level or a population level. One way to see the connection is to think about O'Neill and King's "inescapable conclusion": "If you move far enough across scale, the dominant processes change. It is not just that things get bigger, but the phenomena themselves change. Unstable systems now seem stable . . . Bottom up control turns into top down control . . . Competition becomes less important and climate seems to dominate patterns. These same changes can be observed in aquatic ecosystems . . . terrestrial vegetation . . . and geomorphological dynamics. It is this observation of changing dynamics with scale that formed the basis for the development of hierarchy theory in ecology" (O'Neill and King 1998:5–6). Understanding the connections between hierarchical system organization and scaling and its associated terminology should provide more clarity when scaling issues are addressed. When referring to scale, an overt reference to which aspect (resolution or extent) is being discussed is essential.

Disjunctive Connections

Many observers do not distinguish between the *characteristic*, *phenomenon*, or *intrinsic scale* (the resolution and extent that the physical problems or processes under consideration exist in nature) and the *observation scale* or *analytical scale* (the scale of measurement and sampling or the scale that a problem is analyzed or modeled). The disjunct, quite simply, is that problems in nature exist at a characteristic scale extent and resolution, and the way we measure or sample is often at a different scale extent and resolution. What makes this important is that the scale of our observation (our sampling) fundamentally determines our description and explanation of the natural world (O'Neill and King 1998:7) and should coincide with the scales inherent in the patterns or processes we are measuring. Animals select habitats, forage, migrate, and otherwise fulfill their life requirements within certain domains of scale (Addicott et al. 1987). We observe animals and their processes, and often we measure variables at single or multiple scale resolutions and extents. Consequently, the measurements we make, based on the questions we ask, tend to portray different patterns. To say that animals select habitats at various scales, without explicitly stating if we are referring to resolution or extent, is imprecise and misleading.

MacNally (1999, 2005) described an organism-centric approach designed to deal with the scale issue that incorporated a species-"idiosyncratic" view of the world based on animal size; sensory capabilities were disregarded for simplicity. He used maximum body length and expected length of lifetime to develop a "characteristic measure" that represented the spatial extension of the organism. I interpret this to mean the area over which the animal is expected to move over its existence; that is, the natural scale that informs how the landscape is used by the organism. MacNally's approach represents an interesting and heuristic approach to addressing the scale problem.

Choosing biologically relevant scale measures for investigation is imperative. How to do that is less obvious, but a biological rationale seems necessary. One pertinent question involves whether the patterns that emerge from a study of specified resolution and extent appropriately reflect the resolution and extent that the physical problem or process under consideration exists in nature—or are they simply an artifact of the scale used? It appears that few have explicitly addressed this distinction. A recent review of 79 multiscale wildlife studies from three journals (*Landscape Ecology, Journal of Wildlife Management*, and *Journal of Applied Ecology*) found that "only 29%" of the observational scales examined had a biological rationale for their use. In other words, observational scales tended to be "chosen arbitrarily with no biological connection to the system being studied" (Wheatley and Johnson 2009:152). Could it be that the attributes of scale (i.e., resolution and extent) are insufficient by themselves, to understand animal response to changing resource availability as reflected in changing landscape patterns? Are we missing something?

INCORPORATING LANDSCAPE CONTEXT

Åberg et al. (1995), Gascon et al. (1999), Guisan et al. (2006), and more recently Wheatley and Johnson (2009) focused on spatial context, specifically the landscape structure and composition of an area surrounding a site. De Knegt et al. (2011) argued that the landscape surrounding a used resource provided the environmental context for understanding the animal response to that resource, and proposed the term "range" to represent the ambit radii at which context can be considered. Referring to an area in which something acts or operates or has power or control (http://wordnetweb.princeton.edu /perl/webwn?s=ambit), ambit has antecedents with the idea

of ecological neighborhoods (Addicott et al. 1987), and the term appears in Haury et al. (1978) and Wiens (1989). Since de Knegt et al. (2011) had no a priori knowledge about the scales that elephants (*Loxodonta africana*) responded to environmental heterogeneity, they suggested that within constraints set by a chosen resolution and extent, measuring environmental or landscape context at different ambit radii centered on used resource sites could possibly reveal the ecologically most relevant scales for studying species–environment relationships. They focused on understanding how elephant responses scaled to food and water resources. They varied the range of environmental contexts that they measured (i.e., the ambit radii from zero [no context] to a 40-km radius). They reported that the strength of habitat selection was highly dependent on the size of the ambit radius used to measure the environmental context variables of tree cover, herbaceous biomass, water occurrence, and a proxy measure of vegetation heterogeneity. In other words, elephants appeared to be keying in on resources at a characteristic scale that could be characterized by ambit radius length. They reported that the characteristic scale did not match the scales at which environmental heterogeneity was most dominant (de Knegt et al. 2011). Accounting for the elephants' environmental context revealed relationships that would not have been exposed had it not been considered. If these results are robust for other species, they suggest that examining a range of environmental contexts, given a specified resolution and extent, will provide a clearer understanding of animal–environment relationships.

Franklin and Lindenmayer (2009) argue that many biologists have largely overlooked the pivotal importance of the matrix and the habitat it provides. Brady et al. (2011) reported that matrix development intensity by anthropogenic forces influenced mammal abundance and landscape use, but that the responses tended to be species specific. Interestingly, native species richness peaked at moderate levels of matrix development, but exotic species richness and feral predators increased with increasing development. Clearly, consideration of landscape context is an important influence that can inform our understanding of animal response. De Knegt et al. (2011) have provided a way to incorporate landscape context while at the same time limiting the problems that wildlife ecologists face when trying to conduct scale-sensitive studies.

SUMMARY

It is clear that scale and scaling in wildlife ecology will continue to pose problems. First, one easily resolved difficulty is to gain accuracy and consistency in the terminology we use to describe scale concepts (Dungan et al. 2002). Here context is important. Understanding that scale and scaling are intimately tied to hierarchy theory (i.e., system organization) may account for some of the confusing terms. Having common terms with mutually agreed-upon meanings seems a logical first step. Second, understanding that qualitative changes often occur when measurements taken at finer-scale resolutions and extents are scaled larger is the *sine qua non* for developing scale-sensitive and biologically realistic sampling protocols and analyses. It is important to understand that the field of landscape ecology came into prominence primarily because cross-scale extrapolation was recognized as problematic. Prior to the attention to scale in the early 1980s, ecologists "scaled up," as if the dynamics they were investigating were linear. It is clear that few important relationships in ecology are that simple. Nonlinearity is more often the case, and hence the problem of scaling is much more difficult. Different patterns emerge when the scale resolution or the scale extent of the study is changed; this is the concept of scale sensitivity. Perhaps it is best to note is that there is no single correct scale; the "appropriate" scales depend on the questions one asks and the specific processes of interest. Space–time diagrams illustrate the difficulty with achieving a match between the spatial extent over which the dynamics of a process or pattern are expressed and the time it takes for the dynamics to be expressed. Ideally, one attempts to match space and time to understand the dynamics. This is often not possible, so the standard practice appears to involve a substitution of space for time, coupled with a sometimes-arbitrary selection of the hopefully meaningful scales. If domains of scales exist, then this may be the closest we can come to adjusting our sampling and observational scales to the scales inherent to natural processes and patterns.

De Knegt et al. (2011:271) have stated that "no question in spatial ecology can be answered without explicitly referring to these components" (resolution and extent). However, Wheatley and Johnson (2009) make a clear distinction between a strictly scalar study and a spatial study. The distinction deals primarily with holding at least one scale component (resolution or extent) constant while varying the other. When either grain or extent, or other independent variables, is changed across scales as part of a study design, the resulting study is not strictly scalar, and hence cross-scale extrapolation and generalization are not possible. It might be that, to achieve accurate cross-scale predictions, dealing with adjustment of resolution and extent may be insufficient. Today it is generally accepted that landscape context is an important driver of animal response (Saunders et al. 1991, McIntyre and Barrett 1992, McIntyre et al. 1996, McIntyre and Hobbs 1999, Manning et al. 2004, Fisher and Lindenmayer 2006; but see Prevedello and Vieira 2010). If we are interested in cross-scale prediction of animal response, then understanding the environmental context is necessary. Addressing scale and scaling issues is a necessary part of what wildlife ecologists need to know if they wish to understand the broad-scale management problems facing the profession.

Literature Cited

Åberg, J., G. Jansson, J. E. Swenson, and P. Angelstam. 1995. The effects of matrix on the occurrence of hazel grouse (*Bonasa bonasia*) in isolated habitat fragments. Oecologia 103:235–269.

Addicott, J. F., J. M. Aho, M. F. Antolin, D. K. Padilla, J. S. Richardson, and D. A. Soluk. 1987. Ecological neighborhoods: scaling environmental patterns. Oikos 49:340–346.

Allen, T. F. H. 1998. The landscape "level" is dead: persuading the family to take it off the respirator. Pages 35–54 in D. L. Peterson and V. T. Parker, editors. Ecological scale: theory and application. Columbia University Press, New York, New York, USA.

Allen, T. F. H., and T. W. Hoekstra. 1990. The confusion between scale-defined levels and conventional levels of organization in ecology. Journal of Vegetation Science 1:5–12.

Allen, T. F. H., and T. B. Starr. 1982. Hierarchy: perspectives for ecological complexity. University of Chicago Press, Chicago, Illinois, USA.

Anthony, R. G., and N. S. Smith. 1977. Ecological relationships between mule deer and white-tailed deer in southeastern Arizona. Ecological Monographs 47:255–277.

Antonovics, J., and D. A. Levin. 1980. The ecological and genetic consequences of density-dependent regulation in plants. Annual Review of Ecology and Systematics 11:411–452.

Bannon, I., and P. Collier, editors. 2003. Natural resources and violent conflict: options and actions. The World Bank, Washington, D.C., USA.

Berger, J. 2003. Is it acceptable to let a species go extinct in a national park? Conservation Biology 17:1451–1454.

Berger, K. M. 2007. Conservation implications of food webs involving wolves, coyotes, and pronghorn. Dissertation, Utah State University, Logan, USA.

Berger, K. M., E. M. Gese, and J. Berger. 2008. Indirect effects and traditional trophic cascades: a test involving wolves, coyotes, and pronghorn. Ecology 89:818–828.

Bissonette, J. A. 1997. Scale-sensitive properties: historical context, current meaning. Pages 3–31 in J. A. Bissonette, editor. Wildlife and landscape ecology: effects of pattern and scale. Springer-Verlag, New York, New York, USA.

Bissonette, J. A., and W. A. Adair. 2008. Restoring habitat permeability to roaded landscapes with isometrically-scaled wildlife crossings. Biological Conservation 141:482–488. doi:10.1016/j.biocon.2007.10.019.

Bowman, J., J. A. G. Jaeger, and L. Fahrig. 2002. Dispersal distance of mammals is proportional to home range size. Ecology 83:2049–2055.

Brady, H. J., C. A. McAlpine, C. J. Miller, H. P. Possingham, and G. S. Baxter. 2011. Mammal responses to matrix development intensity. Austral Ecology 36:35–45.

Collins, J. W. 1884. Notes on the habits and methods of capture of various species of sea birds that occur on the fishing banks off the eastern coast of North America, and which are used as bait for catching codfish by New England fisherman. Report of the Commissioner of Fish and Fisheries for 1882 13:311–335.

Conner, M. M., J. E. Gross, P. C. Cross, M. R. Ebinger, R. R. Gillies, M. D. Samuel, and M. W. Miller. 2007. Scale-dependent approaches to modeling spatial epidemiology of chronic wasting disease. Special report 2007. Utah Division of Wildlife Resources, Salt Lake City, USA.

Connor, E. G., and E. D. McCoy. 1979. The statistics and biology of the species-area relationship. American Naturalist 113:791–833.

de Knegt, H. J., F. van Langevelde, A. K. Skidmore, A. Delsink, R. Slotow, et al. 2011. The spatial scaling of habitat selection of African elephants. Journal of Animal Ecology 80:270–281.

Delcourt, H. R., P. A. Delcourt, and T. Webb. 1983. Dynamic plant ecology: the spectrum of vegetational change in space and time. Quaternary Science Review 1:153–175.

Dungan, J. L., J. N. Perry, M. R. T. Dale, P. Legendre, S. Citron-Pousty, M.-J. Fortin, A. Jakomulska, M. Miriti, and M. S. Rosenberg. 2002. A balanced view of scale in spatial statistical analysis. Ecography 25:626–640.

Dyksterhuis, E. J. 1949. Condition and management of range land based on quantitative ecology. Journal of Range Management 2:104–115.

Fischer, J., and D. B. Lindenmayer. 2006. Beyond fragmentation: the continuum model for fauna research and conservation in human-modified landscapes. Oikos 112:473–480.

Franklin, J. F., and D. B. Lindenmayer. 2009. Importance of matrix habitats in maintaining biological diversity. Proceedings of the National Academy of Sciences 106:349–350.

Gascon, C., T. E. Lovejoy, R. O. Bierregaard, J. R. Malcom, P. C. Stouffer, H. L. Vasconcelos, W. F. Laurance, B. Zimmerman, M. Tocher, and S. Borges. 1999. Matrix habitat and species richness in tropical forest remnants. Biological Conservation 91:223–229.

Guisan, A., A. Lehmann, S. Ferrier, M. Austin, J. M. C. Overton, R. Aspinall, and T. Hastie. 2006. Making better biogeographical predictions of species' distributions. Journal of Applied Ecology 43:386–392.

Gustafson, E. J. 1998. Quantifying landscape spatial pattern: what is the state of the art? Ecosystems 1:143–156.

Haury, L. R., J. A. McGowan, and P. H. Wiebe. 1978. Patterns and processes in the time–space scales of plankton distribution. Pages 227–327 in J. H. Steele, editor. Spatial pattern in plankton communities. Plenum, New York, New York, USA.

Holl, K. D., E. E. Crone, and C. B. Schultz. 2003. Landscape restoration: moving from generalities to methodologies. Bioscience 53:491–502.

Holling, C. S. 1992. Cross-scale morphology, geometry, and dynamics of ecosystems. Ecological Monographs 62:447–502.

Humphreys, M. 2005. Natural resources, conflict, and conflict resolution: uncovering the mechanisms. Journal of Conflict Resolution 49:508–537.

Hutchinson, G. E. 1959. Homage to Santa Rosalia. American Naturalist 93:145–159.

Johnson, M. P., N. J. Frost, M. W. J. Mosley, M. F. Roberts, and S. J. Hawkins. 2003. The area-independent effects of habitat complexity on biodiversity vary between regions. Ecology Letters 6:126–132.

King, A. W. 1991. Translating models across scales in the landscape. Pages 479–517 in M. G. Turner and R. H. Gardner, editors. Quantitative methods in landscape ecology. Springer-Verlag, New York, New York, USA.

King, A. W. 1997. Hierarchy theory: a guide to system structure for wildlife biologists. Pages 185–212 in J. A. Bissonette, editor. Wildlife and landscape ecology: effects of pattern and scale. Springer, New York, New York, USA.

King, A. W., A. R. Johnson, and R. V. O'Neill. 1991. Transmutation and functional representation of heterogeneous landscapes. Landscape Ecology 5:239–353.

Kleiber, M. 1947. Body size and metabolic rate. Physiological Reviews 27:511–541.

Krausman, P. R. 1976. Ecology of the Carmen Mountains white-tailed deer. Dissertation, University of Idaho, Moscow, USA.

Levin, S. A. 1992. The problem of pattern and scale in ecology: the Robert H. MacArthur Award Lecture. Ecology 73:1943–1967.

Lindenmayer, D. B., and J. Fischer. 2006. Habitat fragmentation and landscape change. Island Press, Washington, D.C., USA.

MacKenzie, D. I. 2005. What are the issues with presence–absence data for wildlife managers? Journal of Wildlife Management 69:849–860.

MacNally, R. 1999. Dealing with scale in ecology. Pages 10–17 in J. A. Wiens and M. R. Moss, editors. Issues in landscape ecology. International Association for Landscape Ecology, Pioneer Press, Greeley, Colorado, USA.

MacNally, R. 2005. Scale and an organism-centric focus for studying interspecific interactions in landscapes. Pages 52–69 in J. A. Wiens and M. R. Moss, editors. Issues and perspectives in landscape ecology. Cambridge University Press, Cambridge, United Kingdom.

Manning, A. D., D. B. Lindenmayer, and H. A. Nix. 2004. Continua and Umwelt: novel perspectives on viewing landscapes. Oikos 104:621–628.

Martens, G. G. 1972. Censusing mouse populations by means of tracking. Ecology 53:859–867.

Martinez, N. D., and J. A. Dunne. 1998. Time, space, and beyond: scale issues in food-web research. Pages 207–226 in D. L. Peterson and V. T. Parker, editors. Ecological scale: theory and applications. Columbia University Press, New York, New York, USA.

May, R. M. 1988. How many species are there on earth? Science 241:1441–1449.

May, R. M. 1994. The effects of spatial scale on ecological questions and answers. Pages 1–17 in P. J. Edwards, R. M. May, and N. R. Webb, editors. Large-scale ecology and conservation biology. Blackwell Scientific, London, United Kingdom.

McIntyre, S., and G. W. Barrett. 1992. Habitat variegation, an alternative to fragmentation. Conservation Biology 6:146–147.

McIntyre, S., G. W. Barrett, and H. A. Ford. 1996. Communities and ecosystems. Pages 154–170 in I. F. Spellerberg, editor. Conservation biology. Longman Group, Essex, United Kingdom.

McIntyre, S., and R. J. Hobbs. 1999. A framework for conceptualizing human effects on landscapes and its relevance to management and research models. Conservation Biology 13:1282–1292.

McNab, B. K. 1963. Bioenergetics and the determination of home range size. American Naturalist 97:133–140.

Miller, J. R., M. G. Turner, E. A. H. Smithwick, C. L. Dent, and E. H. Stanley. 2004. Spatial extrapolation: the science of predicting ecological patterns and processes. Bioscience 54:310–320.

Montello, D. R. 2001. Scale in geography. Pages 13,501–13,504 in N. J. Smelser and P. B. Baltes, editors. International encyclopedia of the social and behavioral sciences. Pergamon Press, Oxford, United Kingdom.

Murphy, R. C. 1914. Observations on birds of the South Atlantic. Auk 31:439–457.

O'Neill, R. V. 1979. Transmutations across hierarchical levels. Pages 59–78 in G. S. Innis and R. V. O'Neill, editors. Systems analysis of ecosystems. International Cooperative, Fairland, Maryland, USA.

O'Neill, R. V., D. L. De Angelis, J. B. Waide, and T. F. H. Allen. 1986. A hierarchical concept of ecosystems. Monographs in Population Ecology 23. Princeton University Press, Princeton, New Jersey, USA.

O'Neill, R. V., and A. W. King. 1998. Homage to St. Michael: or, why are there so many books on scale? Pages 3–15 in D. L. Peterson and V. T. Parker, editors. Ecological scale: theory and application. Columbia University Press, New York, New York, USA.

Openshaw, S. 1984. The modifiable areal unit problem. Headley Brothers, Kent, United Kingdom.

Pickett, S. T. A. 1989. Space-for-time substitution as an alternative to long term studies. Pages 110–135 in G. E. Likens, editor. Long-term studies in ecology: approaches and alternatives. Springer-Verlag, New York, New York, USA.

Prevedello, J. A., and M. V. Vieira. 2010. Does the type of matrix matter? A quantitative review of the evidence. Biodiversity Conservation 19:1205–1223.

Rastetter, E. B., A. W. King, F. J. Cosby, G. M. Hornberger, R. B. O'Neill, and J. E. Hobbie. 1992. Aggregating fine-scale ecological knowledge to model coarser-scale attributes of ecosystems. Ecological Applications 2:55–70

Saunders, D. A., R. J. Hobbs, and C. R. Margules. 1991. Biological consequences of ecosystem fragmentation: a review. Conservation Biology 5:18–32.

Sawyer, H., F. Lindzey, and D. McWhirter. 2005. Mule deer and prong-horn migration in western Wyoming. Wildlife Society Bulletin 33:1266–1273.

Schneider, D. C. 1991. The role of fluid dynamics in the ecology of marine birds. Oceanography and Marine Biology Annual Review 29:487–521.

Schneider, D. C. 2001. The rise of the concept of scale in ecology. Bioscience 51:545–553

Sherry, T. W., and R. T. Holmes. 1988. Habitat selection by breeding American restarts in response to a dominant competitor, the least flycatcher. Auk 105:350–364.

Siegel, S. 1956. Nonparametric statistics for the behavioral sciences. McGraw-Hill, New York, New York, USA.

Southwood, T. R. E. 1977. Habitat, the template for ecological strategies? Journal of Animal Ecology 46:337–365.

Stevens, S. S. 1946. On the theory of scales of measurement. Science 103:677–680.

Stommel H. 1963. The varieties of oceanographic experience. Science 139:572–557.

Sutherland, G. D., A. S. Harestad, K. Price, and K. P. Lertzman. 2000. Scaling of natal dispersal distance in terrestrial birds and mammals. Conservation Ecology 4:1–16.

Thompson, F. R., III, W. D. Dijak, T. G. Kulowiec, and D. A. Hamilton. 1992. Breeding bird populations in Missouri Ozark forests with and without clearcutting. Journal of Wildlife Management 56:23–30.

Tilman, D. 1989. Ecological experimentation: strengths and conceptional problems. Pages 136–157 in G. E. Likens, editor. Long-term studies in ecology. Springer, New York, New York, USA.

Tilman, D. 2000. Causes, consequences and the ethics of biodiversity. Nature 405:208–211.

Turner, M. G., R. H. Gardner, and R. V. O'Neill. 2001. Landscape ecology in theory and practice: pattern and process. Springer, New York, New York, USA.

Turner, M. G., R. V. O'Neill, R. H. Gardner, and B. T. Milne. 1989. Effects of changing spatial scale on the analysis of landscape pattern. Landscape Ecology 3:153–162.

West, G. B., J. H. Brown, and B. J. Enquist. 1997. A general model for the origin of allometric scaling laws in biology. Science 276:122–126.

Wheatley, M., and C. Johnson. 2009. Factors limiting our understanding of ecological scale. Ecological Complexity 6:150–159.

Wiens, J. A. 1989. Spatial scaling in ecology. Functional Ecology 3:385–397.

Withers, M. A., and V. Meentemeyer. 1999. Concepts of scale in landscape ecology. Pages 205–252 in J. M. Klopatek and R. H. Gardner, editors. Landscape ecological analysis: issues and applications. Springer, New York, New York, USA.

Wolff, J. O. 1999. Behavioral model systems. Pages 11–40 in G. W. Barrett and J. D. Peles, editors. Landscape ecology of small mammals. Springer, New York, New York, USA.

Wright, S. 1943. Isolation by distance. Genetics 28:114–138.

Wright, S. 1946. Isolation by distance under diverse systems of mating. Genetics 31:39–59.

Wu, J., and H. Li. 2006. Concepts of scale and scaling. Pages 3–13 in J. Wu, K. B. Jones, H. Li, and O. L. Loucks, editors. Scaling and uncertainty analysis in ecology: methods and applications. Springer, New York, New York, USA.

WILDLIFE POPULATION DYNAMICS

L. SCOTT MILLS AND HEATHER E. JOHNSON

INTRODUCTION

Are the recovery and recolonization of gray wolves (*Canis lupus*) reducing the number of elk (*Cervus canadensis*) available to hunters? Does oil and gas development influence sage grouse (*Centrocercus urophasianus*) populations, and if so, how great is the impact? How fast are raccoons (*Procyon lotor*) proliferating and spreading diseases that affect humans? These are just a few cases where wildlife management and conservation depend critically on understanding population dynamics. Without this knowledge, wildlifers do not know even the most basic things about populations: whether a population is increasing or decreasing in size, how a particular stressor may affect the persistence of a population, or the quality of the habitat that supports a population.

In short, wildlife cannot be conserved or managed without understanding population dynamics. But to understand population dynamics, biologists must master many fields. They must understand basic ecology, the conceptual foundation upon which our current knowledge has been built. They must know some math, because mathematic models reveal processes underlying population dynamics that cannot be seen with even a keen eye or sharp intuition. They must understand the natural history of different species, to help apply insights from one species to another and to avoid wandering into irrelevance when applying models. And they must grasp the fundamentals of disparate fields including population genetics, quantitative biology, animal behavior, animal physiology, plant ecology, and human dimensions, because the mechanisms influencing population dynamics are best understood by deciphering the interactions animals have with each other and their environment.

Readers will approach this chapter from different backgrounds and have varying levels of training or knowledge in the different pieces needed to understand wildlife population dynamics. Therefore we assume relatively little background, building up from the basics and providing some references to more advanced readings once you master the fundamentals. Expanded coverage of most of these topics is presented by Mills (2007, 2013).

In this chapter we provide an overview of some core concepts, describe exponential growth as the basic foundation for understanding population dynamics, discuss some of the factors that most affect wildlife population dynamics, consider management insights that can be gained from analyzing the dynamics of individual age or stage classes, examine dynamics of multiple populations across a landscape, consider key aspects of monitoring wildlife population dynamics, and close with a case study applying many of the topics in the chapter. Throughout we stress a few key themes: (1) variation is as important as the mean in understanding population dynamics (embrace uncertainty!); (2) some of the most powerful insights into outcomes of wildlife management actions are nonintuitive, revealed by applying data to models; and (3) because different management actions influence population dynamics in different ways, we must understand population processes to identify the most successful actions to meet population objectives.

KEY DEFINITIONS AND CONCEPTS

A discussion of wildlife population dynamics requires precise language, often stated in mathematical terms for clarity. We begin by introducing some of the key definitions and concepts to provide a solid foundation to build on throughout the chapter.

First, what is a population? We prefer a broad and practical definition, and refer to a population as being a collection of individuals of a species occupying a defined area, for which it is meaningful to refer to birth and death rates, sex ratios, abundances, and age structures (Cole 1957). For certain applications, more restrictive definitions of populations are needed (e.g., limited gene flow with other populations or long-term occupancy), but for our purposes this general definition allows us to focus on the wide range of population interactions and dynamics that form the basis of wildlife management and conservation.

Fundamentally, all population dynamics can be determined through births, immigration, deaths, and emigration (BIDE).

The abundance (N) of a population at time $t + 1$ equals the abundance in the previous time step (t), plus the number of animals that have been added to the population through births (B) and immigration (I), and minus those animals that have died (D) or emigrated (E):

$$N_{t+1} = N_t + B + I - D - E$$

Almost everything in this chapter will revolve around using these components to interpret the status of wildlife populations and how management actions are most likely to affect their dynamics.

The pieces of BIDE make up some of the vital rates that drive population dynamics. Births are quantified as litter size or clutch size or, as time passes and some of the newborns die, as the recruitment of juveniles into a population (i.e., the net number of new animals added to a population). Deaths are often described as a mortality rate, and the flip side of mortality is survival (survival rate = [1 – mortality rate]). Immigration refers to individuals arriving into one population from another, and emigration refers to individuals leaving a target population. Both of these interpopulation vital rates are mediated by dispersal, defined as the permanent movement of an individual from one population to another population; dispersal can be different from gene flow, where animals move from one population to another and reproduce successfully.

Other vital rates that influence the dynamics of a population include age structure and sex ratio. Age structure refers to the proportion of individuals of each age in a population; often, age structure is generalized to stage structure, grouping categories of animals based on similar survival and reproductive rates. For example, a male deer might live for ten years, but the stage classes most relevant for management agencies to collect data and evaluate population dynamics might be fawns, yearlings, and adults. These "stages" of deer have different vital rates (i.e., average annual fawn survival may be 0.50 while adult male survival may be 0.75) and are easily distinguished during field surveys. Another key characteristic of populations is the sex ratio, or the proportion of the population that is males versus females. The sex ratio can vary among different stage classes of animals, from birth to weaning to adulthood, particularly in populations with a sex-biased harvest.

Collectively, these different vital rates determine the abundance or density of a population, and allow us to track demographic trends through time. Abundance is the number of individuals in a population; easy to say, but remarkably hard to actually estimate because we almost never count every animal present (we will return to this idea in Monitoring Population Dynamics, below). Density is the abundance scaled to unit area, allowing comparisons on the same scale (e.g., 3 mice/m² in a hay barn compared to 0.1 mice/m² in a forest).

The population growth rate describes the trend in abundance (or density) over time. While this is arguably the most critical parameter when determining the dynamics of a population, the term is a bit misleading, because population growth can refer to an increasing population, or to a decreasing or stationary population (i.e., neither increasing nor decreasing).

We will come back to this concept throughout the chapter, but it is so crucial that we introduce two key descriptors of population growth here.

The first is the geometric growth rate, or the discrete growth rate, referred to as lambda (λ), which describes the proportional change in abundance from one year (or other appropriate time step) to the next. In a simple equation, with abundance this year (N_t) and abundance next year (N_{t+1}):

$$\lambda = N_{t+1}/N_t \tag{1}$$

If $\lambda = 1$, the population is stationary in size; $\lambda < 1$ indicates a declining population, and $\lambda > 1$ an increasing population. Lambda is easy to work with because it easily converts to percentage change per year: percent change = $(\lambda - 1) \times 100$. For example, if $\lambda = 1.25$, the population will increase by 25% next year; if $\lambda = 0.75$, it will decrease by 25%.

Although λ is intuive and easy to understand as a proportional change in population size, the discrete growth represented by λ has some awkward mathematical properties. As a result, a solid understanding of population growth requires a second descriptor, the calculus-based continuous time analog of λ, defined by r and called the exponential growth rate or the instantaneous per-capita growth rate. The two measures, λ and r, are interchangeable after a simple conversion:

$$\lambda = e^r \text{ or } r = \ln \lambda \tag{2}$$

Here ln is the natural logarithm, with the base e (which is about 2.718). A population with an $r > 0$ is increasing, and a population with $r < 0$ is decreasing.

When should we use λ versus r? Typically, biologists use λ when describing population growth to managers or the public because it is intuitive and easy to interpret. Conversely, r is often used when doing mathematical calculations of population growth, as it is independent of a specific time interval, can be easily compared among taxa, and values can be added across time intervals or averaged among them.

Other core concepts related to population dynamics are variation and uncertainty (Mills 2013). Population dynamics are inherently variable over space and time, and biologists must understand and embrace that variation. No understanding of wildlife population dynamics can emerge without understanding how variation affects dynamics, where it comes from, and how managers might use it to their advantage. We provide examples throughout the chapter but start with a lively anecdote, first quoted by Ankney (1996:41), to describe how a clever, if imagined, waterfowl biologist might respond to setting duck population goals based on the average population size, ignoring variation:

> Did someone say that the "average" numerical standing of the North American mallard population over some period of years would be a good standard for management to try to maintain? Don't let the Old Forecaster hear such talk. Not long ago, he got involved in certain philosophical deliberations regarding the "average" condition of dynamite. Which commodity, in its quiescent state, was a small cylinder having a volume of

a few cubic inches, yet at its peak of explosion occupied hundreds of cubic yards. Seemingly, the "average" condition of dynamite could be determined by adding together measurements taken at various levels between these two extremes, and dividing their sum by the number of measurements. The forecaster took one look at the results of all this arithmetic, turned slightly purple, and then decided ruefully that dynamite in its "average condition" must be one helluva thing to crate, ship, and otherwise handle. He has assiduously avoided "averages" ever since that unfortunate experience.

Indeed, variation around a mean can be even more important than the average, or mean, itself. In the broadest sense, there are two main sources of variation in population dynamics: process variation and sampling variation.

Process variance is the real variation in population processes that comes from nature or management. Often we describe process variance as arising from stochastic, or random, events in nature such as those imposed by weather. For example, in barn owls (Tyto alba), harsh winter weather is associated with declines in juvenile and adult survival, which has resulted in population crashes in some years (Altwegg et al. 2006). Process variance can also arise from more predictable factors, called deterministic factors, including those arising from wildlife management activities that attempt to increase or decrease a population (e.g., a change in harvest or a habitat enhancement project).

In addition to real process variance, population dynamics may appear variable given difficulties in accurately estimating population parameters. Sample variation, or observation error, is the inevitable uncertainty that arises from estimates based on incomplete sampling of animals. Animals avoid detection by moving around, making it hard to know who has already been sampled, and by spreading widely across landscapes so that only a portion of the population can be sampled. Sample variance or observation error is a constant challenge in wildlife studies, but can often be accounted for with sound field methods and appropriate statistical tools. Embrace uncertainty from sampling variation as an honest representation of the amount of confidence you have in a population parameter estimate. Conclusions about population dynamics must always be couched in terms of uncertainty about the ecological system, given limitations in our abilities to estimate parameters in the field. In sum, in population dynamics, as with dynamite, the conditions influencing temporal or spatial variation can be at least as important as the average.

PROJECTING POPULATION CHANGE WITH EXPONENTIAL GROWTH

The simplest dynamics a population can exhibit is to change at a constant exponential rate through time. For just one time step, and building off equations (1) and (2), this exponential change would be:

$$N_{t+1} = N_t \lambda \quad \text{or equivalently} \quad N_{t+1} = N_t e^r \qquad (3)$$

Next, extend this constant growth rate (λ or r) for T time steps (say, years) into the future, starting from the initial abundance at time (N_0):

$$N_T = N_0 \lambda^T \quad \text{or} \quad N_T = N_0 e^{rT} \qquad (4)$$

As an example, if a population of 20 black-footed ferrets (Mustela nigripes) experienced a constant exponential change of $\lambda = 1.6$ or $r = 0.47$ (Grenier et al. 2007), how many ferrets would be expected in five years?

$$N_T = 20 \times 1.6^5 = 210 \text{ ferrets}$$

or equivalently

$$N_T = 20 \times e^{0.47 \times 5} = 210 \text{ ferrets}$$

This example illustrates the power exponential growth can have in leading to rapid increases in population size (Fig. 7.1A). At the same time, however, do not make the common mistake of confusing exponential growth with really big growth. For example, after five years, our initial 20 ferrets would only be 21 if they had an exponential growth rate of $\lambda = 1.006$, and only 12 ferrets if $\lambda = 0.9$.

Because population growth is a geometric process (where the growth rate is multiplied by, not added to, N; Fig. 7.1A), a straight line based on the abundances plotted as logarithms can replace a curved line (Fig. 7.1B). In this case, the slope of the line is equal to r, which can be illustrated by rewriting the log-transformed version of equation (4):

$$\ln(N_T) = \ln(N_0) + rT.$$

This is an equation for a straight line of log abundance, with slope r.

Of course, population growth is not necessarily constant over time for any wildlife population, and process variance has two strong implications for population dynamics. First, greater variation in λ over time makes it more likely that a population will stumble toward extinction, regardless of trend (Boyce 1977, Vucetich et al. 2000). Population viability analyses, where long-term count or demographic data underlie estimates of the risk of extinction for a population, incorporate this phenomenon (Dennis et al. 1991, Morris and Doak 2002, Mills 2013).

Second, in addition to the direct effect of increasing extinction probability, process variance can have a more subtle, yet still important, effect on population dynamics by decreasing the future expected population size below that expected from the arithmetic mean λ in the absence of variation (Mills 2013). This interesting phenomenon follows the same laws that cause financial planners to advise investors to choose less volatile stocks, even if they have a lower average net return, because the most likely growth of any multiplicative process (like money in an interest-bearing account, or numbers in a wildlife population) is better described by the geometric mean growth rate than by the arithmetic mean. This is because the probability of a population being any particular size in the future gets more skewed over time, with most populations being relatively small (they cannot go below zero) but a few get-

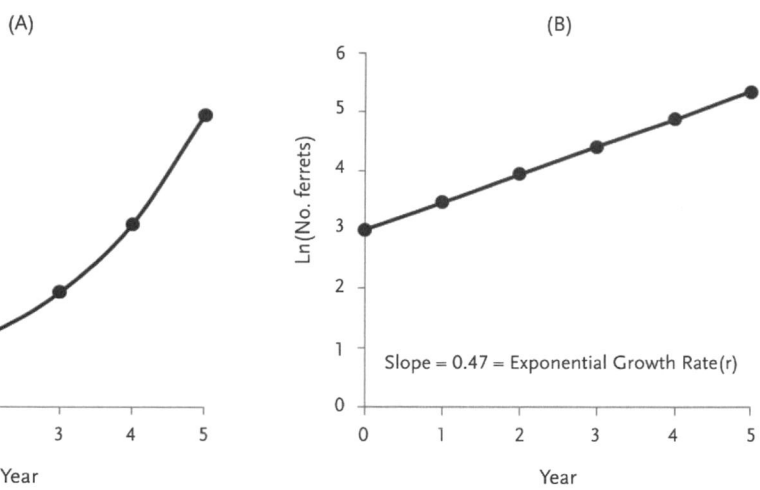

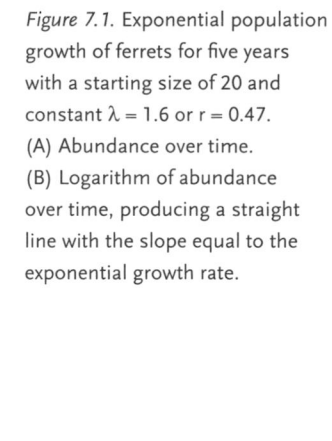

Figure 7.1. Exponential population growth of ferrets for five years with a starting size of 20 and constant λ = 1.6 or r = 0.47. (A) Abundance over time. (B) Logarithm of abundance over time, producing a straight line with the slope equal to the exponential growth rate.

ting incredibly large. Those few large populations can inflate the average arithmetic mean population size, making it larger than the most likely future population size. Without variation, the most likely population size is the same as the average, but with variation, the most likely population size will be less than the average. The geometric mean reflects this reality, becoming smaller than the arithmetic mean when variation is present (Lande et al. 2003, Doak et al. 2005). The geometric mean λ, typically called the stochastic growth rate (λ_S), is calculated by multiplying λ values over T time steps, and taking the T root of the product:

$$\lambda_S = \sqrt[T]{(\lambda_1 * \lambda_2 * \lambda_3 * \ldots \lambda_T)} \tag{5}$$

The stochastic growth rate based on the geometric mean contrasts with the more familiar arithmetic mean (λ_A):

$$\lambda_A = \frac{(\lambda_1 + \lambda_2 + \lambda_3 + \ldots \lambda_T)}{T} \tag{6}$$

For an example of how λ_S becomes smaller than λ_A when λ is variable, consider a population changing across three time intervals: in the first interval, λ = 1.1; in the second, λ = 0.3; and in the third, λ = 1.9. Across all time periods, the arithmetic mean population growth rate is $\lambda_A = 1.1$ (increasing by 10%/yr), but the geometric mean stochastic growth rate is $\lambda_S = 0.86$ (declining by 14%/yr). The implications of this phenomenon are profound; a population whose arithmetic mean average growth rate indicates positive growth might actually be most likely to decrease over time! The effect of stochasticity on population growth rates gets complicated, but this basic and nonintuitive phenomenon is fundamental to understanding population dynamics. Just remember, the most likely population trajectory in a variable environment (as experienced by any wildlife population) is governed by λ_S.

FACTORS AFFECTING POPULATION DYNAMICS

Many factors affect dynamics of wildlife populations, often in complex ways. Some of these are stochastic, or random, while others are deterministic. We first discuss the main sources of stochastic variation in populations (e.g., demographic, genetic, and environmental stochasticity) and then discuss several sources of deterministic variation in populations (e.g., habitat quality, predation, disease, interspecific interactions, anthropogenic factors, and density dependence), realizing that both sources of variation often affect populations in simultaneous and interactive ways, yielding complex population processes.

Stochastic Factors
Demographic Stochasticity
Demographic stochasticity comes from random deviations that arise from sampling whole animals that experience probabilistic birth and death rates (e.g., an animal with a 70% survival rate cannot 70% survive; either it lives or it dies) particularly when population sizes are small. Demographic stochasticity is sometimes called "penny flipping variation," because a coin flip has a 50:50 chance of heads or tails. Toss it three times, and you cannot possibly get 50% heads (you will get 0, 33, 67, or 100% heads); even if you toss it four times, you would not be surprised to get 0, 25, 75, or 100% heads. The effect of demographic stochasticity disappears with more than about 100 animals, just as you would expect to converge closely on 50% heads with 100 coin tosses. For a biological example of demographic stochasticity, in 2009 there were only five caribou (Rangifer tarandus caribou) left in Banff National Park. All five died in a single avalanche, extirpating the herd (Hebblewhite et al. 2009), an outcome that would be highly unlikely in a larger population.

Genetic Stochasticity
Genetic stochasticity, or inbreeding due to genetic drift, is also the consequence of chance variation. In this case, random variation in the allele frequencies of relatively small, isolated populations tends to lead to a loss of genetic variation (i.e., heterozygosity) that in turn can decrease vital rates through inbreeding depression (Mills and Tallmon 1999, Allendorf et al. 2012). Genetic stochasticity can be another driver of stochastic variation in the dynamics of small populations.

Environmental Stochasticity

In contrast to demographic and genetic stochasticity that affect small populations, environmental stochasticity can cause variation in the vital rates of a population of any size. Environmental stochasticity refers to unpredictable changes that affect mean vital rates from year to year. Weather often drives environmental stochasticity (e.g., wet years versus dry, heavy snow versus light snow), but also unexpected events such as disease outbreaks or changes in predator abundance. Chapter 17 further discusses climate change on wildlife populations.

Deterministic Factors
Habitat Quality

All wildlife species need certain food items, cover types, climatic conditions, and other characteristics (e.g., sites for nesting, hibernating, lekking) to survive and to reproduce successfully. As a result, habitat quality is highly species specific, varying across the landscape for each species in accordance with the availability of different resources. This inherent variation in habitat quality can account for differences in the vital rates of populations. For example, Ozgul et al. (2006) reported that juvenile survival in yellow-bellied marmots (*Marmota flaviventris*) was highly variable based on habitat conditions, with survival decreasing at higher elevations and on northeast-facing slopes. Such differences in habitat had large effects on the growth rates of different populations, as λ ranged from 0.96 to 1.09 depending on the conditions of the site. Human-induced changes to habitat, such as loss, degradation, or fragmentation, can also have major effects on wildlife populations. These changes pose some of the greatest threats to wildlife (Wilcove et al. 1998) and have been associated with declining population trends (Wittmer et al. 2007, Harper et al. 2008).

Predation

In their most widely recognized role, predators can affect prey population dynamics through direct consumption, with the degree of their impact depending upon the rate of predation, which individuals are killed, and whether predation is additive or compensatory (Mills 2013). The first piece, the predation rate, is simply expressed as:

$$\frac{\text{number of prey killed}}{\text{prey abundance}} \times 100, \qquad (7)$$

where the number of prey killed is a function of both the number of predators present at a particular prey density (i.e., the numerical response) and the number of prey killed per predator (i.e., the functional response, or kill rate; Holling 1959). The second piece determining how predation may affect prey numbers reflects the fact that individual prey differ in their value to the growth rate of a population and in their susceptibility to predation. For example, a reproductive adult female Kemp's Ridley sea turtle (*Lepidochelys kempii*) that can lay over 100 eggs/yr is worth much more to the growth rate of a population than a newborn hatchling with a very low survival rate (Heppell et al. 1996; see Accounting for Effects

of Ages and Stages on Population Dynamics, below). Finally, the last factor that determines the impact of predation on a prey population is whether those animals that are killed were likely to die anyway as doomed surplus (Errington 1956). Because factors such as disease or the availability of territories, cover, or food can inherently limit the number of animals that could survive in a population, predation in some cases may be compensatory, having little or no effect on overall survival. At the other extreme, predation decreases survival beyond other mortality factors. Compensation of predator-caused mortality can also arise from increased immigration or reproduction.

In addition to the direct effects of predators on prey population dynamics, predators can exert indirect effects on prey behavior and distribution. For example, prey might avoid high-quality habitat, increase their vigilance, or alter their foraging patterns to avoid predators (Fortin et al. 2005, Hamel and Côté 2007, Bourbeau-Lemineux et al. 2011). These behaviors have the potential to decrease survival and reproductive rates of prey populations, ultimately affecting their dynamics.

Disease

Disease can be a major influence of wildlife population dynamics, with outbreaks exacerbated by factors such as population density, specific environmental conditions, or rare pathogen exposure. As with predation, certain stage classes or vital rates may be more or less susceptible to a particular disease. For example, outbreaks of respiratory disease in bighorn sheep (*Ovis canadensis*) typically have the greatest impact on juvenile animals, potentially causing a reduction in juvenile survival of ~60%, while adults might only suffer a reduction of ~5% (Festa-Bianchet 1988, Singer et al. 2000, George et al. 2008). The duration and the magnitude of disease events can also be highly variable, as respiratory disease may chronically affect bighorn sheep populations at low levels (Cassirer and Sinclair 2007) or cause catastrophic die-offs (Martin et al. 1996). The influence of a disease event can be particularly pronounced when a population is exposed to a pathogen that the species did not evolve with. Diseases introduced from domesticated animals or from invasive species have traditionally posed some of the greatest threats to wildlife populations (Pedersen et al. 2007, Smith et al. 2009).

Other Interspecific Interactions

Predation and disease represent interactions where one or more species benefit at the expense of others. Interspecific competition arising from shared use of a finite resource also leads to negative interactions among one or more interacting species. The ways that species can compete varies as much as life itself, and includes consuming a shared resource, preempting space and physically excluding the other species by fights, perhaps to the death (e.g., wolves excluding coyotes [*Canis latrans*]; Berger and Gese 2007).

But of course interspecific interactions are not always negative. Other types of species interactions that can affect population dynamics include mutualisms, where both species benefit each other, and facilitation or commensalisms, where one species benefits and the other is unaffected.

A classic example of a mutualism can be found between whitebark pine (*Pinus albicaulis*) and Clark's nutcracker (*Nucifraga columbiana*; Hutchins and Lanner 1982, Lorenz et al. 2011). The pine has evolved mechanisms that allow only Clark's nutcrackers to disperse the seeds successfully away from the tree, and the nutcracker has evolved spectacular adaptations to harvest pine seeds. The nutcrackers carry up to 50 seeds in a cheek pouch especially suited for the purpose; in a single year, one bird will cache up to 98,000 seeds at distances up to 22 km from the tree of origin! The birds benefit by eating the seeds, although more than 66% of the seeds are never eaten and so remain available for germination.

An example of facilitation or commensalism can be found in the life cycle of hermit crabs (*Paguroidea* family). Most hermit crab species are dependent upon finding and using snail shells to protect themselves from predators. While the crabs do not influence snail populations, the abundance and distribution of different types of snails (and thus their shells) can dramatically influence the dynamics of local hermit crabs, affecting their reproduction, growth, and survival (Williams and McDermott 2004).

Anthropogenic Effects

Of the different deterministic factors that influence wildlife populations, impacts from humans are among the greatest. Many of those impacts are by-products of human population growth, development, and per-capita consumption (see box on p. 92), resulting in habitat loss, degradation and fragmentation, pollution, the introduction of nonnative species, overexploitation, and other such factors. Scientists are realizing that even species that have historically experienced minimal direct impacts from human development are being affected through human-induced climate change, arguably the greatest threat to global biodiversity (Parry et al. 2007, Running and Mills 2009). For example, polar bears (*Ursus maritimus*) in the Arctic are experiencing a reduction in the available sea ice caused by climate change, a key habitat component for successful foraging. Researchers suspect that the continual loss of sea ice will be associated with future population declines (Stirling et al. 2011). The impacts of climate change are not only expected to have singular effects on wildlife populations, but to interact synergistically with many of the other factors that influence populations like disease, overexploitation, and habitat loss (Brook et al. 2008, Smith et al. 2009).

Of course, unlike the unintentional consequences of human population growth on wildlife species, people also intentionally influence wildlife populations through management. Depending on societal goals (e.g., recovery, recreationally viable, ecologically functional), managers may work to increase, decrease, or stabilize populations using a variety of approaches. These approaches include public harvests, habitat enhancement, lethal removal, translocations, captive breeding, contraception, and the enforcement of various regulatory

mechanisms. For some species termed "conservation reliant" (Scott et al. 2005), continuing active management is necessary for persistence. As we stress throughout the chapter, management actions will be most effective when applied with a sound understanding of their influence on the dynamics of a population. Ultimately, however, whether anthropogenic effects on wildlife populations are unintentional (e.g., increase in road density associated with urban development) or intentional (e.g., change in harvest strategy), it is critical to recognize that the human footprint is pervasive across the globe. Managers need to acknowledge and incorporate human factors into wildlife management strategies to be successful at meeting population objectives.

Density Dependence: Positive and Negative

Density dependence arises when a population's density or abundance affects the vital rates (e.g., reproduction, survival) of individuals in the population, which in turn can affect the population growth rate (Fig. 7.2).

Positive density dependence, where vital rates (and population growth) are positively related to density, is often referred to as the Allee effect because it was first described by Allee (1931). Although sometimes referred to as "inverse density dependence" or "depensation," we prefer positive density dependence because it captures the fact that density and vital rates track each other positively; when one goes up or down, so does the other. Positive density dependence is often caused by mechanisms such as cooperative defense, foraging efficiency, or the rearing of offspring (Table 7.1; Kramer et al. 2009, Gregory et al. 2010), as an increase in the number of animals that

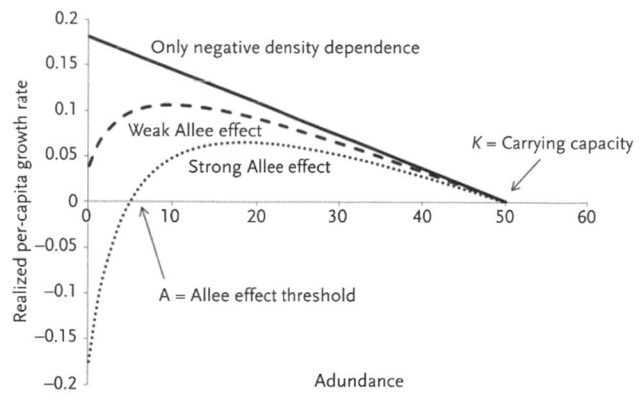

Figure 7.2. The influence of density dependence on a population with an exponential growth rate of $r = 0.18$ and carrying capacity of 50. Changes in realized per-capita population growth rate $[\ln(N_{t+1}/N_t)]$ are plotted. The solid line shows pure logistic (negative density–dependent) dynamics. The dashed and dotted lines show positive and negative density dependence operating simultaneously. Although both of these exhibit decreases in per-capita growth rate at small abundances expected under positive density dependence (Allee effects), only the dotted line shows strong Allee effects that create a threshold abundance below which the population would actually decline toward extinction. From Mills (2013)

Table 7.1. Some mechanisms that can lead to positive density dependence (Allee effects) in wild populations

Mechanism	Example
Minimizing predation	
Increased predator detection and defense	Survival rates in dwarf mongoose (*Helogale parvula*) increase with group size because guard mongoose decrease predator attacks (Rasa 1989).
Greater confusion for predator	Largemouth bass (*Micropterus salmoides*) take longer to capture silvery minnows as minnow school size increases (Landeau and Terborgh 1986).
Predator satiation	Black brant (*Branta bernicla nigricans*) per-capita nest mortality declines as colony size increases and predators become satiated (Raveling 1989).
Foraging advantages	
Access to foods	Small colonies of blind mole rats (*Cryptomys damarensis*) are more likely to fail because they are unable to rapidly extend burrow systems to obtain food during the brief time period when the soil is moist and easily worked (Jarvis et al. 1998).
Increased resource detection	After being reduced in numbers by hunters and habitat alteration, a cause of further decline in passenger pigeons (*Ectopistes migratorius*) may have been that small flocks were compromised in their ability to find their patchy and sporadic food sources (e.g., acorns and nuts; Reed 1999).
Cooperative resource defense	Larger groups of coyotes (*Canis latrans*) are better able to defend carcasses against intruder coyotes (Bekoff and Wells 1986).
Mating and caring for young	
Finding mates	Glanville fritillary butterflies (*Melitaea cinxia*) are less likely to locate mates and successfully reproduce in small populations (Kuussaari et al. 1998).
Caring for young	The number of young fledged per nest increases with group size in white-fronted bee eaters (*Merops bullockoides*) because helpers provision food and assist in nest excavation, nest defense, and egg incubation (Emlen 1990).
Conditioning of environment	
Temperature tolerance	Bobwhite quail (*Colinus virginianus*) in large coveys standing in a circle with their tails toward the center of the circle are better able to survive extreme low temperatures (Allee et al. 1949).

Source: Modified from Mills (2013).

can participate in these activities leads to an increase in individual survival and reproduction that ultimately boosts population growth. Positive density dependence can occur in populations of any size. For example, in house finches (*Carpodacus mexicanus*), positive density dependence has been reported to lead to a higher carrying capacity and faster spread of large populations (Veit and Lewis 1996), a phenomenon that may in general exacerbate rapid increases of invasive species.

Although positive density dependence can occur at any population size, some of the most important instances for wildlife management occur at very small numbers (Fig. 7.2). In these cases, vital rates decrease as densities decrease (still a positive relationship between density and vital rates), which can destabilize a population, sending it spiraling toward extinction. This pattern has been observed in several wildlife species (Kramer et al. 2009), including caribou, where smaller populations have declined faster than larger populations (Wittmer et al. 2005). Appropriate management actions in these instances would emphasize increasing abundance at all costs, potentially through translocations or captive breeding, to raise abundance above the threshold where the population growth rate is declining.

Negative density dependence arises when increases in density lead to decreases in a population's vital rates and growth rate and vice versa (Fig. 7.2). Factors that elicit negative density dependence include intraspecific competition (i.e., competition among individuals within a species, which occurs through many of the same mechanisms discussed above for interspecific competition among different species) and heightened susceptibility to predation, parasites, and disease. Because resources are finite, any population, including humans, must eventually exhibit negative density dependence (see box on p. 92). Sometimes intraspecific competition occurs as direct interference or contests among individuals, as in fights for food, mates, or territories; winners obtain sufficient resources to survive and reproduce, while losers may not. In other instances the competitive interaction is exploitative, or based on a scramble for resources, with simultaneous use of a common resource (often food) lowering the amount of that resource available to each individual.

Whether influenced by intraspecific competition or other processes, negative density dependence tends to regulate abundance around an equilibrium size range, called the carrying capacity (*K*; Fig. 7.3). Theoretically, the carrying capacity is the population size at which reproduction and mortality are equal. That said, it is best to think of the carrying capacity as a range of values, not as a single number. The *K* for any population is determined by different limiting factors that include density-related factors (i.e., limitations in food, nesting sites, territories, as the population increases in size) but also density-independent factors (i.e., weather). For example, the endangered San Joaquin kit fox (*Vulpes macrotis mutica*) population in California is limited by negative density dependence, likely because of a restricted number of available territories, and rainfall, which influences the abundance of small mammals that are prey for foxes (White and Garrott 1999, Dennis and Otten 2000).

The simplest way to represent negative density dependence in a predictive model of population dynamics is to assume a linear decline in per-capita growth rate as density increases. In this case, the per-capita growth rate is positive and exponential at very small densities, zero at the carrying capacity, and negative at densities above *K* (Fig. 7.3). This specific form of negative density dependence leads to an S-shaped logistic

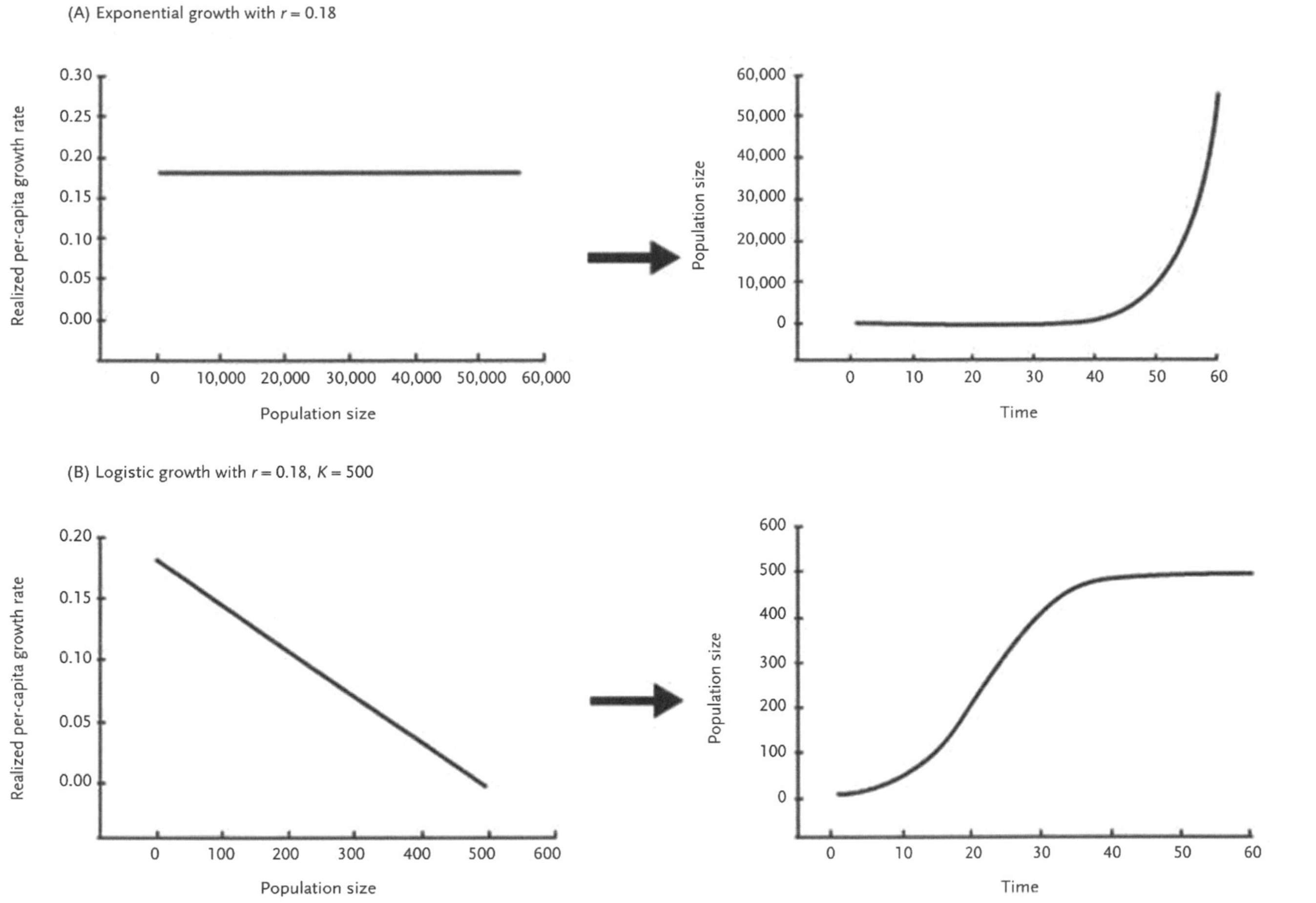

(A) Exponential growth with $r = 0.18$

(B) Logistic growth with $r = 0.18$, $K = 500$

Figure 7.3. The contrast between exponential and logistic growth. (A) Exponential growth. The realized per-capita population growth rate measured from a time series ($\ln[N_{t+1}/N_t]$) is equal to the intrinsic growth rate (r), no matter the population size. This lack of density dependence leads to exponential growth of the population over time. (B) Logistic growth. With a linear decline in the realized per-capita population growth rate as the population size increases, the population increases exponentially at first, then slows its growth as it approaches carrying capacity. From Mills (2013)

curve of abundance plotted against time, with rapid growth at small population sizes (r_0, the exponential growth at small abundance) and dampened growth as the populations size approaches K (Fig. 7.3B). Skipping the math that derives and details this relationship, here is a discrete-time form of the logistic equation, called the Ricker equation, which projects the population one time step forward:

$$N_{t+1} = N_t e^{\left(r_0\left[1 - \frac{N_t}{K}\right]\right)} \tag{8}$$

Compare equation (8) to equation (3), and notice that the only difference is that the exponential r now becomes penalized as density increases. When density is zero, there is no penalty ($1 - 0/K = 1$), so r_0 is unaffected by density and growth is exponential. When abundance reaches the carrying capacity, however, the growth term becomes zero ($[1 - N/K] = [1-1] = 0$, so $r \times 0 = 0$). In short, a linear decline in per-capita growth rate as density goes from zero to K (Fig. 7.3B).

Notice also in equation (8) that a one-year time lag exists between the operation of density dependence and abundance the following year. Biologically, it makes sense that birth and death rates would respond to past population density. Return-

ing to the San Joaquin kit foxes, rainfall affects small mammal abundance with a one-year time lag, which in turn affects kit fox abundance with an additional one-year lag (i.e., rainfall affects foxes with a two-year lag; Dennis and Otten 2000). It turns out that a population following logistic growth with a time lag can exhibit some crazy dynamics that include cycles and even unpredictable chaotic dynamics (Mills 2013). Wildlife cycles are interesting population dynamics phenomena in their own right, and the inherent dynamics from negative density dependence are one contributing factor causing cycles (see box on p. 93).

Realize that the logistic curve is just one of a nearly infinite number of ways that density dependence might manifest in wildlife populations (Mills 2013). The logistic model of population growth assumes that only negative density dependence happens, and that its effects become gradually worse from 0 to K in a linear way. No law dictates this to be so; in fact, most species probably exhibit positive density dependence at small numbers, and some form of negative density dependence at high numbers (Sibly et al. 2005, Peacock and Garshelis 2006).

Although the logistic curve should not be overinterpreted

PAUL EHRLICH (b. 1932) AND HUMAN POPULATION DYNAMICS

By 1800, aided by the Industrial Revolution, the development of global agriculture, and improved nutrition and hygiene, the human population numbered one billion and began a period of strong population growth. By the mid-1960s, human numbers increased to 3.5 billion and reached a peak annual growth rate of just over 2%, which would be expected to double our numbers to seven billion people in about 34 years.

In 1968, Paul Ehrlich published *The Population Bomb*. Although only one of more than 40 books written over an astonishingly productive career—including works on coevolution, population dynamics, genetics of checkerspot butterflies (*Euphydryas editha bayensis*), natural history of birds and insects, demography, human ecology, and evolutionary biology—*The Population Bomb* has been Ehrlich's highest-impact publication. It introduced the public to the fundamental ecological premise that no species, including humans, can sustain exponential growth indefinitely, and that exceeding the earth's carrying capacity will negatively affect humans and other organisms. The book sold more than two million copies, and has been the source of political controversy from both the far Right and the far Left. As Ehrlich and Ehrlich (2009:64) state, "none of those constituencies seemed to understand that the fundamental issue was whether an overpopulated society, capitalist or socialist, sexually repressed or soaked, egalitarian or racist/sexist, religious or atheist, could avoid collapse . . . Perhaps the biggest barrier to acceptance of the central arguments of *The Bomb* was—and still is—an unwillingness of the vast majority of people to do simple math and take seriously the problems of exponential growth."

Ehrlich also popularized another critical point fundamental to population dynamics: the overall impact of any population is a function of both population size and per-capita consumption (Ehrlich 1968, Ehrlich and Holdren 1971). Eventually, this concept—that impact depends on both population size and resource use per person—became the famous IPAT equation, where *impact = population × affluence × technology*.

What about now? Ehrlich was right, that the rapid exponential growth of the mid-1960s could not and would not be sustained. Our growth rate has fallen steadily since that peak, to about 1%, and our numbers reached seven billion in late 2011. So it took about 45 years to double our population size from the mid-1960s, not the 34 years he predicted; even so, the human population continues to increase. Collectively, the ever-increasing product of IPAT continues to manifest as deforestation, climate change, and accelerated extinction rates. We do not know what our carrying capacity will be or how it will be realized, though some credible papers predict that our numbers will stabilize around ten billion by the end of the 21st century (Lutz et al. 2001, Bongaarts 2009). Clearly, the subjects Ehrlich raised—human numbers and resource use confronting a finite global carrying capacity—are still relevant and pressing.

Photo courtesy of Paul Ehrlich

as a law of population growth, as long as the assumptions are recognized, the logistic equation can be useful for generating clear predictions of how negative density dependence may affect populations and management actions. One such prediction of direct relevance to management is that a population following logistic growth will recruit the most new individuals into the population when abundance is at 50% K (Fig. 7.4). Why? Because at small densities, few young can be added to the population because there are fewer females present. At high densities, near K, the per-capita growth rate is severely affected by negative density dependence so that, again, few young are added to the population. But at 50% K, recruitment is maximized because density effects are not yet severe, and the number of females is reasonably high (Fig. 7.4). This phenomenon was used for many years in harvest management, especially in fisheries, to determine the maximum sustained yield. The idea was that if the population was harvested down to 50% K, then the population would add the most new recruits, which meant that humans could harvest the maximum number of animals. The general idea is useful for informing us that under pure negative density dependence, the population will provide the most recruits at some intermediate density. But an appropriate level of humility about the assumptions of the logistic equation (and uncertainties in estimating densities and K) should caution us to not treat 50% K as any sort of magic number of maximum recruitment. To do so is to risk overharvest toward extinction, an all-too-common occurrence for ocean fisheries. For these reasons, the concept of sustained yield has emphasized the need to avoid harvesting at the theoretical maximum sustained yield of 50% K and instead consider multiple scenarios of density dependence and other complexities (Mills 2013; Chapter 5).

ACCOUNTING FOR EFFECTS OF AGES AND STAGES ON POPULATION DYNAMICS

So far we have seen how various outside factors can act on the mean and variance of population growth rates. But for three

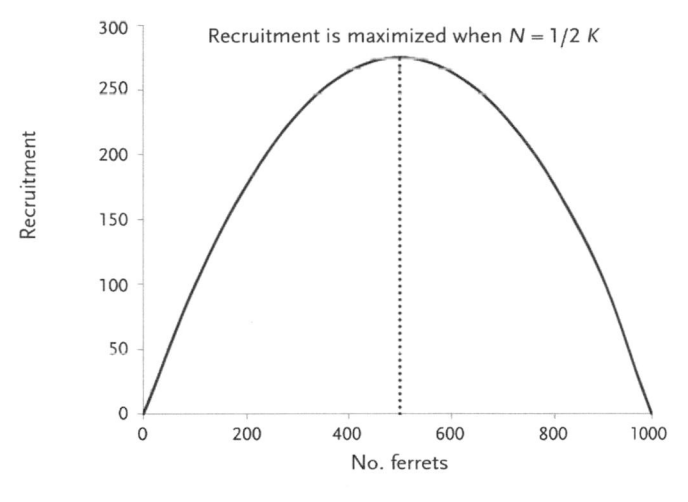

Figure 7.4. Under ideal logistic growth with no variability, recruitment of new individuals into the population is maximized at 50% *K*. Here the logistic curve, with *r* = 1.1, is used to show recruitment relative to population size.

main reasons, wildlife managers are often as interested in the dynamics of specific age classes or stage classes as they are in dynamics of the population as a whole. First, different stages may have different values to the public. Consider big game hunting regulations, where populations are managed based on the harvest of certain sexes or stages (e.g., branch-antlered, spike, or nonantlered animals). Second, stressors and manage-

ment actions affect different stages of animals in unique ways. In woodpeckers, for example, drilling nesting cavities in trees might increase fecundity, while removing invasive predators might improve adult survival. This leads to the third, and most important, reason why wildlifers must consider dynamics of different ages or stages; all stage classes and stage-specific vital rates are not equal in their effects on population growth, which means that management actions that target different stage classes will not be equal in their effectiveness.

A great story to convey these points is based on management of threatened loggerhead sea turtles (*Caretta caretta*) in the Atlantic. Prompted by long-term declines in population numbers, management efforts focused on reducing the stressors that seemed intuitively to be causing the decline: mortality of the eggs and newborn hatchlings that were killed by predators, crushed by vehicles, and disoriented by lights as they tried to make their way from the nest to the ocean. However, in one of the best examples of population models informing management, an analysis by Crouse et al. (1987) demonstrated that even large increases in egg or hatchling survival would do little to reverse the population decline, because hatchling survival had such a minimal effect on the overall population growth rate. A small increase in the survival of young adults, however, was found to lead to substantial increases in population growth. This scientific finding was translated into a management action to reduce young adult mortality via "turtle excluder devices" on shrimp nets to reduce lethal entanglement. Though loggerhead turtles still face many challenges, the identification of vital rates that most affect population growth led to management actions that would most facilitate recovery.

Given that identifying effective management actions requires knowledge of the influence of specific stage or age classes on population growth, we must consider mathematical models. Next we describe the fundamentals of matrix models and then describe their utility for guiding wildlife management decisions (Mills 2013).

Fundamentals of Population Projection Matrices

Recall the BIDE equation, which established that population dynamics arise from vital rates within and among populations. For now, we focus on within-population dynamics. At the simplest level, births and deaths determine dynamics within a population. Because numbers of births and deaths differ among animals of different ages or stages, we need to keep track of them in a specific way. Initially, demographers used a life table (Deevey 1947) to keep track of age- or stage-specific survival rates, but now we tend to use a population projection matrix. A matrix provides a convenient accounting system to track all stage-specific vital rates and abundances, and allows us to project populations through time to determine how different vital rates and stages affect dynamics.

The population matrix is conventionally denoted by *M* (Fig. 7.5), a square of *k* columns and *k* rows, where *k* is the total number of stage classes. Each element (or cell) of the matrix contains a reproduction or survival value that is used to project the individuals in each stage class forward one time step

From This Stage . . .

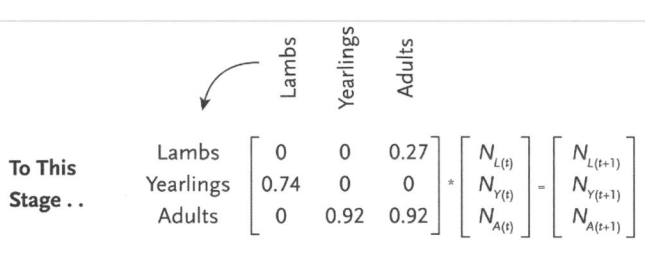

Figure 7.5. Anatomy of a female-based projection matrix, using bighorn sheep as the species of interest (see Recovering Endangered Sierra Nevada Bighorn Sheep case study). This species has three stages that can be identified by wildlife managers in the field and which were used to develop the matrix: lambs (first year of life), yearlings (the second year of life), and adults (individuals over two years of age). The first row of the matrix represents reproduction from each stage class that contributes to the number of lambs in the population the following year. In this example, we assume that bighorn sheep do not consistently reproduce until they are adults, so $a_{1,1}$ (the element in row 1, column 1) and $a_{1,2}$ are both 0, as neither lambs or yearlings contribute to reproduction. Each adult female, however, does contribute an average of 0.27 lambs to the population, as seen in $a_{1,3}$. The diagonal ($a_{1,1} = 0$, $a_{2,2} = 0$, and $a_{3,3} = 0.92$) represents the proportion of individuals in a stage class that will survive and will still be in the same stage next year (all lambs and yearlings transition to the next stage, while adults can remain in that stage class for several years), while the subdiagonal ($a_{2,1} = 0.74$ and $a_{3,2} = 0.92$) represents the proportion of the stage class that survives and advances to the next stage next year. The matrix is then multipled by a vector of abundances of each stage class in year t ($N_{L(t)}$, $N_{Y(t)}$, $N_{A(t)}$) to obtain a vector of abundances of each stage class in year $t + 1$ ($N_{L(t+1)}$, $N_{Y(t+1)}$, $N_{A(t+1)}$).

(for more details on the structure and projection of wildlife population matrices, see Mills et al. 2012). Fig. 7.5 provides a simple example of a matrix and population projection. Notice that this bighorn sheep matrix has three stage classes based on physical characteristics that can be identified by managers in the field: lambs, yearlings, and adults.

If conditions are relatively constant through time, the matrix of vital rates provides several useful population descriptors known as asymptotic properties. First, a population defined by a given matrix will converge on a stable age distribution (or stable stage distribution, or SSD) that describes the expected constant proportion of animals in each stage class. Given the matrix of bighorn sheep vital rates (Fig. 7.5), the SSD would maintain 18% of the population as lambs, 12% as yearlings, and 70% as adults. Second, when the population reaches SSD, it will grow at a constant asymptotic growth rate that is characteristic for that set of vital rates; this asymptotic growth rate at SSD is symbolized by λ_{SSD}. For the vital rates in the bighorn sheep matrix (Fig. 7.5), $\lambda_{SSD} = 1.08$, indicating an expected 8% increase per year once the population achieves SSD. Finally, under asymptotic (constant) conditions, the matrix can reveal reproductive values for each stage (or age) class, which can be interpreted as the relative contribution individuals in each

stage will make to future population growth, values that add to 1 across all stages. From this matrix the reproductive value of the lamb stage class is 0.24, for yearlings it is 0.35, and for adults it is 0.41. In other words, these values indicate that each adult is nearly twice as valuable as each lamb as a seed for future population growth.

Of course, conditions often are not constant through time, and matrix models can easily accommodate vital rates that change either stochastically or deterministically (perhaps through density dependence or other factors described above as affecting population dynamics). For example, an important phenomenon, population inertia (Koons et al. 2006), arises when the actual stage distribution is very different from SSD, perhaps following a translocation, or because a stressor (e.g., a severe disease event, poaching, or predators) killed certain stages disproportionately. The effects of population inertia can be readily seen with the bighorn sheep matrix in Fig. 7.5. If we conducted a translocation with ten bighorn sheep, how would the composition of those introduced sheep affect expected abundance 20 years later? Assuming a constant growth rate, if the initial ten sheep were approximately at SSD (18% lambs, 12% yearlings, and 70% adults rounds to two lambs, one yearling, and seven adults), the population would be expected to grow by the asymptotic growth rate, $\lambda_{SSD} = 1.08$, and after 20 years we would expect 21 bighorn sheep. By contrast, if the initial ten translocated sheep were all lambs (the stage class with the lowest reproductive value), then after 20 years the expectation would be only 14 bighorn sheep in the population, while starting with all adult sheep (the stage class with the highest reproductive value), we would expect 23 bighorn sheep after 20 years. In all cases, given constant vital rates, the population ends up converging on asymptotic SSD and λ_{SSD}, but the different reproductive values of translocated individuals resulted in different initial population changes. These insights can be used to maximize the impact of management to reach population objectives.

Evaluating Vital Rate Importance and Management Actions through Sensitivity Analysis

In a broad sense, population sensitivity analysis refers to how changes in vital rates, due to either a natural change or through management, may change population growth or persistence (Mills 2013). This is one of the most useful contributions of population ecology to wildlife management, because it quantifies how certain management actions, which change certain vital rates, would be expected to affect population growth. As a result, it allows us to assess the relative effects of different management actions for meeting our population goals. We mention several approaches for conducting sensitivity analysis and return to examples at the end of the chapter.

A measure called analytical sensitivity uses reproductive value and stable age distributions to quantify how a small, equal, absolute change in any stage-specific vital rate will change the asymptotic population growth rate (i.e., the λ value calculated from a matrix at SSD). For example, in the bighorn sheep matrix, if each vital rate were raised by 0.01, holding all other vital rates constant, the change in adult fecundity would increase the population growth rate by 0.5%, the same change in yearling survival by 0.2%, and adult survival by 0.9%.

Because a fixed absolute change will lead to a larger proportionate change in vital rates with smaller values than in large ones (e.g., increasing the fecundity rate from 0.27 to 0.28 by an absolute change of 0.01 will be an ~4% proportionate change compared to increasing the adult survival rate from 0.92 to 0.93, an ~1% proportionate change), analytical sensitivity can be rescaled as analytical elasticity. This defines the *proportional* change in λ given a tiny one-at-a-time proportional change in a vital rate. If each bighorn sheep vital rate was changed by a proportional 1%, the increase in the population growth rate would be 0.1% for changes in fecundity and yearling survival, and 0.8% for changes in adult survival.

Analytical sensitivities and elasticities have led to some important generalizations across life history strategies of different species. For example, species such as large mammals and birds, with late maturation, fewer offspring, and higher survival rates, tend to have higher elasticities for adult survival than reproduction. Conversely, species such as small mammals with early maturation and large litters tend to have higher elasticities for reproduction, including litter size and offspring survival (Sæther and Bakke 2000, Crone 2001). These generalizations can be useful as first-cut predictions of the effects of management actions. For example, a management action that could change adult survival by a fixed amount in a long-lived vertebrate would more efficiently increase population growth than an action that changed juvenile survival by the same proportional amount.

That said, for actual applications of predicting which management actions will most efficiently change population growth, analytical sensitivity and elasticity have two serious limitations. First, they say nothing about how much different vital rates can be changed. Rarely in the real world do different management actions change different vital rates by the same absolute or proportionate amount (Mills et al. 1999). For example, for a woodpecker, it may be that fire management increases nest success by 8%, removal of invasive predators improves fledgling survival by 12%, and adult survival by 23%, and so on (see Hartway and Mills 2012 for examples with managing bird populations). Gaillard et al. (1998) recognized this issue in a summary of ungulate population dynamics. They reported that while adult female survival consistently had the highest analytical elasticity, it was a vital rate with little natural variation, allowing little room for management activities to have an appreciable effect. Meanwhile, juvenile survival had low elasticity but high variability that was largely responsible for changes in population size, and thus the key rate to target for management. Analytical sensitivities and elasticities are limited by their reliance on changes being equal and tiny.

The second limitation of analytical sensitivities and elasticities is that in their basic form they rely on asymptotic dynamics. If a population was not at SSD for any reason (e.g., perhaps a natural or anthropogenic stressor killed off a large number

of animals of one stage class), then the sensitivities and elasticities would be invalid (stochastic forms of these analyses are possible but complicated). This might be particularly relevant for small populations of conservation interest, which can often be out of SSD.

Some alternative methods of conducting sensitivity analysis avoid analytical sensitivities and elasticities and directly quantify how changes in vital rates affect population growth. For example, one could simply perturb vital rates in a matrix in a sensible way, and ask how the future growth rate or abundance is affected. Drawing on the sheep example, if adult survival decreases by 10% (from 92% to 82%), to simulate the effects of an increase in harvest or predation or some other mortality factor, we would find that the population growth rate declines from 1.08 to 0.99, going from an increasing to a decreasing trajectory. This manual perturbation method is a quite flexible and powerful approach to sensitivity analysis, quantifying the outcome from any number of "what if" scenarios. Different vital rates can be projected to change by any amount either naturally (i.e., environmental stochasticity) or under management. Additionally, you can start with any age structure, and no assumptions of SSD are necessary.

Another simulation-based sensitivity analysis method is called life-stage simulation analysis (LSA; Wisdom et al. 2000). LSA is in some ways a formalized version of the manual perturbation method. Many plausible matrices are built (perhaps 1,000 or more on the computer), embracing the full range of variation that is possible for that population by randomly building each matrix from the means and variances of vital rates from field data. The population growth rate is then calculated for each matrix, so one can determine what percentage of replicates have positive versus negative population growth. Baseline scenarios can then be compared to those where realistic management alternatives are simulated given changes in different vital rates, and the outcomes of those scenarios can be evaluated. Additionally, λ values from each of the 1,000 matrices can be regressed against each vital rate for that particular matrix. The coefficient of determination (R^2) between λ's and their associated vital rate values represents the proportion of the variation in the population growth rate explained by variation in each vital rate, with all other vital rates varying simultaneously (as expected to occur in nature). When all main effects and interactions are included, the R^2 values sum to 1. Most applications of LSA have been based on asymptotic growth, but nothing in the method restricts it to that.

Although we have described stage-structured matrix models in their simplest form, they are quite flexible and can incorporate many other processes, such as density dependence (described above) and multiple population dynamics (discussed below).

MULTIPLE POPULATION DYNAMICS

Again, recall the BIDE equation: births, immigration, deaths, emigration. So far, we have focused on births and deaths within populations, but a fundamental truth in wildlife population dynamics is that animals move, and they move a lot. In fact, it could be said that the more we study dispersal, through improved tools ranging from GPS radiocollars to sophisticated mark-recapture analyses to genetic sampling (Mills et al. 2003, Crooks and Sanjayan 2006, Lowe and Allendorf 2010), the more we see that long-distance movements among populations are surprisingly common. For example, juvenile Columbia spotted frogs (*Rana luteiventris*) in Montana commonly move among ponds each year, traveling up to 5 km and gaining 750 m in elevation (Funk et al. 2005); a young male wolverine (*Gulo gulo*) radio collared in northern Wyoming traveled over 800 km to central Colorado (www.wcs.org); a female burrowing owl (*Athene cunicularia*) traveled 1,860 km from Arizona to Saskatchewan (Holroyd et al. 2011); and a mountain lion (*Puma concolor*) moved almost 3,200 km from South Dakota to Connecticut (Drajem 2011). In this section we describe how these movements could affect wildlife population dynamics.

Metapopulations

Populations across a landscape can have distinct characteristics, just as individuals making up a population do. At one extreme, populations can be isolated from each other. This might occur in oceanic islands, or for invasive species colonizing a new area, or in habitat fragments or protected areas surrounded by impenetrable hostile habitat. If populations are entirely isolated and small, they may be vulnerable to loss of genetic variation and inbreeding depression, as mentioned above. Such populations are also highly susceptible to other catastrophic events, such as a severe weather episode, a disease outbreak, or the introduction and expansion of an exotic predator or competitor. Persistence of these populations may also be hindered by the inability for recolonization, as isolated populations can easily go extinct.

In contrast to isolates, multiple populations can exist with some level of connectivity, or dispersal and gene flow, among them. The term "metapopulation" was coined by Levins (1970) to refer to a population of populations, where the dynamics of multiple populations depend on vital rates within— and also among—populations (immigration and emigration). Connectivity between populations can therefore influence local population dynamics and overall metapopulation persistence. The concept highlights the importance of connectivity as a dominant force in population ecology, and captures the nonintuitive fact that sometimes management on areas where a species does not even live (but does travel through) may be as important to persistence as occupied habitat. For example, connectivity and high gene flow from Canada into the rather small and scattered lynx populations in the western United States may be as important for recovering this federally threatened species as management to improve the local habitat conditions of the U.S. populations (Schwartz et al. 2002). Other key concepts in wildlife population ecology, such as sources and sinks and ecological traps, spin off of the metapopulation concept.

Sources and Sinks

A form of metapopulation dynamics of special interest to wildlife biologists is captured by the metaphor of sources and sinks (Lidicker 1975, Pulliam 1988). If some populations are strong contributors to the metapopulation (sources which contribute to positive overall population growth) while others are drains on the system (sinks whose dynamics can decrease overall population growth), then identifying these areas becomes a priority for managers. Sinks cannot be identified based only on abundance or density, because abundances vary in different seasons, and poor habitats may have high numbers of animals for a variety of reasons (e.g., adults may exclude juveniles from high-quality habitat areas; Van Horne 1983).

To identify from field data the sources contributing to metapopulation growth and sinks that drain the metapopulation, we must account for the two ways that a population can contribute to the greater metapopulation (Runge et al. 2006, Mills 2013). First, a population can contribute to its own growth (call this self-recruitment of population x, or R^x) via births. Second, a local population x can contribute by providing successful emigrants to other subpopulations (E^x). Jumping over a fair bit of details, a source and sink can be defined by a contribution metric that is like multipopulation λ. A population where $R^x + E^x > 1$ is a source, while a population where $R^x + E^x < 1$ is a sink.

Snowshoe hares (*Lepus americanus*) provide an example of how this concept can be extended to heterogeneous landscapes. The hares exhibit remarkable population cycles in northern latitudes but dampened cycles in the southern portion of their range (see box on p. 97), a pattern attributed to the role of different vegetation patch types in driving source–sink dynamics. For northern populations the landscape primarily consists of relatively continuous boreal forest, while the landscape becomes increasingly patchy in the South, with many detrimental openings in the forest (natural and because of forest management). In Montana, source–sink dynamics were quantified for hares in dense versus open forest stands (Griffin and Mills 2009). Dense stands operated as sources having positive contributions to the metapopulation through self-recruitment and emigration, while open stands were sinks where immigrating hares were much more susceptible to predation. The maintenance of hares across a landscape will require an appropriate threshold of high-producing source stands to maintain the overall metapopulation (Wirsing et al. 2002, Griffin and Mills 2003).

A key point to recognize when identifying population sources and sinks is that one cannot use λ alone to make the determination, because it does not account for how much of a population's growth rate comes from immigrants, or how much a local population contributes to the metapopulation via emigration (see box p. 97). For example, headwater salamanders (*Gyrinophilus porphyriticus*) in upper reaches of first-order streams in New Hampshire have positive population growth rates only because downstream salamanders have high reproduction and preferentially move upstream (Lowe 2003). To truly discriminate between a population source and sink, all

QUANTIFYING SOURCE–SINK DYNAMICS FOR HARVEST MANAGEMENT

In the western United States, many wildlife management agencies have promoted a source–sink metapopulation approach in the harvest management of mountain lions (Sweanor et al. 2000, Logan and Sweanor 2001, Laundre and Clark 2003, Stoner et al. 2006). The approach assumes that areas with little to no harvest will serve as sources for the overall metapopulation, supplying emigrants to sink areas that are more heavily harvested. Hunting has been found to induce source–sink dynamics in other carnivore species (Boyd and Pletscher 1999, Loveridge et al. 2007, Adams et al. 2008), and in mountain lions it seemed likely that immigration might sustain populations that were heavily harvested (Robinson et al. 2008). To quantify the effects of harvest on connectivity among populations, and to quantify how source–sink dynamics might be created, researchers used vital rate data across multiple years from hundreds of radio-collared mountain lions in hunted and unhunted areas in Wyoming and Montana (Newby et al. 2013, Robinson and DeSimone 2011). Mountain lions were captured and treed with hounds, immobilized, fitted with telemetry collars, and monitored for survival and movements. Field crews then tracked the dispersal of subadults to quantify survival, emigration rates, dispersal distance, and disperser success.

Hunter harvest was by far the leading cause of mountain lion mortality, affecting not only within-population dynamics but also the dynamics of the greater metapopulation (Fig. 7.6). Hunting led to decreased emigration, dispersal distance, and disperser success, which in turn affected whether hunted populations acted as sources or sinks in the landscape (Newby et al. 2013). The heavily hunted Garnet population before harvest closures (1997–2000) was strongly declining with low survival and little emigration, resulting in a population sink ($R^x + E^x$ = contribution metric < 1). However, after hunting closures were enacted in that area (2000–2006), the population exhibited positive population growth and increased emigration, making it a source providing a positive contribution to overall metapopulation growth (see also Robinson and DeSimone 2011).

The influence of connectivity was also clear in the region around Yellowstone National Park studied during two different time periods. For example, during 1987–1993, the population was a source, because the sum of its self-recruitment and emigration to other areas ($R^x + E^x$) exceeded 1 (Fig 7.6). Interestingly, these populations would actually decline if they did not also receive immigrants from other areas; for example, the self-recruitment (R^x) in phase 1 was 0.98, and the positive population growth ($\lambda = 1.11$) only occurred because the population was subsidized by a high average per-capita annual immigration rate of 0.14. Thus connectivity both supports the Yellowstone population and allows it to support other surrounding populations (Newby et al. 2013), underscoring the wisdom of conserving multiple, mutually supportive source areas across a landscape.

Overall, these results show how harvest can affect not only local population dynamics but also dispersal and dynamics across populations. To manage wildlife populations, we need to understand these between-population dynamics (immigration and emigration).

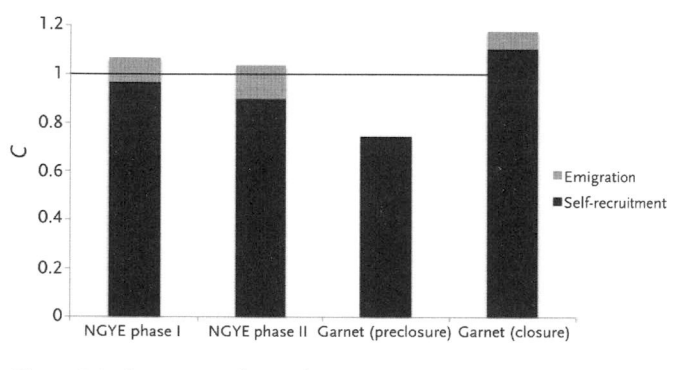

Figure 7.6. The estimated contribution metric (C) based on self-recruitment into a population and on successful emigration (dispersal) from that population to others. The solid horizontal line indicates $C = 1$; a population with $C > 1$ is a source, and one with $C < 1$ is a sink. The contribution is estimated from field data on mountain lions in the northern Greater Yellowstone Ecosystem (NGYE) in 1987–1993 (phase I) and 1998–2005 (phase II), and in the Garnet, Montana, region before and after closure of a 915 km² area to hunting. From Newby et al. (2013)

four of the components of population growth—births, immigration, deaths, emigration—must be accounted for.

Ecological Traps

Sometimes a sink habitat is preferred over better habitats because formerly reliable cues for habitat selection become mismatched with current fitness consequences in a modified landscape. This phenomenon creates a metapopulation dynamic known as an ecological trap, often associated with human-modified habitats. Ecological traps can arise when habitat modification amplifies the cues for making a poor habitat choice (making an area more attractive even when it decreases fitness), or when the quality of a preferred habitat is decreased, or both (Robertson and Hutto 2006). The drain created when good animals love bad habitats (Battin 2004) can overwhelm even strong sources and cause entire metapopulations to decline toward extinction. As a result, the identification of ecological traps can lead to critical management actions for recovering species of concern (Mills 2013). For example, if a grassland bird preferentially nests in agricultural fields with abundant fences or poles as perches (attractive habitat cues) but mowing of the fields destroys nests, management solutions might include adjusting mowing schedules to allow chicks to fledge, or eliminating perches and scaring birds to reduce attraction to the trap (Battin 2004). Indigo buntings (*Passerina cyanea*) provide another example as a species that prefers to nest along man-made forest edges. These edges are often ecological traps because they support a large number of subsidized and introduced predators (Weldon and Haddad 2005). The attractiveness of such a trap could be reduced by making clear-cut edges straight, minimizing the edge-to-area ratio, and thus reducing the area affected by the traps.

How Much Connectivity Is Needed among Multiple Populations?

Connectivity among populations, whether through a landscape porous to movement or via particular movement corridors, plays multiple critical roles in population dynamics across a species' range (Haddad et al. 2011, Mills 2013). These roles include recolonizing extinct patches, colonizing new patches as the environment changes, minimizing inbreeding depression, and maintaining small populations via demographic rescue. On the other hand, issues that could arise from too much connectivity (typically less of a problem than too little connectivity) may include swamping of local adaptation and transmission of disease.

A common question in wildlife management is how connected populations need to be (Mills et al. 2003). One answer for identifying appropriate levels of connectivity might be to mimic predisturbance, or historic levels. Historic levels of connectivity might be inferred by estimating movement rates using radiotelemetry or genetic tools (Lowe and Allendorf 2010) across an undisturbed or at least unfragmented natural area. An example of this approach is the use of genetic connectivity indices in areas with and without major highway development to infer how highway corridors may fracture grizzly bear (*Ursus arctos*) movements (Proctor et al. 2012). Another way to infer historic connectivity takes advantage of the ability of genetic sampling to compare current levels of connectivity to those in the past based on genetic signatures from museum or other historic specimens. For example, a break in gene flow due to habitat fragmentation and loss was quantified for greater prairie chickens (*Tympanuchus cupido pinnatus*) and Attwater's prairie chickens (*T. c. attwateri*) by comparing samples from 12 contemporary populations to historic samples collected between 1936 and 1970 (Bouzat et al. 2009). While these approaches have been useful, one must recognize that intact habitat areas are increasingly hard to find for many species and that natural fragmentation may still be present because of fire, disease, or other factors.

In cases where there is no knowledge of background or historic levels of connectivity, a genetic rule of thumb does exist. One migrant per generation, where a migrant in the genetic sense is a breeding disperser, allows local adaptation to proceed in populations while minimizing the negative effects of inbreeding depression (Mills and Allendorf 1996). One migrant per generation may translate to ten or more actual dispersers per generation, depending on the reproductive success of individual animals. While this guideline has been useful, the ecological and demographic needs of different populations will often mandate higher or lower levels of connectivity (Vucetich and Waite 2000).

MONITORING POPULATION DYNAMICS

We have seen some of the complications in understanding wildlife population dynamics, but also the considerable body of science that can be harnessed to deepen our insights. The full power of wildlife population ecology can be realized when monitoring wildlife population dynamics over time. Wildlife monitoring can increase our understanding of population processes, allow us to assess the factors influencing those processes, and determine how populations will change in response to management actions. Some of the management-based motivations for wildlife monitoring include: evaluating effects of a stressor on one or more species; developing regulations for the sustainable harvest of a game species; assessing the recovery of a threatened species; evaluating the outcome of a reintroduction program; determining the status of an invasive or pest species; or quantifying changes in biodiversity over time. Often, monitoring is a government mandate or policy. For example, many governmental agencies must conduct monitoring as part of the legal requirements outlined by legislation, such as the U.S. Endangered Species Act or the National Environmental Policy Act (Schultz 2010).

If approached casually, wildlife monitoring can suffer from poorly defined objectives and deficient statistical design and analysis, which in turn lead to inferences that are impossible to defend in court or in the eyes of the public. Happily, in the last decade, monitoring has been the subject of intense attention in mainstream wildlife population ecology, resulting in a rigorous scientific basis for its application. Here we describe some keys to a successful and efficient program to monitor wildlife population dynamics and the influence of management actions. In the process, we draw on some of the fundamental concepts of wildlife population dynamics already discussed in this chapter.

To place wildlife monitoring on the rigorous scientific footing it requires and deserves, there must be thorough consideration of why the monitoring is occurring, what will be monitored, and how it will be done. To help develop these goals, we consider a couple of different forms of monitoring and describe the field and analytical techniques most commonly employed.

Why Monitor? Targeted versus Surveillance Monitoring

The first step in monitoring a wildlife population is to define the purpose for a project. By setting up a priori (before the start) hypotheses related to the mechanisms influencing population dynamics and the effects of management on a species of interest, and including (when possible) natural or induced manipulations, targeted monitoring programs can efficiently lead to a deep understanding of biological and management questions (Nichols and Williams 2006). A classic example of this approach is demonstrated by the monitoring objectives of midcontinent mallard ducks (*Anas platyrhynchos*), a species for which managers want to maximize hunting opportunities while maintaining a sustainable population. To meet this objective, numerous federal, state, and provincial agencies in the United States and Canada annually survey the duck population, and use that survey data and harvest information to evaluate a suite of hypotheses describing how harvest influences mallards (i.e., does harvest have additive or compensatory effects on mallard survival?). By applying monitoring data to a specific set of ecological and management related questions, biologists have gained valuable information about the dynamics of the system and about sustainable harvest regulations (Nichols et al. 1995, Nichols and Williams 2006).

In some cases where little is known about the species, the relevant spatial scale is unknown, and management effects are particularly complex, all that can be done is omnibus surveillance monitoring (Nichols and Williams 2006) or cumulative effects monitoring (Boutin et al. 2009). In these cases, numerous attributes of a system are measured over time, usually without manipulations directed at testing specific mechanistic hypotheses. Multiple management effects (e.g., harvest levels, patch size, road density, fire suppression, human access, invasive species), occurring cumulatively and simultaneously across several spatial scales and species, are monitored concurrently to track numerous processes. In developing countries where the capacity for wildlife research is limited, surveillance monitoring can play a critical role in raising public interest in biology (and in some cases getting biologists out of their offices and into the field), empowering local people in natural resource decision making, and may be the only practical form of monitoring if knowledge of population ecology and study design is limited (Danielsen et al. 2009).

While omnibus surveillance monitoring is routinely conducted, biologists should strive to initiate targeted monitoring as much as possible to answer mechanistic, hypothesis-driven questions about the factors (natural and management induced) that affect population dynamics (Yoccoz et al. 2001, Nichols and Williams 2006). A study designed around causal hypotheses will be better situated to separate multiple population influences or stressors and to determine the effects of management actions. Similarly, after many years of omnibus surveillance monitoring, one might detect a problem (e.g., a decline in the abundance of focal species or in species richness) but might not be able to pinpoint the mechanism responsible or the best management action. In short, even though it is harder to do, targeted monitoring based on multiple competing hypotheses is much more efficient for identifying problems and solutions. The key is to try to articulate the biological and management-based hypotheses you want to test before the monitoring begins, instead of struggling to tease apart different effects after data have already been collected. Targeted monitoring is a central component of adaptive management (Lindenmayer and Likens 2009), a synthetic framework to understand population dynamics and effects of management actions; Chapter 5 covers this topic in detail.

Field and Analytical Techniques for Monitoring

After identifying the objectives for wildlife population monitoring, one must decide how to collect and analyze field data. These issues of study design and statistical analysis of carefully specified variables are too often ignored in wildlife monitoring studies (Marsh and Trenham 2008). Although details of study design and analysis are beyond the scope of this chapter, some key points warrant mention. First, the study should be at an appropriate scale and pay attention to the basic principles of statistical sampling (Yoccoz et al. 2001). Typically, inferences for wildlife populations are desired across some large area (e.g., a national park, a state, a province, a country), but it is possible to sample only a small portion of the area. There are many ways to link limited sampling frames to broader areas of inference, but all methods are rooted in the basic principle of random sampling, the most powerful approach for strong statistical inference. A pitfall to avoid is the temptation to initiate convenience or subjective sampling. A monitoring program where field sites are chosen for convenience or because they are subjectively felt to be representative of the larger area will always compromise inferences from that study to a larger area.

A second point to emphasize is the need to consider, before the study begins, whether statistical power will be sufficient to estimate accurately a population parameter of interest (e.g., abundance) or to detect whatever change in a population you are interested in (e.g., a change in density or distribution; Mills 2013). Statistical power will be reduced when sample sizes are small, variance in the data is large, or a trend is subtle (e.g., a 5% decline in population size). Unfortunately, these are exactly the conditions that inevitably occur when monitoring most species of concern. While we want to detect subtle changes in population dynamics, the small population sizes and elusive habits of rare species inevitably reduce sample sizes and increase variability of the data. As a result, without a priori attention to statistical power, many wildlife monitoring programs will be doomed from the start, unable to estimate population parameters with desired precision or to detect meaningful trends in the dynamics of populations (Taylor and Gerrodette 1993, Gibbs et al. 1998).

What are the solutions? One is to initiate discussions between scientists and managers about the desired precision in a population estimate and the costs and benefits of detecting, or failing to detect, a change in population parameters through monitoring. This can be formalized in a decision theory framework. An example: koalas (*Phascolarctos cinereus*) are declining in much of eastern Australia despite being a major tourist attraction. Field et al. (2004) conducted power analyses incorporating the financial impact of lost tourism if an undetected decline of koalas occurred. The researchers reported that classical approaches to monitor trends would always fail to detect economically important declines of koalas due to low statistical power. In fact, $5 million would be saved in this case if monitoring were abandoned in favor of direct management action to prevent a decline from occurring. The decision theory framework was critical to point out in this case where statistical power was inadequate given the economic consequences of koala decline.

Another way to increase statistical power and to improve inferences from a monitoring program is to carefully choose the monitoring field methodology. Although traditional monitoring techniques to capture and mark animals continue to be useful (e.g., trapping or sighting transects for mammals, cover boards for amphibians, mist netting for birds and bats), many powerful new approaches have emerged in noninvasive sampling (no capture or handling of animals; Long et al. 2008, Kelly et al. 2012, Mills 2013). Noninvasive sampling can include photos from remote cameras and genetic sampling. Noninvasive collection of genetic samples might include scats, eggshells, and hair obtained via snags, rub devices, transects, or snow tracking. Especially for species that are rare, elusive, and hard to monitor with conventional approaches, noninvasive sampling can provide data on the species present, individual identity, gender, relatedness, and genetic distinctiveness (Oyler-McCance and Leberg 2012). These data can also feed into the monitoring of state variables described below.

The final decisions to make before implementing monitoring include choosing the specific measures that summarize the status of a population of interest (state variable) and other variables or covariates of interest to understanding the population's dynamics (auxiliary information). State variables include those addressing the presence, distribution, abundance, growth rate, or viability of a population(s). At the same time, auxiliary information is collected to facilitate tests of how the system works or on the influence of management. For example, harvest management of North American mallard ducks measures population size as the state variable, but also includes mallard harvest and number of wetlands on key breeding areas to help inform the models that connect the monitoring data to an understanding of harvest and mallard population dynamics (Yoccoz et al. 2001). In the next few sections we describe the most common state variables: abundance, trends in population growth, presence/absence of the target species, and risk of decline.

Monitoring Abundance or Density

One of the most common targets of wildlife studies in general and monitoring in particular is an estimate of abundance or density (abundance per unit area). Although abundance may seem to be an obvious and simple population attribute to measure, it is actually quite challenging because of the simple fact that animals move, hide, and generally avoid being detected. This means that a simple count of animals in an area will be less than the number of animals present. Said more formally, detection probabilities are typically less than one. In fact, the case where detection probability equals 1, so that we detect all

individuals in an area, is so unusual that we use a special word to describe it: census.

In addition to accounting for incomplete detection, animal abundance estimates must also account for incomplete sampling. Neither the time nor resources are always available to sample the entire population of interest (e.g., Big Bend National Park, state of Wisconsin). If the study area represents only a portion of the population, then the proportion of area sampled (a) is less than 1. Both a and the detection probability ($\hat{p}$; the hat over the p is a standard convention indicating an estimate) are incorporated in this generalized equation for estimating abundance ($\hat{N}$):

$$\text{estimate of abundance} = \hat{N} = \frac{\text{count of animals}}{\hat{p} \times a} \qquad (9)$$

For example, suppose 40 rabbits on an island are counted on a spotlight transect that includes the entire island ($a = 1.0$), but detection probability is 0.5; the estimate of abundance ($\hat{N}$) would be 80 (40/0.5). If the transect included only 25% of the area occupied by the rabbit population, the abundance estimate would be:

$$\hat{N} = \frac{40}{0.5 \times 0.25} = 320$$

Counting animals can be achieved through a variety of methods, including transect sampling (e.g., sighting animals by plane, foot, vehicle, horseback), live trapping, and noninvasive genetic and camera sampling. In all cases, detection probability must be estimated, requiring thoughtful methodologies (e.g., mark-recapture, double sampling) and oftentimes sophisticated statistical analyses. Although the rich analytical details of estimating detection probability (and abundance) with different methods are beyond the scope this chapter (see overviews in Williams et al. 2002, Mills 2013), an example using the simplest possible approach follows (see box on p. 101).

In some cases, the study design prohibits an estimate of detection probability. For example, perhaps only raw counts of animals are available, or observations of animal signs such as numbers of pellets, tracks, calls, nests, scrapes, or burrow entrances. These are index counts, which in general should be avoided in monitoring studies because they can give misleading signals of changes in abundance. The problem is that an index is a function of both the abundance of animals and the relationship between the index and true abundance. Unless a biologist can be sure that the relationship between the index and abundance stays constant across space and time, any change in the index could be caused by a change in that relationship *or* by a change in abundance. For example, suppose in one year you record an average of 60 pheasants (*Phasianus colchicus*) at calling stations. The next year you record 80. You cannot assume the population increased, because the change in the index from 60 to 80 could have been caused by an increase in call rate or call detection (maybe each bird was calling more, or observers were better trained to hear them, or weather conditions were more favorable for sound to travel). The bottom line is that indices are best avoided for monitoring studies, but if they must be used, one should do everything

ESTIMATING ABUNDANCE WITH THE LINCOLN-PETERSEN METHOD

The Lincoln-Petersen (LP) estimator is based on two sampling occasions where animals are marked on the first occasion. The population is assumed to be closed to births, deaths, immigration, and emigration, and all individuals are assumed to be equally detected (captured) and unaffected by the capture or marking procedure (these assumptions can be relaxed in more complex abundance estimators). The capture and marking could be physical live capturing or alternatives such as mark-resight or noninvasive sampling with cameras or genetic sampling.

Suppose you are estimating abundance from two sampling occasions that occur on consecutive days. Denote the number of animals marked and released on day one as n_1. During the second occasion (day two), a total of n_2 animals are captured, of which m_2 have marks from the previous day. The capture probability, $\hat{p}$, is simply the proportion of the animals marked the first day that are captured on the second day (m_2/n_1). Taking the total captures on the second day (n_2) as the count, the estimate of abundance is:

$$\hat{N} = \frac{n_2}{\hat{p}} = \frac{n_2}{m_2/n_1} = \frac{n_1 n_2}{m_2}.$$

Applying this equation to an example, suppose we wanted to estimate the number of flying squirrels (*Glaucomys sabrinus*) in a forest patch. In one night of trapping, 32 squirrels were captured, ear-tagged, and released. One week later, the traps were set again, and a total of 44 squirrels were captured, 20 of which had been ear-tagged during the first trapping occasion.

$$\hat{N} = \frac{n_1 n_2}{m_2} = \frac{32 \times 44}{20} = 70.4$$

Given these numbers, the population size of flying squirrels would be estimated to be 70.4. It turns out that this simple and intuitive equation needs to be corrected for statistical reasons, so the actual LP estimator (and its variance) for actual application are:

$$\hat{N} = \left[\frac{(n_1 + 1)(n_2 + 1)}{(m_2 + 1)} \right] - 1$$

and

$$\text{var}(\hat{N}) = \frac{(n_1 + 1)(n_2 + 1)(n_1 - m_2)(n_2 - m_2)}{(m_2 + 1)^2 (m_2 + 2)}.$$

possible to keep constant the relationship between the index and abundance.

Finally, we will not discuss monitoring based on abundance indices arising from casual perceptions or field notes. Although these may have advantages in being inexpensive and involving local people, the quality of the data is typically unknown, as is the link between perceptions and actual population dynamics, leading to weak scientific value for either changes in abundance or the mechanisms underlying perceived changes.

Monitoring Trends in Abundance (Population Growth over Time)

If the monitoring objective includes evaluating population growth over time, two primary approaches can be used. One way uses vital rates and population projection models; for example, following the methods described above in the section on stage-structured population dynamics. The second and more common way to estimate trend uses abundance estimates over time from a monitoring program. It is this abundance-based trend estimator approach that we cover briefly below (for more details, see Humbert et al. 2009, Mills 2013). Although trend estimators can incorporate real-world complications such as density dependence, observer effects and other covariates, for simplicity we focus only on exponential growth estimation without covariates or other auxiliary information.

The most commonly used method of estimating trend from a time series of abundance values is a simple linear regression of natural log (ln) of abundances against time (the natural log accounts for the fact that birth and death processes cause wildlife populations to change geometrically, not arithmetically). The slope of the regression represents the estimated average rate of change ($\hat{r}$). The simplicity of the method explains its popularity, but the method has a major limitation: it assumes that all variation in the trend arises only from the uncertainty in estimating N (i.e., pure observation error or sample variance; Humbert et al. 2009). That is, this method assumes that population growth is completely constant, unaffected by process variance arising from weather, predators, or other environmental conditions, so that all deviations in abundances from the trend line arise solely from the uncertainty of estimating abundances. A suite of other widely used methods make the opposite assumption, that no observation error exists and that all variation in the trend arises from process variance or process noise (e.g., the diffusion approximation; Dennis et al. 1991).

Recent developments permit exponential trend estimation with both process and observation error occurring simultaneously. These new methods include state-space statistical models containing a component to account for the stochastic fluctuations due to process noise and a component to accommodate the observation error in abundance estimates (Humbert et al. 2009, Mills 2013). Other approaches may also be used to estimate trends of wildlife populations over time using count or abundance data. For example, Bayesian analyses

(Taylor et al. 1996) are becoming more popular. Also, if mark-recapture data have been collected, both λ and its variance can be estimated directly (Nichols and Hines 2002).

How long must a biologist collect time series data to estimate population trends reliably? The answer depends on many factors, of course, but for most wildlife species, the state-space model should use a bare minimum of ten years, with at least five samples of abundance during that time (Humbert et al. 2009). Of course, unusual events that affect process variance (e.g., 20-year floods or 15-year fire events) will only be picked up with longer sampling.

Monitoring Species Distribution or Community Composition

Another set of state variables in wildlife population monitoring is the distribution of a single species, or the diversity of a suite of species. Instead of estimating abundance or vital rates for assessing population growth, these approaches estimate changes in occupancy—or presence versus absence—of target species in a sampled area.

Although distribution or occupancy of a species in an area would seem to be straightforward, it is (like abundance) complicated by incomplete detection probability. Just as a raw count of animals does not describe abundance when individuals are undetected, estimates of occupancy are also biased low if species presence is detected imperfectly. If a species is detected, then of course it is present. But if it is not detected, it is not necessarily absent. Therefore a single presence/absence survey should be called a present/not detected survey. The challenge, then, is to adjust counts of presences of the species to account for the probability that they were present but not detected.

The scientific framework for such estimates is called occupancy modeling. Just as raw counts can be adjusted by detection probability to derive estimates of abundance, with occupancy modeling, the raw count of where a species was detected can be adjusted by the estimated probability of detecting the species, derived from the pattern of detections and nondetections in the data. Space prevents us from detailing the methods of occupancy modeling (MacKenzie and Royle 2005, MacKenzie et al. 2006). In brief, however, data collection for occupancy surveys involves searching for evidence of presence in sampling units, with detection methods ranging from visual observations or captures to photographs and indirect evidence of presence such as hair, feces, or tracks. Detections and nondetections at multiple sample units form the detection history used to estimate the probability of the species being present but not detected.

Monitoring Extinction Risk and Population Viability Management

Monitoring abundance, trend, or distribution can be important for many reasons in applied wildlife population studies. But in other cases the ultimate goal centers on risk assessment, or potential vulnerability of the population to extinction or deep decline. Applied population ecology has a variety of tools for as-sessing viability of wildlife populations, including quantitative approaches of population viability analysis (PVA; Mills et al. 2012). PVA can be defined as "the application of data and models to estimate likelihoods of a population crossing thresholds of viability within various time spans, and to give insights into factors that constitute the biggest threats" (Mills 2013:227). The two main ways to conduct PVAs with monitoring data are with population projection models and time series analysis. A useful example of PVA linked to abundance-based time series data has been called risk-based viable population monitoring (Staples et al. 2005), which uses the trend and variance in trend in monitoring data to provide early warnings of risk by iteratively updating the probabilities of persistence or decline each year (for software, see http://www.cnr.uidaho.edu/population_ecology).

An even more comprehensive and integrative approach is population viability management (PVM; Bakker and Doak 2009), where population dynamics and risk assessment become part of an integrated adaptive management process. Researchers used PVM in the management of the endangered island fox (*Urocyon littoralis*) on the Channel Islands (Bakker and Doak 2009, Bakker et al. 2009). The first step in the management process was to set recovery criteria in terms of acceptable risk in a public forum, accounting for the inevitably complex sociopolitical and biological considerations. Next, readily monitored population attributes were selected (e.g., adult population size and adult mortality) and linked to extinction thresholds (where extinction could be a number greater than zero; for example, it may be the number of animals that would trigger a captive breeding program to be initiated). Finally, PVM incorporated management actions predicted to affect population attributes; for example, determining how decreasing predation on foxes by removing golden eagles (*Aquila chrysaetos*) would increase adult survival, given the practical constraints of monitoring. PVM provides a platform for integrated adaptive decision making, whereby increasing knowledge and changing management outcomes can be accounted for in data-based decisions. Also, this approach embraces various types of parameter and process uncertainty, while presenting managers with straightforward results that do not bury the findings in distracting complexity.

CASE STUDY: RECOVERING ENDANGERED SIERRA NEVADA BIGHORN SHEEP

We use a case study to demonstrate how a mechanistic understanding of population dynamics can inform on-the-ground management decisions. It highlights how monitoring data can be used to develop stage-structured models that in turn help assess population responses to different management scenarios, and how those responses can be used to prioritize conservation actions. This example also illustrates the complexity of deciphering the underlying factors driving populations, a constant challenge for wildlife managers.

Sierra Nevada bighorn sheep (*Ovis canadensis sierrae*) are the rarest subspecies of mountain sheep in North America (Fig.

Figure 7.7. Federally endangered female Sierra Nevada bighorn sheep. The ear tag is used for estimating population size using mark-resight field surveys. Photo courtesy of Art Lawrence

7.7). There are approximately 400 individuals in the subspecies, and they are endemic only to the Sierra Nevada mountain range that spans eastern California (Fig. 7.8). The subspecies was listed as endangered under the U.S. Endangered Species Act in 1999, when biologists could locate only about 100 adults in the wild (U.S. Fish and Wildlife Service 2007). Since then, government agencies, nonprofit organizations, and university scientists have been collecting detailed monitoring data on the dynamics of this endangered subspecies. To identify and prioritize management actions that would be most effective at increasing the size and growth rates of Sierra Nevada bighorn sheep populations, researchers outlined three objectives:

1. Identify which stage-specific vital rates are most important for influencing the growth rates of bighorn sheep populations.
2. Determine the stochastic and deterministic factors responsible for spatial and temporal variation in those key vital rates influencing population growth.
3. Use information from objectives 1 and 2 to develop effective management strategies to increase population sizes and to reach recovery goals.

To meet these objectives, researchers used a suite of different field methods to obtain data on Sierra Nevada bighorn sheep populations. They used telemetry collars to mark over 180 animals, tracking their survival and annual reproduction. Since 1980, they also conducted ground counts of each population each year, with attempts to annually census the number of bighorn sheep in individual herds. Additionally, in 2006,

biologists started conducting annual mark-resight surveys (analogous to mark-recapture, except animals are resighted as opposed to recaptured) to obtain estimates of population sizes with detection probabilities and associated measurements of error. Using information from these different data types, biologists estimated annual stage-specific vital rates for different populations of bighorn sheep (Johnson et al. 2010*a,b*); here we will focus specifically on data from the Mono Basin, Wheeler, and Langley populations (Fig. 7.9).

Objective 1: Identify Which Stage-specific Vital Rates Are Most Important for Influencing the Growth Rates of Bighorn Sheep Populations

To develop effective management strategies for the recovery of endangered species, it is necessary to identify those vital rates responsible for poor population performance and those vital rates that can be increased to most efficiently change a population's trajectory. To do that, biologists tracked three vital rates that captured the basic life cycle and population dynamics of bighorn sheep and were relevant to stage classes that could be reliably distinguished in the field: adult female survival (S_A, the number of adults that survived to year t given the number of adults and yearlings in year $t-1$); yearling female survival (S_Y, the number of lambs in year $t-1$ that survived to be yearlings in t); and adult female fecundity (F_A, the number of lambs born per adult female in year t). These vital rates were estimated from marked individuals and census counts of lambs, yearlings, and adults.

The field data revealed significant differences in the means and variances of these vital rates in different populations, and even within the same population for different phases of population growth (Johnson et al. 2010*a*). For example, the Mono Basin population experienced two very different phases of growth. The herd dramatically increased after it was initially reintroduced in 1986, but it then declined precipitously around 1994 (Fig. 7.9). Below are two matrices based on the mean vital rates measured during the increasing phase (from 1986 to 1993) and the decreasing phase (from 1994 to 2007). Although we will not delve into the details of the population matrix, we point out that fecundity (matrix element $a_{1,3}$) is multiplied by adult survival because field surveys occurred just after lambs were born, so that the number of lambs, in this case, is also a function of the number of adult females that survived to reproduce (see Mills 2013 for details about postbirth pulse matrices).

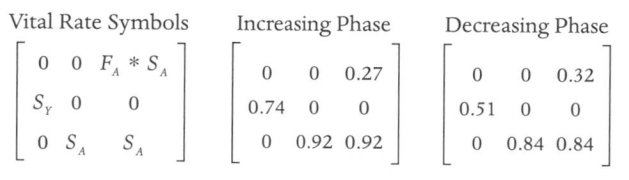

During the increasing phase in Mono Basin, the asymptotic growth rate was 1.08 (the population grew at 8%/yr), while during the decreasing phase it was 0.98 (declining by 2%/yr). Notice that the fecundity rate was actually higher during the decreasing phase, but yearling (element $a_{2,1}$) and adult survival rates (elements $a_{3,2}$ and $a_{3,3}$) were much lower. Which stage-specific vital

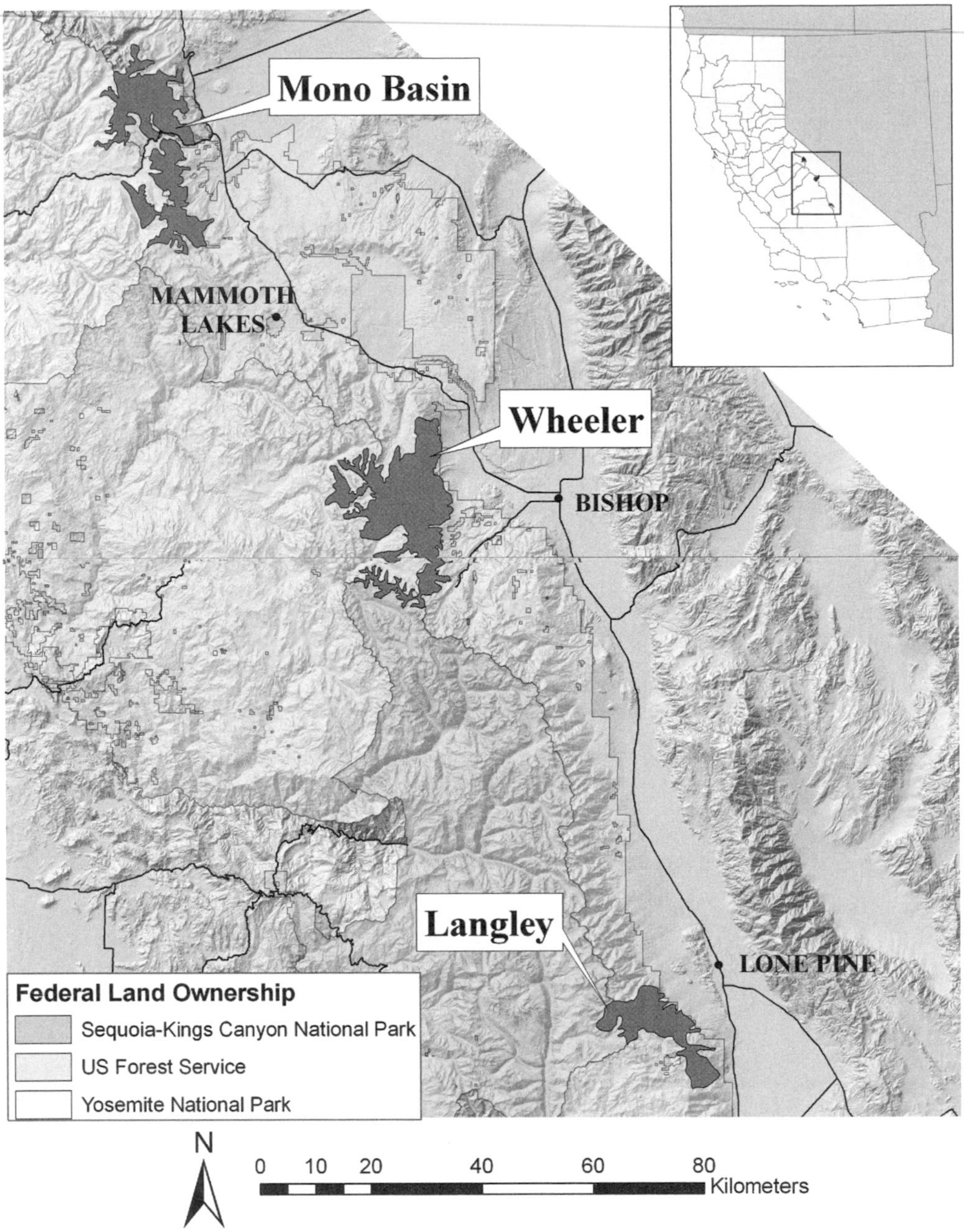

Figure 7.8. Locations of the Mono Basin, Wheeler, and Langley populations of Sierra Nevada bighorn sheep in California.

rates were responsible for these disparate population trends, and which rates were associated with the period of population decline? Which vital rate(s) should managers try hardest to improve in order to boost the size of this population?

To answer these questions, researchers turned to sensitivity analyses. First, they calculated the analytical elasticities of each vital rate and found that the vital rate with the highest elasticity, across all populations and phases of growth, was adult female survival (Fig. 7.10), which corresponds to general patterns from long-lived species (Sæther and Bakke 2000, Crone 2001). Indeed, given that adult females have the highest reproductive value and comprise a large proportion of the population (>65% of the females are adults), it is not surprising that a small change in adult female survival would yield a

proportionately greater change in the population growth rate than a small change in either yearling survival or fecundity.

While this pattern has been well recognized in the scientific literature, it has limitations for wildlife managers. For example, the temptation to conclude that adult female survival should be the vital rate targeted for efficient population increase must be tempered by the fact that elasticity does not take into account how much that rate can realistically be manipulated through management. During the increasing phase in Mono Basin, adult survival was 92%. This is close to the expected biological maximum for adult survival, with little room for improvement. If managers wanted to further increase population growth but could not easily increase adult survival, what other vital rate should be the focus?

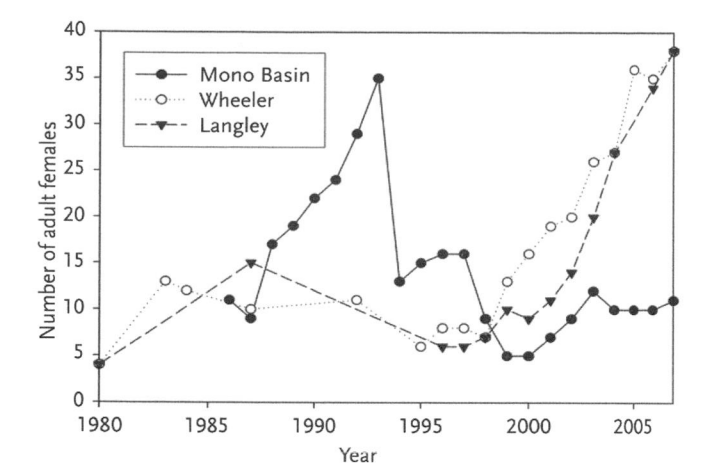

Figure 7.9. Number of adult females in the Mono Basin, Wheeler, and Langley populations, 1980–2007. From Johnson et al. (2010*a*)

To incorporate the realistic range of variation that is observed in different vital rates, researchers turned to LSA. This analysis determines how observed variation in different vital rates is associated with observed variation in population growth rates. In other words, which vital rates are really responsible for variation in population growth? During the increasing phase, researchers reported that variation in yearling survival at Mono Basin explained most of the variation in λ (63%), even though adult female survival had the highest elasticity (Fig. 7.10). If managers wanted to increase the population growth rate during this phase, they would have been most successful by working to boost yearling survival. During the decreasing phase, however, variation in adult survival explained 98% of the variation in λ (Fig. 7.10), as the decrease in adult survival from 92% to 84% was primarily responsible for the drop in population growth from 1.08 to 0.98. During this kind of declining phase, managers should work to increase adult female survival to increase overall growth rates.

Across the study herds, researchers reported that LSA patterns were highly population specific, and each of the three vital rates was found to be the primary population driver in different herds and for different phases of growth (Johnson et al. 2010*a*). Wildlife managers often focus on improving the same vital rate in different populations of the same species, or in populations of similar species. This analysis of the dynamics of different endangered populations, however, demonstrates that appropriate management targets may often be idiosyncratic in space and time.

Objective 2: Determine the Stochastic and Deterministic Factors Responsible for Spatial and Temporal Variation in Those Key Vital Rates Influencing Population Growth

Once biologists identified which vital rates were most important in the dynamics of endangered bighorn sheep populations, particularly during phases of population decline, they wanted to identify the factors influencing the variation in

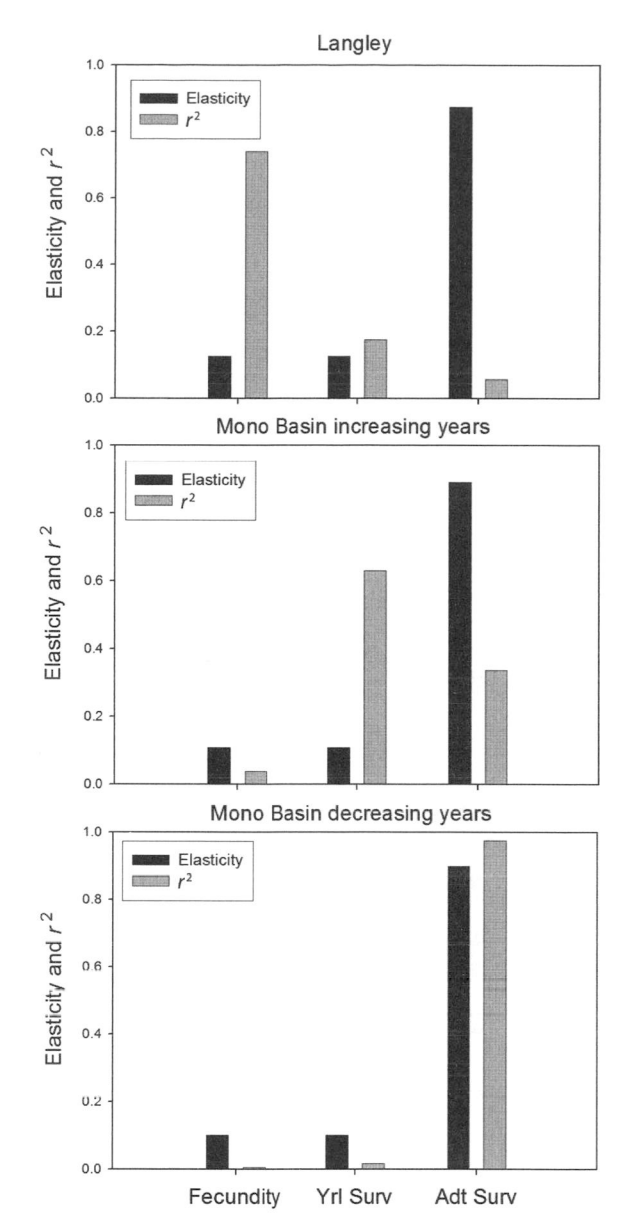

Figure 7.10. Elasticity and r^2 values (the association between variation in population growth and variation in different vital rates) for fecundity, yearling survival, and adult survival of different bighorn sheep populations and different phases of growth.

those rates. Some of the factors hypothesized to influence bighorn sheep population dynamics were density dependence, predation, inbreeding depression, and weather (i.e., winter severity). A suite of analyses revealed that all these factors were significant, but their degree of influence varied among the different populations and for different vital rates (Johnson et al. 2010*b*, 2011, 2013).

For example, adult female survival had the greatest influence on population growth in the Wheeler herd and during the decline in Mono Basin, but the factors that affected this rate were dramatically different between them. In Mono Basin, adult female survival was positively associated with population density (positive density dependence) and summer

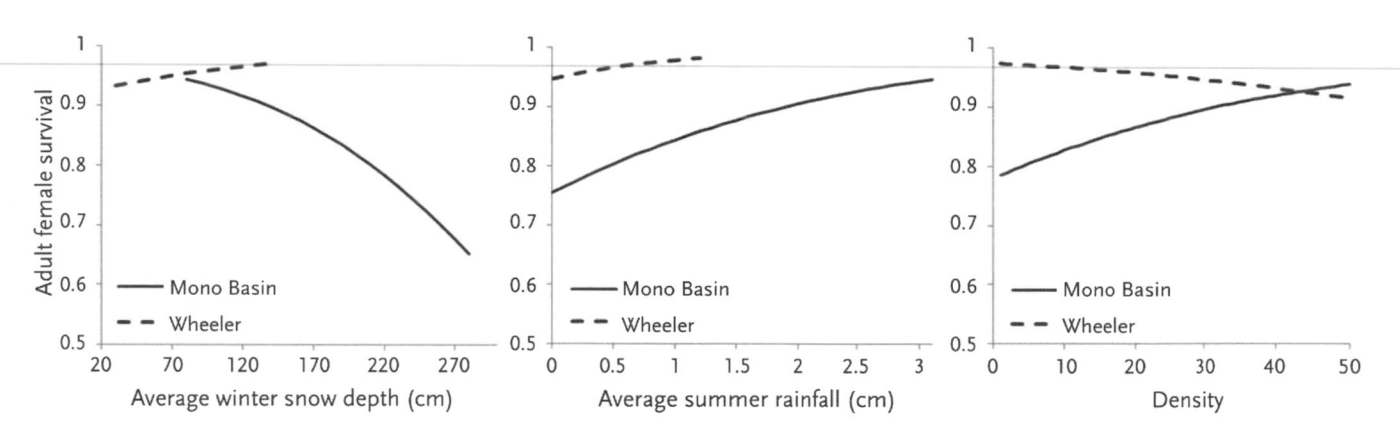

Figure 7.11. Predicted effects of winter snow depth, summer rainfall, and density on adult female survival for the Mono Basin and Wheeler populations. Predictions for weather covariates were only modeled for the observed range of variation in each population.

rainfall (an index of summer forage quality), and negatively associated with winter snow depth (an index of winter severity). Meanwhile, adult survival in Wheeler was negatively associated with population density (negative density dependence) and mountain lion predation, and slightly positively associated with winter snow depth and summer rainfall (Fig. 7.11).

Why would adult female survival of bighorn sheep in two populations be affected so differently by the same set of factors? One explanation is that habitat conditions were highly variable between the herds, particularly during winter. For example, the Mono Basin population persisted at high elevations (~3,400 m) during winter, relying on high plateaus that were free of snow. Meanwhile, most bighorn sheep in the Wheeler population spent the winter at lower elevations (~2,550 m), on south-facing slopes below snowline. This difference likely explains the negative effect of winter snow depth on survival at Mono Basin compared to Wheeler. That same distinction in winter ranges likely also accounts for the observed disparity in mountain lion predation. The winter range of bighorn sheep in the Wheeler population overlaps with the winter range of thousands of mule deer that support a healthy mountain lion population, posing a significant threat to endangered bighorn sheep. Meanwhile, the Mono Basin population spent winter at high elevations, away from deer, where there was little evidence of mountain lion predation (Johnson et al. 2013). As for density dependence, bighorn sheep are a species that probably experiences both positive and negative density dependence at different population sizes. As a gregarious species, foraging efficiency and predator detection are likely enhanced above a threshold population size (Mooring et al. 2004), and the very small size of the Mono Basin population (about ten adult females in recent years) may have induced Allee effects. For larger populations like Wheeler, limitations in food resources or increased predation may induce patterns associated with negative density dependence.

As with populations of Sierra Nevada bighorn sheep, biologists often find that different factors have variable effects on vital rates in distinct populations. The key is to maintain focus on those factors that influence the vital rates most consequential to population growth for the group of animals of interest.

Factors that influence vital rates that are largely inconsequential to λ should not be a priority for managers, as they will not be able to significantly influence the trajectory of a population. For example, researchers found evidence of inbreeding depression in fecundity rates of Sierra Nevada bighorn sheep, but have not initiated direct actions to address this threat. While genetic management had been proposed to counteract inbreeding, analyses with population models illustrated that such efforts would do little to benefit the population in the near term, because fecundity was not an important driver of growth rates (Johnson et al. 2011). In this case, managers focused on actions that would yield greater short-term benefits to populations while planning for increased gene flow among herds in the future.

Objective 3: Use Information from Objectives 1 and 2 to Develop Effective Management Strategies to Increase Population Sizes and to Reach Recovery Goals

Vital rates can either change naturally (i.e., the influence of winter severity on adult survival in Mono Basin) or as a result of management. Managers cannot change the weather (although knowing the importance of winter severity may be critical for identifying future reintroduction sites), but they often have a suite of potential actions that can be used to manipulate the trajectories of populations. The key, however, is to identify which of those potential actions is most effective in influencing the dynamics of populations, and thus reaching management objectives. Often, well-intentioned management actions have wasted critical resources because they had little to no influence on population dynamics (see the sea turtle example above).

For Sierra Nevada bighorn sheep, the initial laundry list of potential management options included mountain lion removal, augmentation (i.e., increasing the size of an existing population), genetic management (i.e., increasing gene flow among herds to counteract inbreeding depression), disease prevention, and prescribed fire to enhance forage quality. To determine which of those activities would be most effective for stimulating population growth, researchers simulated their

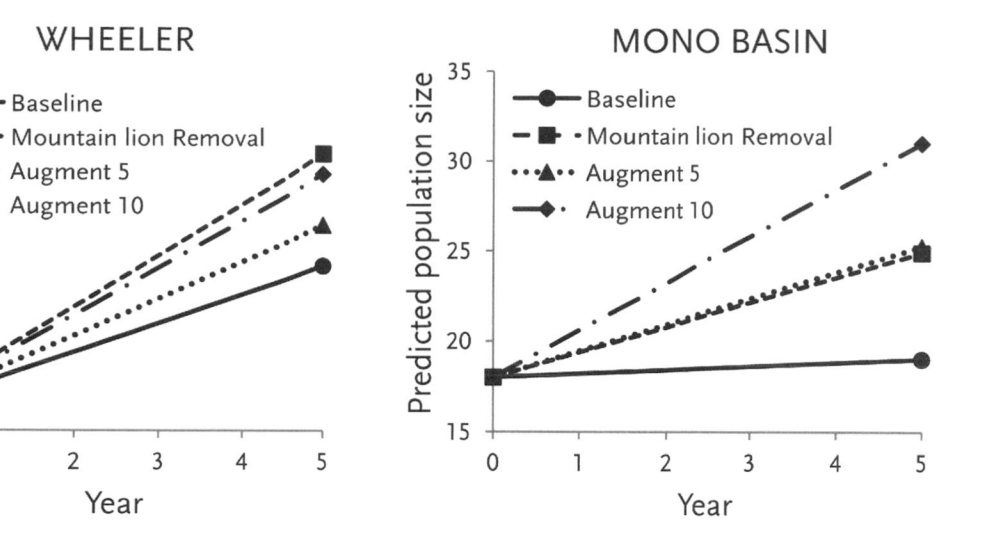

Figure 7.12. Predicted increase in population size resulting from mountain lion removal and augmentation (adding five and ten adult females to the population) for the Wheeler and Mono Basin populations of Sierra Nevada bighorn sheep over a five-year period.

effects using stage-structured population models. To simulate these "what if" scenarios, managers must have information about how different activities influence key vital rates, as well as the magnitude of their effects. For Sierra Nevada bighorn sheep, managers used models to compare the effects of two specific actions, mountain lion removals and population augmentations. Both actions could be readily implemented in the field and had clear and quantifiable influences on vital rates or numbers of adult bighorn sheep.

Researchers evaluated the relative influence of these two actions over a five-year period on Wheeler and Mono Basin, given their unique dynamics (Johnson et al. 2010a). In this analysis, different vital rates were manipulated in a population model to reflect expected management outcomes, and their anticipated effects on population growth were compared. Biologists simulated mountain lion removal by conservatively modeling a 5% increase in each vital rate, and an augmentation by adding either five or ten adult females to each population, a realistic number given limited source stock for translocations. These kinds of demographic modifications can be simulated from actual data on a study system or, when data are not available, from values published in the peer-reviewed scientific literature. For Sierra Nevada bighorn sheep, over the modeled time period, investigators reported that predator control would be most effective for boosting population growth in the Wheeler herd, while an augmentation would have a greater influence on population growth in Mono Basin (Fig. 7.12). Effective management actions were unique to each herd based on their individual dynamics.

Even once the best management action has been identified and implemented, the job of a biologist is not finished. They must carefully monitor the demographic impacts of those actions on populations over time. This information can then be used within the iterative process of monitoring and adaptive management to improve future modeling efforts, so that bet-

ter predictions of the responses of populations to management can be used for sound decision making in the future (Williams 2001, Lyons et al. 2008; Chapter 4).

SUMMARY

We have covered a lot of ground, from the basics of population growth and dynamics, to the factors that influence populations, to stage-structured models, to considerations for monitoring population dynamics. The field of wildlife population dynamics is broad and complex. While we could only briefly introduce many of the key concepts here, we aim to instill an appreciation for the importance of population dynamics in understanding and managing wildlife populations. Before taking any management action intended to change the size or distribution of a population, one should ask, what is our specific population objective? Which vital rate(s) should be altered to achieve that objective? And which management action(s) can most effectively change the vital rate(s) of interest? In those cases where we do not have a clear understanding of how a management action will influence the dynamics of a population, biologists can use targeted monitoring and adaptive management as powerful tools for elucidating such information.

The application of carefully collected field data to population models can yield critical and nonintuitive insights about the mechanisms driving populations. As human development, climate change, the expansion of nonnative species, and other factors continue to affect wildlife populations, understanding the processes that govern population dynamics is increasingly critical for preserving biodiversity. Given the available suite of powerful field, statistical, and modeling tools, biologists can apply information about the dynamics of populations to maximize their management efforts and to achieve population objectives.

Literature Cited

Adams, L. G., R. O. Stephenson, B. W. Dale, R. T. Ahgook, and D. J. Demma. 2008. Population dynamics and harvest characteristics of wolves in the Central Brooks Range, Alaska. Wildlife Monographs 170:1–25.

Allee, W. C. 1931. Animal aggregations: a study in general sociology. University of Chicago Press, Chicago, Illinois, USA.

Allee, W. C., O. Park, A. E. Emerson, T. Park, and K. P. Schmidt. 1949. Principles of animal ecology. W. B. Saunders, Philadelphia, Pennsylvania, USA.

Allendorf, F. W., G. Luikart, and S. N. Aitken. 2012. Conservation and the genetics of populations. Second edition. Wiley-Blackwell, Cambridge, Massachussetts, USA.

Altwegg, R., A. Roulin, M. Kestenholz, and L. Jenni. 2006. Demographic effects of extreme winter weather in the barn owl. Oecologia 149:44–51.

Ankney, C. D. 1996. Why did the ducks come back in 1994 and 1995: was Johnny Lynch right? Proceedings of the Seventh International Waterfowl Symposium 7:40–44.

Bakker, V. J., and D. F. Doak. 2009. Population viability management: ecological standards to guide adaptive management for rare species. Frontiers in Ecology and Environment 7:158–165.

Bakker, V. J., D. F. Doak, G. W. Roemer, D. K. Garcelon, T. J. Coonan, S. A. Morrison, C. Lynch, K. Ralls, and R. Shaw. 2009. Estimating and incorporating ecological drivers and parameter uncertainty into a demographic population viability analysis for the Island Fox (*Urocyon littoralis*). Ecological Monographs 79:77–108.

Battin, J. 2004. When good animals love bad habitats: ecological traps and the conservation of animal populations. Conservation Biology 18:1482–1491.

Bekoff, M., and M. C. Wells. 1986. Social ecology and behavior of coyotes. Advances in the Study of Behavior 16:251–338.

Berger, K. M., and E. M. Gese. 2007. Does interference competition with wolves limit the distribution and abundance of coyotes? Journal of Animal Ecology 76:1075–1085.

Bongaarts, J. 2009. Human population growth and the demographic transition. Philosophical Transactions of the Royal Society B 364:2985–2990.

Bourbeau-Lemieux, A., M. Festa-Bianchet, J.-M. Gaillard, and F. Pelletier. 2011. Predator-driven component Allee effects in a wild ungulate. Ecology Letters 14:358–363.

Boutin, S., D. L. Haughland, J. Schieck, J. Herbers, and E. Bayne. 2009. A new approach to forest biodiversity monitoring in Canada. Forest Ecology and Management 258:S168–S175.

Bouzat, J. L., J. A. Johnson, J. E. Toepfer, S. A. Simpson, T. L. Esker, and R. L. Westemeier. 2009. Beyond the beneficial effects of translocations as an effective tool for the genetic restoration of isolated populations. Conservation Genetics 10:191–201.

Boyce, M. S. 1977. Population growth with stochastic fluctuations in the life table. Theoretical Population Biology 12:366–373.

Boyd, D. K., and D. H. Pletscher. 1999. Characteristics of dispersal in a colonizing wolf population in the central Rocky Mountains. Journal of Wildlife Management 63:1094–1108.

Brook, B. W., N. S. Sodhi, and C. J. A. Bradshaw. 2008. Synergies among extinction drivers under global change. Trends in Ecology and Evolution 23:453–460.

Cassirer, E. F., and A. R. E. Sinclair. 2007. Dynamics of pneumonia in a bighorn sheep population. Journal of Wildlife Management 71:1080–1088.

Cole, L. C. 1957. Sketches of general and comparative demography. Cold Spring Harbor Symposia on Quantitative Biology 22:1–15.

Crooks, K. R., and M. Sanjayan, editors. 2006. Connectivity conservation. Cambridge University Press, Cambridge, United Kingdom.

Crone, E. 2001. Is survivorship a better fitness surrogate than fecundity? Evolution 55:2611–2614.

Crouse, D. T., L. B. Crowder, and H. Caswell. 1987. A stage-based population model for loggerhead sea turtles and implications for conservation. Ecology 68:1412–1423.

Danielsen, F., N. D. Burgess, A. Balmford, P. F. Donald, M. Funder, et al. 2009. Local participation in natural resource monitoring: a characterization of approaches. Conservation Biology 23:31–42.

Deevey, E. S., Jr. 1947. Life tables for natural populations of animals. Quarterly Review of Biology 22:283–314.

Dennis, B., P. L. Munholland, and J. M. Scott. 1991. Estimation of growth and extinction parameters for endangered species. Ecological Monographs 61:115–143.

Dennis, B., and M. R. M. Otten. 2000. Joint effects of density dependence and rainfall on abundance of San Joaquin kit fox. Journal of Wildlife Management 64:388–400.

Doak, D. F., W. F. Morris, C. Pfister, B. E. Kendall, and E. M. Bruna. 2005. Correctly estimating how environmental stochasticity influences fitness and population growth. American Naturalist 166:E14–E21.

Drajem, B. 2011. A cougar in Connecticut. Science News. Web Edition 2:August.

Ehrlich, P. R. 1968. The population bomb. Ballantine Books, New York, New York, USA.

Ehrlich, P. R., and H. H. Ehrlich. 2009. The population bomb revisited. Electronic Journal of Sustainable Development 1:63–71.

Ehrlich, P. R., and J. P. Holdren. 1971. Impact of population growth. Science 171:1212–1217.

Elton, C. 1924. Periodic fluctuations in the numbers of animals: their causes and effects. British Journal of Experimental Biology 2:119–163.

Emlen, S. T. 1990. White-fronted bee-eaters: helping in a colonially nesting species. Pages 489–526 *in* R. B. Stacey, and W. D. Koenig, editors. Cooperative breeding in birds: long-term studies of ecology and behavior. Cambridge University Press, Cambridge, United Kingdom.

Errington, P. L. 1956. Factors limiting higher vertebrate populations. Science 124:304–307.

Festa-Bianchet, M. 1988. A pneumonia epizootic in bighorn sheep, with comments on preventative management. Proceedings of Biennial Symposium of the Northern Wild Sheep and Goat Council 6:66–76.

Field, S. A., A. J. Tyre, N. Jonzén, J. R. Rhodes, and H. P. Possingham. 2004. Minimizing the cost of environmental management decisions by optimizing statistical threshold. Ecology Letters 7:669–675.

Fortin, D., H. L. Beyer, M. S. Boyce, D. W. Smith, T. Duchesne, and J. S. Mao. 2005. Wolves influence elk movements: behavior shapes a trophic cascade in Yellowstone National Park. Ecology 86:1320–1330.

Funk, W. C., A. E. Greene, P. S. Corn, and F. W. Allendorf. 2005. High dispersal in a frog species suggests that it is vulnerable to habitat fragmentation. Biology Letters 1:13–16.

Gaillard, J.-M., M. Festa-Bianchet, and N. G. Yoccoz. 1998. Population dynamics of large herbivores: variable recruitment with constant adult survival. Trends in Ecology and Evolution 13:58–63.

George, J. L., D. J. Martin, P. M. Lukacs, and M. W. Miller. 2008. Epidemic pasteurellosis in a bighorn sheep population coinciding with the appearance of a domestic sheep. Journal of Wildlife Diseases 44:388–403.

Gibbs, J. P., S. Droege, and P. Eagle. 1998. Monitoring populations of plants and animals. Bioscience 48:935–940.

Gregory, S. D., C. J. A. Bradshaw, B. W. Brook, and F. Courchamp. 2010. Limited evidence for the demographic Allee effect from numerous species across taxa. Ecology 91:2151–2161.

Grenier, M. B., D. B. McDonald, and S. W. Buskirk. 2007. Rapid population growth of a critically endangered carnivore. Science 317:779.

Griffin, P. C., and L. S. Mills. 2003. Snowshoe hares in a dynamic managed landscape. Pages 438–449 in H. R. Akcakaya, M. A. Burgman, O. Kindvall, C. Wood, P. Sjogren-Gulve, J. Hatfield, and M. A. McCarthy, editors. Species conservation and management: case studies. Oxford University Press, Oxford, United Kingdom.

Griffin, P. C., and L. S. Mills. 2009. Sinks without borders: snowshoe hare dynamics in a complex landscape. Oikos 118:1487–1498.

Haddad, N. M., B. Hudgens, E. I. Damschen, D. J. Levey, J. L. Orrock, J. J. Tewksbury, and A. J. Weldon. 2011. Assessing positive and negative ecological effects of corridors. Pages 475–503 in J. Liu, V. Hull, A. Morzillo, and J. Wiens, editors. Sources, sinks, and sustainability across landscapes. Cambridge University Press, Cambridge, United Kingdom.

Hamel, S., and S. D. Côté. 2007. Habitat use patterns in relation to escape terrain: are alpine ungulate females trading off better foraging sites for safety? Canadian Journal of Zoology 85:933–943.

Harper, E. B., T. A. G. Rittenhouse, and R. D. Semlitsch. 2008. Demographic consequences to terrestrial habitat loss for pool-breeding amphibians: predicting extinction risks associated with inadequate size of buffer zones. Conservation Biology 22:1205–1215.

Hartway, C., and L. S. Mills. 2012. A meta-analysis assessing the effects of common management actions on the nest success of North American birds. Conservation Biology 26:657–666.

Hebblewhite, M., C. White, and M. Musiani. 2009. Revisiting extinction in national parks: mountain caribou in Banff. Conservation Biology 24:341–344.

Heppell, S. S., L. B. Crowder, and D. T. Crouse. 1996. Models to evaluate headstarting as a management tool for long-lived turtles. Ecological Applications 6:556–565.

Holling, C. S. 1959. The components of predation as revealed by a study of small-mammal predation of the European pie sawfly. Canadian Entomologist 91:293–320.

Holroyd, G. L., C. J. Conway, and H. E. Trefry. 2011. Breeding dispersal of a burrowing owl from Arizona to Saskatchewan. Wilson Journal of Ornithology 123:378–381.

Humbert, J.-Y., L. S. Mills, J. S. Horne, and B. Dennis. 2009. A better way to estimate population trend. Oikos 118:1487–1498.

Hutchins, H. E., and R. M. Lanner. 1982. The central role of Clark's nutcracker in the dispersal and establishment of whitebark pine. Oecologia 55:192–201.

Jarvis, J. U. M., N. C. Bennett, and A. Spinks. 1998. Food availability and foraging by wild colonies of Damaraland mole-rats (Cryptomys amarensis): implications for sociality. Oecologia 113:290–298.

Johnson, H. E., M. Hebblewhite, T. R. Stephenson, D. W. German, B. M. Pierce, and V. C. Bleich. 2013. Evaluating apparent competition in limiting the recovery of an endangered ungulate. Oecologia 171:295–307.

Johnson, H. E., L. S. Mills, J. Wehausen, and T. R. Stephenson. 2010a. Population-specific vital rate contributions influence management of an endangered ungulate. Ecological Applications 20:1753–1765.

Johnson, H. E., L. S. Mills, J. Wehausen, and T. R. Stephenson. 2010b. Combining ground count, telemetry, and mark-resight data to infer population dynamics in an endangered species. Journal of Applied Ecology 47:1083–1093.

Johnson, H. E., L. S. Mills, J. Wehausen, T. R. Stephenson, and G. Luikart. 2011. Translating inbreeding depression on component vital rates to overall population growth in endangered bighorn sheep. Conservation Biology 25:1240–1249.

Keith, L. B. 1990. Dynamics of snowshoe hare populations. Pages 119–195 in H. H. Genoways, editor. Current mammalogy. Plenum, New York, New York, USA.

Kelly, M. A., J. Betsch, C. Wultsch, B. Mesa, and L. S. Mills. 2012. Noninvasive sampling for carnivores. Pages 47–69 in L. Boitani, and R. Powell, editors. Carnivore ecology and conservation. Oxford University Press, Oxford, United Kingdom.

Koons, D. N., R. F. Rockwell, and J. B. Grand. 2006. Population momentum: implications for wildlife management. Journal of Wildlife Management 70:19–26.

Kramer, A. M., B. Dennis, A. M. Liebhold, and J. M. Drake. 2009. The evidence for Allee effects. Society of Population Ecology 51:341–354.

Krebs, C. J., R. Boonstra, S. Boutin, and A. R. E. Sinclair. 2001. What drives the 10-year cycle of snowshoe hares? BioScience 51:25–35.

Krebs, C. J., S. Boutin, R. Boonstra, A. R. E. Sinclair, J. N. M. Smith, M. R. T. Dale, K. Martin, and R. Turkington. 1995. Impact of food and predation on the snowshoe hare cycle. Science 269:1112–1115.

Kuussaari, M., I. Saccheri, M. Camara, and I. Hanski. 1998. Allee effects and population dynamics in the Glanville fritillary butterfly. Oikos 82:384–392.

Lande, R., S. Engen, and B.-E. Sæther. 2003. Stochastic population dynamics in ecology and conservation. Oxford University Press, Oxford, United Kingdom.

Landeau, L., and J. Terborgh. 1986. Oddity and the "confusion effect" in predation. Animal Behaviour 34:1372–1380.

Laundre, J., and T. W. Clark. 2003. Managing puma hunting in the western United States: through a metapopulation approach. Animal Conservation 6:159–170.

Levins, R. 1970. Extinction. Lectures on Mathematics in the Life Sciences 2:75–107.

Lidicker, W. Z., Jr. 1975. The role of dispersal in the demography of small mammals. Pages 103–128 in B. Golley, K. Petruscwicz, and L. Ryszkowski, editors. Small mammals: their productivity and population dynamics. Cambridge University Press, Cambridge, United Kingdom.

Lindenmayer, D. B., and G. E. Likens. 2009. Adaptive monitoring: a new paradigm for long-term research and monitoring. Trends in Ecology and Evolution 24:482–486.

Logan, K. A., and L. L. Sweanor. 2001. Desert puma: evolutionary ecology and conservation of an enduring carnivore. Island Press, Washington, D.C., USA.

Long, R. A., P. MacKay, W. J. Zielinski, and J. C. Ray. 2008. Noninvasive survey methods for carnivores. Island Press, Washington D.C., USA.

Lorenz, T. J., K. A. Sullivan, A. V. Bakian, and C. A. Aubry. 2011. Cachesite selection in Clark's nutcracker (Nucifraga columbiana). Auk 128:237–247.

Loveridge, A. J., A. W. Searle, F. Murindagomo, and D. W. Macdonald. 2007. The impact of sport-hunting on the population dynamics of an African lion population in a protected area. Biological Conservation 134:548–558.

Lowe, W. H. 2003. Linking dispersal to local population dynamics: a case study using a headwater salamander system. Ecology 84:2145–2154.

Lowe, W. H., and F. W. Allendorf. 2010. What can genetics tell us about population connectivity? Molecular Ecology 19:3038–3051.

Lutz, W., W. Sanderson, and S. Scherbov. 2001. The end of world population growth. Nature 412:543–545.

Lyons, J. E., M. C. Runge, H. P. Laskowski, and W. L. Kendall. 2008. Monitoring in the context of structured decision-making and adaptive management. Journal of Wildlife Management 72:1683–1692.

MacKenzie, D. I., J. D. Nichols, J. A. Royle, K. H. Pollock, L. L. Bailey, and J. E. Hines. 2006. Occupancy estimation and modeling. Elsevier, Oxford, United Kingdom.

MacKenzie, D. I., and J. A. Royle. 2005. Designing occupancy studies: general advice and allocating survey efforts. Journal of Applied Ecology 42:1105–1114.

Marsh, D. M., and P. C. Trenham. 2008. Current trends in plant and animal population monitoring. Conservation Biology 22:647–655.

Martin, K. D., T. Schommer, and V. L. Coggins. 1996. Literature review regarding the compatibility between bighorn and domestic sheep. Biennal Symposium of the Northern Wild Sheep and Goat Council 10:72–77.

Mills, L. S. 2007. Conservation of wildlife populations: demography, genetics, and management. Wiley-Blackwell, Malden, Massachusetts, USA.

Mills, L. S. 2013. Conservation of wildlife populations: demography, genetics, and management. Second edition. Wiley-Blackwell, Malden, Massachusetts, USA.

Mills, L. S., and F. W. Allendorf. 1996. The one-migrant-per-generation rule in conservation and management. Conservation Biology 10:1509–1518.

Mills, L. S., D. F. Doak, and M. J. Wisdom. 1999. The reliability of conservation actions based on sensitivity analysis of matrix models. Conservation Biology 13:815–829.

Mills, L. S., M. K. Schwartz, D. A. Tallmon, and K. P. Lair. 2003. Measuring and interpreting connectivity for mammals in coniferous forests. Pages 587–613 in C. J. Zabel and R. G. Anthony, editors. Mammal community dynamics: management and conservation in the coniferous rorests of western North America. Cambridge University Press, Cambridge, United Kingdom.

Mills, L. S., J. M. Scott, K. M. Strickler, and S. A. Temple. 2012. Ecology and management of small populations. Pages 270–292 in N. J. Silvy, editor. The wildlife techniques manual: management. Seventh edition. John Hopkins University Press, Baltimore, Maryland, USA.

Mills, L. S., and D. A. Tallmon. 1999. The role of genetics in understanding forest fragmentation. Pages 171–184 in J. A. Rochelle, L. A. Lehmann, and J. Wisniewski, editors. Forest fragmentation: wildlife and management implications. Brill Academic, Leiden, Netherlands.

Mooring, M. S., T. A. Fitzpatrick, T. T. Nishihira, and D. D. Reisig. 2004. Vigilance, predation risk, and the Allee effect in desert bighorn sheep. Journal of Wildlife Management 68:519–532.

Morris, W. F., and D. F. Doak. 2002. Quantitative conservation biology: theory and practice of population viability analysis. Sinauer Associates, Sunderland, Massachussetts, USA.

Newby, J. R., L. S. Mills, T. K. Ruth, D. H. Pletscher, M. S. Mitchell, H. B. Quigly, K. M. Murphy, and R. DeSimone. 2013. Human-caused mortality influences spatial population dynamics: pumas in landscapes with varying mortality risks. Biological Conservation. In press.

Nichols, J. D., and J. E. Hines. 2002. Approaches for the direct estimation of λ, and demographic contributions to λ, using capture-recapture data. Journal of Applied Statistics 29:539–568.

Nichols, J. D., F. A. Johnson, and B. K. Williams. 1995. Managing North American waterfowl in the face of uncertainty. Annual Review of Ecology and Systematics 26:177–199.

Nichols, J. D., and B. K. Williams. 2006. Monitoring for conservation. Trends in Ecology and Evolution 21:668–673.

Oyler-McCance, S. J., and P. L. Leberg. 2012. Conservation genetics and molecular ecology in wildlife management. Pages 526–546 in N. J. Silvy, editor. The wildlife techniques manual: research. Seventh edition. John Hopkins University Press, Baltimore, Maryland, USA.

Ozgul, A., K. B. Armitage, D. T. Blumstein, and M. K. Oli. 2006. Spatiotemporal variation in survival rates: implications for population dynamics of yellow-bellied marmots. Ecology 87:1027–1037.

Parry, M. L., N. Adger, P. Aggarwal, S. Agrawala, J. Alcamo, et al. 2007. Technical summary. Pages 23–78 in M. L. Parry, O. F. Canziani, J. P. Palutikof, P. J. van der Linden, and C. E. Hanson, editors. Climate change 2007: impacts, adaptation and vulnerability. Contribution of Working Group II to the fourth assessment report of the Intergovernmental Panel on Climate Change. Cambridge University Press, Cambridge, United Kingdom.

Peacock, E., and D. L. Garshelis. 2006. Comment on "On the regulation of populations of mammals, birds, fish, and insects" IV. Science 313:45.

Pedersen, A. B., K. E. Jones, C. L. Nunn, and S. A. Altizer. 2007. Infectious disease and mammalian extinction risk. Conservation Biology 21:1269–1279.

Post, E., N. C. Stenseth, R. O. Peterson, J. A. Vucetich, and A. M. Ellis. 2002. Phase dependence and population cycles in a large-mammal predator-prey system. Ecology 83:2997–3002.

Proctor, M. F., D. Paetkau, B. N. McLellan, G. B. Stenhouse, K. C. Kendall, et al. 2012. Population fragmentation and inter-ecosystem movements of grizzly bears in western Canada and the northern United States. Wildlife Monographs 180:1–46.

Pulliam, H. R. 1988. Sources, sinks, and population regulation. American Naturalist 132:652–661.

Rasa, O. A. E. 1989. The costs and effectiveness of vigilance behavior in the Dwarf Mongoose: implications for fitness and optimal group size. Ethology Ecology and Evolution 1:265–282.

Raveling, D. G. 1989. Nest-predation rates in relation to colony size of black brant. Journal of Wildlife Management 53:87–90.

Reed, J. M. 1999. The role of behavior in recent avian extinctions and endangerments. Conservation Biology 13:232–241.

Robertson, B. A., and R. L. Hutto. 2006. A framework for understanding ecological traps and an evaluation of existing evidence. Ecology 87:1075–1085.

Robinson, H. S., R. B. Wielgus, H. S. Cooley, and S. W. Cooley. 2008. Sink populations in carnivore management: cougar demography and immigration in a hunted population. Ecological Applications 18:1028–1037.

Robinson, H. S., and R. M. DeSimone. 2011. The Garnet Range Mountain Lion Study: characteristics of a hunted population in west-central Montana. Final report. Montana Department of Fish, Wildlife and Parks, Helena, USA.

Royama, T. 1992. Analytical population dynamics. Chapman and Hall, London, United Kingdom.

Runge, J. P., M. C. Runge, and J. D. Nichols. 2006. The role of local populations within a landscape context: defining and classifying sources and sinks. American Naturalist 167:925–938.

Running, S., and L. S. Mills. 2009. Terrestrial ecosystem adaptation. Resources for the Future, Washington, D.C., USA. http://www.rff.org/News/Features/Pages/09-07-08-Managing-for-Resilience.aspx. Accessed 14 November 2012.

Sæther, B.-E., and Ø. Bakke. 2000. Avian life history variation and contribution of demographic traits to the population growth rate. Ecology 81:642–653.

Schultz, C. 2010. Challenges in connecting cumulative effects analysis to effective wildlife conservation planning. BioScience 60:545–551.

Schwartz, M. K., L. S. Mills, K. S. McKelvey, L. F. Ruggiero, and F. W. Allendorf. 2002. DNA reveals high dispersal synchronizing the population dynamics of Canada lynx. Nature 415:520–522.

Scott, J. M., D. D. Goble, J. A. Wiens, D. S. Wilcove, M. Bean, and T. Male. 2005. Recovery of imperiled species under the Endangered Species Act: the need for a new approach. Frontiers in Ecology and the Environment 3:383–389.

Sibly, R. M., D. Barker, M. C. Denham, J. Hone, and M. Pagel. 2005. On the regulation of populations of mammals, birds, fish and insects. Science 309:607–610.

Singer, F. J., E. Williams, M. W. Miller, and L. C. Zeigenfuss. 2000. Population growth, fecundity, and survivorship in recovering populations of bighorn sheep. Restoration Ecology 8:75–84.

Smith, K. F., K. Acevedo-Whitehouse, and A. B. Pedersen. 2009. The role of infectious diseases in biological conservation. Animal Conservation 12:1–12.

Staples, D. F., M. L. Taper, and B. B. Shepard. 2005. Risk-based viable population monitoring. Conservation Biology 19:1908–1916.

Stirling, I., T. L. McDonald, E. S. Richardson, E. V. Regehr, and S. C. Amstrup. 2011. Polar bear population status in the northern Beaufort Sea, Canada, 1971–2006. Ecological Applications 21:859–876.

Stoner, D. C., M. L. Wolfe, and D. M. Choate. 2006. Cougar exploitation levels in Utah: implications for demographic structure, population recovery, and metapopulation dynamics. Journal of Wildlife Management 70:1588–1600.

Sweanor, L. L., K. A. Logan, and M. G. Hornocker. 2000. Cougar dispersal patterns, metapopulation dynamics, and conservation. Conservation Biology 14:798–808.

Taylor, B. L., and T. Gerrodette. 1993. The uses of statistical power in conservation biology: the vaquita and northern spotted owl. Conservation Biology 7:489–500.

Taylor, B. L., P. R. Wade, R. A. Stehn, and J. F. Cochrane. 1996. A Bayesian approach to classification criteria for spectacled eiders. Ecological Applications 6:1077–1089.

U.S. Fish and Wildlife Service. 2007. Recovery plan for the Sierra Nevada bigorn sheep, Sacramento, California, USA.

Van Horne, B. 1983. Density as a misleading indicator of habitat quality. Journal of Wildlife Management 47:893–901.

Veit, R. R., and M. A. Lewis. 1996. Dispersal, population growth, and the allee effect: dynamics of the house finch invasion of eastern North America. American Naturalist 148:255–274.

Vucetich, J. A., and T. A. Waite. 2000. Is one migrant per generation sufficient for the genetic management of fluctuating populations? Animal Conservation 3:261–266.

Vucetich, J. A., T. A. Waite, L. Qvarnemark, and S. Ibargüen. 2000. Population variability and extinction risk. Conservation Biology 14:1704–1714.

Weldon, A. J., and N. M. Haddad. 2005. The effects of patch shape on indigo buntings: evidence for an ecological trap. Ecology 86:1422–1431.

White, P. J., and R. A. Garrott. 1999. Population dynamics of kit foxes. Canadian Journal of Zoology 77:486–493.

Wilcove, D. S., D. Rothstein, J. Dubow, A. Phillips, and E. Losos. 1998. Quantifying threats to imperiled species in the United States. Bioscience 48:607–615.

Williams, B. K. 2001. Uncertainty, learning, and optimization in wildlife management. Environmental and Ecological Statistics 8:269–288.

Williams, J. D., and J. J. McDermott. 2004. Hermit crab biocoenoses: a worldwide review of the diversity and natural history of hermit crab associates. Journal of Experimental Marine Biology and Ecology 305:1–128.

Williams, B. K., J. D. Nichols, and M. J. Conroy. 2002. Analysis and management of animal populations. Academic Press, San Diego, California, USA.

Wirsing, A. J., T. D. Steury, and D. L. Murray. 2002. A demographic analysis of a southern snowshoe hare population in a fragmented habitat: evaluating the refugium model. Canadian Journal of Zoology 80:169–177.

Wisdom, M. J., L. S. Mills, and D. F. Doak. 2000. Life-stage simulation analysis: estimating vital rate effects on population growth for conservation. Ecology 81:628–641.

Wittmer, H. U., B. N. McLellan, R. Serrouya, and C. D. Apps. 2007. Changes in landscape composition influence the decline of a threatened woodland caribou population. Journal of Animal Ecology 76:568–579.

Wittmer, H. U., A. R. E. Sinclair, and B. N. McLellan. 2005. The role of predation in the decline and extirpation of woodland caribou. Oecologia 144:257–262.

Yoccoz, N. G., J. D. Nichols, and T. Boulinier. 2001. Monitoring of biological diversity in space and time. Trends in Ecology and Evolution 16:446–453.

8 DISEASES AND PARASITES

DAVID A. JESSUP

INTRODUCTION

"The role of disease in wildlife conservation has probably been radically underestimated" (Leopold 1933:325). Eighty years later, the specter of disease still exerts a major influence on state, federal, and private wildlife management activities. It occurs against a background of increasing human population densities and habitat loss and degradation. New diseases have emerged, and old ones have unfortunately been moved around. Not only has there been an apparent increase in the incidence of wildlife disease, but there has been an awareness raised of the implications for domestic animal and human health, as well as a push for social, political, and legal means to control and limit wildlife diseases.

A more traditional view is that wildlife populations can compensate for disease losses by increased reproduction and survival of the young; thus disease losses are of little importance to wildlife management. There are some notable historical exceptions. Rinderpest virus, introduced into Africa from Asia around the turn of the 20th century, caused such massive mortality in cattle and across wild ungulate species, that the ecology and economy of southern and eastern Africa were altered for nearly 100 years. There have been several examples in North America of major wildlife die-offs or chronic disease problems that have had serious impacts on the harvest of wildlife. In the latter half of the 20th century, type C botulism and avian cholera deaths were of such a magnitude in some areas that they curtailed the harvest of some species of waterfowl. For a few years in the 1950s and 60s, bluetongue (actually two viral diseases) in white-tailed deer (*Odocoileus virginianus*) was believed to be a serious threat to deer population recovery in the southeastern United States, and even today, major outbreaks can result in altered hunting regulations.

The most striking recent examples of the potential impacts of wildlife disease at the population level do not involve harvested species. These examples are the global mass mortality—and in some cases extinction of many amphibian species because of chythrid (*Batrachochytrium dendrobatidis*) fungus infections (Berger et al. 1998)—and massive bat mortality in North America caused by white-nose syndrome (Fig. 8.1), another fungal (*Geomyces destructans*) disease (Frick et al. 2010). Ironically and somewhat mysteriously, fungal organisms are generally considered neither highly infectious nor highly pathogenic, so the emergence and rapid spread of both of these diseases may be related to larger phenomena (e.g., immune function effects of climate change or contaminants) that compromise disease resistance.

With disease—as with other areas of conservation—controversy, public perception, and social values often override biological realities. For example, there has never been a single case of chronic wasting disease (CWD; Fig. 8.2) diagnosed in humans, yet the perception that deer (*Odocoileus* spp.), elk (*Cervus canadensis*), moose (*Alces alces*), and other cervids might harbor an untreatable prion organism similar to mad cow disease, and early-onset Creutzfeldt-Jakob disease in humans, has

Figure 8.1. Bats showing signs of white-nose syndrome, a highly fatal fungal infection that has spread rapidly across the United States. The ecological consequences of losing large numbers of insectivorous and plant-pollinating bats may be enormous. Provided by the U.S. Geological Survey National Wildlife Health Center

affected big game hunting across North America. The emergence and persistence of this disease has also greatly retarded the development of the cervid game ranching industry. The harvest of wildlife is big business, in part because people view wild meat as a healthy and desirable luxury. If that perception is significantly altered, and harvested wildlife are viewed as unhealthy or diseased, the financial consequences could be dire for recreation-based communities, hunting and fishing equipment manufacturers and suppliers, and government wildlife management agencies whose budgets are supported by hunting, licenses, and taxes.

The interface between public health and wildlife disease is another area where perception is reality. Rabies in human beings is a terrible and almost uniformly fatal viral disease, but it is exceedingly rare in North America. Vaccination greatly diminished the role domestic animals held in spreading rabies

Figure 8.2. Captive elk showing signs of chronic wasting disease: depression, emaciation, drooling, incoordination. Even liberal and either-sex hunting seasons in affected areas has not slowed the spread of this prion disease. Provided by Beth Williams

in the 1960s and 70s, and wildlife inherited the weight of social and political commitment to stamp it out. The result was decades of studies on rabies ecology, development of a vaccine and bait technologies (Fig. 8.3), laboratory and field trials, and field applications that yielded positive results in controlling rabies in wildlife. Government agencies (i.e., U.S. Department of Agriculture, or USDA; Centers for Disease Control and Prevention; and state public health agencies), private foundations, and universities headed these efforts. Because rabies is a public health issue, traditional wildlife management agencies have had limited involvement. Strictly from a traditional wildlife management perspective, rabies is not of significance to wildlife populations, but it is a clear example that wildlife management cannot, and should not, exist in a vacuum.

Two bacterial diseases that were once common in cattle—bovine tuberculosis (*Mycobacterium bovis*, or TB) and bovine brucellosis (*Brucella abortus*)—are now endemic in a few populations of free-ranging ungulates in North America. In an effort to make the meat and milk supply safe, and to gain a favorable export market position at considerable expense, these diseases were essentially eliminated from livestock in North America in the 21st century. But several deer populations in the upper midwestern United States are persistently infected with TB. Elk and bison (*Bison bison*) populations in the greater Yellowstone region are infected with brucellosis. Elk, deer, and bison in two parks in Canada have brucellosis and TB. These remaining pockets of infection in wildlife have become an important management issue, even though they pose little or no threat to populations of deer, elk, or bison.

There is great financial and political pressure to eliminate these diseases from wildlife populations. One result has been animosity between wildlife and agriculture agencies, particularly as the involvement of the latter in wildlife management programs has increased. State, national, and regional efforts to involve stakeholders have required consideration of various

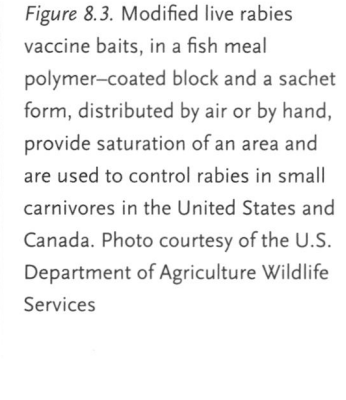

Figure 8.3. Modified live rabies vaccine baits, in a fish meal polymer–coated block and a sachet form, distributed by air or by hand, provide saturation of an area and are used to control rabies in small carnivores in the United States and Canada. Photo courtesy of the U.S. Department of Agriculture Wildlife Services

perspectives, including those of ecologists, wildlife managers, hunters, and recreation industry representatives, economists, the livestock industry, large corporations, and international trade representatives. These stakeholders have proposed—and even tried—some expensive, aggressive, and controversial proposals for elimination of TB and brucellosis. For example, the capture, testing, and quarantine of all bison in Yellowstone National Park have been offered as a way to eliminate brucellosis. Elimination of these diseases from wild ungulate populations across vast landscapes, however, might not be possible with current technology, applications, and attitudes. Reducing risks of transmission and disruption of business may be more viable options.

The efforts to manage these diseases have revealed serious problems in existing wildlife management practices. At Riding Mountain National Park in Canada, focused hunting and culling to reduce numbers and the proportion of older animals and to reduce contact with livestock have reduced the prevalence of TB in deer and elk and the incidence of infection in adjacent cattle. In contrast, wildlife managers have been unable to halt baiting of deer (with TB) in Michigan and the supplementary winter feeding of elk (with brucellosis) in Wyoming. Feeding wildlife fosters artificially high densities and concentrations of animals and increases the risk of contact with livestock. Both of these factors increase potential for disease transmission. Although increased hunting pressure has been used in Michigan and Wyoming, culling has not been widely implemented, and neither has proven very effective. In fairness, the large areas and complexities of ownership and access are at least an order of magnitude greater in Michigan and Wyoming than in Riding Mountain National Park. Vegetation succession, loss of winter range and migration corridors to ranching and development, urban sprawl, and balancing predator control and the potential culling benefits of predation are additional major management challenges that can affect disease control efforts. Brucellosis and TB have forced wildlifers to question some fundamental management assumptions and programs, and they have required that wildlife professionals deal with financial, political, and social realities when seeking to limit or eliminate undesirable effects of wildlife diseases.

CASE STUDIES

I present six case studies of wildlife diseases to illustrate agent–host–environment relationships, emergence or recognition of disease problems, how they are diagnosed and managed, and which agencies or organizations in North America are responsible for their management.

Avian Cholera

Avian cholera (*Pasteurella multocida* septicemia) possibly existed in North American waterfowl before it was described in California in the 1940s and 50s, and then in the Rainwater Basin of Nebraska in the 1950s and 60s (Friend 2006). However, it is unlikely that an epidemic that killed thousands of waterfowl and that had such recognizable, gross lesions would

have been missed (Jessup 1986). Most current evidence suggests that the *P. multocida* of waterfowl came from domestic turkeys or chickens, where it had been a recognized cause of disease for many years. *P. multocida* bacteria colonize the upper respiratory tract and invade the bloodstream, where they reproduce in large numbers and are spread to the body's organs (Jessup 1986). Clots of bacteria plug blood vessels, and the toxins produced by dying bacteria cause tissue damage and septic shock. Infected waterfowl die very quickly, sometimes literally falling from the sky (Friend 2006).

How and why did this virulent strain of bacteria emerge, spreading steadily over 50 years to become one of the scourges of North American waterfowl? Avian cholera is transmitted from bird to bird via respiratory droplets, mucus from dead and dying birds, and in feces. As such, it is a disease of confinement and population concentration. Wintering waterfowl have always occurred in dense flocks, but the losses of marshlands (90–95%) by midcentury in California and elsewhere led to much denser concentrations. Heavy rains, cold weather, and storms have been posited as potential stressing factors, but avian cholera occurs in good and bad weather. Reduced immune function in a population of waterfowl might explain increased susceptibility to an upper respiratory bacterial pathogen. The temporal emergence of avian cholera coincides reasonably well with the widespread use of dichlorodiphenyltrichloroethane (DDT) and other chlorinated hydrocarbon insecticides to control mosquito-borne diseases and crop pests. Mallard ducks (*Anas platyrhynchos*) dosed with petrochemicals are more susceptible to fatal infection with *P. multocida* (Friend 2006), but little evidence exists to support chemical-induced immune suppression as a factor in most outbreaks in free-flying waterfowl. Some evidence suggests that environmental conditions and water characteristics prolong bacterial survival outside the host. And some species, notably snow geese (*Chen caerulescens*), appear to be capable of carrying the bacteria back to summer nesting areas and are often the first species to show mortality when winter outbreaks occur (Friend 2006). In the mid-20th century, fundamental changes occurred in the parasite–host–environment relationship regarding waterfowl, *P. multocida*, and North American waterfowl habitat; that legacy remains today.

Ducks, geese, and swans with avian cholera die acutely and peracutely and are almost always in good to excellent body condition (Jessup 1986). The gross lesions consist of multiple small (i.e., petechial to ecchymotic) hemorrhages across the surface of the heart and on great vessels, and multiple small white spots (septic infarcts) across the liver (Fig. 8.4; Davidson 2006). A simple blood smear stained with Wright's or Giemsa can reveal large numbers of bipolar staining, rod-shaped bacteria in the blood of infected birds. A presumptive diagnosis of fowl cholera is warranted for waterfowl in good body condition with characteristic lesions and with positive blood smear stains (Davidson 2006). Although it is beneficial to do full postmortem examination, bacterial culture, and perhaps histopathology, these tests are impractical when hundreds to thousands of birds are dying during an outbreak. One problem with

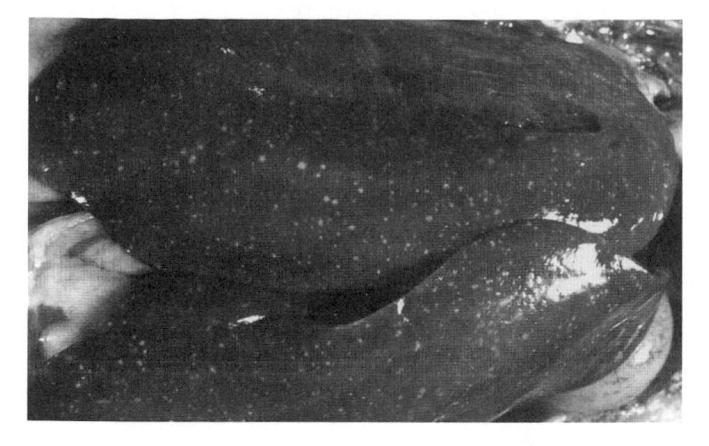

Figure 8.4. The lesions of fowl cholera (*Pasteurella multocida* septicemia) are relatively obvious on gross postmortem examination, but similar lesions can be caused by avian influenza and duck virus enteritis. Affected ducks, geese, and swans usually die peracutely, in good body condition, have small (petechial to ecchymotic) hemorrhages on the heart and great vessels, and have multiple small white spots (septic infarcts) on the liver (shown). A blood smear stained with Wrights or Geimsa reveals many bipolar staining rod-shaped bacteria. Provided by Cornell University

taking shortcuts when diagnosing avian cholera is that two dangerous viral diseases (i.e., avian influenza and duck virus enteritis) cause similar, often indistinguishable gross lesions (Jessup 1986, Friend 2006) and can occur under the same temporal and environmental circumstances. A complete postmortem examination of a subset of fresh dead birds conducted at a state, federal, or university diagnostic laboratory is recommended to verify the cause of disease-related waterfowl die-offs.

Avian cholera often occurs in public and private waterfowl hunting areas during hunting season. Efforts to manage avian cholera have largely consisted of picking up carcasses to prevent seeding the environment with more bacteria and so that dead birds do not act as decoys attracting live birds (Wobeser 1994). State, provincial, and federal agency personnel, often assisted by volunteers and waterfowl advocacy groups, coordinate carcass removal. Avian cholera is a relatively easy disease to diagnose, but management has sometimes proven frustrating and unrewarding. There is no strong body of evidence to show that carcass collection is even effective (Wobeser 1994). Currently, various constituencies prefer action over inaction when large outbreaks occur. Keeping waterfowl populations dispersed across a flyway may also be of some benefit.

Brucellosis

In the 19th and early 20th centuries, cattle infections with the bacteria *Brucella abortus* were relatively common, as was the resulting human disease, undulant fever. As the bacteria's Latin name suggests, the most common sign of infection in cattle was abortion. Brucellosis was considered an occupational hazard for stockmen and for those who consumed unpasteurized milk. Efforts to eradicate brucellosis focused on test and quarantine, test and slaughter, farm sanitation,

and milk pasteurization. Because the focus of the disease was livestock and public health, the agencies in charge of brucellosis programs were within the USDA and the various state agriculture and livestock agencies. Sometime in the late 19th century, elk and bison became infected, possibly in a number of locations. Since that time, brucellosis has persisted in Yellowstone National Park and surrounding areas (Williams and Barker 2001). As eradication efforts in livestock progressed, the remaining infections in wildlife became the focus of research and control efforts. In the 1970s and 80s, the Wyoming Game and Fish Department (WGFD) began a series of landmark experiments at its Sybille Wildlife Research Center. These experiments revealed the infectivity of aborted tissues, how long the bacteria could last under environmental conditions, and the relative susceptibility of other species (Williams and Barker 2001). Extensive vaccination trials were conducted using the cattle product known as Strain 19 vaccine on captive elk and for nearly a dozen years on free-ranging elk on winter feeding grounds. When Strain 19 proved ineffective at protecting elk from infection, another vaccine (RB 51) was tried experimentally, but it proved neither protective nor effective. The National Park Service and the governors of Wyoming, Idaho, and Montana have cooperated in the formation of an interagency working group to coordinate the management of brucellosis in the greater Yellowstone ecosystem with tribal, advocacy groups, and nongovernmental organizations.

Brucellosis management involves a suite of management actions that together might accomplish what no single action has to date (i.e., reduce the prevalence of infection in elk and reduce or eliminate the likelihood of elk transmitting brucellosis to cattle). Disease transmission occurs when a domestic cow or elk comes into contact with the fetal membranes from an aborted calf (Fig. 8.5), or from an infected normal birthing event, which are heavily contaminated with the *B. abortus* bacteria (Williams and Barker 2001). The bacteria can survive outside the body for a few hours to a few days, depending on environmental conditions. Transmission from elk to cattle appears to be rare, but it does occur, often in winter, when elk may enter ranches to feed off haystacks. One approach to reducing rates of brucellosis has been reducing elk populations by liberal hunting of female elk. But implementing heavy professional culling, or elimination of winter feeding of elk to reduce animal concentrations and numbers, is considered by agency managers to be ineffective, costly, and unpopular with hunters and many communities. Much has been written on this subject, but the dense aggregation of elk on government-run winter feeding grounds (Fig. 8.6), which seriously exacerbates disease transmission problems, was initially made necessary by loss of natural winter ranges to ranching. As noted above, predation may have an influence on some elk populations, reducing numbers and the potential for transmission. So the translocation of wolves (*Canis lupus*) and the recovery of grizzly bear (*Ursus arctos*) could help reduce brucellosis. Although vaccination has not proven effective at reducing disease prevalence, recent experiments report that dispersing feed across feed sites for elk does help reduce infection rates. Another

Figure 8.5. When brucellosis causes cow elk to abort, the fetus and membranes contain millions of bacteria that can remain infectious for days under winter feed and ground conditions. Provided by the Wyoming Game and Fish Department

Figure 8.6. Artificial feeding and baiting of wildlife, or other manipulations that result in crowding, can greatly exacerbate the transmission of wildlife diseases. These diseases include tuberculosis, brucellosis, fowl cholera, rabies, distemper, and many others. Provided by the Wyoming Game and Fish Department

potential strategy is to develop a more efficacious *B. abortus* vaccine for livestock that could reduce the potential for disease transmission at the livestock–wildlife interface. However, this strategy is not currently feasible, because *B. abortus* is listed as a potential biological warfare agent through the Department of Homeland Security, which makes further vaccine development extremely difficult and expensive.

Although experimental conditions suggest that bison can transmit brucellosis to livestock, it has never been proven definitively to occur in free-ranging conditions. Even so, state agriculture authorities in Montana and Idaho have been adamant that bison wandering out of Yellowstone must be shot. The spread of brucellosis has a political component, as well. The disease control programs mandated by state and federal governments have historically made eradication of the disease the only goal, placing extreme financial pressure on stockmen

in states where the disease occurs, making them very risk averse. Wildlife interest groups point out that brucellosis is not a threat to the viability of wildlife populations, only rarely gets transmitted to livestock, and that wildlife culling and other proposed draconian measures are unlikely to have beneficial effects. It is becoming clear that eradication or management of a disease like brucellosis in large, free-ranging ungulate populations, over vast and rugged geographic areas, and with varying ownership patterns, is a daunting task. This situation remains unresolved, but more emphasis is now being placed on risk management and reduction, and less on disease eradication.

Distemper

Canine distemper has been common in Europe for almost 200 years (Williams and Barker 2001). Caused by a morbillivirus that infects domestic and wild canids, and all mustelids and procyonids, canine distemper is capable of infecting other species, including wild felids (Williams and Barker 2001). The incidence of distemper declined considerably in dogs in North America with the implementation of widespread and effective vaccination in the 1950s and 60s. Transmitted by aerosol or contact with fluids of infected and dying animals, the virus cannot survive outside a living host for long, so its survival depends on repeated transmission to and between animals in susceptible populations. Periodic outbreaks in suburban and rural red and gray fox (*Vulpes vulpes* and *Urocyon cinereoargenteus*) and raccoon (*Procyon lotor*) occur in various parts of North America, often in summer and when populations are relatively high (Davidson 2006). The absence of domestic dog outbreaks suggests it has become endemic in wildlife.

Once inhaled or ingested, the virus invades local mucous membranes and is picked up by white blood cells (i.e., lymphocytes). It multiplies in lymphoid organs and spreads to the lymph nodes throughout the body, eventually lodging in the central nervous system (Williams and Barker 2001). Although some of the signs and lesions of distemper (Fig. 8.7)—oculonasal discharge, diarrhea, hard pad—are fairly characteristic (Davidson 2006), the most common signs are not exclusive to distemper. Depression, seizures, and partial paralysis are also potential signs of rabies. Coinfections of distemper and rabies can occur. Professional postmortem examination is the only way to differentiate between the two diseases with confidence.

When black-tailed prairie dog (*Cynomys ludovicianus*) ecosystems covered 20% of the western rangelands, the black-footed ferret (*Mustela nigripes*) was common in 12 western states and two Canadian provinces. Ferrets are exquisitely sensitive to distemper virus, and infections are fatal in 95–100% of cases. On shortgrass prairies, distemper outbreaks naturally occur every three to seven years, affecting coyote (*Canis latrans*), swift fox (*Vulpes velox*), badger (*Taxidea taxus*), skunk (*Mephitis* spp.), and ferrets (Williams and Barker 2001). Black-footed ferrets were placed on the endangered species list in 1967, but were thought to have become extinct in the 1970s when the last known captive individuals died after receiving the modified live canine distemper vaccine. But in 1981, a rem-

Figure 8.7. Raccoon showing signs of depression, emaciation, and occulonasal discharge characteristic of canine distemper. This viral disease can affect a wide variety of small and large carnivores but can also co-occur with plague, rabies, and other diseases. Professional postmortem examination is strongly recommended for reaching a diagnosis. Provided by Kevin Keel, Southeastern Cooperative Wildlife Disease Study

nant wild population of black-footed ferrets was discovered near Meeteetse, Wyoming.

Narrowly avoiding a distemper outbreak, 18 wild ferrets were taken into captivity between 1985 and 1987 by the U.S. Fish and Wildlife Service (USFWS) and WGFD. Subsequently, the National Black-footed Ferret Conservation Center was established, and additional breeding programs at the Cheyenne Mountain, Louisville, Phoenix, Toronto, and National (Front Royal facility) zoos produced hundreds of kits. Eventually, an effective ferret-safe distemper vaccine was developed that made captive management easier, and release into infected environments safer (Williams and Thorne 1999). These successes made possible several successful translocations of black-footed ferrets to historic habitat in the northern Great Plains. Currently, there are more than 500 adult black-footed ferrets in self-sustaining wild populations in Montana and Wyoming. A successful translocation occurred to the Grasslands National Park, Canada, and the success of another in northern Mexico is still being assessed. The management goal is to establish 3,000 adults scattered across historic ferret habitat.

Highly infectious and highly lethal in many species, canine distemper has been blamed for catastrophic die-offs of African wild dogs (*Lycaon pictus*) and lions (*Panthera leo*). It is one of the diseases with potential to destroy whole populations of susceptible species and, because asymptomatic animals can be carriers, it has proven difficult to predict or limit in wildlife. When canine distemper destroys or seriously limits large carnivore populations, it can alter predator–prey relationships and cause serious ecological ripple effects.

The survival of wild ferret populations is intimately connected to the presence of prairie dog colonies. Another disease, plague (caused by the bacteria *Yersinia pestis*), threatens the survival of both prairie dogs and ferrets. It can be lethal to ferrets but also has the ability to eliminate whole prairie dog colonies and to starve species that are dependent on them. Plague was introduced into North America in the late 1800s (Friend 2006). In the last decade, an effective plague vaccine and delivery system for prairie dogs has been developed (Rocke et al. 2008). Although vaccination has significant limitations as a general wildlife treatment, the successful recovery of black-footed ferrets is dependent on vaccination of these animals for plague, distemper, and other intensive wildlife management strategies. Oral immunization of prairie dogs through distribution of vaccine-laden baits would significantly enhance ferret recovery by protecting their prey base (Rocke et al. 2008).

Brain Worm

Brain worm (*Parelaphostrongylus tenuis*) is a relatively common round worm parasite often found in the meninges of white-tailed deer, where it generally causes little damage in the host to which it is adapted (Davidson 2006). Eggs laid by adult worms are carried in the bloodstream to the lungs, where they hatch. The larvae migrate up the trachea and are swallowed and eventually shed in the feces (Davidson 2006). The larvae penetrate the foot of slugs and snails, which are ingested along with vegetation by various grazing species. If reindeer or caribou (*Rangifer tarandus*), moose, mule or black-tailed deer (*Odocoileus hemionus* and *O. h. columbianus*), llama (*Lama glama*), or sheep ingest the larvae, they invade the spinal column and cord, causing severe and often fatal neurologic disease (Wobeser 1994). Elk and fallow deer (*Dama dama*) are also affected but are somewhat resistant to clinical disease. Elk have even been reported to shed the parasite without signs of disease. In many ways, this pattern of a complex life cycle with a fairly high degree of host specificity and natural host parasite tolerance, but severe pathology in related but unadapted species, is fairly common for parasitic organisms.

Incoordination, circling, and partial paralysis are characteristics of infection. Postmortem examination may show accumulation of a brownish exudate in the meninges and lesions in the spinal cord (Davidson 2006). Definitive diagnosis requires recovery and identification of the adult worm, as the first-stage larvae of the related muscle worm (*P. andersoni*) in feces cannot be distinguished reliably from the brain worm (Wobeser 1994).

ELIZABETH S. WILLIAMS (1951–2004)

Elizabeth S. Williams graduated from Purdue University with a doctorate in veterinary medicine in 1977 and completed a residency and Ph.D. in veterinary pathology from Colorado State University in 1981. While writing her Ph.D. thesis on *Mycobacterium paratuberculosis* in wildlife, Williams astutely recognized chronic wasting disease of captive mule deer and elk as a spongiform encephalopathy. This discovery guided her subsequent research interests and culminated in her recognition as the foremost expert on chronic wasting disease in deer and elk in the United States. Williams was active and skilled in research, diagnostic veterinary pathology, and teaching. A professor in veterinary sciences and an adjunct professor in zoology and physiology at the University of Wyoming, as well as an adjunct professor in veterinary pathology at Colorado State University, she was recognized as an outstanding mentor of students. Williams was a diplomate of the American College of Veterinary Pathologists, coedited the latest edition of *Infectious Diseases of Wild Mammals*, and was editor of the *Journal of Wildlife Diseases* at the time of her death.

E. TOM THORNE (1941–2004)

In 1967, Tom Thorne received his doctor of veterinary medicine degree from Oklahoma State University. He then started a career as a wildlife veterinarian by supervising research projects and by providing veterinary care for the Wyoming Game and Fish Department. As a prominent researcher of brucellosis and chronic wasting disease in wild ungulates, Thorne was involved in the management of bighorn sheep and initiated the successful captive breeding program for endangered black-footed ferrets. He progressed to services division chief and was acting director of the department prior to his retirement. He authored and coauthored many publications. Over the years he was also vice president of the Wildlife Disease Association, president and chairman of the Advisory Council for the American Association of Wildlife Veterinarians, and chairman of the U.S. Health Association's Wildlife Diseases Committee. In 2003, Thorne retired from the Wyoming Game and Fish Department after 35 years of service.

Both Williams and Thorne were renowned for their collaborative work in infectious diseases and management of wildlife, as well as for important contributions to conservation of the black footed-ferret, Wyoming toad, and other sensitive species. They died together in a car accident in late 2004 and are remembered as two thoughtful, generous, and productive scientists who loved wildlife and the people who work with them.

Photo originally published in *Wildlife Society Bulletin* 33:392–393. Used with permission

Brain worm is so lethal for moose and caribou that they are at a significant survival disadvantage in the presence of infected deer populations. As a consequence, dense populations of infected white-tailed deer essentially limit the ability of moose or caribou populations to expand their range into otherwise suitable areas. Several efforts to repatriate caribou, moose, and elk into historic ranges have failed or been limited by mortality from meningeal worm and muscle worm infestation. Treating wildlife populations for these parasites is considered impractical, but treatment of limited areas for the snail intermediate host is one solution that has been tried.

Multicausality of Pneumonia of Bighorn Sheep

Bighorn sheep (*Ovis canadensis*) populations across the western United States and Canada have declined by more than 90% from historic population numbers. There were likely many causes of the decline, but the most prominent appear to have been market hunting, habitat loss and degradation, and diseases contracted from domestic sheep (Jessup 2011). The most severe, persistent, and deadly of these diseases are

those that cause pneumonia (Williams and Barker 2001). Historic observations of massive die-offs of bighorn sheep of all ages following contact with domestic sheep date back almost a century, but causes were poorly documented. Beginning in the early 1980s, die-off investigations and experimental exposures by contact and inoculation affirmed that bighorn frequently died of bacterial pneumonia within days to weeks following contact with domestic sheep (Foreyt and Jessup 1982). Several different *Pasteurella* species could be isolated from sick and dying animals, and investigators noted its close parallel to shipping fever in cattle (Williams and Barker 2001, Jessup 2011). Although wildlife agencies sought only to develop agreements allowing better geographic separation for protection of wild sheep, domestic sheep advocates responded first with denial, then with alternative explanations exculpating domestic sheep, and then with demands for absolute proof, beyond even that attainable for cattle after over 40 years of research and observation on shipping fever (Jessup 2011).

An indigenous cause of pneumonia in bighorn is lungworm (*Protostrongylus stilesi*; Wobeser 1994). This parasite coexisted with bighorn sheep in parts of North America and requires a snail as an intermediate host. It generally exists only where bighorn populations occupy habitat with sufficient vegetation and duff to retain the soil moisture required by snails (Wobeser 1994). It is uncommon in desert bighorn populations. As there is an evolved host–parasite relationship, infected adults are fairly resistant to primary or secondary pneumonias. When population numbers peak, lambs may become heavily infested and die, and these spring lamb pneumonias caused by lungworm act as a brake on population growth (Wobeser 1994). Provision of anthelminthics in fermented apple mash, feed, or salt blocks has been used to reduce lungworm infection levels in bighorn sheep herds. However, this tool is best combined with population reduction and other management programs, because increasing lungworm loads may be an indicator of limited carrying capacity, and lungworms may develop drug resistance.

Several bacteria of the *Pasteurellacea* family (*Pasteurella multocida*, *Pasteurella trehalosi* and *Mannheimia*—formerly *Pasteurella*—*haemolytica*) can cause pneumonia in domestic sheep, goats, and bighorn sheep (Williams and Barker 2001, Wehausen et al. 2011). Bacterial infection may be secondary to lungworm infestation, but it also may be a primary cause of pneumonia epidemics, including all-age die-offs. Bighorn sheep are extremely sensitive, and researchers have reported that the white blood cells (neutrophils) of some bighorn sheep are less capable of killing these bacteria than those of domestic sheep (Jessup 2011). When apparently healthy bighorn sheep are experimentally placed in contact with domestic sheep, bighorn contract pneumonia and die, while the domestic sheep remain healthy (Foreyt and Jessup 1982, Wehausen et al. 2011). This does not occur when bighorn are allowed contact with cattle, deer, elk, horses, or llamas. Contact with domestic goats is less fatal but can result in serious diseases, including pneumonia, bacterial keratoconjunctivitis (e.g., eye infections; Jansen et al. 2006), and others.

Mannheimia haemolytica originally isolated from healthy domestic sheep has been genetically marked and labeled with a fluorescent dye, and inoculated back into the oropharynx of domestic sheep. They remained healthy. Fence line contact allowed pneumonia to spread from domestic sheep to bighorn sheep, and nose-to-nose contact proved uniformly fatal for bighorn (Lawrence et al. 2010). The dyed and genetically marked bacteria were subsequently isolated from the bighorn that died of pneumonia (Lawrence et al. 2010). Infection did not occur when separation barriers of 3 m or more were maintained. This conclusive proof, that domestic sheep can carry respiratory bacteria that are nearly always fatal to bighorn sheep, should end the three-decade-long battle over whether contact with domestic sheep may be responsible for many bighorn sheep die-offs. But the problem of what to do about it remains.

Today, when domestic sheep graze on public rangelands where there is a goal of conservation of bighorn sheep populations, geographic separation seems to hold the greatest promise for protecting bighorn. Unfortunately, the propensity of

bighorn to wander, and for stray or lost domestic sheep and goats to seek out the company of bighorn, makes separation somewhat difficult when either species strays. For land-use planning purposes, 40–50 km is generally seen as an optimal separation distance. But a number of domestic sheep management changes, like trucking them from the high-elevation summer grazing ranges instead of trailing them, can also help reduce risk.

Many efforts have been made to develop vaccines that can elicit sufficient immune response to *Pasteurella* spp. to protect domestic livestock and bighorn sheep. After many failures, a candidate *Mannheimia haemolytica* vaccine has emerged. But four vaccinations were required to confer immunity, and the logistics of vaccinating free-ranging bighorn over vast and rugged terrain, at exactly the right time of life, perhaps repeatedly, and without stress and injury remain a serious limitation to potential use on bighorn (Wehausen et al. 2011). Such a vaccine would provide no protection from other factors (like mycoplasma; see below) that could help initiate bacterial pneumonia. Further, vaccination of bighorn would place the responsibility of preventing disease transmission on public trust agencies rather than on those who would use public land for their financial gain. Perhaps a better use for this technology would be to require that all domestic sheep grazing on public lands anywhere near bighorn sheep habitat be vaccinated, so that they do not carry potentially lethal bacteria.

A further complication is that a primitive bacterial organism (*Mycoplasma ovis ovipneumoniae*) appears to be capable of facilitating or exacerbating pneumonia in bighorn. It has been isolated from bighorn suffering from pneumonia, and epidemiologic studies implicate it in some large bighorn pneumonia die-offs. Winter penning of Dall's sheep (*Ovis dalli*) with domestic sheep in a zoo resulted in all the Dall's sheep dying of *Mycoplasma ovis ovipneumoniae* pneumonia (Jessup 2011, Wehausen et al. 2011). It can readily be isolated from healthy domestic sheep and goats. Mycoplasma infection alone does not consistently cause fatal respiratory disease in experimental exposure trials of bighorn, but infections often occur in concert with aggressive respiratory bacterial flora like *Pasteurella* (Jessup 2011, Wehausen et al. 2011). Surviving bighorn may also harbor and carry it for many months or years. When mycoplasma is present during pneumonia outbreaks in wild sheep populations, the disease process is often highly infectious and fatal, and it tends to persist in the recovered female population and to cause pneumonia and death in lambs in subsequent years. Respiratory viruses, severe weather, malnutrition, and other stressors may contribute to pneumonia in bighorn populations.

That bighorn sheep can harbor mycoplasma, and (rarely) virulent forms of *Pasteurella*, should call into question the common practice of relocating of surplus bighorn sheep into existing herds or populations. Relocated animals always carry with them a flora of bacteria, viruses, mycoplasma, and parasites. The end result of these supplementation efforts might not be increased genetic diversity, but disease-related morbidity and mortality, and occasionally loss of founder stock. This bargain should be weighed carefully in light of the fact that the amount of genetic diversity desired, and whether it already exists in a herd, is unmeasured, and that the genes for which diversity might be beneficial are unknown. The risk of disease transmission seems likely to outweigh any perceived or real value of increased genetic diversity under these circumstances.

Diagnosis of pneumonia is reasonably straightforward; affected sheep often cough and have nasal discharge. But coughing in bighorn sheep does not reveal either the cause or the likely outcome for the individual or the populations. Acute cases can result in death within days with few outward signs. Professional postmortem examination should help reveal the cause or causes, but as the disease progresses, the initiating organisms may become harder to identify. Isolation and identification of the bacterial and mycoplasmal causes of pneumonia in bighorn sheep require typing of the organisms to determine the origin of infection.

Perhaps one of the best summary statements regarding what is known about pneumonia in bighorn sheep and domestic sheep, and general management options, came out of a meeting hosted by biologists of the Payette National Forest in 2008. The meeting had six noteworthy recommendations (Jessup 2011):

1a. Scientific observation and field studies demonstrate that contact between domestic sheep and bighorn sheep is possible under range conditions. This contact increases risk of subsequent bighorn sheep mortality and reduced recruitment, primarily because of respiratory disease.

1b. The complete range of mechanisms and causal agents that lead to epizootic disease has not been conclusively proven.

1c. Given the previous two statements, it is prudent to undertake management to prevent contact between these species.

2. Not all bighorn sheep epizootic disease events can be attributed to contact with domestic sheep.

3. Gregarious behavior of bighorn sheep and domestic sheep may exacerbate potential for disease introduction and transmission.

4. Dispersal, migratory, and exploratory behaviors of individual bighorn sheep traveling between populations may exacerbate potential for disease introduction and transmission.

5. There are factors (e.g., translocation, habitat improvement, harvest, weather, nutrition, fire, interspecies competition, and predation) that can be managed, and some that cannot, that can influence bighorn sheep population viability.

6. *Pasteurellaceae*, other bacteria, viruses, and other agents may occur in healthy, free-ranging bighorn sheep.

The controversy over the need to enforce separation between domestic and bighorn sheep has not been resolved.

Sea Otters, Diseases, and Habitat Degradation

Failure of a wildlife population to recover when wildlife health and disease is part of the problem can involve more than a single disease process or causal agent. Despite decades of legal protection, the recovery of California's southern sea otter (*Enhydra lutris nereis*) population has been hindered by high mortality from the deaths of prime-aged adult animals. Up to 50% of sea otter mortality has been attributed most notably to protozoal and bacterial infections (Thomas and Cole 1996, Kreuder et al. 2003). The connections between many infections and various sources and types of pollution have been documented (Jessup et al. 2004, 2007; Miller et al. 2010a). Additional southern sea otter mortality caused by ingestion of marine biotoxins appears to be influenced by nutrient loading coming from coastal freshwater and estuarian sources, and directly from biotoxins originating from freshwater sources (Kudela et al. 2008, Miller et al. 2010c). Most persistent organic pollutants (POPs) and contaminants of ecological concern (COECs) are found in the blood of live (and in liver samples of dead) southern sea otters at levels 20–50 times higher than those of sea otters from more pristine areas (Kannan et al. 2006, 2007, 2008; Jessup et al. 2010). Some sources of POPs and COECs are known, and most come from nonpoint sources on land. These disease and health problems all appear to be more prevalent along urbanized coastlines and near river mouths, suggesting that land–sea pathogen and toxin flow is an important component of exposure to these causes of morbidity and mortality (Miller et al. 2002, Jessup et al. 2004).

Some pathogens appear to be new. One cause of fatal systemic and neurologic protozoal infections in sea otters is *Sarcocystis* (i.e., *Sarcocystis neurona*; Miller et al. 2010b), which was inadvertently introduced to California by its nonnative invasive species host, the Virginia opossum (*Didelphis virginiana*; Dubey et al. 2001). A second, similar protozoal parasite is *Toxoplasma gondii* (Fig. 8.8). Although native felid species, bobcat (*Lynx rufus*) and mountain lion (*Puma concolor*), coevolved with sea otters and are competent hosts of *T. gondii*, the relatively recent introduction of millions of domestic cats into California occurred in the last 150 years, and at a time when sea otters

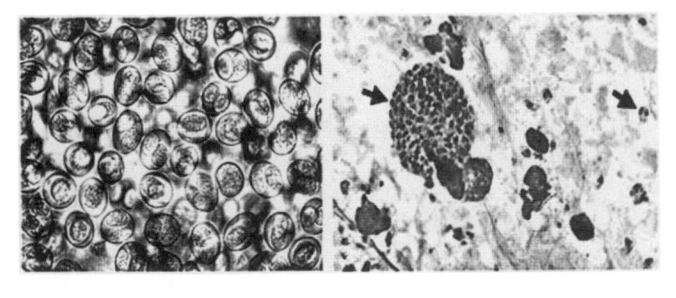

Figure 8.8. Oocyst (*left*) and tissue cyst (*right*) phases of *Toxoplasma gondii*. The oocysts are shed only in the feces of cat species, and have been shown to bioaccumulate in filter-feeding shellfish, which appears to be the route by which sea otters become infected. The tissue cysts can lay dormant for years before other stressors allow it to break out, causing fatal neurologic and systemic disease in sea otters. Provided by the California Department of Fish and Game

were recovering from the brink of extinction. Most municipal sewage generated along coastal California is released into the ocean after treatment, but primary and secondary treatment does not kill many pathogens, notably *Toxoplasma* and *Sarcocystis* (Conrad et al. 2005).

Bacterial infections, many of them caused by organisms commonly associated with feces, are another significant cause of sea otter illness and death (Thomas and Cole 1996, Kreuder et al. 2003, Jessup et al. 2007, Miller et al. 2010a). Old or inadequate sewage infrastructure, seasonally heavy precipitation, and accidents can result in uncontrolled releases of sewage. Many residences in rural portions of California's coastal counties where sea otters live are connected to private septic systems, and some boats discharge untreated sewage directly into the ocean. Untreated feces from pets and wildlife are periodically flushed into the ocean with storm runoff. This fecal matter, plus human and livestock feces, all may be sources of bacterial and protozoal pathogens (Sercu et al. 2009).

Some of the most productive and intensively farmed land in the United States is adjacent to southern sea otter habitats. The heavy use of nitrogen fertilizers, phosphates, and other nutrients; light and porous soils; and seasonally heavy rainfall result in significant nutrient pulses in embayment areas. Higher levels of urea promote domoic acid toxin production by *Pseudo-nitzschia australis*, the marine diatom that causes amnesic shellfish poisoning in humans and major mortality events in various marine birds and mammals (Kudela et al. 2008). Recently, cyanotoxins produced by *Microcystis* spp. in nutrient- (phosphate-) rich freshwater reached the ocean, killing sea otters (Miller et al. 2010a). The toxin is 1,000 times more potent than domoic acid and kills fish, dogs, people, and most other living organisms. More than 30 sea otter deaths have now been traced back to it.

A growing body of evidence suggests that the cumulative effects of all of these forms of pollution (i.e., protozoa, bacteria, biotoxins, POPs, COECs) and habitat degradation contribute to southern sea otter morbidity and mortality and are limiting population recovery. Southern sea otters also show some evidence of being food limited. They respond to limited abundance of preferred prey like abalone (*Haliotis* spp.) by diversifying their diets. Unfortunately, many of the less preferred prey items are filter feeders or pseudo–filter feeders and are capable of harboring or concentrating protozoa, pathogenic bacteria, biotoxins, and organic pollutants (Johnson et al. 2009).

Southern sea otter problems are a classic example of the concept that human, animal, and environmental health is inexorably linked (i.e., One Health; Jessup et al. 2007). Most of the organisms and toxins that kill sea otters also sicken and kill other marine mammals, domestic species, and humans. Sea otters are a keystone species for kelp forests, protecting them from kelp stipe grazers like sea urchins (*Stronglyocentrotus* spp.). Kelp forests and the biological diversity they support are healthier with sea otters than without them (Estes 2005). Kelp forests reduce storm damage and beach erosion. In the eastern Pacific Ocean, kelp forests are one of the major marine carbon sequestering macro-organisms. Diseases and intoxica-

tions that reduce sea otter health and abundance also have serious environmental health effects.

The efforts to understand the causes of southern sea otter mortality and potential management options to foster recovery have taken over 15 years to develop and have required the resources of the U.S. Geological Survey (USGS); U.S. Fish and Wildlife Service; California Department of Fish and Game; Monterey Bay Aquarium; Marine Mammal Center; University of California (UC), Davis; Wildlife Health Center, UC Santa Cruz Institute for Marine Science; Defenders of Wildlife; and many other organizations. Over 20 years of consistent population surveys, carcass pickup and examination, and numerous focused studies on particular pathogens and toxins have provided an understanding of sea otter health and population recovery challenges. This work has required the merging of ecological and epidemiological data sets with those developed by pathologists and clinical veterinarians. It has benefited from data collected on live, dead and stranded, captive, and free-ranging sea otters. The overall effort is a classic example of what widely collaborative and cooperative wildlife health programs can accomplish.

METHODS OF INVESTIGATION

Any wildlife disease investigation must ask three primary questions. Is disease present? What is causing it? What effect is it having on the population? Wildlife population surveys and demographic data can be indirect measures of health as well as a means to investigate whether a disease might be present and to measure a disease's effects on populations. Wildlife management often requires conducting population surveys, determining age class distribution, and measuring changes in those data over time using many different methods. Individual animals and age classes can be determined directly by counting small populations at a point in time to establish a minimum number (e.g., southern sea otters). Surveys involving standardized methods and routes and a multiplication or correction factor are used for very large populations (e.g., waterfowl). Mark-recapture methods are often used when a known number of marked animals are in a population (e.g., collared bighorn sheep, deer, or elk; Wobeser 1994). When populations are very large and remote, and when disease processes are subtle, a decline in population numbers or changes in age distribution may be one of the best or only indicators of disease.

Field observation of die-offs and on-the-ground investigations often provide information that cannot be gained by other means. These include the age of carcasses (the approximate time of death), the population structure of affected individuals, involvement of more than one species, proximity to water, water quality, presence of vectors, and many other vital observations (Wobeser 1994). Investigators familiar with the terrain, ecology, and general management of the area best perform such observations. A team made up of biologists and a veterinarian or disease specialist is recommended. Local ranchers and farmers can be quite helpful if properly motivated.

Postmortem examination (i.e., necropsy) is the primary method of establishing a cause of mortality. Fresh carcasses representative of the population and event are needed, particularly if several diseases may be present or if several factors compromise health at the same time. Each carcass must be examined methodically, opening body cavities and examining all organs. Frequently, tissue samples are taken and preserved in formalin or placed in culture media or frozen for further analysis. Veterinary pathologists specialize in this type of work. Under most circumstances, submitting wildlife carcasses to a veterinary diagnostic laboratory or submitting samples taken from field necropsies conducted under professional supervision are the most effective ways to determine the cause of illness or death.

In some cases, gross lesions, or those apparent by visual observation, and basic on-site testing may be characteristic enough to establish a diagnosis. Clinical pathology including the rapid analysis of blood counts, chemistries, and stains may be a useful diagnostic adjunct. More often, definitive diagnosis requires laboratory culture for bacterial, fungal, viral, or protozoal organisms; microscopic examination of sectioned and stained tissues; and possibly immunohistochemistry, toxicological, or genetic sequence analysis. To better understand population and management implications, the genus and species of parasites may need to be identified. For cases or die-offs involving complicated or multiple disease processes, identification can take days to weeks or months to complete.

Epidemiology, the study of disease and health in populations, often takes over after the diagnostic process has established a cause or causes. Spatial and temporal patterns of disease or mortality over time are useful in establishing answers to the primary questions of presence, cause, and population effect. Incidence (i.e., the number of new cases per unit time) and prevalence (i.e., number of cases at any one point in time) rates, and their change over time, are key measures of the progress of a disease epidemic (i.e., major outbreak). Epidemiology may involve a field component (shoe leather epidemiology) and a modeling (mathematical and statistical) component.

There are a variety of models that can be used to predict the course of a given disease in a population through time, and these can be used to develop intervention and treatment strategies (e.g., when and how many susceptible individuals would need to be vaccinated to stop the spread of the disease). The susceptible, infectious, recovered (SIR) compartmental model is one of the more fundamental ones. It relies on the fact that individuals (animals or people) move from one stage of disease to the next through time at a predictable rate. Measles is a classic example that shows how the number of susceptible individuals in a naive population drops quickly and predictably as the number of infected begins to peak, followed by a rise and peak of recovered individuals, and the end of the epidemic. The rate at which individuals transition from susceptible to infected is called the force of infection (F), but for many infectious diseases, to calculate F, it is more realistic to consider the fraction of individuals in the population that are susceptible, rather than the absolute number. A more sophisticated version of SIR takes into account the fact that, with many diseases, susceptible individuals are exposed and might or might not become infected (SEIR). Another model factors in the effects of maternal-derived immunity on susceptibility (MSIR). Modeling is one of the most powerful tools available to shape and scale disease intervention strategies, and to indicate when and which interventions might or might not be successful.

At some point along the continuum of disciplines that are used in wildlife health investigations, veterinary epidemiology begins to blend with disease ecology. A thorough understanding of various aspects of the science of ecology is vital to understand wildlife diseases and the ecosystems in which they occur. Disease ecology is in turn a special area of ecology that has grown and greatly matured in the last 20 years.

MANAGEMENT OF HEALTH AND DISEASE

Theoretically, wildlife populations in pristine environments at or below carrying capacity should not have significant disease problems beyond a low level of well-adapted parasites. This idyllic set of circumstances is rare to nonexistent. The general principles of wildlife management—providing sufficient food, water, cover, and space—are generally good methods for preventing or reducing health problems. More than any other basic factor, crowding and concentrating wildlife contributes to the occurrence, incidence, and severity of many wildlife diseases (i.e., avian cholera, TB, brucellosis, CWD, distemper, parasitism in general, and pneumonias).

Agent and Vector Reduction

One of the most important contributions of wildlife disease research to wildlife conservation is that, by knowing the intimate details of the biology of disease agents and the vectors that transmit them, cycles can be broken and the impacts of disease lessened or managed. One classic example is the extensive effort to manage cattle trypanosomiasis and human sleeping sickness in southern and eastern Africa, which are transmitted by the tsetse fly (*Glossina* spp.). Tsetse flies can be attracted to large, dark-colored cloth targets that pivot and move in the breeze and smell like cow's breath, mimicking the host, such as cow, Cape buffalo (*Syncerus caffer*), or other large ruminants. But insecticide permeates the targets, and local tsetse fly populations and transmission of trypanosomes (vectors and disease agent) can be reduced.

Other examples are the switches in the United States to steel shot for waterfowl hunting and to copper rifle bullets in California condor (*Gymnogyps californianus*) range. Both manage disease (i.e., lead poisoning) by removal of the agent from the environment. Banning the use of several persistent organic pollutants (e.g., DDT, chlordane) is another example. Interestingly, the loss of DDT has made it harder to control fleas that transmit wildlife plague in the United States. However, the eventual development of insecticide resistance is an expected consequence of repeated and continual use of pesticides, which can have unintended side effects or be unpredictable.

Some states also try to reduce wildlife disease by informing the public and enlisting their support. California State law requires a warning on cat litter that cat feces should not be flushed down the toilet or placed where it can be washed into the ocean in an attempt to reduce the levels of *Toxoplasma* oocysts ingested by shellfish and sea otters.

Host Management

In some cases the behavior, spatial or temporal abundance, or other critical features of affected wildlife or other hosts can be managed. Historically, efforts to manage outbreaks of avian cholera, botulism, and duck virus enteritis have all relied on dispersal of birds to reduce contact and transmission (Friend 2006). Pyro techniques, carbide cannons, airboats, and aerial hazing have been used to disperse waterfowl and are classic examples of host management. Dispersal can reduce exposure and mortality, but animals often adapt to disturbance and threat, and it is very difficult to permanently haze off birds or many other animals (Wobeser 1994).

Permanent separation by 3 m between domestic sheep and bighorn can prevent transmission of pneumonia. Potential for contact with domestic sheep can be significantly reduced if they are not allowed to graze within 40 km of desert bighorn populations on unfenced ranges. Bighorn sheep that have suffered pneumonia outbreaks should not be used to augment other bighorn populations, as they may carry lethal respiratory disease agents. These are further examples of host management to prevent disease.

Many wildlife disease management efforts have relied on population reduction as a form of host management (Table 8.1). These have included focal population reduction, zonal reduction (i.e., creation of barriers), and general population reduction (Wobeser 1994). Once popular, these types of programs are used less commonly today.

General population reduction is a blunt tool for disease management. An example of general population reduction was the widespread and extensive effort to kill skunk, fox, raccoon, and coyote by shooting, poisoning, and trapping as a means of combating rabies. It was carried out for decades until it was finally determined that killing mesopredators on a massive scale had little or no effect on rabies prevalence.

Recent efforts to reduce the prevalence and spread of CWD in Wisconsin by encouraging and focusing high levels of hunter harvest have also not proven effective. An example of a more successful program might be the widespread culling of brushtail possum (*Trichosurus vulpecula*) that harbor TB in Australia. Because these possum are not native to the habitats from which they are being culled, this program is more popular than culling of native species. The problem with host reduction as a disease management strategy is that, to be effective, the diseased animal must be relatively easy to identify and remove, or the disease must be relatively host specific, not very environmentally resistant, or not very infectious. Where this is the case, diseases often die out spontaneously without much culling.

Immunization

The use of vaccines to create or boost an animals' immune response to a disease is a specific case or type of host management. Treatment of individual animals has not been a traditional tool of wildlife management, but that is changing. Rabies vaccines for dogs were developed before the virus itself was even identified, and modern killed or modified live rabies vaccines are very effective at stimulating an immune response. Louis Pasteur developed a vaccine for rabies decades before viruses were known to exist. When years of large-scale population reduction efforts failed to reduce the incidence of red fox rabies in Europe, vaccination was tried. The first vaccines were delivered by putting them in chicken heads and feeding them to foxes. Subsequently better vaccines and more stable artificial baits were developed for fox, skunk, and raccoon. Splicing portions of the rabies gene into other viruses has improved both safety and efficacy of vaccination. Wildlife rabies control and outbreak response is a major program of the USDA, Wildlife Services, and a major success of wildlife management from a public perspective (Slate et al. 2009).

As discussed above, the restoration of the black-footed ferret to its native ranges would probably have failed if ferret-safe canine distemper vaccines had not been developed. By the 1980s, wildlife managers recognized that commercially produced killed vaccines were not antigenic enough to offer much protection for many species and that modified live vaccines were lethal to ferrets and some other species. Extensive efforts led to development of ferret-safe gene-spliced vaccines in use today. Vaccination, along with large-scale captive breeding programs, appears to have secured the future of black footed ferrets.

The emergence of plague as a threat to remnant prairie dog communities and the species that are dependent on them is a further, and perhaps even more profound, example of how vaccination can contribute to wildlife conservation. It is also worth noting that by controlling plague, conservation is served, and human and domestic animal health are protected. This is another example of the One Health approach (Rocke et al. 2008).

Some disease organisms do not stimulate sufficient host immune response so that vaccination will reliably provide immunity. No effective or widely usable vaccine for protection of human or animals populations against tuberculosis has been developed, but not for lack of effort. Some host species are less responsive to vaccines than others. Brucellosis vaccination of elk and bison has not proven effective even though Strain 19 and RB 51 vaccines for cattle are effective.

Rinderpest virus infection in wildlife and livestock has now been eliminated from Africa through large-scale and uniform vaccination campaigns. First, communities had to be educated, diseased populations reduced or isolated from healthy ones, and an effective, long-lasting heat stable vaccine developed. Although it took nearly a decade, the vaccination campaign provided sufficient numbers of immune populations so that the virus could not continue to move between naive populations. Every susceptible animal, wild and domestic, did not

Table 8.1. Examples of the use of depopulation on wild animals as a method of disease management

Species	Disease	Method[a]	Location	Year reported
Depopulation of focal area(s)				
Striped skunk (*Mephitis mephitis*)	Rabies	T	Wisconsin	1966
		P	New Mexico	1970
		T, P, S	Alberta	1986
Vampire bat (*Desmodus rotundus*)	Rabies	P	Nicaragua	1976
Ground squirrel	Plague	S	Colorado	1982
American coot (*Fulica americana*)	Avian cholera	W	Virginia	1976
European badger (*Meles meles*)	Tuberculosis	T, G	England	1974
				1982
				1984
Blackbirds	Histoplasmosis	W	Tennessee	1985
Creation of a barrier or control zone				
Striped skunk	Rabies	P, G, S, T	Alberta	1978
Red fox	Rabies	T	New York	1960
		G, S	Europe	1974
Vampire bat	Rabies	G	Argentina	1974
		P	Latin America	1980
Cape buffalo (*Syncerus caffer*)	Rinderpest	S	Uganda	1953
General depopulation over a large area				
Coyote, red fox, wolf, bear	Rabies	P, T, S	Alberta	1954
Red fox	Rabies	T	Virginia	1966
		G, S	Germany	1981
Mongoose	Rabies	T	Puerto Rico	1966
Vampire bat	Rabies	S, P	Trinidad	1960
Ungulates	Rinderpest	S	Africa	1982
Deer	FMD[b]	S	California	1921

Source: Modified from Wobeser (1994).

[a]T, trapping; S, shooting; P, poisoning; G, gassing of dens or caves; W, wetting with surfactants.

[b]FMD, Foot and mouth disease.

need to be vaccinated, just enough so that the virus could not survive. Rinderpest is only the second virus scourge (smallpox was the first), and the first one in wildlife, to ever have been eliminated. Wildlife vaccination programs are inherently expensive and risky, but when properly designed and carried out, they can be effective where other ecologically damaging efforts like large-scale, long-lasting depopulation or cordon fencing (discussed below) have failed. Vaccination is now also being tried for fertility control (Liu et al. 1999).

Environmental Management

Diseases occur or become epidemic when conditions of the host animal, agent, and environment allow it. Manipulation of the environment is a tool used frequently for control or management of disease outbreaks.

One example is type C botulism poisoning of waterfowl. In the western United States and Canada, conditions often favor the conversion of the nearly ubiquitous spore form of *Clostridium botulinum* type C into the vegetative state (Jessup 1986, Wobeser 1994). A warm, moist alkaline environment and a source of decomposing protein are key ingredients in the recipe. Under these conditions the spores germinate, and the vegetative form replicates rapidly, producing an extremely potent neurotoxin (Jessup 1986, Friend 2006). Under natural

conditions this process occurs when alkaline lake beds become flooded and small mammals, reptiles, or insects die, providing a protein source. Flooding of agricultural fields creates the same conditions.

Fly maggots feed on the toxic carcasses and uptake high levels of botulinum toxin. In feeding trials, as few as two maggots contain enough toxin to paralyze and kill a duck in a few hours (Jessup 1986). Flies lay eggs on the fresh duck carcasses, which are steeped in the warm alkaline waters of botulism soup, and the resulting maggots are eaten by more ducks. In a few days, mortality can increase from a few ducks to a few thousands ducks. The toxin will also paralyze and kill geese, swans, and wading and shore birds (Friend 2006). Hawks are somewhat resistant; vultures are extremely resistant and nearly impossible to kill with type C toxin. Humans and other mammals, including dogs used for retrieving, are also very resistant to ingested type C toxin.

In what were once alkaline semidesert areas of much of the West, irrigation has brought vast wealth via farming. California is a classic example. But practices like flood irrigating in summer or early fall create the perfect conditions for type C botulism die-offs. Sprinkler irrigation, periodic ditch irrigation, and flood irrigating when temperatures are cooler all help reduce the conditions favoring botulism die-offs. When flood irrigation is necessary for crops like rice, the steeper the slope of the check or earth wall, the less area for dead animals to collect that start die-offs. Irrigation practices are not the only human activity that can exacerbate potential for botulism die-offs. Anything that kills ducks and leaves their dead carcasses in warm, alkaline water can cause a die-off that can expand if conditions are right. Power lines, windmills, or towers near shallow, warm water sources can be the initial source of bird mortality (Friend 2006). Adjusting farming practices and design of infrastructure can help keep yearly die-offs from occurring.

Perhaps the most infamous effort to control or contain disease by manipulation of the environment has been the attempt to keep foot and mouth disease out of Botswana and other southern African countries by constructing huge cordon fences across much of the country. This expensive effort helped protect a cattle industry that was highly subsidized and conducted in an arid land largely unsuited to cattle grazing. The practice is continued at the cost of millions of dead wildlife, severe disturbance of normal ecological processes, and the ending of some nomadic people's traditional lifestyles. It has been fairly effective in protecting those who depend on the cattle export economy, but at a high environmental price.

In California, native tule elk (*Cervus canadensis nannodes*) were repatriated to the swamps of the Sacramento–San Joaquin River Delta when a population was put back onto the Grizzly Island Wildlife Area between Sacramento and San Francisco. The herd grew and expanded from a few dozen to several hundred. Their grazing impact was noticeable, and plans for hunts and translocations to reduce and manage the population were being developed when a die-off killed about 20% of the herd, primarily younger animals. Investigation revealed that lack of fire or tilling to remove the dead grass thatch, and a long, wet, foggy winter had not allowed the emergence of green grasses or forbs (Jessup et al. 1986). The primary green feed available in many locations was poison hemlock (*Conium maculatum*), which contains the neurotoxin conine (Jessup et al. 1986). This case is an example of the failure to foresee the need for—and subsequent use of—environmental management. The emergency response included provision of alfalfa hay, discing and spraying of hemlock patches, and hazing of elk from hemlock-rich areas. This initial step stopped the die-off and was then followed by population reduction by relocation and either-sex hunts to control numbers. These measures have kept large-scale hemlock poisoning from reoccurring, even though the plant is still very common on the wildlife area.

As noted above, one of the more dangerous environmental management practices, unless one takes the perspective of the disease organism, is concentrating wild animals. Doing so results in greatly increased animal-to-animal contact, ingestion of unnatural feeds, and unnatural feeding behaviors that exacerbate disease transmission. Tuberculosis and brucellosis have already been mentioned. Both organisms can survive outside the host, particularly in feces or fluids. It should be no surprise that these diseases, along with paratuberculosis (Johne's disease) and CWD, which are also environmentally persistent, are a serious concern in captive settings such as in zoos and on game farms. Even under the most stringent conditions of observation, treatment, and sanitation, these diseases are very difficult to manage or eliminate in captive or crowded animal populations.

Another dramatic example is *Mycoplasma gallisepticum* infections in finches and seed-eating migratory birds. This mycoplasma is highly infectious and causes conjunctivitis; the bacterial discharge from the eyes, mouth, and nares is the source of infection (Friend 2006). Although bird feeding is extremely popular, the concentration of birds around feeders—and repeated contact of the eye and nares of infected birds with contaminated feeder surfaces—is partially responsible for the toll it takes on finches and other species (Davidson 2006). Often, dozens to hundreds of birds sicken and die in less than a week around a few adjacent feeders. Only removal and sanitation of the feeders can help control the infection. The inherent characteristics of mycoplasmas make development of effective treatment or vaccines extremely unlikely.

Like dispersion of feed and natural feed sources, dispersion of water and clean water sources can reduce prevalence of wildlife diseases and parasites. Irrigated pastures exacerbate infestations of lungworm (*Dictyocaulus* spp.) in deer and the giant liver fluke (*Fascioloides magna*; Wobeser 1994). Providing multiple small water sources like spring boxes or drinkers instead of one large muddy stock pond appears to reduce the prevalence and severity of hemorrhagic diseases of deer, particularly where water is scarce and must be shared with livestock.

Fire, properly managed, can improve feed conditions for grazing wildlife, disperse them, and in some habitats reduce tick infestations and the diseases they transmit. When parasites and chronic diseases reoccur on lands managed for wildlife, it is sometimes valuable to review how food, water, cover, and habitat are being managed, including carrying capacity, and predator–prey balance.

Some wildlife management activities may strongly influence the manifestation, emergence, or severity of disease. A case in point is wildlife translocation. The translocated host animal may harbor disease agents and allow their introduction to new environments (Jessup 1993). Conversely, they may encounter new agents or vectors in a new environment. There may be health implications for the translocated animals, conspecifics or other species in the new location, or for domestic species or humans (Jessup 1993). When tule elk were relocated to historic ranges at Point Reyes National Seashore, California, the soils of the peninsula where they were released had a very high molybdenum content that led to fatal and debilitating copper deficiency (Gogan et al. 1989). The presence of Johne's disease (i.e., paratuberculosis) in the adjacent dairy cattle resulted in infection of these elk. Temporary diet supplementation of copper, separating elk and cattle, and extensive disease screening and culling of infected elk herds resulted in effective control but not eradication of these two diseases (Gogan et al. 1989, Jessup and Williams 1999).

Many other examples of transplantation and disease could be cited: the emergence of raccoon rabies in the eastern seaboard states owing to the relocation of raccoons for hunting (Davidson 2006); potential transplantation of *Mycoplasma gallisepticum* and *M. meleagridis* along with wild turkeys in the West (Jessup et al. 1983); *Brucella suis*, *Leptospira* spp., and pseudorabies virus spread with movement and stocking of feral hogs (Clark et al. 1983); the emergence of neurologic disease in horses and sea otters caused by *Sarcocystis neurona* shed by Virginia opossum transplanted to California; and the withering syndrome rickettsia to abalone off California via release of infected ship ballast water or possibly aquaculture practices (Haaker et al. 1992). Disease analysis and health evaluation should be part of any major stocking or translocation effort. Significant wildlife disease outbreaks and problems have convinced many wildlife management agencies to incorporate disease investigation and control into their core programs. Especially with regard to free-ranging wildlife, Benjamin Franklin's quote rings true: "An ounce of prevention is worth a pound of cure."

Translocation is only one example of many human activities that can upset the host–agent–environment relationship with regard to wildlife disease. More pervasive and profound are common agricultural or resource extraction practices that transform environments and bring people, domestic animals, and disease agents into contact with wildlife in novel ways (Friend 2006). The emergence of Hendra and Nipah viruses, fatal epidemics in pigs and people, largely resulted from the development of swine feeding facilities in fruit bat (*Epomophorus* spp.) habitats where the diseases were endemic and where bat feces and carcasses could get into pig sheds. Other examples include trapping and consumption of civets (*Paradoxurus* spp.) and the outbreak of severe acute respiratory syndrome (SARS) in China and its rapid spread to North America via airlines; the trade in African bush meat and potential for outbreaks of Ebola and Marburg virus in major North American cities; and live bird importations and avian influenza. At least 30 viruses appear to have been introduced to North America by the exotic bird trade.

Integrated Approach

Much like integrated pest management has proven effective in dealing with increasing threats to more intractable agricultural disease problems, integrated approaches hold promise for managing wildlife diseases, where efforts are made to manage various host, agent, and environmental factors at the same time. Rinderpest vaccination was only effective after infected populations of wildlife and livestock were identified, reduced, or eliminated—or limited in movement by traditional regulatory, testing, and population reduction methods. The comprehensive brucellosis control efforts in the greater Yellowstone region provide an excellent example of coordinated and integrated disease control efforts. Management of distemper and plague in prairie dog and black-footed ferret habitats also include host, agent, and environmental components (Williams and Barker 2001). The examples given of tule elk manage-

ment to mitigate hemlock poisoning (Jessup et al. 1986), molybdenum mediated copper deficiency (Gogan et al. 1989), and Johne's disease (Jessup and Williams 1999) were more primitive integrated efforts. Outbreaks of such waterfowl diseases as avian cholera and botulism usually involve efforts to manage the environment and host, but less so the agent.

One Health

Although what is now called the One Health concept has been around under various names and in various forms for about 50 years, it seems to have recently gained serious traction and acceptance in the general field of veterinary medicine, and at least the public health sector of human medicine. The concept is that we need to seek and implement optimal solutions to health problems that maximize benefits to human health, animal health, environmental health, and sustainability. Political, social, and financial realities cause wildlife managers to seek One Health solutions to wildlife health problems. The mission statement of the Wildlife Disease Associations is "to acquire, disseminate, and apply knowledge of the health and diseases of wild animals in relation to their biology, conservation, and ecology, including interaction with humans and domestic animals." Although written more than 30 years ago, this is clearly One Health (Jessup et al. 2007).

Considering wildlife and the environment when human and animal health programs are developed and implemented exemplifies the One Health concept. From an environmental perspective this represents a significant advance over the days of widespread use of DDT, which did not consider the ecological and vector resistance consequences. Consideration at the policy level of government that outbreaks of human and domestic animal disease could have roots in a disturbed ecosystems or disrupted environments may result in health programs that are more ecologically sound and sustainable. We hope One Health will be kept in mind should we see outbreaks of highly pathogenic avian influenza in waterfowl, or foot and mouth disease in wild ungulates in North America. Because it is still being developed, wildlife management and conservation participation in One Health has not been very pervasive. But One Health represents potential positive change, and should be incorporated into wildlife health, management, and conservation programs.

SUMMARY

The more rapidly we move people, animals, diseases, and disease vectors around the world, and the deeper we penetrate once-pristine ecosystems, the more likely exotic and zoonotic wildlife diseases are to emerge. Seventy percent of emergent diseases in the last five decades have been zoonotic (i.e., of animal origin), and the majority of these are associated with wildlife, not domestic, species (Friend 2006). This situation is unlikely to change in the coming five decades.

Some of the larger centers for wildlife disease diagnosis and research (Table 8.2) and other locations provide excellent wildlife health diagnostic work, research, and conservation.

Table 8.2. Selected large centers of wildlife health/disease research and diagnosis in North America

Group	Location(s)	Services
California Department of Fish and Game	Rancho Cordova, CA	State management (terrestrial)
	Santa Cruz, CA	Diagnosis, state management (marine)
Canadian Cooperative Wildlife Health Centres	Guelph, Quebec; Saskatoon, Calgary; Prince Edward Island	Diagnosis, epidemiology, research management advice; provides all services for Canada
Colorado State University Veterinary School and Labs	Fort Collins, CO	Diagnosis, epidemiology, state management for Colorado and cooperating states
Cornell University	Ithaca, NY	Diagnosis, epidemiology, research
Michigan Department of Natural Resources and Michigan State University Diagnostic Lab	East Lansing, MI	Diagnosis, epidemiology, research, state management
National Animal Disease Center (USDA—Animal and Plant Health Inspection Service, Agricultural Research Service, Veterinary Services)	Ames, IA	Diagnosis, epidemiology, research; foreign animal and program diseases
National Wildlife Research Center (USDA—Animal and Plant Health Inspection Service and Wildlife Services)	Fort Collins, CO	Research, disease and damage management, epidemiology
National Wildlife Health Center (U.S. Department of the Interior-U.S. Geological Survey/Biological Resources Division), University of Wisconsin	Madison, WI	Diagnosis, epidemiology, research, information technology services, U.S. Department of the Interior agency management advice
National Park Service, Veterinary Services	Fort Collins, CO	All national parks
National Oceanic and Atmospheric Association, National Marine Fisheries Service	Beltsville, MD	Marine animal diseases, diagnosis, epidemiology, management
Washington State University Animal Disease Diagnostic Laboratory, Global Programs	Pullman, WA	Diagnosis, epidemiology, research
Southeastern Cooperative Wildlife Disease Study	Athens, GA	Diagnosis, epidemiology, research, management advice for 20 southeastern and midwestern states
University of Florida Florida Game and Fish	Gainesville, FL	Diagnosis, epidemiology, research
University of Georgia	Athens, GA	Diagnosis and research
University of California, Davis, Veterinary School, Wildlife Health Center, Animal Health and Food Safety, and other labs	Davis, CA	Diagnosis, epidemiology, research, OWCN, Sea Doc, mountain gorilla management advice
Tufts Cummings School of Veterinary Medicine	North Grafton, MA	Diagnosis, epidemiology, research, Center for Conservation Medicine, wildlife clinic, international programs
Wyoming Department of Game and Fish with Wyoming State Veterinary Lab, Sybille Research Unit	Wheatland and Laramie, WY	Diagnosis, epidemiology, research
Colorado Division of Wildlife	Fort Collins, CO	Diagnosis, epidemiology, research

The discovery of the emerging West Nile virus epidemic ten years ago resulted from the workup of diseased exotic birds and crows simultaneously at a major zoo and a state wildlife health investigation facility (McLean et al. 2002).

The most appropriate group to assist with investigation of a wildlife health problem will largely be determined by the species involved and the location. In general, all native wildlife is managed under the authority of that state's wildlife or natural resource department. Twenty states in the southeastern and midwestern United States use the Southeastern Wildlife Cooperative Disease Study at the University of Georgia for some or most of their wildlife health investigations. Endangered and migratory species and those from lands under the U.S. Department of the Interior (e.g., national wildlife refuges) are the responsibility of the USFWS. Much of the USFWS and other Department of the Interior work is done at the National Wildlife Health Center (NWHC). National parks have their own wildlife disease programs but often cooperate with USGS/NWHC. Marine mammals and marine fish are under the National Oceanic and Atmospheric Administration's Na-

tional Marine Fisheries Service. Canadian wildlife falls under the purview of Fish and Wildlife Canada, but the cooperative formed by all the veterinary schools and provincial agencies in Canada (Canadian Cooperative Wildlife Health Centres) do disease investigations.

Increasingly rapid, worldwide movement of people and animals has resulted in widespread and accelerating introduction of potential diseases, bringing organisms into wildlife habitats and exposing people to once-rare disease organisms. Wildlife pathogens are twice as likely to become emerging diseases of humans as pathogens without wildlife hosts. Since the 1980s, there has been increasing recognition that the ecology of wildlife diseases is a complex and significant field of study within the field of ecology. During this same period, key wildlife management programs—including big game hunting, wildlife translocations, sensitive species recovery, upland game bird stocking, waterfowl hunting, and many endangered species programs—have been shaped by economic, legal, financial, political, and biological aspects of wildlife disease. Wildlife diseases can have profound ecological and wildlife management implications. Optimizing the health of wildlife populations requires the cooperation of many aspects of society if the goals of One Health and the protection of public and ecosystem health—and healthy and sustainable wildlife and domestic animal populations—are to be met.

Literature Cited

Berger, L. R., R. Spear, P. Daszak, D. E. Green, A. A. Cunningham, et al. 1998. Chytridiomycosis causes amphibian mortality associated with population declines in rain forests of Australia and Central America. Proceedings of the National Academy of Science 95:9031–9036.

Clark, R. K., D. A. Jessup, D. W. Hird, R. Ruppanner, and M. E. Meyer. 1983. Serologic survey of California wild hogs for antibodies against selected zoonotic disease agents. Journal of the American Veterinary Medical Association 183:1248–1251.

Conrad, P. A., M. A. Miller, C. Kreuder, E. R. James, J. Mazet, H. Dabritz, D. A. Jessup, F. Gulland, and M. E. Grigg. 2005. Transmission of Toxoplasma: clues from the study of sea otters as sentinels of Toxoplasma gondii flow into the marine environment. International Journal of Parasitology 35:1125–1168.

Davidson, W. R. 2006. Field manual of wildlife diseases in the southeastern United States. Third edition. Southeastern Cooperative Wildlife Disease Study, Athens, Georgia, USA.

Dubey, J. P., D. S. Lindsay, W. J. Saville, S. M. Reed, D. E. Granstrom, and C. A. Speer. 2001. A review of Sarcocystis neurona and equine protozoal myeloencephalitis (EPM). Veterinary Parasitology 95:89–131.

Estes, J. A. 2005. Carnivory and trophic connectivity in kelp forests. Pages 61–81 in J. C. Ray, K. H. Redford, R. S. Steneck, and J. Berger, editors. Large carnivores and the conservation of biodiversity. Island Press, Washington, D.C., USA.

Foreyt, W. J., and D. A. Jessup. 1982. Fatal pneumonia of bighorn sheep following association with domestic sheep. Journal of Wildlife Diseases 18:163–168.

Frick, W. F., J. F. Pollock, A. C. Hicks, K. E. Langwig, D. S. Reynolds, G. R. Turner, C. M. Butchkoski, and T. H. Kunz. 2010. An emerging disease causes regional population collapse of a common North American bat species. Science 329:679–682.

Friend, M. 2006. Disease emergence and resurgence: the wildlife human connection. Circular 1285. U.S. Geological Survey, Reston, Virginia, USA. http://www.nwhc.usgs.gov/publications/disease_emergence/index.jsp. Accessed 16 November 2012.

Gogan, P. J. P., D. A. Jessup, and M. Akeson. 1989. Copper deficiency in tule elk at Point Reyes, California. Journal of Range Management 42:233–238.

Haaker, P. L., D. O. Parker, H. Hogstad, D. V. Richards, G. E. Davis, and C. S. Friedman. 1992. Mass mortality and withering syndrome in Black Abalone, Haliotis cracherodii, in California. Pages 214–224 in S. A. Shepherd, M. J. Tegner, and S. Guzman del Proo, editors. Abalone of the world: biology, fisheries and culture. Blackwell Scientific, Oxford, United Kingdom.

Jansen, B. D., J. R. Hefelfinger, T. H. Noon, P. R. Krausman, and J. C. deVos. 2006. Infectious keratoconjunctivitis in bighorn sheep, Silver Bell Mountains, Arizona, USA. Journal of Wildlife Diseases 42:407–411.

Jessup, D. A. 1986. Anseriformes: avian cholera, waterfowl botulism, duck virus enteritis. Pages 342–353 in M. E. Fowler, editor. Zoo and wildlife medicine. Second edition. Iowa State University Press, Ames, USA.

Jessup, D. A. 1993. Translocation of wildlife. Pages 493–499 in M. E. Fowler, editor. Zoo and wild animal medicine. Third edition. W. B. Saunders. Philadelphia, Pennsylvania, USA.

Jessup, D. A. 2011. Wild and domestic sheep disease information. American Association of Wildlife Veterinarians. http://aawv.net/bighorn/. Accessed 14 October 2011.

Jessup, D. A., J. H. Boermans, and N. D. Kock. 1986. Toxicosis in tule elk caused by ingestion of poison hemlock. Journal of the American Veterinary Medical Association 189:1173–1175.

Jessup, D. A., A. J. DaMassa, R. Lewis, and K. R. Jones. 1983. Mycoplasma gallisepticum infection in wild-type turkeys living in close contact with domestic fowl. Journal of the American Veterinary Medical Association 183:1245–1247.

Jessup, D. A., C. K. Johnson, J. Estes, D. Carlson-Bremer, W. Jarman, S. Reese, E. Dodd, M. T. Tinker, and M. H. Ziccardi. 2010. Persistent organic pollutants and other contaminants of concern in the blood of free ranging sea otters (Enhydra lutris) in Alaska and California. Journal of Wildlife Diseases 46:1214–1233.

Jessup, D. A., M. Miller, J. Ames, M. Harris, P. Conrad, C. Kreuder, and J. A. K. Mazet. 2004. The southern sea otter (Enhydra lutris nereis) as a sentinel of marine ecosystem health. Ecohealth 1:239–245.

Jessup, D. A., M. A. Miller, C. Kreuder-Johnson, P. Conrad, T. Tinker, J. Estes, and J. Mazet. 2007. Sea otters in a dirty ocean. Journal of the American Veterinary Medical Association 231:1648–1652.

Jessup, D. A., and E. S. Williams 1999. Paratuberculosis in free-ranging wildlife in North America. Pages 616–620 in M. E. Fowler and R. E. Miller, editors. Zoo and wildlife medicine. Fourth edition. W. B. Saunders, Philadelphia, Pennsylvania, USA.

Johnson, C. K., M. T. Tinker, J. A. Estes, P. A. Conrad, M. Staedler, M. A. Miller, D. A. Jessup, and J. K. Mazet. 2009. Prey choice and habitat use drive sea otter pathogen exposure in a resource-limited coastal system. Proceedings of the National Academy of Science 106:2242–2247.

Kannan, K., H. B. Moon, S. H. Yun, T. Agusa, N. J. Thomas, and S. Tanabe. 2008. Chlorinated, brominated, and perfluorinated compounds, polycyclic aromatic hydrocarbons and trace elements in livers of sea otters from California, Washington, and Alaska (USA), and Kamchatka (Russia). Journal of Environmental Monitoring 10:552–558.

Kannan, K., E. Perrotta, and N. J. Thomas. 2006. Association between perfluorinated compounds and pathological conditions in southern sea otters. Environmental Science and Technology 40:4943–4948.

Kannan, K., E. Perrotta, N. Thomas, and K. Aldous. 2007. A comparative analysis of polybrominated diphenyl ethers and polychlorinated biphenyls in southern sea otters that died of infectious diseases and noninfectious causes. Archives of Environmental Contamination and Toxicology 53:293–302.

Kreuder, C., M. A. Miller, D. A. Jessup, L. J. Lowenstein, M. D. Harris, J. A. Ames, T. E. Carpenter, P. A. Conrad, and J. A. K. Mazet. 2003. Patterns of mortality in southern sea otters (Enhydra lutris nereis) from 1998–2001. Journal of Wildlife Diseases 39:495–509.

Kudela, R. M., J. Q. Lane, and W. P. Cochlan. 2008. The potential role of anthropogenically derived nitrogen in the growth of harmful algae in California, USA. Harmful Algae 8:103–110.

Lawrence, P. K., S. Shanthalingham, R. Dassanayake, R. Subramaniam, C. N. Herndon, et al. 2010. Transmission of Mannheimia haemolytica from domestic sheep (Ovis aries) to bighorn sheep (Ovis canadensis): unequivocal demonstration with green fluorescent protein tagged organisms. Journal of Wildlife Diseases 46:706–717.

Leopold, A. 1933. Game management. Charles Scribner's Sons, New York, New York, USA.

Liu, I. K. M., J. W. Turner, and J. F. Kirkpatrick. 1999. Contraception of artiodactylids, using zona 46ellucid vaccine. Pages 628–630 in

M. E. Fowler and R. E. Miller, editors. Zoo and wildlife medicine. Fourth edition. W. B. Saunders, Philadelphia, Pennsylvania, USA.

McLean, R. G., S. R. Ubico, D. Bourne, and N. Komar. 2002. West Nile virus in livestock and wildlife. Current Topics in Microbiology and Immunology 267:271–308.

Miller, M. A., B. A. Byrne, S. S. Jang, E. M. Dodd, E. Dorfmeier, et al. 2010a. Enteric bacterial pathogen detection in southern sea otters (Enhydra lutris nereis) is associated with coastal urbanization and freshwater runoff. Veterinary Research 41:01. doi:10.1051/vetres /2009049.

Miller, M. A., P. A. Conrad, M. Harris, B. Hatfield, G. Langlois, et al. 2010b. Localized epizootic of meningoencephalitis in southern sea otters (Enhydra lutris nereis) caused by Sarcocystis neurona. Veterinary Parasitology 172:183–194.

Miller, M. A., R. M. Kudela, A. Mekebri, D. Crane, S. C. Oates, et al. 2010c. Evidence for a novel marine harmful algal bloom: Cyanotoxin (Microcystin) transfer from land to sea otters. PLoS ONE 5:e12576. doi:10.1371/journal.pone.0012576.

Rocke, T. E., S. Smith, P. Marinari, J. Kreeger, J. T. Enama, and B. S. Powell. 2008. Vaccination with the F1-V fusion protein protects black-footed ferrets (Mustela nigripes) against plague upon oral challenge. Journal of Wildlife Diseases 44:1–7.

Sercu, B., L. C. Van De Werfhorst, J. Murray, and D. Holden. 2009. Storm drains are sources of human fecal pollution during dry weather in three urban Southern California watersheds. Environmental Science and Technology 43:293–298.

Slate, D., T. P. Aldeo, K. M. Nelson, R. B. Chipman, D. Donavan, J. D. Blanton, M. Niezgoda, and C. C. Rupprecht. 2009. Oral rabies vaccination in North America: opportunities, complexities and challenges. PLoS Neglected Tropical Diseases 3:e549. doi:10.1371 /journal.pntd.0000549.

Thomas, N. J., and R. A. Cole. 1996. The risk of disease and threats to the wild population. Endangered Species Update 13:23–27.

Wehausen, J. D., S. T. Kelly, and R. R. Ramey. 2011. Domestic sheep, bighorn sheep and respiratory disease: a review of experimental evidence. California Fish and Game Quarterly 97:7–24.

Williams, E. S., and I. K. Barker, 2001. Infectious diseases of wild mammals. Third edition. Iowa State University Press, Ames, USA.

Williams, E. S., and E. T. Thorne. 1999. Veterinary contributions to the black-footed ferret conservation program. Pages 460–463 in M. E. Fowler and R. E. Miller, editors. Zoo and wildlife medicine. Fourth edition. W. B. Saunders, Philadelphia, Pennsylvania, USA.

Wobeser, G. A., 1994. Investigation and management of diseases in wild animals. Plenum Press, New York, New York, USA.

9

HUNTING AND TRAPPING

JAMES R. HEFFELFINGER

INTRODUCTION

Hunting and trapping in all their various forms provide the very foundation of a system of wildlife conservation developed in North America to stem the loss of native wildlife species caused by overexploitation and habitat alteration. As restrictive regulations halted wildlife population declines, the field of wildlife management shifted from a protective paradigm to a focus on consumptive use of wildlife to manage population abundance and demography within desired goals. Beyond actual population management, hunting provides the financial and social support for continued sustainable conservation of habitat and nonhunted species that benefits all members of the public.

The Human Hunter

Hunting and trapping have been integral to human existence since early hominids emerged from the forest and began carrying weapons to facilitate intake of an increasingly protein-rich diet (Ardrey 1976). The reduction in the size of canine teeth throughout human evolution (Haile-Selassie et al. 2004) corresponds to an increasing use of tools to harvest and process animal protein. The incredible development of the human brain was the result of an increased need to communicate and coordinate abstract plans associated with hunting animals (Watson 1971). The high-protein diet that hunting provided would have also allowed for rapid gains in physical size, strength, and intelligence (Leonard and Robertson 1994). Through time, cultures differentiated and specialized in hunting and trapping the wildlife species with which they coexisted.

As technology developed more effective weapons and trapping techniques, humans were able to exploit a wider variety of quarry. Several sites in southeastern Arizona contain juvenile mammoth fossils with Clovis spear points and butchering tools in association with the bones (Haury et al. 1959). Some researchers believe these cases explain the megafaunal extinctions at the close of the Pleistocene as being caused by humans (Martin and Klein 1984). Although no single cause easily explains these mass extinctions, it is unlikely that Pleistocene humans armed only with spears were able to drive that many species of large mammals to extinction (Grayson and Meltzer 2002, 2003).

Technological advances were not limited to hunting and gathering as humans eventually learned they could grow some of their own plants and even domesticate some animals for food. With the genesis of primitive farming and animal husbandry, humans, edible plants, and animals inevitably began to domesticate one another. Domestication began the quiet and nearly imperceptible shift of hunting and trapping being a survival necessity to an activity that, for most cultures, merely supplemented food obtained more easily elsewhere.

From Sustenance to Conservation

At the early stages of Western civilization in North America, hunting and trapping were still important survival skills that supplemented supplies, but the harvest of wildlife was also one of the leading contributors to the economy of the young, growing nations. However, with a complete lack of limits to harvest, wildlife populations began to diminish rapidly. Unlike their spear-wielding predecessors, Europeans armed with rifles easily overharvested some species of wildlife that had the misfortune to be edible, wrapped in fur and leather, adorned by ornate feathers, or otherwise desirable. Out of these early years of unmitigated slaughter for market and subsistence use came the realization by the young nations that something had to change. Concerned citizenry, led by some of the most prominent hunters and other conservationists of the day, developed a collective sense of stewardship and conservation (Roosevelt et al. 1902, Grinnell 1913, Leopold 1933, Reiger 1986, Bradley 1995, Thomas 2010). The desire to maintain wildlife populations evolved into an incredibly successful system whereby consumptive use of wildlife in the form of hunting and trapping became the cornerstone of what is now known as the North American Model of Wildlife Conservation.

This conservation paradigm built around scientifically regulated harvest has benefited nearly all species of native wildlife because as wildlifers conserved, restored, and managed

CLINTON HART MERRIAM (1855–1942)

Clinton Hart Merriam was born the son of a U.S. congressman in 1855 and began studying animals as a small boy. By the time he was 17 years old, he took part in the Hayden Survey exploring the Yellowstone area. He entered medical school at Yale and Columbia, graduating with his M.D. in 1879. Merriam practiced medicine for a few years, but by the age of 30 he had turned to the study of wildlife and was leading the Division of Economic Ornithology and Mammalogy, which later became the U.S. Fish and Wildlife Service. He was a founding member of the National Geographic Society and was president of many scientific organizations of the time.

For most of his career, Merriam focused on collecting and describing mammals, including approximately 660 he named as previously unknown and new to science. However, he was known as a taxonomic "splitter," and many of these "new" species were invalidated through further study. Still, Merriam's lasting contribution to science was in perfecting study methods that relied on large specimen samples with accurate geographic data and a combination of field and laboratory studies. His published writings number nearly 500, but he may be best known for developing the concept of "life zones" showing that altitudinal changes in local vegetation associations corresponded to latitudinal zones from the equator to the poles. He retired from government service in 1910 and was given an annual pension of $12,000 to study anything he desired. By then his interests were shifting from mammalogy to ethnology, and he studied and published on Native tribes in California well into his 80s. Merriam died in 1942 at the age of 87.

habitat for game species, they also conserved a landscape that provided habitat for nonhunted species. The North American Model of Wildlife Conservation has been so successful that recognition is growing worldwide that hunting and trapping are proven and valuable tools for maintaining the sustainable use of wildlife in perpetuity. Management in other countries is quite different from the system that evolved in North America (Putnam et al. 2011). At recent worldwide symposia in London and Namibia, participants reaffirmed the importance of hunting to wildlife conservation and urged other countries to

model their programs after the North American model (Geist 2006, Patterson 2009, Duda et al. 2010b).

The support for wildlife by hunters and trappers arises partly out of self-interest because of the importance of harvest to them personally (Geist 2006). Participants accrue many benefits—psychological, physical, sociological, and nutritional—from hunting and trapping, but the real benefit in North America is collective stewardship of wildlife and the habitats on which they rely (Leopold 1943). Consumptive use of wildlife provides several important functions beyond the personal benefits to participants (Duda et al. 1998). Hunters and trappers have voluntarily and willingly contributed billions of dollars to support conservation for all wildlife species, not just those that are hunted (Southwick and Allen 2010). They have been the central pillars of this conservation paradigm and thus are responsible for supporting a wide variety of conservation activities that the general public values.

CONSERVATION THROUGH HUNTING AND TRAPPING

Conservation by Consumption

Both the need and success of the North American Model of Wildlife Conservation have revolved around sustainability of harvest (Mahoney 2009). Depletion of wildlife resources in the late 1800s spurred the invention of a unique system of conservation in which wildlife could be used in a sustainable manner that was closely regulated by law and based on the best available science. In its infancy, wildlife management began as a system of harvest limits to stop the rapid decline of wildlife populations during the era of overexploitation (Baughman and King 2008). As the laws gained in effectiveness, management programs evolved slowly to restore and manage hunted species as they responded positively to early game laws. It is a common misconception that "hunting" caused the extinction or extirpation of some species. However, when unregulated killing was controlled in concert with growing restoration efforts, these populations rebounded vigorously only after regulated hunting was used to influence a bold, experimental system of comprehensive wildlife conservation.

This system of conservation has been so undeniably successful, it has been recognized as a model to emulate worldwide. The World Conservation Strategy of the International Union for Conservation of Nature (1980:18) defines conservation as "the management of human use of the biosphere so that it may yield the greatest sustainable benefit to present generations while maintaining its potential to meet the needs and aspirations of future generations." A recent symposium in Namibia, South Africa, on the ecological and economic benefits of hunting reaffirmed the superiority of managed conservation over strictly protection. Participants recognized that "sound scientific information demonstrates the importance of hunting to the future of wildlife" (Rowe 2009:392).

Legal Basis

There is sometimes considerable confusion about the function and purpose of the various natural resource agencies as it relates to conservation. The National Park Service, U.S. Forest Service, and Bureau of Land Management are all federal agencies and are referred to as land management agencies because they are responsible for managing land-based resources (e.g., timber, vegetation) rather than wildlife populations. The U.S. Fish and Wildlife Service (USFWS) has land management (national wildlife refuges) and wildlife management (endangered species and migratory birds) responsibilities. The Canadian Wildlife Service (CWS) has wildlife and land management responsibilities almost identical to the USFWS. Having a federal agency in charge of migratory birds makes sense because they can be managed in large flyways that span a large portion of the continent. However, states and provinces have agencies responsible for managing resident wildlife in their jurisdictions (excluding tribal or First Nation lands). The authority of the state and provincial agencies to manage native wildlife evolved because wildlife belongs to the public and is held in public trust. Additionally, because of the complex mosaic of land ownership patterns, it would make no sense for this authority to change as wildlife moved through parcels of land owned by different agencies and entities.

State and provincial wildlife agencies use research, population monitoring, adaptive management, emerging technologies, and experience to develop management guidelines and protocols to determine the appropriate level of harvest for each population. The allowable level of harvest might simply be that which is sustainable with no ill effects to the population, or it might be a prescription to reach specific management goals of animal abundance and demography. Societal pressures play a role in some cases, when such issues as vehicle collisions, agricultural crop damage, or simply hunter preferences affect management goals. Management goals are normally set through an open and transparent process where the public has ample opportunity to provide input. In addition, agencies are now using surveys to obtain the opinions and desires of hunters, anglers, and the general public (Decker et al. 2001, 2010b). In most wildlife management scenarios, there is a wide range of management goals that are appropriate ecologically and biologically. Because wildlife are held in public trust and managed on behalf of all citizens, it is wholly appropriate that the public have a voice in how wildlife are managed within this wide range of possible goals.

Once goals are determined, proper management involves monitoring wildlife population abundance and demography and harvest-related parameters. Both survey (e.g., abundance, distribution, sex and age ratios) and harvest (e.g., number harvested, age structure, harvest per unit effort) information represent vital data that help biologists manage wildlife populations. No single piece of information alone is sufficient; managers must use all available information in concert to make informed choices about how to achieve management goals. Managers look at the current conditions and, more importantly, the recent trends in population parameters. Trend data from male:female ratios, fawn:female ratios, age structure, population estimates or density, and abundance should all be monitored to measure current performance against the intended goals of the agency. Using these points of information, the manager can learn from and predict how certain management actions will affect the population.

The management goals and guidelines, and the actual data collected, vary greatly by species throughout North America because of a diversity of financial, logistical, and social constraints. However, all state and provincial agencies monitor populations of hunted species and apply monitoring data to management goals. Each jurisdiction has a commission or wildlife board made up of citizens or political appointees. These commissions receive recommendations from agency biologists and consider research and input from the public to make wildlife management decisions, to direct agencies, and to set the annual hunting and trapping rules and regulations.

Managing Animal Abundance

Today, hunters and trappers are the cornerstone of North American wildlife conservation because of the funds and advocacy they bring to the table, and because they remain the most effective logistical agents of actual population management. The early days of North American wildlife management were spent stopping declines of species that humans deemed worth saving, as well as encouraging population growth with limited seasons, male-only hunting for some species, daily bag limits, and other restrictions. As successful law enforcement, habitat restoration and conservation, and wildlife management programs grew, so did most wildlife populations. After the initial protections resulted in overpopulations of some species, biologists, hunters, and the public realized that their efforts were too successful in some cases, and that local populations had been allowed to exceed the carrying capacity of the habitat (Meine 1991).

When some large mammal populations were too abundant for the amount of habitat available, reproduction decreased and mortality increased because of intraspecific competition for resources available (McCullough 1979, Dusek et al. 1989). Reducing densities lessened competition and increased the population growth rate by improving reproduction and survival. Early biologists saw this compensatory effect of harvest as evidence that game populations could be managed as a renewable natural resource where the population could replace the portion removed by hunters or trappers.

It is often necessary to maintain wildlife populations below environmental and social carrying capacity to reduce die-offs, to provide for more productive populations, to protect habitat, to reduce the spread of disease, or to reduce conflicts with humans (Conover 2001). In cases where population reduction is the management goal, managers must implement harvest beyond the level at which the population can replace itself in the short term (Carpenter 2000). Population reductions or maintenance at appropriate levels are examples of hunters acting as partners in wildlife management. Conflicts with humans in the form of vehicle collisions, nuisance wildlife, livestock depredation, conflicts with agricultural production, as well as human and domestic animal health and safety may result in a goal to manage at a social carrying capacity lower than the biological limit of the habitat (Conover 2002, Duda et al. 2010b). For example, the number of deer–vehicle collisions is estimated to exceed 1.5 million every year on U.S. roadways (Conover et al. 1995). This tremendous loss of life and property illustrates the importance to all society of effectively managing wildlife abundance to appropriate levels.

In recent decades, there are many examples of hunting alone being ineffective in controlling a few species, most obviously white-tailed deer (*Odocoileus virginianus*; Brown et al. 2000) and white geese (*Chen* spp.). White-tailed deer are a highly adaptable and prolific species benefiting from habitat disturbance (including agriculture) and urban refugia. Declining access to hunting opportunities, reluctance to harvest females, and a counterproductive protectionist attitude toward wildlife exacerbate the problems related to managing their abundance. This is not an indictment of the failure of hunting as a wildlife management tool in general, but an example of how socially and biologically complex wildlife management can be.

Overabundant big game populations can dramatically alter the habitat to the detriment of many other sympatric species (Horsley et al. 2003, Rawinski 2008), underlining the importance of hunting in controlling wildlife populations. In some areas, deer overpopulation is exacerbated by trends in landowners charging access fees or otherwise restricting hunting access and thereby greatly reducing the number of hunters on the landscape (Carpenter 2000). Hunters paying for access expect lower hunter densities and high deer densities, which makes it more difficult for agencies to control game populations without full cooperation of the private landowner.

Like hunting, trapping is scientifically regulated to ensure it does not impact populations, unless that is the management goal (Todd 1981). Furbearer species are renewable natural resources, as are other harvested species, and trapping contributes to supporting the North American model. When furbearer or predator population reduction is needed, trapping provides a different tool to manage some species of wildlife that are difficult to harvest or capture any other way (Payne 1980). Participation in trapping has historically varied with changes in fur prices, but more recent societal changes have resulted in a steady decline in the number of trappers across the continent (Armstrong and Rossi 2000).

For many years, hunting and trapping were partially justified as necessary management actions to save animals from a lingering death by starvation. That is certainly true in many cases, but also not true in many more. If prey species have to be hunted because predator populations were reduced, why are predators still being hunted? The truth is more complicated than the simplistic idea of wildlife overpopulation. In reality, the importance of hunting to conservation in the broad sense is not tied simply to population control. One has to understand that a simple deer season or duck season might seem like an isolated activity, but it is merely a component—a critical one—of a much larger wildlife conservation model. Game populations are renewable resources that literally pay the bills for a far-reaching, comprehensive system of sustainable wildlife conservation that has proven itself superior to any other widely implemented model.

Demographic Effects of Consumptive Use

Disproportionate harvesting of certain sex or age classes can affect population demographics. Heavily hunted populations might have age structures and sex ratios that are very different from unexploited populations. Males of many species naturally have a higher mortality rate, resulting in more females in the population even when not hunted, which becomes exaggerated in populations with a predominately male harvest. This is much less of an issue with trapping, because trapping is not as sex biased.

Many ungulate populations are managed for high hunter opportunity, which often results in more females than males in the population. Also, heavy exploitation of the male segment will lower the average age of that part of the population. Low male:female ratios affect reproductive behavior but do not significantly affect productivity (Desimone et al. 1993, Noyes et al. 1996, Bender and Miller 1999). In heavily hunted populations of white-tailed and mule deer (*Odocoileus hemionus*), there is no indication a low number of males negatively affects reproductive rates or overall population robustness (Ozoga and Verme 1985, White 2001).

Changes to population demographics may alter social structure and breeding behavior. Studies have suggested that white-tailed deer populations with a young male age structure and low male:female ratios experience a longer, later, and less intensive breeding season in the southeastern United States, where photoperiod changes less than northern latitudes (Guynn et al. 1988, Jacobson 1992). When hunting occurs prior to the breeding season, this effect may be more pronounced. Concerns have been raised that delaying breeding dates in northern climates may produce younger, smaller offspring entering the harsh winter, resulting in lower winter survival rates. Although more research is needed, there is currently very little empirical evidence to support this concern (Bender 2002). In fact, some of the populations with the heaviest male exploitation also have the highest reproductive rates (McCaffery et al. 1998). Low male:female ratios in ungulates appear to be less of a biological concern and much more of a social concern in terms of hunter perception and satisfaction.

Hunting older ungulate males appears to be increasing in popularity, encouraged by a segment of society that places heavy emphasis on large horns and antlers. Individuals who prefer hunting large, mature animals sometimes pressure wildlife agencies to manage the entire state or province more conservatively to provide for a mature male age structure and higher hunt success. However, such requests by a minority of hunters present a dilemma for agencies because conservative management would dramatically reduce hunter opportunity, which in turn negatively affects the financial support and advocacy for wildlife and their habitat. For example, states like

Arizona issue hunting permits through a lottery-style drawing where 61,818 applicants competed for only 43,993 deer permits in 2010. Increasing the amount of conservative hunting over large areas runs contrary to the foundational success of the North American model, whereby everyone has access to hunting opportunity.

Genetic Effects of Hunting

Harvest strategies that remove a high percentage of males have the potential to affect the genetic composition of a population. Changes to the gene pool could simply occur from the outright reduction in the number of male individuals (e.g., reduced effective population size) that might result in the loss of genetic material in the population. If certain traits (e.g., horn or antler size) are heritable and hunters are exerting an intensive directional selective pressure, it could change the frequency of alleles affecting that phenotypic trait. Although many speculative papers have been written (Festa-Bianchet 2003, 2008; Darimont et al. 2009; Mysterud and Bischof 2010, Mysterud 2011), studies showing an effect of selective harvest on the size of horns or antlers are very rare. One fact that is frequently overlooked is that for hunter selection to reduce horn or antler size, animals with poor genetic potential for horn or antler growth must be given a fitness advantage and pass on more genes than those with superior genetic potential. Hunters removing the largest males are usually simply removing the oldest males in the population, not removing genetically superior animals.

Garel et al. (2007) studied a population of mixed subspecies of mouflon sheep (Ovis orientalis) in southern France that was exposed to unrestricted hunting. Through 28 years of habitat loss and intensive selection against a particular horn shape, they documented a change in horn size and shape characteristics. They attributed this change to selective hunter harvest of a particular phenotype. With a mixture of taxonomically different phenotypes and unrestricted harvest, selective harvest for the larger phenotype in this case may have provided a more obvious opportunity for detectible selective changes to take place.

Coltman et al. (2003) reported that intense selection of large male bighorn sheep (Ovis canadensis) resulting from a specific hunting regime contributed to a reduction in horn size in an isolated bighorn sheep population in Alberta. This finding triggered a chain reaction of reports in the news media and several speculative pieces in the scientific literature that inappropriately extrapolated these findings to other sheep populations and even to all hunting in general (Allendorf and Hard 2009, Minard 2009). This is clearly not the case, as this small and isolated study population and the hunt structure practiced there was unique. Also, further work on this same population indicated that the selective effect reported by Coltman et al. (2003) might have been overestimated, because they were not able to account for the confounding nutritional effects related to changes in population density (Coltman 2008). Recent work by Hengeveld and Festa-Bianchet (2011) compared completely different subspecies of sheep in different environmental condi-

tions and speculated selective pressures from trophy hunting explained the changes in horn size for one of the subspecies.

Despite all the speculative discussion papers, no other studies have shown significant phenotypic changes in ungulates that could be attributed to selective hunting, while several show no effect or contradictory trends (Brown et al. 2010, Loehr et al. 2010, Rughetti and Festa-Bianchet 2010). Singer and Zeigenfuss (2002:695) offer an alternative view regarding genetic diversity and removal of old males: "trophy hunting permits more subdominant and smaller-horned rams to obtain copulations, and thus may increase the ratio of effective population size to census population size (Ne:N) and thus increase total genetic diversity."

FUNDING THE NORTH AMERICAN MODEL

Nearly everyone enjoys wildlife, but most people are not aware of the financial contributions made by the hunters, trappers, anglers, and recreational shooters to support sustainable conservation. Neither do they realize the fundamental role hunters and trappers play in preserving the wildlife and wild places they enjoy (Duda et al. 1998). Although many consumptive users know that their financial contributions from licenses and some equipment helps pay for wildlife management, they do not always fully appreciate their key role in wildlife conservation on the broader scale, far beyond the game they pursue.

During 2006, the last year for which data are available, $233 million was distributed to state wildlife agencies in the United States from the excise tax collected on hunting, fishing, and shooting purchases (Federal Aid in Wildlife Restoration Funds, also known as the Pittman-Robertson Act; Southwick and Allen 2010). The sale of hunting and trapping licenses ($612 million) and private donations by hunters for conservation efforts ($313 million) also contributed, bringing the total to nearly $1.2 billion in 2006 alone (Southwick and Allen 2010). Some of these contributions are voluntary, but most are a requirement of participation. However, hunters who saw the value of a conservation model that protects nature through collective public stewardship built these requirements into the system (Geist 1994).

There are about 12.5 million hunters in the United States alone (USFWS 2007), and their annual expenditures provide significant support to rural communities in the United States, Mexico, and Canada. Overall, hunting and trapping voluntarily redistributes wealth from urban centers to smaller rural communities, where it is multiplied through the local economy (Duda et al. 2010b). Economic multipliers are commonly used to estimate this compounding ripple effect. In 2006, it was estimated that $24.7 billion spent in America by 12.5 million hunters had an economic impact of $66.7 billion, supporting almost 660,000 jobs in the United States (Southwick and Allen 2010). If similar figures were available for Canada, the total contribution would be staggering.

In addition to institutionalized programs, nongovernmental organizations raise and contribute additional money for specific research projects, habitat acquisition and enhancement, and population monitoring. For example, the Wild Sheep Foundation (WSF) has raised and contributed more than $70 million in the last 30 years to activities that benefit wild sheep and other wildlife (R. Lee, former chief executive officer, WSF, personal communication). Similarly, the Rocky Mountain Elk Foundation has funded 7,400 individual projects protecting or enhancing more than 2,428,114 ha of wildlife habitat (Rocky Mountain Elk Foundation 2011). Since 1973, the National Wild Turkey Federation (NWTF), in cooperation with state and federal partners, has spent more than $331 million to restore wild turkeys and to conserve more than 6,474,970 ha of habitat (J. E. Kennamer, chief conservation officer, NWTF, personal communication).

Agency Infrastructure

In recent decades, some state and provincial wildlife agencies became creative in their ability to garner additional funding sources to supplement the long-standing contribution of hunters, anglers, and shooting enthusiasts. Lottery sources, state income tax, special stamps, and similar funds are sometimes channeled to wildlife agencies and ear-marked for programs that have not received adequate financial support in the past (e.g., nongame or habitat acquisition and management). These recent supplemental funds are an important addition to the budgets of wildlife agencies, but they are vulnerable to legislative meddling and do not replace or negate the importance of the base funds from consumptive activities. Supplemental funds are only effective because there is an agency infrastructure in place that can take additional money and apply it directly to a specific program area. Funds from consumptive wildlife support law enforcement, personnel resources, and all other day-to-day agency operations, allowing supplemental dollars to be effective. The agencies most effective in conserving all resources, whether hunted or not, are those with a solid financial foundation provided by the well-regulated consumption of a few wildlife species.

Law Enforcement

One of the important contributions to wildlife conservation that hunters, trappers, and anglers have made is the creation and maintenance of law enforcement officers to uphold the massive amount of legal restrictions to wildlife harvest. Regulated hunting is only regulated if the laws are obeyed. Surveys consistently report that about 50% of hunters and anglers have had recent contact with these enforcement personnel in the field and hold them in high regard (Duda et al. 1998).

Currently, thousands of wildlife conservation law enforcement officers are actively working in the United States and Canada, and most are paid with income from the sale of hunting, trapping, and fishing licenses. This ubiquitous wildlife law enforcement presence is almost always independent of other law enforcement agencies and allows for the majority of their time and energy to be devoted to protecting natural resources. Besides policing hunters and anglers, they also perform duties related to water quality, habitat protection, public safety,

search and rescue, littering, vandalism, trade in threatened and endangered species, and providing backup to other local law enforcement agencies. Opponents of hunting rarely offer alternatives for funding trained officers to protect wildlife against exploitation. If hunting were made illegal in North America, we would immediately lose this massive protection force and likely degrade into the unregulated destruction that was common before hunting was institutionalized as the basis (rather than the bane) of conservation.

Population Restoration

The restoration of wildlife populations across North America is the greatest wildlife success story in the history of conservation anywhere. Conservation has restored nearly all of the badly overexploited populations before the development and implementation of the North American model. Species like Canada geese (*Branta canadensis*), wood ducks (*Aix sponsa*), white-tailed deer, pronghorns (*Antilocapra americana*), bighorn sheep, and wild turkeys (*Meleagris gallopavo*) all represent important species whose restoration was made possible by funding and advocacy generated by hunting.

The state of Arizona began restoring desert bighorn sheep in 1955 with translocations to historical habitat. Since then, more than 100 translocations of at least 1,800 bighorn sheep have restored this iconic species in all previously occupied habitat and many other areas of suitable habitat (O'Dell 2007). Across North America, wild sheep populations have been restored with over 1,500 translocations involving about 25,000 wild sheep since 1922 (K. Hurley, Wild Sheep Foundation, personal communication), exemplifying the type of restoration activity that has occurred for decades throughout all states and provinces in North America with the funds generated from the regulated consumption of a few species.

North America has a nearly full complement of native wildlife living in habitat that has changed remarkably little in the last 300 years, compared to other continents. Restoration of large mammal populations continues today, with elk (*Cervus canadensis*) being successfully translocated into historical ranges in the East for the enjoyment of all residents. Work also continues for other species, such as bison (*Bison bison*) and large predators whose restoration comes with significant societal controversy. Individual hunters or even some organizations might not be supportive of the restoration of some large predators, but they support the system of collective stewardship that works to restore native species. With the restoration success of hunted species, focus has shifted to restoring nonhunted species, with threatened and endangered animals receiving the highest priority.

Monitoring and Management of Wildlife Populations

Monitoring wildlife populations and accumulating baseline trend data are the bases of well-informed, science-based decisions that are foundational to the North American model (Baughman and King 2008). Hunted species are not the only ones monitored, but they generally do receive the most attention. State, provincial, and federal agencies have a history of monitoring wildlife populations, beginning at the very genesis of wildlife conservation in North America. Many agencies have examples of monitoring programs that have remained relatively consistent for decades and provide valuable trend data. The CWS and USFWS have conducted continent-wide aerial waterfowl surveys since 1955. This cooperative survey effort involves flying more than 128,000 km of survey each year throughout waterfowl areas from southern Mexico to northern Canada.

It is usually not necessary to monitor intensively the abundance and demographics of populations that are not being annually harvested, unless such species are causing significant damage, are negatively impacting habitats or other species, or are threatened or endangered. Under the current system, wildlife species are monitored and managed fairly intensively by agencies financially solvent enough to use resources for any other species as the need arises.

Habitat Acquisition, Protection, Restoration, and Enhancement

Land management agencies manage wildlife habitat on millions of hectares of federal and Crown land. Many states and provinces have also purchased wildlife habitat with the proceeds from hunting licenses and taxes on some specific hunting, fishing, and shooting equipment. Funding sources for habitat vary among Canadian provinces. For example, in British Columbia, surcharges collected on hunting, angling, trapping, and guide-outfitting licenses go into a trust fund managed by an independent organization, the Habitat Conservation Trust Foundation (HCTF). The HCTF annually invests $5–6 million in habitat acquisition, restoration, and enhancement and other priority conservation projects throughout the province.

During a five-year period (2005–2009) in the United States, $58.5 million from Federal Aid in Wildlife Restoration funds was apportioned to the states for the acquisition of more than 12.2 million ha of wildlife habitat (USFWS 2010b). In addition, such wildlife conservation organizations as the Rocky Mountain Elk Foundation, Wildlife Habitat Canada, Mule Deer Foundation, The Nature Conservancy, Ducks Unlimited, Canadian Wildlife Federation, National Wild Turkey Federation, Pheasants Forever, the Wild Sheep Foundation, and myriad state and provincial wildlife organizations used private donations to purchase land or conservation easements on large tracts of wildlife habitat. Most of these areas are purchased with game animals in mind, but wetlands acquired for waterfowl, forests purchased for deer or turkeys, mountainous areas protected for wild sheep, and grasslands restored for quail and pronghorn have benefited countless nongame and endangered species that rely on those habitat associations. Recent estimates indicate about 70% of users in these areas do not hunt, and in some properties the percentage may be as high as 95% (USFWS 2010c). Ironically, there are sometimes conflicts when nonconsumptive users express concern about seeing hunters on these properties during the few days or weeks each year the hunting seasons are open.

Research

One of the foundations of the North American model is that management decisions are based in science. In the United States, about $57 million was apportioned in 2009 to state wildlife agencies from the Federal Aid in Wildlife Restoration Program for conducting more than 10,000 wildlife research projects (USFWS 2010b). In the early years of the wildlife management profession, money was spent exclusively on learning more about species that were at low levels because of the lack of a comprehensive system of wildlife conservation. As more was learned about managing those species back to abundance, focus shifted to all species and their habitats.

Canadian wildlife researchers obtain funding from a wide variety of sources. All revenue from hunting and fishing in Canada is placed into the general revenue and then distributed to each fish and wildlife agency. Most provinces have an association or trust fund that funds research and land acquisition. Wildlife conservation organizations (e.g., Rocky Mountain Elk Foundation, Alberta Conservation Association, WSF) and extractive industries (e.g., oil companies) also contribute money for wildlife research.

To facilitate meaningful research in the United States, a series of cooperative research units were established in 1935 at universities to provide an opportunity for the federal government, state wildlife agencies, universities, and nongovernmental organizations to work together. These units receive federal funds to employ two to five scientists, but most of the annual baseline operating budget comes from hunters and anglers through the state wildlife agencies. This system began with seven "co-op units" and has grown to 44 distributed across 40 states. Canada has also established a Cooperative Fish and Wildlife Research unit in New Brunswick modeled after this U.S. system. Through the last 75 years, funding from hunters has provided information that influences intelligent management decisions to better understand and conserve hunted and nonhunted species.

Hunter and Trapper Education Programs

Hunter education programs are important to wildlife conservation because illegal acts by a few hunters can cast all hunters in a bad light and erode public support for this successful conservation paradigm. North America has a network of hunter education programs delivering information and coursework on wildlife management, hunter ethics, firearms safety, and hunting and trapping techniques. Each year a volunteer hunter education instructor force of more than 70,000 trains about 650,000 hunters (W. East, International Hunter Education Association, personal communication). An annual apportionment of Federal Aid in Wildlife Restoration provides funding for such training in the United States, funds that exceeded $472 million in 2010. Each state receives between $2 million and $20 million, depending on their size and need (USFWS 2010a). In Canada, these programs are either paid for by users or funded through the provincial wildlife agencies. The successful completion of a hunter education course is mandatory for certain age classes and certain kinds of hunting in all 50 states and ten Canadian provinces, and has resulted in more than 35 million students be-

OLOF CHARLES WALLMO (1919–1982)

O. C. "Charlie" Wallmo was born in Iowa in 1919 and studied forestry and wildlife at the Universities of Wisconsin and Montana before completing his bachelor's degree at Utah State University in 1947. He returned to the University of Wisconsin for his masters degree and then attended Texas A&M University, where he earned a Ph.D. Through his work in Texas, Arizona, Alaska, and the Rocky Mountains, Wallmo pioneered research that resulted in many of the fundamental and foundational concepts in wildlife management.

He conducted the first comprehensive study of scaled quail ecology early in his career. He was also one of the first to use free-ranging tame deer as research tools to elucidate diet, behavior, and metabolism of mule deer. Wallmo was sought-after for his knowledge of mule deer nutrition and the effects of habitat manipulations on deer population dynamics. His work in the central Rockies showed the benefits of small forest clear-cuts to deer nutrition, and early work on deer survey methodology formed the basis for improved management of deer populations. His efforts in southeast Alaska demonstrated the value of overstory cover for black-tailed deer during winter.

Wallmo published more than 50 significant publications, and his edited tome *Mule and Black-tailed Deer of North America* still serves as the primary source of basic information about that species. Even though he was known for his dedication to science and the scientific process, his lasting legacy is not volumes of esoteric scientific publications or reams of data analysis, but important contributions to the body of knowledge that wildlife managers used for decades as the foundation for improved management. In addition, many of his former graduate students have become known for their work with cervids across North America. Finally, every two years, North America's leading black-tailed and mule deer biologist is honored with the Wallmo Award.

Photo courtesy of Joe Wallmo

ing trained since the beginning of the program (International Hunter Education Association 2010a,b).

Nongame Including Threatened or Endangered Species

A preponderance of hunter-generated money is still expended on the management and protection of hunted species. This is appropriate because populations of species that are being annually hunted generally require a greater intensity of monitoring, law enforcement, research, and management.

Various sources, including income tax check-offs, special stamps, independent grants, donations, lottery or gambling revenue, some sales tax, and hunters' dollars from the Federal Aid in Wildlife Restoration Program fund nongame activities. In Canada, funding comes from provincial and federal sources with the management responsibility remaining with provinces (except for migratory birds). A portion of Wildlife Restoration funds in the United States is available to the states for conservation of birds and mammals that are not hunted. This funding mechanism provides millions of dollars annually for the conservation of nonhunted species. However, this is a small percentage of what is needed and does not begin to address fully the needs of all other taxonomic groups such as native fish, songbirds, amphibians, and reptiles. As a consequence, wildlife agencies must be creative to find and maintain additional funds for management of species that were not obviously exploited historically and therefore are not the immediate focus of the Wildlife Restoration Program.

Through the current authorization of the Wildlife Restoration Program, hunters' dollars contribute to the restoration of many threatened and endangered species such as the California condor (*Gymnogyps californianus*), Mexican gray wolf (*Canis lupus baileyi*), black-tailed prairie dog (*Cynomys ludovicianus*), and the black-footed ferret (*Mustela nigripes*). Under this system, conservation actions have the potential to protect other wildlife species before they become threatened or endangered (Baughman and King 2008). The future of conservation in North America will have to include the existing model of using hunters' dollars to conserve all wildlife for all people.

Information and Public Relations

Communicating with the public and considering human dimensions in wildlife management have become vital to the effectiveness of management agencies (Decker et al. 2001). All wildlife agencies have some public information officers on staff to disseminate wildlife information and to update stakeholders on agency activities through press releases, websites, social networking media, radio, television, and a multitude of publications for diverse audiences. Some wildlife agencies use Federal Aid in Wildlife Restoration funds, but most simply use money garnered from the sale of hunting and fishing licenses. In this way, the entire public benefits from the baseline funding provided by regulated hunting. In Canada, much of the funding for these kinds of activities comes from taxes, which poses a problem for wildlife agencies as funding cuts and downsizing continually erodes their ability to be effective.

ADVOCACY FOR WILDLIFE AND WILDLANDS

Management guidelines control hunting and trapping efforts and law enforcement officers enforce regulations, but much of North America's conservation success is due to incentives based on self-interest and personal ethics (Leopold 1933). Many outdoorspeople in North America go far beyond what the law requires. Historically, hunting has been the greatest force in assuring wildlife a place on the landscape. Consumptive use fosters attention, and wildlife thrives with attention and withers from neglect (Geist 2006). The strong desire to hunt wildlife appears to be deeply primordial. Most commonly, the passion to hunt expresses itself as a deep, lifelong interest in and devotion to wildlife, often accompanied by considerable work, even sacrifice, by the hunter on behalf of wildlife. Witness the many organizations dedicated to the conservation of wildlife in North America. Arizona's first desert bighorn sheep hunt occurred in 1953, and it was probably no coincidence that two years later the state wildlife agency began its aggressive translocation program that has now restored all of the state's historical sheep populations. There are endless examples of hunter advocacy for being instrumental in implementing game management and wildlife conservation on a broad scale. It comes as no surprise that Aldo Leopold, considered the founder of the wildlife management profession, was an avid, lifelong hunter (Bradley 1995, Peyton 2000).

No collective group is composed entirely of active leaders, and hunting is no different. A small percentage of the hunting community rises to the position of spokespersons or leaders of conservation organizations. As in any group, the majority is happy to follow leaders and follow the rules. For this reason, it may be unreasonable to expect every hunter to act as a steward of broad-scale biodiversity at the individual level (Holsman 2000). The hunting community, acting as collective stewards for the greater good of wildlife and their habitat, has influenced this successful conservation paradigm.

Political Support

Early groups of organized hunters were instrumental in providing the political support necessary to implement many of the laws that coalesced into the system of conservation we have today. For example, Theodore Roosevelt organized the Boone and Crockett Club by assembling most of the powerful and influential conservation-minded people of the day, many of them hunters. The Boone and Crockett Club successfully lobbied for the establishment of Yellowstone National Park, the preservation of the bison, cessation of market hunting, and much more in the 20th century.

It is difficult to maintain separation between the sometimes-detrimental world of politics and wildlife management. When political influence threatens proper wildlife conservation efforts, sportsmen at the local and national level have shown themselves willing and able to come together in support of wildlife. There are many examples of wildlife agency funds or commission structure coming under attack by politicians, only to have the organized wildlife groups step up to its de-fense. A survey of outdoor user groups in the southwestern United States asked respondents if they would be willing to write a letter if wildlife conservation funding was threatened with diversion to other uses. Seventy-four percent of hunters responded they would be likely or very likely to write a letter protesting such action (Responsive Management 1995).

Nonhunters are often involved, but it is the organizational infrastructure of hunting organizations that is frequently the vehicle that drives such coordinated defensive activities. This infrastructure is also used for mounting campaigns in defense of crucial wildlife habitat threatened by conflicting interests. With declines in the proportion of the population that hunts, wildlife may have a less organized, less effective voice on their behalf. As wildlife and their habitat are subjected to increasing pressures, hunters and nonhunters will need to focus on common goals and combine their collective resources for the good of the wildlife they both enjoy.

Biological Samples and Information

Hunters and trappers are an important source of biological information for wildlife managers. Harvest data such as total number harvested, sex and age ratios, body weight or condition, and harvest location have been collected at check stations since the early years of wildlife management. Other hunt-related information is routinely collected at check stations, in the field, or with posthunt questionnaires by phone or mail. Whether the hunter was successful, the total number of hunter-days expended, area hunted, and other information can be used to track trends in population parameters or abundance. In some cases, these check stations and questionnaires are mandatory, but in many instances hunters go to great lengths to provide information that might help managers.

Biological samples can help determine prior disease exposure, parasite loads, nutritional status, genetic relationships or diversity, and approximate age. Hunters themselves sometimes collect these samples, requiring a certain amount of cooperation and commitment. Along with other members of the public, hunters routinely provide information on species distribution and sources of unusual mortality. Such input is valuable in tracking changes in wildlife distributions in the face of habitat loss, natural disasters, climate change, and emerging disease issues.

Volunteerism

Hunters and trappers individually, and the organizations to which they belong, have always been active in volunteering for habitat improvement projects, constructing nesting structures or boxes, altering fences to be wildlife friendly, teaching hunting and trapping education courses, conducting wildlife surveys, working check stations, performing routine facility maintenance, cleaning up trash, and many other beneficial activities (Bleich 1990). These volunteer efforts benefit wildlife directly and allow wildlife management agencies to stretch their conservation dollars farther to accomplish additional goals.

Most of these projects benefit more than game species. For example, big game hunters throughout the West have installed water collection and retention devices, but uncounted numbers of bird, mammal, insect, reptile, and amphibian species use them, too. Designs of water catchments for large mammals have been altered through the years to accommodate the needs of bats and smaller terrestrial nongame animals.

Residents in the state of Maryland were asked if they would be interested in volunteering their time to help the state wildlife agency (Responsive Management 1993). Results revealed that 22% of hunters were "very likely" to volunteer, compared to 7% of the nonhunters. This does not imply nonhunters do not care about wildlife, but it does illustrate the level of commitment to collective stewardship of natural resources that is inherent in the hunting community.

THE CONSISTENCY OF CHANGE

One constant throughout the history of wildlife management is the societal change to which managers must adapt. The contribution of hunters and trappers to wildlife and habitat conservation is undeniable. Consumptive users' contributions have been steady and consistent through time, even as society at large has changed. These sociological trends are not likely to abate or reverse themselves, so resource managers must accept these sociological shifts as the new reality.

Foundational Changes and Trends

The number of hunters has decreased for decades, while the overall population continues to grow (Duda et al. 2010a). From 1980 to 1996 the percentage of Americans who hunted dropped from 10% to 5% (Schuett et al. 2009). This decline had many causes, mostly related to changes in society and not directly to the act of hunting. This decline in consumptive use of wildlife mirrored declines in the percentage of people who engage in wildlife watching. Wildlife watching declined from 17% to 10% during the same 1980–1996 period (Schuett et al. 2009).

Southwick (2010) made predictions about future trends in hunting participation in the United States using data from the 1991 and 2006 National Survey of Fishing, Hunting, and Wildlife-Associated Recreation (USFWS 2007). They predicted the number of hunters would rise from 12.7 million in 2010 to 13.9 million by 2050, but the percent of the overall population that hunts would decline from 5.2% to 4% during that same period (Southwick 2010).

Trapping as practiced avocationally has been on the decline for decades. However, trapping conducted for the control of nuisance wildlife is on the rise because of increasing conflicts with urban dwellers (Armstrong and Rossi 2000). Trapping is also an important research and management technique used by wildlife professionals for specific objectives that would be difficult to accomplish any other way. It was estimated there were about 145,000 avocational trappers in the United States in 1999, with only 18 states having more than 1,500 trappers (Armstrong and Rossi 2000). Declines in avocational trapping are the result of many things, including depressed fur prices, less access to lands, and societal changes that have led to trapping bans for many species in California (1998), Colorado

(1996), Arizona (1994), Washington (2000), and Massachusetts (1996; Duda et al. 2010a:197).

The public's perceptions of, and interactions with, nature have changed tremendously in only a few decades. After more than four million years of interacting first hand with animals and nature, humans became detached from the natural world in a geologic blink of an eye. More people moved to urban settings where they no longer hunted for food or butchered their own livestock. An increasing number of people obtain their information about nature from digital media in the form of television and the Internet. Television presentations of nature focus on individual animals and not on the realities of managing populations and executing successful conservation programs. Many television programs depict the rescue of individual animals, sometimes members of extremely abundant species that may be overpopulated at the time. These programs, coupled with incorrect information available online, create a public with very distorted views of traditional wildlife management. This severing of first-hand ties to nature has had a profound impact on how the latest generations perceive the natural world as a whole and specific wildlife management actions in particular.

Television programs in the 1950s depicted hunters as rugged and self-sufficient, skilled to thrive in the outdoors. They were portrayed as admirable individuals who cared about nature. Today, hunters and trappers in television and movie roles are more often portrayed as criminals or unsavory characters. This incremental decline in the status of hunters in popular culture is partially because of the societal changes above, but also because wildlife educators, agencies, and professionals have not done an effective job articulating and illustrating how important hunting and trapping are to the successful North American system of conservation.

As society changed, the profession of wildlife management was changing, too. Wildlife professionals in universities and state, provincial, and federal agencies grew more diversified to meet the challenges of all wildlife, and an increasing percentage of them did not come from farming, ranching, or hunting backgrounds. Individuals who have had no exposure to consumptive use of wildlife or even those who come from an animal protectionist background are ascending into the ranks of tenured university professors, researchers, scientific consultants, and agency administrators. With them comes an unfamiliarity and sometimes bias against some or all forms of consumptive wildlife use. Unfortunately, these biases make their way into inferences made in research and the recommendations made to wildlife management practitioners.

Some are unwilling or reluctant to appropriately credit hunters and trappers for the success this continent has enjoyed in the conservation of wildlife and the habitat upon which they depend (Nelson et al. 2011). Unfortunately, criticisms are often rooted in personal bias rather than scholarly critiques or credible faults in the North American model. For example, Nelson et al. (2011) begin their historical interpretation in the 1960s, decades after the North American model was constructed around the principle of controlled consumptive use of wildlife. They further obfuscate the issue by comparing the historical role of hunters in conservation to child labor and by asking esoteric questions about social justice, human liberty, and ethical reasoning. Other criticisms will follow, and there is certainly room for improvement in some aspects of our current implementation of conservation, but the lack of a superior alternative only supports the framework of the North American model.

Continuing the Hunter's Role in Conservation

Because hunters had a lead role in the development of the most successful system of wildlife conservation does not mean they own the future of this paradigm. Societal changes are already challenging the foundation of conservation through consumption. Those truly interested in perpetuating a realistic and proven conservation model will need to work to preserve it. Future efforts to conserve wildlife and wild places will not succeed without a broad base of public support (Mahoney 2007). History has illustrated the failure of conservation prescribed by the elite and fashioned after the desires of the minority. Stewardship of wildlife has always taken a lot of hard work and that will not change, nor will the need for advocacy backed by science-based research, education, and management programs.

Internationally, the superiority of the North American model has not gone unnoticed. Africa, Asia, Russia, and many other parts of the world are making attempts to implement what has worked so well in North America (Mahoney 2007). Two recent symposia (in London and Namibia) addressed the success and application of hunting-based conservation, recognizing the success of this conservation model (Mahoney 2009, Patterson 2009). Decades of attempting to build conservation programs based on a preservation paradigm without local grassroots support or funding have largely failed.

Regulated hunting currently enjoys a broad base of public support in North America. Several surveys have consistently reported that 75–81% of respondents support hunting and agree it should continue (Duda et al. 2010a:32). Trend data from sequential surveys indicate there may be an increasing proportion of Americans who approve of legal fair-chase hunting when the harvest is utilized (Duda et al. 2010a:44). The re-emergence of positive media portrayals of hunters and the political attention paid to hunters by presidential candidates reflect this trend in subtle ways. Wildlife management agencies and conservation organizations have been discussing and celebrating the continued success of the North American model for the last decade, and perhaps the general public has noticed.

The continued success of hunters, trappers, and anglers supporting conservation will depend on a public that understands how the system works. Everyone enjoys seeing wildlife and beautiful wild places when enjoying the outdoors. A camping, hiking, or hunting trip is much more enjoyable when wildlife are seen. And yet most people do not understand what drives and funds the programs that provide the foundation for the nature-based opportunities they enjoy. The founders of the North American model worked hard to turn the tide of public opinion in support of a widespread conservation ethic (Reiger 1986, Mahoney 2007). Those interested in wildlife must continue that hard work by increasing the awareness of this uniquely effective system among an increasingly disengaged public.

The future success of the North American model will require hunters and trappers to remain relevant to conservation (Mahoney 2007). Hunters and trappers must be recognized for their past, present, and potential future contributions (Geist et al. 2001). Remaining relevant also means they are not seen as degrading or obstructing wildlife conservation efforts. Without an alternative paradigm, no serious wildlife advocate is calling for fundamental changes in the North American model, but there is room for improvement in several areas. Consumptive wildlife users will have to do a better job of policing themselves so as not to show hunters in a bad light, thereby eroding public confidence in a hunter-based system of conservation. Hunters and trappers must continue to demonstrate and articulate that they are truly stewards of all wildlife (Posewitz 2002). How the hunter communicates and positions himself in the minds of the nonhunting public will decide whether hunting will be supported far into the future.

Producing deer, ducks, and turkeys in abundance is not what wildlife conservation has been about in the past, nor will it be in the future. Wildlife managers and their constituent public need to support conservation of all native wildlife, not just those that are hunted. Currently, carnivores are testing the public's support for a full complement of native wildlife (predators and prey), as well as our ability to manage both predators and prey without being crippled by laws, lawsuits, and regulations that can grind proper management efforts to a halt. Completing the recovery of all native wildlife in North America continues, but it will not be possible to restore all species to all former habitat, because humans are now well distributed across the landscape. We should support restoration of large carnivores, but they must be subject to the same scientific management that has been so successful for all other restored species of large mammals.

Human dimensions research to assess the desires of stakeholders and the public at large has become an important part of modern resource management. Gauging public sentiment, helping the public understand the trade-offs, and then determining how the public would like their wildlife resources managed will become even more critical in the future. Most wildlife enthusiasts have a common goal; everyone has an investment in sustainable wildlife resources for the future. Resource managers, consumptive users, nonconsumptive users, and all other stakeholders will have to work together to guide conservation actions that result in abundant and properly managed populations of native wildlife persisting in minimally disturbed habitat for generations to come.

SUMMARY

Hunting and trapping have been an important part of human existence since before we were capable of recognizing it as such. In recent centuries, technology evolved rapidly to the point where humans could have an impact on the numbers of animals around them. On most continents, unsustainable exploitation of wildlife resources was not recognized and halted in time to retain most native species. In North America, however, individuals concerned about natural resources recognized that something had to change in order to protect major species of native birds and mammals. This concern, led by some of the most prominent hunters and other conservationists of the day, developed a collective sense of stewardship and conservation ethic. The desire to maintain wildlife populations into perpetuity led to North America's unique and incredibly successful system, whereby consumptive use of wildlife in the form of hunting and trapping became the cornerstone of what is now known as the North American Model of Wildlife Conservation.

This conservation paradigm, built around scientifically regulated harvest, has benefited nearly all species of native wildlife because advocates conserved, restored, and managed habitat for game species. They also conserved a landscape that provided habitat for nonhunted species. This system of conservation has been so successful, it has been recognized as a model to emulate worldwide. State and provincial wildlife agencies use research, population monitoring, adaptive management, emerging technologies, and experience to develop management guidelines and protocols to determine the appropriate level of harvest for each population. The allowable level of harvest might simply be that which is sustainable to the population, or it might be a prescription to reach specific management goals of animal abundance and demography. All of this is determined through a public process that sets management goals and guidelines. The types of management data collected vary greatly by species throughout North America because of a diversity of financial, logistical, and social constraints; however, all state and provincial agencies monitor populations of hunted species and apply monitoring data to management goals.

Today, hunters and trappers are the bedrock of North American wildlife conservation because of the funds and advocacy they contribute, and because they remain the most effective logistical agents of actual population management. The early days of North American wildlife management were spent stopping declines of those species humans deemed important and worth saving. Early protections resulted in recovery of populations and even overpopulations of some species, and so management agencies switched from recovery actions to management programs. Continual population management through hunting is mindful of genetic effects and strives to maintain demographic and abundance characteristics defined by management goals.

Most people are not aware that the wildlife they enjoy viewing depend on the financial contributions made by hunters to support sustainable conservation. Hunters, trappers, anglers, and recreational shooters play a critical role in preserving wildlife and wild places. The agencies most effective in conserving all resources, whether hunted or not, are those with a solid financial foundation provided by the well-regulated consumption of a few wildlife species. Hundreds of millions of dollars are generated for wildlife conservation as part of a program that levies an excise tax on hunting, fishing, and shooting equipment and distributes these funds to wildlife management agencies. In addition, the sales of hunting and fishing licenses, as well as private contributions, raise more than $1.2 billion each year. This money supports law enforcement officers and agency infrastructure, helps monitor trends in wildlife populations, restores species to historical habitat, and finances research. It also goes toward conserving nonhunted species, both threatened and endangered; informing the public; and acquiring, protecting, and enhancing wildlife habitat.

Hunters and trappers and the organizations to which they belong are strong advocates for wildlife and their habitat. They provide volunteer labor for improving habitat, educating hunters and trappers, and surveying wildlife, among other activities. Consumptive users of wildlife also consistently provide biological samples and information that biologists use to manage wildlife populations. Through these activities emerges a community-based infrastructure advocating for wildlife conservation.

Regulated hunting currently enjoys a broad base of public support in North America. But societal change brings with it new problems, and wildlife managers must be prepared to respond. The contribution of hunters and trappers to wildlife and habitat conservation is undeniable, but sociological trends will likely pose challenges in the future. Human dimensions research to assess the desires of stakeholders and the public at large plays an increasing role in resource management. Most wildlife enthusiasts have a common goal; everyone has an investment in sustainable wildlife resources for the future. As wildlife and their habitat are subjected to increasing pressures, hunters and nonhunters will need to focus on common goals and combine their collective resources for the good of the wildlife they all enjoy.

Literature Cited

Allendorf, F. W., and J. J. Hard. 2009. Human-induced evolution caused by unnatural selection through harvest of wild animals. Proceedings of the National Academy of Sciences 106:9987–9994.

Ardrey, R. 1976. The hunting hypothesis. McClelland and Stewart. Toronto, Ontario, Canada.

Armstrong, J. B., and A. N. Rossi. 2000. Status of avocational trapping based on the perspectives of state furbearer biologists. Wildlife Society Bulletin 28:825–832.

Baughman, J., and M. King. 2008. Funding the North American Model of Wildlife Conservation in the United States. Pages 57–64 in J. Nobile and M. D. Duda, editors. Strengthening America's hunting heritage and wildlife conservation in the 21st century: challenges and opportunities. Responsive Management, Harrisonburg, Virginia, USA.

Bender, L. C. 2002. Effects of bull elk demographics on age categories of harem bulls. Wildlife Society Bulletin 30:193–199.

Bender, L. C., and P. J. Miller. 1999. Effects of elk harvest strategy on bull demographics and herd composition. Wildlife Society Bulletin 27:1032–1037.

Bleich, V. C. 1990. Affiliations of volunteers participating in California wildlife water development projects. Pages 187–192 in G. K. Tsukamoto and S. J. Stiver, editors. Wildlife water development. Nevada Department of Wildlife, Reno, USA.

Bradley, N. L. 1995. How hunting affected Aldo Leopold's thinking and his commitment to a land ethic. Pages 10–13 in Proceedings of the fourth annual governor's symposium on North American hunting. North American Hunting Club, Minnetonka, Minnesota, USA.

Brown, D. E., W. C. Keebler, and C. D. Mitchell. 2010. Hunting and trophy horn size in male pronghorn. Proceedings of the Pronghorn Workshop 24:30–45.

Brown, T. L., D. J. Decker, S. J. Riley, J. W. Enck, T. B. Lauber, P. D. Curtis, and G. F. Mattfeld. 2000. The future of hunting as a mechanism to control white-tailed deer populations. Wildlife Society Bulletin 28:797–807.

Carpenter, L. H. 2000. Harvest management goals. Pages 192–213 in S. Demarais and P. R. Krausman, editors. Ecology and management of large mammals in North America. Prentice-Hall, Upper Saddle River, New Jersey, USA.

Coltman, D. W. 2008. Molecular ecological approached to studying the evolutionary impacts of selective harvesting in wildlife. Molecular Ecology 16:221–235.

Coltman, D. W., P. O'Donoghue, J. T. Jorgenson, J. T. Hogg, C. Strobeck, and M. Festa-Bianchet. 2003. Undesirable evolutionary consequences of trophy hunting. Nature 426:655–658.

Conover, M. R. 2001. Effect of hunting and trapping on wildlife damage. Wildlife Society Bulletin 29:521–532.

Conover, M. R. 2002. Resolving human–wildlife conflicts: the science of wildlife damage management. CRC Press, Boca Raton, Florida, USA.

Conover, M. R., W. C. Pitt, K. K. Kessler, T. J. DuBow, and W. A. Sanborn. 1995. Review of human injuries, illnesses, and economic losses caused by wildlife in the United States. Wildlife Society Bulletin 23:407–414.

Darimont, C. T., S. M. Carlson, M. T. Kinnison, P. C. Paquet, T. E. Reimchen, and C. C. Wilmers. 2009. Human predators outpace other agents of trait change in the wild. Proceedings of the National Academy of Sciences 106:952–954.

Decker, D. J., T. L. Brown, and W. F. Siemer. 2001. Human dimensions of wildlife management in North America. The Wildlife Society, Bethesda, Maryland, USA.

Desimone, R., J. Vore, and T. Carlson. 1993. Older bulls—who needs them? Pages 29–35 in J. D. Cada, J. G. Peterson, and T. N. Lonner, editors. Proceedings of the western states and provinces elk workshop. Montana Wildlife, Fisheries and Parks, Helena, USA.

Duda, M. D., S. J. Bissell, and K. C. Young. 1998. Wildlife and the American mind: public opinions on and attitudes toward fish and wildlife management. Responsive Management, Harrisonburg, Virginia, USA.

Duda, M. D., M. F. Jones, and A. Criscione. 2010a. The sportsman's voice. Venture, State College, Pennsylvania, USA.

Duda, M. D., M. Jones, A. Criscione, and A. Ritchie. 2010b. The importance of hunting and the shooting sports on state, national and global economies. Pages 276–293 in World symposium: ecologic and economic benefits of hunting. World Forum on the Future of Sport Shooting Activities, Windhoek, Namibia.

Dusek, G. L., R. J. Mackie, J. D. Herringes Jr., and B. B. Compton. 1989. Population ecology of white-tailed deer along the lower Yellowstone River. Wildlife Monographs 104:1–68.

Festa-Bianchet, M. 2003. Exploitative wildlife management as a selective pressure for the life history evolution of large mammals. Pages 191–207 in M. Festa-Bianchet and M. Apollonio, editors. Animal behavior and wildlife conservation. Island Press, Washington, D.C., USA.

Festa-Bianchet, M. 2008. Ecology, evolution, economics, and ungulate management. Pages 183–202 in T. E. Fulbright and D. G. Hewitt, editors. Wildlife science: linking theory and management applications. CRC Press, Boca Raton, Florida, USA.

Garel, M., J. M. Cugnasse, D. Maillard, J. M. Gaillard, A. J. M. Hewison, and D. Dubray. 2007. Selective harvesting and habitat loss produce long-term life history changes in a mouflon population. Ecological Applications 17:1607–1618.

Geist, V. 1994. Wildlife conservation as wealth. Nature 368:491–492.

Geist, V. 2006. The North American Model of Wildlife Conservation: a means of creating wealth and protecting public health while generating biodiversity. Pages 285–293 in D. M. Lavigne, editor. Gaining ground: in pursuit of ecological sustainability. International Fund for Animal Welfare, University of Limerick, Limerick, Ireland.

Geist, V., S. P. Mahoney, and J. F. Organ, 2001. Why hunting has defined the North American model of wildlife conservation. Transactions of the North American Wildlife and Natural Resources Conference 66:175–185.

Grayson, D. K., and D. J. Meltzer. 2002. Clovis hunting and large mammal extinction: a critical review of the evidence. Journal of World Prehistory 16:313–359.

Grayson, D. K., and D. J. Meltzer. 2003. A requiem for North American overkill. Journal of Archaeological Science 30:585–593.

Grinnell, G. B. 1913. The game preservation committee. Pages 421–432 in G. B. Grinnell, editor. Hunting at high altitudes. Harper and Brothers, New York, New York, USA.

Guynn, D. C., Jr., J. R. Sweeney, R. J. Hamilton, and R. L. Marchington. 1988. A case study in quality deer management. South Carolina White-tailed Deer Management Workshop 2:72–79.

Haile-Selassie, Y., G. Suwa, and T. D. White. 2004. Late Miocene teeth from Middle Awash, Ethiopia, and early hominid dental evolution. Science 303:1503–1505.

Haury, E. W., E. B. Sayles, and W. W. Wasley. 1959. The Lehner mammoth site, southeastern Arizona. American Antiquity 25:2–32.

Hengeveld, P. E., and M. Festa-Bianchet. 2011. Harvest regulations and artificial selection on horn size in male bighorn sheep. Journal of Wildlife Management 75:189–197.

Holsman, R. H. 2000. Goodwill hunting? Exploring the role of hunters as ecosystem stewards. Wildlife Society Bulletin 28:808–816.

Horsley, S. B., S. L. Stout, and D. S. deCalesta, 2003. White-tailed deer impact on the vegetation dynamics of a northern hardwood forest. Ecological Applications 13:98–118.

International Hunter Education Association. 2010a. Who we are. http://ihea-usa.org/about-ihea/usa/who-we-are. Accessed 11 October 2012.

International Hunter Education Association. 2010b. Hunter education requirements. http://ihea-usa.org/hunting-and-shooting/hunter-education. Accessed 11 October 2012.

International Union for Conservation of Nature. 1980. World conservation strategy. Gland, Switzerland.

Jacobson, H. A. 1992. Deer condition response to changing harvest strategy, Davis Island, Mississippi. Pages 48–55 in R. D. Brown, editor. The biology of deer. Springer-Verlag, New York, New York, USA.

Leonard, W. R., and M. L. Robertson. 1994. Evolutionary perspectives on human nutrition: the influence of brain and body size on diet and metabolism. American Journal of Human Biology 6:77–88.

Leopold, R. A. 1933. Game management. Charles Scribner's Sons, New York, New York, USA.

Leopold, R. A. 1943. Wildlife in American culture. Journal of Wildlife Management 7:1–6.

Loehr, J., J. Carey, R. B. O'Hara, and D. S. Hik. 2010. The role of phenotypic plasticity in responses of hunted thinhorn sheep ram horn growth to changing climate conditions. Journal of Evolutionary Biology 23:783–790.

Mahoney, S. P. 2007. The importance of how society views hunting. Pages 62–67 in Proceedings of the Western Association of Fish and Wildlife Agencies. Arizona Game and Fish Department, Flagstaff, USA.

Mahoney, S. P. 2009. Recreational hunting and sustainable wildlife use in North America. Pages 266–281 in B. Dickson, J. Hutton, and W. M. Adams, editors. Recreational hunting, conservation and rural livelihoods: science and practice. Blackwell, Chichester, West Sussex, United Kingdom.

Martin, P. S., and R. G. Klein. 1984. Quaternary extinctions. University of Arizona Press, Tucson, USA.

McCaffery, K. R., J. E. Ashbrenner, and R. E. Rolley. 1998. Deer reproduction in Wisconsin. Transactions of the Wisconsin Academy of Sciences, Arts and Letters 86:249–261.

McCullough, D. R. 1979. The George Reserve deer herd: population ecology of a k-selected species. University of Michigan Press, Ann Arbor, USA.

Meine, C. D. 1991. Aldo Leopold: his life and work. University of Wisconsin Press, Madison, USA.

Minard, A. 2009. Hunters speeding up evolution of trophy prey? National Geographic News. http://news.nationalgeographic.com/news/2009/01/090112 trophy-hunting.html. Accessed 11 October 2012.

Mysterud, A. 2011. Selective harvesting of large mammals: how often does it result in directional selection? Journal of Applied Ecology 48:827–834.

Mysterud, A., and R. Bischof. 2010. Can compensatory culling offset undesirable evolutionary consequences of trophy hunting? Journal of Animal Ecology 79:148–160.

Nelson, M. P., J. A. Vucetich, P. C. Paquet, and J. K. Bump. 2011. An inadequate construct? The wildlife professional. The Wildlife Society, Bethesda, Maryland, USA.

Noyes, J. H., B. K. Johnson, L. D. Bryant, S. L. Findholt, and J. W. Thomas. 1996. Effects of bull age on conception dates and pregnancy rates of cow elk. Journal of Wildlife Management 60:508–517.

O'Dell, J. 2007. 50 years and a lot of sheep later. Arizona Wildlife Views November–December. Arizona Game and Fish Department, Phoenix, USA.

Ozoga, J. J., and L. J. Verme. 1985. Comparative breeding behavior and performance of yearling vs. prime-age white-tailed bucks. Journal of Wildlife Management 49:364–372.

Patterson, R. 2009. Executive summary. Pages 391–392 in World symposium: ecologic and economic benefits of hunting. World Forum on the Future of Sport Shooting Activities, Windhoek, Namibia.

Payne, N. F. 1980. Furbearer management and trapping. Wildlife Society Bulletin 8:345–348.

Peyton, R. B. 2000. Wildlife management: cropping to manage or managing to crop? Wildlife Society Bulletin 28:774–779.

Posewitz, J. 2002. Beyond fair chase: the ethics and tradition of hunting. Falcon, Kingwood, Texas, USA.

Putnam, R., M. Apollonio, and R. Andersen. 2011. Ungulate management in Europe: problems and practices. Cambridge University Press, Cambridge, United Kingdom.

Rawinski, T. J. 2008. Impacts of white-tailed deer overabundance in forested ecosystem: an overview. Northeastern Area State and Private Forestry, U.S. Forest Service, U.S. Department of Agriculture. http://www.na.fs.fed.us/fhp/special_interests/white_tailed_deer.pdf. Accessed 11 October 2012.

Reiger, J. F. 1986. American sportsmen and the origins of conservation. Revised edition. University of Oklahoma Press, Norman, USA.

Responsive Management. 1993. Wildlife viewing in Maryland: participation, opinions and attitudes of adult Maryland residents towards a watchable wildlife program. Report Prepared for the Maryland Wildlife Division. Responsive Management, Harrisonburg, Virginia, USA.

Responsive Management. 1995. Federal aid outreach survey, region II: Arizona anglers, boaters, and hunters; New Mexico anglers, boaters and hunters; Oklahoma anglers, boaters and hunters; Texas anglers, boaters, hunters and passport holders. Report Prepared for U.S. Fish and Wildlife Service. Responsive Management, Harrisonburg, Virginia, USA.

Rocky Mountain Elk Foundation. 2011. Fast facts. http://www.rmef.org/AboutUs/FoundationFacts/. Accessed 11 October 2012.

Roosevelt, T., T. S. Van Dyke, D. G. Eliot, and A. J. Stone. 1902. The deer family. MacMillan, New York, New York, USA.

Rowe, T. 2009. WFSA President Ted Rowe's closing remarks. Page 392 in World symposium: ecologic and economic benefits of hunting. World Forum on the Future of Sport Shooting Activities, Windhoek, Namibia.

Rughetti, M., and M. Festa-Bianchet. 2010. Compensatory growth limits opportunities for artificial selection in alpine chamois. Journal of Wildlife Management 74:1024–1029.

Schuett, M. A., D. Scott, and J. O'Leary. 2009. Social and demographic trends affecting fish and wildlife management. Pages 18–30 in M. J. Manfredo, J. J. Vaske, P. J. Brown, D. J. Decker, and E. A. Duke, editors. Wildlife and society: the science of human dimensions. Island Press, Washington, D.C., USA.

Singer, F. J., and L. C. Zeigenfuss. 2002. Influence of trophy hunting and horn size on mating behavior and survivorship of mountain sheep. Journal of Mammalogy 83:682–698.

Southwick, R. 2010. Hunter number projections. Southwick Associates, Fernandina Beach, Florida, USA.

Southwick, R., and T. Allen. 2010. Expenditures, economic impacts and conservation contributions of hunters in the United States. Pages 308–313 in World symposium: ecologic and economic benefits of hunting. World Forum on the Future of Sport Shooting Activities, Windhoek, Namibia.

Thomas, E. D., Jr. 2010. How sportsmen saved the world: the unsung conservation effort of hunters and anglers. Lyons Press, Guilford, Connecticut, USA.

Todd, A. W. 1981. Ecological arguments for fur-trapping in boreal wilderness regions. Wildlife Society Bulletin 9:116–124.

USFWS. U.S. Fish and Wildlife Service. 2007. 2006 national survey of fishing, hunting, and wildlife-associated recreation. http://library.fws.gov/Pubs/nat_survey2006_final.pdf. Accessed 11 October 2012.

USFWS. U.S. Fish and Wildlife Service. 2010a. Final apportionment of Pittman-Robertson wildlife restoration Funds (CFDA # 15.611) for fiscal year 2010. http://wsfrprograms.fws.gov/subpages/Grant Programs/WR/WRFinalApportionment2010.pdf#page=4. Accessed 11 October 2012.

USFWS. U.S. Fish and Wildlife Service. 2010b. National summary of accomplishments, 2005–2009. Washington, D.C., USA.

USFWS. U.S. Fish and Wildlife Service. 2010c. Federal Aid Division—The Pittman-Robertson Federal Aid in Wildlife Restoration Act. http://www.fws.gov/southeast/federalaid/pittmanrobertson.html. Accessed 11 October 2012.

Watson, L. 1971. The omnivorous ape. Coward, McCann and Geoghegan, New York, New York, USA.

White, G. C. 2001. Effect of adult sex ratio on mule deer and elk productivity in Colorado. Journal of Wildlife Management 65:543–551.

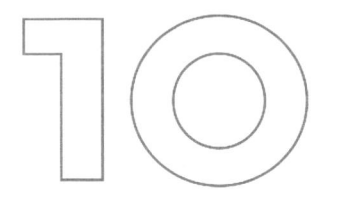

IMPACTS OF WEATHER AND ACCIDENTS ON WILDLIFE

MICHAEL R. CONOVER, JONATHAN B. DINKINS, AND
MICHAEL J. HANEY

INTRODUCTION

Weather is variation in ambient temperature, precipitation, and wind across time and space. A weather event is a specific weather phenomenon that lasts no more than a few days (e.g., thunderstorms, hurricanes, wind storms, and blizzards). A weather pattern is a weather phenomenon that lasts weeks or months, and climate is the prevailing weather at a point on the earth's surface over the course of many years or decades.

Spring weather patterns often cause annual variation in reproductive rates of wildlife, with reduced success during periods of wet and cold weather. For example, snow and ice conditions in June adversely impact reproduction of several species of geese that nest in the Arctic (Reeves et al. 1976). Annual nest success rates of wild turkey (*Meleagris gallopavo*) are correlated with the rainfall amounts during spring (Roberts et al. 1995, Roberts and Porter 1998*a,b*). Precipitation and wind account for over 90% of annual variation in survival of eider ducklings (*Somateria mollissima*; Mendenhall and Milne 1985). In contrast, wet years benefit some species. Annual production of mallards (*Anas platyrhynchos*) in the Mississippi Flyway correlates with the number of ponds containing water in the northern prairies of the United States and Canada, and more ponds fill with water during wet years (Bellrose et al. 1961). In the Mississippi Flyway, winter survival of mallards is highest during years when mild temperatures are combined with heavy winter rains; the heavy rains create more areas where the ducks can forage (Reinecke et al. 1987).

Winter weather patterns account for annual variation in survival of many wildlife species, including caribou (*Rangifer tarandus caribou*; McLoughlin et al. 2003), Dall's sheep (*Ovis dalli*; Burles and Hoefs 1984), and alpine ibex (*Capra ibex*; Jacobson et al. 2004). Ground-feeding birds and mammals can starve when deep snows cover the ground for an extended period of time (Roseberry 1962, Vepsäläinen 1968, Bull and Dawson 1969). In Sweden, the two worst winters recorded in 68 years (>60 days with >5 cm of snow covering) caused a population of barn owls (*Tyto alba*) to decline by more than 70% when their main prey (i.e., rodents) became unavailable under a layer of snow (Altwegg et al. 2006).

In arid regions, the amount and timing of precipitation are critical to reproduction and survival of many wildlife species. In New Mexico, Bender et al. (2011) reported that reproduction and survival of mule deer (*Odocoileus hemionus*) were reduced during droughts. To mitigate the impact of dry periods, some birds and mammals in arid regions delay nesting until the onset of rain (Johnson et al. 1992).

HOW WEATHER EVENTS IMPACT WILDLIFE

Many weather events kill animals but have little impact on the population dynamics of a species unless all individuals of that species are restricted to a small area. One example of a species with a restricted range is the whooping crane (*Grus americana*). All free-ranging whooping cranes winter in a single area along the Texas coast. Even a localized weather event might cause extinction of the species, and the U.S. Fish and Wildlife Service has been trying to establish populations of whooping cranes in other areas to reduce this risk.

Effect of Weather Events on Foraging Efficiency

Weather events can put an animal or species at a competitive disadvantage by changing its foraging efficiency. For instance, rattlesnakes need to keep their body temperatures within a narrow range to hunt small mammals effectively. When it is too cold for them, mammalian predators that hunt the same species have a competitive advantage over rattlesnakes because mammals can better regulate their body temperature. But when ambient temperatures are optimal, rattlesnakes have a competitive advantage over mammalian predators because snakes have a lower metabolic rate and do not need to catch as much prey as do mammalian predators. As another example, sunny days warm the ground and create updrafts, which in turn provide optimal hunting conditions for ferruginous hawks (*Buteo regalis*), a species that hunts for prey while soaring. Their hunting efficiency declines when cloudy skies and the lack of wind prevent the creation of updrafts. But the lack of updrafts does not have the same adverse impact on other hawk species that hunt from perches.

Effect of Weather Events on Risk of Predation

Weather events can alter an animal's risk of predation. Rain and wind increase ambient noise levels, which in turn can muffle the sound of an approaching predator. For this reason, many herbivores remain in secure cover and refrain from foraging under such conditions. Changing ambient light levels can also increase an animal's risk of predation. Leopards (*Panthera pardus*) have better night vision than their prey. This advantage is particularly acute when it is completely dark, such as when clouds block the moonlight. Under nighttime conditions, herbivores cannot see a stalking leopard, but the leopard can still see them.

Other predators such as raccoons (*Procyon lotor*), striped skunks (*Mephitis mephitis*), and feral hogs (*Sus scrofa*) use olfaction to locate prey when weather conditions favor its use. Ideal atmospheric conditions for the use of olfaction include high humidity, a light breeze, and a temperature inversion close to the ground (Conover 2007). This may explain why turkey nests are more likely to be depredated during periods of wet weather (Palmer et al. 1993; Roberts et al. 1995; Roberts and Porter 1998a,b).

Weather can influence depredation rates in unexpected ways. Loss of eider (*Somateria* spp.) ducklings to predators doubles during rainy and windy days because the ducklings do not stay as close to their mother as during calm, dry days (Mendenhall and Milne 1985). Stormy weather also can increase predation rates on ducklings by forcing the broods into calm bays, where they are depredated by gulls (Bergman 1982 as cited by Johnson et al. 1992).

Effect of Weather Events on Reproduction

Weather events have a pronounced impact if they occur when animals are particularly vulnerable, such as when they are only a few days old (i.e., neonates) or when they are migrating. Eggs are especially vulnerable to weather events because of their immobility. A severe snowstorm in the Northwest Territories of Canada that occurred when goose nests were hatching killed 100% of snow goose (*Chen caerulescens*) nests, 75% of white-fronted goose (*Anser albifrons*) nests, and 50% of brant (*Branta bernicla*) nests (Barry 1967, Sargeant and Raveling 1992). Nests built along shorelines, on islands, or in low-lying areas can become submerged during periods of high rainfall, floods, high tides, or storm surges. These calamities may destroy nearly all low-lying waterfowl nests (Hansen 1961, Sargeant and Raveling 1992). Also vulnerable to flooding are birds that build nests over water. In Manitoba, 6% of canvasback duck (*Aythya valisineria*) nests are lost to flooding (Stoudt 1982). Flooding is a problem for ground-nesting birds in upland areas; each year 2% of dabbling duck nests in the prairies and parklands of North America are lost to flooding (Sargeant and Raveling 1992).

While nesting in trees provides protection from floods, high winds can destroy nests or kill young birds. Mourning doves (*Zenaida macroura*) that build structurally sound nests in secure locations (e.g., in a tree fork) are more likely to reproduce successfully than doves that build poor nests or locate them on thin branches where they are vulnerable to being blown out by the wind (Coon et al. 1981).

Neonates have only a limited ability to thermoregulate; rain or cold temperatures can cause mass mortality of young if these conditions occur during periods when the young are vulnerable. Exposure to adverse weather is believed to kill more waterfowl broods than starvation, diseases, or parasites (Johnson et al. 1992).

Effect of Weather Events on Migrating Animals

Common weather events rarely cause mortality of adult wildlife located within their own home range. Migration, however, is a dangerous time for many animals. Headwinds can force migrating birds to fly longer, depleting their energy reserves, while heavy rains can saturate plumage and cause excessive wing loading (Cottam 1929, Williams 1950, Kennedy 1970). These stresses can cause birds to fly lower than they otherwise would, increasing their risk of flying into such obstacles as communication towers, or forcing migrants to land in unsuitable areas.

Animals that have to migrate across inhospitable areas in adverse weather are particularly at risk. Many of the recorded migrating bird mortalities occurred when terrestrial birds encountered storms while crossing large water bodies (Dick and Pienkowski 1979, Morse 1980, Evans and Pienkowski 1984, Spendelow 1985). More than 40,000 migrants representing 45 species were killed during a storm on 8 April 1993 off Louisiana (Wiedenfeld and Wiedenfeld 1995). At least 106 avian species have washed up on beaches following storms over the Gulf of Mexico (James 1956, Webster 1974, King 1976, Wiedenfeld and Wiedenfeld 1995). Small birds are especially vulnerable to storms when migrating across water, yet large birds can succumb to oceanic storms, too. For instance, 4,600 dead rooks (*Corvus frugilegus*) were found in southern Sweden following a storm (Alestam 1988). In April 1980, more than 1,300 birds of prey (including eagles) died off Israel when strong winds blew them over the sea from their usual landward route (Kerlinger 1989).

Likewise, migrating waterbirds may encounter problems when they need to migrate over land. In western North America, eared grebes (*Podiceps nigricollis*) migrate over hundreds of kilometers of desert with few places to land in an emergency. Snowstorms have occasionally brought down thousands of these birds. On 13 December 2011, thousands of eared grebes crash-landed in southern Utah during a snowstorm. Once down, most of the eared grebes were unable to take flight again and died (M. R. Conover, personal communication).

Impact of Temperature Extremes, Hailstorms, and Windstorms on Wildlife

Unusually cold or hot temperatures can kill animals, especially those living along the edge of the geographic range for their species. During 2010, a cold snap killed 431 Florida manatees

(*Trichechus manatus latirostris*)—13% of the population (Segelson 2010)—and stunned over 3,500 sea turtles (McNulty 2010). Waterfowl and shorebirds die when cold temperatures cause their aquatic foraging areas to freeze (Smith 1964, Barry 1968, Fredrickson 1969). High temperatures can cause hyperthermia in wildlife. Hyperthermia may have been the cause of death of four Sonoran pronghorn (*Antilocapra americana sonoriensis*) fawns that died during the three hottest days of 2005 (with highs of 44.1°C, 44.1°C, and 43.6°C; Wilson and Krausman 2008).

Hail and lightning pose a risk to animals (Hochbaum 1955, Glasrud 1976, Roth 1976). One hailstorm killed more than 35 tundra swans (*Cygnus columbianus*; Hochbaum 1955), while another killed more than 100 snow geese (Krause 1959). In July 1953, two hailstorms in Alberta covered 2,500 km² and killed most birds in the area, including over 60,000 waterfowl (Smith and Webster 1955). Sandstorms caused by high winds can be a particular problem for birds moving through arid regions, such as Africa's Sahara Desert. Sandstorms have killed barn swallows (*Hirundo rustica*), northern wheatears (*Oenanthe oenanthe*), and white storks (*Ciconia ciconia*; Moreau 1928, Schüz et al. 1971).

CLIMATE CHANGE AND WILDLIFE

Climate change has occurred at a slow rate throughout earth's history (e.g., ice ages have come and gone). However, humans may be changing the climate at a faster rate than under natural (i.e., no human influence) conditions (Root and Schneider 2002). In response to climate changes, wildlife species adapt by changing phenology, morphology, physiology, genetics, behavior, migration, and competitive ability (Green et al. 2001, Harvell et al. 2002, Root and Schneider 2002, Parmesan and Yohe 2003).

Burning fossil fuels, deforestation, and other land-use practices (e.g., conversion of forested areas to urban or agricultural areas) can affect climate, but burning fossil fuels has the greatest effect by increasing atmospheric levels of carbon dioxide. This gas heats the earth by capturing a large percentage of long-wavelength heat energy near the earth's surface, producing a greenhouse effect (Root and Schneider 2002). The global average temperature has increased 0.6–0.7°C over the past century (Green et al. 2001, Root et al. 2005), and 0.8°C to over 2°C have been projected as possible increases by 2050 (Intergovernmental Panel on Climate Change 2001). The high end of this range is much greater than during most of the Holocene period (the last 10,000 years), when the average global temperature changed by less than 1°C per 1,000 years.

In general, vegetation is expected to move toward the North Pole in the Northern Hemisphere, toward the South Pole in the Southern Hemisphere, and up elevation gradients in response to increased surface temperatures. Wildlife populations are expected to follow (Root et al. 2005). Parmesan and Yohe (2003) verified movement of animals toward the poles; they conducted a meta-analysis involving 99 avian species, 16 butterfly species, and 17 alpine plant species and reported movement of 6.1 km (±2.4 km) per decade of range limits toward the poles during the 20th century.

Not all vegetation and wildlife will be able to move to new areas in response to climate change; some species will be trapped, and especially vulnerable will be species that occupy small, isolated ranges. For example, many species of amphibians and reptiles occupy small ranges located on isolated mountains (ecological islands) and may become extinct when the climate changes (Root and Schneider 2002). Arctic and Antarctic wildlife species, such as polar bears (*Ursus maritimus*) and Adélie penguins (*Pygoscelis adeliae*), have distributions that are at least seasonally tied or restricted to areas with sea ice (Green et al. 2001). As sea ice melts with a warming climate, there might be no suitable habitat left for these species. In contrast, some tropical species will likely benefit from climate change by having a larger area of suitable habitat that they can occupy.

Climate change will produce trophic mismatches affecting many wildlife species. A trophic mismatch occurs when the availability of food and habitat resources are no longer in synchrony with when wildlife needs those resources. Parmesan (2006) suggested that the primary consequence of climate change for wildlife will be trophic mismatches that lower the abundance of many wildlife species. Some birds have responded to climate change by arriving earlier at the breeding grounds and by having earlier hatching dates (Root and Schneider 2002). Many neotropical migratory birds, consisting of approximately 200 species of songbirds, shorebirds, waterfowl, and raptors, are dependent on the timing of food resources at stopover locations (i.e., areas between winter areas and breeding grounds). Vegetation at higher latitudes is expected to start growing and to senesce earlier with warmer temperatures (Root and Schneider 2002); fruits, nectar, and insects will follow similar patterns. The danger is that neotropical birds might not be able to reach their northern breeding grounds with adequate time to reproduce successfully and to build up the energy reserves required to migrate south for the winter.

Warmer temperatures from climate change are expected to increase rainfall worldwide, but the effects of climate changes will vary by location. Some areas will become drier or colder than current conditions, depending on changes in wind patterns. Weather events such as tornadoes, hurricanes, and floods are expected to become more frequent and more severe (Green et al. 2001, Root and Schneider 2002). Chapter 17 addresses climate change and its influence on wildlife in more detail.

IMPACT OF ACCIDENTAL DEATHS ON WILDLIFE

Accidents are a fact of life, and accidental deaths befall every species, but unless animals are tracked using radio collars, it is difficult to determine how frequently accidents kill wildlife. What little is known about accidental deaths comes from species that from a human perspective live in dangerous areas, such as bighorn sheep (*Ovis canadensis*). Jokinen et al. (2008) radio collared 46 female bighorn sheep in Alberta and tracked them from 2003 until 2005. Eleven females died during the study; three were killed in accidents (one fell to its death, one was killed in an avalanche, and one broke its leg), one died from an unknown cause, and seven were killed by predators. In Hells Canyon, 22% of bighorn sheep died from falls or injuries (Cassirer and Sinclair 2007). In the Sierra Nevada Mountains, the cause of death was determined for 105 bighorn sheep—19 were killed in avalanches or accidents, five died of exposure, one was struck by a vehicle, and 80 were killed by predators (U.S. Fish and Wildlife Service 2003). In Arizona, Kamler et al. (2003) reported the cause of death for 46 bighorn sheep; nine of them (20%) died from climbing accidents. Among adults, accidents often peak during the mating season, when animals are distracted or engage in territorial or dominance fights. Among muskoxen (*Ovibos moschatus*) and bighorn sheep, 5–10% of adult males are killed annually from injuries sustained during the rut (Geist 1971, Wilkinson and Shank 1976).

Young animals are particularly vulnerable to accidents because neonates have limited ability to escape the hazards in their environment. In a suburban population of Canada geese (*Branta canadensis*), accidents were the second leading cause of death among young goslings; accidents resulted from being hit by vehicles, falling into holes, being struck by golf balls, and becoming separated from their parents (M. R. Conover, unpublished data). Accidents can even befall eggs inside nests. Parents normally roll the eggs several times a day, because if eggs remain stationary for too long, the membranes surrounding the developing embryo will adhere to the shell, destroying the embryo. But rolling the eggs increases the chance that an egg might tumble out of the nest. When it does, Canada geese will use their bill to roll the eggs back into the nest, a difficult task when the eggs roll downhill, slip over ledges, or roll into the water. M. R. Conover (unpublished data) reported that egg falls were one of the main reasons why some eggs failed to hatch from successful nests (i.e., nests where one or more eggs hatched).

ACCIDENTAL DEATHS CAUSED BY HUMAN ENDEAVORS

One of the main goals of wildlife management is to lessen the impacts of human activities on wildlife. For this reason, accidental deaths that result from human endeavors are a greater source of concern than deaths from the types of accidents that have befallen wildlife for eons. Likewise, most of the research on accidental deaths of wildlife has investigated the impact of human-induced accidents. For instance, dozens of papers have been written about how many birds die from flying into windows or power lines, but we are not aware of any reports about how many birds are killed from flying into trees. For the rest of this chapter, we will focus on accidental deaths for anthropogenic activities and structures.

Oil Ponds

When gas and oil are pumped from the ground, water often comes up with the petroleum. Oil pits, ponds, or tanks are used to separate the oil from the water and then store the contaminated water. Waste oil ponds also result from oil spills, oil drips, or from flaring hydrogen sulfide. In this chapter we refer to these collectively as oil ponds. Birds and mammals are attracted to oil ponds by thirst, struggling insects, or other animals that are caught in the oil. Migratory waterfowl can become trapped by landing in oil ponds before they realize that the ponds are covered by oil. Death results from becoming trapped in the oil and drowning, getting oil on their feathers or fur and ingesting the toxic oil when animals try to clean themselves, or swallowing too much oil when drinking oily water. Over half (62%) of the birds recovered from oil ponds were members of the order Passeriformes; 10% were Anseriformes, 6% Columbiformes, 5% Strigiformes, 4% Ciconiiformes, 3% Charadriiformes, 2% Falconiformes, and 2% Cuculiformes (Trail 2006).

The U.S. Fish and Wildlife Service estimated that oil ponds killed two million migratory birds during 1997 (Ramirez 1999). Since then, oil companies have taken steps to address the problem by replacing many oil ponds with closed tanks or by using nets to keep birds out of open oil ponds. Trail (2006) reported that these measures have decreased annual mortality rates from one million birds to 500,000.

Communication Towers

There are more than 50,000 communication towers in the United States that stand over 66 m above ground level; they are used mostly by the radio, television, and wireless telephone industries. Migratory birds collide with these towers, particularly during inclement weather. Nocturnal migrants, including thrushes, vireos, tanagers, cuckoos, and sparrows, are particularly vulnerable to colliding with communication towers (Avery et al. 1976), and estimates of the number of birds killed annually in the United States range between four and ten million (Kerlinger 2000).

Power Lines

Power lines include transmission lines and distribution lines. The former conduct high-voltage electricity (115–500 kV) over long distances and are typically held aloft by tall wooden or metal towers. Distribution lines deliver electricity from substations to individual buildings and carry up to 70 kV (Manville 2005, Edison Electric Institute 2009). There are more than 1,217,000 km of transmission lines and 9,612,000 km of distribution lines in the United States (Edison Electric Institute 2009).

A bird must complete the circuit of an electrified power line to be electrocuted, which often happens when a large bird perches on a power line and its wings touch another energized line, ground line, or pole (Bevanger 1995, 1998; Harness and Wilson 2001; Lehman 2001; Manville 2005). In the rural western United States, eagles, hawks, and owls accounted for 96% of all electrocutions (Harness and Wilson 2001). Electrocution is the second-largest cause of mortality for golden eagles (*Aquila chrysaetos*) and the fourth-largest cause of death for bald eagles (*Haliaeetus leucocephalus*) in the United States (LaRoe et al. 1995, Lehman 2001). In southwestern Spain, Ferrer et al.

RICHARD A. DOLBEER (b. 1945)

Dr. Richard A. Dolbeer served at the forefront of research, management, and policy development to reduce wildlife hazards to aviation. He was one of first individuals to recognize that the cost in property and lives due to wildlife–aircraft collisions was avoidable. In the mid-1980s, Dolbeer helped create the Aviation Project of the U.S. Department of Agriculture/Wildlife Services (WS) National Wildlife Research Center, and served as its leader until 2002. He then served as the WS national airports coordinator for the Aviation Safety and Assistance Program. His work was international in scope, including collaborative efforts with five foreign governments, and it has produced advances in how airport habitats should be managed to reduce use by wildlife, considerations for the design of turbine-powered engines and airframes to withstand bird strikes, and advances in how wildlife strike data are reported and interpreted.

Dolbeer's work with management of airport habitats to reduce use by wildlife has produced a dramatic reduction in aircraft collisions with birds at John F. Kennedy International Airport (JFKIA). Based on his research findings, JFKIA began clearing large stands of woody vegetation to eliminate food sources, roosting sites, and perching sites for birds that pose bird strike risks and adopted recommendations for maintaining vegetation six to ten inches high to reduce bird and small mammal use of grasslands.

In collaboration with the Federal Aviation Administration, Dolbeer and scientists under his guidance produced a series of publications that culminated in the registration by the U.S. Environmental Protection Agency of two chemical foraging repellents and one investigational new permit by the U.S. Food and Drug Administration for the wildlife-capture drug alpha-chloralose. In addition, Dolbeer and collaborators developed a laser product for bird dispersal and were awarded several patents. Two wildlife foraging repellent applications received patents, as did a collaborative project with Precise Flight, in Bend, Oregon, to develop an aircraft-mounted lighting system. Another patent was issued for a hazard avoidance system to increase the "visibility" of aircraft to birds and to provide hazard data to pilots.

Dolbeer coauthored with Edward Cleary *Wildlife Hazard Management at Airports: A Manual for Airport Personnel*. This manual provided for the first time a detailed course of action for understanding and managing wildlife hazards at airports. In 2001, the manual was translated into Spanish for distribution to airports throughout Mexico and Latin America, and into French for use in Quebec, France, and West Africa.

Since 1995, Dolbeer supervised the editing and entry of over 70,000 strike reports into the national Wildlife Strike Database. These data were critical to understanding the circumstances of wildlife collisions with aircraft. Wildlife biologists who work at airports are now encouraged to increase efforts to detect, remove, and disperse large species of birds from the airport environment. This transition from research findings to applied management is a hallmark of Dolbeer's research career.

Text adapted from Blackwell and DeVault (2009).
Photo courtesy of the Federal Aviation Administration

(1991) estimated that 400 raptors and vultures per year were electrocuted on a 1-km stretch of power line and 1,200 birds of prey were electrocuted annually in Doñana National Park. Between 1974 and 1982, 52% of banded raptors and 69% of eagles found dead in the Doñana area had been electrocuted (Ferrer et al. 1991). Rubolini et al. (2005) estimated that 0.15 birds died per power pole in Italy, 0.21 per pole in Spain, and 0.15–5.2 per pole in the United States.

Birds also are killed by flying into power lines, and the list of birds that are susceptible to collision with power lines is much more diverse than birds susceptible to electrocution; the most impacted groups of birds from a population standpoint are less maneuverable birds, or so-called "poor" flyers, which have small wings relative to their mass, such as ducks and grebes. In prairies, waterfowl are more likely to collide with

power lines than other avian groups (Faanes 1987). Collisions with power lines may be a considerable problem for many threatened or endangered birds. For example, the two largest sources of mortality for California condors (*Gymnogyps californianus*) were collisions with power lines and lead poisoning (Meretsky et al. 2000). In Colorado, collisions with power lines killed 39% of whooping cranes (Brown and Drewien 1995, Sundar and Choudhury 2005).

Several studies have calculated mortality rates per kilometer of power line. Some examples of annual morality estimates include: 124 birds killed per kilometer in North Dakota prairie wetlands (Faanes 1987), 5.3 willow ptarmigan (*Lagopus lagopus*) killed per kilometer in subalpine Norway (Bevanger and Brøseth 2004), 2.3–5.8 common cranes (*Grus grus*) killed per kilometer in Spain (Janss and Ferrer 2000), and 1.6–4.0 great bustards (*Otis tarda*) killed per kilometer in Spain (Janss and Ferrer 1998, 2000). The most thorough studies on avian mortality rates caused by power lines were conducted in the Netherlands. Annual estimates of birds killed in the Netherlands were 113/km of transmission line in grasslands, 58/km in agriculture lands, and 489/km near river crossings (Koops 1987). By combining these fatality rates with the 4,600 km of transmission lines in the Netherlands, Koops (1987) calculated that between 750,000 and one million birds were killed annually by flying into transmission lines in the country. Erickson et al. (2001) made the assumption that the fatality rate per kilometer of transmission line in the United States is similar to the rate in the Netherlands. Considering that there are 800,000 km of transmission lines in the United States, Erickson et al. (2001) and Manville (2005) estimated that 130 to 175 million birds are killed annually by these lines. This mortality rate is higher than we would have expected. However, it excludes mortalities from all of the more than one million kilometers of distribution lines in the United States.

Wind Turbines

Thousands of wind turbines have been built in North America to generate electric power. Almost 5,000 are in the Altamont Pass Wind Resource Area (APWRA) in Contra Costa and Alameda counties in California. Smallwood and Thelander (2008) reported that wind turbines in APWRA collectively kill over 2,700 birds annually, including 67 golden eagles, 188 red-tailed hawks (*Buteo jamaicensis*), 348 American kestrels (*Falco sparverius*), and 440 burrowing owls (*Athene cunicularia*). Johnson et al. (2002) estimated that the more than 350 wind turbines at Buffalo Ridge, Minnesota, killed 72 birds annually. Most of the mortalities were passerines that were migrating through the area. Across the United States, wind turbines kill approximately 33,000 birds each year (Erickson et al. 2001).

Windows

Klem et al. (2004) and Klem (2006) asserted that collisions with windows are the second-largest cause of avian mortality related to human endeavors (after habitat loss). Passerines strike windows more than any other avian group, probably because large numbers of passerines live near human dwellings. The following species strike windows in North America with high frequency: the American robin (*Turdus migratorius*), dark-eyed junco (*Junco hyemalis*), ovenbird (*Seiurus aurocapillus*), northern cardinal (*Cardinalis cardinalis*), downy woodpecker (*Picoides pubescens*), mourning dove, evening grosbeak (*Coccothraustes vespertinus*), house sparrow (*Passer domesticus*), house finch (*Carpodacus mexicanus*), white-throated sparrow (*Zonotrichia albicollis*), Swainson's thrush (*Catharus ustulatus*), northern flicker (*Colaptes auratus*), and yellow-rumped warbler (*Dendroica coronate*; Klem, 1989, Dunn 1993, O'Connell 2001, Gelb and Delacrétaz 2006, Klem 2006). However, mortalities caused by striking windows are a greater concern when rare or endangered birds are involved. Over 1.5% of all endangered swift parrots (*Lathamus discolor*; only 1,000 breeding pairs remain) are killed annually by collisions with windows (Klem 2006). Boal and Mannan (1999) reported that window collisions accounted for 70% of mortalities among banded Cooper's hawks (*Accipiter cooperii*) living in Arizona cities.

Klem (1990, 2006) estimated that each building in the United States kills an average of one to ten birds annually from window collisions. He based his estimate on counts he made of birds killed by striking windows located on residential and commercial buildings in southern Illinois and New York from 1974 to 1986. Dunn (1993) independently verified Klem's estimated fatality rate during "Project Feeder Watch." During the winter of 1989–1990, Project Feeder Watch documented 995 fatal bird–window strikes based on 1,165 reports from participants across the United States and Canada. Dunn (1993) refined her estimate to 0.7–7.7 bird–window strikes per house per year by taking into account possible underestimates caused by scavengers removing bird carcasses before they are counted, and overestimates of bird–window strike frequency owing to observer bias. Based on the number of buildings in North America, Klem (1990) and Dunn (1993) estimated that 100 million to one billion birds are killed annually in North America.

Factors that affect the occurrence and frequency of avian collisions with windows include weather, type of glass, orientation of glass, time of day, time of year, and proximity of bird feeders to windows (Klem 1989). Typically, birds collide with windows when attempting to fly through them, or to reach habitat that they can see through the glass or habitat reflected by the glass (Klem 2006). Territorial birds also may collide with windows because they mistake their reflected image for an intruding bird. Birds that attack their reflection usually collide with the glass at lower velocities than birds that try to fly through the glass and are less likely to suffer fatal injuries (Klem 2006). Migrating birds collide into windows on tall, commercial buildings at a greater frequency than resident birds (O'Connell 2001, Gelb and Delacrétaz 2006), whereas resident birds collide into windows on residential houses with greater frequency than migrating birds (Klem 1990, Dunn 1993).

Cars and Trucks

Conover et al. (1995) estimated that there are more than 700,000 confirmed deer–vehicle collisions each year in the United States; the actual number is probably twice as high, because confirmed collisions exclude those where the deer died away from the road and accidents not reported to state authorities. Being struck by a vehicle is fatal to deer 92% of the time; about 1.5 million deer are killed annually in the United States (Conover 2002). Each year there are about 1,500 collisions between vehicles and moose (*Alces alces*) in Alaska, Maine, and Newfoundland (Rattey and Turner 1991, Pelletier 2006). Data on ungulate–vehicle collisions are available from several European countries (e.g., Austria, Denmark, Finland, Germany, Ireland, Netherlands, Norway, and Sweden). In these countries, there are annually 114,000 collisions involving roe deer (*Capreolus capreolus*), 1,900 red deer (*Cervus canadensis*), 700 white-tailed deer (*Odocoileus virginianus*, which is an exotic species in Europe), 1,900 fallow deer (*Dama dama*), 300 chamois (*Rupicarpra rupicarpra*), 5,700 moose, 3,400 reindeer (*Rangifer tarandus*), and 7,000 wild boar (Groot Bruinderink and Hazelbrook 1996). The total number of wildlife–vehicle collisions in Europe exceeds this number because it does not include collisions in several large countries (e.g., France, Spain, Italy, Portugal, Hungary, and Poland). In Saudi Arabia, 600 free-ranging camels (*Camelus dromedarius*) die annually in vehicle collisions out of a population of over 500,000 camels (Al-Ghamdi and AlGadhi 2003).

It is much harder to estimate the number of small vertebrates that are killed in road accidents, because such collisions often escape the driver's attention and are rarely reported. Data on how frequently small animals are struck by vehicles usually come from investigators counting carcasses on the same stretch of road at regular intervals and are expressed as the number killed per kilometer per day. Vertebrate death rates on a road adjacent to the Big Creek National Wildlife Area in Ontario were estimated at 13/km/day from April through October (Ashley and Robinson 1996). Vertebrate death rate was 5/km/day on a highway through Payne's Prairie State Park, Florida (Smith and Dodd 2003), and 1/km/day on roads in Saguaro National Park in Arizona (Gerow et al. 2010). Most of these reports, however, come from parks or other areas with high wildlife densities and might not be applicable to all highways in the United States.

Banks (1979) estimated that the annual avian mortality rate was 9/km of road in the United States, the same rate Hodson and Snow (1965) estimated for Britain. Since Bank's study, the network of U.S. roads has increased at a rate of 0.2% annually, to a total of 6.7 million km by 2006; vehicle kilometers traveled in the United States have increased at a much more rapid rate of 1.9% annually, to 5.0 trillion in 2006 (U.S. Federal Highway Administration 2008). If we assume that there is a 1:1 relationship between the rate at which birds are struck by vehicles and the number of vehicle kilometers traveled, then during 2006 the avian mortality rate was 14.9/km of road; this means that 100 million birds are killed annually on U.S. roads.

In Saguaro National Park, Arizona, five times as many mammals as birds were killed by vehicle collisions (Gerow et al. 2010). We believe a conservative estimate is that twice as many mammals are killed on roads as birds (i.e., 200 million mammals annually).

Reptiles and amphibians that try to cross a road are more vulnerable to being struck by a vehicle than birds and mammals, because their slower speeds means that it takes them longer to cross the road. Birds and mammals also have a greater ability to evade a vehicle that is heading toward them. In Arizona, Gerow et al. (2010) found 12,264 dead reptiles and 12,208 dead amphibians along the same stretch of road that contained just 759 dead birds and 4,146 dead mammals. During the amphibian breeding season, Hels and Buchwald (2001) estimated that 4–16 amphibians were killed per kilometer per

day along a section of road in Denmark. Rosen and Lowe (1994) estimated that 22.5 snakes were killed annually per kilometer on a stretch of highway running mainly within Organ Pipe Cactus National Monument in Arizona. In Alabama, Dodd et al. (1989) drove 19,000 km over 135 days and found 0.013 dead reptiles per kilometer of road. The authors did not determine how long carcasses remained on roads before they were scavenged or became unrecognizable, but we doubt that the half-life of a reptile or amphibian carcass lying on the road is greater than a week. This would yield an annual mortality rate of 0.68/km/yr. If this rate is extrapolated to the 6.7 million km of roads in the United States, then 4.6 million reptiles are killed annually on U.S. highways. We assume that an equal number of amphibians, mostly frogs and toads, are killed annually on U.S. highways (i.e., 4.6 million).

Fences

One of the major man-made changes to the North American landscape is the construction of fences. Wire fences confine livestock on farms and ranches and line many roads and railroads. These fences pose a barrier to ungulates as they move across their home range, causing a threat to them if they fail to jump successfully over the fence. In Colorado, the annual fatality rate for pronghorns (*Antilocapra americana*) was 0.11/km of fence, for mule deer was 0.08/km, and for elk (*Cervus canadensis*) was 0.06/km (Harrington and Conover 2006). Juveniles were eight times more likely to be killed by fences than adults. Woven-wire fences topped by a single strand of barb were more lethal than woven-wire fences topped with two strands barb or four-strand barbed-wire fences. The reason for this is that when a leg is caught between two stands of barbed wire, there is enough play between the two wires that the ungulate can often pull free, but woven-wire is much more rigid. It is much harder to extract a leg when it is cinched between a stand of barbed wire and woven wire. Considering the millions of kilometers of fences erected in North America, the number of ungulates killed in fences is significant. Most U.S. highways have fences on each side of the road. With 6.7 million km of roads in the United States, this works out to 13 million km of fencing. Given data from Harrington and Conover (2006), we estimate that road fences kill 5,000 elk, 5,000 pronghorns, and 100,000 white-tailed and mule deer each year in the United States. Fences are also used to mark property boundaries and to enclose pastures, but the total length of boundary and pasture fences in the United States is unknown, as is the number of wildlife killed by them.

Aircraft

From 1990 to 2007, aircraft annually struck a mean of 4,700 birds, 102 terrestrial mammals, 15 bats, and six reptiles (Tables 10.1 and 10.2; Dolbeer and Wright 2009). These values are conservative because pilots only report about 25% of all bird strikes (Linnell et al. 1999). Terrestrial mammals were struck when they wandered onto a runway at the wrong time (Table 10.2). Bats involved in bat–aircraft collisions in North America were primarily the Mexican free-tailed bat (*Tadarida*

Table 10.1. Avian species for which ten or more individuals are killed annually in reported bird–aircraft collisions, 1990–2007

Species	Average mortalities per year
Canada goose (*Branta canadensis*)	158
Mourning dove (*Zenaida macroura*)	146
European starling (*Sturnus vulgaris*)	110
American kestrel (*Falco sparverius*)	90
Rock pigeon (*Columba livia*)	86
Killdeer (*Charadrius vociferus*)	65
Red-tailed hawk (*Buteo jamaicensis*)	50
Ring-billed gull (*Larus delawarensis*)	42
Horned lark (*Eremophila alpestris*)	39
Barn swallow (*Hirundo rustica*)	38
Herring gull (*Larus argentatus*)	37
Barn owl (*Tyto alba*)	27
Mallard (*Anas platyrhynchos*)	25
Pacific golden-plover (*Pluvialis fulva*)	24
Eastern meadowlark (*Sturnella magna*)	20
Turkey vulture (*Cathartes aura*)	17
American robin (*Turdus migratorius*)	15
Laughing gull (*Larus articilla*)	13
American crow (*Corvus brachyrhynchos*)	13
Western meadowlark (*Sturnella neglecta*)	13
Cliff swallow (*Hirundo pyrrhonota*)	12
Great blue heron (*Ardea herodias*)	11

Source: Dolbeer and Wright (2009).

Table 10.2. Number of mammals killed annually in strikes with aircraft, 1990–2007

Species	Average mortalities per year
White-tailed deer (*Odocoileus virginianus*)	42
Coyote (*Canis latrans*)	15
Domestic dog (*Canis familiaris*)	14
Woodchuck (*Marmota monax*)	5
Black-tailed jackrabbit (*Lepus californicus*)	4
Virginia opossum (*Didelphis virginiana*)	4
Red fox (*Vulpes vulpes*)	3
Striped skunk (*Mephitis mephitis*)	3
Raccoon (*Procyon lotor*)	3
Mule deer (*Odocoileus hemionus*)	2
Eastern cottontail (*Sylvilagus floridanus*)	2
Total	97

Source: Dolbeer and Wright (2009).

brasiliensis), red bat (*Lasiurus borealis*), horay bat (*Lasiurus cinereus*), Seminole bat (*Lasiurus seminolus*), and silver-haired bat (*Lasionycteris noctivagans*; Peurach et al. 2009). We assume that about 20,000 birds and 400 mammals are killed each year in the United States after being struck by an airplane.

Wildlife Control by Households

There are more than 60 million households in the 100 largest metropolitan centers in the United States; each year 42% of them (25 million) take action to try to solve a pest problem involving wildlife on their property (Conover 1997, 2002). Mice are the most common culprits, followed by squirrels,

Table 10.3. Estimates of the number of animals killed annually in the United States and Canada by human activities and collisions with man-made objects

	Birds	Mammals	Reptiles	Amphibians
Oil ponds	0.5–1 million	Unknown	Unknown	Unknown
Communication towers	7 million	Unknown	Unknown	Unknown
Power lines	135–175 million	Unknown	Unknown	Unknown
Wind turbines	33,000	Unknown	Unknown	Unknown
Windows	0.1–1 billion	Unknown	Unknown	Unknown
Automobiles and trucks	100 million	200 million	4.6 million	4.6 million
Fences	Unknown	110,000	Unknown	Unknown
Aircraft	20,000	400	Unknown	Unknown
Hunting	56 million	7 million	39,000	Unknown
Households	Unknown	78 million	Unknown	Unknown
Total	0.4–1.3 billion	285 million	4.6 million	4.6 million

Data are based on the most recent year for which data are available.

raccoons, moles, pigeons, starlings, and skunks. There are also 34 million households in smaller cities, towns, or rural areas (Conover 2002). Because wildlife populations are higher in towns and in rural areas, we assume that at least 42% of rural households (14 million) take action annually to solve a wildlife pest problem (39 million counting both areas). If we assume that each year 39 million households in metropolitan or rural areas kill two mice or other mammals, the total mortality equals 78 million mammals annually.

SUMMARY

Every year, weather patterns and specific weather events kill millions of birds and mammals; millions more die from accidents. Most of these deaths are unseen by humans and go unreported. The result is that we know little about how often animals die from weather events or accidents.

We estimate that each year 400 million to 1.3 billion birds, 287 million mammals, and nine million amphibians and reptiles are killed accidentally by human endeavors (Table 10.3). Banks (1979) estimated that there are ten billion birds in the United States during the start of the breeding season and probably 20 billion in the fall, with the addition of young that hatch each spring. This implies that less than 10% of all birds in the United States are killed accidently by human endeavors. By way of comparison, U.S. hunters annually harvest over 56 million birds (Table 10.4), 15 million mammals including seven million ungulates (Adams and Hamilton 2011) and six million eastern gray squirrels (*Sciurus carolinensis*), 125,000 eastern diamond rattlesnakes (*Crotalus adamanteus*; Means 2010), and 39,000 free-ranging alligators (*Alligator mississippiensis*; Louisiana Wildlife and Fisheries 2008, Florida Fish and Game Commission 2010).

In contrast to the major sources of mortality listed above, few birds (20,000) or mammals (400) die annually from colliding with aircraft because of a concerted effort by the U.S. Fed-

Table 10.4. Number of birds harvested by hunters in the United States and Canada during 2009

Country	Species	Number killed annually
United States	Duck	13,140,000
	Geese	3,327,000
	Sea duck	70,000
	Brant	38,000
	Mourning dove	17,400,000
	White-winged dove	1,642,000
	Band-tailed pigeon	28,000
	Woodcock	239,000
	Snipe	84,000
	Coot	219,000
	Gallinule	7,000
	Rail	36,000
	Pheasants	20,000,000
	Total	56,230,000
Canada	Ducks	1,019,000
	Geese	945,000
	Total	1,964,000

Source: Migratory bird data are from Raftovich et al. (2010) and pheasant data from Pheasants Unlimited.

eral Aviation Agency, U.S. Department of Agriculture Wildlife Services, and the U.S. Armed Forces to prevent bird–aircraft collisions at airports (Table 10.3). Similar efforts should be undertaken to reduce wildlife mortality rates from other human endeavors.

Literature Cited

Adams, K., and J. Hamilton. 2011. Management history. Pages 355–377 *in* D. G. Hewitt, editor. Biology and management of white-tailed deer. CRC Press, Boca Raton, Florida, USA.

Alestam, T. 1988. Findings of dead birds drifted ashore reveal catastrophic mortality among early spring migrants, especially rooks *Corvus frugilegus*, over southern Baltic Sea. Anser 27:181–218.

Al-Ghamdi, A. S., and S. A. AlGadhi. 2003. Warning signs as countermeasures to camel–vehicle collisions in Saudi Arabia. Accident Analysis and Prevention 36:749–760.

Altwegg, R., A. Roulin, M. Kestenholz, and L. Jenni. 2006. Demographic effects of extreme winter weather in the barn owl. Oecologia 149:44–51.

Ashley, E. P., and J. T. Robinson. 1996. Road mortality of amphibians, reptiles and other wildlife on the Long Point Causeway, Lake Erie, Ontario. Canadian Field Naturalist 110:403–412.

Avery, M., P. F. Springer, and J. F. Chassel. 1976. The effect of a tall tower on nocturnal bird migration—a portable ceilometer study. Auk 93:281–291.

Banks, R. C. 1979. Human related mortality of birds in the United States. Special Scientific Report—Wildlife No. 215. U.S. Fish and Wildlife Service, Department of the Interior, Washington, D.C., USA.

Barry, T. W. 1967. The geese of the Anderson River Delta, Northwest Territories. Dissertation, University of Alberta, Edmonton, Canada.

Barry, T. W. 1968. Observations on natural mortality and native use of eider ducks along the Beaufort Sea coast. Canadian Field Naturalist 82:140–144.

Bellrose, F. C., T. G. Scott, A. S. Hawkins, and J. B. Low. 1961. Sex ratios and age ratios in North American ducks. Illinois Natural History Survey Bulletin 27:391–474.

Bender, L. C., J. C. Boren, H. Halbritter, and S. Cox. 2011. Condition, survival, and productivity of mule deer in semiarid grassland-woodland in east-central New Mexico. Human–Wildlife Interaction 5:276–286.

Bergman, G. 1982. Inter-relationships between ducks and gulls. Pages 241–247 *in* D. A. Scott, editor. Managing wetlands and their birds: a manual of wetland and waterfowl management. International Waterfowl Research Bureau, Slimbridge, England.

Bevanger, K. 1995. Estimates and population consequences of tetraonid mortality caused by collisions with high tension power lines in Norway. Journal of Applied Ecology 32:745–753.

Bevanger. K. 1998. Biological and conservation aspects of bird mortality caused by electricity power lines: a review. Biological Conservation 86:67–76.

Bevanger, K., and H. Brøseth. 2004. Impact of power lines on bird mortality in a subalpine area. Animal Biodiversity and Conservation 27:67–77.

Boal, C. W., and R. W. Mannan. 1999. Comparative breeding ecology of Cooper's hawks in urban and exurban areas of southeastern Arizona. Journal of Wildlife Management 63:77–84.

Brown, W. M., and R. C. Drewien. 1995. Evaluation of two power line markers to reduce crane and waterfowl collision mortality. Wildlife Society Bulletin 23:217–227.

Bull, P. C., and P. G. Dawson. 1969. Mortality and survival of birds during an unseasonable snow storm in South Canterbury, November 1967. Notornis 14:172–179.

Burles, D. W., and M. Hoefs. 1984. Winter mortality of Dall's sheep, *Ovis dalli*, in Kluane National Park, Yukon. Canadian Field Naturalist 98:479–484.

Cassirer, E. F., and A. R. E. Sinclair. 2007. Dynamics of pneumonia in a bighorn sheep metapopulation. Journal of Wildlife Management 71:1080–1088.

Conover, M. R. 1997. Wildlife management by metropolitan residents in the United States: practices, perceptions, costs, and values. Wildlife Society Bulletin 25:306–311.

Conover, M. R. 2002. Resolving human–wildlife conflicts: the science of wildlife damage management. Lewis Brothers, Boca Raton, Florida, USA.

Conover, M. R. 2007. Predator–prey dynamics: the role of olfaction. CRC Press, Boca Raton, Florida, USA.

Conover, M. R., W. C. Pitt, K. K. Kessler, T. J. DuBow, and W. A. Sanborn. 1995. Review of data on human injuries, illnesses, and economic losses caused by wildlife in the United States. Wildlife Society Bulletin 23:407–414.

Coon, R. A., J. D. Nichols, and H. F. Percival. 1981. Importance of structural stability to success in mourning dove nests. Auk 98:389–391.

Cottam, C. 1929. A shower of grebe. Condor 31:80–81.

Dick, W. J. A., and M. W. Pienkowski. 1979. Autumn and early winter weights of waders in north-west Africa. Ornis Scandinavica 10:117–123.

Dodd, C. K., Jr., K. M. Enge, and J. N. Stuart. 1989. Reptiles on highways in north-central Alabama, USA. Journal of Herpetology 23:197–200.

Dolbeer, R. A., and S. E. Wright. 2009. Safety management systems: how useful will the FAA National Wildlife Strike database be? Human–Wildlife Conflicts 3:167–178.

Dunn, E. H. 1993. Bird mortality from striking residential windows in winter. Journal of Field Ornithology 64:302–309.

Edison Electric Institute. 2009. Out of sight, out of mind revisited: an updated study on the undergrounding of overhead power lines. Washington, D.C., USA.

Erickson, W. P., G. D. Johnson, M. D. Strickland, K. J. Sernka, and R. E. Good. 2001. Avian collisions with wind turbines: a summary of existing studies and comparisons to other sources of avian collision mortality in the United States. National Wind Coordinating Committee Resource Document. Western EcoSystems Technology, Cheyenne, Wyoming, USA.

Evans, P. R., and M. W. Pienkowski. 1984. Population dynamics of shorebirds. Behavior of Marine Animals 5:83–123.

Faanes, C. A. 1987. Bird behavior and mortality in relation to power lines in prairie habitats. General Technical Report 7. U.S. Fish and Wildlife Service, Department of the Interior, Washington, D.C., USA.

Ferrer, M., M. de la Riva, and J. Castroviejo. 1991. Electrocution of raptors on power lines in southwestern Spain. Journal of Field Ornithology 62:181–190.

Florida Fish and Game Commission. 2010. Alligator harvest summary for 2009. Florida Fish and Game Conservation Commission, Gainesville, USA.

Fredrickson, L. H. 1969. Mortality of coots during severe spring weather. Wilson Bulletin 81:450–453.

Geist, V. 1971. Mountain sheep: a study in behavior and evolution. University of Chicago, Chicago, Illinois, USA.

Gelb, Y., and N. Delacrétaz. 2006. Avian window strike mortality at an urban office building. Kingbird 56:190–198.

Gerow, K., N. C. Kline, D. E. Swann, and M. Pokorny. 2010. Estimating annual vertebrate mortality on roads at Saguaro National Park, Arizona. Human–Wildlife Conflicts 4:283–292.

Glasrud, R. D. 1976. Canada geese killed during lightning storm. Canadian Field Naturalist 90:503.

Green, R. E., M. Harley, M. Spalding, and C. Zöckler. 2001. Impacts of climate change on wildlife. United Nations Environment Programme, World Conservation Monitoring Centre and the Royal Society for the Protection of Birds, Sandy, United Kingdom.

Groot Bruinderink, G. W. T. A., and E. Hazelbrook. 1996. Ungulate traffic collisions in Europe. Conservation Biology 10.1059–1067.

Hansen, H. A. 1961. Loss of waterfowl production to tide floods. Journal of Wildlife Management 25:242–248.

Harness, R. E., and K. R. Wilson. 2001. Electric-utility structures associated with raptor electrocutions in rural areas. Wildlife Society Bulletin 29:612–623.

Harrington, J. L., and M. R. Conover. 2006. Characteristics of ungulate behavior and mortality associated with wire fences. Wildlife Society Bulletin 34:1295–1305.

Harvell, C. D., C. E. Mitchell, J. R. Ward, S. Altizer, A. P. Dobson, R. S. Ostfeld, and M. D. Samuel. 2002. Climate warming and disease risks for terrestrial and marine biota. Science 296:2158–2162.

Hels, T., and E. Buchwald. 2001. The effect of road kills on amphibian populations. Biological Conservation 99:331–340.

Hochbaum, H. A. 1955. Travels and traditions of waterfowl. University Minnesota Press, Minneapolis, USA.

Hodson, N. L., and D. W. Snow. 1965. The road deaths enquiry, 1960–1961. Bird Study 12:90–99.

Intergovernmental Panel on Climate Change. 2001. Climate change 2001: the scientific basis. J. T. Houghton, Y. Ding, D. J. Griggs, M. Noguer, P. J. van der Linden, and D. Xiaosu, editors. Cambridge University Press, Cambridge, United Kingdom.

Jacobson, A. R., A. Provenzale, A. von Hardenberg, B. Bassano, and M. Festa-Bianchet. 2004. Climate forcing and density dependence in a mountain ungulate population. Ecology 85:1598–1610.

James, P. 1956. Destruction of warblers on Parade Island, Texas, in May 1951. Wilson Bulletin 68:224–227.

Janss, G. F. E., and M. Ferrer. 1998. Rate of bird collision with power lines: effects of conductor-marking and static wire-marking. Journal of Field Ornithology 69:8–17.

Janss, G. F. E., and M. Ferrer. 2000. Common crane and great bustard collision with power lines: collision rate and risk exposure. Wildlife Society Bulletin 28:675–680.

Johnson, D. H., J. D. Nichols, and M. D. Schwartz. 1992. Population dynamics of breeding waterfowl. Pages 446–485 in B. D. J. Batt, A. D. Afton, M. G. Anderson, C. D. Ankney, D. H. Johnson, J. A. Kadlec, and G. L. Krapu, editors. Ecology and management of breeding waterfowl. University of Minnesota Press, Minneapolis, USA.

Johnson, G. D., W. P. Erickson, M. D. Strickland, M. R. Shepherd, D. A. Shepherd, and S. A. Sarappo. 2002. Collision mortality of local and migrant birds at a large-scale wind-power development on Buffalo Ridge, Minnesota. Wildlife Society Bulletin 30:875–879.

Jokinen, M. E., P. F. Jones, and D. Dorge. 2008. Evaluating survival and demography of a bighorn sheep (Ovis canadensis) population. Biennial Symposium of the Northern Wild Sheep and Goat Council 16:138–159.

Kamler, J. F., R. M. Lee, J. C. deVos Jr., W. B. Ballard, and H. A. Whitlaw. 2003. Mortalities from climbing accidents of translocated bighorn sheep in Arizona. Southwestern Naturalist 48:145–147.

Kennedy, R. J. 1970. Direct effects of rain on birds: a review. British Birds 63:401–414.

Kerlinger, P. 1989. Flight strategies of migrating hawks. Chicago University Press, Chicago, Illinois, USA.

Kerlinger, P. 2000. Avian mortality at communication towers: a review of recent literature, research, and methodology. U.S. Fish and Wildlife Service, Office of Migratory Bird Management, Laurel, Maryland, USA.

King, K. A. 1976. Bird mortality, Galveston Island, Texas. Southwest Naturalist 21:414.

Klem, D., Jr. 1989. Bird–window collisions. Wilson Bulletin 101:606–620.

Klem, D., Jr. 1990. Collisions between birds and windows: mortality and prevention. Journal of Field Ornithology 61:120–128.

Klem, D., Jr. 2006. Glass: a deadly conservation issue for birds. Bird Observer 34:73–81.

Klem, D., Jr., D. C. Keck, K. L. Marty, A. J. Miller Ball, E. E. Niciu, and C. T. Platt. 2004. Effects of window angling, feeder placement, and scavengers on avian mortality at plate glass. Wilson Bulletin 116:69–73.

Koops, F. B. J. 1987. Collision victims of high-tension lines in the Netherlands and effects of marking. Report 01282-mob 86-3048. KEMA, Arnheim, Netherlands.

Krause, H. 1959. Northern Great Plains region. Audubon Field Notes 13:380–381.

LaRoe, E. T., G. S. Farris, C. E. Puckett, P. D. Doran, and M. J. Mac. 1995. Our living resources: a report to the nation on the distribution, abundance and health of U.S. plants, animals, and ecosystems. National Biological Service, U.S. Department of the Interior, Washington, D.C., USA.

Lehman, R. N. 2001. Electrocutions on power lines: current issues and outlook. Wildlife Society Bulletin 29:804–813.

Linnell, M. A., M. R. Conover, and T. J. Ohashi. 1999. Biases in bird strike statistics based on pilot reports. Journal of Wildlife Management 63:997–1003.

Louisiana Wildlife and Fisheries. 2008. Louisiana's alligator management program: 2007–2008 annual report. Office of Wildlife, Coastal and Nongame Resources Division, Louisiana Department of Wildlife and Fisheries, Baton Rouge, USA.

Manville, A. M., II. 2005. Bird strike and electrocutions at power lines, communication towers, and wind turbines: state of the art and state of the science-next steps toward mitigation. General Technical Report PSW-GTR-191:1051–1064. U.S. Department of Agriculture Forest Service, Washington, D.C., USA.

McLoughlin, P. D., E. Dzus, B. Wynes, and S. Boutin. 2003. Declines in populations of woodland caribou. Journal of Wildlife Management 67:755–761.

McNulty, S. 2010. Florida sea turtle cold stunning. Wildlife Data Integration Network, Madison, Wisconsin, USA.

Means, D. B. 2010. Time to end rattlesnake roundups. Wildlife Professional 4:64–67.

Mendenhall, V. M., and H. Milne. 1985. Factors affecting duckling survival of eiders Somateria mollissima in northeast Scotland. Ibis 127:148–158.

Meretsky, V. J., N. F. R. Snyder, S. R. Beissinger, D. A. Clendenen, and J. W. Wiley. 2000. Demography of the California condor: implications for reestablishment. Conservation Biology 14:957–967.

Moreau, R. E. 1928. Some further notes from the Egyptian deserts. Ibis 4:453–475.

Morse, D. H. 1980. Population limitation: breeding or wintering grounds? Pages 505–516 in A. Keast and E. S. Morton, editors. Migrant birds in the Neotropics: ecology, behavior and distribution. Smithsonian Institution Press, Washington, D.C., USA.

O'Connell, T. 2001. Avian window strike mortality at a suburban office park. Raven 72:141–149.

Palmer, W. E., S. R. Priest, R. S. Seiss, P. S. Phalen, and G. A. Hurst. 1993. Reproductive effort and success in a declining wild turkey population. Proceedings of the Annual Conference of the Southeastern Association of Fish and Wildlife Agencies 47:138–147.

Parmesan, C. 2006. Ecological and evolutionary responses to recent climate change. Annual Review of Ecology, Evolution, and Systematics 37:637–669.

Parmesan, C., and G. Yohe. 2003. A globally coherent fingerprint of climate change impacts across natural systems. Nature 421:37–42.

Pelletier, A. 2006. Injuries from motor-vehicle collisions with moose—Maine, 2000–2004. Morbidity and Mortality Weekly Report 55:1272–1274.

Peurach, S. C., C. J. Dove, and L. Stepko. 2009. A decade of U.S. Air Force bat strikes. Human–Wildlife Conflicts 3:199–207.

Raftovich, R. V., K. A. Wilkins, K. D. Richkus, S. S. Williams, and H. L. Spriggs. 2010. Migratory bird hunting activity and harvest during the 2008 and 2009 hunting seasons. U.S. Fish and Wildlife Service, Laurel, Maryland, USA.

Ramirez, P., Jr. 1999. Fatal attraction: oil field waste pits. Endangered Species Bulletin 24:10–11.

Rattey, T. E., and N. E. Turner. 1991. Vehicle–moose accidents in Newfoundland. Journal of Bone and Joint Surgery 73:1487–1491.

Reeves, H. M., F. G. Cooch, and R. E. Munro. 1976. Monitoring Arctic habitat and goose production by satellite imagery. Journal of Wildlife Management 40:532–541.

Reinecke, K. J., C. W. Shaiffer, and D. Delnicki. 1987. Winter survival of female mallards in the lower Mississippi Valley. Transactions of the North American Wildlife and Natural Resources Conference 52:258–263.

Roberts, S. D., J. M. Coffey, and W. F. Porter. 1995. Survival and reproduction of female wild turkeys in New York. Journal of Wildlife Management 59:437–447.

Roberts, S. D., and W. F. Porter. 1998a. Influence of temperature and precipitation on survival of wild turkey poults. Journal of Wildlife Management 62:1499–1505.

Roberts, S. D., and W. F. Porter. 1998b. Relationship between weather and survival of wild turkey nests. Journal of Wildlife Management 62:1492–1498.

Root, T. L., J. T. Price, K. R. Hall, S. H. Schneider, C. Rosenzweig, and J. A. Pounds. 2005. The impact of climatic change on wild animals and plants: a meta-analysis. General Technical Report PSW-GTR-191:1115–1118. U.S. Department of Agriculture Forest Service, Washington, D.C., USA.

Root, T. L., and S. H. Schneider. 2002. Climate change: overview and implications for wildlife. Pages 1–56 in S. H. Schneider and T. L. Root, editors. Wildlife responses to climate change: North American case studies. Island Press, Washington D.C., USA.

Roseberry, J. L. 1962. Avian mortality in southern Illinois resulting from severe weather conditions. Ecology 43:739–740.

Rosen, P. C., and C. H. Lowe. 1994. Highway mortality of snakes in the Sonoran Desert of southern Arizona. Biological Conservation 68:143–148.

Roth, R. R. 1976. Effects of a severe thunderstorm on airborne ducks. Wilson Bulletin 88:654–656.

Rubolini, D., M. Gustin, G. Bogliani, and R. Garavaglia. 2005. Birds and power lines in Italy: an assessment. Bird Conservation International 15:131–145.

Sargeant, A. B., and D. G. Raveling. 1992. Mortality during the breeding season. Pages 396–422 in B. D. J. Batt, A. D. Afton, M. G. Anderson, C. D. Ankney, D. H. Johnson, J. A. Kadlec, and G. L. Krapu, editors. Ecology and management of breeding waterfowl. University of Minnesota Press, Minneapolis, USA.

Schüz, E., P. Berthold, E. Gwinner, and H. Oelke. 1971. Grundirss der Voelzugskunde. Paul Parey, Berlin, Germany.

Segelson, C. 2010. Record cold leads to record numbers of manatee deaths. Florida Fish and Wildlife Conservation Commission, Tallahassee, USA.

Smallwood, K. S., and C. Thelander. 2008. Bird mortality in the Altamont Pass Wind Resource Area, California. Journal of Wildlife Management 72:215–223.

Smith, A. G., and H. R. Webster. 1955. Effects of hail storms on waterfowl population in Alberta, Canada—1953. Journal of Wildlife Management 19:368–374.

Smith, L. L., and C. K. Dodd Jr. 2003. Wildlife mortality on U.S. Highway 441 across Payne's Prairie, Alachua County, Florida. Florida Scientist 66:128–140.

Smith, M. A. 1964. Cohoe-Alaska. Audubon Field Notes 18:478–479.

Spendelow, P. 1985. Starvation of a flock of chimney swifts on a very small Caribbean island. Auk 102:387–388.

Stoudt, J. H. 1982. Habitat use and productivity of canvasbacks in southwestern Manitoba, 1961–72. Special Scientific Report—Wildlife 248. U.S. Fish and Wildlife Service, Department of the Interior, Washington, D.C., USA.

Sundar, K. S. G., and B. C. Choudhury. 2005. Mortality of sarus cranes (Grus antigone) due to electricity wires in Uttar Pradesh, India. Environmental Conservation 32:260–269.

Trail, P. 2006. Avian mortality at oil pits in the United States: a review of the problem and efforts for its solution. Environmental Management 38:532–544.

U.S. Federal Highway Administration. 2008. Status of the nation's highways, bridges, and transit: conditions and performance. Report to Congress. U.S. Department of Transportation, Washington, D.C., USA.

U.S. Fish and Wildlife Service. 2003. Draft recovery plan for the Sierra Nevada bighorn sheep. Portland, Oregon, USA.

Vepsäläinen, K. 1968. The effect of the cold spring 1966 upon the lapwing *Vanellus vanellus* in Finland. Ornis Fennica 45:33–47.

Webster, F. S. 1974. The spring migration, April 1–May 31, 1974, South Texas region. American Birds 28:822–825.

Wiedenfeld, D. A., and M. G., Wiedenfeld. 1995. Large kill of neotropical migrants by tornado and storm in Louisiana, April 1993. Journal of Field Ornithology 66:70–80.

Wilkinson, P. F., and C. C. Shank. 1976. Rutting-fight mortality among musk oxen on Banks Island, Northwest Territories, Canada. Animal Behaviour 24:756–758.

Williams, G. C. 1950. Weather and spring migration. Auk 67:52–65.

Wilson, R. R., and P. R. Krausman. 2008. Possibility of heat related mortality in desert ungulates. Journal of the Arizona-Nevada Academy of Science 40:12–15.

11

NUTRITIONAL ECOLOGY

KATHERINE L. PARKER

INTRODUCTION

Nutritional ecology is the science of relating an animal to its environment through nutritional interactions. This field of study is important to wildlife ecologists because nutrition affects animal survival and reproduction. The life histories of animals (e.g., their activities, feeding patterns, movements, and reproductive strategies) are shaped by nutritional requirements and constrained by the foods available to meet them. The goal of this chapter is to provide readers with an overview of basic concepts for nutritional ecology, including examples that help explain why free-ranging animals use their environments as they do.

ENERGY REQUIREMENTS

Energy is the universal currency for survival. It is needed to support basic body functions and activities, to stay warm, to reproduce, and to grow and produce tissues. Animals transform chemical energy from food, with the addition of oxygen, into products that help meet energy demands and into heat that helps maintain an acceptable body temperature. Energy costs (e.g., the energy needed per unit time, or metabolic rate) have been measured indirectly as heat produced (kJ or kcal; 1 kcal = 4.184 kJ) or directly as oxygen consumed for numerous wildlife species. These measurements for specific behaviors and ambient conditions are made on animals in sealed chambers or tanks or wearing respiratory masks.

Basal Metabolism

Small species and young individuals have relatively high energy demands (per unit mass) compared to older or larger ones, in part because the energy needed to support such basal functions as respiration, blood circulation, and muscle tone is higher in small animals. Basal metabolic rate (BMR), as the energy used when animals are lying, calm, fasted, and not thermally stressed, has been measured for numerous wildlife species in captivity. From these measurements, the interspecific relationship commonly used to approximate basal metabolic rate from body mass in placental mammals is BMR =

$293 \times kg^{0.75}$ kJ/day (Kleiber 1947). Birds, especially small passerines, tend to have higher basal metabolism; most marsupials have slightly lower rates (Barboza et al. 2009:220). Deviations from the interspecific relationship can be adaptive. Compared to Arctic canids, desert canids have lower metabolism, which would reduce water loss and risk of dehydration (Careau et al. 2007). The high metabolic rates in Arctic ungulates enable rapid growth in environments with a pulsed forage supply in a short growing season (Lawler and White 2003).

Activity

Average energy costs for daily existence, which incorporate all activities, are usually two to three times higher than BMR for mature birds and mammals (not including reproductive requirements; Robbins 1993:159). Activity-specific costs have been measured for numerous species. Energy costs of standing by game birds and mammals are approximately 17–22% higher than costs for sitting or lying, and costs of feeding by birds and ungulates range from 10% to 60% above perching or standing, depending on the type, composition, and availability of food (Fancy and White 1985, Robbins 1993:143). For all species, energy costs of travel increase at higher speeds. This is generally a linear increase for species that use terrestrial locomotion (Fig. 11.1A). More energy is required to move uphill, and less is needed going downslope. In species such as kangaroos (Macropus spp.), which change from bipedal locomotion to hopping at higher speeds, energy costs actually decline as energy is recovered on the hop. For swimming species, energy costs increase exponentially at higher speeds because of increasing drag in the water (Fig. 11.1C). Leaping out of the water by dolphins (Delphinidae) may allow them to expend less energy because of decreased drag. For birds and bats, energy costs of flight are represented as a U-shaped function relative to speed (Fig. 11.1D). When first taking off, costs are elevated because an individual must support its weight against gravity and overcome drag; with increasing wing-flapping and flying speed, energy costs steadily increase. During energy-saving migration, the optimal speed of flight

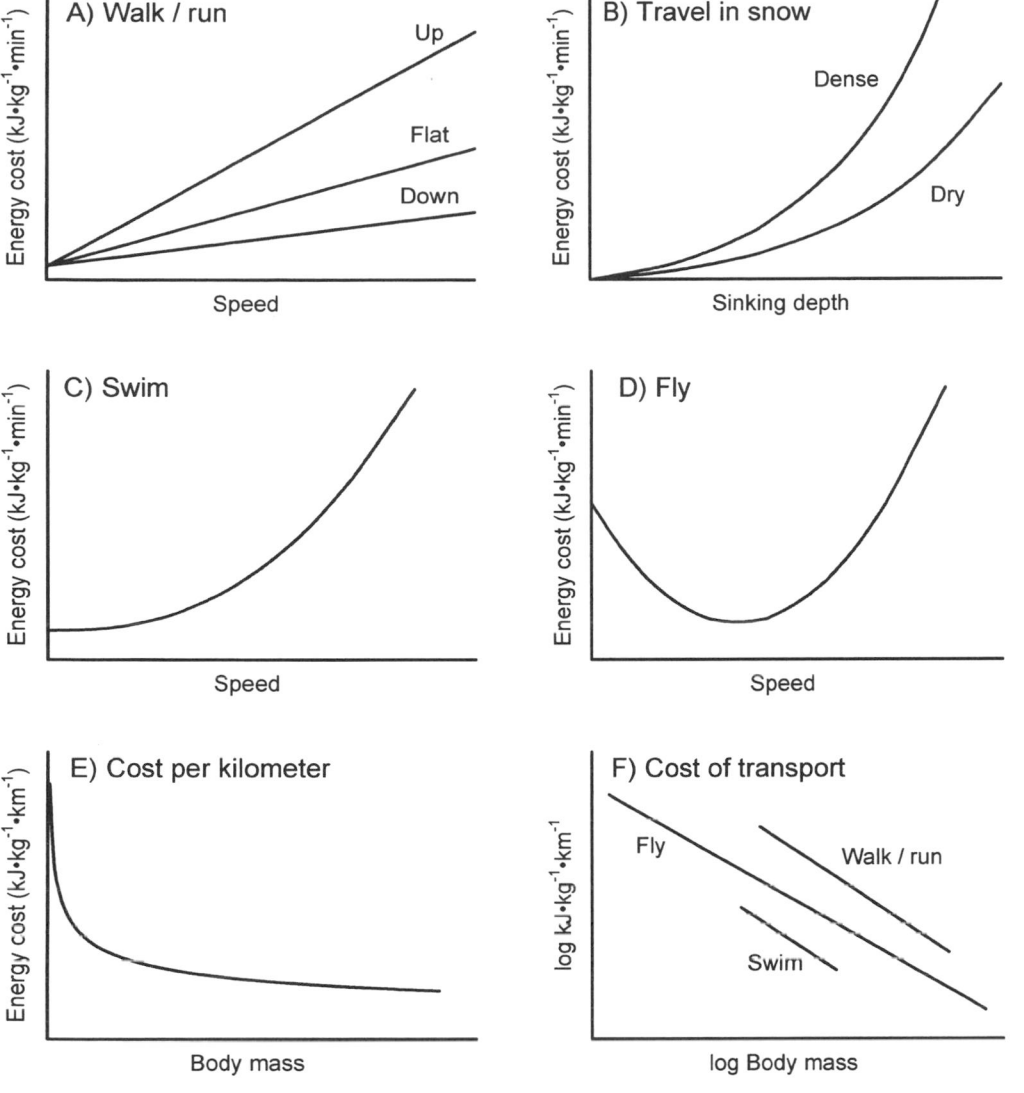

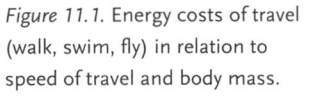

Figure 11.1. Energy costs of travel (walk, swim, fly) in relation to speed of travel and body mass.

for long-distance migrants is assumed to be the speed at which energy costs are lowest. For running, swimming, and flying, relative energy costs of traveling a known distance are highest for small species and individuals (Fig. 11.1E). It takes more energy per kilogram for a neonate to travel a kilometer than it does for its mother. Comparing the three forms of locomotion, net energy costs are least to swim and most to run or walk a kilometer (Fig. 11.1F).

The substrate may exacerbate the energy needed for terrestrial locomotion. Travel by caribou (*Rangifer tarandus*) in bogs can cost 30–40% more energy than when moving on solid footing (Fancy and White 1985). In snow, energy costs of travel increase exponentially as the sinking depth of the animal increases (Fig. 11.1B). Snow depths to front knee height (~60% of leg length) may double energy costs of movement for cervids and bovids. When snow and cooling temperatures trigger autumn migrations of mule deer (*Odocoileus hemionus hemionus*) to areas with less snow, higher energy costs are avoided (Monteith et al. 2011). If animals sink to brisket height (leg length), energy costs are three to seven times higher than

when walking on bare ground, depending on snow density (Robbins 1993:135). Juveniles with shorter leg lengths are compensated to some extent by lower foot loading (i.e., body mass per unit foot area), which can reduce sinking depths (Fancy and White 1985). Nonetheless, snowy winters compounded by reduced food availability take their greatest toll on young animals with relatively high metabolism. Ungulates are disadvantaged compared to such predators as wolves (*Canis lupus*) because of generally heavier foot loading. Delayed snowmelt in spring further increases population mortality rates if energy costs remain high, body reserves have been depleted, and spring food supplies are delayed.

Thermoregulation

Very cold, very hot, and wet, rainy environments increase energy demands on wildlife. These supplemental energy costs for thermoregulation are typically small compared to activity costs but nonetheless contribute to energetic drains. Most birds and mammals exhibit a characteristic thermal response in which energy costs remain relatively constant over a range

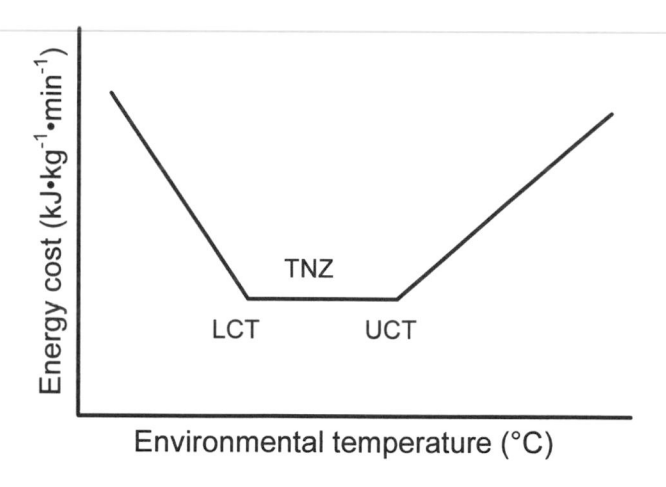

Figure 11.2. Common energetic thermoregulatory response of birds and mammals to environmental temperature. TNZ, thermoneutral zone; LCT, lower critical temperature; UCT, upper critical temperature.

of ambient temperatures (i.e., the thermoneutral zone; Fig. 11.2). Animals piloerect their feathers or hair to increase thermal depth and insulation value at cold temperatures. Below a lower critical temperature, animals shiver to increase metabolic rate and to maintain an acceptable body temperature. Lower critical temperatures are lower in winter, when animals have more thermal insulation from winter pelage and fat depots than in summer. Among north-temperate ungulates, lower critical temperatures in winter range from –10°C for bighorn sheep (*Ovis canadensis*) and Columbian black-tailed deer (*Odocoileus hemionus columbianus*) to below –40°C for moose (*Alces alces*) and reindeer (*Rangifer tarandus*; Parker and Robbins 1985). Fed animals can withstand lower temperatures than animals without food because of heat produced during food digestion. At temperatures above an upper critical temperature, animals usually increase metabolism by panting or sweating to rid the body of excess heat. A potential advantage of sweating in nonarid environments where water is not limited is that animals can continue to forage efficiently at high temperatures.

Some wildlife species avoid increases in energy demands at cold ambient temperatures by going into torpor; they allow body temperature to decline to a regulated low temperature and are capable of arousing back to normal body temperature. Bats (e.g., big brown bat, *Eptesicus fuscus*) and hummingbirds (e.g., *Selasphorus* spp.) use torpor each night; chickadees (*Poecile* spp.) use torpor during extreme weather conditions, and bears (*Ursus* spp.) and ground squirrels (*Spermophilus* spp.) employ extended periods of torpor (e.g., hibernation). At high ambient temperatures, some desert ungulates allow body temperature to rise above normal to minimize the added costs (and inherent water loss) of trying to maintain a stable body temperature.

The ambient thermal environment experienced by free-ranging animals includes more than air temperature. Rather, animals are exposed to the combined effects of air temper-

ature, wind, and solar radiation. High winds decrease the effective temperature that animals experience; sunshine increases it. Mule deer, which shiver violently at –30°C in early morning in winter, cease shivering with incoming direct sunshine (Parker and Robbins 1984). A useful index to quantify the thermal environment that animals experience is standard operative temperature (Bakken et al. 1985), which is similar to a wind chill index but includes the added influence of solar radiation (it does not accommodate rainfall). This index can be calculated for any combination of air temperature, wind, and radiation to compare thermal values of different habitats. Species- and habitat-specific models predict when animals are metabolically stressed outside the thermoneutral zone and therefore when populations could be adversely affected if food resources are limited (Parker and Gillingham 1990, Beaver et al. 1996). For resource managers, modeling exercises are useful to determine the frequency of elevated thermal costs and if suitable cover is available. Supplementary energy costs of thermoregulation are even higher for animals under rainy, wet conditions, depending on temperature and season (Parker 1988), particularly for neonates and nestlings.

Thermal cover is a manageable entity for furbearers and ungulates. Coarse, woody debris on the forest floor and cavities in snags provide added insulation for martens (*Martes americana*), which have relatively high lower critical temperatures. In the deeryards of eastern North America, forest cover is managed to ameliorate the effects of cold temperatures, wind, and deep snow. Thermal cover also provides shelter at high operative temperatures, such as for moose, which are heat stressed above freezing temperatures in winter (Renecker and Hudson 1986). Managing for thermal cover, however, is difficult because animals are not always found in the thermoneutral zone. Animals may remain in thermally stressful environments if the energy gain from the food exceeds the thermal costs. Attempts to validate the influence of thermal cover on the body condition of elk (*Cervus canadensis*) were confounded by variability in small, infrequent weather effects, although access to incoming solar radiation in open areas decreased mortality in winter (Cook et al. 1998).

Reproduction and Production

Reproduction increases energy requirements. For birds and mammals, energy costs for females are relatively higher (per kilogram) for small species than larger ones, because the eggs or offspring are larger in proportion to body mass. In birds, energy costs depend on the number and composition of the eggs. Eggs with large yolks are more costly to produce. Precocial species such as waterfowl have large yolks to support the development of functionally advanced chicks capable of searching for their own food upon hatching (Vleck and Bucher 1998). Depending on the number of eggs laid, energy requirements for egg production by precocial bird species could be more than five times higher than those of altricial species (Robbins 1993:192). During incubation and chick rearing, energy costs vary with the development of the chicks and with uniparental versus biparental care (Tulp et al. 2009). The amount of time spent foraging versus incubating can energetically affect males and females. One or both sexes also may invest energy into the production of plumage color that signals mate quality.

In mammals, energy costs for females increase during fetal development. More than 90% of the energy requirements of gestation occur in the last trimester. Small mammals have shorter gestation periods, but relatively larger young compared to large mammals. Energy costs during lactation are higher than for gestation, to support larger mass gains by larger offspring. Lactation is considered to be the most energetically demanding period for adult females. Daily energy expenditures can rise to four to seven times BMR depending on milk composition and stage of lactation (Robbins 1993:203). Because of the relatively long period of gestation overlapping the environmental constraints of winter, and the subsequent period of lactation, highest total daily energy costs for reproducing female ungulates usually occur from late winter to midsummer. In contrast, highest daily energy costs for male ungulates typically occur when they are most active during the breeding period in autumn, often at the expense of foraging. Males expend some energy and nutrients in secondary characteristics to attract mates (e.g., antlers in cervids).

Free-ranging animals require energy for growth and molt, although these biological processes are largely protein demands. Requirements for energy and nutrients by juveniles are highest when the proportional changes in body mass are highest—when cells are rapidly dividing or expanding (Barboza et al. 2009:249). The energy content of mass gain varies depending on the deposition of fat and protein. Juveniles that are still growing deposit a higher proportion of protein toward body mass than do adults. The highest energy content of gain occurs in marine mammals and species that fatten before hibernation. Some supplemental energy also is needed for seasonal pelage and plumage replacement, and for corresponding thermoregulatory costs that may increase during this period.

Energy Budgets of Free-ranging Animals

Wildlife biologists can multiply time spent by free-ranging wildlife in different behaviors (i.e., the activity budget) by published activity-specific energy costs to calculate an energy budget. These types of calculations provide insights into the effects of disturbance and changing environments. For example, we can quantify the effect of winter recreationists on ungulates by knowing the distances that animals flee and the energy costs of travel in snow. Energy costs in response to snow machine travel are often lower than in response to less predictable cross-country skiers (Reimers et al. 2003). Plowed roadways, which decrease energy costs in winter, may have facilitated the northward expansion of coyotes (Canis latrans) in eastern North America (Crête and Larivière 2003). The energetic effects of aircraft or fishermen on behavior vary among waterfowl species (Conomy et al. 1998).

When researchers cannot monitor daily activities but are able to handle animals for sampling, doubly labeled (isotopic) water can be used to determine energy expenditures (Nagy et al. 1999). This method involves injecting an animal with two labels (i.e., hydrogen—^2H or radioactive ^3H—and oxygen—^{18}O) and measuring the rate of loss of the two isotopes in one or more subsequent water samples (usually blood or urine). Because hydrogen is lost from the body as water (H_2O, via excretion and evaporation), and oxygen (O_2) is lost as both H_2O and carbon dioxide (CO_2), the difference in the rates of these losses is a measure of CO_2 production. The measure of CO_2 produced is used to derive O_2 consumed, a measure of energy cost. Doubly labeled water can accurately quantify energy costs of free-ranging animals, but there are cautions when using the technique. In addition to knowing the time needed for the labels to equilibrate in the body water before sampling, small sampling errors can easily influence the slopes describing rates of isotopic decline and therefore the calculations of energy costs. In spring and summer, if hydrogen is incorporated into milk production or fat deposits (and therefore not lost from the body water pool only as water), the rate of isotopic decline can be affected enough to cause impossible estimates of energy expenditure (Parker et al. 1999). Heart-rate telemetry has also been used to index daily energy costs, because heart rates typically increase with energy expenditures (Butler et al. 2004). This relationship varies widely among individuals and by season, however, making extrapolation to the population difficult.

PROTEIN REQUIREMENTS

Protein is important to wildlife because it is a major constituent of the animal body. Proteins are in pelage and plumage, hooves and claws, the antibodies for disease resistance, blood clotting factors, and all hormones. Proteins enable movement of muscle fibers and biochemical reactions with enzymes. Composed of sequences of amino acids, all proteins contain nitrogen.

For an animal to maintain body protein during routine, nonstressed activity, protein requirements can be approximated from protein losses. Losses caused by the normal breakdown of body proteins are excreted in the urine (i.e., endogenous urinary nitrogen, or EUN) and are related to body mass and metabolism. Losses also occur from the gastrointestinal tract (as metabolic fecal nitrogen, or MFN) because of loss of digestive enzymes, bacterial byproducts, and sloughing of cells along the tract. Protein lost as MFN is of nonfood origin, but it is related to food quality because highly fibrous diets place more wear and tear on the digestive lining. Protein lost as MFN by browsers is generally higher than by grazers (Robbins 1993:177). To estimate protein requirements of free-ranging wildlife, biologists use species-specific equations for EUN and MFN (with some adjustment for dietary nitrogen that is not protein) determined from animals in captivity based on body mass and daily intakes (Spalinger et al. 2010). In birds, nitrogen losses are excreted together.

For all species, relative protein requirements (per unit mass) are typically highest during growth. Game bird chicks require

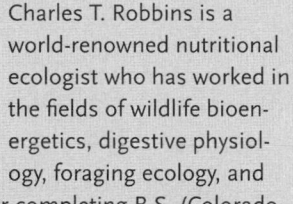

**CHARLES T. ROBBINS
(b. 1946)**

Charles T. Robbins is a world-renowned nutritional ecologist who has worked in the fields of wildlife bioenergetics, digestive physiology, foraging ecology, and nutritional balance. After completing B.S. (Colorado State University) and M.S. (Syracuse University) degrees, he worked on the biological basis of carrying capacity for his doctoral research (Cornell University). Since 1974, Robbins has been at Washington State University.

Robbins developed an internationally recognized research program in wildlife nutritional ecology, which spans specific wildlife requirements and adaptations as well as applied implications and management. His initial emphasis on a variety of wild ruminants has extended to in-depth grizzly and black bear studies. In addition to more than 100 publications, Robbins has brought the science of nutritional ecology to the forefront for many wildlife biologists with his book *Wildlife Feeding and Nutrition*.

Photo courtesy of Darin Watkins

30% dietary protein, compared to the 12% needed for maintenance of adults (Robbins 1993:175). Diets that are more than 75% insects help meet this demand (Johnson and Boyce 1990). Reproduction also increases protein requirements. Many avian species, including hummingbirds, increase the high-protein invertebrate content of their diets during egg laying. The timing of egg laying and the growth of nestlings emphasize the importance of high-protein diets in spring and summer. For mammals, protein requirements needed during lactation are second to juvenile growth. The protein incorporated in milk to support offspring is more than the protein demands of producing a fetus. If the requirements for fetal growth and milk production cannot be met by food intake, species such as caribou must rely on maternal body protein stores (Barboza and Parker 2008). Protein is also important for survival when animals are food stressed, because they mobilize body protein to meet energy requirements if body fat reserves have been significantly depleted.

MINERALS AND VITAMINS

All wildlife species require minerals and vitamins. Deficiencies and toxicities are not that common in wild animals in natural habitats, but they do occur. Concentrations of minerals in particular vary in different areas, and therefore concentra-

tions needed in the diet depend on intake. Requirements vary among species.

Water-soluble vitamins (C, B vitamins) rarely pose deficiencies in free-ranging wildlife and are not usually toxic because any excesses are excreted in the urine. Fat-soluble vitamins (A, D, E, K) are absorbed in the intestinal tract with fat and can be stored in large concentrations. They also do not usually pose nutritional problems for free-ranging species. Vitamin A was assumed to synchronize reproduction in quail (*Colinus* spp.) populations, but more recent findings suggest that plant estrogens are likely to be a regulating factor (Robbins 1993:84). Vitamin D is synthesized in the presence of ultraviolet light; young mammals that are born in burrows without access to sunlight obtain vitamin D through the mother's milk. Deficiencies of vitamin E, which is important in maintaining cell membranes, precipitate tissue damage during capture stress. Vitamin K is important in blood clotting. Warfarin, which is used as a rodent poison, inhibits vitamin K and thus clotting, and poisoned animals bleed to death. Barboza et al. (2009:192) and Robbins (1993:83, 96) give more detailed descriptions of the functions of vitamins as well as signs of deficiency and toxicity.

Twenty-six of the 90 elements in the chemical periodic table are essential to life. These minerals are associated with many essential functions (Barboza et al. 2009:172, Robbins 1993:55). Important macroelements (i.e., elements usually required as ≥1 mg/g of food) include calcium (Ca), phosphorus (P), magnesium (Mg), and sodium (Na). Ratios of 1:1 to 2:1 for Ca:P are required for normal skeletal formation, and increase to 4:1 in laying birds. Calcium appears to be a major determinant of reproductive success in some avian species. Providing supplemental Ca to tree swallows (*Tachycineta bicolor*) in nest boxes resulted in earlier laying dates, larger eggs, larger clutches, and higher growth rates of larger chicks (Bidwell and Dawson 2005). Actively growing plants are usually high in P, and therefore deficiencies are most likely to occur in herbivores in winter. Porcupines (*Erethizon dorsatum*) and bighorn sheep may chew on antlers and bones to balance P requirements. Magnesium is abundant in plants and animals, but very high potassium levels in herbaceous forage can induce a Mg deficiency known as grass tetany, resulting in limited muscle contraction, uncoordinated movements, convulsions, and often death. In contrast, most plants cannot accumulate Na and, when coupled with the leaching of minerals from alpine and mountain soils, may be one reason why ungulates travel to licks to obtain nutrients that cannot be found in forage alone. Besides Na, lick soils often contain elevated levels of Ca, Mg, carbonates, and clay, all of which can help ameliorate gastrointestinal disturbances associated with the transition from winter browse to lush green forage (Ayotte et al. 2006). In addition to movements to licks by northern cervids and bovids and the congregations of tropical parrots at licks, minerals may drive the long migrations of wildebeests (*Connochaetes taurinus*; McNaughton 1990). Aquatic plants that contain high levels of Na (50–500 times higher than most terrestrial plants) reportedly influence the movements and abundance of moose (Belovsky and Jordan 1981).

Trace elements are required in smaller amounts (<1 mg/g of food) than macroelements. Important links to free-ranging wildlife have been made for iron (Fe), copper (Cu), fluoride (F), and selenium (Se). Iron is obtained by geophagia (i.e., consumption of soil) and is adequate in most forages. Copper deficiencies, causing defective keratinization of hair and hooves, occur in moose and deer in areas with Cu-deficient soils and can be precipitated in mine-spoil areas where levels of molybdenum are high. Around hot springs where F accumulates naturally, dental problems associated with tooth enamel have been reported in elk and bison (*Bison bison*). Selenium interacts with vitamin E to maintain the integrity of tissues and cell membranes. When Se is deficient (e.g., in high-rainfall areas), there is increased susceptibility to muscle damage during capture for species such as mountain goats (*Oreamnos americanus*), and white muscle disease may result. In contrast, during periods of poor forage conditions, ingesting some toxic species of milkvetch (*Astragalus* spp.), which contain thousands of times the Se levels needed for maintenance, can cause either chronic (alkali disease) or acute (blind staggers) Se toxicity.

ESSENTIAL FATTY ACIDS

Fatty acids are important to wildlife because they are components of body fat stores and are found in all cell membranes. Although difficult to define structurally, fatty acids are grouped as saturated (single bonds between all carbon atoms) and unsaturated (having one or more double bond). Because of their structure, unsaturated fatty acids do not stack in an orderly manner and remain more fluid especially at low temperatures. The legs of Arctic caribou have high levels of monounsaturated fatty acids, which help to facilitate movement (Blix 2005). Alpine marmots (*Marmota marmota*) store high levels of polyunsaturated fatty acids and can mobilize those fat depots more easily, thereby tolerating longer bouts of hibernation (Geiser et al. 1994). The primary essential fatty acids, which must be obtained in the diet because animals lack the enzymes needed to create specific double bonds, are linoleic, linolenic, arachidonic, eicosapentaenoic, and docosahexanoic acids (Barboza et al. 2009:121). Because plants synthesize fatty acids, however, deficiencies of essential fatty acids are uncommon in free-ranging herbivores and their predators.

WATER

Water is the most essential nutrient for all wildlife and drives the ecological life histories of many desert species. Small and young animals generally require more water per unit mass than larger species because they have higher metabolic rates. Although water can be obtained from freestanding sources, many wild animals rarely, if ever, drink free water. Instead, they acquire preformed water in the tissues of food and metabolic water from the breakdown of carbohydrates, protein, and fats. Small mammals in arid environments are commonly seedeaters, because high carbohydrate content produces the greatest net amount of water for the animal. Nocturnal or fos-

sorial desert species take advantage of lower temperatures and higher humidity. Ungulates may time parturition to avoid the driest periods of the year, thereby having access to higher food abundance and quality. Fawning is delayed by three to four weeks for mule deer in the southwestern United States compared to northern populations (Bowyer 1991), and numbers of white-tailed deer (*Odocoileus virginianus*) track the amount of precipitation in the southern United States (Teer et al. 1965). Huge congregations of bison remained on the North American Great Plains, where the timing of lactation coincided with the highest precipitation levels (supporting high-quality forage abundance), instead of in the drier northwestern river valleys (Mack and Thompson 1982). Chapter 12 discusses water management for wildlife.

FOOD INTAKE TO MEET NUTRITIONAL REQUIREMENTS

Food resources and body reserves are used to meet the nutritional requirements of wildlife. Whether food alone meets animal requirements depends on its availability, as well as the quantity and quality of the food consumed. Food intake drives what is possible in terms of body mass and condition of the animal. Body reserves help meet nutrient demands when food supplies are limited.

Diet

Feeding time is an important measure of nutritional ecology, estimated by watching animals or by using activity sensors attached typically to the head or neck. Defining specifically what animals eat (their diets) can be done through assessments of feeding sites, visual observations of feeding (Table 11.1), and by using cameras mounted on animals or their collars. More often, reporting diets involves the use of hair, bone, skull, and exoskeleton keys for carnivores and insectivores, and requires knowledge of plant epidermal characteristics for herbivores. Samples are obtained from regurgitated pellets in birds, esophageal contents, stomach and rumen contents, and cheek pouches and crops. Diets determined by microhistology of fecal (or scat) samples are limited to plant fragments that are least digested by herbivores and to identifiable remains for carnivores, although DNA methods are useful in distinguishing some prey species. Estimating the proportions of items in diets requires corrections for digestibility, because highly digestible items will be underrepresented in feces.

Recently, stable isotopes have been used to reconstruct diets, particularly for carnivores, for which residues in fecal material are limited if meat is entirely digested. Further, scat samples give a relatively short-term dietary assessment of recent ingestion compared to a longer-term estimate from stable isotopes that index the assimilation of prey material into consumer tissues. Stable isotopes of a chemical element contain the same number of protons but a different number of neutrons. Wildlife ecologists typically use nitrogen ($^{15}N/^{14}N$) and carbon ($^{13}C/^{12}C$) to define diets. Isotopic ratios are expressed in delta notation (δ), comparing the ratio of the heavy:light

Table 11.1. Some commonly used methods to quantify diets, food intake, and food quality for free-ranging wildlife

Index	Premise	Cautions for biologists	Source
Diet and intake			
Feeding site methods	What has been removed by herbivores or what remains after predation can be quantified.	Biased toward foods in which removal is easiest to detect; problems with missing items and regrowth.	Holechek et al. (1982)
Contents of stomach, rumen, crop, cheek pouches	Foods that have recently been consumed will be in the digestive tract.	Biased toward foods least digestible and easiest to identify.	Holechek et al. (1982)
Fecal analyses	Foods can be identified by indigestible residues voided in the feces.	Biased toward foods least digestible and most likely to fragment; detailed microhistological keys needed.	Holechek et al. (1982)
Remote monitoring of feeding behavior	Time spent feeding can be determined using telemetry transmitters and cameras.	Variation around the relation between signal pattern and behavior can be high; difficulties positioning cameras.	Hassall et al. (2001)
Direct observations of feeding by free-ranging animals	Bite counts or time spent feeding reflect food intake.	Biased toward plants easiest to see at a distance; difficult to stay close to wild animals; handpicked bites are assumed to simulate amounts (and quality) of forages consumed.	Holechek et al. (1982)
Direct observations of feeding by captive animals in the wild	Bite-by-bite estimates of intake are determined by visual estimates of plant unit size or handpicked estimates of bite size.	Training needed to recognize sizes of plant units or bite sizes consumed; animals must not be naive to native foods and must allow close observers.	Collins et al. (1978), Parker et al. (1993)
Stable isotopes	Isotopic signatures in consumer tissues reflect assimilation of foods.	All dietary sources must be identified and have isotopic signatures that differ from each other; discrimination between animal and diet is needed; variation around food estimates depends on the mixing model used (IsoSource, SIAR, SISUS, MixSIR).[a]	Karasov and Martinez del Rio (2007:433)
n-alkanes	Alkane profiles in the feces reflect alkane profiles in the diet.	Dietary components are limited by the number of n-alkane markers; dosing with an even-chained alkane helps correct for digestibility.	Bugalho et al. (2005), Dove and Mayes (2005)
Fatty acids	Fatty acids in consumer tissues reflect fatty acids of the prey consumed.	Species with foregut microbial digestion may break down dietary fatty acids so that they are not absorbed intact in the consumer's tissues, biasing results of mixing models (QFASA).[b]	Iverson et al. (2004)
Food quality			
Energy content by bomb calorimetry	Gross energy is released by complete oxidation of food.	Not all gross energy is available to the animal, depending on what is digested and metabolized.	Robbins (1993:9)
Protein content	Kjeldahl procedure or total elemental nitrogen analysis provides a crude estimate of dietary protein.	Nonprotein nitrogen is included in varying amounts; not all protein is available to the animal.	Barboza et al. (2009:138)
Total dietary fiber	Higher fiber decreases digestibility.	Some fiber may be digested by microbial fermentation.	Prosky et al. (1984), Servello et al. (2005)
Digestibility by in vivo digestion trials	Digestion trials with live animals document specific digestibilities of different foods.	Requires relatively tame animals and sufficient time to equilibrate to foods before trials.	Hupp et al. (1996), Mayes and Dove (2000)
Digestibility by internal markers	Marker concentrations in the food and feces can be used to calculate indigestibility.	Food intake and internal marker concentration must be recorded over an extended time period.	Tilley and Terry (1963)
Digestibility by in vitro digestion	Digestion in a test tube simulates digestion in an animal.	Apparent digestibility depends on an adapted donor inoculum (microbes must be habituated to foods).	Robbins (1993:257)
Proximate analysis	Food fiber and nonfiber (water, ash, crude protein, crude fat, nitrogen free extract) components can be quantified.	Components may not be separated completely; outdated procedure.	Robbins (1993:257)
Digestibility by sequential detergent analyses	Nutritive value of plants depends on cell contents and cell wall constituents.	Plant secondary metabolites may confound estimates of the digestible fraction.	Goering and Van Soest (1970), Mould and Robbins (1981)
Tannin content based on bovine serum albumin (BSA) precipitate	Protein-precipitating capacity indexes the reduction in protein availability.	Quantifiable effects of tannins on digestion are species specific.	Hanley et al. (1992), Spalinger et al. (2010)
Fecal protein	Higher fecal nitrogen reflects higher dietary nitrogen.	Limited use for species consuming mostly tannin-containing forages because tannins bind to protein, elevating fecal protein.	Robbins et al. (1987), Leslie et al. (2008)
Fecal 2,6-diaminopimelic acid (DAPA)	DAPA (a component of nondigested rumen microflora) tracks changes in dietary energy.	Should not be used to track dietary nitrogen.	Osborn and Ginnett (2001)
Near-infrared reflectance spectroscopy (NIRS)	Spectral properties index diet components and quality (digestible energy and protein) from fecal samples.	Calibration equations for the relationships between constituents in the sample and NIRS spectral information must be developed; these vary with botanical composition.	Kamler et al. (2003)

[a] SIAR, Stable Isotope Analysis in R; SISUS, Stable Isotope Sourcing Using Sampling; MixSIR, Mixing Sampling Importance Resampling.

[b] QFASA, Quantitative Fatty Acid Signature Analysis.

isotope in a sample with the ratio of a nonvariable standard. Delta values are positive if the ratio is higher than the standard, and negative if the ratio is lower than the standard. Samples collected for isotopic analyses depend on the time frame of interest. Signatures in plasma generally reflect what has been nutritionally absorbed during the previous seven to ten days; red blood cells and muscle are indicative of the last three to four months; bone represents assimilation over a lifetime; and hair, feathers, and nails reflect the time period when they were grown. Segmenting hairs can indicate seasonal changes in diets of grizzly bears (*U. arctos*; Milakovic 2008) and wolves (Darimont et al. 2008). Analyzing eggs can help quantify avian diets during egg laying (Hobson 1995).

Nitrogen isotopes (^{15}N/^{14}N) are particularly important because they help determine tropic position. Animals preferentially excrete the lighter nitrogen isotope (^{14}N) in nitrogenous wastes during normal enzymatic breakdown. Relatively more of the heavier isotope (^{15}N) is incorporated into tissues, and therefore there is an increase in δ^{15}N with each trophic level. Carbon isotopes (^{13}C/^{12}C) distinguish broadly between plants with different photosynthetic pathways. The δ^{13}C of C$_3$ plants (e.g., many herbaceous plants, shrubs, cool-season grasses) usually differs from that of C$_4$ plants (e.g., sedges, warm-season grasses). In tropical areas where grasses are primarily C$_4$ plants and other plants use C$_3$ pathways for photosynthesis, the δ^{13}C of grazers differs from browsers, as do the signatures incorporated into their predators. This distinction among herbivores is less clear in colder environments where there are few C$_4$ plants. Nonetheless, sections of antler and hoof tissue have shown switches between shrub-based, graminoid, and lichen-based diets by reindeer, and from woody browse to leafy and aquatic diets by moose (Kielland and Finstad 2000, Kielland 2001). The contribution to diets by some aquatic plants and cacti, which use a third photosynthetic pathway, can be quantified using δ^{13}C when combined with hydrogen isotopes (δ^2H, ^{2}H/^{1}H; Karasov and Martinez del Rio 2007:449).

As natural markers, stable isotopes provide important information on geographic, temporal, and age-specific variation in diets. Hydrogen signatures in bird feathers are a means of tracking origins of migratory species because of a North American continent-wide gradient in δ^2H in precipitation (Karasov and Martinez del Rio 2007:462). High-sulfur signatures (δ^{34}S, ^{34}S/^{32}S) in whitebark pine (*Pinus albicaulis*) nuts helped quantify the importance of that food resource to grizzly bears in Yellowstone National Park (Felicetti et al. 2003). Strontium (^{87}S/^{86}S) isotopes, as geochemical signatures, were used to track the diets and locations of African elephants (*Loxodonta africana*; Koch et al. 1995). Nitrogen, carbon, oxygen, hydrogen, sulfur, and oxygen (^{18}O/^{16}O) isotopes are all enriched in marine systems. Consequently, they have been used to assess the degree of terrestrial foraging by polar bears (*U. maritimus*) as the sea ice declines (Ramsay and Hobson 1991) as well as the contribution of marine foods when terrestrial sources are limited for species such as Arctic foxes (*Vulpes lagopus*; Angerbjörn et al. 1994).

Fatty acids also are used to reconstruct shifts in diets. They are most useful in simple-stomached carnivores for which fatty acids are deposited in the tissues with minimal modification from the diet. Fatty acids in the blubber of marine mammals and bears reflect the fatty-acid signatures of their major food items (Iverson et al. 2001, 2004). Vibrissae, which continually grow throughout an animal's lifetime, can be a time series of seasonal shifts in fatty acids related to diet composition (Newsome et al. 2010). In non-simple-stomached animals, such as ruminants, fatty acids might not always remain intact because of microbial metabolism, and distinguishing among types of forages eaten might not be possible.

Available Food and Quantity Consumed

Food availability typically changes with season. When one prey species declines, carnivores shift to other prey species, move to other areas, or also decline. Decreases in lemming (*Lemmus* spp.) and ptarmigan (*Lagopus* spp.) populations trigger snowy owls (*Bubo scandiacus*) and other raptors to move southward from Arctic tundra (Gessaman 1972). Canada lynx (*Lynx canadensis*) populations cycle in abundance relative to snowshoe hare (*Lepus americanus*) populations (Stenseth et al. 1997). For herbivores, food availability is constrained by the subset of forage species that they ingest from the total plant biomass. Caribou, for example, consume particular lichen species from the large mats of ground-dwelling lichens (Johnson et al. 2000). Biologists tasked with quantifying food availability for either carnivores or herbivores are furthered challenged by patchy distributions of food and changing availability associated with environmental factors such as snow.

As food abundance increases, food intake by animals per unit time increases asymptotically up to a maximum rate (Fig. 11.3A). This functional response depends on an herbivore's requirements and characteristics of the food (i.e., quality, structure, and spatial arrangement). For predators, per-capita kill rate changes similarly with prey density. At very low food biomass, animals may choose not to feed because the total value of the food obtained over time is less than the costs of obtaining it (Barboza et al. 2009:34). Some studies define thresholds of food abundance (biomass per unit area) below which the environment is not suitable to meet intake requirements.

Physical and physiological factors regulate the quantity of food an animal eats. For species from snowshoe hares to moose, intake rates increase asymptotically with bigger bite sizes (Fig. 11.3B; Shipley and Spalinger 1992). Intake rates by Sitka black-tailed deer (*Odocoileus hemionus sitkensis*) on large skunk cabbage (*Lysichiton americanum*) or devil's club (*Oplopanax horridus*) leaves can be five times higher than when picking individual forb leaves (Gillingham et al. 1997). Fibrous foods lower intake rates because of the processing constraints of chewing and digestion (Fig. 11.3C,D). Ruminants in particular must break down food into small particles before it can move through the ruminant stomach and the remainder of the gastrointestinal tract. Black-tailed deer eating low-quality fibrous food in winter spend up to ten hours each day chewing

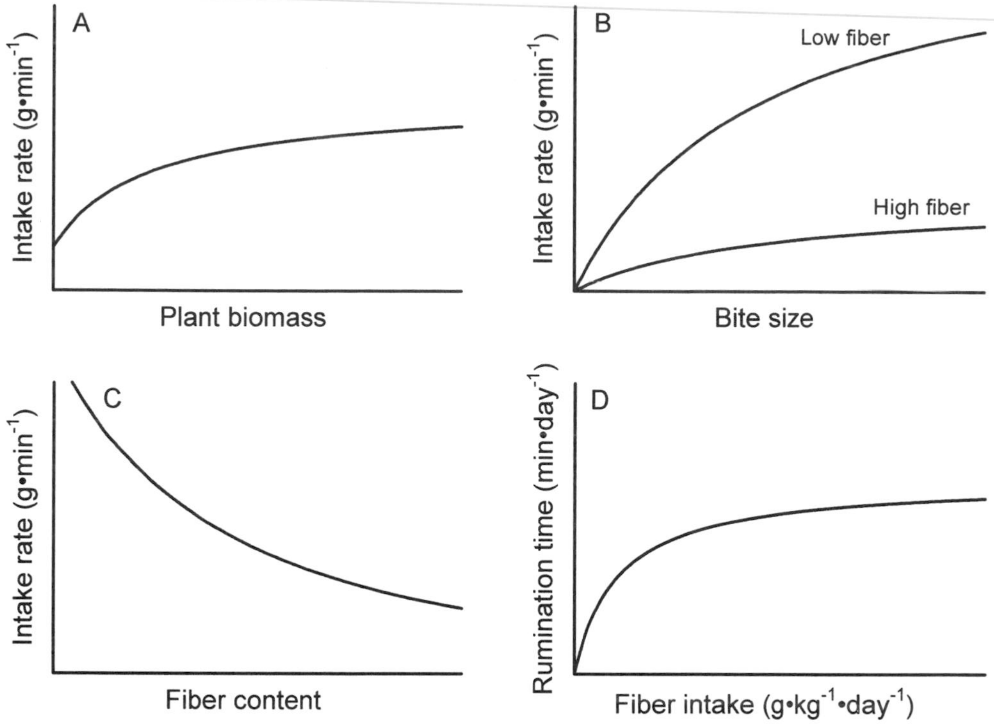

Figure 11.3. Food intake relative to food biomass and bite size, and the influence of plant fiber on intake rate and rumination time.

their cud (Parker et al. 1999). Food quality limits the intake of food quantity under poor forage conditions. The number, size, and nutritional value of bites taken ultimately influence the duration of feeding bouts and time spent feeding each day (Shipley 2007). Some carnivores such as badgers (*Taxidea taxus*) increase digestive efficiency by slowing the passage of prey through the gut, thereby increasing nutrient gain during periods of prey scarcity (Harlow 1981).

Highest voluntary food intakes are typically associated with highest nutrient demands. For mammals, the quantity of food consumed per kilogram by lactating females can be double what it is for males. In birds, intake rates to meet daily energy expenditures during the incubation phase can be twice as high as during the nestling phase for birds that incubate without assistance from their mates (Tinbergen and Williams 2002). Young animals and small species, with relatively higher metabolism, must eat more per unit mass than adults or large species to meet their requirements. Daily intakes when corrected to dry matter generally range from 2% of body mass per day for large mature animals without reproductive demands to 10% for smaller or younger animals, although this is dependent on food quality (Robbins 1993:333). Intakes are even higher when animals put on body stores prior to migration, hibernation, or winter.

Food intakes for free-ranging animals can be approximated from the mass of bites consumed while watching individuals feed (Table 11.1). Typically, these estimates are only for short periods of time (minutes or hours) because of logistical constraints; extrapolations are then made to total daily intakes. Less commonly used to quantify intake by wildlife are the plant wax compounds known as *n*-alkanes. Most of the *n*-alkanes in forage have naturally present odd-numbered

carbon chains with 25–35 carbons (e.g., C_{25}, C_{27}, C_{29}, C_{31}, C_{33}, C_{35}). Because there are differences in the alkane profiles of plants, the proportion of intake from different dietary components can be discriminated by multiple plant wax markers. These compounds are largely indigestible, but not completely recovered in the feces. If published data are not available to adjust for fecal recovery, tame or untamed animals can be dosed orally with a synthetic, slow-release, even-chained alkane similar to the plant alkane of adjacent carbon chain length (e.g., C_{24}, C_{28}, C_{32}, C_{36}). Subsequent fecal samples with both natural and synthetic *n*-alkanes are then used to estimate food intakes.

Food Quality

Estimating food quality is one of the greatest challenges for wildlife ecologists. Nutritive value varies with plant and animal parts and changes with season. Importantly, food quality influences the nutrients that can be extracted from the food. Even when food quantity is sufficient, food quality can limit the growth of individuals and populations. Quality is commonly expressed in terms of energy content per gram and percent protein or nitrogen (N) content on a dry matter basis (Table 11.1). Forages generally range between 18.8 and 20.5 kJ gross energy per gram dry weight (slightly higher for foods containing resins, waxes, or oils), but the digestibility of that energy usually declines from spring to autumn as structural and chemical characteristics of the forage change. Because proteins average 16% N, percent crude protein is calculated as grams N per gram feed × 6.25 (from 100 ÷ 16). Protein is highest in the new growing buds and shoots (20–30%) of grasses, forbs, and shrubs in spring, and declines with plant maturity and senescence (3–4%). Seeds are relatively high in both energy and protein. Animal tissue, which is largely protein, increases in energy content with age because of increasing fat content. For carnivores and simple-stomached species that do not consume highly fibrous foods, available energy in prey can be roughly approximated from average fuel values: 16.7 kJ/g carbohydrate, 16.7 kJ/g protein, and 37.7 kJ/g fat based on dry mass.

Understanding the value of different foods to free-ranging wildlife, however, requires some knowledge of how the animal processes and retains foods. Feeding trials with captive carnivores and herbivores are used to quantify food value specifically from measures of all food consumed minus any digestive (fecal) and metabolic (urine and digestive gas) losses. In birds, digestive and metabolic losses are excreted together. The apparent energy available to a free-ranging animal (metabolizable energy) is calculated by multiplying the gross energy content (kilojoules per dry gram) of the prey or forage by the percent that is digestible (excluding fecal losses) and metabolizable (excluding urinary and gaseous losses). These calculations enable comparison of the values of different food types. One snowshoe hare, for example, is worth the energetic equivalent of 49 small voles (*Microtus* and *Clethrionomys* spp.) to a bobcat (*Lynx rufus*; Powers et al. 1989).

The higher the fiber content of the food, typically the lower is its digestibility. For bears consuming fish, meat, berries, pine nuts, and plants, fiber content varied five-fold, and digestibilities ranged from 90% to 40% (Pritchard and Robbins 1990). Fiber content of forage generally increases from new growth in spring to senescence in winter; stems usually have more fiber than leaves. Consequently, the digestibilities of common forages for herbivores in winter may be only 30–40% of summer forages (Cook 2002). Digestive fecal losses from food not absorbed by the animal tend to be substantially higher and more variable than metabolic urinary losses (i.e., food digested but not metabolized) and are more easily quantified with laboratory measures. Food value is often presented simply as digestible energy or digestible protein, recognizing that animals retain slightly less than these estimates.

Digestion trials that use live animals to determine digestibility by measuring total food intake and fecal output or the concentration of an internal marker in diet and feces are not always possible, and many foods have not been fed during experimental trials. Instead, researchers have simulated digestion in test tubes, quantified fiber and nonfiber components, and chemically digested the plant materials (Table 11.1). The contents of plant cells (i.e., the cell solubles) are highly digestible, but plant cell walls composed of a cellulose matrix provide nutritional value only to wildlife species that can partially digest them using microbial fermentation. These species include all ruminants, and species with microbes in pouches of the hindgut—horses, porcupines and beavers (*Castor canadensis*), giraffes (*Giraffa camelopardalis*), elephants, kangaroos—or microbes in caeca, such as rabbits and hares (Leporidae), as well as grouse (Tetraoninae). Sequential detergent analysis on forage is the most commonly used method to estimate what herbivores with microbes can potentially digest. A neutral detergent digests the cell contents, leaving neutral detergent

fiber (NDF, or the plant cell walls) as the residue. A sequential acid detergent added to the NDF digests a portion of the cell walls that microbes can digest, leaving an acid detergent fiber. Additional protocols are used to quantify other constituents of the fiber (e.g., cellulose, lignin, cutin). Equations that predict digestibility of different forages, validated with digestion trials, incorporate the components of sequential detergent analysis.

Assessing the nutritive value of forages is further complicated by plant defensive compounds, also called plant secondary metabolites (PSMs) or secondary plant compounds because initially they were not known to have primary metabolic function in plants and are presumably produced to deter herbivory. These compounds include toxic alkaloids (found in lupine [*Lupinus* spp.], larkspur [*Delphinium* spp.], nicotine, morphine); flavonoids that alter reproductive patterns; terpenoids (in volatile oils of sagebrush, or *Artemisia* spp., eucalyptus, conifers, and citrus); and tannins. Terpenoids generally have toxic effects on microbes that ferment cellulose. Perhaps uniquely, the habituated microbial populations in pronghorns (*Antilocapra americana*), pygmy rabbits (*Brachylagus idahoensis*), and sage grouse (*Centrocercus urophasianus*) allow those species to consume diets of mostly sagebrush in winter. Much attention has been given to understanding the influence of tannins on forage quality because they are present in the leaves of most deciduous woody species and some forbs. Their levels vary depending on whether they are produced under shade versus sun conditions (Happe et al. 1990). Tannin levels are minimal in grasses and relatively low in winter browse. Tannins bind to forage protein (making it unavailable to the animal) and decrease digestibility. For example, the tannin content of forages consumed by moose increases over the summer, resulting in a similar decrease in protein availability (McArt et al. 2009). Lab procedures using bovine serum albumin (BSA) have been developed to mimic how much protein is bound to tannins; BSA can be incorporated as a correction factor in equations for protein availability and digestibility (Table 11.1). Wildlife ecologists are able to estimate food value from chemical analyses of foods for species including voles, grouse, bears, and ungulates (Servello et al. 2005).

Some animals avoid plants with PSMs; others that frequently consume them have counteradaptations to modify plant defenses (Dearing et al. 2005, Marsh et al. 2006). Moose and deer, for example, produce salivary proteins that bind tannins to reduce their absorption and effects on digestion. Some other species (usually specialists) absorb PSMs and must detoxify them for excretion. The rate of detoxification can determine the rate of feeding. So why do animals eat toxic compounds that can influence food quality, digestion, and potentially reproduction? A uniform threshold for ingestion of toxic PSMs does not appear to exist. Rather, the value of the nutrient content presumably is greater than the cost of detoxification (McArthur et al. 1993). There are important implications of the current increasing levels of CO_2 on the planet to landscape dynamics because some PSM concentrations increase at higher levels of CO_2, and consumers may attempt to avoid elevated defensive compounds. Chickadees, for example, prefer eating caterpillars that feed on leaves with low PSM content. Avoidance by avian predators of gypsy moth (*Lymantria dispar*) caterpillars consuming highly defended leaves may lead to even larger areas of defoliation (Müller et al. 2006).

In the past, resource professionals managed those food species that were most limited during critical times. Enhancing willow (*Salix* spp.) and aspen (*Populus* spp.) stands generally increased or maintained populations of moose and ruffed grouse (*Bonasa umbellus*), respectively. Now there is more emphasis placed on understanding the nutritional components that influence feeding patterns. Ruffed grouse feed in spring on the large male aspen (*P. tremuloides* and *P. grandidentata*) catkins that have the highest energy content (Svoboda and Gullion 1972). Spruce grouse (*Canachites canadensis*) select needles of jack pine (*Pinus banksiana*), which have significantly higher protein and mineral content than other trees (Gurchinoff and Robinson 1972). Pronghorns are adapted to consume big sagebrush (*A. tridentata*) despite its terpenoid content, and may benefit from the higher protein levels that are found in sagebrush compared to other winter browse species (Bailey 1984:96). Mule deer and black-tailed deer eat conifer seedlings and branches blown from treetops, but tend to avoid the medium-aged, low branches of conifers that have higher terpenoid content (Parker et al. 1999). Koalas (*Phascolarctos cinereus*), even as highly specialized foliovores, visit eucalyptus (*Eucalyptus* spp.) trees that are least defended (Moore and Foley 2005). In the subarctic, ptarmigan, snowshoe hares, and beavers select foods to avoid plant defensive compounds (Bryant and Kuropat 1980).

Indexing the quality of the diet that animals consume is a challenge for wildlife ecologists. The use of fecal samples for this purpose has met with mixed success (Table 11.1). Fecal nitrogen can serve as a general index to dietary nitrogen, depending on animal and forage species. Studies that assume that diet quality varies with fecal protein, however, can overestimate digestibility and protein value for wildlife species that consume mostly tanniferous browse leaves in summer. Fecal 2,6-diaminopimelic acid (DAPA) generally tracks dietary energy but not dietary nitrogen. Near-infrared reflectance spectroscopy (NIRS), with extensive calibration, might index various aspects of dietary quality. Recently, fecal chlorophyll has been used to document consumption of green biomass in relation to timing of spring green-up (Christianson and Creel 2009). Values of fecal protein also appear to correspond with changes in large-scale indices of green biomass derived from satellite imagery (e.g., normalized difference vegetation index; Pettorelli et al. 2005, Hamel et al. 2009). This correspondence has important implications for the nutritional understanding of highly mobile species that track plant phenology, selecting for highest digestible energy and protein content. Pregnant barren-ground caribou (*R. t. granti*), for example, move across the Alaskan coastal plain to areas with higher rates of spring green-up (Griffith et al. 2002), and Stone's sheep (*Ovis dalli stonei*) females move up in elevation as the growing season progresses (Walker et al. 2006).

Predicting Food Intake Needed to Meet Requirements and Nutritional Carrying Capacity

Wildlife ecologists estimate the amount of food that an individual needs to consume on the basis of nutritional requirements and food value. For example, energy requirements can be determined after calculating an energy budget derived from observations of all activities, each multiplied by energy-specific costs. Food energy value is its energy content multiplied by digestibility (and metabolizability, if known). From these calculations the amount of food required per unit time can be determined for herbivores or carnivores, recognizing that requirements vary with sex and age. Food needed to meet protein requirements can be assessed in a similar manner. Based on these types of models, long-term changes in benthic communities associated with climate shifts are expected to impact Arctic wintering waterfowl (Richman and Lovvorn 2003). In northern temperate areas the greatest nutritionally limiting factor for coastal black-tailed deer is reported to be digestible energy, whereas inadequate forage protein constrains northern moose populations (Parker et al. 1999, McArt et al. 2009). Digestible energy also appears to regulate elk populations through its influence on body condition and probability of calving (Cook et al. 2004). The ability to predict nutritional value helps define the importance of seasonal habitats. Extrapolation to a nutritional carrying capacity (i.e., the number of individuals supported per unit area) can be made if there are reasonable estimates of individual animal requirements and food biomass and its quality, adjusted for changing food availability (Hobbs and Swift 1985, Hanley and Rogers 1989, Guthery 1999, Hanley et al. 2012). Estimates of the food resources needed to support an individual or group of individuals would be lower if animals are able to mobilize body reserves, particularly during seasons when food availability and quality are low.

BODY CONDITION

Body condition is a general term for an animal's level of energy reserves and results from the integration of nutrient requirements and food intake. Body fat and protein and body mass have direct consequences for reproduction and population dynamics. In ungulates, food resources in summer strongly influence body size and condition in autumn, affecting the timing and probability of conception by females and then carrying a fetus to term (Cook et al. 2004). Also during summer, juveniles must obtain a size that is sufficient for them to survive winter. Juvenile size directly affects adult body mass and age of first reproduction. Food resources and environmental conditions in winter have direct consequences to over-winter survival and body condition in spring, and to timing of parturition and birth mass of young (Parker et al. 2009). Similarly, habitats needed to acquire prebreeding reserves are critical for breeding success and survival of young waterfowl.

The body capital–income continuum helps explain timing of reproduction and the importance of different habitats toward population recruitment (Barboza et al. 2009:246).

Energy and nutrient supplies that birds and mammals use to support reproduction can be obtained from body stores accumulated before breeding (i.e., capital breeders), food supplies available at the time of reproduction (i.e., income breeders), or both. Migrating waterfowl and shorebirds that arrive on northern nesting areas early often use body stores (capital) acquired at the southern wintering areas to support egg production. Income breeders that arrive later rely on food supplies (income) at staging areas and nesting areas. Similarly, reindeer that calve before spring green-up produce calves and the milk to support them by mobilizing maternal tissues deposited the previous fall. Caribou with calves born later can meet more of those nutritional demands with spring forage (Barboza and Parker 2008).

Wildlife biologists estimate body condition using a variety of morphological and physiological indices (Table 11.2; Parker 2003, Servello et al. 2005). In addition to body mass, skeletal measurements provide indirect insights to condition. Linear and circumferential measures (e.g., chest girth, body length, hind foot length) are often measured on ungulates, lagomorphs, and bears. The most commonly used indices describing body condition are associated with body fat, which is the primary energy reserve of the body. Animals with more body fat are assumed to be in better condition. From dead animals, total body fat can be determined based on fat extraction of homogenized tissues. Kidney fat is a general index of body fat of adult ungulates and lagomorphs, but it does not reflect body condition at extremely low body fat levels. Gizzard fat provides a similar index in game birds. Mass of xiphoid fat at the base of the sternum indexes total body fat in canids. Bone marrow fat based on color and consistency is sensitive only for animals in very poor condition. On live animals, total body fat can be predicted from total body water content after injecting animals with hydrogen isotopes (^{2}H or ^{3}H; Speakman et al. 2001). Total body fat also can be indexed reliably in some species using measures of electrical conductivity through the body, because the conductivity of lipids is only 4–5% that of other body tissues (Robbins 1993:236). In ungulates, ultrasound measures of rump fat and body condition scores are used to predict total body fat, often in conjunction with other morphological measures. Visual scoring of avian fat stores and profiles is commonly used for waterfowl and songbirds. Servello et al. (2005) describe many of these techniques.

Physiological indicators when calibrated and validated can be used to infer nutritional deprivation and, indirectly, body condition (Table 11.2). As animals resort to breaking down body protein to meet energy demands when fat reserves are mostly depleted, higher levels of urinary nitrogen are excreted, as are metabolites of muscle breakdown. The use of nitrogen isotopes can distinguish urinary nitrogen that results from breakdown of body protein versus nitrogen from food intake because of the preferential excretion of the lighter nitrogen isotope (^{14}N) in nitrogenous wastes mentioned earlier. A progressive increase in the excretion of the heavier isotope (^{15}N) would indicate greater mobilization of body protein

reserves. Cortisol or corticosterone levels also may increase with increasing nutritional stress, particularly when population densities are high and food becomes more limiting. The excretion of allantoin (a urinary metabolite resulting from microbial fermentation and digestion of nucleic acids in forage) by such herbivores as bison and elk declines as digestible energy intakes decrease when snow accumulates. Higher levels of glucuronic acid are excreted when the consumption of plant secondary metabolites is high, which presumably reflects lower forage quality for such species as deer, snowshoe hares, and grouse. Each of these physiological indicators is obtained noninvasively from urine samples in snow and is commonly presented as a ratio of creatinine. Creatinine is a normal excretory product proportional to muscle mass and serves as a background that avoids differences in dilution or hydration state. Urea nitrogen and glucocorticoid indicators also can be obtained using blood samples, but animal handling is required, and the indicators in blood typically vary more in response to short-term changes. Cook (2002) and Servello et al. (2005) give other blood indices. Whenever possible, wildlife ecologists should use multiple indices of nutritional condition and available food resources to best assess nutritional status and constraints of study animals and study areas.

NUTRITIONAL ECOLOGY IN THE BIG PICTURE

For free-ranging species, life is a trade-off between costs and benefits over the long term. Animals generally use seasonal strategies that minimize the maximum detriment at different times of the year (Parker et al. 2009). In northern winters, ungulates minimize energy expenditures by decreasing movement rates and by avoiding the most thermally stressful environments. At a time when food supplies are usually reduced, minimizing energy costs and maintaining protein stores increase the likelihood of survival and fetal development. Following successful birthing in spring, adult females must regain body mass and replenish body reserves for subsequent breeding in autumn. Males also must regain condition for successful breeding, and juveniles must amass sufficient size to ensure survival though winter. In spring through autumn, animals usually maximize intake of high-quality forage. In species that migrate from summer to winter ranges, strategies to lessen the primary deterrents to fitness depend on body condition. Animals in good nutritional condition may delay migration to benefit from higher food quality at the risk of early winter snows. Animals in poor condition migrate earlier, trading off the potential of higher energy costs for the predictable, albeit lower-quality, food supply (Monteith et al. 2011). Under extreme conditions of hard winters or poor summers, some females may conserve maternal condition at the expense of producing offspring, ultimately enabling higher future reproduction (Clutton-Brock et al. 1989, Festa-Bianchet and Côté 2008). In avian species, food quantity and quality affect body stores for migration, and the ability to support income or restore capital during the reproductive period (Anteau and Afton 2004).

Table 11.2. Common indices used to assess body condition: premise and limitations

Index	Premise	Cautions for biologists	Source
Morphological measurements			
Body mass	Comparative changes in mass or carcass mass among years provide a relative index of changing conditions.	Varies with age and physiology; does not index body fat.	Servello et al. (2005)
Chest girth, hind foot length, femur length, wing length, body length	Structural measures index long-term nutrition.	Considerable variability among individuals.	Servello et al. (2005)
Body length:mass ratio	Size adjusted for mass provides a sound index of condition.	Difficult to interpret biologically because ratio is confounded by effects of body mass.	Servello et al. (2005)
Antler beam width and diameter	Antler size reflects spring and summer range conditions.	Most affected by nutrition just prior to antler growth; confounded by genetics.	Scribner et al. (1989)
Fat indices			
Visual condition scoring or manual palpation	Visual scoring of avian fat depots and profiles indexes subcutaneous fat; ungulate body condition scores index body fat.	Requires extensive training for accuracy.	Krementz and Pendelton (1990), Cook (2002)
Kidney fat:mass ratio	General indicator of total body fat.	Confounded by seasonal variation in kidney mass; not useful in young animals without fat reserves or very fat animals.	Van Vuren and Coblentz (1985)
Bone marrow:fat mass ratio	Body condition based on color and consistency of bone marrow.	Limited use in young animals because bone marrow is active in red blood cell formation or for animals in very good condition.	Harder and Kirkpatrick (1994)
Total body fat by whole body grinding and lipid extraction	Absolute chemical measure of percent fat.	Time-consuming, specialized facilities required.	Servello et al. (2005)
Total body fat by water dilution using hydrogen isotopes	Total fat can be predicted from an estimate of body water after injecting H isotope.	Estimate of the time needed for the isotope to equilibrate in the animal is required.	Torbit et al. (1985), Hildebrand et al. (1998)
Total body fat by total body electrical conductivity or bio-electrical impedance assay	Electrical conductivity through lipids is much lower than through other body components.	Most useful on animals that distribute fat relatively evenly; requires training to consistently position animals for measurements.	Scott et al. (2001)
Ultrasonography	Rump fat through thickness in ungulates is related to total body fat.	Limited use when rump fat is depleted; need for species-specific equations.	Cook et al. (2010)
Dual emission X-ray absorptiometry	X-rays are transmitted differentially depending on tissue type.	Equipment generally not field-portable.	Nagy (2001)
Physiological indices			
Urinary urea nitrogen:creatinine (UN:C) ratio	In winter, when forage protein remains relatively constant, increases in UN:C ratios occur when body protein is broken down.	May be confounded if protein intake is variable; assumes creatinine excretion is constant.	DelGiudice (1995)
Urinary cortisol:creatinine ratio	Cortisol excretion increases during nutritional stress and prior to death.	May be confounded by nonnutritional stressors.	Saltz and White (1991)
Urinary allantoin:creatinine ratio	Excretion of allantoin, a product of microbial fermentation, increases with higher intakes of digestible nutrients.	Levels of allantoin in some species are too low to be used as an index.	Vagnoni et al. (1996)
Urinary glucuronic acid:creatinine ratio	Because glucuronic acid is a by-product of digesting plant secondary metabolites, higher levels indicate reduced dietary quality.	Confounded by level of food intake; not produced from consumption of all plant defensive compounds.	Guglielmo et al. (1996), Servello and Schneider (2000)
Urinary δ^{15}N concentration	Higher concentrations of δ^{15}N are excreted when body protein is being catabolized.	Potentially high variance around model parameters.	Gustine et al. (2011)

For many species, predation risks are highest when young are produced, and reducing this risk coincides with the need to obtain nutrients to support the young (e.g., lactation). Woodland caribou (*R. t. caribou*) choose calving areas that have relatively low wolf risk and select the highest-quality vegetation from what is available. Within two months of parturition, solitary female caribou and their calves move to form large groups in areas with higher food abundance (Gustine et al. 2006), minimizing risk while maximizing intake. Similarly, African ungulates birth when forage quality is high followed by abundant plant biomass. Predation risk is lower during the highly synchronous birthing of large-herd species that swamps predator response and when parturition by small-herd species is spatially and temporally variable (Sinclair et al. 2000). But predation does not negate the importance of nutrition. In some cases, predation may even mask the nutritional value provided (or not provided) by different habitats. In areas where nutritional value is marginal for population recruitment, the good body condition of mothers that have been relieved of the high energetic costs of lactation following neonatal predation might suggest otherwise. In other cases, antipredator behavior may increase the nutritional deficits of winter, most affecting animals with high reproductive demands (Christianson and Creel 2010).

Small differences in food value and distribution compound the effects on animal performance (White 1983). As such, nutrition affects individuals, populations, and ecosystem dynamics. Not surprisingly, diet selection and habitat segregation may differ among females, males, and nonreproductive animals because nutritional requirements of the sexes differ (Barboza and Bowyer 2001). Feeding pressures can modify communities enough to change the complement of herbivores, their predators, and associated species on the landscape (Berger et al. 2001, Allombert et al. 2005).

Ever-increasing rates of environmental change pose challenges for wildlife. The ability to accommodate these changes depends on adaptation through nutritional interactions (i.e., meeting requirements when food resources change in timing, abundance, and quality). Climatic changes in winter (e.g., deep snows, freezing rains, icing events) can increase energy requirements and decrease access to food. Body reserves acquired in summer would help buffer these nutritional deficits. Earlier plant green-up induced by globally warmer temperatures in spring, however, may cause trophic mismatches between nutritional supplies and demands if animals are unable to alter the timing of reproduction (and potentially the timing of migration) so that it corresponds with peak food values. In these cases, life histories would be altered because of subsequent declines in body condition. Anthropogenic habitat changes that alter plant succession also inevitably influence the nutritional ecology of free-ranging wildlife. Wildlife species with high behavioral and physiological plasticity to change will be those that are most likely to survive and reproduce on changing landscapes. The concepts and tools referred to in this chapter can help wildlife ecologists assess nutritional constraints on habitat value by quantifying shifts in diet and condition that accompany environmental change.

SUMMARY

Nutritional requirements shape the activities, feeding patterns, and movements of wildlife. Small and young animals generally require more energy, more protein, and more water per unit mass than larger species because of their higher metabolism. Reproductive demands also are relatively higher for small species than larger ones, because eggs or offspring are larger in proportion to body size. Wildlife species use food resources and body reserves to meet these energetic and protein demands, which vary with life history and seasonally changing environments. Even when food quantity is sufficient, food quality may limit the growth of individuals and populations because it influences the nutrients that can be extracted from food. Food quality varies with plant and animal parts and also changes with season. Both the quantity and quality of food consumed drive what is possible in terms of body mass and condition of the animal, which then have direct consequences for survival and reproduction. The field of nutritional ecology relates wildlife to their environments through such nutritional interactions.

The ability of biologists to predict nutritional value helps define the importance of seasonal habitats. Even small differences in nutrient supplies can have multiplier effects on animal performance, population size, and ecosystem dynamics. Nutrition remains important even in systems with predator influences. Understanding the trade-offs that wildlife species make to minimize deterrents to fitness enables biologists to better gauge the effects of changing environments. Foundational scientists of nutritional ecology, such as C. T. Robbins, R. G. White, and R. J. Hudson in the Northern Hemisphere and I. D. Hume and W. J. Foley in the Southern Hemisphere, have documented an impressive range of adaptations that wild species use to survive and reproduce on different landscapes.

Literature Cited

Allombert, S., A. J. Gaston, and J.-L. Martin. 2005. A natural experiment on the impact of overabundant deer on songbird populations. Biological Conservation 126:1–13.

Angerbjörn, A., P. Hersteinsson, K. Lidén, and E. Nelson. 1994. Dietary variation in Arctic foxes (*Alopex lagopus*)—an analysis of stable carbon isotopes. Oecologia 99:226–232.

Anteau, M. J., and A. D. Afton. 2004. Nutrient reserves of lesser scaup (*Aythya affinis*) during spring migration in the Mississippi Flyway: a test of the spring condition hypothesis. Auk 121:917–929.

Ayotte, J. B., K. L. Parker, J. M. Arocena, and M. P. Gillingham. 2006. Chemical composition of lick soils: functions of soil ingestion by four ungulate species. Journal of Mammalogy 87:878–888.

Bakken, G. S., W. R. Santee, and D. J. Erskine. 1985. Operative and standard operative temperature: tools for thermal energetic studies. American Zoologist 25:933–943.

Bailey, J. A. 1984. Principles of wildlife management. John Wiley and Sons, New York, New York, USA.

Barboza, P. S., and R. T. Bowyer. 2001. Seasonality of sexual segregation in dimorphic deer: extending the gastrocentric model. Alces 37:275–292.

Barboza, P. S., and K. L. Parker. 2008. Allocating protein to reproduction in Arctic reindeer and caribou. Physiological and Biochemical Zoology 79:628–644.

Barboza, P. S., K. L. Parker, and I. D. Hume. 2009. Integrative wildlife nutrition. Springer-Verlag, Heidelberg, Germany.

Beaver, J. M., B. E. Olson, and J. M. Wraith. 1996. A simple index of standard operative temperature for mule deer and cattle in winter. Journal of Thermal Biology 21:345–352.

Belovsky, G. E., and P. A. Jordan. 1981. Sodium dynamics and adaptations of a moose population. Journal of Mammalogy 62:613–621.

Berger, J., P. B. Stacey, L. Bellis, and M. P. Johnson. 2001. A mammalian predator–prey imbalance: grizzly bear and wolf extinction affect avian neotropical migrants. Ecological Applications 11:947–960.

Bidwell, M. T., and R. D. Dawson. 2005. Calcium availability limits reproductive output of tree swallows (Tachycineta bicolor) in a nonacidified landscape. Auk 122:246–254.

Blix, A. S. 2005. Arctic animals and their adaptations to life on the edge. Tapir Academic Press, Trondheim, Norway.

Bowyer, R. T. 1991. Timing of parturition and lactation in southern mule deer. Journal of Mammalogy 72:138–145.

Bryant, J. P., and P. J. Kuropat. 1980. Selection of winter forage by subarctic browsing vertebrates: the role of plant chemistry. Annual Review of Ecology and Systematics 11:261–285.

Bugalho, M. N., J. A. Milne, R. A. Mayes, and F. C. Rego. 2005. Plant-wax alkanes as seasonal markers of red deer dietary components. Canadian Journal of Zoology 83:465–473.

Butler, P. J., J. A. Green, I. L. Boyd, and J. R. Speakman. 2004. Measuring metabolic rate in the field: the pros and cons of the doubly labeled water and heart rate methods. Functional Ecology 18:168–183.

Careau, V., J. Morand-Ferron, and D. Thomas. 2007. Basal metabolic rate of canidae from hot deserts to cold Arctic climates. Journal of Mammalogy 88:394–400.

Christianson, D., and S. Creel. 2009. Fecal chlorophyll describes the link between primary production and consumption in a terrestrial herbivore. Ecological Applications 19:1323–1335.

Christianson, D., and S. Creel. 2010. A nutritionally mediated risk effect of wolves on elk. Ecology 91:1184–1191.

Clutton-Brock, T. H., S. D. Albon, and F. E. Guinness. 1989. Fitness costs of gestation and lactation in wild mammals. Nature 337:260–262.

Collins, W. B., P. J. Urness, and D. D. Austin. 1978. Elk diets and activities on different lodgepole pine habitat segments. Journal of Wildlife Management 42:799–810.

Conomy, J. T., J. A. Collazo, J. A. Dubovsky, and W. J. Fleming. 1998. Dabbling duck behavior and aircraft activity in coastal North Carolina. Journal of Wildlife Management 62:1127–1134.

Cook, J. G. 2002. Nutrition and food. Pages 259–349 in D. E. Toweill and J. W. Thomas, editors. North American elk: ecology and management. Smithsonian Institution Press, Washington, D.C., USA.

Cook, J. G., L. L. Irwin, L. D. Bryant, R. A. Riggs, and J. W. Thomas. 1998. Relations of forest cover and condition of elk: a test of the thermal cover hypothesis in summer and winter. Wildlife Monographs 141:1–61.

Cook, J. G., B. K. Johnson, R. C. Cook, R. A. Riggs, T. Delcurto, L. D. Bryant, and L. L. Irwin. 2004. Effects of summer-autumn nutrition and parturition date on reproduction and survival of elk. Wildlife Monographs 155:1–61.

Cook, R. C., J. G. Cook, T. R. Stephenson, W. L. Myers, S. M. McCorquodale, et al. 2010. Revisions of rump fat and body scoring indices for deer, elk, and moose. Journal of Wildlife Management 74:880–896.

Crête, M., and S. Larivière. 2003. Estimating the costs of locomotion in snow for coyotes. Canadian Journal of Zoology 81:1808–1814.

Darimont, C. T., P. C. Paquet, and T. E. Reimchen. 2008. Spawning salmon disrupt trophic coupling between wolves and ungulate prey in coastal British Columbia. BioMed Central Ecology 8:14.

Dearing, M. D., W. J. Foley, and S. McLean. 2005. The influence of plant secondary metabolites on the nutritional ecology of herbivorous terrestrial vertebrates. Annual Review of Ecology, Evolution, and Systematics 36:169–189.

DelGiudice, G. D. 1995. Assessing winter nutritional restriction of northern deer with urine in snow: considerations, potential and limitations. Wildlife Society Bulletin 23:687–693.

Dove, H., and R. W. Mayes. 2005. Using n-alkanes and other plant wax components to estimate intake, digestibility and diet composition of grazing/browsing sheep and goats. Small Ruminant Research 59:123–139.

Fancy, S. G., and R. G. White. 1985. Incremental cost of activity. Pages 143–159 in R. J. Hudson, and R. G. White, editors. Bioenergetics of wild herbivores. CRC Press, Boca Raton, Florida, USA.

Felicetti, L. A., C. C. Schwartz, R. O. Rye, M. A. Haroldson, K. A. Gunther, D. L. Phillips, and C. T. Robbins. 2003. Use of sulfur and nitrogen stable isotopes to determine the importance of whitebark pine nuts to Yellowstone grizzly. Canadian Journal of Zoology 81:763–770.

Festa-Bianchet, M., and S. D. Côté. 2008. Mountain goats: ecology, behavior, and conservation of an alpine ungulate. Island Press, Washington, D.C., USA.

Geiser, F., B. M. McAllan, and G. J. Kenagy. 1994. The degree of dietary fatty acid unsaturation affects torpor patterns and lipid composition of a hibernator. Journal of Comparative Physiology B 164:299–305.

Gessaman, J. A. 1972. Bioenergetics of the snowy owl (Nyctea scandiaca). Arctic and Alpine Research 4:223–238.

Gillingham, M. P., K. L. Parker, and T. A. Hanley. 1997. Forage intake by large herbivores in a natural environment: bout dynamics. Canadian Journal of Zoology 75:1118–1128.

Goering, H. K., and P. J. Van Soest. 1970. Forage analyses (apparatus, reagents, procedures, and some applications). Agricultural Handbook 379. U.S. Department of Agriculture, Washington, D.C., USA.

Griffith, B., D. C. Douglas, N. E. Walsh, D. D. Young, T. R. McCabe, D. E. Russell, R. G. White, R. D. Cameron, and K. R. Whitten. 2002. The porcupine caribou herd. Pages 8–37 in Arctic refuge coastal plain terrestrial wildlife research summaries. Biological Science Report 2002-0001. Biological Resources Division, U.S. Geological Survey, Anchorage, Alaska, USA.

Guglielmo, C. G., W. H. Karasov, and W. J. Jakubas. 1996. Nutritional costs of a plant secondary metabolite explain selective foraging by ruffed grouse. Ecology 77:1103–1115.

Gurchinoff, S., and W. L. Robinson. 1972. Chemical characteristics of jackpine needles selected by feeding spruce grouse. Journal of Wildlife Management 36:80–87.

Gustine, D. D., P. S. Barboza, L. G. Adams, R. G. Farnell, and K. L. Parker. 2011. An isotopic approach to measuring nitrogen balance in caribou. Journal of Wildlife Management 75:178–188.

Gustine, D. D., K. L. Parker, R. J. Lay, M. P. Gillingham, and D. C. Heard. 2006. Calf survival of woodland caribou in a multi-predator ecosystem. Wildlife Monographs 165:1–32.

Guthery, F. S. 1999. Energy-based carrying capacity for quails. Journal of Wildlife Management 63:664–674.

Hamel, S., M. Garel, M. Festa-Bianchet, J.-M. Gaillard, and S. D. Côté. 2009. Spring normalized difference vegetation index (NDVI) predicts annual variation in timing of peak faecal crude protein in mountain ungulates. Journal of Applied Ecology 46:582–589.

Hanley, T. A., C. T. Robbins, A. E. Hagerman, and C. McArthur. 1992. Predicting digestible protein and digestible dry matter in tannin-containing forages consumed by ruminants. Ecology 73:537–541.

Hanley, T. A., and J. J. Rogers. 1989. Estimating carrying capacity with simultaneous nutritional constraints. Research Note PNW-RN-485. Pacific Northwest Research Station, U.S. Forest Service, Portland, Oregon, USA.

Hanley, T. A., D. E. Spalinger, K. J. Mock, O. L. Weaver, and G. M. Harris. 2012. Forage resource evaluation system for habitat–deer: an interactive deer–habitat model. General Technical Report PNW-GTR-858. Pacific Northwest Research Station, U.S. Forest Service, Portland, Oregon, USA. http://treesearch.fs.fed.us/pubs/40300. Accessed 13 October 2012.

Happe, P. J., K. J. Jenkins, E. E. Starkey, and S. H. Sharrow. 1990. Nutritional quality and tannin astringency of browse in clear-cuts and old-growth forests. Journal of Wildlife Management 54:557–566.

Harder, J. D., and R. L. Kirkpatrick. 1994. Physiological indices in wildlife research. Pages 275–306 in T. A. Bookhout, editor. Research and management techniques for wildlife and habitats. Fifth edition. The Wildlife Society, Bethesda, Maryland, USA.

Harlow, H. 1981. Effect of fasting on rate of food passage and assimilation efficiency in badgers. Journal of Mammalogy 62:173–177.

Hassall, M. A., S. J. Lane, M. Stock, S. M. Percival, and B. Pohl. 2001. Monitoring feeding behaviour of Brent geese Branta bernicla using position-sensitive radio transmitters. Wildlife Biology 7:77–86.

Hildebrand, G. V., S. D. Farley, and C. T. Robbins. 1998. Predicting body condition of bears via two field methods. Journal of Wildlife Management 62:406–409.

Hobbs, N. T., and D. M. Swift. 1985. Estimates of habitat carrying capacity incorporating explicit nutritional constraints. Journal of Wildlife Management 49:814–822.

Hobson, K. A. 1995. Reconstructing avian diets using stable-carbon and nitrogen isotope analysis of egg components: patterns of isotopic fractionation and turnover. The Condor 97:752–762.

Holechek, J. L., M. Vavra, and R. D. Pieper. 1982. Botanical composition determination of range herbivore diets: a review. Journal of Range Management 35:309–315.

Hupp, J. W., R. G. White, J. S. Sedinger, and D. G. Robertson. 1996. Forage digestibility and intake by lesser snow geese: effects of dominance and resource heterogeneity. Oecologia 108:232–240.

Iverson, S. J., C. Field, W. D. Bowen, and W. Blanchard. 2004. Quantitative fatty acid signature analysis: a new method of estimating predator diets. Ecological Monographs 74:211–235.

Iverson, S. J., D. E. McDonald, and L. H. Smith. 2001. Changes in the diet of free-ranging black bears in years of contrasting food availability revealed through milk fatty acids. Canadian Journal of Zoology 79:2268–2279.

Johnson, C. J., K. L. Parker, and D. C. Heard. 2000. Feeding site selection by woodland caribou in north-central British Columbia. Rangifer Special Issue 12:159–172.

Johnson, G. D., and M. S. Boyce. 1990. Feeding trials with insects in the diet of sage grouse chicks. Journal of Wildlife Management 54:89–91.

Kamler, J., M. Homolka, and D. Čižmár. 2003. Suitability of NIRS analysis for estimating diet quality of free-living red deer Cervus elaphus and roe deer Capreolus capreolus. Wildlife Biology 10:235–240.

Karasov, W. H., and C. Martinez del Rio. 2007. Physiological ecology: how animal process energy, nutrients, and toxins. Princeton University Press, Princeton, New Jersey, USA.

Kielland, K. 2001. Stable isotope signatures of moose in relation to seasonal forage composition: a hypothesis. Alces 37:329–337.

Kielland, K., and G. Finstad. 2000. Differences in ^{15}N natural abundance reveal seasonal shifts in diet choice of reindeer and caribou. Rangifer 12:145.

Kleiber, M. 1947. Body size and metabolic rate. Physiological Reviews 27:511–541.

Koch, P. L., J. Heisinger, C. Moss, R. W. Carlson, M. L. Fogel, and A. K. Behrensmeyer. 1995. Isotope tracking of change in diet and habitat use in African elephants. Science 267:1340–1343.

Krementz, D. G., and G. W. Pendelton. 1990. Fat scoring: sources of variability. Condor 92:500–507.

Lawler, J. P., and R. G. White. 2003. Temporal responses in energy expenditure and respiratory quotient following feeding in muskox: influence of season on energy costs of eating and standing and an endogenous heat increment. Canadian Journal of Zoology 81:1524–1538.

Leslie, D. M., R. T. Bowyer, and J. A. Jenks. 2008. Facts from feces: nitrogen still measures up as a nutritional index for mammalian herbivores. Journal of Wildlife Management 72:1420–1433.

Mack, R. N., and J. N. Thompson. 1982. Evolution in steppe with few large, hooved mammals. The American Naturalist 119:757–773.

Marsh, K. J., I. R. Wallis, R. L. Andrew, and W. J. Foley. 2006. The detoxification limitation hypothesis: where did it come from and where is it going? Journal of Chemical Ecology 32:1247–1266.

Mayes, R. W., and H. Dove. 2000. Measurement of dietary nutrient intake in free-ranging mammalian herbivores. Nutrition Research Reviews 13:107–138.

McArt, S. H., D. E. Spalinger, W. B. Collins, E. R. Schoen, T. Stevenson, and M. Bucho. 2009. Summer dietary nitrogen availability as a potential bottom-up constraint on moose in south-central Alaska. Ecology 90:1400–1411

McArthur, C., C. T. Robbins, A. E. Hagerman, and T. A. Hanley. 1993. Diet selection by a ruminant generalist browser in relation to plant chemistry. Canadian Journal of Zoology 71:2236–2243.

McNaughton, S. J. 1990. Mineral nutrition and seasonal movements of African migratory ungulates. Nature 345:613–615.

Milakovic, B. 2008. Defining the predator landscape of northeastern British Columbia. Dissertation, University of Northern British Columbia, Prince George, Canada.

Monteith, K. L., V. C. Bleich, T. R. Stephenson, B. M. Pierce, M. M. Conner, R. W. Klaver, and R. T. Bowyer. 2011. Timing of seasonal migration in mule deer: effects of climate, plant phenology, and life-history characteristics. Ecosphere 2:1–34.

Moore, B. D., and W. J. Foley. 2005. Tree use by koalas in a chemically complex landscape. Nature 435:488–490.

Mould, E. D., and C. T. Robbins. 1981. Evaluation of detergent analysis in estimating nutritional value of browse. Journal of Wildlife Management 45:323–334.

Müller, M. S., S. R. McWilliams, D. W. Podlesak, J. R. Donaldson, H. M. Bothwell, and R. L. Lindroth. 2006. Tri-trophic effects of plant defenses: chickadees consume caterpillars based on host leaf chemistry. Oikos 114:507–517.

Nagy, K. A., I. A. Girard, and T. K. Brown. 1999. Energetics of free-ranging mammals, reptiles, and birds. Annual Review of Nutrition 19:247–277.

Nagy, T. R. 2001. The use of dual-energy X-ray absorptiometry for the measurement of body composition. Pages 211–229 in J. R. Speakman, editor. Body composition analysis of animals: a handbook of non-destructive methods. Cambridge University Press, Cambridge, United Kingdom.

Newsome, S. D., M. T. Clementz, and P. L. Koch. 2010. Using stable isotope biogeochemistry to study marine mammal ecology. Marine Mammal Science 26:509–572.

Osborn, R. G., and T. F. Ginnett. 2001. Fecal nitrogen and 2,6-diaminopimelic acid as indices to dietary nitrogen in white-tailed deer. Wildlife Society Bulletin 29:1131–1139.

Parker, K. L. 1988. Effects of heat, cold, and rain on black-tailed deer. Canadian Journal of Zoology 66:2475–2483.

Parker, K. L. 2003. Advances in the nutritional ecology of cervids at different scales. Ecoscience 10:395–411.

Parker, K. L., P. S. Barboza, and M. P. Gillingham. 2009. Nutrition integrates environmental responses of ungulates. Functional Ecology 23:57–69.

Parker, K. L., and M. P. Gillingham. 1990. Estimates of critical thermal environments for mule deer. Journal of Range Management 43:73–81.

Parker, K. L., M. P. Gillingham, and T. A. Hanley. 1993. An accurate technique for estimating forage intake of tractable animals. Canadian Journal of Zoology 71:1462–1465.

Parker, K. L., M. P. Gillingham, T. A. Hanley, and C. T. Robbins. 1999. Energy and protein balance of free-ranging black-tailed deer in a natural forest environment. Wildlife Monographs 143:1–48.

Parker, K. L., and C. T. Robbins. 1984. Thermoregulation in mule deer and elk. Canadian Journal of Zoology 62:1409–1422.

Parker, K. L., and C. T. Robbins. 1985. Thermoregulation in ungulates. Pages 161–182 in R. J. Hudson and R. G. White, editors. Bioenergetics of wild herbivores. CRC Press, Boca Raton, Florida, USA.

Pettorelli, N., J. O. Vik, A. Mysterud, J.-M. Gaillard, C. J. Tucker, and N. C. Stenseth. 2005. Using the satellite-derived NDVI to assess ecological responses to environmental change. Trends in Ecology and Evolution 20:503–510.

Powers, J. G., W. W. Mautz, and P. J. Pekins. 1989. Nutrient and energy assimilation of prey by bobcats. Journal of Wildlife Management 53:1004–1008.

Pritchard, G. T., and C. T. Robbins. 1990. Digestive and metabolic efficiencies of grizzly and black bears. Canadian Journal of Zoology 68:1645–1651.

Prosky, L., N. Asp, I. Furda, J. W. Devries, T. F. Schweizer, and B. F. Harland. 1984. Determination of total dietary fiber in foods, food products, and total diets: interlaboratory study. Journal of the Association of Official Agricultural Chemists 67:1044–1052.

Ramsay, M. A., and K. A. Hobson. 1991. Polar bears make little use of terrestrial food webs: evidence from stable-carbon isotope analysis. Oecologia 86:598–600.

Reimers, E., S. Eftestøl, and J. E. Colman. 2003. Behavior responses of wild reindeer to direct provocation by a snowmobile or skier. Journal of Wildlife Management 67:747–754.

Renecker, L. A., and R. J. Hudson. 1986. Seasonal energy expenditures and thermoregulatory responses of moose. Canadian Journal of Zoology 64:322–327.

Richman, S. E., and J. R. Lovvorn. 2003. Effects of clam species dominance on nutrient and energy acquisition by spectacled eiders in the Bering Sea. Marine Ecology Progress Series 261:283–297.

Robbins, C. T. 1993. Wildlife feeding and nutrition. Second edition. Academic Press, San Diego, California, USA.

Robbins, C. T., T. A. Hanley, A. E. Hagerman, O. Hjeljord, D. L. Baker, C. C. Schwartz, and W. W. Mautz. 1987. Role of tannins in defending plants against ruminants: reduction in protein availability. Ecology 68:98–107.

Saltz, D., and G. C. White. 1991. Urinary cortisol and urea nitrogen responses in irreversibly undernourished mule deer fawns. Journal of Wildlife Diseases 27:41–46.

Scott, I., C. Selman, P. I. Mitchell, and P. R. Evans. 2001. The use of total body electrical conductivity (TOBEC) to determine body condition in vertebrates. Pages 127–160 in J. R. Speakman, editor. Body composition analysis of animals: a handbook of non-destructive methods. Cambridge University Press, Cambridge, United Kingdom.

Scribner, K. T., M. H. Smith, and P. E. Jones. 1989. Environmental and genetic components of antler growth in white-tailed deer. Journal of Mammalogy 70:284–291.

Servello, F. A., E. C. Hellgren, and S. R. McWilliams. 2005. Techniques for wildlife nutritional ecology. Pages 554–590 in C. E. Braun, editor. Techniques for wildlife investigations and management. Sixth edition. The Wildlife Society, Bethesda, Maryland, USA.

Servello, F. A., and J. W. Schneider. 2000. Evaluation of urinary indices of nutritional status for white-tailed deer: test with winter browse diets. Journal of Wildlife Management 64:137–145.

Shipley, L. A. 2007. The influence of bite size on foraging at larger spatial and temporal scales by mammalian herbivores. Oikos 116:1964–1974.

Shipley, L. A., and D. E. Spalinger 1992. Mechanics of browsing in dense food patches: effects of plant and animal morphology on intake rate. Canadian Journal of Zoology 70:1743–1752.

Sinclair, A. R. E., S. A. R. Mduma, and P. Arcese. 2000. What determines phenology and synchrony of ungulate breeding in Serengeti? Ecology 81:2100–2111.

Spalinger, D. E., W. B. Collins, T. A. Hanley, N. E. Cassara, and A. M. Carnahan. 2010. The impact of tannins on protein, dry matter, and energy digestion in moose (Alces alces). Canadian Journal of Zoology 88:977–987.

Speakman, J. R., G. H. Visser, S. Ward, and E. Krol. 2001. The isotope dilution method for the evaluation of body composition. Pages 56–98 in J. R. Speakman, editor. Body composition analysis of animals: a handbook of non-destructive methods. Cambridge University Press, Cambridge, United Kingdom.

Stenseth, C. C., W. Falck, O. N. Bjørnstad, and C. J. Krebs. 1997. Population regulation in snowshoe hare and Canadian lynx: asymmetric food web configurations between hare and lynx. Proceedings of the National Academy of Sciences 94:5147–5152.

Svoboda, F. J., and G. W. Gullion. 1972. Preferential use of aspen by ruffed grouse in northern Minnesota. Journal of Wildlife Management 36:1166–1180.

Teer, J. G., J. W. Thomas, and E. A. Walker. 1965. Ecology and management of white-tailed deer in the Llano Basin of Texas. Wildlife Monographs 15:1–62.

Tilley, J. M. A., and R. A. Terry. 1963. A two-stage technique for in vitro digestion of forage crops. Journal of the British Grassland Society 18:104–111.

Tinbergen, J. M., and J. B. Williams. 2002. Energetics of incubation. Pages 299–313 in D. C. Deeming, editor. Avian incubation: behaviour, environment and evolution. Oxford University Press, Oxford, United Kingdom.

Torbit, S. C., L. H. Carpenter, A. W. Alldredge, and D. M. Swift. 1985. Mule deer body composition—a comparison of methods. Journal of Wildlife Management 49:86–91.

Tulp, I., H. Schekkerman, L. W. Bruinzeel, J. Jukema, G. H. Visser, and T. Piersma. 2009. Energetic demands during incubation and chick rearing in a uniparental and a biparental shorebird breeding in the high Arctic. Auk 126:155–164.

Vagnoni, D. B., R. A. Garrott, J. G. Cook, and P. J. White, 1996. Urinary allantoin:creatinine ratios as a dietary index for elk. Journal of Wildlife Management 60:728–734.

Van Vuren, D., and B. E. Coblentz. 1985. Kidney weight variation and the kidney fat index: an evaluation. Journal of Wildlife Management 49:177–179.

Vleck, C. M., and T. L. Bucher. 1998. Energy metabolism, gas exchange, and ventilation. Pages 89–116 in J. M. Starck and R. E. Ricklefs, editors. Avian growth and development—evolution within the altricial–precocial spectrum. Oxford University Press, New York, New York, USA.

Walker, A. B. D., K. L. Parker, and M. P. Gillingham. 2006. Behaviour, habitat associations and intrasexual differences of female Stone's sheep. Canadian Journal of Zoology 84:1187–1201.

White, R. G. 1983. Foraging patterns and their multiplier effects on productivity of northern ungulates. Oikos 40:377–384.

12 WATER AND OTHER WELFARE FACTORS

JAMES W. CAIN III, PAUL R. KRAUSMAN, AND
STEVEN S. ROSENSTOCK

INTRODUCTION

Wildlife agencies indirectly manage populations by manipulating resources upon which animals depend for survival. These resources are called welfare factors and include food, water, cover, space, and special factors (e.g., essential minerals, dust baths). Because food is discussed more thoroughly in Chapter 11, this chapter will focus on water, cover, and special factors. We use water as a model to show how managers utilize required resources to manipulate populations. We restrict the discussion of water to its role as a resource consumed by animals to meet physiological requirements rather than as a component of habitat (e.g., for waterfowl).

WATER

Life processes evolved around water (Louw 1993). The thermal capacity, high heat of vaporization, and action as a solvent make water crucial for all organisms (Edney 1977). Water constitutes the majority of molecules within an animal's body and plays a critical role in physiological processes (e.g., hydrolytic reactions, joint lubrication, digestion, thermoregulation, transport of nutrients, excretion of waste products; Robinson 1957).

Animals obtain water from three sources: free water in the environment (e.g., lakes, streams, rain, snow, and dew), preformed water in food, and metabolic water produced during oxidation of organic compounds containing hydrogen. Preformed water content of plants varies by species, season, and plant parts, ranging from under 10% in dried seeds and senescent grasses (Jarman 1973), to 30–60% in browse, and over 70% in succulent plants (cacti; Cain et al. 2008a); water content of animal tissues consumed by predators commonly exceeds 70% (Robbins 1983). Metabolic water is derived from the oxidation of fats, proteins, and carbohydrates (Robbins 1983). Oxidation of 1 g of fat, protein, or carbohydrate produces 1.07, 0.4, and 0.56 g of water, respectively (Gill 1994).

WATER REQUIREMENTS OF WILDLIFE

Water requirements vary among species and depend on thermal load (i.e., ambient temperature, solar radiation, thermal radiation, vapor pressure deficits), activity patterns, morphology, diet, metabolic processes, reproductive state, and adaptations for water conservation. Determining water requirements requires measurement of intake from free, preformed, and metabolic water (Robbins 1983), and when possible should be conducted on free-ranging animals. Birds and mammals with diets composed mainly of fruit, nectar, insects, or animal tissue often meet their water requirements from preformed and metabolic water (Bartholomew and Cade 1963, Golightly and Ohmart 1984, Nagy and Gruchacz 1994). Similarly, large herbivores may meet their water requirements during all or part of the year from preformed and metabolic water (Zervanos and Day 1977, Jhala et al. 1992, Fox et al. 2000). When unable to meet their water requirements from these sources, animals must have access to free water.

OVERVIEW OF WATER BALANCE AND THERMOREGULATION

Wildlife inhabiting arid and semiarid environments must contend with factors affecting thermoregulation and water balance, including high solar radiation, high ambient temperatures, and limited water and food resources. These species cope by either avoiding or tolerating environmental conditions causing heat stress and dehydration (Schmidt-Nielsen 1979, Louw and Seely 1982). Animals in these environments have evolved a variety of behavioral, morphological, and physiological adaptations. Behavioral adaptations typically facilitate avoidance of stressful environmental conditions, whereas tolerance involves morphological and physiological adaptations (Louw and Seely 1982, Louw 1984). Evaporative cooling is one mechanism by which animals maintain homeothermy. Because high temperatures and scarce free water often characterize arid areas, animals face conflicting needs to maintain body temperature and to minimize water loss. Animals commonly use multiple adaptations simultaneously to avoid heat stress and dehydration. We briefly review some of the physiological, morphological, and behavioral mechanisms involved in water balance and thermoregulation.

Physiological Mechanisms and Water Balance

The total body water (i.e., the total amount of water in the body) of animals has four compartments: intracellular fluid (i.e., fluid within cells), extracellular fluid within the digestive tract (e.g., rumen), blood plasma, and interstitial fluid. The distribution of water among these compartments varies by species and state of hydration (Louw 1993). Species able to withstand severe dehydration typically maintain blood plasma volume and prevent circulatory failure by moving fluid from other compartments (Carmi et al. 1993, Silanikove 1994).

Water turnover rate (i.e., the period of time required for all water molecules within an animal's body to be replaced) varies among species and is lower in animals adapted to arid environments (Nagy and Peterson 1988, Tieleman et al. 2002). In normally hydrated animals, water turnover rate increases with increasing temperature (Longhurst et al. 1970, Degen et al. 1983). When dehydrated, water turnover rates decrease regardless of ambient temperature (McNabb 1969, Maloiy 1973). Water turnover rates are affected by reproductive status and are 40–50% higher in pregnant and lactating mammals (Maloiy et al. 1979) and increase during egg production in birds (Bartholomew and Cade 1963).

There are four primary routes of water loss in animals: feces, urine, cutaneous evaporation, and pulmonary evaporation. Mammalian females experience additional water loss during lactation. Desert-adapted species evolved physiological adaptations that reduce the amount of water lost through each of these routes or are able to tolerate significant amounts of water loss (Dawson 1984, McNab 2002).

Water Loss in Feces and Urine

Animals can minimize water loss by reducing water content of feces and urine volume, and by increasing urine concentration (osmolality; Maloiy et al. 1979, Maclean 1996). A major difference between mammals and birds is the excretion of nitrogenous waste as urea and uric acid, respectively. Mammals are less efficient than birds, using 20–40 times more water to excrete an equivalent amount of urea (Bartholomew and Cade 1963, Robbins 1983).

Fecal moisture content in normally hydrated animals varies widely among species. Arid-adapted ungulates such as gemsbok (*Oryx gazella*) typically range from 40% to 50%, compared to 70% to 80% for ungulates adapted to mesic areas, such as waterbuck (*Kobus ellipsiprymnus*; Maloiy et al. 1979, Woodall and Skinner 1993). When dehydrated, reductions in fecal moisture content of 17–50% have been reported in some ungulates (Maloiy and Hopcraft 1971, Turner 1973). Similarly, arid-adapted birds typically have lower fecal moisture content (Gill 1994), and dehydrated birds can reduce fecal moisture content (Ohmart and Smith 1971). Arid-adapted mammals and birds typically have reduced urine output and greater urine concentration than do temperate species (Louw and Seely 1982, Williams et al. 1991).

Thermoregulation

Thermoregulation is the maintenance of body temperature within specific boundaries under varying environmental temperatures. The environmental heat load plus the metabolic heat produced by the animal compose the overall heat load (Porter and Gates 1969). Environmental heat load is a function of thermal radiation absorbed by the animal, air temperature, wind speed, and vapor pressure deficit.

When body temperature exceeds ambient temperature, animals passively lose heat to the environment by radiation, convection, and conduction (Porter and Gates 1969, Mitchell 1977), commonly accomplished by behavioral or morphological mechanisms (see below). However, once the total heat load exceeds levels that can be dissipated by nonevaporative means, animals must either increase body temperature or use other mechanisms to remove heat from the body.

Evaporative Heat Loss

Evaporative heat loss (i.e., evaporative cooling) occurs when heat is transferred from the animal to the environment via evaporation of water from body surfaces. Evaporative heat loss is accomplished passively via diffusion of water through the skin and respiratory tract, or actively by sweating, panting, and gular fluttering (in birds). The magnitude of heat loss by evaporation is related to rates of sweating or panting, wind speed, and vapor pressure gradient.

EVAPORATIVE HEAT LOSS IN MAMMALS

The number of sweat glands and rate of sweating vary among species (Robertshaw and Taylor 1969, Sokolov 1982). Rates of cutaneous water loss are affected by season, age, nutritional status, and state of hydration (Taylor 1970a, Maloiy and Hopcraft 1971, Parker and Robbins 1984). To minimize water loss, some desert-adapted species use facultative cutaneous evaporation; sweating rates increase with increasing body temperature when hydrated, but they decrease when dehydrated (Maloiy 1970, Taylor 1970b). The body temperature at which dehydrated animals begin to sweat is often higher than when hydrated normally (Schmidt-Nielsen et al. 1957, Taylor 1970a).

The respiratory tract also performs a thermoregulatory function. As body temperature increases, mammals increase respiratory frequency and volume to maximize the movement of air across the evaporative surfaces of the upper respiratory tract, increasing evaporative heat loss (Finch 1972). In normally hydrated mammals, respiratory rates increase with increasing ambient temperature. Dehydrated mammals minimize respiratory water loss by maintaining lower respiratory rates and by panting at higher temperatures than normally hydrated animals (Taylor 1969b, Maloiy and Hopcraft 1971, Finch 1972).

The use of cutaneous or respiratory evaporation is related to body size. Small mammals (≤40 kg) typically pant. Larger mammals rely more on sweating because they have low rates of nonevaporative heat loss (because of low surface area:volume ratio; see below) and respiratory evaporation is less efficient than cutaneous evaporation, which is often insufficient to dissipate heat generated during activity in large bodied animals (Robertshaw and Taylor 1969, Maloiy et al. 1979).

KNUT SCHMIDT-NIELSEN (1915–2007)

Knut Schmidt-Nielsen was born in Trondheim, Norway, in 1915. He was educated in Oslo and Copenhagen before coming to the United States and studying at Swarthmore College, Stanford University, and the University of Cincinnati College of Medicine.

He became a U.S. citizen and a professor at Duke University in 1952. He conducted pioneering work in comparative physiology and ecophysiology, and his research interests included, among other topics, thermoregulation and water balance of camels, kangaroo rats, and birds; avian respiration; energetics of locomotion; and allometric relationships. He published over 270 scientific publications and five books: *How Animals Work* (1972), *Desert Animals: Physiological Problems of Heat and Water* (1979), *Scaling: Why Is Animal Size So Important?* (1984), *Animal Physiology: Adaptation and Environment* (1997), and his autobiography, *The Camel's Nose: Memoirs of a Curious Scientist* (1998). He was a member of the U.S. National Academy of Sciences, the Royal Society of London, and the French Academy of Sciences, as well as other organizations. Schmidt-Nielsen received numerous awards, including the International Prize for Biology.

Photo reproduced with permission from Hoppeler and Weibel (2005)

EVAPORATIVE HEAT LOSS IN BIRDS

Birds lose excess body heat by evaporation of water from the respiratory tract and skin (Dawson 1982, 1984; Williams and Tieleman 2002). Panting and gular fluttering are commonly used to cope with excessive environmental heat loads by evaporative water loss from the nasal, buccal, tracheal, and upper pharyngeal regions (Dawson 1984). Rates of panting and gular fluttering in birds tend to increase with temperature rises (Weathers 1972). Similar to mammals, dehydrated birds begin panting at a higher body temperature compared with normally hydrated birds (Kleinhaus et al. 1985).

Despite lacking sweat glands, cutaneous water loss represents a large proportion (e.g., 40–70%) of total evaporative water loss in birds (Dawson 1982, Williams and Tieleman 2005). The relative contribution of respiratory and cutaneous water loss in birds varies with species and environmental heat load (Dawson 1982, Wolf and Walsberg 1996). For example, in passerines, cutaneous water loss, as a proportion of total evaporative water loss, decreases with increasing temperature, whereas in pigeons and doves, cutaneous water loss increases

(Wolf and Walsberg 1996, Tieleman et al. 2003). State of hydration can also influence the rate of cutaneous water loss in birds. Cutaneous evaporation increased with increasing temperature in hydrated pigeons (*Columba livia*), and decreased in dehydrated birds that relied on panting as their main evaporative cooling mechanism (Arad et al. 1987).

Physiological Adaptations, Thermoregulation, and Water Balance

Metabolic Rate

Reduction in metabolic rate can minimize metabolic heat gain, reducing the need for evaporative cooling. After accounting for body size, species in arid climates typically have lower metabolic rates than those in mesic environments (Macfarlane et al. 1971; Nagy 1987; Tieleman et al. 2002, 2003; White et al. 2007). The lower metabolic rates of desert-adapted species are associated with lower water turnover rates in ungulates (Maloiy et al. 1979) and lower total evaporative water loss in birds (Tieleman et al. 2002, 2003). Furthermore, dehydrated ungulates have lower metabolic rates than when hydrated normally (Schmidt-Nielsen et al. 1967).

Adaptive Heterothermy

Adaptive heterothermy can minimize water loss from evaporative cooling. Body temperature rises during the day, reducing the thermal gradient and heat gain from the environment; the body then releases excess heat at night, when ambient temperature falls below the body temperature (Taylor and Lyman 1972, Dawson 1984). Similarly, antelope ground squirrels (*Ammospermophilus leucurus*) allow their body temperature to rise when foraging during the day in summer, and then periodically return to their cooler burrows to dissipate excess body heat (Chappell and Bartholomew 1981).

Selective brain cooling is a commonly cited mechanism to maintain brain temperature below critical values when body temperature increases (Brinnel et al. 1987). Recent research suggests that the rise in core body temperature observed in thermal studies may have resulted from studying captive animals that were prevented from using behavioral or other thermoregulatory mechanisms (Mitchell et al. 2002, Fuller et al. 2005). Some studies of free-ranging ungulates have not shown evidence of adaptive heterothermy, suggesting that increased daily fluctuations in body temperature in dehydrated animals were actually dehydration-induced hyperthermia (Mitchell et al. 2002). Dehydration-induced hyperthermia increases the temperature at which animals begin to thermoregulate via evaporative cooling, thus conserving water (Mitchell et al. 2002, Fuller et al. 2005). Some investigators have also questioned the use of selective brain cooling to protect the brain when body temperature rises (Fuller et al. 2005).

Behavioral Regulation of Body Temperature and Water Balance

Behavioral adaptations function in combination with physiological and morphological mechanisms and are important for the maintenance of body temperature and water balance

(Bartholomew 1964). Diet selection, timing of activity, use of microhabitats, and body orientation are common behaviors used by animals to aid in thermoregulation (Berry et al. 1984, Gill 1994, Sargeant et al. 1994).

Diet and Water Balance

Preformed water provides a significant portion of total water intake and allows some species (e.g., heteromyid rodents) to survive long periods without access to free water (Taylor 1968, 1969a; Schemnitz 1994). Lower moisture content of forage used by grazing versus browsing animals makes the former more dependent on free water (Maloiy 1973, Kay 1997). Furthermore, foraging on succulent plants (e.g., cacti), underground storage organs (e.g., tubers), and browse with higher moisture content reduces the amount of free water needed to maintain water balance (Taylor 1969a, Williamson 1987). Air-dried seeds typically have very low moisture content (e.g., <10%); many granivorous birds must supplement their diet with insects and succulent vegetation to meet their water requirements by preformed and metabolic water alone (Goldstein and Nagy 1985, Guthery and Koerth 1992).

Timing of Activity

Animals can reduce heat loads by adjusting the duration or timing of activities. During thermally stressful periods, desert-adapted animals spend less time being active and display crepuscular or nocturnal activity patterns, foraging and moving during the cooler periods of the day (Carmi-Winkler et al. 1987, Alderman et al. 1989). Nocturnal feeding has been documented during summer for largely diurnal species such as desert bighorn sheep (*Ovis canadensis*; Miller et al. 1984). Nocturnal feeding may also increase the intake of preformed water. In areas where relative humidity increases at night, water content of dry vegetation may increase because of the hygroscopic uptake of moisture from the air (Taylor 1968).

Use of Microclimates

Animals commonly spend inactive periods using cooler microclimates (Wolf et al. 1996, Tull et al. 2001) or forage in shaded areas (Owen-Smith 1998). Use of cooler microclimates reduces environmental heat loads and helps maintain a temperature gradient that facilitates nonevaporative heat loss. Bedding down on cool substrates can increase conductive heat loss. Small mammals like kangaroo rats (*Dipodomys* spp.) and midsized carnivores living in arid environments, including badger (*Taxidea taxus*) and kit fox (*Vulpes macrotis*), create their own microclimate by burrowing into the soil. However, species largely incapable of creating their own microclimate (e.g., large mammals and birds) seek shade provided by vegetation or other environmental features during the heat of midday (Goldstein 1984, Wolf et al. 1996, Tull et al. 2001). Bats commonly use caves, but so do some birds, small mammals, and large mammals, including desert bighorn sheep, mountain lions (*Puma concolor*), and mule deer (*Odocoileus hemionus*; Krausman 1979, Abeloe and Hardy 1997, Cain et al. 2008b).

Body Orientation

Animals active during hotter periods of the day or in areas lacking shade reduce absorption of solar radiation and increase convective heat loss by adjusting body position relative to the sun and wind. Ungulates stand with the long axis of the body parallel to the sun or wind (Jarman 1977, Berry et al. 1984). Birds likewise minimize exposure to solar radiation by similarly positioning their bodies and increase convective heat loss by lowering their legs during flight (Luskick et al. 1978, Martineau and Larochelle 1988). Small mammals like the antelope ground squirrel actually use their tail like an umbrella, shading their body and reducing thermal load while foraging during the day (Chappell and Bartholomew 1981).

Morphological Adaptations for Thermoregulation and Water Balance

Desert-adapted species possess a variety of morphological adaptations that aid reduction of heat load and minimize water loss. These include body size and shape as well as pelage, feather, and horn characteristics.

Body Size and Shape

Large-bodied animals have a lower surface area:volume ratio, which reduces the proportion of the body exposed to solar radiation, resulting in lower rates of heat gain from the environment (Phillips and Heath 1995). This characteristic can be disadvantageous because it also reduces the rate of heat loss. Shaded microclimates of sufficient size to benefit large animals are more limited where vegetation is sparse and other types of cover are unavailable. Conversely, smaller species, with their larger surface area:volume ratio, gain heat from the environment at a faster rate, but lose excess body heat at a faster rate and can have greater access to shaded microclimates. The shape of the body and appendages influences the rates of heat transfer between the animal and the environment; long and narrow appendages minimize radiant heat gain and maximize convective heat loss. Desert-adapted species characteristically have longer, thinner appendages with a higher surface area:volume ratio, facilitating heat loss compared to similar species inhabiting cooler environments (Phillips and Heath 1995).

Pelage, Plumage, and Horn Characteristics

The thickness and color of pelage and plumage affect heat transfer between animals and their environment. Thick pelage provides increased insulation but limits effectiveness of evaporative cooling. Conversely, thin pelage maximizes heat loss but provides little insulation. Pelage thickness tends to decrease as body size increases in desert-adapted ungulates, facilitating heat loss (Hofmeyr 1985). Lack of pelage or plumage on ventral surfaces may facilitate nonevaporative and evaporative heat loss. Erection of pelage and feathers can also facilitate heat transfer to the environment.

Desert-adapted ungulates typically have glossy, light-colored pelage, which reflects more radiation than dark colored pelage. However, dark-colored pelage or plumage may absorb more solar radiation than light-colored pelage or plumage, reducing the amount of radiation that reaches the skin (Walsberg et al. 1978, Walsberg 1983). Increased absorption of solar radiation by dark-colored pelage lowers the thermal gradient between the surface of the pelage and the environment, reducing heat gain (Louw 1993). Highly vascularized horns of bovids may contribute to thermoregulation, and species from arid areas have relatively larger horn cores and thinner keratin sheaths than temperate species, which facilitates heat loss from the horns (Picard et al. 1999).

TYPES OF WILDLIFE WATER DEVELOPMENTS

Precipitation Catchments

Precipitation catchments harvest rain or snow, storing water for future use by wildlife. These systems can be particularly effective in desert environments, where net annual precipitation is low, but often occurs as brief, intense thunderstorms that generate substantial runoff from impervious surfaces. A brief overview of major designs is presented here; additional information can be found in Halloran and Deming (1958), Lesicka and Hervert (1995), Bleich et al. (2005), and references therein.

Guzzlers

Guzzlers are the most common precipitation catchment used to supply water for wildlife. Numerous designs have been developed over more than 50 years of use in western North America, with management agencies adapting them to accommodate varying project objectives and site characteristics (e.g., accessibility for construction equipment, annual rainfall, topography). Guzzlers capture rainwater on a natural surface or constructed apron (Fig. 12.1). Aboveground storage tanks require a float valve to control flow of water to the drinking

trough (Bleich et al. 2005). When placed below ground, water can be piped into the trough without a float valve, reducing maintenance needs and mechanical failures (Lesicka and Hervert 1995, Bleich et al. 2005). Small guzzlers for game birds may hold only a few hundred liters; units for large ungulates can have capacities of more than 60,000 L. To reduce the potential for animals becoming trapped and drowning, drinking troughs include a ramp that allows animals to escape. Concerns over visual impacts, susceptibility to vandalism, and high costs associated with maintenance and water hauling have fostered new guzzler designs that have few aboveground components and a minimal disturbance footprint (Lesicka and Hervert 1995; Fig. 12.2.).

Figure 12.1. A 1960s-era guzzler near Yuma, Arizona. The system has a metal collection apron and buried concrete storage tank/trough. Photo courtesy of S. S. Rosenstock

Figure 12.2. (A) Modern, low-maintenance/low-visual impact guzzler in Kofa National Wildlife Refuge, Arizona. All components except collection points (see Fig. 12.2B) are below ground; the trough is in the foreground at lower right. Disturbed area has been recontoured and revegetated with native plants. This water development was the subject of litigation asserting adverse impacts on wilderness values. (B) Collection point for buried guzzler system. The concrete and rock diversion dam diverted water flowing in the shallow rill (*right foreground*), which was then piped into a buried storage tank. Photos courtesy of S. S. Rosenstock

Tinajas

Tinajas occur in many desert mountain ranges and can represent the primary source of free water for wildlife. These natural rock tanks collect and hold water for varying lengths of time after rainstorms; capacity ranges from small amounts to over 100,000 L (Bleich et al. 2005). Because many tinajas have insufficient capacity to provide perennial water, many have been modified to increase storage capacity and reduce evaporation. An impermeable dam is commonly built on the downstream side to increase the capacity (Fig. 12.3) and sunshades added to minimize evaporation (Halloran and Deming 1958). Rock gabions can be constructed upstream, trapping sediment, reducing water velocity, and increasing flow into the tinaja. Inclusion of an escape ramp can minimize risk of mortality in steep-sided tinajas (Mensch 1969, Bleich et al. 2005).

Adits

Adits are short (e.g., 4–5 m) tunnels 2–3 m in diameter with a downward sloping floor, constructed into solid rock, typically located immediately adjacent to a wash (Fig. 12.4; Bleich et al. 2005). A diversion dam, perforated by pipes, is often constructed near the entrance to prevent the adit from filling with sand and other debris while allowing water to enter (Bleich et al. 2005). Gabions can be used, as well.

Retention Dams and Sand Tanks

Retention dams with sand tanks or sand dams (Fig. 12.5) were a common method to harvest and store runoff in desert washes (Bleich and Weaver 1983); however, their construction is now less common and has largely been replaced by recent catchment designs. Concrete retention dams are usually built across narrow washes, and sand and gravel then fill the upstream side of the dam during flooding. During subsequent flood events, water seeps into the accumulated material stored behind the dam. The addition of rock gabions upstream of the dam slows the flow of floodwater, increasing infiltration into the sand tank. Pipes through the bottom of the dam transport water to a drinking trough or tinaja.

Figure 12.4. Buck Peak Tank, an adit located in the Cabeza Prieta Mountains, Cabeza Prieta National Wildlife Refuge, Arizona. Photo courtesy of J. W. Cain III

Figure 12.3. Heart Tank, a modified tinaja located in the Sierra Pinta, Cabeza Prieta National Wildlife Refuge, Arizona. Photo courtesy of J. W. Cain III

Figure 12.5. Drainage dam with sand tank in the Kofa National Wildlife Refuge, Arizona. Subsurface water stored behind the dam is flowing into the tinaja immediately below. Photo courtesy of S. S. Rosenstock

Figure 12.6. Water formerly used for livestock converted for use by wildlife in the Kofa National Wildlife Refuge, Arizona. The aboveground metal trough was replaced with a buried concrete trough with escape ramp, placed under new shade cover. The water is fenced to exclude feral ass (*Equus asinus*). Photo courtesy of S. S. Rosenstock

Wells and Windmills

In areas where precipitation catchments are not practical, wells equipped with windmills can provide water for wildlife. To avoid increasing cover of vegetation and predation risk near the well, overflow from the trough should be redirected back down the casing. Many wells originally constructed for livestock have been converted for use by wildlife, typically by installing a more wildlife-friendly drinking trough (Fig. 12.6).

Spring and Seep Development

Making water more readily available from existing springs or seeps may be as simple as removing phreatophytes (Bleich et al. 2005). However, in some situations, extensive modifications may be required to provide a reliable water source, including fencing to exclude domestic livestock and feral equines, construction of access ramps, basins, or pools, or piping water to a nearby drinking trough.

Horizontal Wells

Horizontal wells are a viable option for providing water for wildlife in arid regions and overcome many disadvantages of spring development (Bleich et al. 1982, 2005). Horizontal wells have been successfully developed in areas with historical springs or seeps and the presence of an appropriate geologic formation, such as an impervious dike. Wells are drilled through the impervious formation into the water table. Water flow can then be controlled using a float valve and water distributed to a nearby drinking trough.

Livestock Water Developments

Many livestock waters are used by mule deer, pronghorns (*Antilocapra americana*), elk (*Cervus canadensis*), small mammals, birds, and breeding amphibians, including some species of conservation concern (e.g., ranid frogs; Rosen and Schwalbe 1998). The northern leopard frog (*R. pipiens*) has been extirpated from most of its historic range in Arizona. On the central Mogollon Rim, where the last known remaining metapopulation occurs, 23 of 25 currently known breeding sites are earthen tanks developed for livestock (S. MacVean, Arizona Game and Fish Department, unpublished data). However, the location of livestock waters sometimes renders them inaccessible to some species (e.g., desert bighorn sheep, pronghorn; Ockenfels et al. 1994). Steep-walled drinkers and storage tanks may be unavailable to smaller species and present a risk of drowning. Wire fencing, water surface area, and other features of livestock waters affect use by bats (Tuttle et al. 2006). Taylor and Tuttle (2007) provide guidelines on making livestock water developments more wildlife-friendly.

WATER MANAGEMENT FOR WILDLIFE

Liebig's law of the minimum states that growth is determined by the scarcest resource, not the total amount of all resources. Every population needs a variety of resources (e.g., food, water, cover, space) to grow. When supplies of these resources exceed needs, populations can increase until some resource is exhausted or depleted and becomes a "limiting factor." In arid regions, water is often assumed to be a primary factor limiting the distribution and productivity of desert ungulates and upland game birds. Beginning in the 1940s, resource management agencies and sportsmen's organizations began investing in the construction and maintenance of water sources in areas where natural sources were scarce or unreliable. These water developments were first constructed to benefit quail (*Callipepla* spp.; Glading 1943) and later chukar (*Alectoris chukar*), mule deer, bighorn sheep, pronghorn, and other game species. Water developments have also been used in mitigation efforts to offset loss or inaccessibility of naturally occurring water sources, habitat loss, and other impacts from urban, agricultural, transportation, and industrial development (Rosenstock et al. 1999, Krausman et al. 2006).

Water developments are widely used by wildlife managers in arid regions. At the end of the last century, ten of 11 western state wildlife agencies had ongoing programs that included approximately 6,000 catchments, guzzlers, tinajas, developed springs, and wells providing perennial water for wildlife. Overall, more than $1,000,000/yr was allocated to these efforts (Rosenstock et al. 1999), yet comparatively few resources have been dedicated toward monitoring the influence of water developments on wildlife. Recently, however, more resources are being dedicated to assess the influence of water developments than had been the case in the past.

Some investigators have questioned the efficacy of water developments (Broyles and Cutler 1999; Rosenstock et al. 2001), and their use is controversial, especially in some protected areas (Czech and Krausman 1999) including wilderness areas (e.g., the Arizona Desert Wilderness Act of 1990, or P. L. 101-628, and the California Desert Protection Act of 1994, or P. L. 103-433) and national monuments (e.g., Agua

Fria, Grand Canyon-Parashant, Sonoran Desert). The question of whether water developments are effective in meeting population objectives largely reflects the lack of experimental studies assessing their efficacy. Some populations of desert ungulates and upland game birds occupy areas without perennial water sources (Krausman and Leopold 1986, Alderman et al. 1989), suggesting that perennial water sources might not be required, and some studies have produced conflicting results (Mendoza 1976, Leslie and Douglas 1979, Deblinger and Alldredge 1991). Furthermore, some sections of the public have argued that the presence and maintenance of man-made water sources compromise wilderness values of remote, protected areas, creating unacceptable signs of human intrusion and altering some presumed balance of "undisturbed nature." In response, wildlife managers contend that the failure to build or maintain these facilities puts vulnerable species and other important wildlife at risk (Bleich 2005).

POTENTIAL BENEFITS OF PROVIDING WATER SOURCES FOR WILDLIFE

Wildlife management agencies believed that augmentation of water sources would increase survival and recruitment, expand animal distribution (Rosenstock et al. 1999), and buffer populations during drought periods. Anticipated benefits of water developments typically encompass one or more interrelated population management objectives, including increasing abundance, increasing available habitat (thereby increasing population size and distribution), improving habitat quality, and benefitting species of conservation concern (e.g., sensitive, threatened, or endangered).

Increase Animal Abundance
The abundance of some populations has been associated with the availability of water sources, with populations having higher abundance in areas with perennial water sources. For example, the presence of water developments can affect the population densities of pronghorn, elk, and mule deer. The high densities of some pronghorn populations are associated with water sources (Yoakum 1994). However, water developments are not the sole determinant of pronghorn abundance (Deblinger and Alldredge 1991), and some studies of desert bighorn sheep did not find an influence of water sources on population size (Krausman and Etchberger 1995).

Increase Available Habitat
If all necessary resources with the exception of water are available, construction of new water sources may increase animal distribution and abundance by increasing available habitat. For example, mule deer (Wright 1959) and desert bighorn sheep (Leslie and Douglas 1979) began year-round occupation of previously seasonal ranges following development of water sources. Elk (Rosenstock et al. 1999) and mule deer (Marshal et al. 2006a) expanded their distribution in arid habitats, possibly in association with water developments. The addition of water sources may increase abundance, but density may

remain unchanged because of the increase in available habitat (Bleich et al. 2010). Increases in distribution have not been found in some instances (e.g., desert bighorn sheep; Krausman and Leopold 1986), and the availability of water did not influence habitat use of adult pronghorns in other regions (Deblinger and Alldredge 1991, Ockenfels et al. 1994).

Improve Habitat Quality
Developed water sources may also improve habitat quality. Survival and reproductive success of some wildlife populations in arid areas fluctuate widely from year to year. If this variability is due to a lack of reliable water sources, then water developments could reduce variability in survival and reproductive rates. For example, growth rates influence survival and recruitment of neonatal ungulates (Cook et al. 2004), which in turn are influenced by the quantity and quality of milk, which can be adversely affected by water deprivation (Hossaini-Hilali et al. 1994), thereby influencing juvenile survival. In areas where forage resources are sufficient but water is limiting, wildlife water developments may result in improved reproductive success and juvenile recruitment.

Recovery and Management of Special Status Species
Although most water developments are constructed to benefit specific game species, a wide variety of other species use them, too, including species of conservation concern (Kuenzi 2001, Lynn et al. 2006, O'Brien et al. 2006). The Sonoran pronghorn (*A. a. sonoriensis*) was listed as endangered in 1967, with the distribution in the United States currently limited to southwestern Arizona. Declines in forage quality during recurring droughts reduce fawn recruitment and adult survival and are believed to limit recovery efforts (Hervert and Bright 2005). During dry periods, Sonoran pronghorns selected areas with abundant chain fruit cholla (*Cylindropuntia fulgida*) and other succulent vegetation, comprising over 40% of their diet (Hughes and Smith 1990). However, areas with abundant chain fruit cholla also had higher predation risk (Hervert and Bright 2005, Hervert et al. 2005). Because chain fruit cholla and other cacti have high moisture content (e.g., 85%; Hughes and Smith 1990) but low nutritional content (e.g., <4% protein), pronghorns may choose forage based on moisture rather than nutritional content when water is limiting. Constructing water sources in areas with lower predation risk may benefit Sonoran pronghorns, allowing them to select areas with higher-quality forage but lower moisture content among the forage available in those areas (Morgart et al. 2005).

POTENTIAL NEGATIVE IMPACTS OF WATER DEVELOPMENTS

Critics of wildlife water developments have cited a number of potential adverse impacts to wildlife, wildlife habitat, and other resources, including disease transmission, water quality, increased predation or competition, direct mortality, and changes to plant communities. Recent studies have generally failed to validate these concerns.

Increased Potential for Disease Transmission

Because wildlife water developments receive heavy use by numerous wildlife species, some have suggested that they may contribute directly or indirectly to spread of wildlife diseases (Broyles 1995). One pathogen of concern for direct transmission is the protozoan parasite *Trichomonas gallinae*, which has been found in urban birdbaths and causes epizootic outbreaks in doves (*Zenaida* spp.) and other birds. Rosenstock et al. (2004) collected water samples from a variety of water development types in Arizona and cultured them for *Trichomonas*; all samples were negative. Water samples from other developments proximate to a trichomoniasis outbreak were also negative. They suggested that exposure to ultraviolet radiation and predation by other micro-organisms may limit persistence of *Trichomonas* in water developments. This hypothesis was supported by a follow-up experiment in which replicated "mini-exclosures" in a catchment trough were inoculated with live *Trichomonas* and then sampled over time for presence of the protozoan. Within 24 hours, all samples were negative (S. Rosenstock, Arizona Game and Fish Department, unpublished data). Another water-mediated disease is botulism, caused by toxins produced by the bacterium *Clostridium botulinum*. There is one published occurrence of botulism-related mortality at a wildlife water development in California. Desert bighorn seeking water became entrapped in the storage tank, drowned, and decomposed, and sheep later consumed the contaminated water. A total of 45 animals died during this event, which was subsequently attributed to type C botulism (Swift et al. 2000).

Broyles (1995) suggested that water developments provided nurseries for hematophagous gnat (genus *Culicoides*) larvae, the adults of which are vectors of viral hemorrhagic diseases (bluetongue, epizootic hemorrhagic disease) affecting ungulates (e.g., mule deer) that also use these facilities. Rosenstock et al. (2004) trapped *Culicoides* at sites adjacent to wildlife water developments and unwatered controls. Several species of *Culicoides*, including the known vector *C. sonorensis*, were widely distributed and common at both watered and control sites. They also sampled potential substrates (sand, mud, gravel) used by larvae at natural and modified tinajas (these materials are generally absent in troughs at other types of constructed waters). Larval *C. sonorensis* were absent, except for a few individuals found at one site. Typically, wildlife water developments do not provide optimal conditions for larval *Culicoides*—fine silt or mud at the margins of water that is brackish or heavily enriched with animal manure (Mullens 1991).

Interactions between conspecifics and other species are common at wildlife water developments (O'Brien et al. 2006), creating the possibility of animal–animal pathogen transmission. While this topic has not yet been studied, it seems unlikely that animal interactions at wildlife water developments would be different from those occurring at natural water sources.

Water Quality

During hot summer months, wildlife water developments in desert environments can have high water temperatures and evaporation rates, inputs of organic material—commonly drowned honeybees (*Apis mellifera*)—and infrequent flushing, characteristics that are also common to natural surface water sources. Nevertheless, some have suggested that poor water quality in developed waters can adversely impact health and or survival of wildlife (Kubly 1990, Broyles 1995). Parameters and constituents of potential concern include dissolved solids, pH, heavy metals, toxic chemicals, and toxins produced by blue-green algae. Rosenstock et al. (2004) and Bleich et al. (2006) examined water quality at water developments in Arizona and California, respectively, at sites that included several types of precipitation catchments, improved springs, modified and natural tinajas, and wells. At sites in both states, water-quality parameters were typically within established guidelines for domestic livestock (standards for wildlife have not been published). Elevated levels of pH, alkalinity, and fluoride occurred at some sites but were deemed of negligible impact to wildlife health. Arizona waters were also screened for cyanobacterial toxins; all tests were negative (Rosenstock et al. 2004). Wildlife water developments do not appear to provide appropriate conditions for toxic algal blooms, which typically occur on lakes and other large water bodies (Schwimmer and Schwimmer 1968).

Predation

Because a variety of avian and mammalian predators are regular visitors to water developments and are assumed to expand home ranges to include them, some researchers have suggested that these facilities increase predation rates on desert bighorn and other species (Broyles 1995). This speculation may reflect observations in African savannas, where ungulates and predation events can be concentrated around waterholes during the dry season (Senzota and Mtahko 1990). Evidence supporting a similar pattern at water developments in North American deserts is scant. DeStefano et al. (2000) found higher amounts of predator sign (feces, tracks) around water developments than at unwatered comparison sites, but little evidence of predation events. They suggested that water developments could decrease predation pressure by reducing concentration around the few existing natural water sources. O'Brien et al. (2006) documented only eight attempted or successful predation events in over 37,000 hours of video observations at three water developments in Arizona, the majority involving bobcats or raptors taking birds or small mammals. Camera monitoring of water developments on the Sevilleta National Wildlife Refuge, New Mexico, have documented interactions between coyote and pronghorn and mountain lion and mule deer, but no mortalities (J. Erz, U.S. Fish and Wildlife Service, unpublished data).

Competition

Broyles (1995) suggested that water developments for desert bighorn sheep could attract potential competitors such as mule deer or feral asses (*Equus asinus*). This is unlikely, because deer and desert bighorn typically use spatially discrete habitats, and wildlife water developments in areas occupied by feral equines are often fenced to exclude them (Andrew et al. 1997).

Direct Mortality

Wildlife water developments have been hypothesized to cause direct mortality of wildlife, an assertion likely rooted in anecdotal reports of drowned animals found at these facilities (Baber 1983) and mortalities at natural tinajas (Mensch 1969). One species of particular concern is the desert tortoise (*Gopherus aggassizii*), a federally listed endangered species. Andrew et al. (2001) identified faunal remains from catchments in the California range of the tortoise. They found remains at six sites—primarily small mammals, birds, and reptiles—but none belonging to desert tortoise. Most skeletal materials showed evidence of predation, and likely originated from pellets cast by raptors or scats of mammalian predators. O'Brien et al. (2006) video-documented another source of remains found in or at water developments, those brought to the site by predators or scavengers. During more than 600 visits to various types of wildlife water developments in Arizona, Rosenstock et al. (2004) documented 19 presumed drowning incidents that included several mule deer, a coyote (*Canis latrans*), a desert bighorn sheep, and various small mammals and passerine birds. Overall, the information in hand does not support the notion that developed waters are an important cause of mortality for wildlife.

Detrimental Impacts on Vegetation

Concentrations of animals at water sources can result in trampling and foraging pressure that impact surrounding plant communities. Studies on the African savanna have documented gradients of decreasing forage use and plant community change with increasing distance from man-made waterholes, with a zone of severe impact ("piosphere") immediately around the water source (Thrash 1998, Parker and Witkowski 1999). Krausman and Czech (1998) hypothesized similar effects around water developments in North American deserts. Marshal et al. (2006b) sampled forage biomass and mule deer use in paired desert washes with and without precipitation catchments. During the hot, dry season, deer sign was more abundant in washes with catchments and increased with proximity to water in washes with catchments. However, they found no differences in forage biomass attributable to foraging by deer. This was not an unexpected finding, because desert ungulates occur at relatively low densities and do not linger around catchments after watering (O'Brien et al. 2006, Waddell et al. 2007).

INFLUENCE OF WATER DEVELOPMENTS ON SELECTED WILDLIFE SPECIES

Extensive literature reviews on the effects of wildlife water developments on game and nongame species were conducted by Payne and Bryant (1998) and Rosenstock et al. (1999), and were later updated by Krausman et al. (2006). Key findings of these reviews are presented here, along with subsequent research on the influence of water sources on game and nongame wildlife species. Collectively, these data suggest that water developments may have resulted in increased wildlife distribution and abundance in some instances, but not in others.

Mammals
Desert Bighorn Sheep

Development of water sources is central to management of desert bighorn sheep and can interact with other factors influencing sheep populations, such as forage quality and availability, escape terrain, thermal cover, and fire history (Bleich et al. 2010). Desert bighorn require approximately 4% of their body weight in water per day (Turner 1973); the proportion met by preformed water in forage is unknown, but likely varies seasonally and among populations depending on availability of succulent forage. Many observational studies have documented that desert bighorn use water developments when available (Blong and Pollard 1968, Wilson 1971) and use increases during summer months (Bleich et al. 1997). Desert bighorn reportedly use water every two to three days (Bradley 1963, Knudsen 1963) and consume 2–4 L of water per visit (Knudsen 1963, Hailey 1967). Others did not document use of newly constructed water sources for at least three years following construction (Krausman and Etchberger 1995), however, and Krausman et al. (1985) reported that two females did not visit water sources for ten or more days during the summer. The use of water sources is likely related to preformed moisture content in forage (Krausman et al. 1985, Warrick and Krausman 1989).

Perennial water sources are a key habitat component for desert bighorn sheep (Douglas and Leslie 1999), and their presence on the landscape commonly defines habitat quality for the species (Bleich et al. 2010). Desert bighorn are commonly located less than 4 km from water sources during the hottest and driest times of the year (Wilson 1971, Turner et al. 2004). Because of their smaller body size and the increased water demands during lactation, females with lambs are commonly found closer to water sources than males or females without lambs (Bleich et al. 1997, Krausman 2002). Although numerous studies have shown summer distribution of desert bighorn sheep to be associated with water sources, other populations occupy ranges devoid of perennial water sources (Watts 1979, Krausman et al. 1985, Alderman et al. 1989) and may obtain sufficient metabolic and preformed water from cacti and other succulent vegetation (Warrick and Krausman 1989). Distribution and habitat use of desert bighorn are not limited by the availability of water sources in all cases (Krausman and Leopold 1986, Krausman and Etchberger 1995).

Although the influence of water sources on seasonal distribution of desert bighorn has been reported for numerous populations across the desert Southwest, documentation of population-level responses (i.e., increases in abundance or survival) is limited. In some cases, population increases were attributed to addition of water sources (e.g., Leslie and Douglas 1979), and population declines (Douglas 1988), and mortalities were attributed to loss of water sources (Allen 1980). Others have concluded that addition of water sources does not always increase survival or productivity (Ballard et al. 1998).

The equivocal results in the scientific literature regarding influence of water sources on desert bighorn sheep may in part reflect the fact that most studies were either observational, anecdotal, or derived from research designed to address other objectives (Rosenstock et al. 1999, Krausman 2002). Few studies have experimentally examined the influence of water developments on desert bighorn sheep. Cain et al. (2008a) conducted a manipulative study documenting the responses of desert bighorn sheep to removal of man-made water sources in southwestern Arizona; response variables included diet selection, home range size, movement, survival, and recruitment. Water sources were maintained in both ranges during 2002–2003, then drained in one mountain range and maintained in the other during 2004–2005. Water removal did not result in the predicted changes in the response variables. When water sources were maintained in both mountain ranges, the area experienced the worst drought on record, and high adult mortality was observed. Providing water sources during drought may do little to prevent the adverse consequences of scarce or poor-quality forage; it is unknown if mortality would have been higher in the absence of water sources during the drought. Conversely, increased precipitation during the posttreatment period resulted in better forage conditions, increased forage moisture content, and availability of ephemeral water sources. The lack of significant changes after water removal suggested that during years with above-normal precipitation, water might not be limiting to desert bighorn sheep. However, because of the climatic conditions observed during this study, it is unknown what impact the removal of water catchments would have had during periods with average precipitation.

Additional experimental studies are needed to determine if and under what conditions water developments have population-level influences on desert bighorn sheep. Broader questions also remain on how climate change will impact individual populations and metapopulation structure (Epps et al. 2004). In the southwestern United States, average annual temperatures are predicted to increase while precipitation is expected to decrease, resulting in lower snowpack levels and decreased surface runoff (Seager et al. 2007). Decreased river flow is likely to result in more reliance on groundwater sources to meet agricultural and municipal needs (Alexander et al. 2011). Increasing reliance on groundwater and decreased groundwater recharge rates are likely to result in lower water tables and loss of natural surface water sources (e.g., springs) used by wildlife (Zektser et al. 2005). Epps et al. (2004) reported that desert bighorn populations in Southern California occupying higher-elevation desert mountain ranges with more precipitation and the presence of natural springs were less likely to go extinct than populations in lower-elevation mountain ranges without perennial springs. It seems likely that water developments will play an increasingly important role in the conservation and management of desert bighorn sheep and other wildlife in the future.

Mule Deer

Constructed water sources commonly benefit desert mule deer. Mule deer often use water developments, particularly during the summer months. However, mule deer tend to use natural water sources when available (Krausman 2002), and newly constructed water sources may receive relatively little initial use compared with established water sources. Water sources available for longer than three years have higher use than newly constructed water sources (Marshal et al. 2006a,b). Mule deer visit water sources at all times, but most visits occur at sunset or at night (Hervert and Krausman 1986, Hazam and Krausman 1988). Mule deer visit water sources every one to four days; the frequency of water use is higher for females than males (Hervert and Krausman 1986, Hazam and Krausman 1988). Water consumption rates range from 1 to 6 L per visit and vary seasonally (Hazam and Krausman 1988). Females consume an average of 3.3 L per visit in early summer and 4.2 L per visit in late summer (Hazam and Krausman 1988); consumption by males averaged 2.7 and 3.6 L per visit during early and late summer, respectively (Hazam and Krausman 1988).

Mule deer tend to be located closer to water sources during the summer dry season than other periods of the year (Ordway and Krausman 1986) and may travel outside their normal home range when water sources become unavailable (Hervert and Krausman 1986). Some populations of desert mule deer may migrate during the dry season to areas with perennial water sources (Rautenstrauch and Krausman 1989). Krausman and Etchberger (1995) reported that during summer, females were located closer to water than males; however, all deer were typically within 5 km of water. In addition to influencing the seasonal distribution of individual mule deer, seasonal ranges became occupied year-round after construction of water developments (Wright 1959).

The addition of water sources has also been associated with increases in abundance in some areas. Mule deer numbers increased for at least a five-year period following the construction of water sources near Fort Stanton, New Mexico (Wood et al. 1970). The increase, however, may have been associated with a redistribution of deer to areas near water sources rather than an increase in population size. Current research lacks conclusive information on population-level influences of water developments on mule deer (Heffelfinger 2006:118).

White-tailed Deer

White-tailed deer are typically found only in association with water sources (Krausman and Ables 1981, Maghini and Smith 1990) and appear dependent on free water in the southwestern United States. Coues white-tailed deer (*O.v. couesi*) in Arizona select areas within 0.4 km of water sources and avoid areas located more than 1.2 km from water sources. At the western edge of the distribution of Coues white-tailed deer, animals drank every one to four days (Henry and Sowls 1980). White-tailed deer subjected to water restriction decrease forage intake and lose more weight than those with access to water (Lautier et al. 1988). When forage moisture is low in arid environments, water developments may influence fawn survival and recruitment (Ockenfels et al. 1991).

Elk

Elk habitat is commonly restricted to areas with surface water. In the spring and summer, elk tend to be distributed within 0.4–0.8 km from a water source (Jeffrey 1963, Marcum 1975, Nelson and Burnell 1976, Delgiudice and Rodiek 1984). Water needs of elk most likely vary seasonally and annually, depending on precipitation, forage conditions, and reproductive status (Mackie 1970, Marcum 1975, McCorquodale et al. 1986). In many areas of the western United States, elk inhabit mesic areas with relatively abundant natural sources of surface water (e.g., streams), and water is often not a limiting factor, making water developments unnecessary. The development of water sources has most likely contributed to the expansion of elk populations in arid areas (Strohmeyer and Peek 1996). In areas with variable surface water availability, elk may benefit from water developments, particularly during calving, lactation, or during dry periods (Delgiudice and Rodiek 1984, McCorquodale et al. 1986).

Pronghorn

Much of the research on the dependence of pronghorns on free water is equivocal (Yoakum 2004). Some researchers contend that pronghorns need drinking water to sustain healthy populations (Yoakum 1978), whereas others suggest that some populations obtain sufficient moisture from succulent forage. When succulent vegetation is unavailable, water developments are heavily used by pronghorns (O'Gara and Yoakum 1992, Yoakum 1994), and many pronghorn populations use water sources, particularly during dry periods (Sundstrom 1968, Beale and Smith 1970, Yoakum 1994). Frequency of use ranges from daily to several days or weeks in areas with succulent forage (Buechner 1950). In western Utah, pronghorns did not use water sources when forage moisture content was over 75% (Beale and Smith 1970). When forage moisture content is inadequate to meet water needs, consumption of free water ranges from 0.95 to 3.8 L/day (Sundstrom 1968, Beale and Smith 1970, Beale and Holmgren 1975), depending on forage moisture content and air temperature. Some populations of pronghorn previously thought not to use free water have subsequently been detected using it (e.g., Sonoran pronghorn; Morgart et al. 2005).

When forage moisture content declines, pronghorns in Wyoming were reported within 5–6.5 km of water sources (Sundstrom 1968). In Arizona, Hughes and Smith (1990) reported that Sonoran pronghorns were typically within 6 km of water sources; however, deVos and Miller (2005) concluded that Sonoran pronghorns preferentially used areas within 2 km of water sources. During fawning, pronghorns in areas with limited surface water availability may occur closer to water, particularly females with fawns. Fawns in Arizona selected bed sites within 0.4–0.8 km of water sources (Ticer and Miller 1994), and yearling pronghorns in New Mexico tended to be closer to water sources than adults (Clemente et al. 1995).

The influence of water sources on population density is less clear. High-density populations are often associated with free water (Yoakum 1994). In Wyoming, pronghorn densities are highest in areas with water sources (Sundstrom 1968, Boyle and Alldredge 1984). Deblinger and Alldredge (1991) also reported that pronghorns in Wyoming had higher densities in areas with water sources, but the distribution of pronghorns did not change when water became unavailable. Whether these results reflect an influence of water sources on population abundance rather than simply an influence on the distribution is unclear.

Water sources obviously influence the distribution of pronghorns, but they are not the sole determinant of the distribution or abundance of pronghorns (Deblinger and Alldredge 1991). Fawn:female ratios in Arizona related more strongly to previous winter precipitation than water availability, suggesting that forage availability was more important for fawn recruitment (Bristow et al. 2006). Based on 27 years of aerial survey and precipitation data in the Trans-Pecos area of Texas, fawn production strongly correlated with precipitation in the year (August–July) prior to surveys, whereas abundance more strongly correlated with precipitation indices (e.g., Palmer Drought Severity Index) incorporating other climatic variables, suggesting that fawn production was related to immediate precipitation conditions, and that abundance was influenced more by long-term climatic trends (Simpson et al. 2007).

Predators

Research results have been equivocal regarding free water requirements of mammalian predators (Rosenstock et al. 1999). Many predators have been documented using water developments (Cutler 1996, O'Brien et al. 2006). Most predators obtain sufficient moisture from their prey, and some like the ringtail (*Bassariscus astutus*) and kit fox are believed to be independent of free water (Chevalier 1984, Golightly and Ohmart 1984). O'Brien et al. (2006), however, documented use of man-made water sources by kit fox in southwestern Arizona. Predators commonly use water developments, but their need to drink free water for survival or reproduction has not been examined (Ballard et al. 1998, DeStefano et al. 2000).

Small Mammals

Small mammals inhabiting arid environments typically employ physiological or behavioral mechanisms that minimize the need for free water, and obtain sufficient water from preformed water in food and metabolic water (Mares 1983, Walsberg 2000, Nagy 2004). Some small mammals have been documented at water developments, including black-tailed jackrabbit (*Lepus californicus*), desert cottontail (*Sylvilagus audubonii*), and ground squirrels (O'Brien et al. 2006). Of the ten species of rodents captured near water sources in southwestern Arizona, Cutler (1996) reported drinking only by the round-tailed ground squirrel (*Spermophilus tereticaudus*).

Bats commonly use water developments for drinking and as foraging areas (Kuenzi 2001, Rabe and Rosenstock 2005).

Schmidt and DeStefano (1999) documented higher bat activity at desert water developments compared with control sites; bat visitation rates of over 1,000 passes per hour were documented at several sites. Some bat species are dependent upon surface water, and water developments may influence the distribution of these species (Schmidt and Dalton 1995). Recent research using isotopically labeled water sources in southwestern Arizona (Wolf 2010) reported that developments provided 6–43% (with an average of 12%) of the total body water pool for eight species. Water developments are important tools for conservation of some bat species, particularly in areas where natural water sources are no longer available (Tuttle et al. 2006, Taylor and Tuttle 2007).

Birds

Upland game birds received the initial focus of wildlife water developments (Rosenstock et al. 1999). Water developments have been constructed for dove, quail, chukar, and turkey, and responses of game birds and other avian species to them have varied.

Upland Game Birds

White-winged (*Zenaida asiatica*) and mourning dove (*Z. macroura*) frequently use water developments (O'Brien et al. 2006), increasingly during the summer. When deprived of water, mourning doves lose almost 5% of body weight per day and therefore require access to drinking water (Bartholomew and MacMillen 1960). Both species have probably benefited from water developments in arid areas (Rosenstock et al. 1999). Hyde (2011) used stable-isotope techniques to assess avian use of water developments in southwestern Arizona. After accounting for preformed and metabolic water, over 60% of the total body water pool of mourning and white-winged doves came from water developments (Hyde 2011). Movement and habitat use data in Idaho (Howe and Flake 1988) suggest that construction of water sources may enhance mourning dove populations. However, because doves can travel long distances to water (>11 km; Howe and Flake 1988), water might not be limiting in many areas.

Studies on the influence of water developments on quail populations have yielded equivocal results. Gambel's (*C. gambellii*), scaled (*C. squamata*), and bobwhite quail (*Colinus virginianus*) use them during summer months, particularly when succulent vegetation is not available (Hungerford 1962, Campbell et al. 1973, Guthery and Koerth 1992). Many quail populations obtain sufficient moisture from succulent forage; however, quail populations in areas that commonly experience drought during the breeding season may benefit from the addition of water developments (Campbell 1961). Chronic water deprivation of bobwhite can lead to reproductive failure (Giuliano et al. 1994). Hyde (2011) reported that over 43% of the total body water pool of Gambel's quail came from water developments.

Development of water sources is the primary habitat management activity for introduced chukar in Utah, Nevada, and Idaho (Benolkin and Benolkin 1994). In Nevada, addition of water sources increased existing chukar populations and facilitated new populations (Benolkin and Benolkin 1994). The use of water sources by chukar is inversely related to forage moisture content (Larsen et al. 2007). During the cooler seasons, chukar may meet water requirements from preformed water in forage, but may need drinking water when succulent vegetation is unavailable (Degen et al. 1984). Distribution of chukar during summer is largely dependent on the availability of water sources (Christensen 1996), but it is also affected by forage moisture content. Summer distribution of chukar in western Utah was associated with water sources in two areas when forage moisture content was under 45%, but unassociated with water sources when forage moisture content was over 55% (Larsen et al. 2010). Chukar do not use all water sources built specifically for them; guzzlers in areas without adequate hiding cover were less likely to be used (Larsen et al. 2007).

Some researchers consider water sources to be an essential habitat component for Merriam's turkey (*Meleagris gallopavo merriami*) in the Southwest. In areas where natural sources of water are lacking or seasonally unavailable, development of water sources may benefit turkey populations (Shaw and Mollohan 1992). Management guidelines for Merriam's turkeys suggest that water sources should be regularly available throughout suitable turkey habitat (Hoffman et al. 1993).

Nongame Birds

A variety of nongame birds, including passerines, raptors, shorebirds, and waterfowl, take advantage of water developments (Lynn et al. 2006, 2008; O'Brien et al. 2006). In Arizona, Cutler (1996) documented over 150 bird species near water developments, 60 of which were observed drinking. The use of water developments by birds tends to be highest during the summer, but use has been observed during spring and autumn migration, suggesting that these water sources may be important as stopover locations during migration, either as a source of drinking water or because associated vegetation served as foraging and resting areas (Rosenstock et al. 1999). Lynn et al. (2006, 2008) documented relatively low use by migrants, however, and suggested that the benefits to nongame birds were primarily to resident species. These results were supported by Hyde (2011), who reported that most neotropical migrants derived less than 10% of their body water pool from water developments.

Resident birds may benefit from water sources to a greater extent than migratory species, and they use them most often during the summer (O'Brien et al. 2006). Cutler (1996) reported that avian abundance and species richness were inversely related to the distance from water development in Arizona, but other investigators (Burkett and Thompson 1994) reported no difference in species richness or abundance at water developments compared with control sites. In southwestern Arizona, the proportion of body water among resident species (excluding doves and quail) derived from water developments ranged from 10% to 58%; nine of 15 species derived more than 25% of their water pool from the developed waters, and four species derived more than 45% from these sources (Hyde 2011). The influence of water developments on survival or reproductive success of most passerines is unknown. However, availability of supplemental water has been experimentally shown to increase clutch size of black-throated sparrows (*Amphispiza bilineata*), although it did not influence nest survival rates or the probability of fledging (Coe and Rotenberry 2003).

Raptors commonly use water developments in arid environments (O'Brien et al. 2006, Lynn et al. 2008), with sign of use higher at water developments than control sites (Burkett and Thompson 1994, Schmidt and DeStefano 1996). Raptors use water sources for drinking and bathing and may also capture birds and small mammals gathered there (O'Brien et al. 2006). Harris's hawks (*Parabuteo unicinctus*) require surface water during the breeding season and may have expanded their distribution following development of water sources (Dawson and Mannan 1991).

Reptiles and Amphibians

Desert-dwelling amphibians may benefit from water developments, too. In Arizona, researchers have observed the Colorado River toad (*Bufo alvarius*), red-spotted toad (*B. punctatus*), leopard frog, and Sonoran tiger salamander (*Ambystoma mavortium stebbinsi*) breeding in various water developments, including tinajas, precipitation catchments, and developed springs (S. Rosenstock, Arizona Game and Fish Department, unpublished data). Some amphibians use stock tanks and other water developments as breeding habitat, including some species of conservation concern (e.g., leopard frogs in stock tanks; Rosen and Schwalbe 1998). Reptiles inhabiting arid environments occasionally use water sources (O'Brien et al. 2006), although it is believed that most reptiles do not require drinking water (Mayhew 1968).

COVER

Cover provides shelter for wildlife. Cover can be vegetation and topographic features that provide places for wildlife to feed, hide, sleep, play, and raise young (Leopold 1933). All of these activities are important to the life history of wildlife. Winter cover consists of vegetation or other features that provide an animal protection from snow and cold. Leopold (1933) defined refuge cover as vegetation from which game cannot be driven by hunters, and we expand that to include other pursuers. Loafing cover offers animals protection and rest from the elements, and resting, roosting, and bedding cover provide just that. Thermal cover has different meanings but generally applies to vegetation that protects animals from cold (e.g., synonymous with winter cover) or heat. Escape cover includes features (e.g., rough landscapes) that allow prey to out-maneuver predators (Gionfriddo and Krausman 1986). These generalized categories are sometimes useful, but they gloss over the complexity of the general and specific cover needs of wildlife.

Cover management becomes even more complex because the management for one species' resource may alter other resources needed by other species. For example, the structural diversity of forests provides cover for species that use the edge of forests and forest interior, as well as cover for ubiquitous species. Species with higher densities along ecotones (e.g., the forest edge) would benefit from some degree of fragmentation that would create more of the required ecotone. However, that increase would be at the expense of species that require forest interior and have more homogeneous cover requirements (e.g., salamanders; Herbeck and Larsen 1999). Reduction of old-growth forest reduces microhabitat availability; by cutting forests, habitat for edge species could be enhanced, reduced for interior species, and likely only have minimal influences on ubiquitous species because, as generalists, they can use the edge or the interior of the forest (Whitcomb et al. 1981).

Anthropogenic Influences on Cover

Human use of landscapes has a direct influence on cover for wildlife. Birds that breed in forests are more successful when forest cover measures in 10 × 10 km blocks or larger (Trzcinski et al. 1999). Likewise, in Arizona, grasslands that are not grazed by livestock are more productive sites for birds than those that are, and the ungrazed sites support a higher abundance of granivorous birds (Bock and Bock 1999).

Agriculture that alters cover also affects small mammals. Some rodents are more abundant with dense nesting cover, have intermediate densities in areas where haying is delayed and confined to rights-of-way, and are at their lowest numbers in idle pastures where cover is lowest (Pasitschniak-Arts and Messier 1998).

In some situations, considerable expenditures of time, effort, and money are allocated to the management of cover for ungulates and other wildlife. Over the past six decades, wildlife biologists managed vegetation for ungulates as a mechanism to enhance survival in cold climates. "Thermal cover has been credited widely with moderating the effects of harsh weather, and therefore, may improve overall performance of populations (i.e., survival and reproduction) by reducing energy expenditures required for thermostasis" (Cook et al. 1998:6). This concern for adequate winter cover led to the widespread belief that winter thermal cover was a key component of ungulate habitat in the western United States. As a result, winter thermal cover was incorporated into the development of large-scale national forest plans, and management agencies made numerous site-specific, case-by-case decisions as to how to harvest forests or to prescribe fire based on the perceived view that winter cover was required by ungulates (Edge et al. 1990, Cook et al. 1998). In reality, winter thermal cover had little influence on the herd productivity and demographics of elk, as determined from a series of controlled experiments that demonstrated the importance of understanding how cover is actually used by wildlife (Cook et al. 1998). Prior to discovering the true extent of wildlife use of winter thermal cover, it was managed for elk on millions of hectares and at a high cost. More effort should be placed on forage quality and quantity, as well as the ability to evaluate forage conditions across the landscape, instead of managing for winter thermal cover for elk (Cook et al. 1996, 1998).

Management of Cover

Cover management is the management of vegetation succession. For example, farming essentially establishes plant communities in early successional stages, and range management attempts to maintain rangelands in grassland stages. If succession is not interrupted, shade-intolerant plant communities will be replaced by more shade-tolerant plant communities until a climax community is reached (i.e., the last stage of plant succession in a biome). Climax flora include specific dominant plants in each biome, and the vegetation and related cover at the climax stage are different from successional stages. Specific animal communities associate with specific successional stages (i.e., sere) and all related ecotones. Seres are often replaced by silvicultural treatments or brush control that subsequently replaces one animal community with another (Payne and Bryant 1998).

Succession varies with soil type, moisture, microclimate, climate, topography, slope, and degrees of interference (e.g., autogenic forces caused by flora and fauna or allogenic forces caused by such outside influences as fire and wind). Common successional stages in grasslands go from bare soil, followed by annual grasses and forbs, perennial forbs, and short-lived perennial grasses, until they are replaced with sod-forming grasses or bunch grasses that maintain a relatively stable equilibrium (Coupland 1992). Shrubland and woodland vegetation advances from bare soil to grasses and forbs, followed by dominant shrubs and woody plants. These are only a few of the numerous seres that have been documented.

All successional stages from each type in each biome should be represented in various shapes, sizes, and distribution to enhance species diversity and richness. These landscape characteristics can be advanced or retarded by natural or human means (e.g., mechanical, chemical, fire, wind, flooding, insects, disease, grazing, irrigation, planting, fertilizing; Thomas et al. 1979). Leopold (1933) referred to planting, fencing against livestock, and protecting areas from fire as tools to speed up succession. Plowing, burning, grazing, and cutting were tools to set back succession. These tools are not unique to wildlife management because they are used in agriculture, forestry, and range management, too. When considering any type of cover management, one must coordinate activities with other land management organizations that influence habitats. These concerns are as important today as they were 80 years ago.

SPECIAL FACTORS

Welfare factors that are important to a particular species, usually in small quantities and for short periods of time, have been called special factors (Leopold 1933). Most special factors could be grouped with other welfare factors (i.e., food, water, cover) but are singled out because of their unique importance.

Well-known special factors include "gravel for gallinaceous birds and waterfowl, salt licks for herbivores and some birds, mineral springs for pigeons, dust baths for various birds, mud baths and hibernation places for bear, caves or dense shade for desert bighorn sheep and quail to reduce water loss during the heat of the day in arid climates, open wind-swept parks or deep water for the relief of moose and deer in fly season, and sandy knolls for 'booming grounds' of prairie chickens" (Leopold 1933:27). Other special factors include lambing areas with adequate protection for desert bighorn sheep and breeding knolls for swamp deer (*Cervus duvauceli*). Minerals and vitamins are required in minute amounts but are critical to life. Their presence or absence likely influences the geographic distribution of species (Leopold 1933). There are likely numerous special factors that have not yet been identified, as well as some with greater importance than we currently understand.

SUMMARY

Water and cover are essential for all wildlife species, and many species require special factors. Wildlife inhabiting arid areas use a combination of physiological, morphological, and behavioral adaptations to avoid or cope with limited water availability. Animals obtain water from three sources: preformed, metabolic, and free water. When preformed and metabolic water are insufficient to meet water demands, animals must have access to sources of free water, either naturally occurring or developed. Wildlife managers have devoted substantial time and resources for the development of wildlife water sources in areas where natural sources are scarce or unreliable. The impacts of water developments on wildlife vary among species, and construction of these water sources has not always resulted in the predicted outcomes. Some species have likely benefitted from water developments, whereas questions remain for others. In particular, information on population-level impacts resulting from water developments is still lacking for the vast majority of species.

Vegetation and topographic features provide cover for wildlife. Refuge cover, nesting cover, roosting cover, and escape cover are important to the life history of most species of wildlife. In addition, winter thermal cover provides protection from the snow and cold, whereas summer thermal cover (e.g., shade) provides protection from high summer temperatures and solar radiation, particularly in deserts. Management agencies have spent considerable time, effort, and money for the management of cover for ungulates and other wildlife.

Special factors are welfare factors that are important for wildlife, typically in small quantities and for short periods of time. Well-known special factors include gravel for gallinaceous birds and waterfowl, salt licks for herbivores, and hibernacula. Others include parturition sites for ungulates.

The assumption that water developments are essential to wildlife in arid regions needs to be assessed rigorously using experimental studies, especially as funds for environmental enhancement or conservation become scarce. Studies that exemplify the importance of cover or identify special factors necessary for wildlife populations also would contribute to the wise use of available resources aimed at sustaining wildlife populations.

Literature Cited

Abeloe, T. N., and P. C. Hardy. 1997. Western screech-owls diurnally roosting in a cave. Southwestern Naturalist 42:349–351.

Alderman, J. A., P. R. Krausman, and B. D. Leopold. 1989. Diel activity of female desert bighorn sheep in western Arizona. Journal of Wildlife Management 53:264–271.

Alexander, P., L. Brekke, G. Davis, S. Gangopadhyay, K. Grantz, et al. 2011. SECURE Water Act Section 9503(c)—reclamation climate change and water. Report to Congress. U.S. Bureau of Reclamation, Denver, Colorado, USA.

Allen, R. W. 1980. Natural mortality and debility. Pages 172–185 in G. Monson and L. Sumner, editors. The desert bighorn. University of Arizona Press, Tucson, USA.

Andrew, N. G., V. C. Bleich, A. D. Morrison, L. M. Lesica, and P. J. Cooley. 2001. Wildlife mortalities associated with artificial water sources. Wildlife Society Bulletin 29:175–280.

Andrew, N. G., L. M. Lesica, and V. C. Bleich. 1997. An improved fence design to protect water sources for native ungulates. Wildlife Society Bulletin 25:823–825.

Arad, Z., I. Gavrieli-Levin, U. Eylath, and G. Marder. 1987. Effect of dehydration on cutaneous water evaporation in heat-exposed pigeons (Columba livia). Physiological Zoology 60:623–630.

Baber, D. W. 1983. Mortality in California mule deer at a drying reservoir: the problem of siltation at water catchments. California Fish and Game 70:248–251.

Ballard, W. B., S. S. Rosenstock, and J. C. DeVos Jr. 1998. The effects of artificial water developments on ungulates and large carnivores in the Southwest. Pages 64–105 in Proceedings of a symposium on environmental, economic, and legal issues related to rangeland water developments, 13–15 November 1997, Tempe, Arizona. Center for Law, Science, and Technology, Arizona State University, Tempe, USA.

Bartholomew, G. A., 1964. The roles of physiology and behavior in the maintenance of homeostasis in desert environments. Symposium of the Society of Experimental Biology 18:7–29.

Bartholomew, G. A., and T. J. Cade. 1963. The water economy of land birds. Auk 80:504–539.

Bartholomew, G. A., and R. E. MacMillen. 1960. The water requirements of mourning doves and their use of sea water and NaCl solutions. Physiological Journal 33:171–178.

Beale, D. M., and R. C. Holmgren. 1975. Water requirements for pronghorn antelope fawn survival and growth. Utah Division of Wildlife Resources, Salt Lake City, USA.

Beale, D. M., and A. D. Smith. 1970. Forage use, water consumption, and productivity of pronghorn antelope in western Utah. Journal of Wildlife Management 34:570–582.

Benolkin, P. J., and A. C. Benolkin. 1994. Determination of a cost–benefit relationship between chukar populations, hunter utilization and the cost of artificial watering devices. Nevada Department of Wildlife, Reno, USA.

Berry, H. H., W. R. Siegfried, and T. M. Crowe. 1984. Orientation of wildebeest in relation to sun angle and wind direction. Madoqua 13:297–301.

Bleich, V. C. 2005. Politics, promises, and illogical legislation confound wildlife conservation. Wildlife Society Bulletin 33:66 73.

Bleich, V. C., N. G. Andrew, M. J. Martin, G. P. Mulcahy, A. M. Pauli, and S. S. Rosenstock. 2006. Quality of water available to wildlife in desert environments: comparisons among anthropogenic and natural sources in southeastern California. Wildlife Society Bulletin 34:625–630.

Bleich, V. C., R. T. Bowyer, and J. D. Wehausen. 1997. Sexual segregation in mountain sheep: resources or predation? Wildlife Monographs 134:3–50.

Bleich, V. C., J. Coombes, and J. H. Davis. 1982. Horizontal wells as a wildlife habitat improvement technique. Wildlife Society Bulletin 10:324–328.

Bleich, V. C., J. G. Kie, E. R. Loft, T. R. Stephenson, M. W. Oehler Sr., and A. L. Medina. 2005. Managing rangelands for wildlife. Pages 873–897 in C. E. Braun, editor. Techniques for wildlife investigations and management. The Wildlife Society, Bethesda, Maryland, USA.

Bleich, V. C., J. P. Marshal, and N. G. Andrews. 2010. Habitat use by a desert ungulate: predicting effects of water availability on mountain sheep. Journal of Arid Environments 74:638–645.

Bleich, V. C., and R. A. Weaver. 1983. "Improved" sand dams for wildlife habitat management. Journal of Range Management 36:133.

Blong, B., and W. Pollard. 1968. Summer water requirements of desert bighorn in the Santa Rose Mountains, California, in 1965. California Fish and Game Journal 54:289–296.

Bock, C. E., and J. H. Bock. 1999. Response of winter birds to drought and short-duration grazing in southeastern Arizona. Conservation Biology 13:1117–1123.

Boyle, S. A., and A. W. Alldredge. 1984. Pronghorn summer distribution and water availability in the Red Desert, Wyoming. Proceedings of the Biennial Pronghorn Antelope Workshop 11:103–104.

Bradley, W. G. 1963. Water metabolism in desert mammals with special reference to desert bighorn sheep. Desert Bighorn Council Transactions 7:26–39.

Brinnel, H., M. Cabanac, and J. R. S. Hales. 1987. Critical upper levels of body temperature, tissue thermosensitivity, and selective brain cooling in hyperthermia. Pages 209–240 in J. R. S. Hales, and D. A. B. Richards, editors. Heat stress: physical exertion and environment. Excerpta Medica, Amsterdam, Netherlands.

Bristow, K. D., S. A. Dubay, and R. A. Okenfels. 2006. Correlation between free water availability and pronghorn recruitment. Pages 55–62 in J. W. Cain III and P. R. Krausman, editors. Managing wildlife in the Southwest. The Wildlife Society, Tucson, Arizona, USA.

Broyles, B. 1995. Desert wildlife water developments: questioning use in the Southwest. Wildlife Society Bulletin 23:663–675.

Broyles, B., and T. L. Cutler. 1999. Effect of surface water on desert bighorn sheep in the Cabeza Prieta National Wildlife Refuge, southwestern Arizona. Wildlife Society Bulletin 27:1082–1088.

Buechner, H. K. 1950. Life history, ecology, and range use of the pronghorn antelope in Trans-Pecos, Texas. American Midland Naturalist 43:257–354.

Burkett, D. W., and B. C. Thompson. 1994. Wildlife association with human-altered water sources in semiarid vegetation communities. Conservation Biology 8:682–690.

Cain, J. W., III, B. D. Jansen, R. R. Wilson, and P. R. Krausman. 2008b. Potential thermoregulatory advantages of shade use by desert bighorn sheep. Journal of Arid Environments 72:1518–1525.

Cain, J. W., III, P. R. Krausman, J. R. Morgart, B. D. Jansen, and M. P. Pepper. 2008a. Responses of desert bighorn sheep to removal of water sources. Wildlife Monographs 171:1–32.

Campbell, H. 1961. An evaluation of gallinaceous guzzlers for quail in New Mexico. Journal of Wildlife Management 24:21–26.

Campbell, H., D. K. Martin, P. E. Ferkovich, and B. K. Harris. 1973. Effects of hunting and some other environmental factors on scaled quail in New Mexico. Wildlife Monographs 34:1–49.

Carmi, N., B. Pinshow, M. Horowitz, and M. H. Bernstein. 1993. Birds conserve plasma volume during thermal and flight-incurred dehydration. Physiological Zoology 66:829–846.

Carmi-Winkler, N., A. A. Degen, and B. Pinshow. 1987. Seasonal time-energy budgets of free-living chukars in the Negev Desert. Condor 89:594–601.

Chappell, M. A., and G. A. Bartholomew. 1981. Standard operative temperatures and thermal energetics of the antelope ground squirrel *Ammospermophilus leucurus*. Physiological Zoology 54:81–93.

Chevalier, C. D. 1984. Water requirements of free-ranging and captive ringtail cats in the Sonoran Desert. Ph.D. Thesis, Arizona State University, Tempe, USA.

Christensen, G. C. 1996. Chukar (*Alectoris chukar*). The birds of North America no. 258. Academy of Natural Sciences, Philadelphia, Pennsylvania, and the American Ornithologists Union, Washington, D.C., USA.

Clemente, F., R. Valdez, J. L. Holechek, P. J. Zwank, and M. Cardenas. 1995. Pronghorn home range relative to permanent water in southern New Mexico. Southwestern Naturalist 40:38–41.

Coe, S. J., and J. T. Rotenberry. 2003. Water availability affects clutch size in a desert sparrow. Ecology 84:3240–3249.

Cook, J. G., L. L. Irwin, L. D. Bryant, R. A. Riggs, and J. W. Thomas. 1998. Relations of forest cover and condition of elk: a test of the thermal cover hypothesis in summer and winter. Wildlife Monographs 141:1–61.

Cook, J. G., B. K. Johnson, R. C. Cook, R. A. Riggs, T. Delcurto, L. D. Bryant, and L. L. Irwin. 2004. Effects of summer-autumn nutrition and parturition date on reproduction and survival of elk. Wildlife Monographs 155:1–61.

Cook, J. G., L. J. Quinlan, L. L. Irwin, L. D. Bryant, R. A. Riggs, and J. W. Thomas. 1996. Nutrition–growth relations of elk calves during late summer and fall. Journal of Wildlife Management 60:528–541.

Coupland, R. T., editor. 1992. Ecosystems of the world: 8A. Natural grasslands—introduction and Western Hemisphere. Elsevier, New York, New York, USA.

Cutler, T. L. 1996. Water use of two artificial water developments on the Cabeza Prieta National Wildlife Refuge, southwestern Arizona. Thesis, University of Arizona, Tucson, USA.

Czech, B., and P. R. Krausman. 1999. Controversial wildlife management issues in southwestern U.S. wilderness. International Journal of Wilderness 5:22–28.

Dawson, J. W., and R. W. Mannan. 1991. The role of territoriality in the social organization of Harris' hawks. Auk 108:661–672.

Dawson, W. R. 1982. Evaporative losses of water by birds. Comparative Biochemistry and Physiology A 71:495–509.

Dawson, W. R. 1984. Physiological studies of desert birds: present and future considerations. Journal of Arid Environments 7:133–155.

Deblinger, R. D., and A. W. Alldredge. 1991. Influence of free water on pronghorn distribution in a sagebrush/steppe grassland. Wildlife Society Bulletin 19:321–326.

Degen, A. A., B. Pinshow, and P. U. Alkon. 1983. Summer water turnover rates in free-living chukars and sand partridges in the Negev Desert. Condor 85:333–337.

Degen, A. A., B. Pinshow, and P. J. Shaw. 1984. Must desert chukars (*Alectoris chukar sinaica*) drink water? Water influx and body mass changes in response to dietary water content. Auk 101:47–52.

Delgiudice, G. D., and J. E. Rodiek. 1984. Do elk need free water in Arizona? Wildlife Society Bulletin 12:142–146.

DeStefano, S., S. L. Schmidt, and J. C. deVos Jr. 2000. Observations of predator activity at wildlife water developments in southern Arizona. Journal of Range Management 53:255–258.

deVos, J. C., Jr., and W. H. Miller. 2005. Habitat use and survival of Sonoran pronghorn in years with above-average precipitation. Wildlife Society Bulletin 33:35–42.

Douglas, C. L. 1988. Decline of desert bighorn sheep in the Black Mountains of Death Valley. Desert Bighorn Council Transactions 25:36–38.

Douglas, C. L., and D. M. Leslie Jr. 1999. Management of bighorn sheep. Pages 238–262 *in* R. Valdez and P. R. Krausman, editors. Mountain sheep of North America. University of Arizona Press, Tucson, USA.

Edge, W. D., S. L. Olson-Edge, and L. L. Irwin. 1990. Planning for wildlife in national forests: elk and mule deer habitats as an example. Wildlife Society Bulletin 18:87–98.

Edney, E. B. 1977. Water balance in land arthropods. Springer-Verlag, Berlin, Germany.

Epps, C. W., D. R. McCullough, J. D. Wehausen, V. C. Bleich, and J. L. Rechel. 2004. Effects of climate change on population persistence of desert-dwelling mountain sheep in California. Conservation Biology 18:102–113.

Finch, V. A. 1972. Thermoregulation and heat balance of the East African eland and hartebeest. American Journal of Physiology 222:1374–1379.

Fox, L. M., P. R. Krausman, M. L. Morrison, and R. M. Kattnig. 2000. Water and nutrient content of forage in Sonoran pronghorn habitat, Arizona. California Fish and Game 86:216–232.

Fuller, A., P. R. Kamerman, S. K. Maloney, A. Matthee, G. Mitchell, and D. Mitchell. 2005. A year in the thermal life of a free-ranging herd of springbok *Antidorcas marsupialis*. Journal of Experimental Biology 208:2855–2864.

Glading, B. 1943. A self-filling quail watering device. California Fish and Game Journal 29:157–164.

Gill, F. B. 1994. Ornithology. Second edition. W. H. Freeman, New York, New York, USA.

Gionfriddo, J. P., and P. R. Krausman. 1986. Summer habitat use by mountain sheep. Journal of Wildlife Management 50:331–336.

Giuliano, W. M., R. S. Lutz, and R. Patiño. 1994. Physiological responses of northern bobwhite (*Colinus virginianus*) to chronic water deprivation. Physiological Zoology 68:262–276.

Goldstein, D. L. 1984. The thermal environment and its constraint on activity of desert quail in summer. Auk 101:542–550.

Goldstein, D. L., and K. A. Nagy. 1985. Resource utilization by desert quail: time and energy, food and water. Ecology 66:378–387.

Golightly, R. T., and R. D. Ohmart. 1984. Water economy of two desert canids: coyote and kit fox. Journal of Mammalogy 65:51–58.

Guthery, F. S., and N. E. Koerth. 1992. Substandard water intake and inhibition of bobwhite reproduction during drought. Journal of Wildlife Management 56:760–768.

Hailey, T. L. 1967. Reproduction and water utilization of Texas transplanted desert bighorn sheep. Desert Bighorn Council Transactions 11:53–58.

Halloran, A. F., and O. V. Deming. 1958. Water development for desert bighorn sheep. Journal of Wildlife Management 22:1–9.

Hazam, J. E., and P. R. Krausman. 1988. Measuring water consumption of desert mule deer. Journal of Wildlife Management 52:528–534.

Heffelfinger, J. 2006. Deer of the Southwest. Texas A&M University Press, College Station, Texas, USA.

Henry, R. S., and L. K. Sowls. 1980. White-tailed deer of the Organ Pipe Cactus National Monument, Arizona. Technical Report No. 6. National Park Service and University of Arizona, Tucson, Arizona.

Herbeck, L. A., and D. A. Larsen. 1999. Plethodontid salamander response to silvicultural practices in Missouri Ozark Forests. Conservation Biology 13:623–632.

Hervert, J. J., and J. L. Bright. 2005. Adult and fawn mortality of Sonoran pronghorn. Wildlife Society Bulletin 22:43–50.

Hervert, J. J., J. L. Bright, R. S. Henry, L. A. Piest, and M. T. Brown. 2005. Home range and habitat-use patterns of Sonoran pronghorn in Arizona. Wildlife Society Bulletin 33:8–15.

Hervert, J. J., and P. R. Krausman. 1986. Desert mule deer use of water developments in Arizona. Journal of Wildlife Management 50:670–676.

Hoffman, R. W., H. G. Shaw, M. A. Rumble, B. F. Wakeling, C. M. Mollohan, S. D. Schmnitz, R. Engel-Willson, and D. A. Hengel. 1993. Management guidelines for Merriam's wild turkeys. Wildlife Report 18. Colorado Division of Wildlife, Fort Collins, USA.

Hofmeyr, M. D. 1985. Thermal properties of the pelages of selected African ungulates. South African Journal of Zoology 20:179–189.

Hoppeler, H., and E. R. Weibel. 2005. Scaling functions to body size: theories and facts. Journal of Experimental Biology 208:1573–1574.

Hossaini-Hilali, J., S. Benlamlih, and K. Dahlborn. 1994. Effects of dehydration, rehydration, and hyperhydration in the lactating and nonlactating black Moroccan goat. Comparative Biochemistry and Physiology A 109:1017–1026.

Howe, F. P., and L. D. Flake. 1988. Mourning dove movements during the reproductive season in southeastern Idaho. Journal of Wildlife Management 52:477–480.

Hughes, K. S., and N. S. Smith. 1990. Sonoran pronghorn use of habitat in southwest Arizona. Final Report 14-16-009-1564 RWO #6. Arizona Cooperative Fish and Wildlife Research Unit, Tucson, USA.

Hungerford, C. R. 1962. Adaptation shown in selection of food by Gambel's quail. Condor 64:213–219.

Hyde, T. C. 2011. Stable isotopes provide insight into the use of wildlife water developments by resident and migrant birds in the Sonoran Desert of Arizona. M.S. Thesis, University of New Mexico, Albuquerque, USA.

Jarman, P. J. 1973. The free water intake of impala in relation to the water content of their food. East African Agricultural and Forestry Journal 38:343–351.

Jarman, P. J. 1977. Behaviour of topi in a shadeless environment. Zoologica Africana 12:101–111.

Jeffrey, D. E. 1963. Factors influencing elk and cattle distribution on the Willow Creek summer range, Utah. Thesis, Utah State University, Logan, USA.

Jhala, Y. V., R. H. Giles Jr., and A. M. Bhagwat. 1992. Water in the ecophysiology in blackbuck. Journal of Arid Environments 22:261–269.

Kay, R. N. B. 1997. Responses of African livestock and wild herbivores to drought. Journal of Arid Environments 37:683–694.

Kleinhaus, S., B. Pinshow, M. H. Bernstein, and A. A. Degen. 1985. Brain temperature in heat-stressed, water-deprived desert phasianids: sand partridge (*Ammoperdix heyi*) and chukar (*Alectoris chukar sinaica*). Physiological Zoology 58:105–116.

Knudsen, M. F. 1963. A summer waterhole study at Carrizo Spring, Santa Rosa Mountains of Southern California. Desert Bighorn Council Transactions 7:185–192.

Krausman, P. R. 1979. Use of caves by white-tailed deer. Southwestern Naturalist 24:203.

Krausman, P. R., 2002. Introduction to wildlife management: the basics. Prentice Hall, Upper Saddle River, New Jersey, USA.

Krausman, P. R., and E. D. Ables. 1981. Ecology of the Carmen Mountains white-tailed deer. Scientific Monograph Series No. 15. U.S. Department of the Interior, National Park Service, Washington, D.C., USA.

Krausman, P. R., and B. Czech. 1998. Water developments and desert ungulates. Pages 138–154 in Proceedings of a symposium on environmental, economic, and legal issues related to rangeland water developments, 13–15 November 1997, Tempe, Arizona. Center for Law, Science, and Technology, Arizona State University, Tempe, USA.

Krausman, P. R., and R. C. Etchberger. 1995. Response of desert ungulates to a water project in Arizona. Journal of Wildlife Management 59:292–300.

Krausman, P. R., and B. D. Leopold. 1986. Habitat components for desert bighorn sheep in the Harquahala Mountains, Arizona. Journal of Wildlife Management 50:504–508.

Krausman, P. R., S. S. Rosenstock, and J. W. Cain III. 2006. Developed waters for wildlife: science, perception, values, and controversy. Wildlife Society Bulletin 34:563–569.

Krausman, P. R., S. Torres, L. L. Ordway, J. J. Hervert, and M. Brown. 1985. Diel activity of ewes in the Little Harquahala Mountains, Arizona. Desert Bighorn Council Transactions 29:24–26.

Kubly, D. M. 1990. Limnological features of desert mountain rock pools. Pages 103–120 in G. K. Tsukamoto and S. J. Stiver, editors. Proceedings of the wildlife water development symposium. Nevada Chapter, The Wildlife Society, U.S. Department of the Interior, Bureau of Land Management, Washington, D.C., and Nevada Department of Wildlife, Reno, USA.

Kuenzi, A. J. 2001. Spatial and temporal patterns of bat use of water developments in southern Arizona. Dissertation, University of Arizona, Tucson, USA.

Larsen, R. T., J. A. Bissonette, J. T. Flinders, M. B. Hooten, and T. L. Wilson. 2010. Summer spatial patterning of chukars in relation to free water in western Utah. Landscape Ecology 25:135–145.

Larsen, R. T., J. T. Flinders, D. L. Mitchell, E. R. Perkins, and D. G. Whiting. 2007. Chukar watering patterns and water site selection. Journal of Range Management 60:559–565.

Lautier, J. K., T. V. Dailey, and R. D. Brown. 1988. Effect of water restriction on feed intake of white-tailed deer. Journal of Wildlife Management 52:602–606.

Leopold, A. 1933. Game management. Charles Scribner's Sons, New York, New York, USA.

Lesicka, L. M., and J. J. Hervert. 1995. Low maintenance water developments for arid environments: concepts, materials, and techniques. Pages 52–57 in D. P. Young Jr., R. Vinzant, and M. D. Strickland, editors. Wildlife water development. Water for Wildlife Foundation, Lander, Wyoming, USA.

Leslie, D. M., Jr., and C. L. Douglas. 1979. Desert bighorn sheep of the River Mountains, Nevada. Wildlife Monographs 66:1–56.

Longhurst, W. M., N. F. Baker, G. E. Connolly, and R. A. Fisk. 1970. Total body water and water turnover in sheep and deer. American Journal of Veterinary Research 31:673–677.

Louw, G. N. 1984. Water deprivation in herbivores under arid conditions. Pages 106–126 in F. M. C. Gilchrist and R. I. Mackie, editors. Herbivore nutrition in the subtropics and tropics. Science Press, Craighall, South Africa.

Louw, G. N. 1993. Physiological animal ecology. Longman Scientific and Technical, Burnt Mill, United Kingdom.

Louw, G. N., and M. Seely. 1982. Ecology of desert organisms. Longman Group, Burnt Mill, United Kingdom.

Luskick, S., B. Battersby, and M. Kelty. 1978. Behavioral thermoregulation: orientation toward the sun in herring gulls. Science 200:81–83.

Lynn, J. C., C. L. Chambers, and S. S. Rosenstock. 2006. Use of wildlife water developments by birds in southwest Arizona during migration. Wildlife Society Bulletin 34:592–601.

Lynn, J. C., S. S. Rosenstock, and C. L. Chambers. 2008. Avian use of desert wildlife water developments as determined by remote videography. Western North American Naturalist 68:107–112.

Macfarlane, W. V., B. Howard, H. Haines, P. J. Kennedy, and C. M. Sharpe. 1971. Hierarchy of water and energy turnover of desert mammals. Nature 234:483–484.

Mackie, R. J. 1970. Range ecology and relations of mule deer, elk, and cattle in the Missouri River breaks, Montana. Wildlife Monographs 20:1–79.

Maclean, G. L. 1996. Ecophysiology of desert birds. Springer-Verlag, Berlin, Germany.

Maghini, M. T., and N. S. Smith. 1990. Water use and diurnal seasonal ranges of Coues white-tailed deer. Pages 21–34 in P. R. Krausman and N. S. Smith, editors. Managing wildlife in the Southwest: a symposium. Arizona Cooperative Wildlife Research Unit and School of Renewable Natural Resources, University of Arizona, Tucson, USA.

Maloiy, G. M. O. 1970. Water economy of the Somali donkey. American Journal of Physiology 219:1522–1527.

Maloiy, G. M. O. 1973. Water metabolism of East African ruminants in arid and semi-arid regions. Journal of Animal Breeding and Genetics 90:219–228.

Maloiy, G. M. O., and D. Hopcraft. 1971. Thermoregulation and water relations of two East African antelopes: the hartebeest and impala. Comparative Biochemistry and Physiology A 38:525–534.

Maloiy, G. M. O., W. V. Macfarlane, and A. Shkolnik. 1979. Mammalian herbivores. Pages 185–209 in G. M. O. Maloiy, editor. Comparative physiology of osmoregulation in animals. Volume 2. Academic, London, United Kingdom.

Marcum, C. L. 1975. Summer-fall habitat selection and use by a western Montana elk herd. Ph.D. Dissertation, University of Montana, Missoula, USA.

Mares, M. A. 1983. Desert rodent adaptation and community structure. Great Basin Naturalist Memoirs 7:30–43.

Marshal, J. P., V. C. Bleich, P. R. Krausman, M. L. Reed, and N. G. Andrew. 2006a. Factors affecting habitat use and distribution of desert mule deer in an arid environment. Wildlife Society Bulletin 34:609–619.

Marshal, J. P., P. R. Krausman, V. C. Bleich, S. S. Rosenstock, and W. B. Ballard. 2006b. Gradients of forage biomass and ungulate use near wildlife water developments. Wildlife Society Bulletin 34:620–626.

Martineau, L., and J. Larochelle. 1988. The cooling power of pigeon legs. Journal of Experimental Biology 136:193–208.

Mayhew, W. W. 1968. Biology of desert amphibians and reptiles. Pages 195–365 in G. W. Brown, editor. Desert biology. Academic Press, New York, New York, USA.

McCorquodale, S. M., K. J. Raedeke, and R. D. Taber. 1986. Elk habitat use patterns in the shrub-steppe of Washington. Journal of Wildlife Management 50:664–669.

McNab, B. K. 2002. The physiological ecology of vertebrates. Cornell University Press, Ithaca, New York, USA.

McNabb, F. M. A. 1969. A comparative study of water balance in three species of quail. I. Water turnover in the absence of temperature stress. Comparative Biochemistry and Physiology 28:1045–1058.

Mendoza, V. J. 1976. The bighorn sheep of the state of Sonora. Desert Bighorn Council Transactions 20:25–26.

Mensch, J. L. 1969. Desert bighorn sheep (Ovis canadensis nelsoni) losses in a natural trap tank. California Fish and Game 55:237–238.

Miller, G. D., M. H. Cochran, and E. L. Smith. 1984. Nighttime activity of desert bighorn sheep. Desert Bighorn Council Transactions 28:23–25.

Mitchell, D. 1977. Physical basis of thermoregulation. Pages 1–27 in D. Robertshaw, editor. International review of physiology. Volume 15. Environmental physiology II. University Park Press, Baltimore, Maryland, USA.

Mitchell, D., S. K. Maloney, C. Jessen, H. P. Laburn, P. R. Kamerman, G. Mitchell, and A. Fuller. 2002. Adaptive heterothermy and selective brain cooling in arid-zone mammals. Comparative Biochemistry and Physiology B 131:571–585.

Morgart, J. R., J. J. Hervert, P. R. Krausman, J. L. Bright, and R. S. Henry. 2005. Sonoran pronghorn use of anthropogenic and natural water sources. Wildlife Society Bulletin 33:51–60.

Mullens, B. A. 1991. Integrated management of Culicoides variipennis: a problem of applied ecology. Pages 896–905 in T. E. Walton and B. I. Osburn, editors. Bluetongue, African horse sickness, and related orbiviruses. CRC Press, Boca Raton, Florida, USA.

Nagy, K. A. 1987. Field metabolic rate and food requirement scaling in mammals and birds. Ecological Monographs 57:111–128.

Nagy, K. A. 2004. Water economy of free-living desert animals. Pages 291–297 in Animals and environments. International Congress Series 1275. Elsevier, Amsterdam, Netherlands.

Nagy, K. A., and M. J. Gruchacz. 1994. Seasonal water and energy metabolism of the desert-dwelling kangaroo rat (Dipodomys merriami). Physiological Ecology 67:1461–1478.

Nagy, K. A., and C. C. Peterson. 1988. Scaling of water flux rates in animals. University of California Press, Berkeley, USA.

Nelson, J. R., and D. G. Burnell. 1976. Elk-cattle competition in central Washington. Pages 71–83 in B. F. Roche, editor. Range multiple use management. University of Idaho, Moscow, USA.

O'Brien, C. S., R. B. Waddell, S. S. Rosenstock, and M. J. Rabe. 2006. Wildlife use of water catchments in southwestern Arizona. Wildlife Society Bulletin 34:582–591.

Ockenfels, R. A., A. Alexander, C. L. D. Ticer, and W. K. Carrel. 1994. Home ranges, movement patterns, and habitat selection of pronghorn in central Arizona. Technical Report No. 13. Arizona Game and Fish Department, Phoenix, USA.

Ockenfels, R. A., D. E. Brooks, and C. H. Lewis. 1991. General ecology of Coues white-tailed deer in the Santa Rita Mountains. Technical Report No. 6. Arizona Game and Fish Department, Phoenix, USA.

O'Gara, B. W., and J. D. Yoakum. 1992. Pronghorn management guidelines: a compendium of biological and management principles and practices to sustain pronghorn populations and habitat from Canada to Mexico. Proceedings of the Biennial Pronghorn Antelope Workshop 15:1–101.

Ohmart, R. D., and E. L. Smith. 1971. Water deprivation and use of sodium chloride solutions by Vesper sparrows (Pooecetes gramineus). Condor 73:364–366.

Ordway, L. L., and P. R. Krausman. 1986. Habitat use by desert mule deer. Journal of Wildlife Management 50:677–683.

Owen-Smith, N. 1998. How high ambient temperature affects the daily activity and foraging time of a subtropical ungulate, the greater kudu (Tragelaphus strepsiceros). Journal of Zoology 246:183–192.

Parker, A. H., and E. T. F. Witkowski. 1999. Long-term impacts of abundant perennial water provision for game on herbaceous vegetation in a semi-arid African savannah woodland. Journal of Arid Environments 41:309–321.

Parker, K. L., and C. T. Robbins. 1984. Thermoregulation in mule deer and elk. Canadian Journal of Zoology 62:1409–1422.

Pasitschniak-Arts, M., and F. Messier. 1998. Effects of edges and habitats on small mammals in a prairie ecosystem. Canadian Journal of Zoology 76:2020–2025.

Payne, N. F., and F. C. Bryant. 1998. Wildlife habitat management of forestlands, rangelands, and farmlands. Krieger, Malabar, Florida, USA.

Phillips, P. K., and J. E. Heath. 1995. Dependency of surface temperature regulation on body size in terrestrial mammals. Journal of Thermal Biology 20:281–289.

Picard, K., D. W. Thomas, M. Festa-Bianchet, F. Belleville, and A. Laneville. 1999. Differences in the thermal conductance of tropical and temperate bovid horns. Ecoscience 6:148–158.

Porter, W. P., and D. M. Gates. 1969. Thermodynamic equilibria of animals with environment. Ecological Monographs 39:227–244.

Rabe, M. J., and S. S. Rosenstock. 2005. Effects of water size and type on bat captures in the lower Sonoran Desert. Western North American Naturalist 65:87–90.

Rautenstrauch, K. R., and P. R. Krausman. 1989. Influence of water availability on movements of desert mule deer. Journal of Mammalogy 70:197–201.

Robbins, C. T. 1983. Wildlife feeding and nutrition. Academic Press, Orlando, Florida, USA.

Robertshaw, D., and C. R. Taylor. 1969. A comparison of sweat gland activity in eight species of East African bovids. Journal of Physiology 203:135–143.

Robinson, J. R. 1957. Functions of water in the body. Proceedings of the Nutritional Society 16:108–112.

Rosen, P. C., and C. R. Schwalbe. 1998. Using managed waters for conservation of threatened frogs. Pages 180–202 in Proceedings of a symposium on environmental, economic, and legal issues related to rangeland water developments, 13–15 November 1997, Tempe, Arizona. Center for Law, Science, and Technology, Arizona State University, Tempe, USA.

Rosenstock, S. S., W. B. Ballard, and J. C. deVos. 1999. Benefits and impacts of wildlife water developments. Journal of Range Management 52:302–311.

Rosenstock, S. S., J. J. Hervert, V. C. Bleich, and P. R. Krausman. 2001. Muddying the water with poor science: a reply to Broyles and Cutler. Wildlife Society Bulletin 29:734–743.

Rosenstock, S. S., M. J. Rabe, C. S. O'Brien, and R. B. Waddell. 2004. Studies of wildlife water developments in southwestern Arizona: wildlife use, water quality, wildlife diseases, wildlife mortalities, and influences in native pollinators. Technical Guidance Bulletin No. 8. Arizona Game and Fish Department, Phoenix, USA.

Sargeant, G. A., L. E. Eberhardt, and J. M. Peek. 1994. Thermoregulation by mule deer (Odocoileus hemionus) in arid rangelands of southcentral Washington. Journal of Mammalogy 75:536–544.

Schemnitz, S. D. 1994. Scaled quail. Birds of North America 106. Academy of Natural Sciences, Philadelphia, Pennsylvania, and the American Ornithologists Union, Washington, D.C., USA.

Schmidt, S. L., and D. C. Dalton. 1995. Bats of the Madrean Archipelago (Sky Islands): current knowledge, future directions. Pages 274–287 in L. F. Debano, G. J. Gottfried, R. H. Hamre, C. B. Edminster, P. F. Ffolliot, and A. Ortega-Rubio, technical coordinators. Biodiversity of the Madrean Archipelago: Sky Islands of the southwestern United States and northwestern Mexico. General Technical Report RM-264. U.S. Department of Agriculture Forest Service, Fort Collins, Colorado, USA.

Schmidt, S. L., and S. DeStefano. 1999. Use of water developments by nongame wildlife in the Sonoran Desert of Arizona. Arizona Cooperative Fish and Wildlife Research Unit, Tucson, USA.

Schmidt-Nielsen, K. 1979. Desert animals: physiological problems of heat and water. Dover, New York, New York, USA.

Schmidt-Nielsen, K., E. C. Crawford, A. E. Newsome, K. S. Rawson, and H. T. Hammel. 1967. Metabolic rate of camels: effect of body temperature and dehydration. American Journal of Physiology 212:341–346.

Schmidt-Nielsen, K., B. Schmidt-Nielsen, S. Jarnum, and T. R. Houpt. 1957. Body temperature of the camel and its relation to water economy. American Journal of Physiology 188:103–112.

Schwimmer, M., and D. Schwimmer. 1968. Medical aspects of phycology. Pages 279–358 in D. F. Jackson, editor. Algae, man, and the environment. Syracuse University Press, Syracuse, New York, USA.

Seager, R., M. Ting, I. Held, Y. Kushnir, J. Lu, et al. 2007. Model predictions of an imminent transition to a more arid climate in southwestern North America. Science 316:1181–1184.

Senzota, R. B. M., and G. Mtahko. 1990. Effect on wildlife of a water-hole in Mikumi National Park, Tanzania. African Journal of Ecology 28:147–151.

Shaw, H. G., and C. Mallohan. 1992. Merriam's turkey. Pages 331–349 in J. O. Dickson, editor. The wild turkey: biology and its management. Stackpole Books, Harrisburg, Pennsylvania, USA.

Silanikove, N. 1994. The struggle to maintain hydration and osmoregulation in animals experiencing severe dehydration and rapid rehydration: the story of ruminants. Experimental Physiology 79:281–300.

Simpson, D. C., L. A. Harveson, C. E. Brewer, R. E. Walser, and A. R. Sides. 2007. Influence of precipitation on pronghorn demography in Texas. Journal of Wildlife Management 71:906–910.

Sokolov, V. E. 1982. Mammal skin. University of California Press, Berkeley, USA.

Strohmeyer, D. C., and J. M. Peek. 1996. Wapiti home range and movement patterns in a sagebrush desert. Northwest Science 70:79–87.

Sundstrom, C. 1968. Water consumption by pronghorn antelope and distribution related to water in Wyoming's Red Desert. Biennial Pronghorn Antelope States Workshop 3:39–46.

Swift, P. K., J. D. Wehausen, H. B. Ernest, R. S. Singer, A. M. Pauli, H. Kinde, T. E. Rocke, and V. C. Bleich. 2000. Desert bighorn sheep mortality due to presumptive type C botulism in California. Journal of Wildlife Diseases 36:184–189.

Taylor, C. R. 1968. Hygroscopic food: a source of water for desert antelopes. Nature 219:181–182.

Taylor, C. R. 1969a. The eland and the oryx. Scientific American 220:88–95.

Taylor, C. R. 1969b. Metabolism, respiratory changes and water balance of an antelope, the eland. American Journal of Physiology 217:317–320.

Taylor, C. R. 1970a. Dehydration and heat: effects on temperature regulation of East African ungulates. American Journal Physiology 219:1136–1139.

Taylor, C. R. 1970b. Strategies of temperature regulation: effects on evaporation in East African ungulates. American Journal Physiology 219:1131–1135.

Taylor, C. R., and C. P. Lyman. 1972. Heat storage in running antelopes: independence of brain and body temperatures. American Journal of Physiology 222:114–117.

Taylor, D. A. R., and M. D. Tuttle. 2007. Water for wildlife: a handbook for ranchers and range managers. Bat Conservation International, Austin, Texas, USA.

Thomas, J. W., R. J. Miller, C. Maser, R. G. Anderson, and B. E. Carter. 1979. Plant communities and successional states. Pages 22–39 in J. W. Thomas, editor. Wildlife habitats in managed forests: the Blue Mountains of Oregon and Washington. Agricultural Handbook 533. U.S. Department of Agriculture Forest Service, Washington, D.C., USA.

Thrash, I. 1998. Impact of water provision on herbaceous vegetation in Kruger National Park, South Africa. Journal of Arid Environments 38:437–450.

Ticer, C. L. D., and W. H. Miller. 1994. Pronghorn fawn bed site selection in a semidesert grassland community of central Arizona. Pronghorn Workshop Proceedings 6:86–103.

Tieleman, I. B., J. B. Williams, and P. Bloomer. 2003. Adaptation of metabolism and evaporative water loss along an aridity gradient. Proceedings of the Royal Society B 270:207–214.

Tieleman, I. B., J. B. Williams, and M. E. Buschur. 2002. Physiological adjustments to arid and mesic environments in larks (*Alaudidae*). Physiological and Biochemical Zoology 75:305–313.

Trzcinski, M. K., L. Fahrig, and G. Merriam. 1999. Independent effects of forest cover and fragmentation on the distribution of forest breeding birds. Ecological Applications 9:586–593.

Tull, J. C., P. R. Krausman, and R. J. Steidl. 2001. Bed-site selection by desert mule deer in southern Arizona. Southwestern Naturalist 46:354–357.

Turner, J. C., Jr. 1973. Water, energy, and electrolyte balance in the desert bighorn sheep, *Ovis canadensis*. Dissertation, University of California, Riverside, USA.

Turner, J. C., C. L. Douglas, C. R. Hallum, P. R. Krausman, and R. R. Ramey. 2004. Determination of critical habitat for the endangered Nelson's bighorn sheep in southern California. Wildlife Society Bulletin 32:427–448.

Tuttle, S. R., C. L. Chambers, and T. L. Theimer. 2006. Effects of livestock water trough modifications on bat use in northern Arizona. Wildlife Society Bulletin 34:602–608.

Waddell, R. B., C. S. O'Brien, and S. S. Rosenstock. 2007. Bighorn use of a developed water in southwestern Arizona. Desert Bighorn Council Transactions 49:8–17.

Walsberg, G. E. 1983. Coat color and solar heat gain in animals. BioScience 33:88–91.

Walsberg, G. E. 2000. Small mammals in hot deserts: some generalizations revisited. BioScience 50:109–120.

Walsberg, G. E., G. S. Campbell, and J. R. King. 1978. Animal coat color and radiative heat gain: a re-evaluation. Journal of Comparative Physiology 126:211–222.

Warrick, G. D., and P. R. Krausman. 1989. Barrel cactus consumption by desert bighorn sheep. Southwestern Naturalist 34:483–486.

Watts, T. J. 1979. Status of the Big Hatchet desert sheep population, New Mexico. Desert Bighorn Council Transactions 23:92–94.

Weathers, W. W. 1972. Thermal panting in domestic pigeons, *Columba livia*, and the barn owl, *Tylo alba*. Journal of Comparative Physiology A 79:79–84.

Whitcomb, R. F., C. S. Robbins, J. F. Lynch, B. L. Whitcomb, M. K. Klimbiewicz, and D. Bystrak. 1981. Effects of forest fragmentation on avifauna of the eastern deciduous forest. Pages 125–205 *in* R. L. Burgess and D. M. Sharpe, editors. Forest island dynamics in man-dominated landscapes. Springer-Verlag, New York, New York, USA.

White, C. R., T. M. Blackburn, G. R. Martin, and P. J. Butler. 2007. Basal metabolic rate of birds is associated with habitat temperature and precipitation, not primary productivity. Proceedings of the Royal Society B 274:287–293.

Williams, J. B., M. M. Pacelli, and E. J. Braun. 1991. The effect of water deprivation on renal function in conscious unrestrained Gambel's quail (*Callipepla gambelii*). Physiological Zoology 64:1200–1216.

Williams, J. B., and B. I. Tieleman. 2002. Ecological and evolutionary physiology of desert birds: a progress report. Integrative and Comparative Biology 42:68–75.

Williams, J. B., and B. I. Tieleman. 2005. Physiological adaptation in desert birds. BioScience 55:416–425.

Williamson, D. T. 1987. Plant underground storage organs as a source of moisture for Kalahari wildlife. African Journal of Ecology 25:63–64.

Wilson, L. O. 1971. The effect of free water on desert bighorn home range. Desert Bighorn Council Transactions 15:82–89.

Wolf, B. O. 2010. The use of water developments by the bird and bat communities on the Kofa National Wildlife Refuge, Arizona. Report to Arizona Game and Fish Department. University of New Mexico, Albuquerque, USA.

Wolf, B. O., and G. E. Walsberg. 1996. Respiratory and cutaneous evaporative water loss at high environmental temperatures in a small bird. Journal of Experimental Biology 199:451–457.

Wolf, B. O., K. M. Wooden, and G. E. Walsberg. 1996. The use of thermal refugia by two small desert birds. Condor 98:424–428.

Wood, J. E., T. S. Bickle, W. Evans, J. C. Germany, and V. W. Howard Jr. 1970. The Fort Stanton mule deer herd: some ecological and life history characteristics with special emphasis on the use of water. Agricultural Experiment Station Bulletin 567. New Mexico State University, Las Cruces, USA.

Woodall, P. F., and J. D. Skinner. 1993. Dimensions of the intestine, diet, and faecal water loss in some African antelope. Journal of Zoology 229:457–471.

Wright, J. T. 1959. Desert wildlife. Wildlife Bulletin No. 6. Arizona Game and Fish Department, Phoenix, USA.

Yoakum, J. D. 1978. Pronghorn. Pages 103–121 *in* J. L. Schmidt and D. L. Gilbert, editors. Big game of North America. Stackpole Books, Harrisburg, Pennsylvania, USA.

Yoakum, J. D. 1994. Water requirements for pronghorn. Proceedings of the Pronghorn Antelope Workshop 16:143–157.

Yoakum, J. D. 2004. Habitat characteristics and requirements. Pages 409–445 *in* B. W. O'Gara and J. D. Yoakum, editors. Pronghorn ecology and management. University Press of Colorado, Boulder, and Wildlife Management Institute, Washington, D.C., USA.

Zektser, S., H. A. Loáiciga, and J. T. Wolf. 2005. Environmental impacts of groundwater overdraft: selected case studies in the southwestern United States. Environmental Geology 47:396–404.

Zervanos, S. M., and G. I. Day. 1977. Water and energy requirements of captive and free-living collared peccaries. Journal of Wildlife Management 41:527–532.

PREDATOR–PREY RELATIONSHIPS AND MANAGEMENT

CLINT BOAL AND WARREN B. BALLARD

INTRODUCTION

Predator–prey interactions are a primary mechanism of evolutionary change. As predators develop adaptations to increase their success in capturing prey, prey species correspondingly develop adaptations to avoid capture, leading to new adaptations by predators, and so on. Dawkins and Krebs (1979) likened this process to an evolutionary arms race. Additional work on prey–predator relationships suggests that prey adapt to predator strategies more rapidly, through a mechanism known as the life-dinner principle (Dawkins and Krebs 1979). When predators are successful, the individual prey animal's genetic information is lost forever, so by removing vulnerable individuals from a prey species population, predators have an influence on the behavior and appearance of that species (Newton 1998). Selective pressure is greater on prey species. During any given predation attempt, the individual prey's failure results in death. In contrast, failure by the predator (over the short term) results only in missing a meal (Dawkins and Krebs 1979). But repeated failures by a predator will lead to starvation and death. Therefore repeated failures similarly result in selection against inferior predators, but the evolutionary penalty is substantially greater for the prey species (Newton 1998).

Although direct predation is an important influence of evolutionary processes and biological community dynamics, the function of predator–prey relationships is more complex than just influences of direct mortality (Lima 1998, Lima and Steury 2005). Predators can also influence prey behavior and trophic dynamics through indirect (i.e., disturbance) effects, a phenomenon known as the ecology of fear (Brown et al. 2001, Preisser et al. 2005). The fear of being captured and consumed can dictate behaviors of prey animals. Ecology of fear can lead to indirect reductions in the overall well-being of the prey, which can subsequently reduce the number of offspring that prey produce. For example, Navarrete (2011) noted that sandhill cranes (*Grus canadensis*) in west Texas spent less time feeding and more time being watchful in agriculture fields close to wind farms. She deduced that the cranes were vigilant around the farm trucks, which they associated with a predator (i.e., hunters).

Although food and other resources ultimately limit prey populations (Leopold 1933), in some situations, predators can limit prey populations below levels that the resources would otherwise allow (Craighead and Craighead 1956, Newton 1993). For this to occur, some predation mortality must be additive to, not a replacement of, other mortality factors. When a prey population is near the maximum size the environment can support (i.e., its carrying capacity), competition occurs between individuals of the same species (i.e., intraspecific competition) for such resources as food or nest sites. In these situations, predation mortality is likely to replace other forms of mortality, such as disease or starvation. In contrast, when a prey population is below carrying capacity with little intraspecific competition, predation can add to other sources of mortality and further reduce the prey population. Conservationists are especially concerned when prey species are part of small, fragmented populations (Macdonald et al. 1999).

In this chapter we provide a general overview of predator–prey relationships. We begin with a brief history of predator–prey theory. We then discuss the current understanding of these relationships and provide examples where predation has, and has not, been important in population dynamics of predators and prey. We focus on large carnivore–ungulate and avian predator–avian prey relationships, because these are the primary focus of most predator- and prey-related wildlife management efforts.

PREDATORS

Many animals engage in opportunistic carnivory. For example, traditional herbivores such as white-tailed deer (*Odocoileus virginianus*) and squirrels (Sciuridae) have been reported to consume songbird eggs and nestlings. The grasshopper mouse (*Onychomys leucogaster*) is a well-known predator of insects, other mice, and reptiles. Most birds are to some extent predacious. Common nighthawks (*Chordeiles minor*) and northern mockingbirds (*Mimus polyglottos*) consume insects, and American robins (*Turdus migratorius*) forage extensively on worms. By definition, these are all acts of predation. However, when we discuss predatory animals and predator–prey relationships, we usually focus on those species that capture and consume other vertebrates. Among mammals, these are primarily members of the order Carnivora, such as the canids, or wolves (*Canis lupus*) and coyotes (*C. latrans*); felids, or bobcats (*Lynx rufus*) and mountain lions (*Puma concolor*); mustelids, or martens (*Martes americana*) and badgers (*Taxidea taxus*); and ursids, or bears (*Ursus* spp.). Among birds, raptors are the primary predators, which include hawks, falcons, eagles (Falconiformes), and owls (Strigiformes). However, predatory birds also include the hawk-like skuas and jaegers (*Stercorarius* spp.), the fish-eating species such as herons and egrets (i.e., Aredeidae), cormorants (*Phalacrocorax* spp.), terns (e.g., *Sterna* spp.), many species of cuckoos—for example, greater roadrunners (*Geococcyx californianus*)—and even songbirds such as shrikes (*Lanius* spp.). Among the reptiles, virtually every snake and lizard is predatory.

Changing Views on Predation

Historically, the principles influencing wildlife management were that higher numbers of game species were inevitably desirable, and that predators limited populations of game species. Managers undertook regular efforts toward predator control without completely understanding predator–prey relationships. This approach began to change in the mid-1930s, when researchers began scientifically investigating predator–prey relationships. Leopold (1933) proposed that ecosystem processes produced surpluses of game animals, and implicitly referenced density-dependent and density-independent factors in population limitations, all of which are important in understanding predator–prey relationships. Leopold (1933) considered the main components of predation to be density of prey, density of predators, predator food preference, physical condition of prey, and abundance of alternative prey species. Errington (1934) suggested that predation limited a prey species only when its population size exceeded what the local habitat could support in terms of cover and food (i.e., carrying capacity). He later elaborated with the idea of a doomed surplus of individual prey animals that were fated to die (by predation, starvation, or other causes) because of limited resources (Errington 1946). The key component of his premise was that predation occurred—and could occur at high rates—but could hypothetically have no influence on the subsequent breeding population. Essentially, the prevailing view of researchers was that increased productivity compensated for predation. However, few studies directly examined predator–prey relationships. Murie's (1940) study of coyotes in Yellowstone was one of the first to provide fact-based justification for termination of the predator eradication program conducted by the National Park Service. Murie's (1944) subsequent research of wolves at Mount McKinley, Alaska, involved the first detailed observational studies of wolves and again provided data supporting termination of eradication—but not control—programs. In a landmark study, Craighead and Craighead (1956) reported strong evidence that raptor predation on prey populations was typically continuous, but occurred in proportion to prey densities. They also reported evidence of situations where predators can regulate prey populations at levels below what would be allowable by habitat conditions, such as when prey populations were low in numbers for other reasons. Simply counting the number of prey animals killed would not provide a satisfactory understanding of predator–prey relationships; rather, one must understand what factors make an individual vulnerable to predation in the first place.

In the context of applied wildlife management, the study of predator–prey relationships usually focuses on the influences of predator populations on game species—for example, wolves as predators of elk (*Cervus canadensis*)—on species of conservation concern—for example, rat snake (*Elaphe* spp.) predation on red-cockaded woodpeckers (*Picoides borealis*)—or economic concern—for example, tern predation of salmon (*Oncorhynchus* spp.) smolt—and depredation issues—for example, herons at aquaculture facilities. More recently, there has been an increased awareness of the importance of the interactive role of all members in biotic communities, and of efforts for conservation at the ecosystem level rather than just for a particular species of interest. This shift included conservation of predators and prey as part of functioning ecosystems (Fascione et al. 2004, Reynolds et al. 2006).

The complex interactions among predators and their prey are a fascinating aspect of wildlife ecology. The secretive and elusive behaviors of most predators, and their often-low population densities, make it challenging to study their role in ecological communities. It is uncommon to witness actual predation events; much of the data on prey use are compiled indirectly by analyzing prey remains at kill sites, feces (scat), and prey delivered to nests, and more recently by isotopic analysis of predator tissues. Such approaches are subject to substantial bias and to limitations in the interpretation of predator influences on prey at the population level. Furthermore, such data are based only on what the predator succeeded in capturing. Predators are often logistically challenging (and expensive) to study, so few studies examine the range of prey animals that escaped capture or predator success and failure rates (Temple 1987, Roth et al. 2006). For these reasons, understanding the role of predation in ecosystem processes has often been elusive and incomplete. Furthermore, as the complexity of predator–prey relationships increases, our capacity to fully understand them decreases. In such cases, ecologists often resort to complex mathematical models (Boyce 2000, Eberhardt et al. 2003). Although models are valuable in helping wildlife biologists attempt to make sense of complex systems, they tend to be simplistic surrogates for what actually occurs in nature.

Predator Behavior

An important aspect of predator–prey relationships is territoriality among predators. Also known as intraspecific tolerance thresholds, territoriality can keep predator numbers below what the local prey population can support, which can result in no increase in the predator population beyond a level

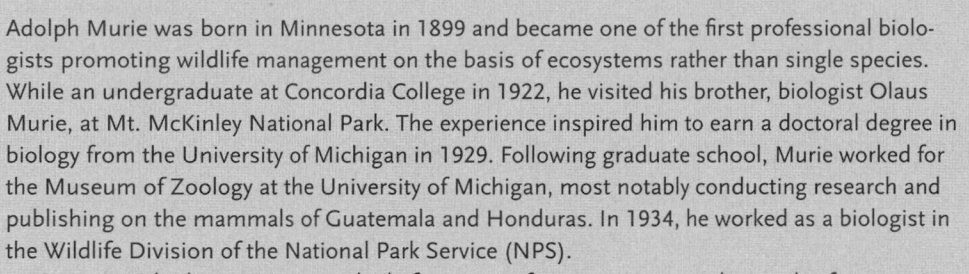

ADOLPH MURIE (1899–1974)

Adolph Murie was born in Minnesota in 1899 and became one of the first professional biologists promoting wildlife management on the basis of ecosystems rather than single species. While an undergraduate at Concordia College in 1922, he visited his brother, biologist Olaus Murie, at Mt. McKinley National Park. The experience inspired him to earn a doctoral degree in biology from the University of Michigan in 1929. Following graduate school, Murie worked for the Museum of Zoology at the University of Michigan, most notably conducting research and publishing on the mammals of Guatemala and Honduras. In 1934, he worked as a biologist in the Wildlife Division of the National Park Service (NPS).

As an NPS biologist, Murie studied of a variety of species. However, his study of coyotes in the Yellowstone region was groundbreaking as one of the first science-based examinations of predator ecology in natural ecosystems. Assigned the task of determining the importance of coyote predation as a limiting factor of deer and elk populations, and after detailed food habitat analysis of coyotes, Murie concluded that it was inadequate winter range, not coyote predation, that limited big game species. Recognizing coyotes as important components of ecosystems, his resulting book, *Ecology of the Coyote in Yellowstone*, published in 1940, met with controversy by going against popularly held beliefs. However, it was one of the first science-based arguments against the practice of predator control and eradication efforts commonplace within the NPS at the time.

Murie was reassigned to Mt. McKinley (now Denali) National Park in 1939, and became the first biologist to study wolves in their natural habitat. Murie's detailed field observations revealed that wolves, weather, and disease were all significant influences on declines of the Dall sheep population. Murie recommended that wolf control continue until the sheep population began to recover. However, rather than broad-based control, Murie led efforts targeted to specific wolves and packs to meet management objectives. His subsequent book, *The Wolves of Mt. McKinley*, published in 1944, is considered a classic and was foundational in changing the commonly held perspectives of predators in natural systems. He was ultimately presented with the NPS's Distinguished Service Award, and received the John Burroughs Medal for his book *A Naturalist in Alaska*. In recognition of his role in both understanding and managing park ecosystems, the NPS dedicated the Murie Science and Learning Center at Denali National Park to Adolph Murie in 2004.

Photo courtesy of the Murie Center Archives

functional response asymptote. For example, Jedrzejewski and Jedrzejewski (1996) reported that tawny owls (*Strix aluco*) were dependent upon small mammals for successful reproduction, but increases in the prey population did not increase owl nesting density because of intraspecific spacing set by the owl's territorial behavior.

Mammalian predators are typically territorial year-round. Wolf packs are social, but they have strongly defined and defended territorial boundaries. Mountain lions are solitary predators that hold territories year-round, but male mountain lions hold a larger territory that can encompass multiple female territories. In contrast, territoriality among avian predators occurs primarily during the breeding season; many avian predators congregate during migration and winter. Broad-winged hawks (*Buteo platypterus*) aggregate in groupings of thousands to migrate to their wintering grounds in South America, and wintering ferruginous hawks (*B. regalis*) congregate at prairie dog (*Cynomys* spp.) colonies. Furthermore, some avian predators (e.g., herons and cormorants) are colonial nesters, but such high nesting densities do not result in intraspecific territorial behavior that would keep predation rates low. The presence of multiple predator species in a given area can result in interspecific competition for prey, which can reduce the foraging success of individuals. For example, Furness et al. (1992) reported that foraging success of herring gulls (*Larus argentatus*) was negatively correlated with the presence and abundance of northern gannets (*Morus bassanus*).

Prey Selection

Predators use a combination of senses to locate prey. Most predatory birds are dependent on sight and sound, whereas mammals are most dependent on sight and scent, which has led to the evolution of distinctive traits. Owls have evolved structural asymmetry of their ears such that they can calculate the exact location of prey in the dark (Norberg 1978). In contrast, diurnal raptors depend on vision to locate prey. They have such a high density of rods and cones (i.e., the photoreceptor tissue in the eye) that their eyes occupy about 66% of the skull. Additionally, each eye has two focal spots as compared to one in mammals, allowing raptors to focus on multiple objects. Canids have well-developed senses of vision and hearing, but their sense of smell is especially highly evolved. Perhaps the most unique evolved trait is the ability of pit vipers (Crotalinae) to detect prey by identifying infrared radiation (i.e., body heat) or the ability of bats (Chiroptera) to detect prey by using sonar.

Once prey has been located, predators employ different tactics to capture them. Mountain lions, Cooper's hawks (*Accipiter cooperii*), and rattlesnakes (*Crotalus* spp.) are classic examples of ambush predators; they wait for prey to approach or stealthily approach their prey, and then make a rapid, surprise attack. Species such as a wolves or gyrfalcons (*Falco rusticolus*) are pursuit hunters, built for endurance and speed to chase down prey. Others systematically search areas until they locate prey: a rat snake or raccoon (*Procyon lotor*) uses olfaction to locate bird nests, whereas a red fox (*Vulpes vulpes*) uses hearing to detect voles moving under the snow. Finally, most predators are solitary hunters, but some—wolves, lions (*Panthera leo*), and Harris's hawks (*Parabuteo unicinctus*)—hunt cooperatively in groups. Others, such as brown boobies (*Sula leucogaster*), will forage in free-for-all groups when chasing schools of fish.

A variety of factors influence a predator's prey use, including the presence of alternate prey, size of prey populations, age- and sex-specific vulnerabilities of prey, specializations of the predator, and environmental conditions that influence the vulnerability of prey or effectiveness of the predator. Predation usually occurs between trophic levels, such as when sharp-shinned hawks (*Accipiter striatus*) prey upon small songbirds (e.g., warblers, sparrows). Within-trophic-level predation, however, can occur to obtain food or as competitive exclusion behavior. For example, northern goshawks (*Accipiter gentilis*) will capture and consume Cooper's hawks, broad-winged hawks, and barred owls (*Strix varia*; Smithers et al. 2005). In an example of competitive exclusion, Kamler et al. (2003) reported that coyotes killed (but did not consume) swift foxes (*Vulpes velox*), to an extent that they suppressed the local swift fox population.

Generalist Predators

Most predators are generalists, regularly capturing a wide variety of prey. Coyotes and Swainson's hawks (*Buteo swainsoni*) are mammalian and avian epitomes, respectively, of generalist, opportunistic predators. Generalist predators do not necessarily take prey in direct relation to their abundance, but in relation to their availability (which might or might not be related to numbers). For example, intraspecific behaviors of prey at high densities (e.g., territoriality, competitive exclusion) could make some individuals more vulnerable to predation. Alternatively, younger or older age class animals could be more susceptible because of inexperience or poorer physiological condition. Generalist predators are able to switch among prey species as they become more or less available; generalist predator populations tend to remain relatively stable because they are less likely to be influenced by a decrease in any one prey species. However, a given prey species could experience increased levels of predation because of decreases in a different prey species (Steenhof and Kochert 1988) or abundance of one prey species having led to an increase in predators in an area.

Some predators concentrate their efforts in areas of high resource availability. Lariviere and Messier (1998) simulated waterfowl nests to examine density-dependent nest predation rates. They reported that such predators as striped skunks (*Mephitis mephitis*) targeted areas with high nest density. They also reported that the proximity of waterfowl nests to dens or nests of predators had a larger influence on waterfowl nesting success. Generalist predators can suppress prey cycles in populations if predator densities are unrelated to any single prey species and if the predation rate is density dependent (Hanski and Korpimaki 1995, Thirgood et al. 2000). Throughout the 1900s, raptor populations on the Scottish moors were kept low through control efforts; raptor numbers cycled with those of the red grouse (*Lagopus lagopus scotica*), a major prey species. Following legal protection in 1990, raptor numbers increased, and when the predicted cyclic peak of the grouse population did not occur, Thirgood et al. (2000) attributed it to the influence of increased raptor predation on grouse.

Specialist Predators

In contrast to generalists, some predators are specialists and, although they can and do capture other prey, they would likely not persist without their key prey species. True specialists are rare; the classic North American example is the black-footed ferret (*Mustela nigripes*), which is dependent upon prairie dogs (Dobson and Lyles 2000). Canada lynx (*Lynx canadensis*) prey almost exclusively on snowshoe hare (*Lepus americanus*), and their cyclic relationship has been the topic of much study (Keith 1983, Stenseth et al. 1998). However, lynx can become an important cause of mortality of neonate caribou (*Rangifer tarandus*) when snowshoe hare populations reach cyclic lows (Bergerud et al. 1983). Similarly, the gyrfalcon is considered a specialist upon ptarmigan (*Lagopus* spp.) throughout much of its Arctic range (Mossop 2011). Hypothetically, other species could specialize under local circumstances. For example, wolves can capture and consume a variety of large ungulate prey, but on Isle Royale they specialize on moose (*Alces alces*) because it is the only ungulate species available (Mech and Peterson 2003). East of the Continental Divide, ferruginous hawks are tied to Richardson's ground squirrels (*Spermophilus richardsonii*) in the northern Great Plains (Schmutz et al. 2008) and to prairie dogs in the southern Great Plains (Giovanni et al. 2007).

Prey Vulnerability

Vulnerability of prey plays a substantive role in predators' prey selection and predation success. Prey can be more or less vulnerable to predation for a variety of reasons, including age, physical condition, environmental conditions, activity, and group size. Black bears (*Ursus americanus*) and grizzly bears (*U. arctos*) have been documented as a significant cause of mortality among elk, mule deer (*Odocoileus hemionus*), white-tailed deer, and moose, but predation focuses almost entirely on calves and fawns (Boertje et al. 1988, Ballard and Miller 1990, Ballard 1992, Zager and Beecham 2006). Urban-nesting Cooper's hawks in Arizona regularly captured Inca (*Columbina inca*) and mourning (*Zenaida macroura*) doves infected with the protozoa *Trichomonas gallinae* (Boal et al. 1998). Bald eagles (*Haliaeetus leucocephalus*) will prey on waterfowl that have been shot but not killed by hunters, or suffering from lead toxicosis after ingesting lead shot (Miller et al. 1998). Wolves sometimes kill prey in excess of their immediate needs (i.e., surplus killing), a behavior that appears to be linked to environmental (i.e., deep snow hindering escape and food access of prey) and physiological (i.e., poor nutrition of prey because of lack of food access) conditions during winter (DelGiudice 1998). Winter had a significant, progressive influence on ungulate vulnerability in the Greater Yellowstone area; wolves' hunting success rose from an estimated 1.6 kills per wolf per month in early winter to 2.2 kills per wolf per month during late winter, presumably because of prey's loss of condition caused by limited food access (Smith et al. 2004). Roth et al. (2006) reported that sharp-shinned hawks had significantly greater success when attacking small birds that were distracted by feeding compared to attacks on birds that were perched or flying. There is also truth in the old adage that there is safety in numbers; Cooper's hawks are significantly more successful attacking solitary birds than those in groups (Roth and Lima 2003; Fig. 13.1).

Management efforts to increase prey populations can result in unintended increased predation, especially if the method of increasing prey makes them more vulnerable to predation. For example, red-tailed hawks (*Buteo jamaicensis*; Turner et al. 2008) and bobcats (Godbois et al. 2004) were detected three and ten times closer, respectively, to feeders provided for northern bobwhite (*Colinus virginianus*) than would be expected by chance. Furthermore, many urban residents are dismayed to find that attracting songbirds in their backyard feeders can result in high predation rates by hawks. Another unintended consequence is when one prey species is increased such that a predator responds numerically, to the point that it negatively influences other prey. For example, feral swine introductions to the Channel Islands of California resulted in a substantial

increase in the previously small golden eagle (*Aquila chrysaetos*) population by increasing the eagles' food resources (Roemer et al. 2001). However, the increased number of eagles led to subsequent predation—and near extirpation—of island foxes (*Urocyon littoralis*).

Habitat Interactions

Wildlife live in complex ecosystems, which are made more so by human actions. Using mites as a study animal, Huffaker (1958) demonstrated that environmental heterogeneity (i.e., different types of patches and resources) was necessary for coexistence of predators and prey. Although he studied invertebrates, the results are just as relevant to wildlife communities and ecosystems. As habitat is lost directly and indirectly through fragmentation, and habitat quality reduced by human activities, there is increasing risk of imbalance in wildlife community structure. For example, Thirgood et al. (2000) reported that habitat loss was the influencing factor behind long-term declines in grouse, but that high levels of raptor predation could subsequently limit the cyclic growth of the grouse population. Predation can be a proximal cause for low numbers, but the ultimate cause is long-term patterns of anthropogenic landscape change and habitat loss. Similarly, Zager and Beecham (2006) indicated the uncertainty of bear predation as the proximate or ultimate cause of ungulate mortality, and concluded that the success of attempts to reduce bear predation likely depends on the prey population's relationship to habitat carrying capacity.

Factors such as weather conditions and human influences on the environment are other important aspects when considering predation. In Ontario, snow depth and snow density were interpreted as factors causing deer to move into areas of shallow snow where they could access food and shelter, but the trade-off was increased risk of predation by wolves (Kittle et al. 2008). Logan et al. (1996) concluded that mountain lion predation was a major mortality factor of mule deer, but drought, habitat quality, and habitat quantity were the main influencing forces on the deer population. Drought in particular can lead to increased predation by concentrating animals at water resources. Management practices in arid landscapes have often included providing water sources for wildlife. DeStefano et al. (2000) reported that predator sign was seven times greater at manmade water sources in the Arizona desert compared with nonwater sites. However, they speculated that the presence of predators likely increased more in response to the presence of the water itself than to the potential for hunting opportunities. Still, evidence from elsewhere suggests increased predation can occur at water sources. Hopcraft et al. (2005) reported that African lions focused their foraging efforts in areas with greater accessibility to prey (which included water sources), not necessarily in areas of prey abundance. Similarly, in southern India, Karanth and Sunquist (2000) reported that sympatric tigers (*Panthera tigris*) and leopards (*P. pardus*) coexist by partitioning prey rather than by activity periods, but both species focused capture attempts at or near habitat features that attracted ungulates.

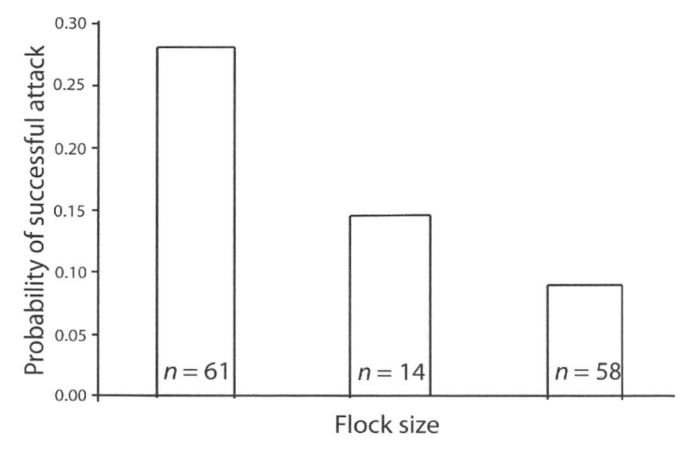

Figure 13.1. Probability of a successful attack by Cooper's hawks on three categories of flock size: solitary individuals (left bar), small flocks (two to four individuals; center bar), and large flocks (more than five individuals; right bar). Cooper's hawks were tracked during two winters in Terre Haute, Indiana; all attacks with known prey species are included (*n* = 133). Adapted with permission from Roth and Lima (2003); copyright the Cooper Ornithological Society

Predator–prey Relationships in Theory

The relationship of predators and prey in natural systems is quite complex; prey populations are dependent upon various components that constitute their habitat (e.g., food, cover, water), and they are usually subject to predation by a suite of predator species. Likewise, factors other than just prey (e.g., territoriality, nest sites) influence predator populations, which will usually prey on a variety of species. Modeling efforts tend to simplify these relationships to focus on single predator–single prey interactions. These models, which we discuss below, have substantially advanced our understanding of predator–prey relationships. However, we reiterate that these are simplified theoretical examinations with numerous assumptions; more complex models will include community or even ecosystem levels of complexity.

Efforts to model predator–prey relationships originated with Lotka (1925) and Volterra (1926), and their approach forms the basis of contemporary models. In general, this modeling approach makes the simplifying assumptions that prey populations (1) experience exponential or logistic growth and (2) mortalities are due to the overall predation rate that is influenced by the functional and numerical response of the predator. A functional response of predators is the changing rate at which an individual predator removes prey individuals from the population. The response reflects how a given predator adjusts its foraging patterns in response to changes in a prey population's density, and how it is affected by behavior of those prey individuals (e.g., spacing). The numerical response is how the predator changes in density (by increasing or decreasing reproduction and survival or by immigrating to or dispersing from the area) in response to changes in prey availability. The predator population within the model is also simplified to assume that population growth occurs as (1) a function of the overall predation rate and (2) density-dependent and density-independent mortalities.

The original Lotka–Volterra model was further simplified by assuming the only limiting factor for predators was the availability of prey, there was no age structure to either species, reproduction was continuous, predation was proportional to the rate at which predators encounter prey, movement was random, and predators experienced a constant density-dependent mortality rate. The simplest Lotka–Voltera model is one in which an absence of predators allows exponential population growth of the prey (H). This is modeled as:

$$\Delta N / \Delta t = rN,$$

in which Δ indicates change in the prey population size (N) and in time (t) given the instantaneous growth rate (r); recall from Chapter 7 that $r = \ln(N_{t+1}/N_t)$. Predators can be incorporated into the equation by estimating the efficiency with which predators capture prey; this is essentially a functional response term (α) indicating the predation rate per individual predator per unit of time ($\alpha = N/NPt$). Otherwise stated, this is the rate at which individual predators capture prey as a function of prey abundance. A larger α indicates individual predators have a greater effect on the prey population, which allows

calculation of prey population trend given the intrinsic rate of increase and the mortality due to predators with the equation:

$$\Delta N / \Delta t = rN - \alpha NP.$$

In the absence of prey, a predator population (P) will decline according to its intrinsic density-dependent mortality rate (m_p):

$$\Delta P / \Delta t = -m_p P.$$

Growth of the predator population can only occur when prey are available. However, this is more complex than just factoring in prey numbers; we must also consider the efficiency with which predators use their food to reproduce—resulting in more predators—referred to as a conversion rate (β). We estimate the conversion rate as P/PNt; thus β will be larger when a single prey item has greater influence on the predator population growth. Population growth of the predator population can then be estimated with the equation:

$$\Delta P / \Delta t = \beta NP - m_p P.$$

The equilibrium for prey and predator populations can be estimated by setting both $\Delta N / \Delta t$ and $\Delta P / \Delta t$ at zero growth. For the prey population, the result is:

$$0 = rN - \alpha NP,$$
$$rN = \alpha NP,$$
$$r = \alpha P,$$
$$P^* = r/\alpha.$$

Although this equation identifies the equilibrium of the prey population, the solution is actually in terms of the specific number of predators (P^*) that will maintain the prey population at zero growth. This can be viewed graphically by charting the prey (r/α) isoclines. If predator numbers are below the isocline at which prey are held steady ($\Delta N / \Delta t = 0$), prey populations increase; however, if $\Delta N / \Delta t > 0$, prey populations will decline (Fig. 13.2). Essentially, this equation indicates that a faster growth rate of the prey population requires more predators to keep the prey population at equilibrium. However, the more efficient predators are in capturing prey, the fewer predators are needed.

Similarly, for the predator population we have:

$$0 = \beta NP - m_p P,$$
$$\beta NP = m_p P,$$
$$\beta N = m_p,$$
$$N^* = m_p / \beta.$$

Here the result is an expression of predator population size in terms of the number of prey (N^*) necessary to keep predator population growth at zero. If prey numbers are below the isocline that holds predators in balance ($\Delta P / \Delta t = 0$), predators will decrease; if $\Delta P / \Delta t > 0$, predators will increase (Fig. 13.3). This equation indicates that the greater the death rate of the predators, the more prey are required to maintain the preda-

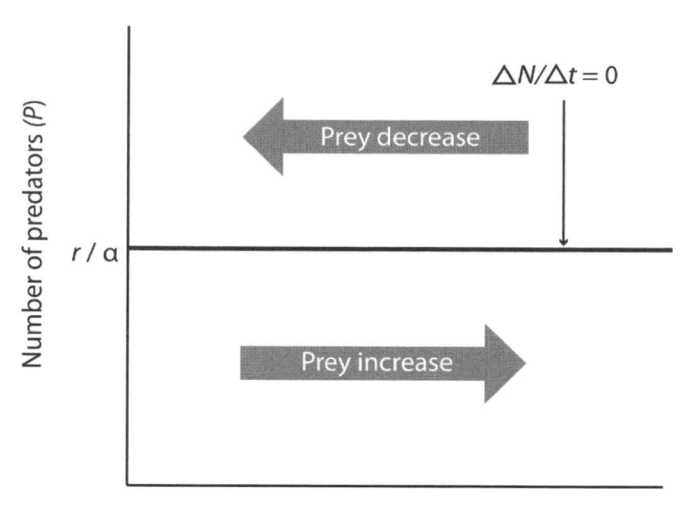

$\Delta N/\Delta t = 0$

Prey decrease

Prey increase

Number of predators (*P*)

r/α

Number of prey (*N*)

Figure 13.2. The Lotka-Volterra predation model predicting a critical number of predators (r/α) that control the prey population at zero growth ($\Delta N/\Delta t = 0$). Above the isocline, the prey population decreases in relation to a greater number of predators. Below the isocline, the prey population increases due to a reduced number of predators.

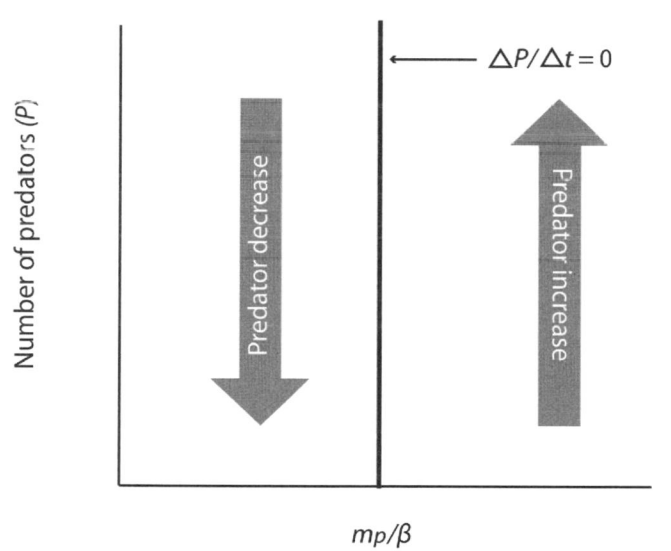

$\Delta P/\Delta t = 0$

Predator decrease

Predator increase

Number of predators (*P*)

mp/β

Number of prey (*N*)

Figure 13.3. The Lotka-Volterra predation model predicting a critical number of prey (mp/β) that control the predator at zero growth ($\Delta P/\Delta t = 0$). Left of the isocline, the predator population decreases in relation to a reducing prey population size. Right of the isocline, the predator population increases due to increasing prey population size.

tors at equilibrium. However, the number of prey needed reduces in accordance with efficiency in converting prey (i.e., food) into more predators.

Under the predation model, these isoclines can be viewed graphically by charting predator (r/α) and prey (m_p/β) isoclines that will intersect at 90° angles (Fig. 13.4). When looking at the upper right quadrant created by the intersecting isoclines, the predator population increases because of high prey numbers, but the increase in predators simultaneously results in a decline of prey. The result is a shift to the upper left quadrant, where the downward trend in prey numbers results in a similar downward trend in predators. The populations then move to the lower left quadrant, where the reduced number of predators results in reduced predation and an increasing trend in prey. This eventually leads to the lower right quadrant, which indicates that predator populations are increasing in response to increased prey. Ultimately, predator and prey populations oscillate in a counterclockwise elliptical path that is stable but without any trend toward equilibrium (Fig. 13.4). The ellipse can be viewed as population growth curves for predators and prey that cycle periodically in relation to each other (Fig. 13.5). However, these models are overly simplistic in that a variety of resources may limit predator and prey populations. For example, we can further refine models of prey population growth by incorporating the density-dependent aspect of habitat carrying capacity (K_n) of the prey species with the equation:

$$\Delta N/\Delta t = rN(K_n - N/K_n) - \alpha NP.$$

This density-dependent influence departs from stable oscillating interaction trajectory to one in which the trajectory spirals into a point of stable equilibrium (Fig. 13.6). This trajectory to a stable equilibrium makes ecological sense; if factors other than just predators limit prey populations, there is less likelihood of the populations cycling. Rosenzweig and MacArthur (1963) used isoclines in this fashion to examine predator–prey interactions under several different scenarios, such as efficient compared to inefficient predators.

Functional Response

Most contemporary predator–prey models are based on the Lotka–Volterra framework, but modified to incorporate how predator populations may respond numerically and functionally to prey population densities. Holling (1959a,b) was the first to describe three types of functional responses of predators to prey (Fig. 13.7). The Type I functional response is the default model already established in the Lotka–Volterra equations, given the assumption that predators engage in random but constant searching for prey and that they have an unlimited appetite. The number of prey consumed by a predator increases linearly as prey population numbers increase, but only to a point. Beyond a certain threshold, there is no increase in per-capita predation because of limitations imposed by handling and digestion requirements. Although this model has been consistent with some invertebrate predator–prey systems, it is a simplistic model with unrealistic assumptions. For example, other biological activities (e.g., sleep, self-maintenance, re-

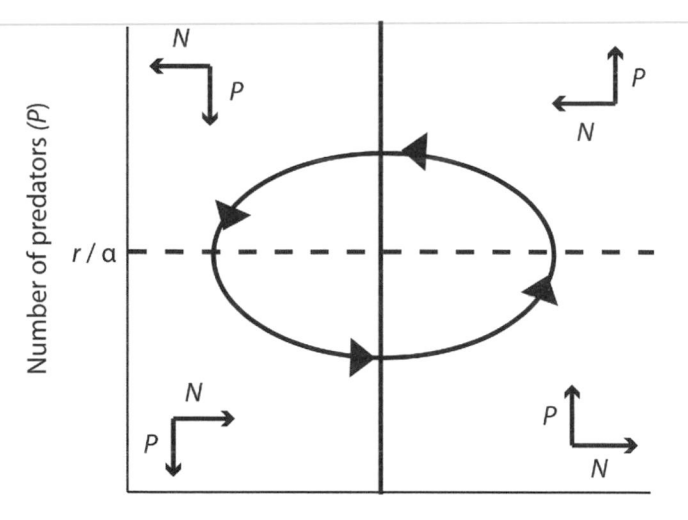

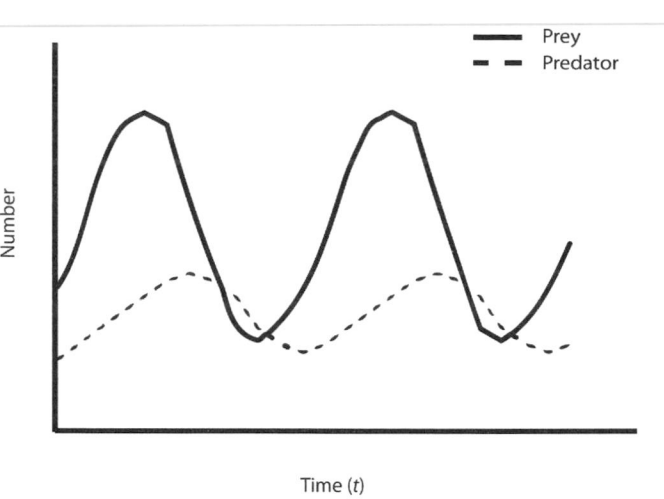

Figure 13.5. The cyclic dynamic of predator and prey populations as predicted by the Lotka-Volterra model.

Figure 13.4. Overlaying the predator and prey isoclines results in four quadrants created by the intersecting isoclines. In the upper right quadrant, predator populations increase and prey populations simultaneously decrease. This results in a shift to the upper left quadrant, in which the decrease in prey results in a decrease in predators. This moves the populations to the lower left quadrant, where the reduction in predators results in prey population increases. This eventually leads to the lower right quadrant, in which predator populations begin to rebound in response to increased prey, ultimately resulting in a predator–prey population dynamic that oscillates in a counterclockwise elliptical path. The dynamic is stable but without any trend toward equilibrium.

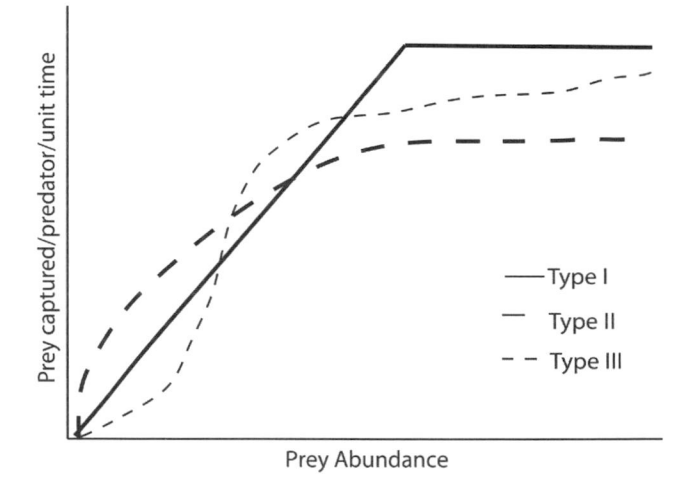

Figure 13.7. Hypothetical Type I, II, and III functional response curves.

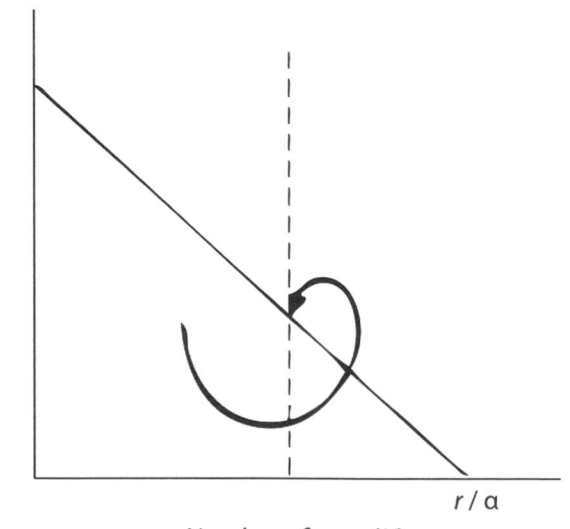

Figure 13.6. Example of the influence of prey population's carrying capacity on the cyclic dynamic of predator and prey populations. The prey population slopes downward because of the density-dependent influence of carrying capacity. The population dynamic results in a stable equilibrium point at the intersection of the isoclines.

production) and handling time (e.g., time required to capture and consume prey when it is encountered) prevent constant searching.

Holling's (1959a,b) second model incorporated constraints that result in a threshold of the number of prey animals consumed (e.g., an asymptote) as prey densities increase (Fig. 13.7); this has been referred to as a Type II functional response. For example, it would be rare for vertebrate predators to search constantly for prey. When prey are encountered, the predator must engage in attempting to capture it and, if successful, consuming and digesting it. This necessitates consideration of the handling time (h), which is the time it takes to kill, consume, and digest prey. The total time (T_t) employed by a predator is the sum of the time spent searching (T_s) for prey and the total handling time (T_h) given by:

$$T_h = hN_a,$$

in which N_a is the number of prey consumed per unit of time calculated by:

$$N_a = cT_sN,$$

where the capture rate (c) is the product of the area searched and the probability of a successful capture attempt. The number of prey consumed per unit of time (N_a) can be calculated for a Type II response rate by:

$$N_a = (cT_tN)/(1 + chN).$$

When the system involves multiple prey species, the formula can be modified to

$$N_a = (c_iT_tN_i)/(1 + \Sigma_j c_j h_j N_j)$$

for prey type (i) consumed per predator summarized across all prey types (j).

Ultimately, a Type II functional response reveals that a decreasing proportion of the prey population is consumed as the prey density rises. Although a Type II response is primarily relevant to invertebrate predator–prey relationships, it does appear to apply to some simple vertebrate systems characterized by one primary predator species and one primary prey species. Korpimaki and Norrdahl (1991) reported that European kestrels (*Falco tinnunculus*) exhibited a Type II functional response to vole (*Microtus* spp.) densities, and Dale et al. (1994) observed a similar relationship between wolves and caribou.

Most predators, however, capture and consume a variety of prey species. A prey species that is more abundant and presumably more available will be encountered and captured more frequently than less abundant prey species. For example, when salmon return to their spawning grounds, brown bears consume them more frequently than moose or deer. Conse-quently, encounter and capture rates will change as prey abundances change; once the salmon run is over, bears seek other prey. This results in a Type III functional response, in which low predation rates occur while a given prey population is low, but rates increase exponentially when the population reaches a threshold density. Eventually, the predation rate reaches an asymptote and levels out; when modeled, a Type III functional response appears similar to a logistic population growth curve (Fig. 13.7). Redpath and Thirgood (1999) observed hen harrier (*Circus cyaneus*) populations changing in response to red grouse populations in a manner consistent with a Type III response. When studying a more complex predator community, Gilg et al. (2003) reported that snowy owls (*Bubo scandiacus*), long-tailed jaegers (*Stercorarius longicaudus*), and Arctic fox (*Vulpes lagopus*) all exhibited a Type III response (Fig. 13.8) to changes in the lemming (*Dicrostonyx* spp.) density.

The shape of the Type III curve is thought to be associated with prey switching by predators. Sinclair et al. (2006) explain prey switching as a predator having a search image for one abundant prey species (species A) and ignoring a less common prey species (species B). If population densities of the prey species change (e.g., species A becomes scarce while species B becomes abundant), the predator will switch to prey species B. For example, great skua (*Stercorarius skua*) predation on seabirds increased dramatically over several years as the population of the skua's primary prey fish declined (Hamer et al. 1991). Such patterns could be especially applicable to specialist predators with high foraging rates. However, it is counterintuitive that a generalist predator, such as a coyote or red-tailed hawk, would forgo capturing an unwary cottontail

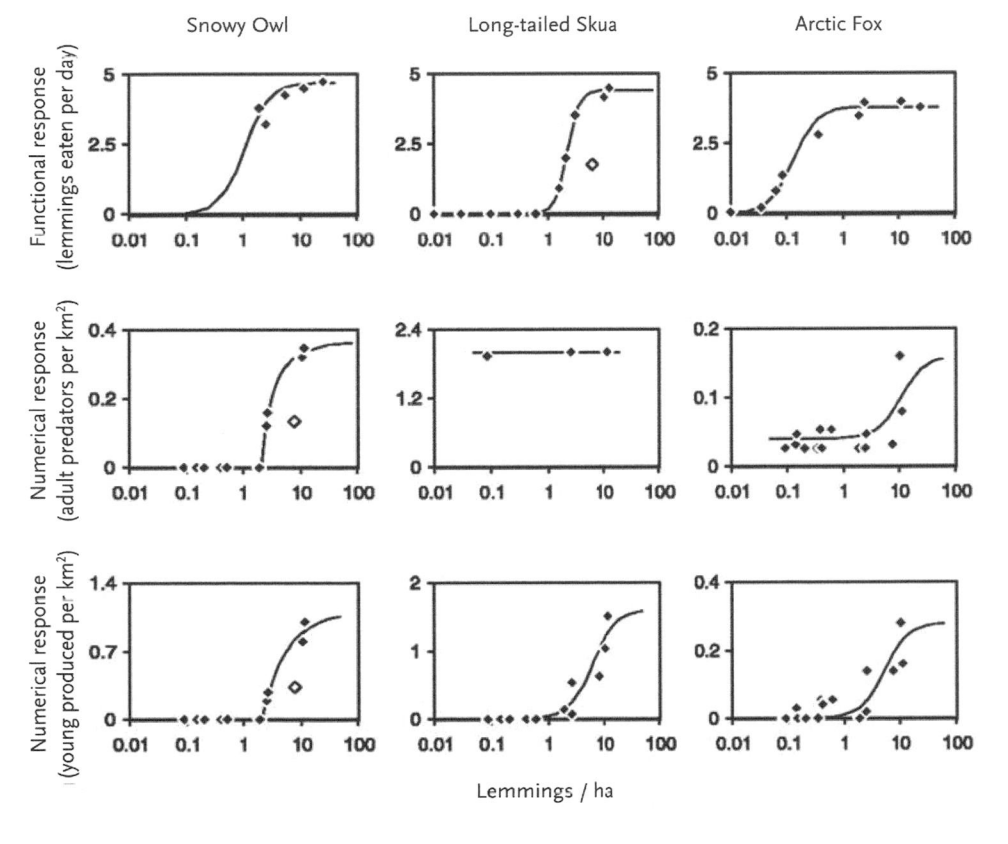

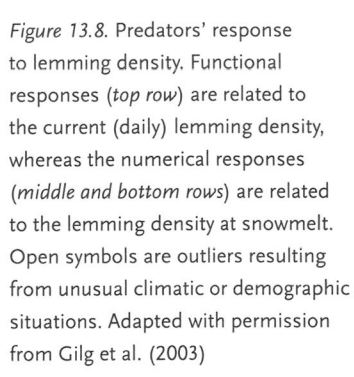

Figure 13.8. Predators' response to lemming density. Functional responses (*top row*) are related to the current (daily) lemming density, whereas the numerical responses (*middle and bottom rows*) are related to the lemming density at snowmelt. Open symbols are outliers resulting from unusual climatic or demographic situations. Adapted with permission from Gilg et al. (2003)

(*Sylvilagus* spp.) only because Richardson's ground squirrels happen to be locally more abundant.

Assessing Predation Rates

Estimating predation rates is difficult because of the inherent mobility of predators and their tendency to avoid humans. Wildlife managers can estimate mammalian predation rates by finding kill sites (especially in snow during winter), by following radio-tagged predators to find killed or cached prey, and most commonly by monitoring survival and causes of mortality of radio-tagged prey animals. For example, to assess predation mortality on the endangered Hutton's shearwater (*Puffinus huttoni*) in New Zealand, Cuthbert and Davis (2002) radio-tracked stoats (*Mustela erminea*) to recover remains and cached carcasses.

It is more common to use data from radio-tagged prey animals to estimate their survival rates and identify sources of mortality. However, radio attachments can predispose some prey animals to predation by negatively influencing their behavior or mobility (Murray and Fuller 2000). Predation rate estimates derived from radio-tagged prey animals might or might not be biased, and caution should be used when extrapolating these estimates to prey populations at large.

Diet of specific raptors can be partially understood by identifying prey (or their remains) at nests. However, this limits data only to the breeding season diets; very few quantitative studies have attempted to determine wintering raptor diets (Roth and Lima 2003, Roth et al. 2006). For this reason, knowledge of raptor diets is primarily for the summer months.

Mammalian predator species, and sometimes specific individuals (e.g., wolf pack, territorial mountain lions), can often be identified by interpreting evidence (e.g., location, tracks, prey condition) at the kill site. For example, mountain lions typically kill their prey by bites to the back of the neck and head, usually feed on the shoulders of the prey first, and then cover the remains with leaves, soil, and other debris until returning to feed again (VerCauteren et al. 2005). In contrast, wolves typically bring down prey by attacking the hind legs and flanks until it is disabled, and they then disembowel it and eat the viscera and hindquarters first (VerCauteren et al. 2005). In contrast, biologists are rarely able to assign a kill to a specific raptor species, much less an individual raptor.

A common problem in assessing predator–prey relationships when game birds are involved is that predators are typically identified only as avian or mammalian. For example, during an intensive three-year study of ruffed grouse (*Bonasa umbellus*) dispersal in southeastern Ohio, Yoder et al. (2004) reported that avian predators caused 23.8% of ruffed grouse mortalities. Similarly, Burger et al. (1995) reported that avian predators accounted for 28% of the natural mortality of northern bobwhite in Missouri. Individuals in either game bird population could have been killed by one raptor species or by any of a diverse community of raptor species. Additionally, the cause of death for some individuals might be misclassified because of subsequent scavenging by other animals. These types of data provide insights to relative mortality rates, but

not to the influence of predation at the population level, complicating the assessment of predator–prey relationships, the influence a given species has on another, and the development of sound management strategies.

The interpretation of a given predator species' influence on a given prey population can be highly variable and influenced by the methods used. In Minnesota, Eng and Guillion (1962) reported that northern goshawks killed 97 (42%) of 232 ruffed grouse in their study area, but they had problems determining the predator. In contrast, Smithers et al. (2005) video-recorded prey deliveries to 13 goshawk nests during three breeding seasons in Minnesota; they reported that goshawks consumed a variety of prey, but ruffed grouse accounted for only 5% of the diet. Similarly low predation rates on game birds were documented in other studies using video monitoring. Giovanni et al. (2007) reported no quail or pheasants (*Phasianus colchicus*) among 937 prey deliveries at ferruginous hawk nests and three quail among 1,057 prey deliveries at Swainson's hawk nests in the Texas panhandle. Elsewhere in Texas, video monitoring revealed three quail (0.9%) and no lesser prairie chickens (*Tympanuchus pallidicinctus*) among 352 prey deliveries to Swainson's hawk nests (Behney et al. 2010). These predation rates on game birds are quite divergent from most mortality estimates derived from radio-tagged quail. These findings could be attributable to different study locations or seasons, or to bias in methodologies. The predation rate on game birds, however, is likely associated with the diversity of prey in the study area. For example, in Greenland, where prey diversity is low compared with temperate and tropical zones, Booms and Fuller (2003) reported that gyrfalcons preyed heavily upon rock ptarmigan (*Lagopus muta*); combined with Arctic hares (*Lepus arcticus*), the two prey species accounted for 79–91% of the diet.

Some studies examined avian predation during the non-breeding season using direct observations of predation attempts by raptors that reside in open areas. Behney et al. (2011) attempted to assess raptor predation on lesser prairie chickens engaged in lekking activities. They reported a diverse community of raptors in the study area but observed only 0.02 attack attempts per hour. However, most direct observation assessments of raptor predation have been focused on falcons because they typically hunt over open areas. Buchanan et al. (1988) reported a 22.5% success rate for merlins (*Falco columbarius*) engaging in 111 capture attempts of dunlins (*Calidris alpina*) on the Washington coast. In British Columbia, Dekker and Ydenberg (2004) reported that 14.4% of peregrine falcon (*Falco peregrinus*) hunts on dunlins were successful. Cresswell and Whitfield (1994) observed predation attempts on wintering wading birds, primarily dunlins and redshanks (*Tringa totanus*), in Scotland. They reported that Eurasian sparrowhawks (*Accipiter nisus*), merlins, and peregrine falcons experienced success rates of 11.6%, 8.8%, and 6.8%, respectively.

These are only examples of methodologies used to identify which and how many prey are killed, and which predator was responsible. These types of data can be used for estimating kill rates. Kenward (2006) proposed estimates of kill rates as a function of the daily food requirements of a given predator, the biomass available from each prey, and the proportion of given prey in the diet (assuming that which is not consumed is subtracted from prey biomass). This included material not consumed by the predator, and the amount wasted if a sated predator abandons prey (such as when a scavenger consumes cached prey before the predator retrieves it; Kenward et al. 1981). This estimate of waste is highly variable among predator species. For example, there would be no wastage of fish swallowed whole by great blue herons (*Ardea herodias*), but wastage could occur by osprey (*Pandion haliaetus*), which tear apart their fish prey. Raptors will cache prey, as will mountain lions and bobcats. In contrast, wolf packs will gorge themselves and remain close by the kill until it is consumed.

Confounding factors in Kenward's approach include accounting for how daily food requirements vary with environmental conditions and behavior. For example, a male Cooper's hawk is smaller than a female and requires less prey biomass. However, prior to egg laying, the male hunts for both himself and the female; his kill rate thereby increases while hers drops. He also subsequently hunts for himself, the female, and the brood of nestlings. Additionally, the behavioral drive to hunt and capture food during the breeding period can result in surplus killing. Such situations are difficult to document and, given the difficulty of capturing prey, probably rare; however, they do occur. Boal (1997) documented male Cooper's hawks in urban settings killing excess prey for themselves, their mates, and their nestlings during the breeding season. Many cached prey were subsequently wasted by spoilage. Such behaviors are not limited to the breeding season. Eurasian pygmy owls (*Glaucidium passerinum*) exhibited a functional response to abundant prey by surplus killing and caching during winter (Solheim 1984). Regardless, such estimates of kill rates are still based on knowledge of the proportion of prey in the diet, and understanding the daily food requirements of the given predator species, both of which are often lacking. Furthermore, understanding the kill rate is only one component of more sophisticated models required to estimate the population-level impact of predators on prey (Eberhardt et al. 2003).

Predator Control

The difficulty in understanding predator–prey relationships, and how to manage them, is compounded by human influence on natural systems. Habitat loss, habitat degradation, altered faunal communities, and altered predator–prey ratios contribute to an environment where predation losses could have implications for species conservation (Reynolds and Tapper 1996). Perhaps no other wildlife management practice generates more controversy than predator control (Van Ballenberghe 2006). Originally directed at reducing livestock losses, predator control became commonplace during the 19th century and allowed an increase in game populations. Predator control to conserve threatened or endangered nongame species is a more contemporary practice.

If predator control is warranted, it must be clear what one hopes to achieve, and what strategy will best achieve those goals. There are numerous approaches to controlling predators, including nonlethal (e.g., deterrents, translocation) and lethal means. Regardless of method, reliably predicting results of predator control on game species is problematic. The dynamic nature of natural systems results in changing interactions among biological, environmental, and practical factors (Boertje et al. 2010). When examining predator control for moose management, Boertje et al. (2010) made excellent recommendations for criteria allowing for virtually any predator control program. When met, these recommended criteria work to ensure that predator control is justified and scientifically defensible. First, it should be established that predators kill substantial numbers of the species of interest that would usually otherwise survive if the predator was removed. Second, that reduced predation can facilitate reliably higher harvests or abundance of the species of interest. Third, that given less predation, habitats can sustain more of, but also be protected from, the species of interest. Fourth, that sustainable populations of predator species will persist in and out of control areas. Ballard et al. (2001) and Ballard (2011) presented criteria for predator reduction programs that might result in increased numbers of mule, black-tailed deer (*O. hemionus columbianus*), and white-tailed deer. However, they emphasized that predator control programs are not likely to influence an ungulate population that is near carrying capacity.

In situations where carrying capacity has not been attained, predator control can have positive influences on prey populations. Keech (2005) concluded that relocation of black and grizzly bears (*U. arctos horribilis*) resulted in an increase in moose calf survival in interior Alaska. Similarly, relocation of black bears from Great Smoky Mountains National Park improved calf elk recruitment from 0.306 (pretreatment) to 0.544 (posttreatment; Yarkovich et al. 2011). However, removal and relocation of predators might be short-term or partial solutions: although moose calf proportions increased 5–24% in Saskatchewan following bear removal, they later returned to preremoval levels when bears were not removed (Stewart et al. 1985). Brown and Conover (2011) reported that coyote control in Utah and Wyoming resulted in increased pronghorn (*Antilocapra americana*) fawn survival but had no effect on mule deer.

Predator control could have unintended consequences, such as increasing undesirable small mammal and lagomorph populations (Henke and Bryant 1999). Dion et al. (2000) reported high predation rates on grassland bird nests by ground squirrels in areas where duck nest predators had been removed. Another unintended consequence is a negative influence on species that would initially appear unrelated to the predator. For example, Berger et al. (2001) reported correlations between wolf and grizzly bear removal and increases in moose populations, which led to subsequent alteration of riparian vegetation and decreases in avian abundance and richness. A more compelling case for unintended consequences of predator removal is the mounting evidence in Yellowstone National Park that wolf removal resulted in loss of woody plants caused by overgrazing by ungulates (Ripple and Beschta 2003). Olechnowski and Debinski (2008) suggested that overgrazing

WARREN BAXTER BALLARD JR. (1947–2012)

Warren Ballard was born in Boston, Massachusetts, in 1947, but soon moved and grew up in Albuquerque, New Mexico. After earning a B.S. in fish and wildlife management from New Mexico State University in 1969, and an M.S. in environmental biology from Kansas State University in 1971, he went to work for the Alaska Department of Fish and Game as a wildlife biologist and research scientist. Over 18 years he conducted in-depth, groundbreaking research on predator–prey relationships among wolves, bears, and ungulate populations. He went on to earn a Ph.D. from the University of Arizona in 1993 with the dissertation "Demographics, Movements, and Predation Rates of Wolves in Northwest Alaska." His research contributed significantly to the 1997 National Research Council report "Wolves, Bears and Their Prey in Alaska: Biological and Social Challenges in Wildlife Management." Ballard went on to serve as director and associate professor with the New Brunswick Cooperative Fish and Wildlife Research Unit at the University of New Brunswick, and then as a research supervisor with the Arizona Game and Fish Department. In 1998, he joined the faculty at Texas Tech University, where he supervised graduate students in his lab as they conducted studies of a variety of species and issues, including white-tailed deer, mule deer, wild turkey, lesser prairie chickens, and songbirds.

But it was his lifelong interest in the ecology and conservation of predators and their relationships with prey that he pursued with passion. Ballard not only studied wolf and bear ecology in Alaska, but he also expanded this research to Mexican gray wolves and black bears in the arid Southwest, and conducted groundbreaking studies of the ecology and interactions of coyote and swift fox. Throughout his career, Ballard authored or coauthored more than 200 peer-reviewed journal articles. He twice served as editor-in-chief of *Wildlife Society Bulletin* and was named Wildlife Society Fellow by The Wildlife Society in 2005. In 2007, he received the Outstanding Achievement Award from the Texas Chapter of The Wildlife Society. Ballard was recognized academically with the Texas Tech University Chancellor's Council Distinguished Research Award in 2002 and the Outstanding Research Award from College of Agricultural Sciences and Natural Resources in 2009, held the Bricker Chair in Wildlife Management, and was named a Horn Professor (the highest honor a faculty member can receive from Texas Tech University) in 2008. Despite all these achievements, his true legacy lies with the more than 60 students who received graduate degrees under his supervision.

Photo courtesy of Heather Whitlaw

---

by elk was the ultimate cause for substantially reduced songbird species richness and abundance in Yellowstone compared with other areas. Following wolf reintroductions, woody cover has increased (Ripple and Beschta 2012) and, presumably in time, so will avian species.

Control of nest predators is a common method used in attempting to enhance waterfowl populations. Pearse and Ratti (2004) reported that duckling survival at 30 days was 60% higher in plots where foxes, skunks, raccoons, coyotes, and badgers were removed. In an analysis of duck nesting success among multiple sites spanning more than 50 years, Beauchamp et al. (1996) reported that using methods to exclude mammalian predators resulted in increased nest success, but lethal control or removal of predators did not reduce nest failure compared with areas where predators were not controlled.

In contrast to mammalian and reptilian predators, predatory birds in the United States are protected at the federal level by the Migratory Bird Treaty Act and, depending on species and status, other federal legislation; they usually receive similar protection at the state level. Avian predators are therefore primarily controlled via nonlethal methods (VerCauteren et al. 2005). Legal lethal control of predatory birds is an exceptional occurrence and requires issuance of permits by

the U.S. Fish and Wildlife Service. When lethal control is initiated, it is more often associated with human–wildlife conflict than predator–prey issues. For example, some raptors such as Mississippi kites (*Ictinia mississippiensis*), Cooper's hawks, and great horned owls (*Bubo virginianus*) have become common nesting species in urban areas and can become aggressive toward humans who approach their nests or young (Mannan and Boal 2004). Similarly, piscivorous birds can have substantial influences on aquaculture facilities (Taylor and Strickland 2008). Some form of predator control could potentially be necessary to reduce human–wildlife conflict and for conservation of threatened and endangered species. Although various nonlethal methods have been developed, lethal control of problem species may be allowed when other deterrents are not available or are ineffective.

Lethal control continues to be controversial, exemplified by the ongoing conflict between fishermen and nonconsumptive wildlife users over control of cormorants (Schusler and Decker 2002). Societal and cultural values often drive these debates, and science used to support differing positions can be misunderstood, misinterpreted, or inherently biased (Boertje et al. 2010). Lethal control can also be challenging when the predator is a species of concern. For example, lethal control of mammalian and avian predators promoted nesting success by the endangered California least tern (*Sternula antillarum browni*; Butchko and Small 1992). However, one of the predators (i.e., northern harrier) was also a species of concern (Butchko and Small 1992), posing a potential conservation conflict.

Development of methods for nonlethal control of predators has been directed at protection of livestock (Bodenchuk and Hayes 2007), primarily in the form of fencing, netting, or other predator exclusion devices. Few effective methods have been developed to reduce predation on wildlife prey species. Habitat management can result in landscapes less attractive to a given predator species. While this does not result in direct mortality, and hence lethal control, mortality may result because of decreased habitat value or dispersal and competition with individuals already residents in occupied areas.

Translocation has become a frequent tool in lieu of lethal control, and also for population introduction or enhancement; results have been mixed. Ruth et al. (1998) reported that two of 14 translocated mountain lions returned to their home ranges despite being translocated an average of 477 km away. Eight others headed in a general direction that led the authors to suspect they potentially could have been homing. Additionally, nine of the 14 died during the study period, but the mortality rate was not different from that of mountain lions in a reference area. The authors reported that translocations of mountain lions would be most successful with younger (12- to 27-month-old) individuals. Bradley et al. (2005) translocated wolves with mixed success. Most translocated wolves (67%) never established or joined a pack and experienced lower survival than reference animals. The primary cause of mortality was lethal control as a result of over 25% of translocated wolves preying upon livestock. Translocated wolves also demonstrated a homing tendency, either returning to or moving in the direction of their original area. Bradley et al. (2005) summarized that those individuals that returned tended to be adults, had been translocated comparatively shorter distances, and had gone through hard rather than soft releases. A hard release is when animals are essentially captured, transported, and released. A soft release is when the animal is relocated to the area but held in an enclosure for a period of time to allow adjustment and acclimation to the new area. Bradley et al. (2005) suggested that translocation efforts would likely be more successful if soft releases were used. Homing behavior is often thought of as a trait of higher, more sophisticated vertebrates (e.g., mammals, birds). However, it appears it could be a concern in virtually any translocation project. For example, Nowak et al. (2002) reported translocated western diamondbacked rattlesnakes (*Crotalus atrox*) experienced decreased survival, and over 50% returned to their original locations.

Moehrenschlager and Macdonald (2003) reported that translocated swift foxes went through three stages following translocation. The first was an acclimation stage characterized by erratic movements and rapid distancing from the release sites. During the second (establishment) phase, distances from the release site did not change significantly, but daily movements were more wide ranging than those of resident foxes. The third and final stage, considered the settlement phase, was characterized by movements similar to those of resident foxes. More importantly, Moehrenschlager and Macdonald (2003) learned that survival rates and litter sizes of translocated foxes were similar to those of resident animals, indicating translocation can be an effective management tool.

Another consideration of translocations or introductions is the influence the introduced predators will have on the local area. For example, prior to reintroduction of wolves into Yellowstone National Park, substantial study of the anticipated impact of wolves on prey populations was made. In contrast, endangered aplomado falcons (*Falco femoralis*) were successfully introduced to Matagorda Island National Wildlife Refuge, Texas, but pre- and postintroduction assessments of the falcons' impact on the biotic community were not conducted. Whether introducing or reintroducing predators to an area, enhancing existing populations, or just translocating nuisance animals to other locations, it is prudent to evaluate the potential effects.

Population Limitation, Regulation, and Trophic Cascades

We have presented examples of predator–prey relationships and management issues. There remains a great deal that is unknown about the interactions of predators and prey, especially at the community level. Do predators control prey, or do prey control predators? As with so many questions regarding wildlife, the answers do not come easily.

In wildlife management, we often focus our efforts on identifying and reducing the impact of limiting factors, the density-independent and density-dependent environmental factors that impede growth to, or maintenance of, the species' equilibrium carrying capacity (i.e., the population size allowed by available resources). These factors influence the population through mortality or reduced reproduction. For example, over 90% of radio-tagged lesser prairie chicken hens initiated nests over a three-year period (2008–2010) in west Texas, but only 20% nested during severe drought in 2011 (Grisham 2012). The population was thus limited by a density-independent factor that reduced reproduction. An important subset of the limiting factors, however, is the regulating factor, which operates only in a density-dependent fashion.

Debate continues as to whether predation is a regulating or limiting factor of wildlife populations. The number of predator species also has large implications for whether predation is solely a limiting factor or a regulating factor. Undoubtedly, the effects of a single predator species are different than those from multiple predator species (Craighead and Craighead 1956). A number of studies have provided data, arguments, and models of how predation could limit or regulate ungulate prey numbers. The debated models focused on four basic hypotheses: recurrent fluctuations, low-density equilibria, multiple stable states, and stable limit cycles (Van Ballenberghe and Ballard 1994, Ballard and Van Ballenberghe 1998a,b; Ballard et al. 2001). Under a recurrent fluctuation hypothesis,

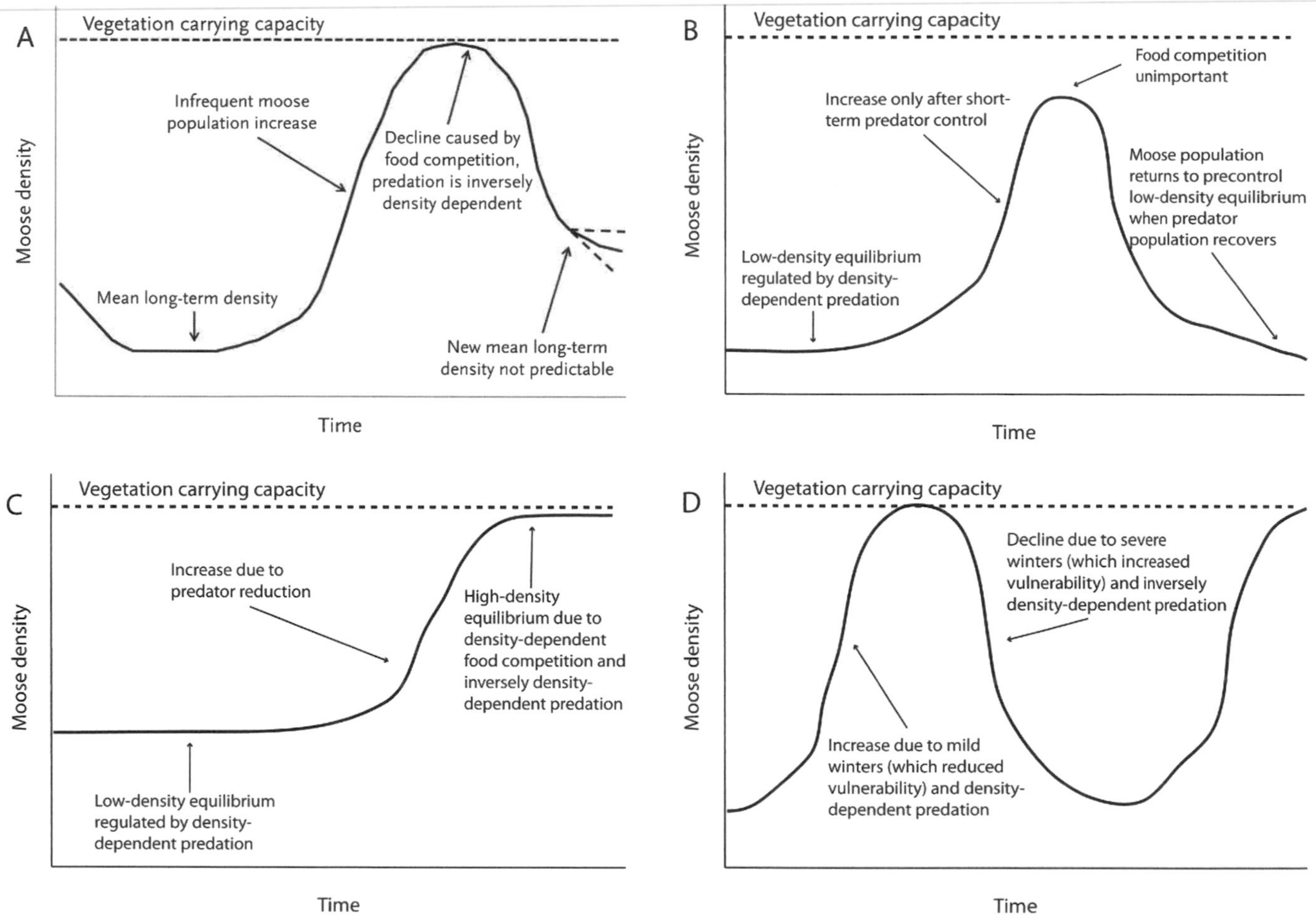

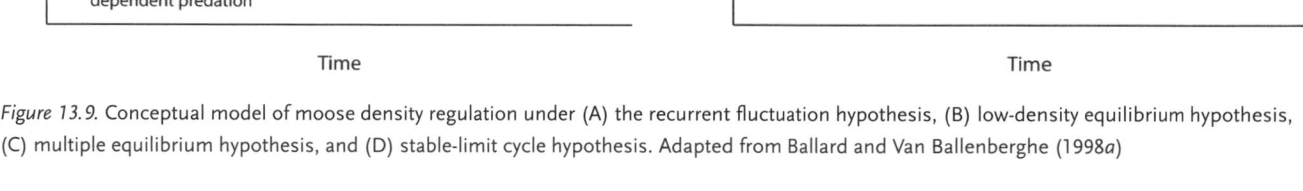

Figure 13.9. Conceptual model of moose density regulation under (A) the recurrent fluctuation hypothesis, (B) low-density equilibrium hypothesis, (C) multiple equilibrium hypothesis, and (D) stable-limit cycle hypothesis. Adapted from Ballard and Van Ballenberghe (1998*a*)

prey densities fluctuate and can change because of numerous factors. Predation is inversely density dependent at high prey densities and is not regulatory (Fig. 13.9A). A low-density equilibrium hypothesis suggests that density-dependent predation occurs at densities lower than what the carrying capacity of an area would allow, and the prey species competition for resources is unimportant because the carrying capacity is not reached. The prey persists at a low-density equilibrium unless predation is alleviated (Fig. 13.9B). A multiple-stable-state hypothesis suggests that the prey is regulated at low density by predators and at high density by intraspecific competition for resources (Fig. 13.9C). The stable-limit hypothesis proposes that prey populations experience regular cyclic oscillations (Fig. 13.9D). In all of these hypotheses, predation is a limiting factor and in some cases could be regulatory. Evidence from northern latitudes involving wolves, bears, and moose support the low-density equilibrium model, where ungulate densities would be held at low levels for extended periods of time, with little surplus available for hunters (Boertje et al. 2010).

In lower latitudes the effects of predation are more complicated. One reason is that human activities have greatly changed the relationships between predators and prey and

their habitats (Ballard et al. 2001). Large predator species have been extirpated from many areas of their former range, allowing increases of smaller predators (i.e., mesopredator release); facultative carnivores, such as bears and coyotes, have numerically increased and expanded their distributions, and humans have greatly altered habitat and grazing schemes. In such situations, any of the proposed models are possible and warrant further study. In any case, regardless of whether predation is a limiting or regulating factor, it is clear that predation must be accounted for when developing management strategies.

Attempting to determine whether predators limit or regulate prey is a challenging endeavor. Most attempts to assess prey limitation or regulation have focused on some form of predator control followed by monitoring predation rates of the species of interest. This measure alone, however, does not provide evidence of predation as a regulating factor regardless of how many prey animals are killed; what is important is how the population responds demographically. Alternatively, well-designed studies of demographic responses of prey to habitat manipulations can provide similar evidence. For example, Bishop et al. (2009) provided supplemental feed (effectively raising the potential carrying capacity) to mule deer in large experimental units in Colorado. By comparing survival among treated and untreated areas, they reported evidence that the deer were food limited; enhanced nutrition substantively increased survival and reduced female and fawn mortality by coyotes and mountain lions. Their findings suggested coyote predation was likely compensatory in removing fawns in poor condition that would have likely succumbed to other causes. Their results revealed that observed rates of coyote predation alone would not have been sufficient for evaluating impact on the deer population. In Canada, Krebs et al. (1995) conducted predator exclusion and food provisioning to assess their influence on snowshoe hare cycles. They reported that while excluding mammalian predators resulted in a doubling of hare densities during cyclic peaks and declines, the addition of food alone tripled hare density, and a combination of predator exclosure and food addition resulted in an 11-fold increase in density. In contrast, Newton and Perrins (in Newton 1998) examined a 50-year data set of nesting blue tits (*Cyanistes caeruleus*) and great tits (*Parus major*) in an area where sparrowhawks (the primary predator of the songbirds) had been present until 1960, but then absent because of pesticides until about 1973, when population recovery led to residency of six to nine nesting pairs per year. If sparrowhawks had historically depressed the songbird population through regulation, songbird-nesting numbers should have increased during the sparrowhawk absence, and declined again when sparrowhawks recovered. Evidence from survey data suggests this was not the case (Fig. 13.10).

Trophic Cascades

Given the multitude of factors that can limit populations, the trophic structure of biotic communities has been an intriguing topic for decades. Ecologically, energy flows from the lowest level to the top level of the trophic hierarchy. Photosynthesis drives primary productivity at the lowest trophic level, feeding consumers at midtrophic levels, which are in turn eaten by predators at top trophic levels. However, concomitant with this energy flow are the top-down influences of predators and the question of whether they eat enough consumers to influence conditions at the producer level (Terborgh and Estes 2010). These two currents have essentially led to two hypotheses explaining how ecosystems are structured. The first contends that food and resources (i.e., bottom-up regulation) structure ecosystems, and the second argues that predation does (i.e., top-down regulation; Kay 1998).

Terborgh and Estes (2010) succinctly summarize the argument. If predators limit herbivorous animals sufficiently (e.g., pressure from top down), then production at the bottom of the food chain is abundant because of the reduced number of herbivores (Hairston et al. 1960). In contrast, if predators only take the infirm individuals of prey populations, healthy prey populations must be regulated by productivity of their food resources (e.g., bottom up). Bottom-up processes have traditionally been accepted, whereas top-down processes, owing to significant challenges in study of large predator–prey relationships, have been viewed skeptically (Terborgh and Estes 2010). Paine (1966) presented compelling evidence of top-down regulation, however; within one year of removal of sea stars (*Pisaster ochraceus*), an important predator of mussels (*Mytilus californianus*) in rocky intertidal zones, mussels had expanded and crowded out other species. Ultimately, species richness dramatically decreased from a diverse community to one dominated by mussels. Estes et al. (1998) later convincingly demonstrated that loss of sea otters (*Enhydra lutris*) in coastal California waters resulted in population expansion of their primary prey, sea urchins (*Stronglyocentrotus* spp.), which in turn denuded the sea floor of its kelp forest, negatively affecting the entire ecosystem.

These top-down interactions have become termed "trophic cascades," and there is now little question that they occur. In addition to direct predation, Kauffman et al. (2010) discussed behaviorally mediated trophic cascades, in which fear of pre-

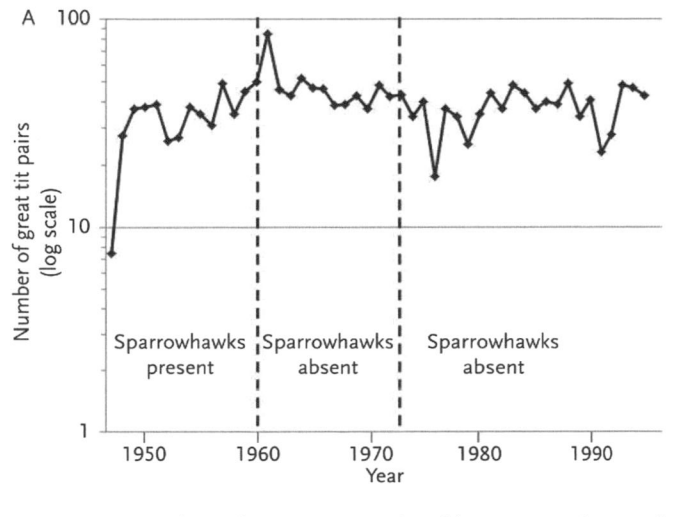
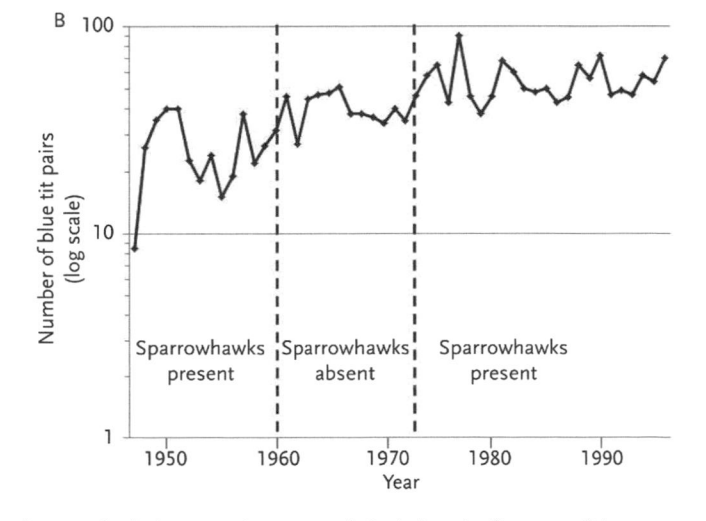

Figure 13.10. Numbers of (A) great tits and (B) blue tits in Marley Wood, Wytham, Oxford, during a 50-year period. Excluding the first year of the study, when tit nest boxes were first provided, nesting numbers were not conspicuously higher in years when sparrowhawks were absent than in years when they were present. Adapted with permission from Newton (1998)

dation alone can lead to reduction in grazing pressures by ungulates, and subsequent increases in plant productivity. For example, debate among ecologists has occurred over the benefits of wolf restoration on riparian woody vegetation and subsequent benefits to neotropical migrant birds in Yellowstone National Park (Berger et al. 2001, Ripple et al. 2001, Kauffman et al. 2010). Certainly, more study of the phenomenon is necessary and warranted to better understand ecosystem dynamics, especially as pertains to where, and under what conditions, trophic cascades occur (Estes 2005).

SUMMARY

The wildlife profession, traditionally associated with game species management, has made great strides in understanding predator ecology and predator–prey relationships. We have progressed from primarily viewing predators as simply a threat to game animals, to considering them a form of compensatory mortality, and finally to recognizing their potential role in regulating prey populations. We have developed models to better predict and understand how predators respond numerically and functionally to prey densities.

Wildlife management has a history of identifying and reducing the influence of limiting factors, especially density-dependent population regulation. The debate over predation as a limiting or regulating factor on prey species is likely specious; the reality is that both effects likely occur in variable, situation-dependent contexts. The key point is the need to reliably identify which is occurring in a given management situation through carefully designed field studies. While it is difficult enough with simple systems, it is an even greater challenge in multipredator systems. Even more challenging is identifying when, where, and why trophic cascades occur; making reliable predictions of the magnitude of impact of trophic cascades across an ecosystem; and devising appropriate, informed management strategies to reduce negative environmental effects. The more we understand predator–prey relationships, the truer the words of John and Frank Craighead, written over half a century ago, ring: "predation is a powerful and complex natural force that should be visualized in its ecological entirety" (Craighead and Craighead 1956:352).

Literature Cited

Ballard, W. B. 1992. Bear predation on moose: a review of recent North American studies and their management implications. Alces Supplement 1:162–176.

Ballard, W. B. 2011. Predator–prey relationships. Pages 251–285 in D. G. Hewitt, editor. Biology and management of white-tailed deer. CRC Press, Boca Raton, Florida, USA.

Ballard, W. B., D. Lutz, T. W. Keegan, L. H. Carpenter, and J. C. deVos Jr. 2001. Deer–predator relationships: a review of recent North American studies with emphasis on mule and black-tailed deer. Wildlife Society Bulletin 29:99–115.

Ballard, W. B., and S. D. Miller. 1990. Effects of reducing brown bear density on moose calf survival in south-central Alaska. Alces 26:9–13.

Ballard, W. B., and V. Van Ballenberghe. 1998a. Predator–prey relationships. Pages 247–273 in A. W. Franzmann and C. C. Schwartz, editors. Ecology and management of North American moose. Smithsonian Institution Press, Washington, D.C., USA.

Ballard, W. B., and V. Van Ballenberghe. 1998b. Moose–predator relationships: research and management needs. Alces 34:91–105.

Beauchamp, W. D., T. D. Nudds, and R. G. Clark. 1996. Duck nest success declines with and without predator management. Journal of Wildlife Management 60:258–264.

Behney, A. C., C. W. Boal, H. A. Whitlaw, and D. R. Lucia. 2010. Prey use by Swainson's hawks in the lesser prairie-chicken range of the Southern High Plains. Journal of Raptor Research 44:317–322.

Behney, A. C., C. W. Boal, H. A. Whitlaw, and D. R. Lucia. 2011. Interactions of raptors and lesser prairie-chickens in the Texas Southern High Plains. Wilson Journal of Ornithology 123:332–338.

Berger, J., P. B. Stacey, L. Bellis, and M. P. Johnson. 2001. A mammalian predator–prey imbalance: grizzly and wolf extinction affect avian neotropical migrants. Ecological Applications 11:947–960.

Bergerud, A. T., M. J. Nolan, K. Curnew, and W. E. Mercer. 1983. Growth of the Avalon Peninsula, Newfoundland, caribou herd. Journal of Wildlife Management 47:989–998.

Bishop, C. J., G. C. White, D. J. Freddy, B. E. Watkins, and T. R. Stephenson. 2009. Effect of enhanced nutrition on mule deer population rate of change. Wildlife Monographs 172:1–28.

Boal, C. W. 1997. An urban environment as an ecological trap for Cooper's hawks. Dissertation, University of Arizona, Tucson, USA.

Boal, C. W., R. W. Mannan, and K. S. Hudelson. 1998. Trichomoniasis in Cooper's hawks from Arizona. Journal of Wildlife Diseases 34:590–593.

Bodenchuk, M. J., and D. J. Hayes. 2007. Predation impacts and management strategies for wildlife protection. Pages 221–263 in A. M. T. Elewa, editor. Predation in organisms: a distinct phenomenon. Springer-Verlag, Berlin, Germany.

Boertje, R. D., W. C. Gasaway, D. V. Grangaard, and D. G. Kellyhouse. 1988. Predation on moose and caribou by radio-collared grizzly bears in east central Alaska. Canadian Journal of Zoology 66:2492–2499.

Boertje, R. D., M. A. Keech, and T. F. Paragi. 2010. Science and values influencing predator control for Alaska moose management. Journal of Wildlife Management 74:917–928.

Booms, T. L., and M. R. Fuller. 2003. Gyrfalcon diet in central west Greenland during the nesting period. Condor 105:528–537.

Boyce, M. S. 2000. Modeling predator–prey dynamics. Pages 253–287 in L. Boitani and T. K. Fuller, editors. Research techniques in animal ecology: controversies and consequences. Columbia University Press, New York, New York, USA.

Bradley, E. H., D. H. Pletscher, E. E. Bangs, K. E. Kunkel, D. W. Smith, C. M. Mack, T. J. Meier, J. A. Fontaine, C. C. Niemeyer, and M. D. Jimenez. 2005. Evaluating wolf translocation as a nonlethal method to reduce livestock conflicts in the northwestern United States. Conservation Biology 19:1498–1508.

Brown, D. E., and M. R. Conover. 2011. Effects of large-scale removal of coyotes on pronghorn and mule deer productivity and abundance. Journal of Wildlife Management 75:876–882.

Brown, J. S., B. P. Kotler, and A. Bouskila. 2001. Ecology of fear: foraging games between predators and prey with pulsed resources. Annales Zoologici Fennici 38:71–87.

Buchanan, J. B., C. T. Schick, L. A. Brennan, and S. G. Herman. 1988. Merlin predation on wintering dunlins: hunting success and dunlin escape tactics. Wilson Bulletin 100:108–118.

Burger, L. W., Jr., T. V. Dailey, E. W. Kurzejeski, and M. R. Ryan. 1995. Survival and cause-specific mortality of northern bobwhite in Missouri. Journal of Wildlife Management 59:401–410.

Butchko, P. H., and M. A. Small. 1992. Developing a strategy of predator control for the protection of the California least tern. Proceedings of the Vertebrate Pest Conference 15:29–31.

Craighead, J. J., and F. C. Craighead Jr. 1956. Hawks, owls and wildlife. Wildlife Management Institute, Washington, D.C., USA.

Cresswell, W., and D. P. Whitfield. 1994. The effects of raptor predation on wintering wader populations at the Tyninghame estuary, southeast Scotland. Ibis 136:223–232.

Cuthbert, R., and L. S. Davis. 2002. The impact of predation by introduced stoats on Hutton's shearwaters, New Zealand. Biological Conservation 108:79–92.

Dale, B. W., L. G. Adams, and R. T. Bowyer. 1994. Functional response of wolves preying on barren-ground caribou in a multiple-prey ecosystem. Journal of Animal Ecology 63:644 652.

Dawkins, R., and J. R. Krebs. 1979. Arms races between and within species. Proceedings of the Royal Society of London B 205:489–511.

Dekker, D., and R. Ydenberg. 2004. Raptor predation on wintering dunlins in relation to the tidal cycle. Condor 106:415–419.

DelGiudice, G. D. 1998. Surplus killing of white-tailed deer by wolves in northcentral Minnesota. Journal of Mammalogy 79:227–235.

DeStefano, S., S. L. Schmidt, and J. deVos Jr. 2000. Observations of predator activity at wildlife water developments in southern Arizona. Journal of Range Management 53:255–258.

Dion, N., K. A. Hobson, and S. Lariviere. 2000. Effects of removing duck-nest predators on nesting success of grassland songbirds. Canadian Journal of Zoology 77:1801–1806.

Dobson, A., and A. Lysle. 2000. Black-footed ferret recovery. Science 288:985–988.

Eberhardt, L. L., R. A. Garrott, D. W. Smith, P. J. White, and R. O. Peterson. 2003. Assessing the impact of wolves on ungulate prey. Ecological Applications 13:776–783.

Eng, R. L., and G. W. Gullion. 1962. The predation of goshawks upon ruffed grouse on the Cloquet Forest Research Center, Minnesota. Wilson Bulletin 74:227–242.

Errington, P. L. 1934. Vulnerability of bob-white populations to predation. Ecology 15:110–127.

Errington, P. L. 1946. Predation and vertebrate populations. Quarterly Review Biology 21:144–177; 221–245.

Estes, J. A. 2005. Carnivory and trophic connectivity in kelp forests. Pages 61–80 in J. Ray, K. Redford, R. Steneck, and J. Berger, editors. Large carnivores and the conservation of biodiversity. Island Press, Washington, D.C., USA.

Estes, J. A., M. T. Tinker, T. M. Williams, and D. F. Doak. 1998. Killer whale predation on sea otters linking oceanic and nearshore ecosystems. Science 282:473–476.

Fascione, N., A. Delach, and M. E. Smith. 2004. People and predators: from conflict to coexistence. Island Press, Washington, D.C., USA.

Furness, R. W., K. Ensor, and A. V. Hudson. 1992. The use of fishery waste by gull populations around the British Isles. Aredea 80:105–113.

Gilg, O., I. Hanski, and B. Sittler. 2003. Cyclic dynamics in a simple vertebrate predator–prey community. Science 302:866–868.

Giovanni, M. D., C. W. Boal, and H. A. Whitlaw. 2007. Prey use and provisioning rates of breeding ferruginous and Swainson's hawks on the Southern Great Plains, USA. Wilson Journal of Ornithology 119:558–569.

Godbois, I. A., L. M. Conner, and R. J. Warren. 2004. Space-use patterns of bobcats relative to supplemental feeding of northern bobwhite. Journal of Wildlife Management 68:514–518.

Grisham, B. A. 2012. The ecology of lesser prairie-chickens in shinnery oak-grassland communities in New Mexico and Texas with implications toward habitat management and future climate changes. Dissertation, Texas Tech University, Lubbock, USA.

Hairston, N. G., F. E. Smith, and L. B. Slobodkin. 1960. Community structure, population control, and competition. American Naturalist 94:421–425.

Hamer, K. C., R. W. Furness, and R. W. G. Caldow. 1991. The effects of changes in food availability on the breeding ecology of great skuas Catharacta skua in Shetland. Journal of Zoology 221:175–188.

Hanski, I., and E. Korpimaki. 1995. Microtine rodent dynamics in northern Europe: parameterized models for the predator–prey interaction. Ecology 76:840–850.

Henke, S. E., and F. C. Bryant. 1999. Effects of coyote removal on the faunal community in western Texas. Journal of Wildlife Management 63:1066–1081.

Holling, C. S. 1959a. The components of predation as revealed by a study of small mammal predation of the European pine sawfly. Canadian Entomologist 91:293–320.

Holling, C. S. 1959b. Some characteristics of simple types of predation and parasitism. Canadian Entomologist 91:385–398.

Hopcraft, J. G. C., A. R. E. Sinclair, and C. Packer. 2005. Planning for success: Serengeti lions seek prey accessibility rather than abundance. Journal of Animal Ecology 74:559–566.

Huffaker, C. B. 1958. Experimental studies on predation: dispersion factors and predator–prey oscillations. Hilgardia 27:795–835.

Jedrzejewski, W., and B. Jedrzejewski. 1996. Tawny owl (Strix aluco) predation in a pristine deciduous forest (Bialowieza National Park, Poland). Journal of Animal Ecology 65:105–120.

Kamler, J. F., W. B. Ballard, R. L. Gilliland, P. R. Lemons, and K. Mote. 2003. Impacts of coyotes on swift foxes in northwestern Texas. Journal of Wildlife Management 67:317–323.

Karanth, K. U., and M. E. Sunquist. 2000. Behavioural correlates of predation by tigers (Panthera tigris), leopard (Panthera pardus) and dhole (Cuon alpinus) in Nagarahole, India. Journal of Zoology 250:255–265.

Kauffman, M. J., J. F. Brodie, and E. S. Jules. 2010. Are wolves saving Yellowstone's aspen? A landscape-level test of a behaviorally mediated trophic cascade. Ecology 91:2742–2755.

Kay, C. E. 1998. Are ecosystems structured from the top-down or bottom-up? A new look at an old debate. Wildlife Society Bulletin 26:484–498.

Keech, M. A. 2005. Factors limiting moose at low density in Unit 19D East, and response of moose to wolf control. Federal Aid in Wildlife Restoration Final Research Performance Report, Grants W-27-5 and W-33-1 through W-33-3, Project 1.58. Alaska Department of Fish and Game, Juneau, USA.

Keith, L. B. 1983. Role of food in hare population cycles. Oikos 40:385–395.

Kenward, R. E. 2006. The goshawk. T and AD Poyser, London, United Kingdom.

Kenward, R. E., V. Marcstrom, and M. Karlbom. 1981. Goshawk winter ecology in Swedish pheasant habitats. Journal of Wildlife Management 45:397–408.

Kittle, A. M., J. M. Fryxell, G. E. Desy, and J. Hamr. 2008. The scale dependent impact of wolf predation risk on resource selection by three sympatric ungulates. Oecologia 157:163–175.

Korpimaki, E., and K. Norrdahl. 1991. Numerical and functional responses of kestrels, short-eared owls, and long-eared owls to vole densities. Ecology 72:814–826.

Krebs, C. J., S. Boutin, R. Boonstra, A. R. E. Sinclair, J. N. M. Smith, M. R. T. Dale, K. Martin, and R. Turkington. 1995. Impact of food and predation on the snowshoe hare cycle. Science 269:1112–1115.

Lariviere, S., and F. Messier. 1998. Effect of density and nearest neighbours on simulated waterfowl nests: can predators recognize high-density nesting patches? Oikos 83:12–20.

Leopold, A. 1933. Game management. Charles Scribner's Sons, New York, New York, USA.

Lima, S. L. 1998. Stress and decision making under the risk of predation: recent developments from behavioral, reproductive, and ecological perspectives. Advanced Study in Behavior 27:215–290.

Lima, S. L., and T. D. Steury. 2005. Perception of predation risk: the foundation of nonlethal predator–prey interactions. Pages 166–188 in P. Barbosa and I. Castellanos, editors. Ecology of predator–prey interactions. Oxford University Press, Oxford, United Kingdom.

Logan, K. A., L. L. Sweanor, T. K. Ruth, and M. G. Hornocker. 1996. Cougars of the San Andres Mountains, New Mexico. Final report to the New Mexico Department of Fish and Game. Hornocker Wildlife Institute, Moscow, Idaho, USA.

Lotka, A. J. 1925. Elements of physical biology. Williams and Wilkins, Baltimore, Maryland, USA.

Macdonald, D. W., G. M. Mace, and G. R. Barretto. 1999. The effects of predators on fragmented prey populations: a case study for the conservation of endangered prey. Journal of Zoology 247:487–506.

Mannan, R. W., and C. W. Boal. 2004. Birds of prey in urban landscapes. Pages 105–117 in N. Fascione, A. Delach, and M. E. Smith, editors. People and predators: from conflict to coexistence. Island Press, Washington, D.C., USA.

Mech, L. D., and R. O. Peterson. 2003. Wolf–prey relations. Pages 131–160 in L. D. Mech and L. Boitani, editors. Wolves: behavior, ecology, and conservation. University of Chicago Press, Chicago, Illinois, USA.

Miller, M. J., M. Restani, A. R. Harmata, G. R. Bortolotti, and M. E. Wayland. 1998. A comparison of blood lead levels in bald eagles from two regions on the great plains of North America. Journal of Wildlife Diseases 34:704–714.

Moehrenschlager, A., and D. W. Macdonald. 2003. Movement and survival parameters of translocated and resident swift foxes Vulpes velox. Animal Conservation 6:199–206.

Mossop, D. H. 2011. Long-term studies of willow ptarmigan and gyrfalcon in the Yukon Territory: a collapsing 10-year cycle and its apparent effect on the top predator. Pages 1–13 in R. T. Watson, T. J. Cade, M. Fuller, G. Hunt, and E. Potapov, editors. Gyrfalcons and ptarmigan in a changing world. Peregrine Fund, Boise, Idaho, USA.

Murie, A. 1940. Ecology of the coyote in the Yellowstone. Fauna of the National Parks of the United States 4. U.S. Department of the Interior, Washington, D.C., USA.

Murie, A. 1944. The wolves of Mount McKinley. University of Michigan, Ann Arbor, USA.

Murray, D. L., and M. R. Fuller. 2000. A critical review of the effects of marking on the biology of vertebrates. Pages 15–64 in L. Boitani and T. K. Fuller, editors. Research techniques in animal ecology: controversies and consequences. Columbia University Press, New York, New York, USA.

National Research Council. 1997. Wolves, bears, and their prey in Alaska: biological and social challenges in wildlife management. National Academy Press, Washington, D.C., USA.

Navarrete, L. 2011. Behavioral effects of wind farms on wintering sandhill cranes (Grus canadensis) in the Texas High Plains. Thesis, Texas Tech University, Lubbock, USA.

Newton, I. 1993. Predation and limitation of bird numbers. Current Ornithology 11:143–198.

Newton, I. 1998. Population limitation in birds. Academic Press, San Diego, California, USA.

Norberg, R. A. 1978. Skull asymmetry, ear structure and function, and auditory localization in Tengmalm's owl, Aegolius funereus (Linne). Philosophical Transactions of the Royal Society of London B 282:325–410.

Nowak, E. M., T. Hare, and J. McNally. 2002. Management of "nuisance" vipers: effects of translocation on western diamond-backed rattlesnakes (Crotalus atrox). Pages 533–560 in G. W. Schuett, M. Haggren, M. E. Douglas, and H. W. Greene, editors. Biology of the vipers. Eagle Mountain, Eagle Mountain, Utah, USA.

Olechnowski, B. F. M., and D. M. Debinski. 2008. Response of songbirds to riparian willow habitat structure in the Greater Yellowstone Ecosystem. Wilson Journal of Ornithology 120:830–839.

Paine, R. T. 1966. Food web complexity and species diversity. American Naturalist 100:65–75.

Pearse, A. T., and J. T. Ratti. 2004. Effects of predator removal on mallard duckling survival. Journal Wildlife Management 68:342–350.

Preisser, E. L., D. I. Bolnick, and M. F. Benard. 2005. Scared to death? The effects of intimidation and consumption in predator–prey interactions. Ecology 86:501–509.

Redpath, S. M., and S. J. Thirgood. 1999. Numerical and functional responses in generalist predators: hen harriers and peregrines on Scottish grouse moors. Journal of Animal Ecology 68:879–892.

Reynolds, J. C., and S. C. Tapper. 1996. Control of mammalian predators in game management and conservation. Mammal Review 26:127–155.

Reynolds, R. T., R. T. Graham, and D. A. Boyce Jr. 2006. An ecosystem-based conservation strategy for the northern goshawk. Studies in Avian Biology 31:299–311.

Ripple, W. J., and R. L. Beschta. 2003. Wolf reintroduction, predation risk, and cottonwood recovery in Yellowstone National Park. Forest Ecology and Management 184:299–313.

Ripple, W. J., and R. L. Beschta. 2012. Trophic cascades in Yellowstone: the first 15 years after wolf reintroduction. Biological Conservation 145:205–213.

Ripple, W. J., E. J. Larsen, R. A. Renkin, and D. W. Smith. 2001. Trophic cascades among wolves, elk and aspen on Yellowstone National Park's northern range. Biological Conservation 102:227–234.

Roemer, G. W., T. J. Coonan, D. K. Garcelon, J. Bascompte, and L. Laughrin. 2001. Feral pigs facilitate hyperpredation by golden eagles and indirectly cause the decline of the island fox. Animal Conservation 4:307–318.

Rosenzweig, M. L., and R. H. MacArthur. 1963. Graphical representation and stabilization conditions of predator–prey interactions. American Naturalist 47:209–223.

Roth, T. C., II, and S. L. Lima. 2003. Hunting behavior and diet of Cooper's hawks: an urban view of the small-bird-in-winter paradigm. Condor 105:474–483.

Roth, T. C., II, S. L. Lima, and W. E. Vetter. 2006. Determinants of predation risk in small wintering birds: the hawk's perspective. Behavioral Ecology and Sociobiology 60:195–204.

Ruth, T. K., K. A. Logan, L. L. Sweanor, M. G. Hornocker, and L. J. Temple. 1998. Evaluating cougar translocation in New Mexico. Journal Wildlife Management 62:1264–1275.

Schmutz, J. K., D. T. T. Flockhart, C. S. Houston, and P. D. McLoughlin. 2008. Demography of ferruginous hawks breeding in western Canada. Journal of Wildlife Management 72:1352–1360.

Schusler, T. M., and D. J. Decker. 2002. Engaging local communities in wildlife management area planning: an evaluation of the Lake Ontario Islands search conference. Wildlife Society Bulletin 30:1226–1237.

Sinclair, A. R. E., J. M. Fryxell, and G. Caughley. 2006. Wildlife ecology, conservation, and management. Blackwell, Malden, Massachusetts, USA.

Smith, D. W., T. D. Drummedr, K. M. Murphy, D. S. Guernsey, and S. B. Evans. 2004. Winter prey selection and estimation of wolf kill rates in Yellowstone National Park, 1995–2000. Journal of Wildlife Management 68:153–166.

Smithers, B. L., C. W. Boal, and D. E. Andersen. 2005. Northern goshawk diet in Minnesota: an analysis using video recording systems. Journal of Raptor Research 39:264–273.

Solheim, R. 1984. Caching behavior, prey choice and surplus killing by pygmy owls *Glaucidium passerinum* during winter, a functional response of a generalist predator. Annals Zoologica Fennici 21:301–308.

Steenhof, K., and M. N. Kochert. 1988. Dietary responses of three raptor species to changing prey densities in a natural environment. Journal of Animal Ecology 57:37–48.

Stenseth, N. C., W. Falck, K.-S. Chan, O. N. Bjornstad, M. O'Donoghue, H. Tong, R. Boonstra, S. Boutin, C. J. Krebs, and N. G. Yoccoz. 1998. From patterns to processes: phase and density dependencies in the Canadian lynx cycle. Proceedings of the National Academy of Science 95:15,430–15,435.

Stewart, R. R., E. H. Kowal, R. Beaulieu, and T. W. Rock. 1985. The impact of black bear removal on moose calf survival in east-central Saskatchewan. Alces 21:403–418.

Taylor, J., and B. Strickland. 2008. Effects of roost shooting on double-crested cormorant use of catfish ponds—preliminary results. Proceedings of the Vertebrate Pest Conference 23:98–102.

Temple, S. A. 1987. Do predators always capture substandard individuals disproportionately from prey populations? Ecology 68:669–674.

Terborgh, J., and J. A. Estes. 2010. Trophic cascades: predators, prey, and the changing dynamics of nature. Island Press, Washington, D.C., USA.

Thirgood, S. J., S. M. Redpath, D. T. Haydon, P. Rothery, I. Newton, and P. J. Hudson. 2000. Habitat loss and raptor predation: disentangling long- and short-term causes of red grouse declines. Proceedings of the Royal Society of London B 267:651–656.

Turner, A. S., L. M. Conner, and R. J. Cooper. 2008. Supplemental feeding of northern bobwhite affects red-tailed hawk spatial distribution. Journal of Wildlife Management 72:428–432.

Van Ballenberghe, B. 2006. Predator control, politics, and wildlife conservation in Alaska. Alces 42:1–11.

Van Ballenberghe, V., and W. B. Ballard. 1994. Limitation and regulation of moose populations: the role of predation. Canadian Journal of Zoology 72:2071–2077.

VerCauteren, K. C., R. A. Dolbeer, and E. M. Gese. 2005. Identification and management of wildlife damage. Pages 740–778 *in* C. E. Braun, editor. Techniques for wildlife investigations and management. Sixth edition. The Wildlife Society, Bethesda, Maryland, USA.

Volterra, V. 1926. Fluctuation in the abundance of species considered mathematically. Nature 118:558–560.

Yarkovich, J., J. D. Clark, and J. L. Morrow. 2011. Effect of black bear relocation on elk calf recruitment at Great Smoky Mountains National Park. Journal of Wildlife Management 75:1145–1154.

Yoder, J. M., E. A. Marschall, and D. A. Swanson. 2004. The cost of dispersal: predation as a function of movement and site familiarity in ruffed grouse. Behavioral Ecology 15:469–476.

Zager, P., and J. Beecham. 2006. The role of American black bears and brown bears as predators of ungulates in North America. Ursus 17:95–108.

ANIMAL BEHAVIOR

JOHN L. KOPROWSKI AND W. SUE FAIRBANKS

INTRODUCTION

The behavior of wildlife is an integral component for developing strategies to conserve and manage wildlife. Leopold (1933:123) recognized this in *Game Management*, in which he stated, "the game manager who observes, appraises, and manipulates these half-known properties of mobility, tolerance, and sex habits of wild creatures, is playing a game of chess with nature. He but dimly sees the board, the men, or the rules. He can be sure of only two things: for intricacy and interest, any other game pales into insignificance; he must win if wild life is to be restored. If any braver challenge inheres in any human vocation, it takes something more than a sportsman to see it."

Why do pronghorn (*Antilocapra americana*) fawns conceal themselves in cover for their initial days of life? Why do eastern gray squirrels (*Sciurus carolinensis*) nest in groups within the cavity of a tree? What are the consequences of sage grouse (*Centrocercus urophasianus*) mating in leks? The answers to questions about behavior have important ramifications for management decisions. For example, Beck et al. (2006) calculated the carrying capacity for elk of five different cover types on summer range in northeastern Nevada. Carrying capacity was calculated on the basis of abundance and digestibility of 15 key forage species in the five cover types, as well as the energy needs of a lactating female. The authors also studied habitat selection behavior of female elk on this range, and of the models they tested, the best model was based on distance to the nearest perennial stream and the ratio of aspen:conifer cover. Based on the animals' selection of areas close to streams and with high aspen:conifer cover, the authors redistributed among the cover types the total number of elk that the carrying capacity analysis determined could be supported on the range based on the quantity and quality of forage. The results indicated that certain cover types, those with high aspen:conifer and nearby perennial streams, would be selected by more elk than the forage could support, while other cover types would be underused despite the abundance of forage. Thus a central goal of wildlife management (i.e., assessing carrying capacity of an area for specific wildlife species) could

lead to damaging management decisions if behavior is not taken into account.

Behavior is often thought of as a conscious or instinctive choice. Innate behaviors are those instinctive actions that have been honed over many generations by the process of natural selection, the differential reproductive success of individuals differing in one or more genetic traits. Recall that for natural selection to operate, three components are required. A behavior has to be variable, heritable, and result in differential reproductive success. The popular label for natural selection—survival of the fittest—is a misnomer, because differential reproductive success is the key to the process. Learned behaviors are accrued through individual experience over the course of an individual's lifetime, though there are also genetic components to the ability to learn.

WHAT IS THE NAME OF THE GAME? FITNESS

A useful way to think about behavior is from an economic perspective. Behaviors have certain associated costs (e.g., energetic, lost time that could be spent in other ways, predation risk, lost mating opportunities) but also provide benefits (e.g., energetic savings, more time to feed, decreased predation, enhanced mating opportunities, more young surviving to join the population) to the actor. Ultimately, we measure the costs and benefits in terms of fitness. Fitness has a number of connotations in biology. A population geneticist measures the fitness of a trait with respect to the average contribution to the gene pool of the average individual with the trait, whereas a physiologist might measure physical fitness. One must be careful in how the term is used. When examining behavior, we consider fitness as individual fitness, where the propensity to contribute offspring to the next generation is the key measure. An individual with high fitness has a high likelihood of reproducing, whereas an individual with low fitness has a low likelihood of contributing offspring to the next generation. As a result, individual fitness correlates with lifetime reproductive success, which is often used as a proxy in behavioral research (Clutton-Brock 1988, Brommer et al. 2005). If we think more broadly, beyond production of offspring, the fitness of individuals may also benefit from the success of relatives with whom they share genes, the concept of inclusive fitness. Production of offspring and direct descendants, such as grand-offspring, contributes to the direct component of fitness; the indirect component of inclusive fitness comes from helping nondescendant relatives (e.g., offspring of siblings or parents) produce offspring that would not have survived without assistance. The direct and indirect components collectively result in the inclusive fitness of an individual, which is especially important in social animals.

HOW IS BEHAVIOR STUDIED?

Wildlife behavioral studies are used in numerous ways to inform conservation and management strategies. Direct observation of individuals is commonly used to assess diet and foraging ecology, habitat use, space use, dominance and social interactions, nesting ecology, reproductive behavior, and anthropogenic influence, to name a few applications. The most common techniques used to sample behavior are as follows:

1. Focal animal sampling. Following individual animals and recording behaviors over a set period of time. Often used to create a catalog of all behaviors (i.e., ethogram) or to determine how individuals apportion their time to specific behaviors (i.e., activity or time budgets).
2. Scan sampling. Scanning a study population at predetermined intervals and recording all behaviors and locations of individuals. Often used to examine space or habitat use, interindividual distances, and population-wide activity-time budgets.
3. All-occurrence or ad libitum sampling. Recording all observations of a specific behavior or behaviors within a study population. Often used to assess the frequency of rare behaviors, such as direct social interactions or matings.

Indirect observations are used in quantifying wildlife behavior. Traditional techniques, such as using sign (e.g., tracks, scat, food remains, nests) to provide data on behavior, can provide important insight into the ecology of wildlife. Increasingly, technology permits wildlifers to use indirect methods with considerable effectiveness. Radiotelemetry and satellite telemetry permit remote monitoring of individuals, trail cameras record time and location of individuals on the landscape, radioisotopic analyses permit assessment of diet, and molecular genetics provide insight into mating patterns and reproductive success.

MATING SYSTEMS

An integral part of wildlife behavior is the mating system in which individuals must operate to achieve reproductive success. The mating system is integrated with the social system of most species, and it is profoundly influential to the ecology

Figure 14.1. Male elk tending his harem of female elk. Note the telemetry collars attached to the male and several females. These uniquely colored and numbered collars permit individual identification at a distance. As in nearly all ungulates, the male is considerably larger than females and is the only sex adorned with weaponry—antlers in this case. Photo courtesy of Eric Godoy

of wildlife species because of its immediate impact on reproduction. We know that elk form harems of females that are tended by a large male (Fig. 14.1). Males compete for the opportunity to maintain a harem and must be able to attract females and defend them from other males. Traits such as large body size and large antlers are the result of sexual selection for success in such competitions. Sexual selection acts solely upon traits that are necessary to maximize access to mates and reproductive success and can result in what may seem like rather bizarre traits, such as the massive antlers (record lengths of main beam exceed 1.4 m) and extreme body size of adult male elk (>450 kg). Humans similarly value these traits for food and trophies; as a result, mating systems influence management and conservation efforts. The number of mates garnered by each sex defines mating systems (Fig. 14.2).

Monogamy

Monogamy occurs when a single male and female are paired. At least 90% of avian species are classified as monogamous, whereas less than 5% of mammals form pair bonds with a single mate. Serial monogamy is a specific case of monogamy where individuals pair with only a single mate during a breeding season, but that mate changes during each season. Many passerine birds are examples of this type of monogamy. In lifelong monogamy, individuals pair with a single mate for that breeding season and for all subsequent breeding seasons, as long as both individuals are alive. Canada geese (*Branta canadensis*) practice lifelong monogamy.

Polygamy

Polygamy is a mating system where only one sex has multiple mates. We can further classify polygamy depending on the sex that obtains multiple mates. Polyandry occurs when one

Males   Females

Monogamy

Polygamy

Polyandry

Polygyny

Polygynandry

Promiscuity

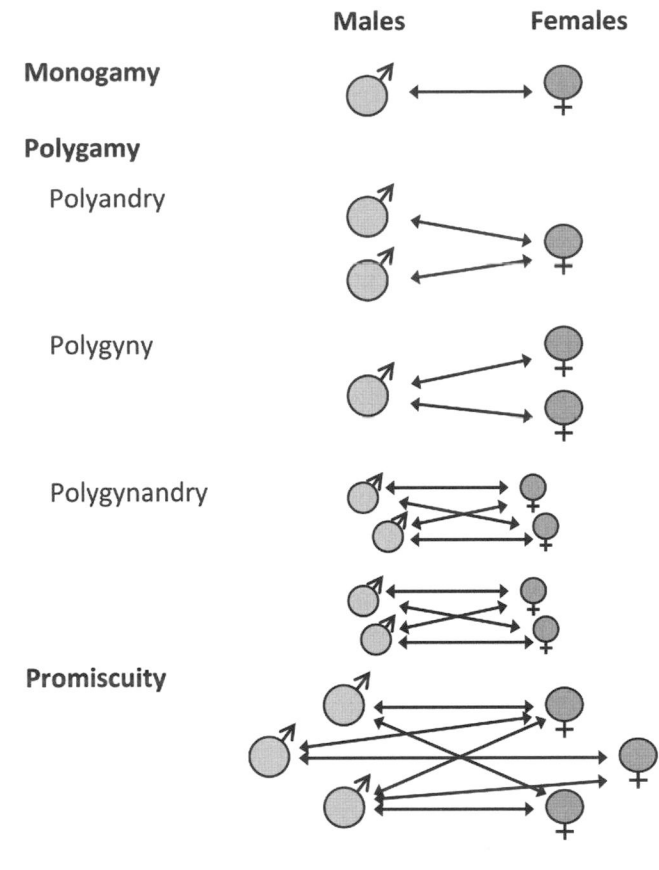

Figure 14.2. Schematic diagram of wildlife mating systems. Arrows that connect individuals symbolize mating between the individuals.

Figure 14.3. Sage grouse lek with males displaying for females. Note the proximity of males to each other and the lack of resources at the lekking grounds, a traditional site used only for mating. The exaggerated plumage and esophageal air sacs for sound production are the products of sexual selection to demonstrate vigor in courtship. Top photo courtesy of Dave M. Shumway; bottom photo courtesy of James Ownby

female has multiple male mates. Many sandpipers, jacanas, and raptors practice this type of polygamy. Polygyny is when one male has multiple female mates. Polygyny is common in many mammals, including white-tailed (*Odocoileus virginianus*) and mule (*O. hemionus*) deer. Polygynandry occurs when two or more males have an exclusive relationship with two or more females; the numbers of males and females need not be equal, and in vertebrate species studied so far, there are usually fewer males. Red foxes (*Vulpes vulpes*), bonobos (*Pan paniscus*), and avian species, including alpine accentors (*Prunella collaris*) and acorn woodpeckers (*Melanerpes formicivorus*), are examples of species exhibiting polygynandry.

Polygamy is often further categorized on the basis of the nature of the resource that is defended. Mate-defense polygamy is a form of polygamy in which the sex that has multiple mates defends against potential competitors. Many ungulates practice this type of polygamy. In resource-defense polygamy, the sex with multiple mates defends a resource that attracts mates from potential competitors. A number of Phocidae, such as elephant seals (*Mirounga angustirostris*), defend protected haul-out areas that females visit. Wildlife that practice lek-based polygamy defend a position at a lek, a traditional site where one sex gathers to display and the other sex visits to select mates. The lek contains no resources (other than po-

tential mates) of value to the visiting sex, and the only value to the displaying sex is that of location in the lek. Central locations often are more successful than peripheral locations. Sage grouse (Fig. 14.3) and greater prairie chickens (*Tympanuchus cupido*) are excellent examples of this mating system.

Promiscuity

A promiscuous mating system is where males and females have multiple mates and do not maintain an exclusive relationship with any individuals. This is a common mating system among mammals, especially in many small- and medium-sized species such as squirrels (*Sciurus* and *Tamiasciurus* spp.), rabbits (*Sylvilagus* spp.), hares (*Lepus* spp.), and many furbearers.

Through the use of molecular genetics, we know that extrapair fertilizations (EPFs) are more common than previously believed and that the observable and apparent mating system might not tell the whole story. For instance, within some species classified as monogamous, more than 70% of young are the result of EPFs (Griffith et al. 2002). As a result, it is common to refer to the behavioral or social mating system and the genetic mating system of a species to distinguish between the potential differences that EPFs create.

Mating systems also have an important impact on the ecology of wildlife in more subtle ways through the process of sexual selection. Sexual dimorphism, or the differences between the sexes as expressed in one or more traits, is often the result. In species where mates must be actively defended, intrasexual competition (i.e., competition between members of the same sex) results in intrasexual selection for characteristics that enhance success in physical combat such as large body size, horns and antlers, and teeth and claws. Additionally, where one sex actively chooses mates, mate choice can be extremely influential owing to intersexual selection. The bright colors and elaborate plumage of many male birds are results of the response of traits to mate choice by females. Sexual dimorphism also has an impact on how we manage wildlife populations, and characteristics attractive to hunters must be incorporated into management plans. The differences in physical appearance also permit more refined management, because we can plan harvests to target specific sex and age classes. For species without obvious sexual dimorphism (e.g., many small game species such as rabbits, squirrels, doves, and most carnivores), we are not able to target specific segments of the population and are likely to have a reduced ability to manipulate population growth through selective harvest.

SPACE AND HABITAT USE

How animals use space is important to wildlife management. It is relevant to habitat use, carrying capacity, and even population estimation. When we think about space use by animals, we often classify animals in terms of their fidelity for an area along a continuum (Fig. 14.4). Truly nomadic species with no fidelity to space are not known, likely because knowledge of resources in an area is beneficial. Home ranges result from fidelity to an area without exclusive use. Home ranges are defined as the area traversed by an individual in which their daily needs are met; home ranges are not defended and can be used by other individuals of the same species. Many species maintain nearly exclusive access to core areas within a home range, whereas exclusive use, or territoriality, occurs when an individual or social group defends the entire area from intruders. The key determinant in the development of areas of exclusive use is the concept of economic defensibility. Benefits of defending a site include access to resources (e.g., food, limited nutrients, dens, nests, mates, space). Costs that accrue include lost time, lost opportunities elsewhere, energetic costs, and catastrophic risk of territory loss (Fig. 14.5).

Territoriality is important to wildlife populations in numerous ways. Territoriality often results in a relatively uniform distribution of individuals on the landscape as territory owners defend resources. The behaviors of defense limit access to resources and ultimately can control population size in a given area. Such behavior also makes it challenging to translocate individuals to new locations where resident animals already exist. Defense of territories can be through direct interactions that can escalate to conflict between individuals. The weaponry among many male ungulates and carnivores often is attributed to such territorial squabbles; however, other subtler behaviors—such as scent marking by urine, feces, or other body secretions, visual displays by territory owners, and vocalizations—define territory boundaries and settle differences without repeated combat. Indeed, wildlife managers capitalize on many of these territorial behaviors to develop indices of the abundance of wildlife populations.

Body size energetics and the economics of energy likely set the size of home ranges and territories (Harestad and Bunnell 1979). Once the effect of body size on home range size is removed from consideration, significant variation remains

Continuum of Space Use

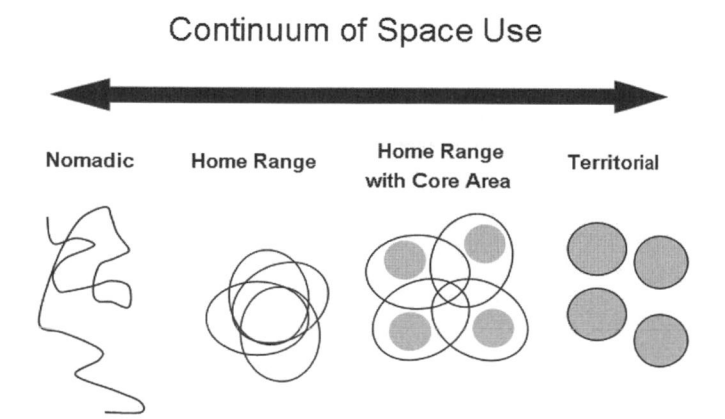

Figure 14.4. Patterns of space use based on affinity for space and level of defense.

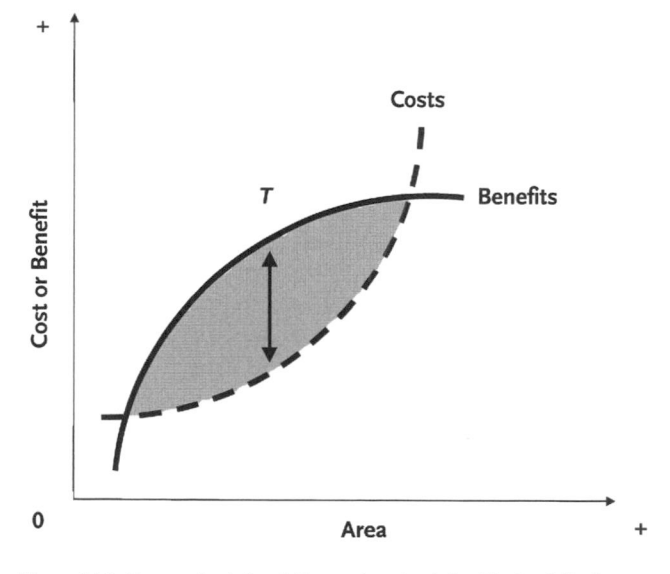

Figure 14.5. Economic defensibility and territoriality. Territoriality is possible wherever the benefits exceed the costs (shading); however, we would expect the most common territory size to be found where the benefits exceed the costs by the greatest amount (point *T*).

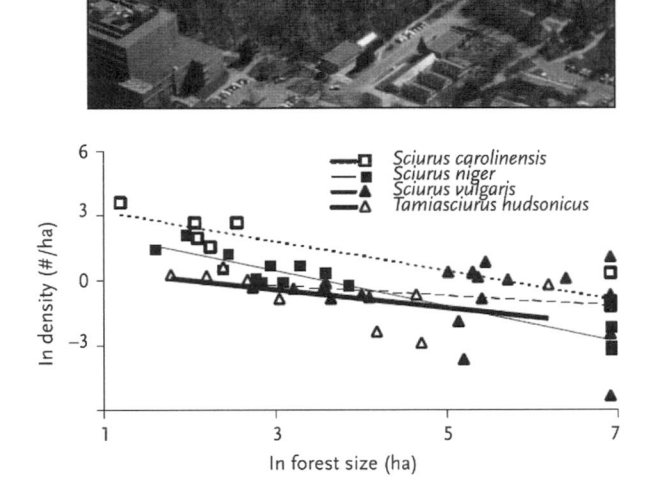

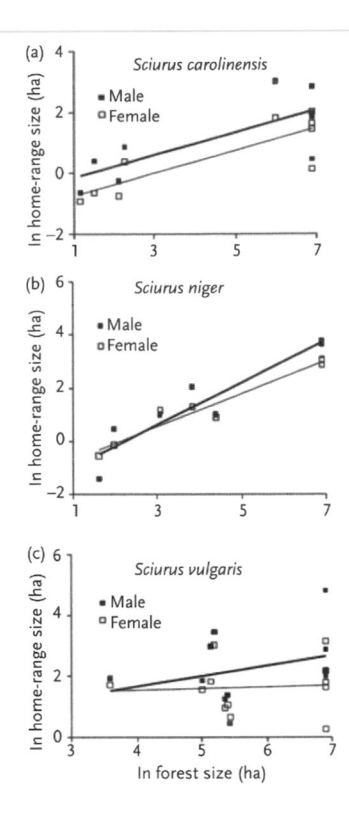

Figure 14.6. Forest fragmentation results in high densities of tree squirrels and a concomitant reduction in home range size as animals are relegated to the remaining fragment of quality of habitat. From Koprowski (2005)

and is related to resource availability, predation risk, interspecific competition, diet, and climate. We might expect variation in the home range size of a species across habitats that differ in quality. Habitat fragmentation increases population density and can result in severe compaction of home ranges (Fig. 14.6; Koprowski 2005). Some additional patterns are clear among wildlife species. Males often traverse larger areas than females, except in monogamous systems where the ranges of the sexes are typically equivalent, suggesting that mating systems have a strong influence on the space use of a majority of mammals and some birds (Clutton-Brock 1989). Other factors like climate can also have a great impact on the extent of sex differences in home range size and may be the dominant factor in such taxa as the large marsupials of Australia, where large home ranges are in poor-quality habitats in which both males and females must traverse vast areas to find resources (Fisher and Owens 2008). Seasonal variation in the availability of mates, food, cover, climate, and predation risk induces seasonal changes in home range size. Sexual dimorphism in home range size often is reduced outside of the breeding season when searching for mates declines as males may not need to roam large home ranges (Cudworth and Koprowski 2010).

SEXUAL SEGREGATION

Sexual segregation (i.e., the sexes living apart during certain seasons) is an often-overlooked component of space use. During segregation, the sexes may use different habitats (habitat segregation), or they may select similar habitats in different places (spatial segregation). Social segregation, defined as individuals of different sex, age, or size classes living in separate social groups, may or may not be associated with differences in habitat or space use by the social groups (Conradt 2005).

Sexual segregation occurs in many migratory species of birds and marine mammals when one sex migrates farther than the other, or the sexes migrate to different locations for specific needs. For example, male sperm whales (*Physeter macrocephalus*) migrate to high latitudes following breeding, while groups of females and offspring remain closer to tropical breeding grounds (Lyrholm et al. 1999). Among migrating birds, males in numerous species winter at higher latitudes than females (Cristol et al. 1999). Many aquatic reptiles exhibit habitat segregation with respect to foraging habitats (Shine and Wall 2005). In Galapagos marine iguanas (*Amblyrhynchus cristatus*), only the males forage in the sea, diving to feed on algae. Females and smaller males feed at the edges of the ocean (Buttemer and Dawson 1993). While many animal taxa exhibit sexual segregation, it has been most studied and is particularly widespread in ungulates (Bowyer 2004).

Why is it important to recognize the differences in grouping or spatial patterns in wildlife species? Studies designed to investigate habitat relationships of particular species commonly focus on only one sex, typically the sex that is easiest to observe or fit with radio transmitters. In mammals, females are easier to track; in birds, males may be. Depending on the objectives of the study, this sex bias may be defensible. In grazing mammals, females often live in larger groups than males and may have greater impacts on the habitat. Alternatively, females may have specific needs for rearing offspring, and knowing their use of habitat would be critical to the dynamics of the entire population. During the period of sexual segregation, however, the sexes may have different habitat requirements,

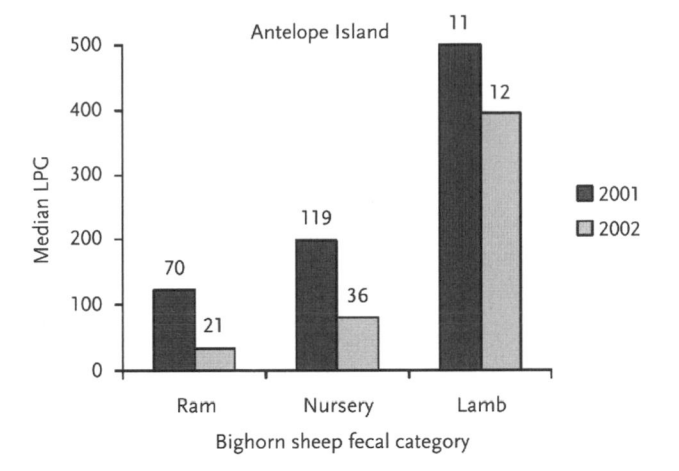

Figure 14.7. Median lungworm larvae per gram (LPG) of feces from male bighorn sheep, nursery groups (predominantly female), and lambs on Antelope Island in Great Salt Lake, Utah, in May–August 2001 and 2002. Numbers above the bars indicate sample size (number of pellet groups sampled). Modified with permission from Rogerson et al. (2008)

2005). Differences in movement patterns and habitat use may also result in differential exposure of the sexes to disease and parasites (Rubin and Bleich 2005). Counts of lungworm larvae (*Protostrongylus* spp.) were higher in fecal samples from nursery groups (predominantly female) of bighorn sheep than in samples from male groups on Antelope Island in Great Salt Lake, Utah (Fig. 14.7; Rogerson et al. 2008). Abundance of the intermediate hosts (terrestrial gastropods) of the parasite was higher around water sources, and the only gastropods infected with lungworm were collected near water sources, leading Rogerson et al. (2008) to suggest that the higher larvae counts in female bighorn sheep may be due, in part, to different patterns of water use during sexual segregation. Not only did males use a larger area than females during segregation, males spent less time at water sources than females, and males and females used different water sources during segregation (Fig. 14.8; Whiting et al. 2010). The potential for differential exposure of the sexes to parasites must be considered as an explanation for differential parasite loads or infection rates.

Most of the hypotheses offered to explain sexual segregation result from studies of sexually dimorphic ungulates, in which males are larger than females (Bowyer 2004). For this reason, the following hypotheses may not relate well to other groups (Wearmouth and Sims 2008). The predation-risk hypothesis (or reproductive strategy hypothesis) is based on the idea that males and females use different strategies to maximize their lifetime reproductive success. Outside the breeding season, females must balance the energy needs required for offspring production with the need to select habitats that minimize predation risk to their offspring or themselves (Rachlow and Bowyer 1998). Males, on the other hand, may be less vulnerable to predators during some seasons because of their size or weapons. The best strategy for males might be to select habitats offering abundant or high-quality forage, facilitating an increase in body condition prior to the mating season.

requiring management of multiple habitats or larger areas that accommodate both sexes. Further, the sexes may respond differently to management activities because of the differences underlying sexual segregation (Stewart et al. 2003, Long et al. 2009).

Males and females may be exposed to different mortality factors, such as predation or vehicle collisions, or different rates of mortality during segregation (Bowyer 2004). For example, male bighorn sheep (*Ovis canadensis*) have larger home ranges and move over greater distances, traveling between groups of females, and as such, males may sustain higher mortality rates from vehicle collisions than females (Rubin and Bleich

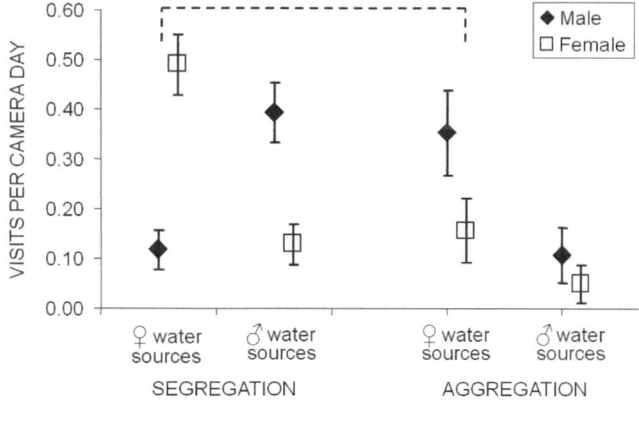

Figure 14.8. Number of visits divided by the number of days in which cameras were set (±95% confidence interval) at water sources located in areas used by male and female bighorn sheep on Antelope Island State Park, Utah, during 2005 and 2006. Males visited water sources used by females almost three times more often during aggregation compared with when bighorns were segregated (dashed line highlights the relationship). Reproduced with permission from Whiting et al. (2010)

Another hypothesis, the forage selection hypothesis (or sexual dimorphism or body-size hypothesis), has several variations. These hypotheses are based on differences in nutritional requirements between the sexes. Larger individuals that require more food are also more efficient at digesting food because of greater gut capacity. The larger sex should choose habitats with high availability of forage, even if it is of lower quality, while the smaller sex must obtain a high-quality diet to make up for their lower digestive efficiency. Other hypotheses fall under the category of social factors hypotheses. While these hypotheses have not received much support as explanations of sexual segregation among ungulates, they may be useful among other taxa, such as marine mammals (Wearmouth and Sims 2008). As sexual segregation is studied in additional groups of animals, other hypotheses may be proposed. For example, the thermal niche-fecundity hypothesis may be particularly relevant to sexual segregation in ectotherms. Under this hypothesis, the sexes segregate to maintain the optimal body temperatures that maximize reproductive output (Sims 2005).

DISPERSAL AND MIGRATION

Dispersal and migration are two important events in the life history strategies of many wildlife species. They are similar because both involve movement of individuals away from their current home ranges, but the causes and consequences of these behaviors are very different. Dispersal is the one-way movement of an individual from one home range to another, in which it establishes a new home range. Migration is round-trip movement between different seasonal areas. Dispersal is more or less permanent; migration is seasonal shuttling between two locations. It is important to establish the definitions used in any discussion, as the same terms in some fields are defined differently. For example, population geneticists may use the term "migration" to refer to the movement of alleles from one gene pool to another. In an ecological or behavioral sense, this movement is usually accomplished by dispersal of individuals.

Dispersal and migration are behaviors that are relevant to conservation and management. Dispersal is an individual behavior that is at the heart of metapopulation dynamics. Metapopulations are sets of subpopulations that are linked by some level of dispersal among the subpopulations. These connections can result in the rescue of declining subpopulations by increasing the local population size and genetic variation. If local subpopulations become extirpated, dispersers from other subpopulations can recolonize the vacant habitat. Persistence of the metapopulation as a whole is increased because of the interconnections resulting from dispersal. Without dispersal, the recolonization of subpopulations within the metapopulation could not occur by natural processes and would require an active management intervention through translocation. Dispersal may have significant impacts on local demographics of subpopulations, as well, with large numbers of immigrants (dispersers moving in) or emigrants (residents dispersing out) affecting population growth or decline, respectively. Dispersal can have major impacts on population genetics, evolution, and persistence of metapopulations. Dispersal processes impact predator–prey dynamics and disease transmission and epidemiology, too. An understanding of dispersal will be important in the new field of invasion ecology and prediction or control of invasive species. Dispersal and migration affect species distribution and will be highly relevant to the consequences of land-use change and climate change for wildlife. Migration is closely related to habitat selection and necessitates large-scale conservation and management strategies. Migration may also have important effects on the communities that they move between, with respect to what they may carry with them: contaminants (e.g., DDT, or other pesticides), pathogens, and parasites.

Dispersal

Several different types of dispersal are often recognized. The two major types are natal dispersal and breeding dispersal. Natal dispersal is the movement of an animal from the home range where it was born or hatched to the home range where it first reproduces. Breeding dispersal is movement to a new home range between reproductive events.

Dispersal has been difficult to study in the past, because it is a behavior of an individual. However, new technologies are providing unprecedented opportunities to conduct dispersal studies. Advances in global positioning system telemetry and satellite tracking enable biologists to track remotely the movements of individual animals over large distances or inhospitable habitats at all times of the day or night. Indirect genetic methods facilitate study of the genetic consequences of dispersal, and genetic markers may even allow identification of dispersers or distinction between immigrants and residents in a population. While stable isotope methods are becoming widely used for study of migration, they might also be useful in dispersal studies (Caudill 2003).

In birds and mammals, there is a strong bias as to which sex disperses (Greenwood 1980). In birds, females are the primary dispersers; in mammals, the males typically disperse. However, a smaller proportion of the opposite sex may disperse, as well. In species in which both sexes disperse, one sex often disperses farther than the other.

Dispersal can be divided into three basic stages: emigration (leaving the current home range), transfer (moving through unfamiliar territory), and immigration (settlement into a new home range). The proximate mechanism triggering an individual to emigrate may be fixed (genetically hardwired) or condition dependent. If current population or environmental conditions (e.g., sex ratio, population density, habitat quality, body condition, competitive ability) are at least somewhat reliable indicators of future conditions in the natal home range, selection may favor sensitivity to these indicators with respect to whether an individual disperses (i.e., condition-dependent dispersal). In an experimental field study of prairie voles (*Microtus ochrogaster*), Lin et al. (2006) manipulated food and cover resources in habitat patches created by mowing intervening vegetation (Fig. 14.9), released a mated pair and their two to three weaning-aged offspring into each patch, and observed dispersal. Males were significantly more likely to disperse overall, but both sexes were significantly less likely to disperse from the highest-quality patches (i.e., those with supplemental food and cover) than from other patches. Voles were more likely to disperse to habitat patches of similar or better quality than the one into which they were released. The sample size of dispersing females was small, but females only dispersed

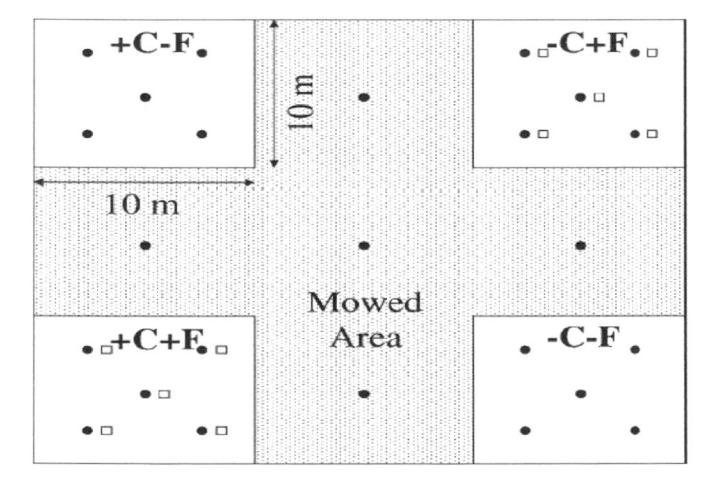

Figure 14.9. Diagram of habitat patches within each of four 0.1-ha enclosures in a field experimental study of dispersal in prairie voles (*Microtus ochrogaster*). Locations of habitats were randomly assigned within each enclosure. The four habitats are coded as +C+F, supplemental cover and supplemental food; −C+F, no supplemental cover and supplemental food; +C−F, supplemental cover and no supplemental food; and −C−F, no supplemental cover and no supplemental food. Locations of trapping stations (circles) and feeding stations (squares) are indicated. Reproduced with permission from Lin et al. (2006)

from patches without supplemental food; all females released in patches with supplemental food were philopatric (i.e., remained in the original home range). Thus local conditions impacted dispersal by both sexes.

Less is known about the transfer stage of dispersal for logistic reasons, but also because interpretation of some behaviors is difficult. Are short excursions from the home range failed dispersal attempts? Or are they fact-finding trips, so that when dispersal occurs, the individual already knows where it is going? For simplicity, in many metapopulation and other types of models, dispersers are often assumed to take a straight path in a random direction. In reality, of course, many factors in the landscape and biological community likely affect the path of dispersal and depend on species' perceptual abilities (i.e., how far away can the animal detect suitable areas) and tolerances to different microhabitats. Another difficult question to answer is whether survival probabilities of dispersers differ from survival of philopatric individuals. If dispersers differ in body condition, competitive ability, and sex, simply comparing survival of dispersing individuals to that of philopatric individuals will not be an appropriate test.

Settling into a new home range (immigration) is largely a function of habitat selection, but for some species the ability to integrate into a new social group may also play a role. Conspecifics already inhabiting the area may act as a "social fence," preventing the disperser from establishing a home range. In other cases, the presence of conspecifics may be a factor in the disperser recognizing suitable areas (i.e., conspecific attraction). In some species, especially generalist species (Davis 2008), choice of settlement habitat may be influenced by the habitat into which the individual was born (i.e., habitat imprinting). Dispersers may also respond differently to cues about the quality of the habitats they encounter during transfer, depending on time since leaving their natal home range (Stamps et al. 2007). For example, many insects and some other taxa initially exhibit refractory periods after leaving home, during which they do not respond to cues from even high-quality settlement habitats. Alternatively, dispersers may be limited in terms of time or resources to support dispersal, and so they may exhibit lower acceptance thresholds for settlement habitat as time since emigration increases (Stamps et al. 2007). In other words, the longer a disperser has been searching, the more willing it might be to accept a lower-quality habitat. A refractory period or declining acceptance thresholds (or both) may explain why dispersers sometimes move through high-quality habitats and later settle in similar- or lower-quality habitats.

Two major hypotheses have been suggested as evolutionary explanations of dispersal: inbreeding avoidance and avoidance of local competition. The inbreeding avoidance hypothesis suggests that dispersal reduces the likelihood of inbreeding depression, the reduced reproductive success that can result from breeding with close relatives. The strong sex bias in avian and mammalian dispersal seems to support this hypothesis. However, the hypothesis does not predict which sex should disperse. The avoidance of local competition hy-

MARDY MURIE (1902–2003)

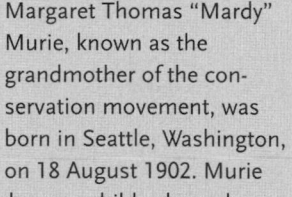

Margaret Thomas "Mardy" Murie, known as the grandmother of the conservation movement, was born in Seattle, Washington, on 18 August 1902. Murie moved to Fairbanks, Alaska, as a child, where she became the first woman to graduate from what would later become the University of Alaska Fairbanks. There she met and married Olaus Murie, a biologist and eventual president of The Wildlife Society; their honeymoon was spent following caribou through the Brooks Range of Alaska, one of many exciting wilderness adventures by this wildlife biologist, author, and avid conservationist. After completing the caribou research, the couple moved to Moose, Wyoming, to conduct research on elk. The Muries called Moose their home for more than 75 years.

Beyond wildlife research, Murie was a tireless proponent of conservation and collaborated with various agencies and nongovernmental organizations to lobby for legislation that set aside conservation and wilderness lands. She was instrumental in the passage of the Wilderness Act of 1964 and the establishment and enlargement of the Arctic National Wildlife Refuge, among many other similar accomplishments. For her efforts, Murie received the Audubon Medal, the John Muir Award, and the highest civilian honor awarded by the United States government, the Presidential Medal of Freedom. Mardy Murie passed away at the age of 101 on 19 October 2003. Her lifelong dedication to conservation is chronicled in a book and documentary entitled *Arctic Dance*.

Photo courtesy of the U.S. Fish and Wildlife Service

pothesis has two parts and offers suggestions as to why males are the dispersing sex in mammals, and why females are the dispersing sex in birds. Local mate competition suggests that young males should disperse to avoid competing for mating opportunities with their fathers or other closely related males. Of course, dispersers will likely compete with unrelated males for mates, but by avoiding competition with close relatives, they avoid decreasing their own inclusive fitness. Alternatively, one sex may disperse to avoid competition with close relatives for food or other resources. Polygyny is the primary mating system in mammals, typically resulting in greater competition among males for mates than among females, which could explain the strong bias toward dispersal by males in mammals. The majority of bird species are socially monogamous, and philopatric males may be more likely to obtain a territory, cru-

cial to their reproductive success, in a familiar area. Females, on the other hand, may benefit from searching widely for high-quality territories and high-quality mates, and may be selected to disperse.

Migration

Migration represents a critical, repeated phase in the life cycle of migratory animals that has major impacts on the physiology and behavior of individuals. The choice of whether to migrate varies even among closely related species. In fact, within a species, some populations might be migratory, whereas other populations remain in an area as year-round residents. Understanding migration, its requirements, and consequences is important to the management and conservation of these species, but also difficult because of the often-large geographic range, numerous habitats, and diverse political and land-use backdrops across which migration occurs.

Migration can be studied as four different stages: preparation, movement, stopovers, and arrival (Ramenofsky et al. 1999). During the preparation stage, both physiological and behavioral changes occur to support migration. Changes in metabolism and foraging assist with fuel loading to cover the energetic costs of sustained movement and periods of fasting during migration. Before migration, animals will often exhibit increased activity with an orientation in the direction that they will eventually take when they migrate. The movement stage may occur over land, through the air, or through water, and each mode has its advantages and disadvantages. In general, migration is more costly over land than by air or water, so there are many more examples of extreme, long-distance migration in birds and marine animals than in terrestrial wildlife. During stopovers, migrants settle in a particular habitat to rest and feed. Some species use many stopovers during migration, some very few. Stopover habitats may be used for a few hours to several weeks, depending on the species. In some species, animals store fuel for the entire journey before the start of migration. In others, feeding may occur during stopovers to replace energy stores used up to that point, which can be necessary if carrying large fuel loads is detrimental (e.g., with respect to predator evasion) or if unpredictable conditions occur during movement that increase the use of stored energy (e.g., storms). Upon arrival in wintering habitat, animals must insert themselves into a community of competitors (and predators) made up of migrants and year-round residents. For migrants, this may mean different diets or foraging strategies, as well as different social settings (flocks versus territories), than experienced on their breeding grounds. But the seasonal influx of migratory species may also have profound effects on the existing community of resident species. Upon arrival back on their breeding grounds, migrants may need to fight for and set up exclusive territories, choose mates, and build up energy stores quickly to support reproduction.

Managing and conserving migratory species involves many challenges. In addition to the wide range of habitats used (e.g., winter, breeding, and stopover) and the geographic scale at which their life history plays out, numerous threats face migratory species. As with all species, habitat loss and degradation are threats, but for migratory species a single population may face these threats in numerous countries or even continents. The political challenges of managing a species that is dependent on regulations and land-use changes in multiple countries can be daunting. Even within a country, terrestrial migrants are in danger of having their migratory routes cut off by development or such human activities as increased oil and gas extraction efforts (Berger et al. 2006). Climate change has already impacted the timing of migration in numerous species (McCarty 2001). However, for species that rely on invariant cues like photoperiod to trigger migration, a concern is that changes in habitats caused by climate change will result in arrival at times when the area is no longer suitable. Ultimately, the more we can learn about migration, the better we will be able to manage migratory species in the face of rapid changes in ecosystems.

ANTIPREDATOR BEHAVIOR

For prey species, the correct behavioral response to predation risk can mean the difference between life and death. Antipredator behavior can enable an animal to avoid detection or evade capture by the predator. Animals that rely on crypsis (coloration or patterns that reduce a predator's ability to distinguish prey from the background) may enhance this strategy with behaviors such as immobility or orientation. For example, when an American bittern (*Botaurus lentiginosus*) points its beak upward, the striping on the head, neck, and body matches the reeds in which it lives. Use of refuges can help a prey individual avoid detection by a predator or evade capture. Vigilance—scanning the surroundings for danger—is critical to avoid detection or capture. Early detection of a predator may increase the likelihood of escape or provide more options for predator avoidance. While time spent being vigilant may reduce the time available for other activities, such as foraging, living in groups may reduce this cost. In numerous species, time spent being vigilant by an individual decreases with group size. Other antipredator advantages to living in groups include a dilution effect (less likely that a given individual will be the one attacked) and, in some species, group defense against a predator by mobbing or counterattack.

In addition to the direct demographic effects of predators on prey populations by removing individuals, there has been much recent interest in the indirect, or nonlethal, effects of predators on prey. Laundré et al. (2001) suggested that prey species may perceive different levels of predation risk in different locations within the physical landscape in which they live, creating a virtual "landscape of fear." Prey may adjust their habitat selection based on perceived risk, or adjust their behavior as they move through areas with higher and lower risk of predation. Alternatively, prey may shift their activity schedules to avoid predators. Such nonlethal effects of predators on prey populations will have important implications for managing prey populations in the face of changing predator populations. In fact, the need to understand the impacts on

prey populations resulting from the translocation of wolves (*Canis lupus*) in Yellowstone National Park, and large predators in other parts of the world, was the driving force stimulating this area of research.

In places where top predators have been extirpated for as long as a century, translocation occurs actively by wildlife managers or passively through protections that have allowed the predators to return on their own. This brings up an important question: how will naive prey respond to increased predation risk? Numerous prey species adjust their behavior to different levels of predation risk, being more cautious in areas of high risk or when cues indicate predator presence, and increasing activity or performing more risky behavior when and where predation risk is low. The latter suggests that antipredator behavior is costly or limits other important activities. However, when predators have been absent for many generations, selection for antipredator behavior may be relaxed. Of concern would be changes in recognition of predators, culturally transmitted patterns of predator avoidance, group size, vigilance behavior, active defense, and shifts in habitat use. Often, studies attempting to address changes in behavior associated with the loss of predators have compared behavior in populations with and without predators. However, whether differences in prey behavior are associated with differences in predation risk or differences in habitat between the two sites can be difficult to determine. Experimental methods can compare the behavior of the same individuals with and without predators using playback or presenting olfactory signals of predators. In an experimental study in South Africa, an internal fence bisected the Phinda Resource Preserve, and a perimeter fence surrounded the entire preserve (Hunter and Skinner 1998). Following an absence of at least four decades, lions (*Panthera leo*) and cheetahs (*Acinonyx jubatus*) were translocated to one side of the preserve but not the other. The proportion of time spent being vigilant by impala (*Aepyceros melampus*) and wildebeest (*Connochaetes taurinus*) increased significantly over time in the presence of translocated lions and cheetahs, but not on the side that lacked large predators. In addition, wildebeest group sizes were higher with exposure to the large predators. Impala and wildebeest responded quickly to the return of large predators despite the fact that those individuals had no prior exposure to lions or cheetahs. Although antipredator behavior can be costly, it might not be lost from a species' behavioral repertoire entirely.

WHY DO ANIMALS LIVE IN GROUPS? SOCIAL SYSTEMS

Aldo Leopold realized the importance of understanding group dynamics and the dearth of knowledge on these behaviors in wildlife species that are often secretive and wary of humans: "Cutting across many of these properties is the habit in many species of forming gregarious units. The existing literature tells which species form coveys, herds, and packs, and which do not, but it seldom suggests what these units consist of . . . a brief summary of what little is known about this question is a necessary basis for an understanding of management technique" (Leopold 1933:48).

Although our knowledge of wildlife groups has increased dramatically during the 80 years since these words were written, we still have much to learn. Social systems influence how animals use space. Most mammals and birds are not highly social and live either a solitary life or life paired with a mate; however, where resources are dense and aggregated, assemblages may form. Where groups form, more elaborate social interactions and interindividual bonds may occur. Most species form aggregations at some time during the year, even if they are only ephemeral feeding aggregations at a seasonal food source (Table 14.1).

Social organization in birds and mammals tends to follow a continuum from solitary individuals to increasing levels of overlap and shared duties. The incipient unit of groups is often a pair, or dyad, of individuals. In monogamous species, this may be a male and female; however, in birds, males are the fundamental unit of a group, and in mammals the fundamental unit of a group is the female–female dyad. The distribution of males in birds and females in mammals are often determined by the distribution of resources because of their high levels of parental investment. Male birds hold territories in which nest and food resources provision offspring; female mammals are the only sex able to nurse their young. The distribution of male birds and female mammals often determines the distribution of the opposite sex. For example, male ungulates and Phocidae are distributed most often in ways that maximize their likelihood of encountering females. Most male carnivores, rodents, and lagomorphs expand their home ranges to overlap female groups. Most birds are monogamous, but in the relatively few species that are polygamous or promiscuous (principally game birds), females settle in male territories or select males by visiting lekking grounds.

Most wildlife species are either solitary or form groups with significant spatial overlap but with no shared duties. White-tailed and mule deer and most ungulates, coyotes (*Canis latrans*), raccoons (*Procyon lotor*), eastern fox squirrel (*Sciurus niger*), eastern gray squirrels, most quail (family Odontophoridae), turkeys (*Meleagris gallopavo*), and blackbirds (family Icteridae) are examples (Fig. 14.10). Communal groups share space and duties, most commonly in the care and provisioning of offspring. Examples of this advanced level of sociality include black-tailed prairie dogs (*Cynomys ludovicianus*), gray wolves, a number of primates, banded mongooses (*Mungos mungo*), red-cockaded (*Picoides borealis*) and acorn woodpeckers, and Harris's hawks (*Parabuteo unicinctus*). Eusociality, with extreme division of labor and reproductive castes like the social bees, ants, and termites (order Hymenoptera), is considered to be the highest level of sociality but is only known from one vertebrate, naked mole rats (*Heterocephalus glaber*). Because only one mammalian species practices eusociality, it is of little consequence in wildlife management and conservation.

Why do animals live in groups? Weighing costs and benefits is a fruitful approach to understanding sociality (Waterman

1997). Survival and reproduction tend to increase with population size (small populations experience Allee effects because of reduced ability to find mates), the avoidance of inbreeding, the detection of predators, and the recruitment of nest mates (Stephens and Sutherland 1999). That most wildlife species are solitary suggests that the costs of group living must be substantial in most cases. The costs are especially evident when considering the lengthy list of potential benefits of group living that have been reported in birds and mammals (Table 14.2). Wildlife species that do live in groups tend to gain great benefit from reduced predation risk, location of food, and energetic advantage through drafting off or huddling with other individuals or, when relatives are present, through inclusive fitness gains. Inclusive fitness benefits can be considerable when relatives live in proximity. W. D. Hamilton, the scientist who conceived of the idea, hypothesized that the inclusive fitness benefits accrued by assisting relatives must be devalued by the level of relatedness (r) shared between the two relatives. The simple equation that is Hamilton's rule predicts that cooperation or assistance should be favored when the adjusted benefits (B) exceed the costs (C):

$$C < rB.$$

Birds and mammals that live in groups often include related individuals, and such inclusive benefits may be significant. White-tailed deer, elk, gray wolves, eastern gray squirrels, white-nosed coati (*Nasua narica*), California quail (*Callipepla californica*), wood ducks (*Aix sponsa*), and acorn woodpeckers appear to capitalize on relatedness within groups and demonstrate amicable behaviors toward related individuals.

In some species such as acorn woodpeckers, Florida scrub jays (*Aphelocoma coerulescens*), naked mole rats, beavers (*Castor canadensis*), African wild dogs (*Lycaon pictus*), and gray wolves, individuals of the same species, but who are not the biological parents, provide care in rearing young from a single litter or nest; these individuals are known as alloparents, helpers-at-the-nest, or auxiliaries. This collaborative rearing system is known as cooperative breeding. As many as 9% of all bird species (Cockburn 2006) and less than 3% of all mammal species are considered cooperative breeders (Lukas and Clutton-Brock 2012). In a smaller number of birds, including greater anis (*Crotophaga major*), and mammals, including banded mongooses and white-nosed coati, young from multiple breeders are pooled, and all breeders share care in a system known as communal breeding. Cooperative breeding often occurs where opportunities for alloparents to survive and breed elsewhere are severely limited by habitat availability, habitat quality, and environmental uncertainty (Jetz and Rubenstein 2011). No universal explanation exists for all cases of cooperative breeding. Current evidence suggests that cooperative breeding in birds appears to have evolved most frequently through reciprocity and mutualism in such situations, where helpers gain directly through immediate or delayed benefits of feeding young, although indirect fitness gains through kin selection are also influential. For example, individual help-

Table 14.1. Aggregations of birds and mammals

	Birds			Mammals	
Species	Aggregation		Species	Aggregation	
Adult birds	A flight (in air), flock (on ground), volary, brace (game birds or waterfowl, referring to a pair killed by a hunter)		Apes	A shrewdness	
Bitterns	A sedge		Asses	A pace	
Bobolinks	A chain		Badgers	A cete	
Buzzards	A wake		Bats	A colony	
Chicks	A brood; clutch		Bears	A sloth, sleuth	
Coots	A cover		Bison	A herd	
Cormorants	A gulp		Buffalo	A gang, an obstinacy	
Cranes	A sedge		Cats	A clowder, a pounce; for kittens, a kindle, litter, an intrigue	
Crows	A murder, horde		Cattle	A drove, herd	
Dotterel	A trip		Deer	A herd, bevy (roe deer)	
Doves	A dule, pitying (specific to turtle doves)		Dogs	A litter (young), pack (wild), cowardice (of curs); specific to hounds, a cry, mute, pack, kennel	
Ducks	A brace, flock (in flight), raft (on water) team, paddling (on water), badling		Elephants	A herd	
Eagles	A convocation		Elk	A gang	
Finches	A charm		Ferrets	A business	
Flamingos	A stand		Foxes	A leash, skulk, earth	
Geese	A flock, gaggle (on the ground), skein (in flight)		Giraffes	A tower	
Grouse	A pack (in late season)		Goats	A tribe, trip	
Gulls	A colony		Gorillas	A band	
Hawks	A cast, kettle (flying in large numbers), boil (two or more spiraling in flight)		Hippopotamuses	A bloat	
Herons	A sedge, a siege		Horses	A herd, team, harras, rag (colts), stud (a group of horses of a single owner), string (ponies)	
Jays	A party, scold		Hyenas	A cackle	
Lapwings	A deceit		Kangaroos	A troop	
Larks	An exaltation		Leopards	A leap	
Magpies	A tiding, gulp, murder, charm		Lions	A pride	
Mallards	A sord (in flight), brace		Martens	A richness	
Nightingales	A watch		Moles	A labor	
Owls	A parliament		Monkeys	A troop, barrel	
Parrots	A company		Mules	A pack, span, barren	
Partridge	A covey		Otters	A romp	
Peacocks	A muster, an ostentation		Oxen	A team, yoke	
Penguins	A colony		Pigs	A drift, drove, litter (young), sounder (of swine), team, passel (of hogs), singular (refers to a group of boars)	
Pheasant	A nest, nide (a brood), nye, bouquet		Porcupines	A prickle	
Plovers	A congregation, wing (in flight)		Rabbits and hares	A colony, warren, nest, herd (domestic only), litter (young); specific to hares, a down, husk	
Ptarmigans	A covey		Rhinoceroses	A crash	
Quail	A bevy, covey		Seals	A pod, herd	
Ravens	An unkindness		Sheep	A drove, flock, herd	
Rooks	A building		Squirrels	A dray, scurry	
Snipe	A walk, a wisp		Tigers	A streak	
Sparrows	A host		Whales	A pod, gam, herd	
Starlings	A murmuration		Wolves	A pack, rout or route (when in motion)	
Storks	A mustering				
Swallows	A flight				
Swans	A bevy, wedge (in flight)				
Teal	A spring				
Turkeys	A rafter, gang				
Widgeons	A company				
Woodcocks	A fall				

Source: Modified from Fellows (2012).

Figure 14.10. Assemblages of wildlife species. (A) Aggregation of pintails at a pond. (B) Covey of bobwhite quail. (C) Nesting group of eastern gray squirrels. (D) Band of white-nosed coati consisting of adult females and yearling offspring. (E) Group of female pronghorns, likely related, with territorial buck nearby. Photo credits: A, U.S. Fish and Wildlife Service Migratory Bird Program; B, copyright Joe Coelho; C, Ken Atkinson; D, Greg Boreham; E, James Ownby

ers gain experience, time, grooming partners, and perhaps a future mate, nest, or territory by caring for young that are not their offspring. In mammals, cooperative breeders are strongly associated with monogamy and with a high degree of relatedness among group members that suggests kin selection plays a stronger role among the rodents, canids, mongooses, and primates that display this behavior (Lukas and Clutton-Brock 2012).

Dominance rank is a key aspect of group-living wildlife species. A near-universal truth is that dominant individuals have greater reproductive success than subordinate individuals (i.e., those with low dominance rank), because they are able to control access to mates and resources required by mates. Low-ranking individuals often adopt alternative reproductive tactics (Gross 1996, Koprowski 2007) to enhance their reproductive success. Subordinate male tree squirrels do not challenge the dominant males in pursuit of an estrous female, but they do adopt a satellite tactic that capitalizes on the female's avoidance behavior and garners some mating success

(Koprowski 2007). Similarly, dominant bighorn males tend females and rebuff subordinate males that use a coursing tactic to achieve some reproductive success. Coursing involves initiating a distraction of the tending male to enable the female to race away on a coursing run, during which the subordinate male may successfully breed with the female (Hogg 1984). Subordinate animals are often young or very old, small, in poor health, or unrelated to individuals within the group. The consequences of low rank can include reduced access to resources, resulting in poor condition, reduced survival, low reproductive success, and overall poorer fitness. High-ranking female elk tend to produce more male young, nurse young more substantially, and wean young that are more dominant than low-ranking females (Clutton-Brock et al. 1986). Dominance among pronghorn females results in high-ranking females having better access to food and higher levels of aggression (Fairbanks 1994). Dominant Harris's hawk females have greater access to food, mates, and nest sites (Dawson and Mannan 1991).

Table 14.2. Generalized costs and benefits of living in groups for wildlife species

Costs	Benefits
Increased competition	Dilution effect: decreased per-capita risk of predation
Increased spread of parasites and disease	Confusion effect: predator success rates decrease due to inability to focus on an individual
Increased incidence of cuckoldry	Increased number individuals vigilant
Increased visibility to predators	Increased potential for group defense
Increased reproductive suppression	Increased success in locating food due to group foraging
Increased risk of injury during combat	Increase variety of and proximity to mates
	Aerodynamic or hydrodynamic advantages during locomotion
	Thermodynamic advantage of huddling
	Information transfer between individuals about resources
	Increased cooperation in social interaction
	Increased inclusive fitness

SUMMARY

The field of wildlife management has progressed considerably since Leopold's (1933) observations about the value of understanding animal behavior. Our knowledge has increased with theoretical advances that enrich our understanding of wildlife behavior. Most importantly, applying acquired knowledge to develop informed management strategies has greatly improved. Yet research needs remain considerable and will ensure a plethora of exciting opportunities for wildlife biologists in the future. Pressing future issues involve the response of wildlife species to habitat fragmentation, increased road development, urban sprawl, alternative energy development, and restoration of predators to ecosystems. What will be the behavioral responses to habitat changes induced by invasive plant and animal species and climate change? These management challenges will require scientists with a well-rounded education in the many facets of the field, including a firm grounding in wildlife behavior.

Literature Cited

Beck, J. L., J. M. Peek, and E. K. Strand. 2006. Estimates of elk summer range nutritional carrying capacity constrained by probabilities of habitat selection. Journal of Wildlife Management 70:283–294.

Berger, J., S. L. Cain, and K. M. Berger. 2006. Connecting the dots: an invariant migration corridor links the Holocene to the present. Biology Letters 2:528–531.

Bowyer, R. T. 2004. Sexual segregation in ruminants: definitions, hypotheses, and implications for conservation and management. Journal of Mammalogy 85:1039–1052.

Brommer, J. E., K. Ahola, and T. Karstinen. 2005. The colour of fitness: plumage coloration and lifetime reproductive success. Proceedings of the Royal Society B 272:935–940.

Buttemer, W. A., and W. R. Dawson. 1993. Temporal pattern of foraging and microhabitat use by Galapagos marine iguanas, *Amblyrhynchus cristatus*. Oecologia 96:56–64.

Caudill, C. C. 2003. Measuring dispersal in a metapopulation using stable isotope enrichment: high rates of sex-biased dispersal between patches in a mayfly metapopulation. Oikos 101:624–630.

Clutton-Brock, T. H. 1988. Reproductive success. University of Chicago Press, Chicago, Illinois, USA.

Clutton-Brock, T. H. 1989. Mammalian mating systems. Proceedings of the Royal Society of London B 236:339–372.

Clutton-Brock, T. H., S. D. Albon, and F. E. Guinness. 1986. Great expectations: dominance, breeding success and offspring sex ratios in red deer. Animal Behaviour 34:460–471.

Cockburn, A. 2006. Prevalence of different modes of parental care in birds. Proceedings of the Royal Society B 273:1375–1383.

Conradt, L. 2005. Definitions, hypotheses, models and measures in the study of animal segregation. Pages 11–32 *in* K. E. Ruckstuhl and P. Neuhaus, editors. Sexual segregation in vertebrates: ecology of the two sexes. Cambridge University Press, Cambridge, United Kingdom.

Cristol, D. A., M. B. Baker, and C. Carbone. 1999. Differential migration revisited: latitudinal segregation by age and sex class. Pages 33–88 *in* V. Nolan Jr., E. D. Ketterson, and C. F. Thompson, editors. Current ornithology. Volume 15. Plenum Press, New York, New York, USA.

Cudworth, N. L., and J. L. Koprowski. 2010. Influences of mating strategy on space use of Arizona gray squirrels. Journal of Mammalogy 91:1235–1241.

Davis, J. 2008. Patterns of variation in the influence of natal experience on habitat choice. Quarterly Review of Biology 83:363–380.

Dawson, J. R., and R. W. Mannan. 1991. Dominance hierarchies and helper contributions in Harris' hawks. Auk 108:649–660.

Fairbanks, W. S. 1994. Dominance, age and aggression among female pronghorn, *Antilocapra americana* (family: Antilocapridae). Ethology 97:278–293.

Fellows, D. 2012. Animal congregations, or what do you call a group of . . . ? Northern Prairie Wildlife Research Center, Jamestown, North Dakota, USA. http://www.npwrc.usgs.gov/about/faqs/animals/names.htm. Accessed 15 October 2012.

Fisher, D. O., and I. P. F. Owens. 2008. Female home range size and the evolution of social organization in macropod marsupials. Journal of Animal Ecology 69:1083–1098.

Greenwood, P. J. 1980. Mating systems, philopatry and dispersal in birds and mammals. Animal Behaviour 28:1140–1162.

Griffith, S. C., I. P. F. Owens, and K. A. Thuman. 2002. Extra-pair paternity in birds: a review of interspecific variation and adaptive function. Molecular Ecology 11:2195–2212.

Gross, M. R. 1996. Alternative reproductive strategies and tactics: diversity within the sexes. Trends in Ecology and Evolution 11:92–98.

Harestad, A. S., and F. L. Bunnell. 1979. Home range and body weight—a reevaluation. Ecology 60:389–402.

Hogg, J. T. 1984. Mating in bighorn sheep: multiple creative male strategies. Science 225:526–529.

Hunter, L. T. B., and J. D. Skinner. 1998. Vigilance behaviour in African ungulates: the role of predation pressure. Behaviour 135:195–211.

Jetz, W., and D. R. Rubenstein. 2011. Environmental uncertainty and the global biogeography of cooperative breeding in birds. Current Biology 21:1–7.

Koprowski, J. L. 2005. The response of tree squirrels to fragmentation: a review and synthesis. Animal Conservation 8:369–376.

Koprowski, J. L. 2007. Reproductive strategies and alternative reproductive tactics of tree squirrels. Pages 86–95 in J. Wolff and P. Sherman, editors. Rodent societies: an ecological and evolutionary perspective. University of Chicago Press, Chicago, Illinois, USA.

Laundré, J. W., L. Hernandez, and K. B. Altendorf. 2001. Wolves, elk, and bison: reestablishing the "landscape of fear" in Yellowstone National Park, USA. Canadian Journal of Zoology 79:1401–1409.

Leopold, A. 1933. Game management. University of Wisconsin Press, Madison, USA.

Lin, Y. K., B. Keane, A. Isenhour, and N. G. Solomon. 2006. Effects of patch quality on dispersal and social organization of prairie voles: an experimental approach. Journal of Mammalogy 87:446–453.

Long, R. A., J. L. Rachlow, and J. G. Kie. 2009. Sex-specific responses of North American elk to habitat manipulation. Journal of Mammalogy 90:423–432.

Lukas, D., and T. Clutton-Brock. 2012. Cooperative breeding and monogamy in mammalian societies. Proceedings of the Royal Society B 279:2151–2156.

Lyrholm, T., O. Leimar, B. Johanneson, and U. Gyllensten. 1999. Sex-biased dispersal in sperm whales: contrasting mitochondrial and nuclear genetic structure of global populations. Proceedings of the Royal Society B 266:347–354.

McCarty, J. P. 2001. Ecological consequences of recent climate change. Conservation Biology 15:320–331.

Rachlow, J. L., and R. T. Bowyer. 1998. Habitat selection by Dall's sheep (Ovis dalli): maternal trade-offs. Journal of Zoology 245:457–465.

Ramenofsky, M., R. Savard, and M. R. C. Greenwood. 1999. Seasonal and diet transitions in physiology and behavior in the migratory dark-eyed junco. Comparative Biochemistry and Physiology A 122:385–397.

Rogerson, J. D., W. S. Fairbanks, and L. Cornicelli. 2008. Ecology of gastropod and bighorn sheep hosts of lungworm on isolated, semiarid mountain ranges in Utah, USA. Journal of Wildlife Diseases 44:28–44.

Rubin, E. S., and V. C. Bleich. 2005. Sexual segregation: a necessary consideration in wildlife conservation. Pages 379–391 in K. E. Ruckstuhl and P. Neuhaus, editors. Sexual segregation in vertebrates: ecology of the two sexes. Cambridge University Press, Cambridge, United Kingdom.

Shine, R., and M. Wall. 2005. Ecological divergence between the sexes in reptiles. Pages 221–253 in K. E. Ruckstuhl and P. Neuhaus, editors. Sexual segregation in vertebrates: ecology of the two sexes. Cambridge University Press, Cambridge, United Kingdom.

Sims, D. W. 2005. Differences in habitat selection and reproductive strategies of male and female sharks. Pages 127–147 in K. E. Ruckstuhl and P. Neuhaus, editors. Sexual segregation in vertebrates: ecology of the two sexes. Cambridge University Press, Cambridge, United Kingdom.

Stamps, J. A., J. M. Davis, S. A. Blozis, and K. L. Boundy-Mills. 2007. Genotypic variation in refractory periods and habitat selection by natal disperser. Animal Behaviour 74:599–610.

Stephens, P. A., and W. J. Sutherland. 1999. Consequences of the Allee effect for behaviour, ecology and conservation. Trends in Ecology and Evolution 14:401–405.

Stewart, K. M., T. E. Fulbright, D. L. Drawe, and R. T. Bowyer. 2003. Sexual segregation in white-tailed deer: responses to habitat manipulations. Wildlife Society Bulletin 31:1210–1217.

Waterman, J. M. 1997. Why do male Cape ground squirrels live in groups? Animal Behaviour 53:809–817.

Wearmouth, V. J., and D. W. Sims. 2008. Sexual segregation in marine fish, reptiles, birds and mammals: behaviour patterns, mechanisms and conservation implications. Advances in Marine Biology 54:107–170.

Whiting, J. C., R. T. Bowyer, J. T. Flinders, V. C. Bleich, and J. G. Kie. 2010. Sexual segregation and use of water by bighorn sheep implications for conservation. Animal Conservation 13:541–548.

HABITAT

R. WILLIAM MANNAN AND ROBERT J. STEIDL

INTRODUCTION

Managing habitat is fundamental to wildlife conservation because individual animals and the populations they comprise cannot persist without an appropriate place to live. Populations decimated temporarily by environmental contaminants or overharvest, for example, can recover if adequate habitat is available. In contrast, a reduction in the amount of habitat for a species will lead to permanent decreases in abundance and distribution unless habitat is restored. Manipulating habitat, therefore, is a primary strategy that biologists use to alter the distribution and abundance of species of conservation or management concern. In this chapter we provide a theoretical and practical overview of the concept of habitat, including defining habitat and reviewing ideas about how animals identify and select habitat. We then introduce concepts and methods important for studying habitat of terrestrial vertebrates, including issues in the design and analysis of studies to characterize habitat, describe habitat use, and assess habitat selection. We conclude with examples that illustrate strategies for managing habitat for single and multiple species.

DEFINITION

Habitat is an area with the combination of resources (e.g., food, cover, water) and environmental conditions (e.g., temperature, precipitation) that promotes residency by individuals of a given species and allows them to survive and reproduce (Morrison et al. 2006). Habitat is inherently a species-specific concept because each species has a unique set of physiological, morphological, and behavioral adaptations that are suited to a particular suite of resources and conditions. Although two or more species may inhabit the same general area, their specific habitat needs are likely to be unique; therefore environmental changes will affect them differently, even if they are sympatric. Qualifiers of habitat—breeding habitat, nonbreeding habitat, and foraging habitat—are useful ways to identify areas or sets of conditions required by an animal during some part of its life cycle or to meet a specific need, but should not be miscon-strued to represent all the places or resources an animal needs to survive and reproduce.

Use of the term "habitat" can differ markedly from the species-specific concept outlined above (Morrison et al. 2006:10). For example, habitat is sometimes used to denote a particular vegetation community, such as "ponderosa pine (*Pinus ponderosa*) habitat." Use of the term in this context probably grew from the phrase "habitat type," coined by Daubenmire (1976:125) to refer to "land units having approximately the same capacity to produce vegetation." Habitat also is sometimes used to describe a general physical environment, such as riparian habitat or mountain habitat. When biologists refer to an area as "suitable habitat" or "unsuitable habitat," they are most likely noting whether an area will (or will not) support a given species. Because habitat refers implicitly to those areas that are suitable for a given species, the phrase "suitable habitat" is redundant, and "unsuitable habitat" is an oxymoron. Although these phrases are commonplace, they dilute the species-specific concept of habitat and can hinder communication, especially when mixed together in the same document; we therefore suggest that they be avoided.

HABITAT SELECTION

Inherent in the definition of habitat provided above is the idea that individuals of a species will settle and establish residency in areas that contain the set of physical and biological resources necessary for their survival. The duration of time an individual resides in an area can vary. Animals that do not migrate might, after natal dispersal, remain in the same area throughout their lives, whereas migratory animals might inhabit one area for weeks or months during the breeding period, inhabit another area for weeks or months during the nonbreeding period, and inhabit multiple areas, each for a short time, during migration. Presumably, animals seeking an area to inhabit recognize suitable sites by the presence of environmental cues that are either directly or indirectly associated with resources in an area (Lack 1933, Svardson 1949,

Hilden 1965). The link between environmental cues and the resources they indicate is essential to the process of habitat selection. Over evolutionary time, individuals that recognized and selected areas with the appropriate set of resources had higher survival and reproductive success than individuals that selected areas with fewer resources. Over many generations, individuals with genotypes that allowed them to recognize appropriate areas have come to dominate populations. Habitat selection, therefore, is a behavioral process that has been honed by natural selection to ensure that individuals settle in those areas that contain resources necessary for their survival and reproduction (Jaenike and Holt 1991). The patterns of residency that we observe in nature are manifestations of that long-term evolutionary process. And although we might expect some systematic variation in habitat of a species across its geographical range, the suite of resources necessary for an area to provide habitat for individuals of a species is likely to be reasonably consistent within a species.

Environmental cues that trigger an animal to settle in an area might include those associated with food, nest sites, and cover from predators or inclement weather, but also those associated with the presence of conspecifics (Stamps 1987, Citta and Lindberg 2007) and interspecific competitors (Williams and Batzli 1979). Collectively, cues should indicate areas that have all of the resources needed by an animal and that competitors do not inhabit. In some circumstances, learning can play a role in aspects of habitat selection, such as when an animal's experiences in its natal environment (Davis and Stamps 2004) and prior adult experiences (Baker 2005) influence its selection of an area. Importantly, absence of a single essential resource or its associated cue could prevent an animal from settling in an area. Resources that are absent or in short supply and that limit the number of residents are called limiting factors. If biologists can identify these features and manage to increase their abundance, then the number of residents an area supports can be increased. In some instances, adding an essential habitat feature that is missing can even create new areas of habitat (Hamerstrom et al. 1973).

Habitat selection by many animals is likely a complex process that involves a set of behavioral responses to environmental cues that span a range of spatial and temporal scales. One set of cues, for example, might trigger an animal to initiate a settling response, but other cues might be necessary for an animal to remain resident throughout its life cycle, breeding cycle, or nonbreeding period. This process is often considered to be hierarchical, because cues expressed at large geographical extents (e.g., presence of a particular plant community) can trigger an animal to initiate a localized search (e.g., for a particular species of nest tree), and more specific cues (e.g., a structural feature of a tree that serves as a place to locate a nest) might narrow the residency response to a particular site. There also might be instances where this hierarchy could be inverted, with animals first identifying important resources at a particular site (e.g., a specific feature that would support a nest) before ensuring the presence of necessary resources at larger spatial extents (e.g., presence of a particular plant com-

munity; Flesch and Steidl 2010). For highly mobile animals, the spatial scale of the habitat selection process can span the spectrum from broad regional attributes, such as vegetation zones, to particular resources within a home range (Wiens 1985). The process of selection for less mobile animals also could be hierarchical, but confined within smaller spatial extents because of limitations in perception and mobility.

Understanding the behavior of habitat selection is important when providing man-made structures as habitat features. For example, if cavities are a limiting factor for a population of secondary cavity-nesting birds, then providing artificial nest sites might create new areas of habitat and increase the size of an existing population. However, a nest box nailed to a living tree or affixed to a metal pole does not resemble all aspects of a natural cavity in a dead tree. In these situations, biologists rely on the idea that animals sometimes respond only to a subset of features when identifying some important resources. If a nest box mimics appropriately the critical environmental features associated with a natural nest cavity, such as size of the cavity opening and depth of the cavity, and if the box is positioned in an area that includes the other resources needed by a species, then a cavity-nesting bird such as a wood duck (*Aix sponsa*) might settle in the area and use the box for nesting.

Habitat Quality

Habitat selection can be considered a threshold response (Wiens 1985), where settling behavior is triggered when necessary resources in an area reach critical levels of abundance. Not all areas with resources above this threshold will be equal, however; plus, abundance of many habitat-related resources can change over time. Variation in levels of important habitat resources among areas is likely to be expressed as variation in survival and reproduction among resident individuals (Johnson 2007). Although it is plausible that some of the observed differences in demographic performance among animals could be due to variation in the quality of the animals themselves—including inherent variation in their ability to recognize suitable places or to survive and reproduce after settling—we favor the idea that every individual of a species can recognize habitat and will settle in locations that meet some minimum threshold of resources, provided those locations are available. Therefore observed differences in reproduction and survival among areas that provide habitat for a species do not, in our view, often reflect differences in genetic fitness, but instead likely reflect differences in habitat quality. Consequently, demographic performance of a population is likely a good measure of the quality of habitat it occupies (Johnson 2007). We favor this idea because habitat selection is a critical behavioral process and, like all essential behaviors and morphologies, is unlikely to vary markedly among most individuals. Thus areas with sufficient resources to support consistently high densities of residents or high rates of survival and reproduction of individuals can be classified as high-quality habitat. Conversely, areas with resource levels that exceed the threshold for habitat and that trigger settling, but where densities of residents and rates of survival and reproduction are consistently low, can be classified as low-quality habitat. Habitat quality for a given species, therefore, is a product of the types and abundances of important resources in an area, and how those resources allow members of the species to deal with competitors and predators. If the suite of resources necessary for a species is well known, then habitat quality could conceivably be determined by measuring the resources themselves, but knowledge of the resources critical to most species is uncommon. Thus habitat quality might best be gauged by assessing demography of the species in question (i.e., the number of individuals that inhabit an area and their relative rates of survival and reproduction; Van Horne 1983, Johnson 2007) and the duration of time an area is inhabited.

Habitat selection behaviors are implicit in theoretical models that describe how animals might select among patches (i.e., discrete and identifiable areas) of habitat that vary in quality. Some models, such as the preemptive form of ideal free distribution, suggest that within a heterogeneous landscape, patches of the highest quality will be inhabited first (Brown 1969, Fretwell and Lucas 1970), with patches of lesser quality being inhabited only when higher-quality patches are saturated. Models of this kind depend on animals being able to gauge patch quality and to sample widely across numerous patches before settling. Although the highest-quality habitat for a species often is inhabited consistently, predictions of the ideal free distribution model may not be fully expressed in nature because the number of patches of habitat an animal samples depends on its perceptual ability, its mobility relative to the distribution of patches (Lima and Zollner 1996), what patches it encounters first or most often (Kristan 2003, Krishnan 2007), and risks associated with predation and losing a patch to a conspecific competitor while sampling.

Ecological Traps

The process that animals use to select habitat must be reliable. Humans have altered many natural environments, which can sometimes disconnect the environmental cues that animals use to identify habitat and the resources on which animals rely. An ecological trap is an area with cues that it provides habitat resources, but where animals have relatively low rates of survival, reproduction, or both. An ecological trap in essence tricks animals into settling in areas that are not favorable to them (Schlaepfer et al. 2002, Battin 2004, Robertson and Hutto 2006).

Reduced demographic success in ecological traps can arise in at least two ways. First, resources important for survival or reproduction are absent in the area, despite cues indicating their presence. The Arizona cotton rat (*Sigmodon arizonae*), a native rodent common in semidesert grasslands of the southwestern United States, is abundant in areas dominated by the nonnative plant, Lehmann lovegrass (*Eragrostis lehmanniana*). Survival of cotton rats in patches of Lehmann lovegrass is relatively high, but reproduction is well below that of populations in areas of native grasses (Litt 2007). Apparently, some critical resource (such as food) required for successful reproduction is reduced in grasslands dominated by the nonnative

grass. Ecological traps can also reduce demographic success by harboring a novel feature, event, or organism with which a species has not evolved and that reduces habitat quality by increasing mortality or reducing reproduction. Farming operations that destroy nests of ground-nesting birds (Rodenhouse and Best 1983), utility poles that electrocute raptors (Dwyer and Mannan 2007), and cars that kill animals living near roads (Mumme et al. 1999) are examples of events and features that reduce survival, reproduction, or both, and that could create ecological traps (see Chapter 7 for a description of the potential influence that ecological traps may have on populations dynamics). Although ecological traps might exist in natural conditions, we believe they are rare because, over evolutionary time, there has been strong selective pressure against misleading cues. The likelihood of ecological traps is therefore higher in areas where humans have rapidly modified natural landscapes, and where there has been inadequate time for an evolutionary response.

CHARACTERIZING HABITAT

Every wildlife biologist has an intuitive understanding of what constitutes habitat for familiar species, so describing habitat for a species seems like it should be a straightforward exercise. On careful consideration, however, characterizing the suite of biotic and abiotic features that constitute habitat for a species can become surprisingly complex, especially without first constraining the concept of habitat in space or time. The overarching nature of the habitat concept, which can be defined at multiple spatial and temporal scales, measured in a variety of ways and resolutions, considered for different demographic subsets of a population, and analyzed in what seems like an endless variety of ways, can explain much of the complexity. Consequently, although generally intuitive, the habitat concept comprises a rich and intricate set of topics based on fundamental interactions between animals and their environment, any subset of which might be relevant to a particular set of conservation or management questions. Few topics have generated as much thought and discussion among wildlife scientists as strategies to characterize and evaluate habitat.

The ultimate challenge of characterizing habitat and understanding the process of habitat selection involves identifying the environmental features that trigger an individual's decision to settle in an area and the resources required for successful survival and reproduction. Such identification is sometimes difficult because animals respond to resources at different scales; therefore the scale at which we study habitat can affect our perspective on how habitat is used (Wiens 1989). Further, many species are cryptic and secretive, so we cannot always be certain that we have identified all areas used by a species. Lastly, we often wish to establish the relative importance of resources used by animals compared to the availability of those resources in the environment, which requires that we evaluate resource use relative to some reference; the way in which we establish that reference can influence our conclusions about the relative importance of different resources.

Design Considerations for Habitat Studies

There are many effective strategies for assessing habitat and identifying habitat features important to animals, all of which involve assumptions and compromises. In general, a strategy for collecting data should reflect the fundamental question of interest, which in turn will dictate the scope of inference (i.e., how broadly the information gained can be applied).

Scale

Scale is a construct we impose on complex systems to help us better understand patterns in nature, including patterns of habitat use by animals (see Chapter 6 for a review of the concept of scale). Because we can consider and evaluate the ways that animals use habitat across a range of spatial scales (we use the term "scale" to refer to extent and not resolution), it is often helpful to consider habitat use as a spatial hierarchy, with choices made by animals at one spatial scale constraining choices at smaller spatial scales. Although habitat for a species can be considered along a continuum of spatial scales, classifying habitat into a three-level spatial hierarchy has proven useful for many wildlife species, with geographical distribution of the species constraining choices at the broadest level, previous choices at the geographical scale constraining choices at the home-range scale, and previous choices at the home-range scale constraining choices of patches for foraging, nesting, and other activities (Johnson 1980). No single spatial scale is correct for all habitat studies, because animals respond to resources and other aspects of their environment at multiple scales (Wiens 1989, Levin 1992). Nonetheless, the scale (or scales) we use to evaluate habitat is a critical element of a study design, because it constrains the habitat features that we can potentially discern as important to a species. For example, an individual animal may select an area for its home range that is dominated by one vegetation community. If we contrast vegetation at locations used by an individual during its daily activities with the vegetation available within its home range, we could conclude that the vegetation community in which the animal lives out much of its life was not especially important, as it was not used at levels higher than was available. For this reason, the spatial scale of a study can affect its results. The appropriate spatial scale for a study must be based on the scale of the research question and should reflect the scale at which an animal is thought to respond to changes in abundance and distribution of the resources of interest (Wiens 1989). In part because of the uncertainty related to which scale (or scales) is most important to a species, most studies assess habitat use at multiple spatial scales.

In addition to issues of spatial scale, habitat studies must also consider issues of temporal scale, because habitat resources important to individuals can vary throughout the day, such as when foraging or resting, throughout the year, especially between breeding and nonbreeding periods, and in response to environmental changes that can occur over longer time periods. For example, for species that inhabit dynamic systems that are altered frequently, such as the intertidal zone, habitat features on which these species depend are integrally linked to the periodic disturbances that characterize these systems and govern their structure. Over longer time periods, temporal changes in the composition of a plant community that result from the process of ecological succession (see below) have important consequences for the habitat of many animals.

Sampling and Inference

The foundation for inference in habitat studies is similar to any other scientific study where it is usually impractical or impossible to measure every unit in the population we wish to know—inference is based on information gathered from a subset of units selected from all units in the larger population of interest (instances where we study every individual in a population are rare). For example, we might be interested in characterizing habitat features used by grasshopper sparrows (*Ammodramus savannarum*) breeding in a particular geographical region. Because it is unlikely that we would be able to survey all grasslands in the region that might be inhabited by this species, or that we would be able to locate every individual, what we know about this species in this region (i.e., our inference) will be based on the subset of areas we survey and individuals we locate. When we select the subset of units to study from a larger population through a randomization scheme (e.g., a stratified random sample), then inference is design based, which means we justify inference to the larger population of interest by the approach we used to select sample units (i.e., the sampling design).

In habitat studies, two types of sampling units are common: animals and plots of land. If the sampling unit is an individual animal, such as in radiotelemetry studies or studies based on individual features such as nests or den sites, a subset of units is selected from the larger population of animals or features that is the target of inference. In these studies, individual animals or features are the sampling unit, the collection of individuals selected for study is the sample, and the overall collection of individuals from which the sample is drawn is the population. If the sampling unit is a plot, such as in field surveys where identities and locations of individuals are not known in advance, surveys to assess presence or abundance of individuals on each plot establish use of plots by animals. When the entire study area will not be surveyed (i.e., when sampling), the subset of plots to survey is selected from the larger population of plots that is the target of inference. In these studies, plots are the sampling unit, the collection of plots studied is the sample, and the overall collection of plots from which the sample is drawn (the universe or frame) is the population about which we wish to draw an inference. In studies where the plots are the sampling unit, plots are often assigned to different general land-cover classes, usually on the basis of the dominant vegetation community or other features thought to influence habitat use by animals.

Common Types of Habitat Studies

Although there are many approaches for studying animal habitat, the two most common types of studies are those that assess *habitat use* and *habitat selection*. Studies of habitat use usually involve gathering information about characteristics of locations used by animals with the goal of characterizing environmental features associated with these locations. Studies of habitat selection usually evaluate the level of use of habitat resources by animals relative to the general availability of those resources in the environment. In contrast to studies of habitat use, studies of habitat selection span a wider range of possible sampling designs and therefore encompass a wider range of analyses that have become increasingly sophisticated.

Imperfect Observations of Animals and Presence-only Data

Because most animals are impossible to detect with certainty during many field studies, information from studies based solely on descriptions of locations where animals have been observed can be biased when the probability of detecting animals varies across the study area. Specifically, if the probability of detecting an animal is associated or confounded with one or more habitat features, results of habitat studies could be biased. For example, if abundance or presence of a species is relatively consistent across an area, but tree density varies and reduces the ability of surveyors to detect animals that are present, then we might conclude incorrectly that abundance was lower in areas with high tree densities if we failed to adjust for the influence of tree density on detection. Further, if the set of habitat-use locations was not generated from a planned survey, but instead was gathered from incidental observations such as might be available from museum records or Natural Heritage databases, the validity of inferences about habitat use depends on the sample of used locations representing all locations used by the target species in the area of interest (Pearce and Boyce 2006, Phillips et al. 2009). Contemporary approaches to using presence-only data to model species distributions on the basis of environmental features attempt to overcome these potential biases and to make these inferences more reliable, such as those implemented through MaxEnt software (Phillips et al. 2006, 2009). Simply, the strategy is to develop a model that predicts the distribution of a species based on the contrast between environmental features at use locations and the distribution of those features across the landscape of interest (Elith et al. 2011).

Habitat Use

Understanding the specific resources that are important to habitat of a species is often the first step in developing strategies for conserving or managing animal populations, because manipulating habitat is the primary means of influencing the distribution and abundance of animals. Although studies of habitat use often are descriptive, quantitative assessments of important habitat features are critical for developing habitat prescriptions. If all individuals in a population use a particular habitat feature, such as a cavity for nesting, then it seems reasonable to assume that this feature is an important habitat component, regardless of its availability. Further, if all other habitat features required by a species are present in an area where this feature is lacking (i.e., a limiting factor), adding the missing feature could enhance habitat for that species.

All studies designed to identify important habitat features must characterize aspects of the areas used by animals; therefore a fundamental activity of any habitat study is to identify reliably those locations used by the species of interest. This process can focus on locations used by animals when individuals are the sampling unit, whether their identity is known, such as when a bird has been banded or a mammal tagged, or can be inferred, such as when the sample of locations used is determined by observations of unmarked animals. The process can also focus on characterizing plots that have been classified as used by animals, in which case plots, rather than individuals, are the sampling units. The way that use is characterized has implications for analysis, which we explain below.

Habitat Selection

More common than studies to characterize habitat use are studies to assess habitat selection (Thomas and Taylor 2006), where the general goal is to characterize the degree to which animals *use* habitat features relative to their *availability* in the environment. Although these types of studies share a name with the behavioral process of habitat selection—the set of behaviors that animals employ to identify areas with the resources they need to survive and reproduce—habitat selection studies do not often assess this process but instead contrast the characteristics of habitat features in areas used by animals relative to those in locations unused or available to a species in a defined area. In these types of studies, selection implies that a resource is used in greater proportion than is available in the environment.

Even when there is no evidence that a species selects a resource, that resource can still be an essential component of its habitat. In particular, resources that are abundant in the environment are unlikely to be classified as selected even when they are essential to habitat, because selection is not an absolute measure of importance of a resource but a relative measure of use contrasted with availability in the environment. Consequently, resources that animals select in higher proportions than available are not likely to be the only resources important as habitat features (Garshelis 2000). When a habitat resource is essential, such as a cavity for secondary cavity-nesting birds, but is not in limited supply, we will likely find no evidence of selection for that resource, even though it is clearly an integral habitat component. In general, we expect to observe strong evidence of selection for an important habitat resource when the resource is uncommon and to observe no evidence of selection when the feature is common. If evidence of selection for a habitat resource increases as availability increases, however, this resource might be a limiting factor for the species. Therefore evaluating the importance of a resource to a species must be interpreted within the context of how availability is defined, regardless of whether a resource is limiting for a species, and the relative amount of resources in the area considered available to a species.

If absolute use of a particular habitat feature is consistent among individuals, the feature is likely to be important to the species in the region of interest. In contrast, the magnitude of selection can vary among individuals, even when resource use is consistent, because of variation in availability of resources in the reference area. For example, use of several nest-related resources was consistent across multiple vegetation communities inhabited by cactus ferruginous pygmy owls (*Glaucidium brasilianum cactorum*) in Sonora, Mexico. Because the availability of these resources varied geographically, however, the magnitude of selection of these resources varied with availability (Flesch and Steidl 2010).

Habitat selection is a specific topic within the broader discipline of resource selection, and many of the methods and tools relevant to evaluating resource selection are applicable to studies of habitat selection (Manly et al. 2002). There are multiple sampling designs available for assessing habitat selection (Table 15.1), which have been classified into four main types based on the approaches used to quantify resource use and resource availability (Thomas and Taylor 1990, 2006; Manly et al. 2002). In design I studies, animals are not identified uniquely, resource use is quantified on the basis of information gathered from surveys on plots (often presence or abundance), and availability is characterized at the level of the study area. In design II studies, animals are identified uniquely, resource use is quantified on the basis of information gathered from individuals, and availability is characterized at the level of the study area. Design II studies are common when animals are radio-marked. Resource availability in design I and design II studies can be based on either complete surveys of resources in the study area (e.g., with data from remote sensing) or estimates of resource availability generated by sampling a subset of the entire study area, such as might be gathered through a sample of plots established at random across a study area. In design III studies, animals are identified uniquely, and both resource use and resource availability are quantified for each individual. An example design could involve defining use based on specific locations of a radio-marked or territorial animal and defining availability based on resources measured in the home range of that individual (Aebischer et al. 1993). Lastly, in design IV studies, resource use is defined by locations for each individual animal, and resource availability is defined uniquely for every use location. The idea is to assess the choices an individual makes in light of the resources available in the immediate vicinity; the spatial and temporal limits of "immediate" are flexible (Erickson et al. 2001).

Habitat Analyses

The type of sampling unit, the type of data collected, and how, when, and where those data are collected (i.e., the sampling design) govern the types of analyses that are appropriate for all scientific studies, including habitat studies. Some common considerations involved with analyzing data from habitat studies relate to collecting multiple observations from the same animal, accounting for our inability to detect individuals with certainty and the breadth of sampling designs available in

Table 15.1. Characteristics of common statistical methods of resource selection

Characteristics	Neu et al. (1974)	Johnson (1980)	Friedman (1937)	Compositional analysis: Aebischer et al. (1993)	Logistic regression: Hosmer and Lemeshow (2000)	Log-linear modeling: Knoke and Burke (1980)	Discrete choice: Hensher et al. (2005)
Use based on unmarked individuals; availability measured at population level (design I)	Yes	No	No	No	Yes	Yes	Yes
Use based on marked individuals; availability measured at population level (design II)	No[a]	Yes	Yes	Yes	Yes	Yes	Yes
Use based on marked individuals; availability measured for each individual (design III)	No[a]	Yes	Yes	Yes	Yes	Yes	Yes
Assumes temporal independence of locations	Yes	No	No	No	Yes/no[b]	Yes/no[b]	No
Assumes independence among animals	Yes	Yes	Yes	Yes	Yes	Yes	Yes
Assumes sample of animals representative of the population; inferences based on average selection in the population	Yes[c]	Yes	Yes	Yes	Yes[d]	Yes[d]	Yes
Allows for categorical covariates (e.g., sex, age class)	No	No	No	Yes	Yes	Yes	Yes
Allows use of continuous covariates (e.g., distance to roads, body mass)	No	No	No	No	Yes	No	Yes

Source: Adapted from McDonald et al. (2005).

[a]Method can be applied after pooling data across individuals, but this is not recommended.

[b]When data are collected for multiple animals, independence among animals is important; animals must be identified as the unit of analysis.

[c]Assumes that the measure of habitat use represents the measure for the population.

[d]Inference to the population is justified if animals are treated as units of analysis (i.e., replicates).

studies of habitat selection, many of which require careful attention to details of the analysis.

Multiple Observations per Animal

Collecting multiple observations from the same sample unit over time is a common strategy in ecological studies. In habitat studies involving marked animals, this approach can be an efficient strategy for evaluating the range of locations and habitat features used by an individual during different activities or different time periods (or both). Multiple observations collected from the same individual are not independent (Aebischer et al. 1993, Otis and White 1999), and therefore the approach to data analysis must ensure that the number of animals—not the number of locations—are treated as the primary sampling unit.

In general, for habitat studies where individual animals are the sampling unit and habitat use by individuals is characterized by recording a series of locations over time, there are three common alternatives for analysis: (1) analyze separately the data for each individual separately, (2) analyze data for all individuals combined but disregard the identities of individuals sampled, or (3) analyze data for all individuals combined but include the identity of individuals as a factor in the analysis. If data from individuals are analyzed separately (alternative 1), then inference about the population from which they were selected can be based on evaluating consistency in patterns of habitat use or selection among individuals. Although this approach can be reasonable when the number of individuals studied is small, analyses at the individual level are not as statistically powerful or as scientifically compelling as analyzing data for all individuals combined. We do not recommend combining observations across individuals without also identifying the individual as the sample unit in the analysis (alternative 2), because this approach inflates the true sample size in the study. The appropriate sample size is typically the number of sample units studied and not the number of locations recorded (Aebischer et al. 1993, Alldredge et al. 1998); using the number of locations inflates sample size and exaggerates precision of estimates. For example, analysis of habitat use recorded for 20 individuals on five occasions should reflect a sample size of 20 sample units, not 100. Further, when the number of observations recorded varies among individuals, individuals with more locations can have a disproportionate influence on results. Combining observations across individuals while also identifying individuals as the sampling unit in the analysis (alternative 3) overcomes the limitations inherent in the other approaches, reflects the underlying two-level hierarchy inherent in the data (individuals and observations), and reflects explicitly the inferential foundation of individual samples representing the larger population from where they were selected. There are several methods for analyzing data in this way, one of which is to treat individual units as a random effect in the model for analysis (Gillies et al. 2006). Identifying individuals as a random effect recognizes that the particular set of individuals studied is a sample of all individuals within a population, and includes a measure of uncertainty that re-

flects this source of sampling error (i.e., if we had selected a different subset of individuals, the estimates we generated would be slightly different).

Sampling Designs for Habitat Selection

Two common approaches to study habitat selection are to contrast characteristics of habitat features in areas *used* by animals relative to those in locations that are *unused* (use-nonuse studies) or locations thought to be available to the species (use-availability studies). Use-nonuse designs maximize the contrast between locations used by an individual (habitat) and the reference (nonhabitat); the disadvantage is that it is difficult to classify with certainty areas as unused. Use-availability designs have the advantage of not needing to classify sites as unused, although the contrast between used locations and the reference (availability) is weaker than in use-nonuse studies, because measures of availability usually include areas used by animals, which reduces the relative contrast between used and available locations relative to used and unused locations. Classifying with certainty areas as unused, however, can be difficult in many circumstances, because cryptic species that are present in an area can be easily overlooked during surveys or temporarily absent from an area they routinely inhabit. Contrasting used with unused or available locations while accounting for uncertainty in the process of identifying used locations introduced by imperfect detection is a useful and important strategy for overcoming potential biases in the sample of used locations (Gu and Swihart 2004); these studies require additional effort in the form of multiple visits to some or all of the sampling units (MacKenzie 2006).

For most studies of habitat selection, there is usually more than one reasonable strategy for analyzing data. Strategies can vary appreciably in their flexibility to work with different sampling designs, both in their assumptions and in the type of information they provide to facilitate interpretation of results, and ultimately our understanding of habitat selection. In a classic study of moose (*Alces alces*) habitat selection in Minnesota (Neu et al. 1974), available habitat was based on the size of the study area classified in each of four general land-cover classes, and habitat use was based on the number of locations of moose or tracks in each class observed during aerial surveys (Table 15.2). This is an example of a design I study, where use was measured for individual animals that were not identified uniquely, and habitat availability was determined at the level of the study area (Table 15.1). A common analysis for habitat data recorded as counts of animals in different habitat classes is based on contingency tables, which are often used to evaluate the null statistical hypothesis that habitat use is independent of availability (i.e., the null that moose do not use habitat classes more or less than their availability in the study area). A common test statistic for analysis of contingency tables is Pearson's chi-square (Neu et al. 1974, Byers et al. 1984, Alldredge and Griswold 2006):

$$\chi^2 = \sum_{i=1}^{k} \frac{(O_i - E_i)^2}{E_i},$$

where k is the number of habitat classes, O_i is the observed frequency of use for each habitat class i, and E_i is the expected

Table 15.2. Habitat classes and habitat-related statistics for moose in Minnesota

Habitat class	Available		Use		Number of moose expected under null hypothesis	Selection ratio	
	Hectares in study area	Proportion of study area	Number of moose observed	Proportion of use		Estimate	SE[a]
In burn, not near periphery	4,570	0.340	25	0.214	39.8	0.63	0.111
In burn, near periphery	1,355	0.101	22	0.188	11.8	1.86	0.358
Out of burn, near periphery	1,394	0.104	30	0.256	12.2	2.46	0.388
Out of burn, not near periphery	6,128	0.455	40	0.342	53.2	0.75	0.096
Total	13,447		117				

Source: From Neu et al. (1974).

[a]SE, standard error.

frequency of use based on the availability of each habitat class *i* in the study area. For the moose data, the test statistic indicates that we should reject the null hypothesis of independence among habitat use and availability ($\chi^2 = 43.5$, $P < 0.0001$), which suggests that moose use one or more habitat classes disproportional to its availability (Table 15.2).

Although this analysis is appropriate given the study design and the type of data collected, it is not especially informative because it provides no information on which habitat classes were selected, avoided, or used in proportion to availability; additional calculations are necessary to evaluate the specific patterns of habitat selection. This shortcoming is true of all contingency table analyses based on chi-square tests because they are not designed around parameters that link directly to patterns of selection, which should be the primary focus of results of any resource selection study. One parameter that can be computed to facilitate the biological interpretation of these data is a selection ratio, which is available in several forms but is always based on a ratio of habitat use to availability. When use and availability are approximately equal, the basic selection ratio (use:availability) will be near 1.0, indicating no evidence of selection, and when use is higher or lower than availability, the ratio will be appreciably greater than or less than 1.0, respectively, indicating that resource use is more or less than available. In the moose example, the estimated selection ratio for the habitat class "out of burn, near periphery" was approximately equal to 2.5 (standard error, or SE = 0.39), suggesting that moose used these areas more frequently than expected based on availability. Confidence intervals can be computed for selection ratios to standardize decisions about whether use of a resource differs significantly from availability (McDonald et al. 2005).

When used in combination with selection ratios, analyses based on contingency tables can be informative for studies that evaluate few habitat factors, and the data collected are counts of animals or their sign. When study designs become more complex, however, results from contingency table analyses become less informative and are cumbersome to interpret. In most circumstances, we prefer analyses that are designed around parameters that facilitate interpretation of results. Among the available alternatives, many biologists have ad-

opted regression-type approaches to analysis, most of which are based on *generalized linear models*, which provide a comprehensive, flexible, and compelling framework for analyzing data from a wide range of resource selection studies (Manly et al. 2002). Generalized linear models link the mean of the response variable to one or more explanatory variables through a regression function, where the appropriate link function depends on the distribution of the response variable (McCullagh and Nelder 1989). Parameter estimates are straightforward to interpret because they represent the influence of a one-unit change in the level of the explanatory variables on the mean response, which in studies of habitat selection is always some measure of the magnitude of selection. Generalized linear models can be run with nearly all general-purpose statistical software (e.g., SAS, R, JMP).

To illustrate this approach, we assessed the effect of woody plants on habitat selection of two songbirds, verdin (*Auriparus flaviceps*) and eastern meadowlark (*Sturnella magna*), that were sampled on 40 ten-hectare plots in semidesert grasslands of southern Arizona. We surveyed plots for breeding birds six times between May and August 2006 and estimated the combined density of shrubs and trees on each plot, which we transformed with the natural log to normalize the right-skewed distribution (Table 15.3). We classified a plot as used for breeding by each species if the species was detected at least once during surveys, and assumed that the sample of 40 plots represented availability across the entire study area, as plots were established at random. This, too, is a design I study (Table 15.1).

To model the effect of woody plants on selection on each species, where presence or absence was the response variable and log density of woody plants was the explanatory variable, we used a form of the generalized linear model called *logistic regression*:

$$\text{logit}(\pi) = \beta_0 + \beta_1 X_1 + \ldots + \beta_p X_p + \varepsilon,$$

where π is the average proportion or probability of a positive response (i.e., a species present on a plot), β_0 is the *y* intercept (which is frequently excluded; Manly et al. 2002:100), $\beta_1 - \beta_p$ are regression coefficients that describe the influence of the explanatory variables $X_1 - X_p$ on the mean response (in these ex-

Table 15.3. Presence or absence of eastern meadowlarks and verdins and natural log-transformed density of shrubs and trees on 40 plots in southern Arizona, 2006

Eastern meadowlark	Verdin	Log (density of woody plants)
Present	Absent	0.83
Present	Absent	1.19
Absent	Present	1.41
Present	Absent	1.63
Present	Absent	2.02
Present	Absent	2.08
Present	Absent	2.22
Present	Present	2.58
Absent	Absent	2.63
Present	Absent	2.79
Present	Absent	2.85
Absent	Present	2.87
Present	Absent	3.04
Present	Present	3.22
Present	Present	3.25
Present	Present	3.27
Absent	Present	3.27
Absent	Absent	3.29
Present	Absent	3.32
Absent	Present	3.41
Absent	Present	3.42
Present	Present	3.44
Present	Absent	3.47
Present	Present	3.53
Present	Present	3.57
Present	Present	3.61
Present	Present	3.69
Absent	Present	3.75
Absent	Present	3.75
Absent	Present	3.89
Absent	Present	3.91
Absent	Present	4.01
Absent	Present	4.12
Absent	Present	4.20
Absent	Present	4.21
Absent	Present	4.39
Absent	Present	4.42
Present	Present	4.77
Absent	Present	5.05
Absent	Present	5.15

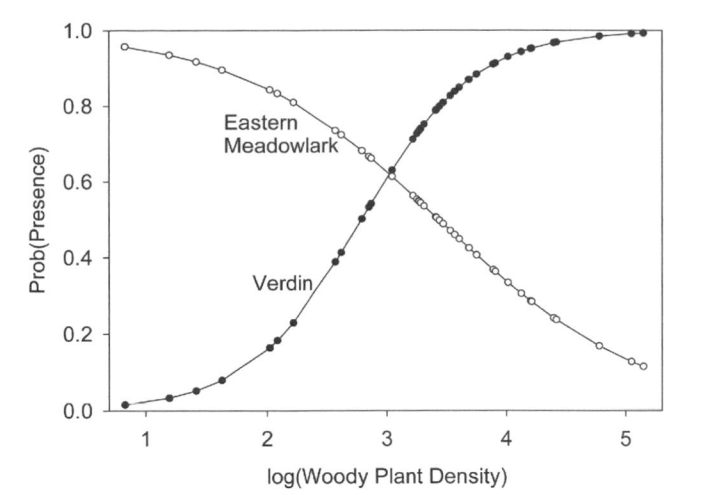

Figure 15.1. Relationship between density of woody vegetation and predicted probability of presence of eastern meadowlarks and verdins in semidesert grasslands of Las Cienegas National Conservation Area, southeastern Arizona, in 2006.

relationship between presence of verdins and woody plants based on these data is described by:

$$\text{logit(presence)} = -5.91 + 2.11 \, \text{log(woody plant density)}.$$

Because the response variable is modeled as the log of the odds of a positive response, we need to back-transform the estimated regression coefficient, $\hat{\beta}_1$, to describe the effect of the explanatory variable on the mean response on the original scale of measure:

$$\text{odds of a positive response} = e^{\hat{\beta}_1},$$

where e is the base of the natural logarithm (2.718 . . .). If an explanatory variable has no effect on the response, $\hat{\beta}_1$ will be near 0, and the estimated odds will be near 1 ($e^0 = 1$). For verdins, we estimated the change in the odds of a positive response for each one-unit increase in the explanatory variable to be $e^{2.11} = 8.3$. That means that for a one-unit increase in the log density of woody plants, the odds of a verdin selecting an area for breeding increased about eight times (Fig. 15.1).

In contrast to verdins, the probability of eastern meadowlarks selecting an area for nesting decreased as the amount of woody vegetation increased ($\chi^2 = 6.11$, $P = 0.013$). We estimated the effect of woody vegetation on presence of meadowlarks to be -1.19 (SE = 0.48), and the change in odds for a one-unit increase in the log density of woody plants to be $e^{-1.19} = 0.30$. That means that for each one-unit increase in log density of woody plants, the odds of a meadowlark selecting an area for breeding decreased about 3-fold (Fig. 15.1). Given a convenient property of odds, by simply changing the sign of the parameter estimate ($e^{1.19} = 3.3$) we can state equivalently that for each one-unit *decrease* in log density of woody plants, the odds of a meadowlark selecting an area for breeding will increase about three times.

Although the examples above are based only on a single explanatory variable, by modeling the probability of habitat

amples, $p = 1$), and ε is the distribution of the errors around the predicted relationship, which is assumed binomial. In logistic regression, the binary response variable, π, is transformed with the logit function, which is the log of the odds, $\log(\pi/1 - \pi)$. Odds represent the ratio of the probability of a positive response (π) to a negative response ($1 - \pi$), and provide an intuitive metric for interpreting results of these analyses.

Habitat use of both species of songbirds changed markedly in response to changes in the amount of woody vegetation (Fig. 15.1). For verdins, as the amount of woody vegetation increased, the probability of selecting an area for breeding increased ($\chi^2 = 8.79$, $P = 0.003$). We estimated the effect of woody vegetation on presence of verdins to be 2.11 (SE = 0.71), and the y intercept to be -5.91 (SE = 2.27). Therefore the

DAVID LAMBERT LACK (1910–1973)

Born in London on 16 July 1910, David Lack was an avid naturalist as a child, studied natural sciences at Magdalene College in Cambridge, and became a well-known evolutionary ecologist. He was influenced early in his professional career by the opportunity to study bird behavior in the Galapagos Islands, and by interactions with Ernst Mayr. Lack became director of the Edward Grey Institute of Field Ornithology in Oxford in 1945 and served in that post until his death in 1973. He is perhaps best known for his 1947 book *Darwin's Finches* and for his research on population biology. Lack was among the first researchers to propose that animals use environmental cues to help them find and settle in places with resources needed for their survival and reproduction, and coined the term "habitat selection" (Lack 1933).

PAUL LESTER ERRINGTON (1902–1962)

Paul Errington was born in Bruce, South Dakota, on 14 June 1902. He grew up working on his family's farm and developed a keen interest in natural history, especially in surrounding marshes and streams, which became evident in his work as a scientist. Errington graduated from South Dakota State College in 1929 and earned a Ph.D. from the University of Wisconsin in 1932. Errington then attended Iowa State University to establish and lead the first cooperative wildlife research unit in the United States, as well as to serve as professor of zoology, which he did through his entire career. Considered a great pioneer of animal ecology, Errington authored more than 200 scholarly papers that focused on predator–prey relationships, population dynamics, population regulation, and the biology and management of muskrats, minks, bobwhite quail, and great horned owls. He also authored four popular books, including *Of Men and Marshes*, that influenced both scientists and the general public. In 1961, Errington was named one of ten outstanding naturalists by *Life* magazine, and in 1962, he was awarded the Aldo Leopold Medal, the highest honor bestowed by The Wildlife Society.

Photo by Alfred Eisendstadt. Courtesy of the U.S. Fish and Wildlife Service

use as a function of different types or levels of one or more resources, these analyses provide examples of *resource selection functions*. This approach of modeling resource use versus availability or nonuse in a regression-type framework extends readily to many types of selection studies, including those that include both categorical and continuous factors, and where multiple observations have been recorded for each sample unit (Manly et al. 2002, Keating and Cherry 2004, Johnson et al. 2006).

One important assumption in these examples is that presence of a species in the area surveyed was determined without error. That is, no areas where the species was truly present were classified as absent (i.e., no false negatives; false positives can also be a problem, but for easily identifiable species, these errors are rare). For many vertebrates, this assumption is unlikely to be met. As mentioned previously, if animals cannot be detected with certainty, and the probability of detection is associated with one or more habitat features being evaluated, there is potential for results to be biased and for study conclusions to be incorrect. The design and analysis of studies where the chance of imperfect detections is likely have been explored in many contexts, including habitat selection (e.g., Gu and Swihart 2004, MacKenzie 2006).

We offered here a brief introduction to the many issues relevant to design and analysis of habitat selection studies. Additional coverage of the topic might begin with the book by Manly et al. (2002), papers and book chapters by Erickson et al. (2001) and McDonald et al. (2005), and the special sec-

tion on resource selection in the *Journal of Wildlife Management* (Strickland and McDonald 2006).

MANAGING HABITAT

Knowledge of the environmental conditions and resources needed to support a species is required before developing a strategy to meet habitat-based conservation or management goals. Frequently, these goals involve maintaining or increasing abundance of a population by protecting existing habitat, creating new habitat by adding resources that are limiting factors, or restoring habitat. Occasionally, the goal is to decrease abundance of a nuisance population, such as Canada geese (*Branta canadensis*) or white-tailed deer (*Odocoileus virginianus*), by reducing the amount or quality of their habitat (Gosser at al. 1997, Ayers et al. 2010). One of the primary ways biologists can either increase or decrease habitat resources for a species is by manipulating vegetation, and among the most effective ways to change vegetation is to control plant succession.

Habitat Management and Plant Succession

Plant succession is the natural process of a plant community on a site changing over time. Patterns of succession vary geographically, but in many areas of the temperate zone, succession begins with annual plants that colonize a site rapidly after a disturbance, then proceeds through a series of different species assemblages called seres or seral stages, and is eventually characterized by large, long-lived tree species. Because requirements of individual species are thought to drive succession (Glenn-Lewin et al. 1992), the assemblage of plant species at any given point along the succession trajectory is a collection of species that thrive in current environmental conditions. Discrete communities, therefore, often cannot be distinguished during succession, although some seral stages are dominated by characteristic plant species and are often named for those species.

Many animal species rely on plants for food and cover. As succession proceeds and the community of plant species on a site changes, animal species that inhabit the site also change. Thus habitat management for a given animal species might involve manipulating succession to ensure that a particular seral stage is maintained in a configuration on the landscape (e.g., size and number of patches) appropriate to maintain populations of that species over time. For example, Kirtland's warbler (*Dendroica kirtlandii*), a rare species found primarily in the northern lower peninsula of Michigan, nests on the ground only in stands of jack pine (*Pinus banksiana*) between seven and 21 years old (Marshall et al. 1998). Natural wildfires historically maintained nesting habitat. Today, a carefully scheduled combination of logging, prescribed burning, and tree planting maintain nesting habitat. Stands of pine are harvested and replanted on a 50-year rotation so that a portion of state and federal lands under management is maintained in a structural condition that provides nesting habitat for warblers (Marshall et al. 1998).

Other animals require habitat elements provided by a combination of seral stages. Ruffed grouse (*Bonasa umbellus*), for example, need brushy areas and stands of young aspen (*Populus tremuloides*) that provide food in summer; stands of mature aspen that provide drumming sites in spring and food in fall, winter, and spring; and stands of dense aspen that provide cover for broods in spring (DeStefano et al. 1984). Thus ideal habitat conditions for ruffed grouse are likely to exist in areas with an equal mix of young (<15 years), intermediate-aged (15–30 years), and old (>30–40 years) stands of aspen trees in small patches so that each seral stage can be incorporated into the home range of individual grouse (Natural Resource Conservation Service 2001). A general prescription for managing forests for ruffed grouse habitat might involve harvesting 1-ha patches of aspen from 4-ha blocks every ten years to maintain a mosaic of different-aged stands in proximity to each other (Natural Resource Conservation Service 2001).

Habitat Management and Human Activities

Human activities affect wildlife habitat in many ways and at many spatial and temporal scales. Activities that are likely to have adverse effects on wildlife populations can be divided into those that function primarily by altering the physical environment and those that affect an animal's behavior (Steidl and Powell 2006). Agriculture, forestry, and urban development are examples of human activities that alter significantly the local, physical environment. In contrast, climate change—also a result of human activities—alters the physical environment more broadly and in ways that can affect profoundly the quality and amount of habitat for wildlife. Some effects of climate change on wildlife habitat likely will include increases in temperature, which could render currently habitable areas as uninhabitable if the limits of thermal tolerance for the species are exceeded (Ward and Mannan 2011). Changes in temperature or rainfall patterns likely will alter abundance and distribution of a wide array of physical and biological resources, many of which provide critical habitat elements for wildlife species. Examples range from sea ice, an important habitat element for polar bears (*Ursus maritimus*; Stirling et al. 2011), to individual plant species crucial for many species (Kerns et al. 2009).

Although changes in land use and climate affect wildlife habitat by altering the biotic and abiotic features of an area, effects of climate change are likely to be more pervasive than those of other human activities (see Chapter 17 for a discussion of climate change). For example, wildlife habitat in a forest that is eliminated as a result of timber harvest or clearing for agriculture will likely recover, given sufficient time. Conversely, if the dominant plant community in an area changes as a result of shifts in temperature and rainfall regimes, eliminated habitat is unlikely to recover within any reasonable time frame. Although changes in climate present unique challenges to the maintenance of biodiversity, a critical element in managing effects of most human activities—understanding the resources that constitute habitat for a species—also will be a significant factor in predicting and managing the effects of climate change.

Perhaps less obvious in their effects on wildlife habitat are nonconsumptive human activities that do not appreciably alter the physical environment but nonetheless can reduce habitat quality. Examples include such recreational activities as hiking, wildlife viewing, and boating. Many factors influence the magnitude of nonconsumptive human activities' effects on wildlife, including the type, duration, frequency, magnitude, location, and timing of the disturbance and the particular species of interest. Although effects of these activities are typically of short duration, they can cumulatively affect wildlife populations adversely in both the short and long term (Steidl and Powell 2006). For example, if recreational activities preclude the use of an area by a species, that area could contain all of the resources necessary for a species but remain uninhabited. In this situation, controlling the number of people and their recreational activities is of obvious importance. However, for the remainder of the chapter, we focus on management activities that affect the resources upon which animals depend.

Human activities that modify natural landscapes inevitably alter habitat resources and lead to declines in abundance and distribution of some species and increases in others. Eliminating these activities entirely is not possible, because many produce resources that humans need to survive. But sound management practices can reduce the negative effects of each of these activities on animal populations, sometimes with relatively minor modifications. On lands under federal or state jurisdiction, improving conditions for target wildlife species often can be accomplished relatively easily, if the modifications help meet mandates of federal and state environmental laws. On lands that are privately owned, cooperation of landowners is critical to the implementation of such modifications, and ultimately to maintenance of wildlife populations on those lands. Financial incentives can encourage landowners to participate in habitat management programs, such as set-aside programs for agricultural lands (Warner et al. 2005). Habitat management to improve conditions for animals in areas being managed primarily for agriculture, forestry, or livestock production are summarized by Warner et al. (2005), Yahner et al. (2005), and Bleich et al. (2005). Below, we outline briefly ways that the negative effects of urbanization on animal populations can sometimes be reduced.

Urban environments, which we define broadly as lands developed for human habitation (e.g., towns and cities), support a variety of wildlife species (Adams 1994) but favor those that can live in close association with people. For example, assemblages of bird species in urban areas tend to have a higher total number of individuals and biomass but fewer species than those in more natural areas (Hansen et al. 2005, Chace and Walsh 2006), primarily because nonnative species—such as house sparrows (*Passer domesticus*) and European starlings (*Sturnus vulgaris*)—thrive, whereas many native species are eliminated. Land-use changes and introduction of many nonnative predators, especially domestic cats and dogs, are examples of challenges to maintaining native species in urban areas. Relatively simple management activities including (1) using native plant species when landscaping, (2) discouraging open lawns on public and private property, (3) maintaining patches of native vegetation in parks, and (4) maintaining connections, potentially via riparian zones, between urban parks and natural areas outside developed areas can enhance the presence of some native species of all taxonomic classes in urban areas (Marzluff and Ewing 2001).

The importance of habitat features in maintaining populations of native species in urban environments depends to some extent on how the features are distributed across the landscape (Daniels and Kirkpatrick 2006). Some native species can take advantage of resources even if they are present in small amounts at small scales, such as a single plant or feeding station in a backyard. For these species, actions of individual homeowners can potentially influence their distribution in urban settings (McCaffrey and Mannan 2012). Other species require that resources be distributed more broadly, and habitat management for these species might require actions of neighborhood groups or city and county planners (McCaffrey and Mannan 2012).

Habitat Management on Large Spatial Scales for Single Species

Managing land primarily for the conservation of a single species is costly and usually motivated by the needs of species listed under the Endangered Species Act. Questions that must be answered when developing a strategy to manage land for one species often focus on how much habitat is needed, how it should be distributed, and how it can be maintained over time. Below we describe how the conservation plan for the northern spotted owl (*Strix occidentalis caurina*) addressed these questions.

Conservation Strategy for the Northern Spotted Owl

The original plan to conserve habitat for the northern spotted owl (Thomas et al. 1990) is among the best examples of a management strategy based on the resource needs and behaviors of a single species over a broad geographical range. Specific areas identified for conservation in that plan have been modified to some extent in subsequent recovery plans (U.S. Fish and Wildlife Service 2008), but the rationale used to develop the plan has remained essentially unchanged and serves as one model for designing habitat reserves for a single species.

The northern spotted owl occupies coniferous forests from southwestern British Columbia through western Washington, western Oregon, and northwestern California (Gutierrez et al. 1995). It primarily inhabits old-growth forests (Forsman et al. 1984) and was listed as threatened under the Endangered Species Act, primarily because of habitat loss. At that time, old-growth forests were being harvested at an unsustainable rate (Parry et al. 1983). Therefore maintaining northern spotted owls required that some old-growth forests be protected from harvest, but deciding how much to protect and where to protect it was controversial, because any protection would result in lost revenues to the timber industry.

How much habitat is needed? The conservation strategy for the northern spotted owl was based on maintaining habitat for roughly 1,500 breeding pairs of owls. A panel of experts convened by the Audubon Society identified this target based on concerns about persistence of spotted owls given demographic and environmental stochasticity. The total protected area would be the amount of habitat capable of supporting about 1,500 breeding pairs (see below).

How big should habitat patches be? The plan called for maintaining patches of forest, called habitat conservation areas (HCAs), large enough to support at least 20 pairs of owls. Maintaining large patches was considered an advantage, because the potential for internal recruitment enhanced the likelihood of persistence of owls inside an HCA (Thomas et al. 1990). The formula used to determine HCA size was:

$$\text{HCA size} = [(\text{median annual home range of pairs}) \times 0.75] \times 20 \text{ pairs.}$$

The size of HCAs varied because the size of annual home ranges of adult owls varied across the geographic range. The multiplier of 0.75 was applied to allow for 25% overlap among home ranges. Current demographic models, developed to be spatially explicit, predict the response of owls to the distribution and quality of habitat and allow for evaluation of different planning strategies (U.S. Fish and Wildlife Service 2010).

How should habitat patches be distributed? HCAs were distributed across the entire geographical range of the northern spotted owl, and individual HCAs were situated so that the distance between them was no farther than the distance that dispersing spotted owls were likely to move. The rationale for this distribution was that resident breeding pairs in an HCA could be replaced by dispersing owls either from within the HCA or from adjacent HCAs.

How should areas between habitat patches be managed? Movement of owls between patches depends not only on the distance between patches, but also on the environmental conditions in areas through which dispersing owls must travel. The plan called for areas between habitat patches (sometimes called *habitat matrix*) to be managed so that half was maintained in forest conditions thought to facilitate movement of dispersing owls.

How can habitat be maintained over time? Old-growth patches identified for conservation could not persist forever, even if vigorously protected. How long patches last would depend on the frequency and intensity of natural disturbances such as fire. Although not described explicitly in the original plan, careful planning would be required to identify, set aside, protect, and allow additional forest stands to develop into old-growth forests. These stands could then be used to replace existing HCAs lost over time. Maintaining habitat reserves for a particular species over long time periods will depend in part on the environmental stability of areas selected as reserves. Changes in resources in response to climate change could affect how well an area functions as a habitat reserve, making management of reserves an increasingly complex challenge (Griffith et al. 2009).

Habitat Management on Large Spatial Scales for Multiple Species

Strategies to conserve habitat for single species are based commonly on identifying and maintaining areas that include the specific set of resources required by that species. As the number of species targeted for conservation increases, however, identifying the habitat resources necessary to support all species becomes much more challenging, as does identifying lands that will provide these resources over long time horizons. Therefore strategies to conserve habitat for multiple species tend to focus on identifying lands that provide the full range of environmental conditions thought necessary to support all target species. Further, given the inevitable changes in landscapes we anticipate in response to natural and anthropogenic processes, conservation strategies must include lands that provide redundancy of these environmental conditions so that all species can persist despite natural and human-caused changes in landscape composition over time.

The Sonoran Desert Conservation Plan

Pima County in southern Arizona initiated a comprehensive land-use planning effort to identify and establish an integrated system of conservation lands, with the goal of maintaining biodiversity while providing a framework to guide future land use. The effort was called the Sonoran Desert Conservation Plan and encompassed an area of nearly 24,000 km². The biological foundation for the plan was based on identifying lands of high conservation value that would form a network of conservation lands around which other regional planning needs would be incorporated. Although several ecological targets could have provided the basis for evaluating the potential conservation value of different areas, the strategy was based on understanding habitat needs of 39 target species (nine mammals, eight birds, seven reptiles, two frogs, six fish, and seven plants) selected to represent the range of biological diversity in the region (Steidl et al. 2009).

A detailed account of the habitat requirements of each species was gathered from the literature and from experts, which were then used as the basis for creating a series of spatially explicit models predicting the distribution of potential habitat across the region for each species. Models of potential habitat were used because published distributions tend to be too general in scope, documented locations are uneven in geographic coverage and often biased toward areas that are traveled commonly, and expert opinion has significant limitations in that on-the-ground knowledge is rarely complete (Steidl et al. 2009). Additionally, habitat can often be identified even if the target species is currently absent from an area, which seems especially likely for many species in jeopardy.

Models of potential habitat were based on values established for four major categories of environmental features represented by 130 variables, including vegetation and land cover (60 variables, including mixed broadleaf forest cover, agriculture), hydrology (11 variables, including perennial stream width, groundwater depth), topography and landform (45 variables, including elevation, slope, aspect), and geology (14 variables, including soil type, presence of carbonates). Scores based on the value of these environmental features to each species were combined to produce a habitat suitability surface that represented the distribution of habitat potential for each species across the landscape.

A geographic information system was used to overlay areas of high potential habitat for all species that resulted in a map that portrayed species richness (i.e., number of species in an area where potential habitat for species was classified as high) across the region; this map illustrated the relative importance of different areas to conservation of species across the region (Fig. 15.2). The landscape was then divided into a collection of discrete polygons representing areas with different levels of species richness on which the conservation lands system was built. To establish boundaries for each contiguous collection of landscape units with the same level of species richness (termed a patch), the guidelines on reserve design provided by the literature were followed to maximize conservation benefits of each patch and the overall network of patches. Specifically, the size of each patch was maximized, distances between adjacent patches minimized, contiguity maximized and fragmentation within and among patches minimized, and connectivity maximized among patches to maintain processes that occur at scales larger than individual patches. Additionally,

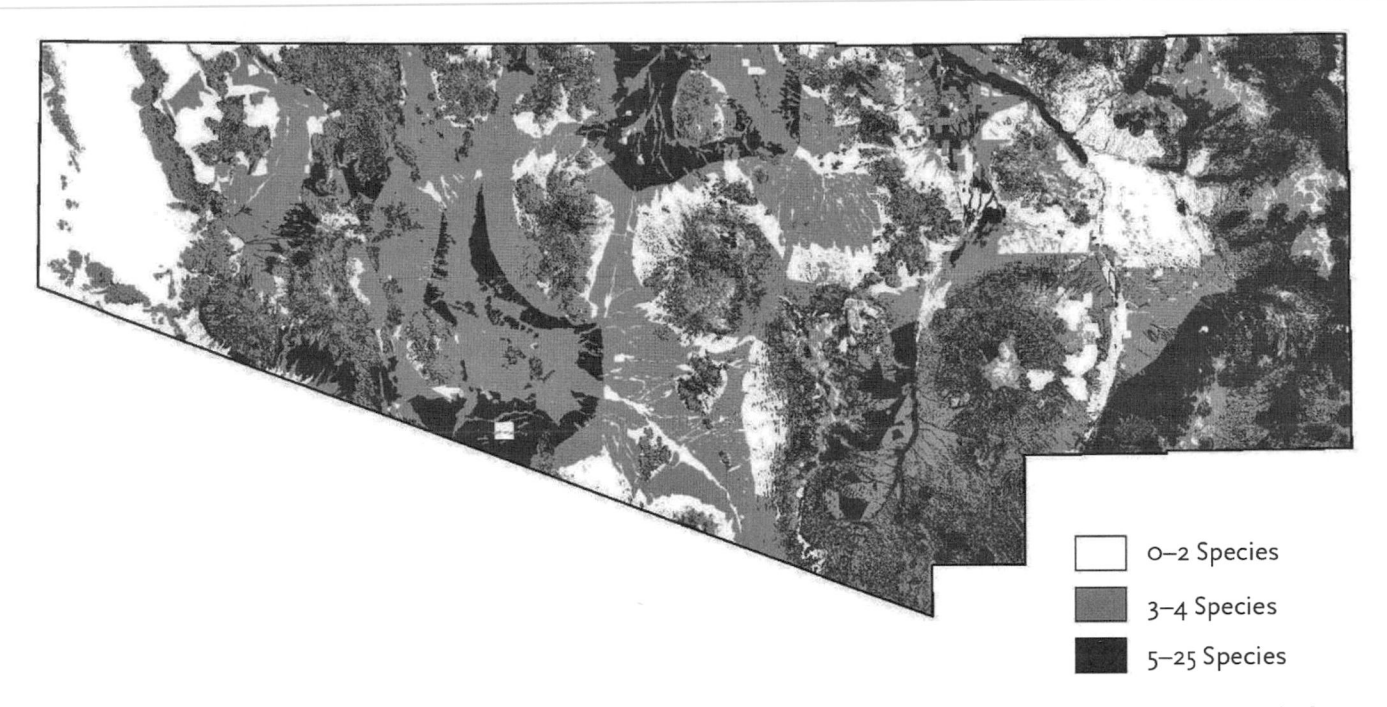

Figure 15.2. Predicted richness of target species across Pima County in southern Arizona, which provided the basis for establishing a network of conservation lands on which the Sonoran Desert Conservation Plan was designed. From Steidl et al. (2009)

boundaries were adjusted to meet a set of specific conservation objectives established to capture ecological processes that manifest at spatial scales broader than an individual patch, and to capture other elements important to conservation of biodiversity in the region that were too small to be captured in landscape-scale analyses (Steidl et al. 2009).

Ultimately, areas within the conservation lands system are predicted to conserve an average of 75% of potential habitat (range = 28–100%) for each of the target species when the region is fully developed. Currently, only about 4% of the area targeted for conservation has been developed, with an additional 4% of these lands predicted to be developed in the future (ESI Corporation 2003). Of the 12,000 km^2 of conservation lands, 57% is federal, 24% state, 14% private, and 5% county and city combined. With such a high percentage of these lands in public ownership, achieving the conservation objectives driving this large-scale process seems tenable.

CONTEMPORARY CHALLENGES

As the number of people occupying our planet continues to increase, the challenge of maintaining habitat for animals and all other forms of life will become increasingly difficult. Although humans will make extraordinary efforts to maintain populations of a few individual species (which almost always means maintaining their habitat), maintaining environmental conditions that support the needs of all forms of life on earth will depend increasingly on efforts designed to provide for a great many species over long time periods in the face of changes we can anticipate, as well as those that will manifest unexpectedly. Some measure of success in this task can be attained only if

biologists continue to strive to understand what constitutes habitat for a wide range of species, and to be innovative and persistent in developing strategies to reduce the adverse effects of anthropocentric activities while enhancing effects that are positive. We suggest that effective strategies include activities that consider the needs of humans and animals across a range of spatial and temporal scales.

SUMMARY

Habitat is a species-specific concept that describes an area with the combination of resources and environmental conditions that promotes residency by individuals of a species and allows those individuals to survive and reproduce. Animals seeking a habitat presumably recognize suitable sites by the presence of environmental cues that are directly or indirectly associated with the resources they require. The process of selecting an area to inhabit may span a range of spatial and temporal scales and is likely a hierarchical process for many species. Variation in levels of important resources among habitats is often expressed as variation in the demographic performance of resident individuals. Sampling strategies for identifying habitat and important habitat features should be influenced by the research question of interest, which in turn will dictate the scope of inference of the study. The two most common types of habitat studies are those that gather information about characteristics of locations used by animals (i.e., habitat use) and those that evaluate the use of habitat resources by animals relative to the general availability of those resources in the environment (i.e., habitat selection). Sampling designs available for assessing habitat selection have been classified into four main types on the basis of the approaches used to quantify resource use and resource availability, and many biologists have adopted regression-type approaches for analyzing resource selection. Biologists can either increase or decrease the amount and quality of habitat for a species by manipulating the quantity of important resources, setting aside key lands in reserve systems, and altering the way humans use land.

Literature Cited

Adams, L. W. 1994. Urban wildlife habitats. University of Minnesota Press, Minneapolis, USA.

Aebischer, N. J., P. A. Robertson, and R. E. Kenward. 1993. Compositional analysis of habitat use from animal radio-tracking data. Ecology 74:1313–1325.

Alldredge, J. R., and J. Griswold. 2006. Design and analysis of resource selection studies for categorical resource variables. Journal of Wildlife Management 70:337–346.

Alldredge, J. R., D. L. Thomas, and L. L. McDonald. 1998. Survey and comparison of methods for study of resource selection. Journal of Agricultural, Biological, and Environmental Statistics 3:237–253.

Ayers, C. R., C. E. Moorman, C. S. Deperno, F. H. Yelverton, and H. J. Wang. 2010. Effects of mowing on anthraquinone for deterrence of Canada geese. Journal of Wildlife Management 74:1863–1868.

Baker, M. B. 2005. Experience influences settling behaviour in desert isopods, *Hemilepistus raumuri*. Animal Behaviour 69:1131–1138.

Battin, J. 2004. When good animals love bad habitats: ecological traps and the conservation of animal populations. Conservation Biology 18:1482–1491.

Bleich, V. C., J. G. Kie, E. R. Loft, T. R. Stephenson, M. W. Oehler, and A. L. Medina. 2005. Managing rangelands for wildlife. Pages 873–897 *in* Techniques for wildlife investigations and management. The Wildlife Society, Bethesda, Maryland, USA.

Brown, J. L. 1969. The buffer effect and productivity in tit populations. American Naturalist 103:1313–1325.

Byers, C. R., R. K. Steinhorst, P. R. Krausman. 1984. Clarification of a technique for analysis of utilization-availability data. Journal of Wildlife Management 48:1050–1053.

Chace, J. F., and J. J. Walsh. 2006. Urban effects on native avifauna: a review. Landscape and Urban Planning 74:46–49.

Citta, J. J., and M. S. Lindberg. 2007. Nest-site selection of passerines: effects of geographic scale and public and personal information. Ecology 88:2034–2046.

Daniels, G. D., and J. B. Kirkpatrick. 2006. Does variation in garden characteristics influence the conservation of birds in suburbia? Biological Conservation 133:326–335.

Daubenmire, R. 1976. The use of vegetation in assessing the productivity of forest lands. Botanical Review 42:115–143.

Davis, J. M., and J. A. Stamps. 2004. The effect of natal experience on habitat preferences. Trends in Ecology and Evolution 19:411–416.

DeStefano, S., S. R. Craven, and R. L. Ruff. 1984. Ecology of the ruffed grouse. University of Wisconsin, Madison, USA.

Dwyer, J. F., and R. W. Mannan. 2007. Preventing raptor electrocutions in an urban environment. Journal of Raptor Research 41:259–267.

Elith, J., S. J. Phillips, T. Hastie, M. Dudik, Y. E. Chee, and C. J. Yates. 2011. A statistical explanation of MaxEnt for ecologists. Diversity and Distributions 17:43–57.

Erickson, W. P., T. L. McDonald, K. G. Gcrow, S. Howlin, and J. W. Kern. 2001. Statistical issues in resource selection studies with radio-marked animals. Pages 211–245 *in* J. J. Millspaugh and J. M. Marzluff, editors. Radio tracking and animal populations. Academic, San Diego, California, USA.

ESI Corporation. 2003. Pima County economic analysis. Report to Pima County. Phoenix, Arizona, USA. http://www.pima.gov/cmo/sdcp/reports.html. Accessed 23 October 2012.

Flesch, A. D., and R. J. Steidl. 2010. Importance of environmental and spatial gradients on patterns and consequences of resource selection. Ecological Applications 20:1021–1039.

Forsman, E. D., E. C. Meslow, and H. M. Wight. 1984. Distribution and biology of the northern spotted owl in Oregon. Wildlife Monographs 87:3–64.

Freidman, M. 1937. The use of ranks to avoid the assumption of normality implicit in the analysis of variance. Journal of the American Statistical Association 32:675–701.

Fretwell, S. D., and H. L. Lucas. 1970. On territorial behaviour and other factors influencing habitat distribution in birds. Acta Biotheoretica 19:16–36.

Garshelis, D. L. 2000. Delusions in habitat evaluation: measuring use, selection, and importance. Pages 111–164 *in* L. Boitani and T. K. Fuller, editors. Research techniques in animal ecology: controversies and consequences. Columbia University, New York, New York, USA.

Gillies, C. S., M. Hebblewhite, S. E. Nielsen, M. A. Krawchuk, C. L. Aldridge, J. L. Frair, D. J. Saher, C. E. Stevens, and C. L. Jerde. 2006. Application of random effects to the study of resource selection by animals. Journal of Animal Ecology 75:887–898.

Glenn-Lewin, D. C., R. K. Peet, and T. T. Veblen, editors. 1992. Plant succession: theory and prediction. Chapman and Hall, New York, New York, USA.

Gosser, A. L., M. R. Conover, and T. A. Messmer. 1997. Managing problems caused by urban Canada geese. Berryman Institute Publication 13. Utah State University, Logan, USA.

Griffith, B., J. M. Scott, R. Adamcik, D. Ashe, B. Czech, R. Fischman, P. Gonzales, J. Lawler, A. D. McGuire, and A. Pidgorna. 2009. Climate change adaptation for the U.S. National Wildlife Refuge System. Environmental Management 44:1043–1052.

Gu, W., and R. K. Swihart. 2004. Absent or undetected? Effects of nondetection of species occurrence on wildlife–habitat models. Biological Conservation 116:195–203.

Gutierrez, R., A. B. Franklin, and W. S. Lahaye. 1995. Spotted owl. Pages 1–28 *in* A. Poole and F. Gill, editors. The birds of North America No. 179. Academy of Natural Sciences, Philadelphia, Pennsylvania, and American Ornithologists' Union, Washington, D.C., USA.

Hamerstrom, F., F. N. Hamerstrom, and J. Hart. 1973. Nest boxes: an effective management tool for kestrels. Journal of Wildlife Management 37:400–403.

Hansen, A. J., R. L. Knight, J. M. Marzluff, S. Powell, K. Brown, P. H. Gude, and K. Jones. 2005. Effects of exurban development on biodiversity: patterns, mechanisms, and research methods. Ecological Applications 15:1893–1905.

Hensher, D. A., J. M. Rose, and W. H. Greene. 2005. Applied choice analysis: a primer. Cambridge University Press, New York, New York, USA.

Hilden, O. 1965. Habitat selection in birds. Annales Zoologici Fennici 2:53–73.

Hosmer, D. W., and S. Lemeshow. 2000. Applied logistic regression. Second edition. John Wiley and Sons, Hoboken, New Jersey, USA.

Jaenike, J., and R. D. Holt. 1991. Genetic variation for habitat preference: evidence and explanations. The American Naturalist 137:S67–S90.

Johnson, C. J., S. E. Nielsen, E. H. Merrill, T. L. McDonald, and M. S. Boyce. 2006. Resource selection functions based on use-availability data: theoretical motivation and evaluation methods. Journal of Wildlife Management 70:347–357.

Johnson, D. H. 1980. The comparison of usage and availability measurements for evaluating resource preference. Ecology 61:65–71.

Johnson, M. D. 2007. Measuring habitat quality: a review. Condor 109:489–504.

Keating, K. A., and S. Cherry. 2004. Use and interpretation of logistic regression in habitat selection studies. Journal of Wildlife Management 68:774–789.

Kerns, B. K., B. J. Naylor, M. Buonopane, C. G. Parks, and B. Rogers. 2009. Modeling tamarisk (*Tamarisk* spp.) habitat and climate change effects in the northwestern United States. Invasive Plant Science and Management 2:200–215.

Knoke, D., and P. J. Burke. 1980. Log-linear models. Sage, Newberry Park, California, USA.

Krishnan, V. V. 2007. Optimal strategy for time-limited sequential search. Computers in Biology and Medicine 37:1042–1049.

Kristan, W. B. I. 2003. The role of habitat selection behaviour in population dynamics: source-sink systems and ecological traps. Oikos 103:457–468.

Lack, D. 1933. Habitat selection in birds with special reference to the effects of afforestation on the Breckland avifauna. Journal of Animal Ecology 2:239–262.

Levin, S. A. 1992. The problem of pattern and scale in ecology: the Robert H. MacArthur Award lecture. Ecology 73:1943–1967.

Lima, S. L., and P. A. Zollner. 1996. Towards a behavioural ecology of ecological landscapes. Trends in Ecology and Evolution 11:131–135.

Litt, A. R. 2007. Effects of experimental fire and nonnative grass invasion on small mammals and insects. Dissertation, University of Arizona, Tucson, USA.

MacKenzie, D. W. 2006. Modeling the probability of resource use: the effect of, and dealing with, detecting a species imperfectly. Journal of Wildlife Management 70:367–374.

Manly, B. F. J., L. L. McDonald, D. L. Thomas, T. L. McDonald, and W. P. Erickson. 2002. Resource selection by animals: statistical analysis and design for field studies. Second edition. Kluwer, Boston, Massachusetts, USA.

Marshall, E., R. Haight, and F. R Homans. 1998. Incorporating environmental uncertainty into species management decisions: Kirtland's warbler habitat management as a case study. Conservation Biology 12:975–985.

Marzluff, J. M., and K. Ewing. 2001. Fragmented landscapes for the conservation of birds: a general framework and specific recommendations for urbanizing landscapes. Restoration Ecology 9:280–292.

McCaffrey, R. E., and R. W. Mannan. 2012. How scale influences birds' responses to habitat features in urban residential areas. Landscape and Urban Planning 105:274–280.

McCullagh, P., and J. A. Nelder. 1989. Generalized linear models. Second edition. Chapman and Hall, Boca Raton, Florida, USA.

McDonald, L. L., J. R. Alldredge, M. S. Boyce, and W. P. Erickson. 2005. Measuring availability and vertebrate use of terrestrial habitats and foods. Pages 465–488 in C. E. Braun, editor. Techniques for wildlife investigations and management. Sixth edition. The Wildlife Society, Bethesda, Maryland, USA.

Morrison, M. L., B. G. Marcot, and R. W. Mannan. 2006. Wildlife–habitat relationships. Third edition. Island Press, Washington, D.C., USA.

Mumme, R. L., S. J. Schoech, G. E. Woolfenden, and J. W. Fitzpatrick. 1999. Life and death in the fast lane: demographic consequences of road mortality in the Florida scrub-jay. Conservation Biology 14:501–512.

Natural Resource Conservation Service. 2001. Ruffed grouse (*Bonasa umbellus*): fish and wildlife habitat management guide sheet. U.S. Department of Agriculture, Washington, D.C., USA.

Neu, C. W., C. R. Byers, and J. M. Peek. 1974. A technique for analysis of utilization-availability data. Journal of Wildlife Management 38:541–545.

Otis, D. L., and G. C. White. 1999. Autocorrelation of location estimates and the analysis of radiotracking data. Journal of Wildlife Management 63:1039–1044.

Parry, B. T., H. J. Vaux, and N. Dennis. 1983. Changing conceptions of yield policy on the national forests. Journal of Forestry 81:150–154.

Pearce, J. L., and M. S. Boyce. 2006. Modelling distribution and abundance with presence-only data. Journal of Applied Ecology 43:405–412.

Phillips, S. J., R. P. Anderson, and R. E. Schapire. 2006. Maximum entropy modeling of species geographic distributions. Ecological Modelling 190:231–259.

Phillips, S. J., M. Dudik, J. Elith, C. H. Graham, A. Lehmann, J. Leathwick, and S. Ferrier. 2009. Sample selection bias and presence-only distribution models: implications for background and pseudo-absence data. Ecological Applications 19:181–197.

Robertson, B. A., and R. L. Hutto. 2006. A framework for understanding ecological traps and an evaluation of existing evidence. Ecology 87:1075–1085.

Rodenhouse, N. L., and L. B. Best. 1983. Breeding ecology of Vesper sparrows in corn and soybean fields. American Midland Naturalist 110:265–275.

Schlaepfer, M. A., M. C. Runge, and P. W. Sherman. 2002. Ecological and evolutionary traps. Trends in Ecology and Evolution 17:474–480.

Stamps, J. A. 1987. Conspecifics as cues to territory quality: a preference of juvenile lizards (*Anolis aeneus*) for previously used territories. American Naturalist 129:629–642.

Steidl, R. J., and B. F. Powell. 2006. Assessing the effects of human activities on wildlife. George Wright Forum 23:50–58.

Steidl, R. J., W. W. Shaw, and P. Fromer. 2009. A science-based approach to regional conservation planning. Pages 217–233 in A. X. Esparza and G. R. McPherson, editors. The planner's guide to natural resource conservation: the science of land development beyond the metropolitan fringe. Springer, New York, New York, USA.

Stirling, I., T. L. McDonald, E. S. Richardson, E. V. Regehr, and S. C. Amstrup. 2011. Polar bear population status in the northern Beaufort Sea, Canada 1971–2006. Ecological Applications 21:859–876.

Strickland, M. D., and L. L. McDonald. 2006. Introduction to the special section on resource selection. Journal of Wildlife Management 70:321–323.

Svardson, G. 1949. Competition and habitat selection in birds. Oikos 1:57–74.

Thomas, D. L., and E. J. Taylor. 1990. Study designs and tests for comparing resource use and availability. Journal of Wildlife Management 54:322–330.

Thomas, D. L., and E. J. Taylor. 2006. Study designs and tests for comparing resource use and availability II. Journal of Wildlife Management 70:324–336.

Thomas, J. W., E. D. Forsman, J. B. Lint, E. C. Meslow, B. R. Noon, and J. Verner. 1990. A conservation strategy for the northern spotted owl. Interagency Scientific Committee to Address the Conservation of the Northern Spotted Owl, U.S. Department of Agriculture Forest Service, Department of the Interior, Bureau of Land Management, U.S. Fish and Wildlife Service, and National Park Service, Portland, Oregon, USA.

U.S. Fish and Wildlife Service. 2008. Recovery plan for the northern spotted owl, *Strix occidentalis caurina*. Portland, Oregon, USA.

U.S. Fish and Wildlife Service. 2010. Draft revised recovery plan for the northern spotted owl, *Strix occidentalis caurina*. Portland, Oregon, USA.

Van Horne, B. 1983. Density as a misleading indicator of habitat quality. Journal of Wildlife Management 47:893–901.

Ward, M. S., and R. W. Mannan. 2011. Habitat model of urban-nesting Cooper's hawks (*Accipiter cooperii*) in southern Arizona. Southwestern Naturalist 51:17–23.

Warner, R. E., J. W. Walk, and C. L. Hoffman. 2005. Managing farmlands for wildlife. Pages 861–872 *in* C. E. Braun, editor. Techniques for wildlife investigations and management. The Wildlife Society, Bethesda, Maryland, USA.

Wiens, J. A. 1985. Habitat selection in variable environments: shrub-steppe birds. Pages 227–251 *in* M. L. Cody, editor. Habitat selection in birds. Academic, San Diego, California, USA.

Wiens, J. A. 1989. Spatial scaling in ecology. Functional Ecology 3:385–397.

Williams, J. B., and G. O. Batzli. 1979. Competition among bark-foraging birds in central Illinois: experimental evidence. Condor 81:122–132.

Yahner, R. H., C. G. Mahan, and A. D. Rodewald. 2005. Managing forestlands for wildlife. Pages 898–919 *in* Techniques for wildlife investigations and management. The Wildlife Society, Bethesda, Maryland, USA.

16

WILDLIFE RESTORATION

MICHAEL L. MORRISON

INTRODUCTION

Studying animals with practical applications (i.e., management) often involves trial and error. Wildlife managers use classroom knowledge, practical experience (including hunting), and various opinions to try and make a desired change in the health and abundance of a target species. Most of the wildlife professional's attention was historically on consumptive species and species that negatively impacted humans either directly (e.g., predators) or indirectly (e.g., crop damage, livestock depredation). Over time, the collective experience expanded and—along with new information from the fields of ecology, physiology, genetics, and others—allowed the wildlife profession to become increasingly rigorous scientifically, leading to the development of management applications based on those data.

The restoration profession has been built on a foundation of plant biology and horticulture. Similar to the wildlife profession, restoration developed into an organized discipline as investigators sought to restore ecologies to previous, more desirable conditions; that is, to manage ecological conditions. Trial and error, as well as projects designed and implemented on the basis of prior experience and expert opinion, characterized restoration. Restoration is also moving toward the use of more rigorous study designs for use in developing practical applications of the available data (Palmer et al. 2006).

In this chapter I review how the principles of wildlife ecology and restoration ecology can be linked to results in comprehensive planning for the management and conservation of natural resources. Elsewhere in the literature is more detailed coverage of wildlife restoration (Morrison 2009). In a condensed manner, herein I give wildlife and restoration professionals—including biologists, managers, and administrators—a basic outline of the fundamental issues of wildlife populations and habitat relationships as related to restoration. I encourage readers to delve deeper into restoration and its applications to wildlife.

Because virtually any restoration activity will change conditions for animal species, restorationists must be aware of how their actions influence the abundance and behavior of animals (if for no other reason than to ensure that the animals do not ruin the restoration measures). Likewise, wildlife ecologists can tap into the vast knowledge of the restoration community on how to manipulate soils, plant species, and vegetative associations to achieve a desired outcome. Restoration plans should be guided by the likely responses of current or desired wildlife species in the project area, which includes data on myriad factors such as the current and historical abundance and distribution of animal species, habitat and food requirements, breeding locations, how plant succession will change species composition through time, space requirements, and necessary links between geographic areas.

CONCEPTUAL FRAMEWORK

Restoration represents a synthesis of many biotic and abiotic concepts, including habitat and niche ecology, populations, genetics, ecosystem dynamics, historical ecology, geology and soils, fire history, and climatic patterns. As such, the practice of restoring ecosystems should be based on an interdisciplinary approach (Halle and Fattorini 2004, Palmer et al. 2006). In this chapter I briefly review key concepts and terms that form the foundation of restoration. Although such topics as economics, sociology, and politics must be considered for restoration projects to be successful, they are beyond my current scope (see Chapter 4). I focus on restoring wildlife, primarily by condensing and revising the more detailed presentation in Morrison (2009). The terminology and concepts reviewed here serve as the foundation for the remainder of this chapter.

DEFINITIONS AND HISTORICAL CONDITIONS

To restore means "to bring back into existence or use"; thus "restoration" is the act of restoring. The simplicity ends here, because difficulty arises once we ask follow-up questions. What exactly do we want to restore something *to*? What is our desired condition? As I have noted previously, automobile restoration is simple (Morrison 2009). We usually know the year and often the exact day a vehicle was produced. Detailed engineering drawings might be available for every part of the vehicle, pictures might show what it originally looked like, and an owner's manual explains how to maintain the vehicle. Nature does not provide us with such a detailed description of the past; rather, we are only provided hints and oftentimes well-hidden clues concerning former conditions and dynamics.

Two key terms to differentiate are restoration ecology and ecological restoration. Restoration ecology is the scientific process of developing theory to guide restoration and of using restoration to advance ecology. Ecological restoration is the practice of restoring degraded ecological systems (Palmer et al. 2006). There is a clear and necessary link between the practice of restoration (i.e., ecological restoration) and the concepts that were used to develop the restoration plan (i.e., restoration ecology).

There have been attempts to establish an overall conceptual framework for restoration ecology (Hobbs and Norton 1996, Halle and Fattorini 2004, Palmer et al. 2006). Briefly, such conceptual frameworks usually attempt to relate the fields of disturbance ecology and succession, where both natural succession and human-induced changes are used to guide development of a system that has been disturbed, including those that are now apart from what is considered normal variation. Although some basic functions and ecosystem processes can be manipulated and changed to a desired condition, it is usually impossible or at least impractical to restore any system to some previous state (Halle and Fattorini 2004).

Historical ecosystems are ecological systems that existed in the past. Historical ecosystems have been a focus of restorationists because they can be used as analogs to guide development of a restoration plan (Egan and Howell 2001). The ecological condition desired as a result of a restoration project is referred to as the reference condition. Reference conditions inform the restoration plan by (1) defining what the original condition was compared to the present, (2) determining what factors caused the degradation, (3) defining what needs to be done to restore the system, and (4) developing criteria for measuring the success of the restoration project (Morrison 2009).

The age of the analog reference condition largely determines the difficulty of describing and then duplicating those conditions and associated processes. If the project is designed to correct a recently damaged area (e.g., natural or human-caused catastrophe), then there should be many readily available reference conditions in the general geographic area. But if the goal is to restore an ecological condition that existed many centuries ago, reference conditions can be highly controversial and also subject to economic and sociological constraints.

First, we must determine what time period will serve as the historical reference, and then decide what ecological conditions existed during that time period. For example, Noss (1985) concluded that whether American Indians should be considered a natural and beneficial component of the environment cannot be answered conclusively, largely because the impact they had varied by time and location. Many restorationists believe that the arrival of European settlers is the point in time when unnatural conditions began to prevail. Willis and Birks (2006) noted that ecosystems change in response to many factors, including climate variability, invasions of species, and wildfires. A central point developed by Willis and Birks (2006) is the difficulty in defining the "natural" features in ecosystems through time. For example, there are over 150 species of plants that have been introduced to the British flora by humans between 500 and 4,000 years ago. Deciding which plants are native would have a substantial impact on efforts to either conserve or potentially eradicate many species.

In my opinion, however, we have little if anything to gain by attempting to categorize human impacts as natural or unnatural. Rather, I believe the role of restorationists is to help quantify the likely consequences of human activities and natural events on the environment so that the public and managers can make informed decisions when developing the desired condition for an area. For example, because most large, terrestrial mammalian predators have been extirpated from most of the contiguous United States and Mexico, ecological systems are missing a dominant selective force. An excellent case study was provided by Berger et al. (2001), who reported how a series of ecological events were triggered by the local extinction of grizzly bears (*Ursus arctos horribilis*) and gray wolves (*Canis lupus*) from the southern greater Yellowstone ecosystem in the United States. Their removal allowed a substantial increase in the moose (*Alces alces*) population to occur, which in turn caused substantial degradation of riparian vegetation through moose browsing, followed by a reduction of neotropical migrant birds that used the riparian zone. Similarly, Hebblewhite et al. (2005) reported that wolf exclusion from Banff National Park, Canada, caused increased browsing by elk (*Cervus canadensis*), which in turn resulted in changes in beaver (*Castor canadensis*) and songbird behavior.

Rather than attempting to return a system to some historical state, Palmer et al. (2006) suggested that a more realistic goal was to move a damaged system to an ecological state that is within some acceptable limits relative to a less disturbed system. The key is to define acceptable limits. Regardless, it is difficult to achieve a condition that is free from human impacts for the following reasons: (1) local plant and animal extinctions, (2) introduced species, (3) migration and dispersal of many plants and animals is no longer possible, and (4) ecological processes have been retarded or prevented. In a practical sense we are left with planning a restoration project based on (1) knowledge of historical conditions, (2) knowledge of current regional conditions, (3) knowledge of species-specific ecological requirements, (4) evaluation of legal requirements, and (5) political reality (Morrison 2009). Although biologists support attempts to place restoration plans in context of historical conditions, ecological reality must guide what can and cannot be achieved.

APPROACHES TO RESTORATION

Planning for restoration and subsequent management of the area requires that managers at least generally understand how the ecosystem in the region was developed and how it functions under differing environmental conditions. Thus we need to examine organisms from different taxonomic levels, along with their interactions with each other and with abiotic conditions and processes. An ecosystem (i.e., an ecological system) can be generally defined as consisting of organisms of various taxonomic designations and levels of biological organization, along with their interactions among each other and among abiotic conditions and processes. Understanding wildlife in an ecosystem context requires understanding of population dynamics; the evolutionary context of organisms, populations, and species; interactions between species that affect their persistence; and the influence of the abiotic environment on the vitality of organisms (Morrison et al. 2006:387). Restoration must similarly be considered within a broad spatial (i.e., landscape) context. Of course, we cannot study an ecosystem and the many species contained therein per se; rather, we must work with the individual animals that comprise the species in an area. Although we should develop restoration plans in a broad-scale context, and some agencies (e.g., U.S. Department of Agriculture Forest Service) are attempting broad-scale projects, most management occurs on small spatial scales (i.e., 1–100 ha).

ANIMAL POPULATIONS AND ASSEMBLAGES

The goal of wildlife restoration is to create conditions that provide for the survival and protection of individual organisms in sufficient numbers and locations to maximize the probability of long-term persistence (Morrison 2009:17). The health (or fitness) of a species is influenced by the dynamics of interactions among individuals within a population, by interactions among populations and other species, and by interactions between organisms and their habitats and environments. Wildlife restoration requires knowledge of population dynamics and behaviors and the processes that regulate population trends. Habitat (discussed below) by itself does not guarantee long-term fitness and viability of populations (Morrison et al. 2006:61).

The restorationist must understand the spatial and geographic factors that influence habitats and environments, population structure, and fitness of organisms, because they relate directly to the size and location of the area needing restoration to ensure survival of a species. For example, providing the proper habitat conditions does not provide for long-term persistence if no allowance has been made for immigration from other geographic locations. I first review the population concept, then discuss the importance of considering exotic species in restoration planning, and finish with an approach for determining which species are likely to occur within a planning area.

Population Concepts

Traditionally, a population has been defined as a collection of organisms of the same species that interbreed. Mills (2007) broadened the concept by defining a population as a collection of individuals of a species in a defined area that might or might not breed with other groups of that species elsewhere. Regardless of the specific definition, all planning for wildlife restoration must include in-depth knowledge of how the distribution and abundance of a species of interest fluctuates across space and time.

There are several other key terms concerning populations that must be understood within a restoration context (Morrison 2009). A deme is defined as individuals of a species with a high probability of interbreeding. A metapopulation structure can arise when there is partial isolation of individuals among populations. A metapopulation is "a species whose range is composed of more or less geographically isolated patches, interconnected through patterns of gene flow, extinction, and recolonization" and has been termed "a population of populations" (Levins 1970:105, Lande and Barrowclough 1987:106). These component populations have also been referred to as subpopulations. Biologically, a subpopulation refers to a deme or to a portion of a population in a specific geographic location. Unfortunately, the term subpopulation has been used on the basis of nonbiological criteria, often for management or legal reasons (e.g., administrative or political boundaries). Metapopulations are linked by a multitude of factors, including dispersal and migration, habitat conditions, genetics, and behavior (Hanski 1996), and they occur frequently in wild animal populations.

If one or more of the species targeted for restoration are metapopulations, the restoration plan must consider the spatial relationship among areas of habitat so that dispersal among subpopulations can occur. Extinction of the species within one or more subpopulations could occur and become permanent if there is no suitable area of habitat close enough to allow dispersal and recolonization. Subpopulations (within a metapopulation) usually vary in abundance of animals. Note the links between subpopulations (Fig. 16.1), where loss of a subpopulation that is located between other subpopulations could lead to further extinctions because the linkage would be broken. Wildlife restoration requires identification of the structure of the populations of species under consideration, followed by planning of restoration sites to enhance this structure (e.g., by promoting dispersal of individuals between subpopulations, which are sometimes called stepping-stone habitat patches; Morrison 2009). Because of metapopulation structure, not all suitable areas will be occupied at any one time. Even if a habitat patch appears unoccupied in any one year, monitoring wildlife use of habitats should proceed for many years.

Populations are distributed in many patterns across the landscape, from those in many isolated groups to those that are linked through movements between groups. For multiple isolated populations, each population is susceptible to extinction, usually without the possibility of natural recolonization (Mills 2007). From a restoration perspective, understanding the dynamics of each population and if movement between populations is occurring is a core component of any plan for conserving a species.

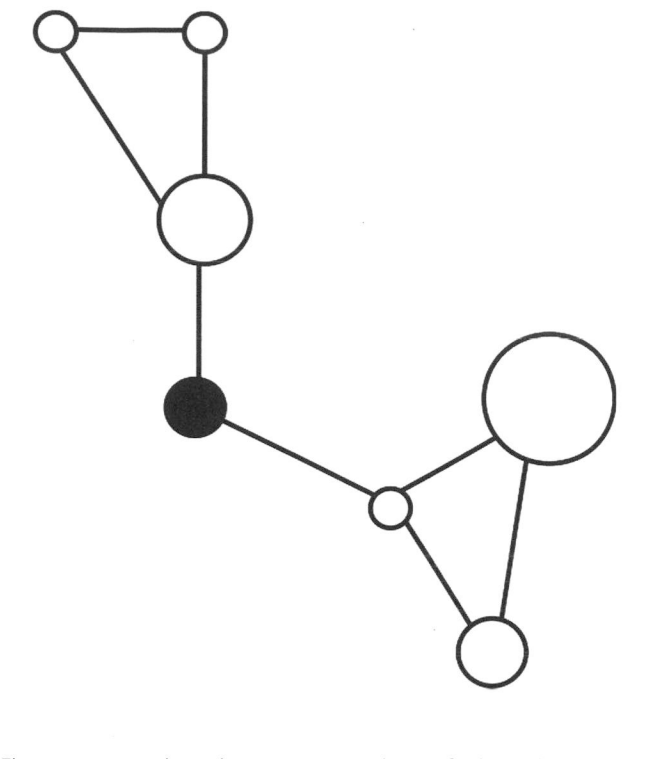

Figure 16.1. Hypothetical arrangement and size of subpopulations (circles) in a metapopulation. The size of the circle reflects population abundance; lines represent dispersal routes. The subpopulation, indicated by the solid circle, would be a key link between other subpopulations and thus a top priority site for restoration. From Morrison (2009:Fig. 3.2)

Exotic Species

The introduction and subsequent range expansion of exotic species are major challenges for conservation and restoration. Exotic plants and animals can affect the health and ultimately the abundance and distribution of native plants and animals (Office of Technology and Assessment 1993). We usually label a species as exotic if humans accidentally or purposefully introduced it. A species is usually considered native if it resides in its presumed area of evolutionary origin or if it appears because of nonhuman-aided dispersal (Willis and Birks 2006). There are many situations where we are uncertain of the cause of range expansions, however; they might be natural, might have been induced or enhanced by human alteration of environments, or might have begun as a minor introduction by humans. The brown-headed cowbird (*Molothrus ater*), for example, evolved in the Great Plains and then spread throughout most of North America during the 1900s, apparently because of forest clearing and agricultural development (Morrison et al. 1999). Humans did not physically capture and move the cowbirds; they moved as a result of human impacts to the environment. Are cowbirds native or exotic within certain portions of their current range? Because cowbirds can negatively impact certain species of (often rare) birds, the answer has a substantial impact on how a restoration project is planned.

Examples of exotic species that have spread throughout the continental United States include the European starling (*Sturnus vulgaris*) and house sparrow (*Passer domesticus*), both introduced from Europe. Both species negatively impact native birds by usurping nest sites and reducing nest productivity. The nonnative house mouse (*Mus musculus*) and several species of rats have spread across North America. Morrison et al. (1994) reported that the house mouse was a dominant species of rodent in disturbed areas in Southern California that were scheduled for restoration.

Restorationists must determine the potential impact that exotic species, and native species whose range has expanded (e.g., cowbird), will have on desired species. Starlings and house sparrows, for example, can occupy the majority of nesting cavities (i.e., woodpecker excavations, bird boxes, eaves in buildings) that could be created as part of restoration for native species (e.g., bluebirds, woodpeckers). Cowbirds can cause a high level of nesting failure in songbirds and have been a specific focus in locations undergoing restoration focusing on endangered species (Kus 1999, Kostecke et al. 2005).

There is not a general prescription that calls for removal of all exotic species, however. Native species can become dependent on an exotic species in severely impacted environments. For example, channelization of rivers has curtailed flooding and lowered water tables, resulting in a loss of native shrub and tree species. In the southwest United States, for example, the exotic plant saltcedar (*Tamarix* spp.) has become established in many riparian areas following loss of native willow (*Salix* spp.)-cottonwood (*Populus* spp.) woodlands. The endangered southwestern willow flycatcher (*Empidonax traillii extimus*) now nests successfully in salt cedar; removal of salt cedar would eliminate their nesting and foraging locations. And because of the altered hydrology due to the channelization, restoring willow-cottonwood is virtually impossible over large areas. Removal of exotics is not always a preferred restoration strategy, especially when an endangered species is involved. As such, restoration per se is not necessarily an attempt to restore past species compositions, but rather an attempt to restore specific parts of the previous system. A major activity in restoration planning is prioritizing what can and cannot be restored.

Assembling Groups of Species

A priority step that must be taken in developing a restoration project is determining which species to explicitly incorporate into the plan. As noted above, a host of factors are responsible for the presence of plant and animal species in a location; the scientist's task becomes more difficult as the physical size of the restoration area decreases. One tool for helping plan and organize a comprehensive restoration plan falls under the general ecological concept termed assemblage rules. Diamond (1975) attempted to develop general rules that predicted broad patterns of species co-occurrence. Diamond reported that species with similar diets would seldom co-occur; he called this pattern a checkerboard distribution and attributed it to competition between species for limited resources.

Assembly rules became controversial because Diamond (1975) and others were largely unsuccessful in showing that competition was the force behind the patterns they reported. Similar patterns as those reported by Diamond (1975) could be created when species were randomly distributed across an area (i.e., a null model); competition was not required (Connor and Simberloff 1979).

Thanks to the pioneering work of Diamond (1975), Connor and Simberloff (1979), and others, the search for assembly rules has gained widespread attention, as evidenced by edited volumes on the subject (Weiher and Keddy 1999, Temperton et al. 2004). Some workers have emphasized biotic interactions (e.g., competition), whereas others have emphasized a more holistic approach that includes interactions of the environment with the organisms. Temperton and Hobbs (2004) concluded, however, that no real consensus existed about which rules should be emphasized. As detailed elsewhere (Morrison 2009), I adhere to a comprehensive view in which biotic and abiotic factors, in combination with other constraints on species, determine the species and their abundance in a specific location. The priority for restoration planning should be identification of the primary factors that can limit the occurrence, abundance, survival, and productivity of a species. The manner in which these limiting factors place boundaries on the distribution of species that can potentially occur in a project area can be captured in the assembly rule process.

Of course, documenting a pattern is not the study of community assembly. Although null models are useful in identifying patterns, they do not identify the mechanisms causing the pattern to exist (Keddy and Weiher 1999). As Temperton and Hobbs (2004) noted, restoration practitioners must try to produce a certain type of ecosystem and need guidelines to follow; identifying the underlying mechanisms substantially enhances what is learned and what can be applied in other locations. Animal ecologists have been able to develop general guidelines for application to large spatial scales and for predicting gross measures of animal performance (e.g., presence–absence). It becomes necessary, however, to identify the mechanisms underlying animal performance as we move into progressively smaller spatial areas. The rules must become more detailed, and more difficult to obtain, as we move to restore more than simple animal presence. Lockwood and Pimm (1999) noted that restoring species composition and diversity was far more difficult than restoring ecosystem function and partial structure.

Terminology

I follow Morrison and Hall (2002) and use the term species assemblage to indicate the group of species that are present and potentially interacting within the area of interest. Such an assemblage could be part of a larger community, but there is no need to even invoke the community concept per se with regard to my treatment of assemblage rules for restoration. Rather, managers should focus on identifying the filters and constraints (see below) that will modify the species present in an area throughout a successional process.

Species Pool

This concept defines a regional species pool as occurring within a biogeographic region and extending over spatial scales of many orders of magnitude larger than those of a local species assemblage. The species pool decreases as we progress to smaller geographic areas, such as moving from the watershed scale down to the stream reach scale (i.e., multiple streams within the larger watershed). Then the local assemblage of species from the larger pool is determined by passing through a series of filters. Van Andel and Grootjans (2006) defined three pools: regional, local, and community. Regional species pools are the set of species occurring in a certain biogeographic or climatic region that are potential members of the target assemblage. Local species pools are the set of species occurring in a subunit of the biogeographic region, such as a valley segment. Community species pools are the set of species present in a site within the target community.

The concept of ecological filters entails a main approach in assembly rule theory. From the total pool of potential colonists, only those that are adapted to the abiotic and biotic conditions present at a site will be able to establish themselves successfully (Hobbs and Norton 2004). A filtering or deletion process takes place that is analogous to a filtering-out of those organisms not adapted to the habitat conditions. This approach focuses on the end product of numerous interactions between a colonist and the ecosystem components. Assembly rules are useful in restoration planning because they allow for a systematic way to determine what factors may be limiting membership in the local pool of species. Some of the rules are obvious and seem trivial (e.g., predators without prey will starve), whereas others are more complicated (e.g., what abundance of prey is necessary for a new predator to enter a community; Temperton and Hobbs 2004). The filters involved that influence the course of system development will determine the approach taken to restore an area (Hobbs and Norton 2004). Filters will also change in type and intensity through space and time. Hobbs and Norton (2004) developed a comprehensive list of potential filters, which Morrison (2009) modified to apply more directly to animals and plants (Table 16.1).

As reviewed elsewhere (Morrison 2009), at least seven figures in Temperton et al. (2004) depicted general pathways from the potential pool of species, through various filters and constraints, to the realized species pool. Morrison (2009:91) synthesized these figures into a single diagram that depicted pathways and filters as well as how species fit into available space throughout the course of succession. The general diagram (Fig. 16.2) shows how filters change in influence on a species and in actual type proceeding through succession and development of the species assemblage.

Various abiotic and biotic factors filter the regional species pool, resulting in the local species pool (Fig. 16.2). Species must be physiologically adapted to occupy a given area based on general climatic conditions; additional abiotic factors will have additional filtering effects. The specific factors that filter or constrain which species will actually occur can change

Table 16.1. Filters that influence the assembly of species within a specified geographic area

Filter type	Description
Abiotic filters	*Climate*: rainfall and temperature gradients
	Substrate: fertility, soil water availability, toxicity
	Landscape structure: landscape position, previous land use, patch size, and isolation
Biotic filters	*Competition*: with preexisting and potentially invading species and between planted or introduced species
	Predation–trophic interactions: from preexisting and potentially invading species, and predation between reintroduced animal species
	Propagule availability (dispersal): bird perches, proximity to seed sources, presence of seed banks
	Mutualisms: mycorrhizae, rhizobia, pollination and dispersal, defense, and so forth
	Disturbance: presence of previous or new disturbance regimes
	Order of species arrival and successional model: facilitation, inhibition, and tolerance
	Current and past composition and structure (biological legacy): how much original biodiversity and original biotic and abiotic structure remains

through time, and different limiting factors will impact each species or perhaps groups of similar species. Further, the habitat or niche space available to species will vary through time as succession proceeds through time. Many factors must be considered when trying to predict and guide the actual species that will be present on the target site (e.g., the "community species pool" of Van Andel and Grootjans 2006). In restoration planning, managers can establish target conditions, such as the structure and floristics of vegetation, and develop lists of desired animal species to be achieved as restoration proceeds. Pools of species change as they proceed through filtering processes that accompany the conditions present within a seral stage (Fig. 16.2). Another interesting feature of this assembly process is that each subsequent community species pool need not be a subset of the pool at a previous seral stage; the community pool is under constant reshuffling as a result of the filtering that is occurring through time from the larger local and regional species pools.

Developing a comprehensive restoration plan using the assembly process can be linked directly with monitoring and especially development and implementation of a valid adaptive management plan (Morrison 2009:Chapter 8). Knowing what should occur during different stages of the restoration process

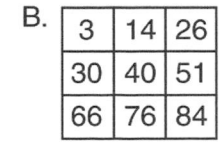

Figure 16.2. The species present in an area (community pool) are those remaining as a result of the filtering process at the local and regional levels. The figure represents two local species pools drawn from the same regional pool and co-occurring in the same two vegetation types. The community pool associated with a seral stage is drawn from the local pool that is specific to the more general vegetation type. The type and number of species present across seral stages will also be a reflection of the size of the target (e.g., project) area and niche space available. The cross-sectional cut (A and B) indicates how species (arbitrarily designated by numbers in the squares) change in type and total number. From Morrison (2009:Fig. 5.2)

FREDERICK LAW OLMSTED (1822–1903)

As the 20th century approached and Americans migrated to cities from rural areas, it became apparent that cities needed to become more hospitable places. Beautification made them more aesthetically pleasing and provided pleasing areas away from the stress of everyday commerce. Frederick Law Olmsted became a pioneer in the city beautiful movement of the post–Civil War generation. Olmsted is acknowledged as the founder of American landscape architecture, even though he had no formal university training and worked a variety of jobs before becoming the superintendent of New York's Central Park, where he became architect-in-chief of construction. He then served as head of the U.S. Sanitary Commission, which was the forerunner of the American Red Cross. Olmsted's goal was to improve American society by designing and promoting recreation in the hearts of cities. He did not envision parks as vast, open spaces but rather as places of harmony—places where people could go to escape city stresses. He wanted these parks to be available to all people no matter their walk of life. He designed trails that curved and flowed with the landscape, screening them with thick plantings along their borders, separating and excluding commercial traffic. He formed his own landscape architecture firm that he ran between 1872 and his retirement in 1895.

to the desired condition allows managers to make midcourse corrections and to improve the opportunity for overall project success.

Implications for Restoration

Restoration projects can actively try to modify the effects of filters to allow desired species in, as well as to prevent the establishment of undesired species (Hobbs and Norton 2004). For example, modifying abiotic filters include providing natural and artificial structures for use as shelter and water sources; modifying biotic filters include controlling exotic species, predator control, and introducing animals.

HABITAT: FOUNDATION OF WILDLIFE RESTORATION

Because restoration requires a synthesis of many disciplines, the restorationist must be well schooled in how a diverse array of scientists and managers—including the community of wildlife professionals—communicate. Virtually everyone

from scientists to the general public understands that habitat describes where something lives. As reviewed elsewhere for wildlife in general (Hall et al. 1997, Morrison and Hall 2002) and for specific application to restoration (Morrison 2009), the term has been used and modified in ways that do not promote understanding and certainly do not easily translate into management actions. This section begins by first providing some brief but critical definitions for key habitat terms (Morrison 2009).

Habitat includes the resources and conditions present in an area that produce occupancy by an animal. If an individual of a species does not occupy a location, then we cannot know for sure that the location is suitable for occupancy. The critical aspect of animal habitat is that it is specific to the presence of a species, population, or individual (animal or plant). Habitat is a synthesis of the specific resources that are needed by the organism. As noted by Miller (2007), an organism-based understanding of habitat is needed to determine appropriate restoration goals.

Habitat and habitat type are not synonymous. Daubenmire (1968) developed the latter term to refer only to the potential climax in an area. We can avoid confusion by referring to the vegetation in an area as the vegetation association or vegetation type instead of the confusing term habitat type (Hall et al. 1997, Morrison and Hall 2002). That habitat is organism specific is a central concept in wildlife ecology and in turn in wildlife restoration, because focusing restoration on vegetation most often will fail to restore the desired assemblage of wildlife.

The definition of habitat has been modified in numerous ways, leading to additional confusion within the wildlife profession (Hall et al. 1997). However, several key terms are widely (and mostly appropriately) used to describe how animals use habitat. Habitat use is the way an animal uses physical and biological resources in an area. Habitat selection refers to the hierarchical and largely innate (as modified by learning) process whereby an animal makes decisions about different scales of the environment (Hutto 1985:458). Johnson (1980) defined selection as a hierarchical process by which an animal chooses which habitat components to use. Habitat preference is restricted to the consequence of the habitat selection process, resulting in the disproportional use of some resources over others.

Habitat availability concerns the ability of an individual to obtain physical and biological components of the environment. In contrast, habitat abundance refers only to the quantity of the resource in the habitat (Wiens 1989:402). Of course, it is challenging to assess resource availability from an animal's perspective (Litvaitis et al. 1994). For example, vegetation that is beyond the reach of an animal is unavailable as a food source. Lastly, habitat quality is the ability of the environment to provide conditions appropriate for individual and population persistence. Although habitat quality occurs along a continuum, it is usually categorized from low- to medium- to high-quality habitats, based on their abilities to provide resources for survival, reproduction, and population persistence, respectively. A critical aspect of habitat quality is that it must be explicitly linked with demographic features if it is to be a useful measure. Indirect measures of quality, such as animal density, often fail to adequately describe actual animal performance (e.g., survival, productivity; Van Horne 1983).

Wildlife ecologists frequently use the terms macrohabitat and microhabitat (Johnson 1980). Macrohabitat refers to large spatial extent (i.e., landscape-scaled) features such as seral stages or vegetation associations (Block and Brennan 1993), which equates to Johnson's (1980) first level ("order") of habitat selection. Alternatively, microhabitat refers to finer-scaled habitat features, such as would-be important factors in levels 2–4 in Johnson's (1980) hierarchy. Macro- and microhabitat are obviously very general categories that have limited usefulness in application, and their use should be minimized except in general conversation.

To advance wildlife ecology and ultimately the transfer of that knowledge to the practice of restoration, we must be sure that the fundamental concepts with which we work are well defined and understood. Peters (1991:76) popularized the term operationalization with regard to ecological concepts, by which concepts such as habitat should have operational definitions that are the practical, measurable specifications of the ranges of specific phenomena the terms represent. But as reviewed above, use of habitat terminology has been imprecise and ambiguous. A lack of explicit definitions leads to ambiguity in what data and results reported in the literature actually mean, which in turn leads to confusion by other workers in the field. As promoted by Hall et al. (1997), Morrison and Hall (2002), and others, there is a strong need for standard definitions or at least a clear understanding of the different uses of the terminology. Without clear definitions of terms, it is unlikely that workers in one field will be able to fully access and to understand information generated by those in another; for our purposes, restorationists and wildlife ecologists must understand one another.

Spatial Scale

Ecologists now recognize that animals go through a series of increasingly refined selection decisions: initial selection of a geographic area; selection of specific combinations of elevation, slope, and vegetation type; selection of specific locations to forage, breed, or rest; and selection of specific items to use, such as types and sizes of food (Fig. 16.3). Moving to an increasingly local spatial scale (i.e., from broad to specific), scientists are able to understand a more detailed amount of information about animals (Fig. 16.3). At broad spatial scales, managers can usually quantify such metrics as presence or absence of a species, but they cannot quantify much, if anything, about survival or reproductive success. However, data obtained about breeding sites inform us about the number of young produced. For application to restoration, this relationship between spatial scale and information content means we must carefully match the goal of a restoration plan with the appropriate scale of study.

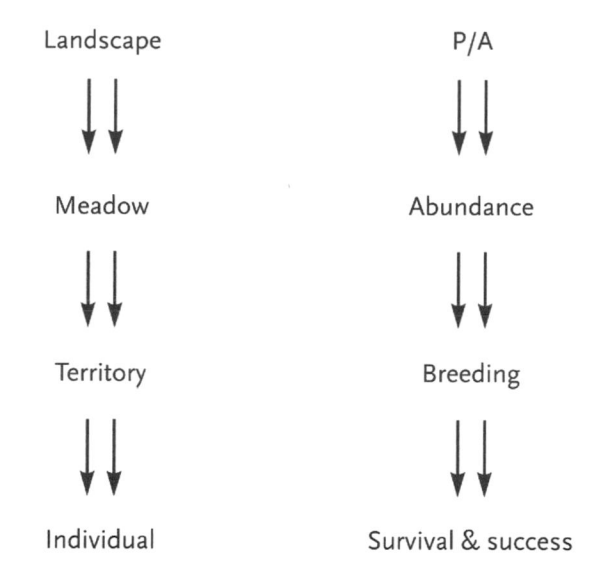

Figure 16.3. Relationship between the measurement of spatial extent and the associated measure of animal performance. P/A is presence or absence. From Morrison (2009:Fig. 4.1)

The Niche

Habitat (in the species-specific sense) provides a general description of the environmental features used by animals and can give initial insight into factors underlying survival and fitness. Other factors, specifically those related to an animals' niche, must be known in order to understand the mechanisms responsible for animal survival and fitness (Morrison et al. 2006). If we do not know ecological mechanisms, we are bound to misinterpret the phenomenon we observed and, in turn, to develop incomplete and likely misguided restoration and management prescriptions. Morrison et al. (2006) provided a detailed account of how the niche can be used to enhance our understanding of wildlife ecology, and Morrison (2009) translated that account into restoration applications. Here I briefly develop the niche concept and how it can be used to improve our ability to restore habitat.

O'Connor (2002) concluded that only one environmental factor is usually limiting in any particular situation for a species, similar to the limiting factor concept of Leopold (1933). The complication, however, is that the same factor is unlikely to always be limiting, because variation in nature results in a continual shifting of resources, which in turn results in shifting among potentially limiting factors. We must understand a good deal about the niche of a species in order to manage and restore its populations.

The potential distribution of a species is defined as the physiological or fundamental niche, which encompasses all of the locations a species could occupy if it was uninhibited by other constraining factors. Within the physiological niche resides the realized niche, which is the actual (observed) distribution of the species. The realized niche does not necessarily contain the full range of conditions under which the species does best, because such factors as competitors and predators exclude it from those conditions. Unless we understand the physiological niche, we cannot know the full range of options available for managing a species. Actions like control of predators or manipulation of plant species may allow some species to thrive in areas where they are rarely found under present conditions (Huston 2002).

The problem with focusing on habitat alone in restoration is that the features measured can stay the same, while use of important resources by an animal within that habitat can change. For example, environmental conditions might change the species composition of arthropods inhabiting an area (niche factor) while not impacting the vegetation (habitat factor) in that area. Changes we observe in habitat use by a species through space and time are often caused by changes in the availability of specific resources that we fail to measure; that is, we do not understand the mechanisms underlying the changes in habitat use. Advancement in knowledge in wildlife ecology and the subsequent translation to restoration entails more than superficial knowledge of natural history and general habitat use. Restoration planning must include these niche factors that have not traditionally been considered, because they influence whether an animal occupies a site and how it performs if occupancy is possible.

DESIRED CONDITIONS FOR RESTORING WILDLIFE

An initial step in designing any natural resource restoration endeavor is to decide upon the desired condition. One must first establish the overall project goal and the specific outcomes desired for plants, animals, and the overall environment. As discussed above, restoring an area to mirror a preexisting condition is difficult unless data on historical conditions are available. And, because of changes in the environment, the historical condition may no longer be attainable.

It is difficult, however, to predict the specific species composition of a locality, especially when dealing with small geographic areas. As reviewed herein, the vegetation type of an area does not directly identify the many species-specific habitats of the locality. The best we can usually do is to assemble lists of probable species occurrences based on the available evidence. The more complete our understanding of the filters and constraints that underpin the process of species colonization, the higher the likelihood we are to establish realistic goals. For most species, however, we have either sporadic, often-vague comments on status in the historical literature, or data too recent, or too incomplete, to be a basis for reconstructing past communities and changes through time. There are a number of tools available that can help develop a reasonable picture of past environmental conditions and associated species occurrences.

Historical Assessments
The primary methods for assessing historical conditions are existing data sets, museum records, fossils, field notes, and literature. As noted by Morrison (2009), the process of assembling the historical view of an area should be seen as a puzzle

whereby one can still identify the picture if some pieces are missing; the number of missing pieces determines the uncertainty in any assessment.

Data Sources
The Biological Resource Division of the U.S. Geological Survey manages the nationwide Breeding Bird Survey (BBS), which began in 1965. The BBS consists of over 2,000 randomly located 40-km-long permanent survey routes established along secondary roads throughout the United States and southern Canada that are surveyed annually during the breeding season (Robbins et al. 1986). Since 1900, the National Audubon Society (NAS) coordinated a bird-counting effort across the United States and Canada during winter known as the Christmas Bird Count (CBC), which is a single-day count conducted by volunteers in a 24-km radius of a chosen location (e.g., a city). The CBC is now a valuable database for long-term monitoring of population trends.

The results of numerous natural history surveys beginning in the late 1800s and continuing to date are available in published reports and scientific papers. A thorough literature review will allow reconstruction of the flora and fauna occurring in the general region of a project site prior to or during the advent of intensive human-induced development. Fleishman et al. (2004) presented a good discussion of the use of historical data in understanding faunal distributions, with special reference to the Great Basin of western North America.

Museum Records
Museums that house natural history collections (e.g., skins, skeletons, eggs, plants) are found within private and public universities, federal and state agencies, and various research organizations. They were developed, in part, to record the historical fauna and flora of a region. Of primary value to our discussion here are the data that accompany each specimen, which usually include the date and location of collection, the collector, the species identity, and perhaps a few natural history notes; much of these data are now available electronically. Of course, a museum specimen only indicates that the species was present at the time of collection. The absence of a species cannot be used to conclude that the species did not occur at the time the collecting was underway. Genetic information available from museum specimens is also being used to reconstruct species' ranges and to identify places where they were extirpated. Morrison (2009) discussed other limitations and cautions associated with museum records.

The Fossil and Subfossil Record
Fossils have been used to reconstruct the former ranges of species. A good example comes from Harris (1993), who reconstructed the succession of microtene rodents from the middle to late Wisconsin period of the Pleistocene in New Mexico. Subfossils, which are unmineralized remains, can also be used to reconstruct more recent environments. Subfossils are found in caves and rocky crevasses, mines, and woodrat (*Neotoma* spp.) middens. Analyzing subfossils requires knowledge of

vertebrate morphology, although only a good undergraduate training in wildlife science or zoology is usually needed. Restorationists can use these types of information to reconstruct the animals, plants, and successional processes that were occurring in the past (Kay 1998).

Literature

Determining the likely environmental changes that occurred through time requires a temporal baseline against which subsequent records can be compared and evaluated. Our baseline of information is improving through time, and has accelerated as we moved through the 20th century. For example, Power (1994) published an example of assembling historical avifaunal records using documents beginning in the 1850s (e.g., from the U.S. Pacific Railroad Survey) and continuing through the early 1900s to reconstruct the distribution and abundance of birds of the coastal islands of California. The NAS has been publishing a compilation of bird observations submitted by the public since the early 1900s. The field journals and other written records of scientists and amateurs alike are housed in their original form in museums, which can help reconstruct the animals present in the region surveyed.

Uncertainty of Observations

Because historical data records will always be incomplete, we must assess the quality of our historical reconstruction by assigning probabilities of certainty to each data source. Such assignments are relative and qualitative. Factors to consider in assigning uncertainty include age of the data source, recorded distance of the source relative to the project site, the number of records available, presence of an actual specimen versus only a visual observation, and (if known) the reputation of the data source. For each restoration plan, the uncertainty of each conclusion and all assumptions must be clearly stated.

Desired Conditions

Conceptual Model

As developed above, a conceptual model of ecosystem processes, functions, filters, and assemblages of species is the initial step for designing a restoration project. This model is a process of feedback between the desired ecological condition and what is ultimately feasible to implement and maintain given current environmental, budgetary, legal, and political constraints. Likewise, the population structure and movements of the target animal species must be incorporated into the plan. As noted by Morrison (2009), there is no reason to restore habitat for a specific suite of animal species if immigration and emigration are impossible, or if predators or competitors cannot be managed either within the project area or on surrounding lands. For example, if fire is an essential component in maintaining the condition of the vegetation, then the frequency and intensity of fires must either be incorporated into the restoration plan, or alternatives to fire must be developed (e.g., cutting trees and thinning shrubs). The model provides a simple visual representation of the ecological processes necessary to maintain the desired condition.

The success of a project will be based in large part on distinguishing between habitat and niche factors, and their relevance to the viability of animals. Because an animal may be absent because the niche factors are inappropriate, models of the presence or viability of animals based on habitat factors often result in poor predictions. Failure to account for niche factors in restoration planning often results in poor success for wildlife diversity and viability. Although evaluation of niche relationships can add substantial time and effort to the planning process, such work substantially improves the efficacy of the final restoration plan and drives anticipated postrestoration management activities. In the previous example of the linkage between wolves, elk, songbirds, beavers, and riparian vegetation (Hebblewhite et al. 2005), failure to incorporate control of elk in the project would likely result in poor plant regeneration and negative responses by the bird assemblage and beavers.

Focal Species

Wildlife management has concentrated time and funds on a restricted set of species, including those regularly hunted and those deemed rare. Many people believe that it is too complicated to address the full potential of the ecological community and that selected species, primarily those of legal concern or consumptive interest, should be the focus of our efforts. Much effort has been expended attempting to simplify how we study and manage species, including a focus on a select list of animal species, which is aptly called a focal species approach.

Lambeck (1997) developed a focal species approach whereby a group of species was selected as a focus of management. These species were the most influenced by specific threatening processes, were area sensitive, and resource and dispersal limited. Related concepts such as indicator, umbrella, and flagship species have generally failed to provide for effective management of multiple species. For example, Lindenmayer et al. (2002) reviewed and criticized the focal species approach in part on the grounds that it incorporated such troublesome concepts as indicator species. Lambeck (2002) countered Lindenmayer et al. (2002) and noted that the focal species approach is multifaceted in that it can incorporate multiple species with many different resource requirements and responses to different threatening factors.

Evaluations of the focal species approach have generally concluded that the approach can be a useful starting point for conservation but with major weaknesses, including the fact that the focal species usually do not serve as surrogates for other species, the selection of species often shows high social bias (i.e., large predators), we do not know enough about every species to correctly choose focal species, and empirical testing of the response of species to management actions is minimal (Lindenmayer et al. 2002, Freudenberger and Brooker 2004).

In contrast to the focal species approach, many management agencies have used a coarse filter approach, which entails managing generally defined habitat conditions such that a majority of the associated animal community will be supported

(Hunter 1991). Morrison et al. (2006) noted, however, that the coarse filter approach usually fails to protect many native species, because the niche requirements of many species are not usually met by this approach.

The most productive way to design a restoration plan involves developing a conceptual ecological framework that will provide suitable general conditions for many species, along with more specific plans for a set of focal species. There will be a multitude of project-specific reasons for selecting species. What we want to avoid are the extremes; namely, only relying on restoration of general conditions in the hope that the species we desire are supported, versus trying to use only a few species as surrogates of how other species will respond to threatening process and management action. An important aspect of a focal species approach is to document the rationale for selecting species.

Implementation Steps

Next I outline specific steps to implement wildlife restoration that are structured in a spatially explicit context (i.e., from broad spatial extent to local-scale applications; Morrison 2009).

Step 1: Planning Area

The essential first step in developing a restoration plan is determining the desired ecological condition of the largest planning area (e.g., a valley composed of multiple creeks). Surrounding areas will influence each management unit (e.g., creek) within the overall planning area. Priority steps to conduct include: develop a conceptual ecosystem model of the planning area that identifies major factors that will lead to desired ecological condition, identify key ecological attributes that influence the ecosystem model, identify the suite of species characteristic of the desired condition, identify focal species, and identify the primary constraints and stressors that inhibit proper functioning of the pathways in the ecological model (and thus prevent attainment of the desired condition). By listing constraints and stressors, one quickly identifies those that can be alleviated and those that cannot. Once identifying these limitations, one can re-evaluate the desired condition and associated conceptual model in light of what is possible. Lastly, a monitoring plan that uses the key ecological attributes at the overall project scale must be developed, including establishing quantitative goals, a time frame for attainment, and thresholds for additional action.

Step 2: Project Area

The restoration plan for each project area (e.g., creek) should follow from the desired ecological condition for the overall planning area (e.g., valley). Each project area's current condition allows an investigator to determine how to allocate target levels of each ecosystem component across the specific project area(s) based on the desired condition for the entire planning area. Certain management actions might not be feasible to include because of the location (e.g., cowbird control adjacent to a residential area). Specific steps include: describe the current vegetation and other environmental features, identify the location and amount of special features (e.g., springs, old trees, caves), identify constraints and stressors that occur spatially and temporally, compare special features relative to the larger planning area, develop species list, identify focal species and evaluate the potential of the site to maintain each species and the constraints on maintenance, develop a management plan for all special features and focal species, and develop the preliminary restoration and monitoring plan.

Step 3: Adaptive Management Implementation

Implementation and management of a restoration plan should follow the tenets of adaptive management (see Chapter 5 for details). Adaptive management requires the specification, during development of the plan, of the potential actions that could be taken if monitoring thresholds are triggered. The plan specifies values for key variables that must be attained by a certain time following restoration. For example, say that a key project goal is to develop 60% cover of riparian vegetation more than 2 m in height within six years of planting. The plan could incorporate a target (threshold) goal for growth of riparian vegetation at three years postrestoration of 40% cover of more than 1 m height, a threshold that was based on knowledge of riparian vegetation growth rate. Failure to attain this threshold by year three would trigger a specific management action (which was clearly elucidated in the original plan). This process helps eliminate (or at least minimize) the trial-and-error approach followed in most projects. Such scenarios or pathways that the project area is likely to follow are an output of the initial ecosystem model upon which the plan was formulated.

Here we see the value in viewing restoration of one area in an overall context of a larger area, rather than as a series of individual, project-level endeavors, which is essential when dealing with populations structured as a metapopulation. To develop an adaptive management approach, one must develop specific management actions if a threshold is triggered at either the overall project or subproject area scale, implement additional management actions, modify vegetation structure, revise treatment schedules, revise the conceptual model on the basis of new information and changes in management actions, revise thresholds and triggers as indicated by revisions in actions, and continue monitoring at appropriate spatial and temporal scales. Chambers and Miller (2004) presented a succinct example of a hierarchical approach to restoration and management using Great Basin riparian systems. Morrison (2009:Chapter 10) provided case studies that developed restoration plans for wildlife.

DESIGNING A WILDLIFE RESTORATION PROJECT

Habitat Heterogeneity

Restoration is perhaps the most integrative of endeavors in the conservation arena. A successful restoration project, regardless of its spatial extent (i.e., how big it is), must consider a plethora of issues that range well outside the physical boundaries of the actual project. Below I review briefly some of the major issues that must be considered in designing a restoration project, including the size and type of patch to be restored, connectivity between patches, fragmentation, and corridors. Broadly, these and other environmental features cause substantial heterogeneity in the habitat available to each animal species.

Habitat heterogeneity is defined as the amount of discontinuity in environmental conditions across a landscape for a particular species. Remember that because animal habitat is a species-specific concept, a particular combination of environmental features may constitute habitat for one species and a barrier for another. Discontinuities in environmental conditions, which lead to heterogeneity, occur naturally with changes in soil type or edges of water bodies, or anthropogenically with agricultural lands or roads. Each species, even those with rather similar general habitat requirements, can show wide differences in tolerance of habitat heterogeneity. After all, at some spatial scale, all locations are heterogeneous (patchy); what may be seen as homogeneous to a large mammal may be heterogeneous at the spatial scale of the amphibian.

Fragmentation

Much has been written, including in the popular press, about fragmentation. Fragmentation is defined as the extent of heterogeneity of habitats across a landscape. The isolation (how far apart) and physical size of resource patches available is used to describe the degree of fragmentation and in turn its impact on various species. As noted by Morrison (2009), fragmentation is necessarily a species-specific concept. The commonly used term "habitat fragmentation" appears frequently in the ecological literature and in the past referred (incorrectly) to any type of heterogeneous condition. Similarly, many authors have used the term "landscape fragmentation," which is strictly incorrect because it is the resources (e.g., habitats for specific species) that become fragmented within landscapes, and not entire landscapes per se. Morrison (2009:119) provided the following truism for designing all restoration projects: "The species-specific concept of habitat selection, which occurs across spatial scales . . . must be the focus of restoration with regard to fragmentation (and the broader issues of habitat heterogeneity). Likewise, the concept of species assembly rules . . . is directly relevant here because fragment size, shape, and the quality of the habitat therein will serve as a filter determining in part if a species can exist in a target area."

Vastly different (and, for specific species, unsuitable conditions) can isolate and surround vegetation patches. Individual animals in these isolated patches can go extinct unless immigrants supplemented their populations. Partial isolation of habitats is also problematic, however, because lowering dispersal rates and thus smaller populations can affect population viability. Other issues involving fragmentation—such as temporal fragmentation and fragmentation caused by subtle discontinuities in environmental conditions within a fragment—are beyond the scope of this chapter but should be reviewed as part of any project (Morrison 2009:119–123).

In planning a restoration project, biologists should consider the following: (1) the absolute loss of habitat area; (2) increased edge; (3) increased distances or permeability for movement of animals between patches; (4) increased penetration of predators, competitors, and nest parasites; and (5) changes in microclimate with changes in patch area and edge. The species-specific nature of animals' response to habitat heterogeneity can be seen in results of a study by Bolger et al. (1997), who studied the response of birds in a matrix of remaining chaparral vegetation in Southern California. They reported that six species showed negative responses, four species showed positive responses, and ten species showed no apparent response to area size. Morrison (2009) reviewed similar results from other studies, concluding that different aspects of fragmentation affect different species. For highly mobile species, total habitat area is probably more important than the area of single patches. In contrast, less mobile animals will be most impacted by the size and proximity of patches.

Corridors

Movements of individuals between populations, especially when a metapopulation is indicated, enhance population viability. Connectivity refers to the extent to which individuals can move among a mosaic of habitat (Hilty et al. 2006). What have become known popularly as corridors are one means of achieving connectivity.

Because species will show differing responses to patchiness, a restoration plan should include a categorization of species by the potential impacts that footpaths, roads, fence lines, canals, structures, changes in vegetation structure, and other features might have on species of interest. Because of the many potential natural and human-placed barriers to animal movement, many conservation biologists advocated the retention of natural corridors or development of artificial corridors in an effort to maintain linkages between patches. Although incorporation of corridors into restoration plans has been frequent, their use is controversial.

As reviewed by Hess (1994), in certain situations corridors can increase the chance of metapopulation extinction by promoting the transmission of disease. The probability of disease and other adverse factors (e.g., parasites) should be considered during the planning stages of all reserve networks. The presence of diseases that are known to infect species likely to use corridors should be evaluated prior to linking reserves. Strategies for containing the disease should be developed prior to establishing the linkages. Contingencies for treating epizootics include vaccination, removal of infected individuals, and temporary severing of the linkage. For example, Simberloff and Cox (1987) and Simberloff et al. (1992) cautioned that too little empirical data were available to warrant wholesale adoption of corridors as a conservation action. In addition, corridors can enhance the spread of disease and fires, and can increase exposure of individuals to predation, domestic animals, and poachers. Although corridors might have great potential, they must be evaluated and planned on a species-specific basis (Cushman et al. 2009). Unfortunately, many restoration plans incorporate corridors with no certitude that they work. If the plan is developed within the context of a true adaptive management plan, then alternatives will be available should corridors fail to provide the necessary linkages between habitat patches to maintain the desired species assemblages. Readers can consult Morrison et al. (2006) and Morrison (2009) for a more thorough review of the empirical evidence for animals dispersing through and otherwise using corridors.

There are many suggested ways to design corridors (Hilty et al. 2006:Chapter 4). I describe several of the typical designs for corridors (Fig. 16.4), beginning with the classic depiction of a corridor (Fig. 16.4A) as a continuous passageway of some width that links two or more habitat areas. When a continuous linkage is not possible or perhaps not needed, the

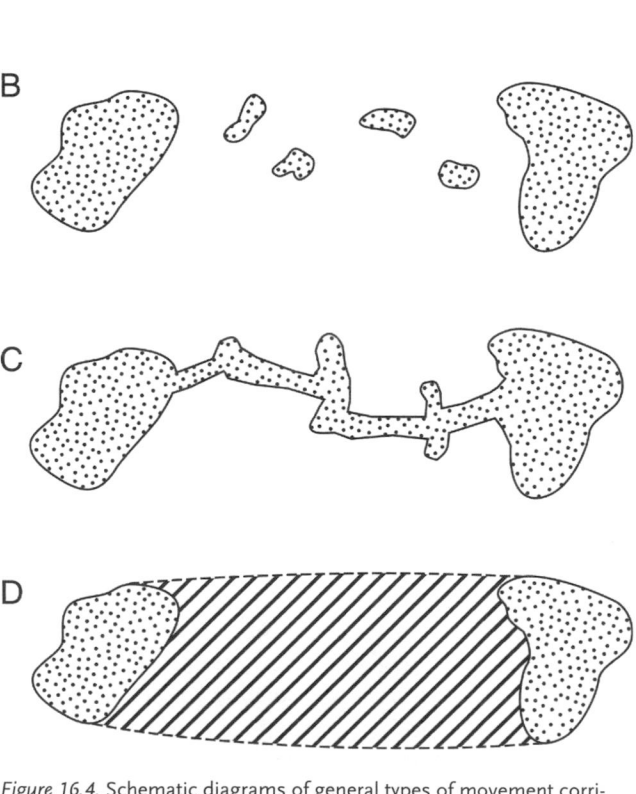

Figure 16.4. Schematic diagrams of general types of movement corridors. (A) The classic design with continuous linkage between core habitat areas. (B) The situation where the linkage is broken into patches of habitat, often called stepping-stones, between core areas. (C) The stepping-stones of Fig. 16.4B have been linked with a narrow but continuous pathway. (D) The core areas are connected by a continuous cover of relatively poor (quality) habitat where animals can move, rest, and perhaps feed but not be sustained for prolonged periods of time. From Morrison (2009:Fig. 7.3)

stepping-stone concept of corridors (Fig. 16.4B) is available. These stepping-stones have been modified to include physical links between the steps (Fig. 16.4C), where the stepping-stone areas might function as a temporary location for resting and feeding. Lastly, I depict a situation where the corridor is actually a large area of lower-quality habitat relative to the core (and thus higher-quality habitat; Fig. 16.4D); individuals are able to move through but not actually breed or survive indefinitely in the corridor. Other combinations of patches and corridors are possible, such as locating the stepping-stones into the matrix of lower-quality habitat (the corridor). Smallwood (2001) suggested that the area surrounding a corridor be classified according to the functions they individually and collectively serve for specific animal species (e.g., on a scale of low to high). Hilty et al. (2006) provided an in-depth accounting of the factors that must be considered when determining if corridors are a viable option within a restoration project; I highly recommend that readers review their work.

Restoration Design

The major principles involved with designing areas for the conservation of species have been developed in the context of reserve design and can be summarized as follows: (1) well-distributed species are less susceptible to extinction than species confined to small locations; (2) larger blocks will allow population persistence longer than small blocks; (3) blocks of habitat close together are better than blocks farther apart; (4) habitat in continuous blocks is better than fragmented habitat; (5) interconnected blocks of habitat are better than isolated blocks; and (6) populations that fluctuate are more vulnerable than populations that are stable (Noss et al. 1997).

Guidelines for habitat management were developed by Morrison et al. (2006) and summarized for restoration in Morrison (2009). Below I summarize these guidelines but encourage readers to pursue the literature in greater depth.

Guidelines for Species Richness and Overall Biodiversity

Although most restoration occurs at the within-patch scale, the planner needs to consider the larger spatial area within which the specific project area resides. Design guidelines should be divided into three spatial scales, and they can be based on how population dynamics are likely to vary as a function of habitat complexity at each scale: (1) within habitat patches, or alpha diversity, such as with the observed correlations between foliar height diversity and bird species diversity; (2) between habitat patches, or beta diversity; and (3) among broader geographic areas, or gamma diversity.

This hierarchy links with the filtering concept that I developed above, where the species present in a location are a result of the limitations imposed by abiotic and biotic factors and other constraints. Moderate disturbance regimes that cause variations across space in vegetation elements, substrates, and abiotic characteristics of the environment usually maintain beta diversity. Of course, beta diversity is reflected at different spatial scales based on body size, size of home range, and movement ability. In contrast, gamma diversity is usually controlled by climate, landform, geographic location, and broad-scale vegetation formations.

Guidelines to Maintain Within-patch Conditions

The size and the topographic location, proximity of other patches, and the susceptibility to disturbances (e.g., floods, fire, human recreation), usually have a strong impact on within-patch conditions. Further, the context of the patch (conditions in adjacent patches); the type and intensity of natural disturbances; and the type, frequency, and intensity of management activities also impact within-patch conditions. Within-patch conditions must be developed and managed to target specific animal species if the outcome of restoration is to meet project goals.

Guidelines to Maintain a Desired Occupancy Rate of Habitat Patches

Specific animal species often come and go within different patches of habitat; occupancy is often not constant. As discussed above, the size and quality of the patches, the proximity and size quality of adjacent patches, the dispersal ability of animals, and other factors determine the dynamics of animal populations within and between patches. Further, the location of patches can change through time because of catastrophic events such as fire, flood, disease, and other disturbances. In restoration planning, habitat patches can often be mapped and links between them identified. Habitat corridors can be retained or developed to provide connections between populations across a landscape.

Planners must consider the range of factors that potentially limit the occurrence of species in a planned restoration area. Such limiting factors and potential ways to mitigate (Table 16.2) these and other factors play a direct role in defining project goals, and in establishing specific management practices for a restoration project.

SUMMARY

This chapter reviewed how the principles of wildlife ecology and restoration ecology can be linked to results in comprehensive planning for the management and conservation of natural resources. Restoration plans should be guided by the likely responses of current or desired wildlife species in the project area, including the current and historical abundance and distribution of animal species, habitat and food requirements, breeding locations, how plant succession will change species composition through time, space requirements, and necessary links between geographic areas. There is a necessary connection between the practice of restoration (i.e., ecological restoration) and the concepts that were used to develop the restoration plan (i.e., restoration ecology). Restorationists must help quantify the likely consequences of human activities and natural events on the environment so that the public and decision makers can make informed choices when restoring an area. No project should be attempted unless it has a

Table 16.2. Potential factors limiting animal populations and potential means of mitigating those factors within a restoration planning context

Limiting factor	Mitigation measure
Disturbance caused by human activities	Control human access or timing of access (e.g., no entry into a sensitive area during breeding season)
	Establish buffers around key, sensitive areas (e.g., roosting areas)
Disease	Do not allow animals to concentrate in small areas
	Consider treatment of the environment or treatment of selected animals
Size of area	Functions as a factor limiting occupancy of certain species; consider possibility of linkages (corridors)
Seasonality	Consider availability of water, roosts, and other resources; such resources might need to be artificially established and maintained
Biotic factors, including predation and competition	Consider direct control of exotic, or in some cases native (e.g., cowbirds), animals

Source: From Morrison (2009:142).

clear conceptual framework with rigor sufficient to allow for continual monitoring and feedback in an adaptive design. Restoration projects are part of an ongoing process that will form the building blocks of a strong discipline of wildlife restoration ecology. Although habitat (in the species-specific sense) provides general knowledge on animal requirements, habitat alone seldom identifies the underlying mechanisms determining occupancy, survival, and fecundity. That is, habitat alone provides a limited explanation of the ecology of an animal. Statistical models of habitat are based largely on surrogates of these mechanisms. If restoration is to successfully restore and maintain wildlife populations, then as a discipline it must take advantage of the best that other disciplines, such as wildlife science, have to offer.

Literature Cited

Berger, J., P. B. Stacey, L. Bellis, and M. P. Johnson. 2001. A mammalian predator–prey imbalance: grizzly bear and wolf extinction affect avian neotropical migrants. Ecological Applications 11:947–960.

Block, W. M., and L. A. Brennan. 1993. The habitat concept in ornithology: theory and applications. Current Ornithology 11:35–91.

Bolger, D. T., T. A. Scott, and J. R. Rotenberry. 1997. Breeding bird abundance in an urbanizing landscape in coastal Southern California. Conservation Biology 11:406–421.

Chambers, J. C., and J. R. Miller. 2004. Restoring and maintaining sustainable riparian ecosystems: the Great Basin ecosystem management project. Pages 1–23 in J. C. Chambers and J. R. Miller, editors. Great Basin riparian ecosystems: ecology, management, and restoration. Island Press, Washington, D.C., USA.

Connor, E. F., and D. Simberloff. 1979. The assembly of species communities: chance or competition? Ecology 60:1132–1340.

Cushman, S. A., K. S. McKelvey, and M. K. Schwartz. 2009. Use of empirically derived source-destination models to map regional conservation corridors. Conservation Biology 23:368–376.

Daubenmire, R. 1968. Plant communities: a textbook of plant synecology. Harper and Row, New York, New York, USA.

Diamond, J. M. 1975. The assembly of species communities. Pages 342–444 in M. L. Cody and J. M. Diamond, editors. Ecology and evolution of communities. Harvard University Press, Cambridge, Massachusetts, USA.

Egan, D., and E. A. Howell. 2001. Introduction. Pages 1–23 in D. Egan and E. A. Howell, editors. The historical ecology handbook. Island Press, Washington, D.C., USA.

Fleishman, E., J. B. Dunham, D. D. Murphy, and P. F. Brussard. 2004. Explanation, prediction, and maintenance of native species richness and composition. Pages 232–260 in J. C. Chambers and J. R. Miller, editors. Great Basin riparian ecosystems: ecology, management, and restoration. Island Press, Washington, D.C., USA.

Freudenberger, D., and L. Brooker. 2004. Development of the focal species approach for biodiversity conservation in the temperate agricultural zones of Australia. Biodiversity and Conservation 13:253–274.

Hall, L. S., P. R. Krausman, and M. L. Morrison. 1997. The habitat concept and a plea for standard terminology. Wildlife Society Bulletin 25:173–182.

Halle, S., and M. Fattorini. 2004. Advances in restoration ecology: insights from aquatic and terrestrial ecosystems. Pages 10–33 in V. M. Temperton, R. J. Hobbs, T. Nuttle, and S. Halle, editors. Assembly rules and restoration ecology: bridging the gap between theory and practice. Island Press, Washington, D.C., USA.

Hanski, I. 1996. Metapopulation ecology. Pages 13–43 in O. E. Rhodes Jr., R. K. Chesser, and M. H. Smith, editors. Population dynamics in ecological space and time. University of Chicago Press, Chicago, Illinois, USA.

Harris, A. H. 1993. Wisconsin and pre-pleniglacial biotic changes in southeastern New Mexico. Quaternary Research 40:127–133.

Hebblewhite, M., C. A. White, C. Nietvelt, J. M. McKenzie, T. E. Hurd, J. M. Fryxell, S. Bayley, and P. C. Paquet. 2005. Human activity mediates a trophic cascade caused by wolves. Ecology 86:2135–2144.

Hess, G. R. 1994. Conservation corridors and contagious disease: a cautionary note. Conservation Biology 8:256–262.

Hilty, J. A., W. Z. Lidicker Jr., and A. M. Merenlender. 2006. Corridor ecology: the science and practice of linking landscapes for biodiversity conservation. Island Press, Washington, D.C., USA.

Hobbs, R. J., and D. A. Norton. 1996. Towards a conceptual framework for restoration ecology. Restoration Ecology 4:93–110.

Hobbs, R. J., and D. A. Norton. 2004. Ecological filters, thresholds, and gradients in resistance to ecosystem reassembly. Pages 72–95 in V. M. Temperton, R. J. Hobbs, T. Nuttle, and S. Halle, editors. Assembly rules and restoration ecology: bridging the gap between theory and practice. Island Press, Washington, D.C., USA.

Hunter, M. L. 1991. Coping with ignorance: the coarse-filter strategy for maintaining biodiversity. Pages 266–281 in K. A. Kohm, editor. Balancing on the brink of extinction. Island Press, Washington, D.C., USA.

Huston, M. A. 2002. Introductory essay: critical issues for improving predictions. Pages 7–21 in J. M. Scott, P. J. Heglund, M. L. Morrison, J. B. Haufler, M. G. Raphael, W. A. Wall, and F. B. Samson, editors. Predicting species occurrences: issues of accuracy and scale. Island Press, Washington, D.C., USA.

Hutto, R. L. 1985. Habitat selection by nonbreeding, migratory land birds. Pages 455–476 in M. L. Cody, editor. Habitat selection in birds. Academic Press, Orlando, Florida, USA.

Johnson, D. H. 1980. The comparison of usage and availability measurements for evaluating resource preference. Ecology 61:65–71.

Kay, C. E. 1998. Are ecosystems structured from the top-down or bottom-up? A new look at an old debate. Wildlife Society Bulletin 26:484–498.

Keddy, P., and E. Weiher. 1999. Introduction: the scope and goals of research on assembly rules. Pages 1–20 in E. Weiher and P. Keddy, editors. Ecological assembly rules: perspectives, advances, and retreats. Cambridge University Press, Cambridge, United Kingdom.

Kostecke, R. M., S. G. Summers, G. H. Eckrich, and D. A. Cimprich. 2005. Effects of brown-headed cowbird (Molothrus ater) removal on black-capped vireo (Vireo atricapilla) nest success and population growth on Fort Hood, Texas. Ornithological Monographs 57:28–37.

Kus, B. E. 1999. Impacts of brown-headed cowbird parasitism on productivity of the endangered Bell's vireo. Studies in Avian Biology 18:160–166.

Lambeck, R. J. 1997. Focal species: a multi-species umbrella for nature conservation. Conservation Biology 11:849–856.

Lambeck, R. J. 2002. Focal species and restoration ecology: response to Lindenmayer et al. Conservation Biology 16:549–551.

Lande, R., and G. F. Barrowclough. 1987. Effective population size, genetic variation, and their use in population management. Pages 87–123 in M. E. Soulé, editor. Viable populations. Cambridge University Press, Cambridge, United Kingdom.

Leopold, A. 1933. Game management. Charles Scribner's Sons, New York, New York, USA.

Levins, R. 1970. Extinction. Lectures on Mathematics in the Life Sciences 2:75–107.

Lindenmayer, D. B., A. D. Manning, P. L. Smith, H. P. Possingham, J. Fischer, I. Oliver, and M. A. McCarthy. 2002. The focal-species approach and landscape restoration: a critique. Conservation Biology 16:338–345.

Litvaitis, J. A., K. Titus, and E. M. Anderson. 1994. Measuring vertebrate use of terrestrial habitats and foods. Pages 254–274 in T. A. Bookhout, editor. Research and management techniques for wildlife and habitats. Fifth edition. The Wildlife Society, Bethesda, Maryland, USA.

Lockwood, J. L., and S. L. Pimm. 1999. When does restoration succeed? Pages 363–392 in E. Weiher and P. Keddy, editors. Ecological assembly rules: perspectives, advances, and retreats. Cambridge University Press, New York, New York, USA.

Miller, J. R. 2007. Habitat and landscape design: concepts, constraints and opportunities. Pages 81–95 in D. B. Lindenmayer and R. J. Hobbs, editors. Managing and designing landscapes for conservation: moving from perspectives to principles. Island Press, Washington, D.C., USA.

Mills, L. S. 2007. Conservation of wildlife populations: demography, genetics, and management. Blackwell, Oxford, United Kingdom.

Morrison, M. L. 2009. Restoring wildlife: ecological concepts and practical applications. Island Press, Washington, D.C., USA.

Morrison, M. L., and L. S. Hall. 2002. Standard terminology: toward a common language to advance ecological understanding and applications. Pages 43–52 in J. M. Scott, P. J. Heglund, M. L. Morrison, J. B. Haufler, M. G. Raphael, W. A. Wall, and F. B. Samson, editors. Predicting species occurrences: issues of accuracy and scale. Island Press, Washington, D.C., USA.

Morrison, M. L., L. S. Hall, S. K. Robinson, S. I. Rothstein, D. C. Hahn, and T. D. Rich. 1999. Research and management of the brown-headed cowbird in western landscapes. Studies in Avian Biology 18:204–217.

Morrison, M. L., B. G. Marcot, and R. W. Mannan. 2006. Wildlife–habitat relationships: concepts and applications. Third edition. Island Press, Washington, D.C., USA.

Morrison, M. L., T. A. Scott, and T. Tennant. 1994. Wildlife–habitat restoration in an urban park in Southern California. Restoration Ecology 2:17–30.

Noss, R. F. 1985. On characterizing presettlement vegetation: how and why. Natural Areas Journal 5:5–19.

Noss, R. F., M. A. O'Connell, and D. M. Murphy. 1997. The science of conservation planning: habitat conservation under the Endangered Species Act. Island Press, Washington, D.C., USA.

O'Connor, R. J. 2002. The conceptual basis of species distribution modeling: time for a paradigm shift? Pages 25–33 in J. M. Scott, P. J. Heglund, M. L. Morrison, J. B. Haufler, M. G. Raphael, W. A. Wall, and F. B. Samson, editors. Predicting species occurrences: issues of accuracy and scale. Island Press, Washington, D.C., USA.

Office of Technology and Assessment. 1993. Harmful non-indigenous species in the United States. OTA-F-565. Two volumes. U.S. Congress, Washington, D.C., USA.

Palmer, M. A., D. A. Falk, and J. B. Zedler. 2006. Ecological theory and restoration ecology. Pages 1–10 in D. A. Falk, M. A. Palmer, and J. B. Zedler, editors. Foundations of restoration ecology. Island Press, Washington, D.C., USA.

Peters, R. H. 1991. A critique for ecology. Cambridge University Press, Cambridge, United Kingdom.

Power, D. M. 1994. Avifaunal change on California's coastal islands. Studies in Avian Biology 15:75–90.

Robbins, C. S., D. Bystrak, and P. H. Geissler. 1986. The Breeding Bird Survey: its first fifteen years, 1965–1979. Research Publication 157. U.S. Fish and Wildlife Service, Washington, D.C., USA.

Simberloff, D., and J. Cox. 1987. Consequences and costs of conservation corridors. Conservation Biology 1:63–71.

Simberloff, D. S., J. A. Farr, J. Cox, and D. W. Mehlman. 1992. Movement corridors: conservation bargains or poor investments. Conservation Biology 6:493–504.

Smallwood, K. S. 2001. Linking habitat restoration to meaningful units of animal demography. Restoration Ecology 9:253–261.

Temperton, V. M., and R. J. Hobbs. 2004. The search for ecological assembly rules and its relevance to restoration ecology. Pages 34–54 in V. M. Temperton, R. J. Hobbs, T. Nuttle, and S. Halle, editors. Assembly rules and restoration ecology: bridging the gap between theory and practice. Island Press, Washington, D.C., USA.

Temperton, V. M., R. J. Hobbs, T. Nuttle, and S. Halle, editors. 2004. Assembly rules and restoration ecology: bridging the gap between theory and practice. Island Press, Washington, D.C., USA.

Van Andel, J., and A. P. Grootjans. 2006. Concepts in restoration ecology. Pages 16–28 in J. van Andel and J. Aronson, editors. Restoration ecology. Blackwell, Oxford, United Kingdom.

Van Horne, B. 1983. Density as a misleading indicator of habitat quality. Journal of Wildlife Management 47:893–901.

Weiher, E., and P. Keddy, editors. 1999. Ecological assembly rules: perspectives, advances, retreats. Cambridge University Press, Cambridge, United Kingdom.

Wiens, J. A. 1989. The ecology of bird communities. Volume 1. Foundations and patterns. Cambridge University Press, Cambridge, United Kingdom.

Willis, K. J., and H. J. B. Birks. 2006. What is natural? The need for a long-term perspective in biodiversity conservation. Science 314:1261–1265.

17

CLIMATE CHANGE AND WILDLIFE

MARTA A. JARZYNA, BENJAMIN ZUCKERBERG, AND
WILLIAM F. PORTER

INTRODUCTION

Climate change is a significant long-term shift of weather patterns. We measure climate change in statistical terms such as long-term averages or variation of temperature or sea level, but to wildlife biologists, climate is a central abiotic component of ecosystems that affects the distribution and abundance of species.

In its Fourth Assessment Report, the Intergovernmental Panel on Climate Change (IPCC) stated that "warming of the climate system is unequivocal, as is now evident from observations of increases in global average air and ocean temperatures, widespread melting of snow and ice and rising global average sea level" (IPCC 2007:30). An organization of scientists from throughout the world and established by the United Nations to provide an objective evaluation of climate research, the IPCC reported that an overwhelming body of evidence suggests recent climate change has been caused mainly by human actions. In fact, the evidence has been accumulating at such a fast pace that it prompted the IPCC to change the wording of their report from "the balance of evidence suggests a discernible human influence on global climate" (IPCC 1995:22) to "human-induced warming of the climate system is widespread" (IPCC 2007:665).

Climate change is not a new phenomenon. The earth's climate has changed repeatedly during its 4.5 billion–year history, and each global climatic change resulted in significant changes in the abundance and distribution of species. However, the climate change we are currently experiencing is distinguished from past events by at least two factors: both the rate and magnitude of change are greater than has been experienced since the advent of civilization, and much of the change is attributable to human activities.

We tend to think of climate change as a simple warming of temperatures and a shift in the distribution of species poleward or higher in elevation. However, the reality is a more complex suite of changes in global-scale processes that lead to profound effects on ecosystems. As such, climate change has potential to alter significantly the species composition and functioning of ecosystems, affecting wildlife populations worldwide. This potential impact is important to wildlife biologists because much of the management of wildlife habitats and populations is based on communities of species and a biogeographic distribution of ecosystems that has been in place for thousands of years. As the climate changes, each species will adapt in different ways, leading to a wholesale reorganization of communities. Wildlife management will need to adapt, as well.

The intent of this chapter is to examine the causes of climate change and to address three central questions. What are the biological implications of climate change to wildlife populations, what will this mean to the functioning of ecosystems, and what does this mean for wildlife managers?

CAUSES OF CLIMATE CHANGE

In this section, we explore the causes of recent climate change. Much of contemporary climate change is caused by increases in carbon dioxide (CO_2) and a handful of other greenhouse gases (GHGs) resulting from use of fossil fuels, land-cover change, and agricultural practices. They principally affect climate by modifying the absorption, scattering, and emission of radiation within the atmosphere and at the earth's surface (IPCC 2007). The increased atmospheric concentrations of four long-lived GHGs have been reported to be the main influence of the recent climate change (IPCC 2007), because they absorb and re-emit the outgoing infrared radiation (heat), while at the same time are transparent to the incoming solar radiation. Those GHGs are carbon dioxide (CO_2), methane (CH_4), nitrous oxide (N_2O), and halocarbons.

Burning fossil fuels and land-use changes are the main causes of recent global increases in concentrations of GHGs. As their name implies, fossil fuels are fossilized remains of dead organisms. Coal is a good example and, because coal is composed of organic matter, it contains high amounts of carbon and some nitrogen, oxygen, and hydrogen. While burning, carbon reacts with oxygen to form CO_2, CH_4, and N_2O are also released while burning fossil fuels. Changes in land cover further contribute to increases in GHGs concentrations. In particular, significant amounts of carbon are released by forest clearing and burning. Agricultural practices are also responsible for increases in GHGs, as CH_4 is a by-product of the fermentative digestion by ruminant livestock (in particular sheep and cattle) and is released from stored manures and inundated rice paddies (Mosier et al. 1998). Soils and manures release nitrous oxide through the microbial transformation of nitrogen (Smith and Conen 2004, Oenema et al. 2005).

Among the GHGs, carbon dioxide has garnered the most attention. The global atmospheric concentration values of CO_2 have increased to 379 ppm in 2005 compared with pre-industrial levels of 280 ppm (IPCC 2007). In March 2013, the Mauna Loa Observatory in Hawaii (the location of the longest record of direct measurements of CO_2 in the atmosphere) recorded a CO_2 concentration of 397 ppm (National Oceanic and Atmospheric Administration 2013).

Methane concentrations increased almost 2.5-fold since preindustrial times, when CH_4 concentration values were about 715 ppb, and 2005, when the concentration values were 1,774 ppb. Methane concentration growth rates stabilized in the late 1990s, but since 2007, global concentration of CH_4 has increased (Fig. 17.1; National Oceanic and Atmospheric Administration 2011).

N_2O continues to grow at a relatively uniform growth rate; its global concentration increased from preindustrial values of about 270 to 324 ppb in 2011 (Fig. 17.1; National Oceanic and Atmospheric Administration 2011). Many halocarbons (i.e., mostly man-made chemical compounds in which carbon atoms are linked with halogens such as fluorine, chlorine, bromine, iodine) have also increased from near-zero preindustrial background concentrations. Most of this increase is due to industrial production. Halocarbon concentrations peaked in the late 1990s (Fig. 17.1; National Oceanic and Atmospheric

Administration 2011) and are now declining as a response to decreased emissions. These reduced emissions are a product of an international treaty known as the Montreal Protocol, which was signed in 1987. It was specifically designed to induce society to phase out halocarbons and other substances responsible for ozone depletion (Montzka et al. 2011).

One of the central contentions in the climate change debate is the role of solar radiation. This contention can be evaluated if we look at the energy balance of the climate system. The energy balance is expressed as radiative forcing and measured in watts per square meter (W/m^2). The combined radiative forcing due to increases in the three main GHGs (CO_2, CH_4, and N_2O) is currently, on average, +2.3 W/m^2. On the other hand, aerosols (i.e., solid or liquid particles suspended in gas) tend to cool the earth by reflecting the incoming solar radiation back to space. The total direct radiative forcing of aerosols is –0.5 W/m^2 (IPCC 2007). Changes in solar radiation have caused a radiative forcing of +0.12 W/m^2 since preindustrial times. Consequently, solar radiation is a relatively small contributor to the overall forcings associated with modern climate change (Fig. 17.2; National Oceanic and Atmospheric Administration 2011).

Observed Changes in Climate

Global warming does not imply that that future changes in weather and climate will be uniform around the globe. Different parts of the world will warm in different ways, and the effects of climate change will vary at local, regional, and global scales. In this section, we discuss the environmental consequences of climate change, including rising average temperatures, sea level rise, melting snow and ice cover, and extreme events such as flood, droughts, and heat waves. The principal message of this section is that changes in the abiotic facets of the ecosystem have already been observed and they

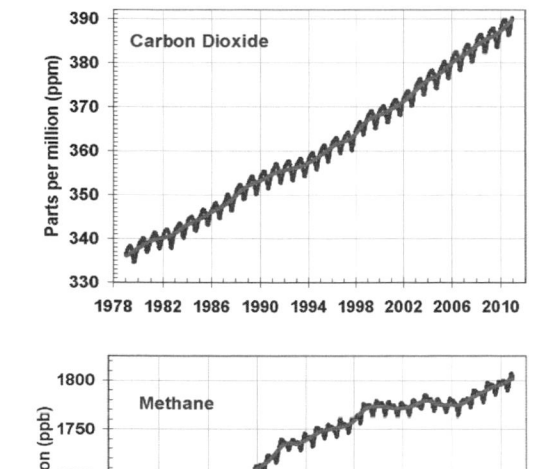
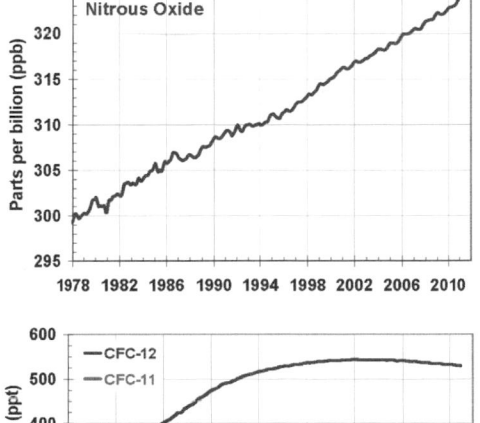
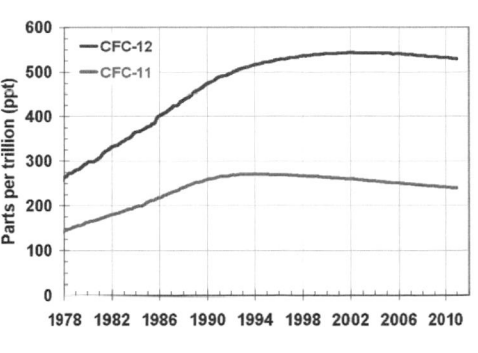

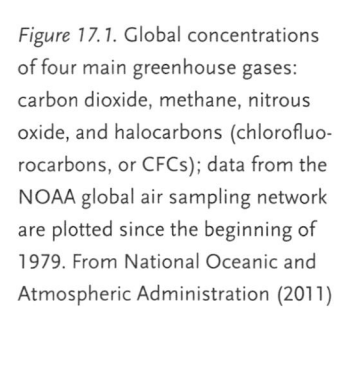

Figure 17.1. Global concentrations of four main greenhouse gases: carbon dioxide, methane, nitrous oxide, and halocarbons (chlorofluorocarbons, or CFCs); data from the NOAA global air sampling network are plotted since the beginning of 1979. From National Oceanic and Atmospheric Administration (2011)

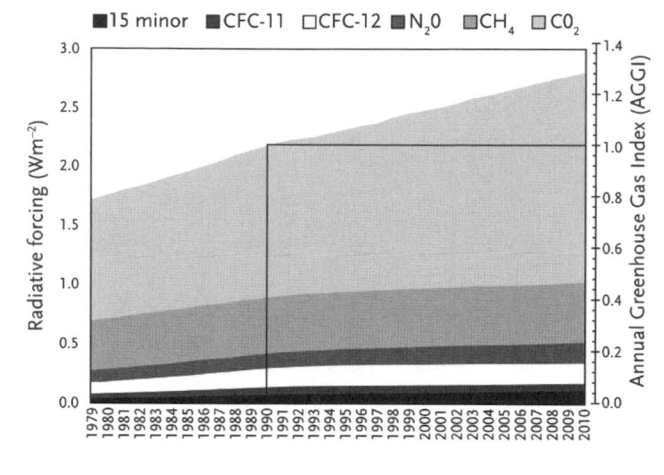

Figure 17.2. Radiative forcing due to increases in the four main greenhouse gases: carbon dioxide (CO_2), methane (CH_4), nitrous oxide (N_2O), and halocarbons (chlorofluorocarbons, or CFCs). Radiative forcing is a measure of the importance of the factor as a potential climate change mechanism. From National Oceanic and Atmospheric Administration (2011)

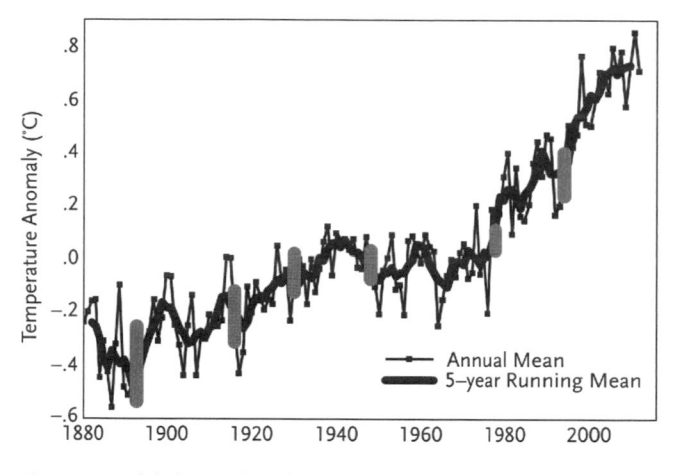

Figure 17.3. Global annual surface temperature increase between 1880 and 2010. In general, global surface temperatures have increased since preindustrial times. Nine of the ten warmest years in the modern meteorological record have occurred since 2000. From National Aeronautics and Space Administration (2010)

are greater in the Northern Hemisphere than in the Southern Hemisphere.

Temperature

The rate of global temperature increase is much faster now than at any other time in the earth's history (IPCC 2007). The 100-year linear trend of global temperature increase was 0.74°C per decade during the time period 1906–2005. Furthermore, the linear warming trend over the second half of this time period (1956–2005) was 0.13°C per decade—nearly twice that for the entire 100 years (Fig. 17.3; National Aeronautics and Space Administration 2010). Projected increases in global average temperature (2090–2099) under different emissions scenarios vary between 1.1–2.9°C and 2.4–6.4°C (Fig. 17.4; IPCC 2007).

Despite being a global pattern, the magnitude of temperature increase varies regionally. Higher northern latitudes have been experiencing greater temperature increases. In the Arctic region, temperatures have risen almost twice as quickly over the past 100 years as the global temperatures. The Northern Hemisphere is warming faster than the Southern Hemisphere because it has more land area; land warms faster than ocean (Folland et al. 2002, Sutton et al. 2007, Dong et al. 2009).

Sea Level

Global average sea level increased at an average annual rate of 1.8 mm between 1961 and 2003 (Bindoff et al. 2007). The rate of increase was almost twice as high for the decade between 1993 and 2003 and averaged 3.1 mm/yr. Scientists are still uncertain about attributing this recent increased rate to climate change, because it is unclear whether it reflects decadal variation or an increase in the longer-term trend (Bindoff et al. 2007).

The forecasts of the IPCC suggest that the annual trend of 1.8 mm is projected to produce average rise in sea level be-

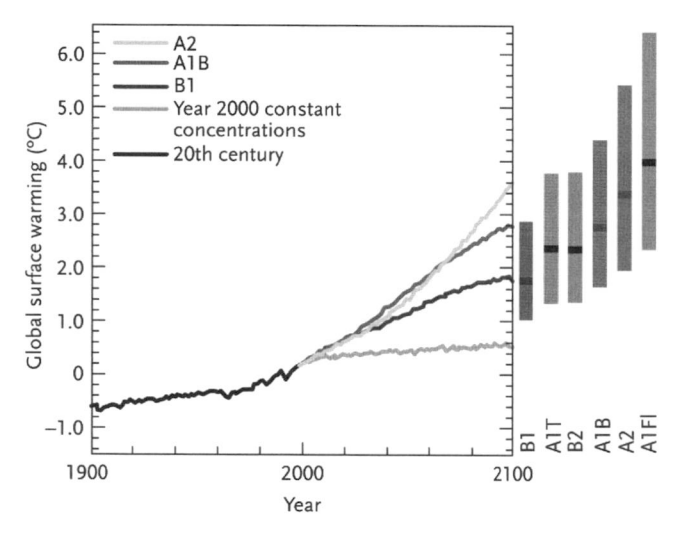

Figure 17.4. Global surface warming for 2000–2100 as predicted by different emissions scenarios. From Intergovernmental Panel on Climate Change (2007)

tween 18–38 and 26–59 cm, depending on emission scenarios, by the end of the century. Sea level rise of such magnitude will cause inundation of coastal areas, flooding, erosion, saltwater intrusion, rising water tables and impeded drainage, and habitat loss (Nicholls et al. 2007). These projections do not account for a possible increase or decrease in the rate of ice flow, because published data on projected ice flow rates are lacking. Therefore the actual sea level rise could potentially exceed the projections (Bindoff et al. 2007).

There are several factors contributing to the sea level rise. Thermal expansion (i.e., the physical increase in volume of water as it warms) of oceans is currently the primary cause. Melting of glaciers and ice caps is important, as well, but a secondary one. Prior to 2004, thermal expansion was responsible for 57% of the sea level rise between 1993 and 2003. It is anticipated that melting glaciers, ice caps, and ice sheets will increasingly contribute to the sea level rise (Bindoff et al. 2007).

Snow Cover, Sea Ice, and Land Ice

To date, melting ice has been secondary to thermal expansion as a cause for rising sea level. Decreases in glaciers and ice caps contributed about 28% of the sea level rise, while melting polar ice sheets contributed the remaining 15%. Data from the U.S. National Snow and Ice Data Center indicate that the average Arctic sea ice extent in September of 2009 was only 70% of what it was at the same time of year in the late 1970s. Not only is the geographic extent of Arctic ice decreasing, the ice is getting shallower. Data on ice thickness collected by U.S. submarines and satellites show decreases of 42% for average midwinter ice depth and 50% during the summer season since the 1970s (Henson 2011). On the other hand, the Antarctic has not shown any clear trend in ice cover (Lemke et al. 2007).

Glaciers and ice sheets of the Arctic and Antarctic are not the only ones threatened by climate change. Glaciers in the tropics and midlatitudes are retreating, and their small size makes them even more vulnerable to climate change. Glaciers of the Andes, which hold 99% of world's tropical ice, are a striking example of the tropical glacier retreat. According to currently available data, the Peruvian Qori Kalis glacier is expected to disappear completely by the end of 2012 (Henson 2011). The Bolivian Chacaltaya glacier disappeared in 2009. Similar losses are evident in Indonesia and Kenya (Klein and Kincaid 2006, Rostom and Hastenrath 2007).

The situation is similar in higher latitudes. The glaciers of the Alps have lost nearly 50% of their surface area and mass since 1850 (Zemp et al. 2006). Nearly 80% of glaciers in Montana's Glacier National Park have retreated since 1850, and predictions indicate that the remaining glaciers will vanish by 2030 (Hall and Fagre 2003, Pederson et al. 2010).

Permafrost is a particularly extensive ecosystem, and the Northern Hemisphere has lost 7–15% of its total permafrost extent since 1900 (Lemke et al. 2007). A maximum temperature increase of 3°C at the top of the Arctic permafrost layer has been observed since 1980s. Decreases in average snow cover have also been observed, and researchers have recorded reduction of the annual duration of northern lake and river ice cover (Lemke et al. 2007).

Under all emissions scenarios, sea ice is projected to decrease in both polar regions. The most drastic scenarios indicate that summer sea ice will almost disappear by the end of the 21st century. Furthermore, snow cover area is projected to decline, and most permafrost regions are projected to experience increased melting (Lemke et al. 2007).

Extreme Events: Heat Waves, Precipitation, and Drought

As climate change intensifies, we can expect an increase in the number of extreme events such as heat waves, floods, and droughts. Cold days, cold nights, and frosts have already become less frequent, while hot days and hot nights have become more frequent (IPCC 2007). The IPCC (2007) also considers it likely that heat waves associated with increasing temperatures have become more frequent over most land areas since the 1970s. For example, in Europe the summer of 2003 was the warmest summer on record since 1540 (World Meteorological Organization 2011), the record broken only by another heat wave seven years later, the 2010 heat wave in Russia (Barriopedro et al. 2011). The likelihood that heat waves will continue to occur is predicted to increase by a factor of five to ten within the next 40 years (Barriopedro et al. 2011).

Increased water vapor and atmospheric moisture have caused higher numbers of precipitation events and increased precipitation intensity, leading to increases in flooding risk in many regions. The risk of flooding is not increasing homogenously around the globe. Regions experiencing increases in the number and intensity of precipitation events are the eastern parts of North and South America, northern Europe, and north-central Asia. On the other hand, some parts of the globe have experienced more intense and longer droughts. Regions with increased droughts included the Sahel, the Mediterranean, southern Africa, and parts of southern Asia.

WHAT ARE THE EFFECTS OF CLIMATE CHANGE ON WILDLIFE POPULATIONS?

The changes in climatic patterns will have far-reaching consequences for wildlife, with impacts already reported for most taxonomic groups (Walther et al. 2002, Parmesan and Yohe 2003, Root et al. 2003, Parmesan 2006). In the following sections, we discuss the observed and predicted effects of climate warming on wildlife. The response of species to long-term environmental changes such as climate change will be determined by their ability (or failure) to adapt. Genetic adaptation will be important over multiple generations and long time intervals, and species with high levels of phenotypic plasticity will be able to adjust more quickly to rapidly changing climate conditions (Dunn and Winkler 2010).

Changes over Time: Evolutionary Consequences

Beyond the most visible aspects of how wildlife can adapt to climate change, many species and populations could adapt over multiple generations and broader periods of time. Shifts in migration and ranges are often attributed to phenotypic plasticity (i.e., the ability of individuals to modify their behavior, morphology, or physiology to altered environmental conditions; Walther et al. 2002). However, many questions remain as to whether these changes have a genetic component leading to evolutionary consequences for species and populations (Bradshaw and Holzapfel 2006). Generally, we have a limited understanding of the evolutionary responses of wildlife populations to climate change when compared to changes in distribution and phenology, but many of these life history traits (e.g., breeding phenology, dormancy) have a strong genetic component that is climate related.

Past climate cycles have left a strong imprint on the historical structure and patterns of plant and animal communities (Hewitt 1996), and they can exert strong influences on biodi-

CAMILLE PARMESAN

Dr. Camille Parmesan is a professor of integrative biology at the University of Texas at Austin. She also holds the National Marine Aquarium Chair in the Public Understanding of Oceans and Human Health at the University of Plymouth in the United Kingdom. Parmesan is currently one of the leading authorities on the implications of climate change for wildlife populations and natural resources. Her research in the mid-1990s on Edith's Checkerspot butterfly (*Euphydryas editha*) was the first to demonstrate that species are shifting their natural ranges in response to climate change. Her subsequent research demonstrated that a number of different taxa have already responded to human-induced climate change.

Parmesan is actively working at the interface of science, policy, and climate change communication. She has been asked to present for the White House and to provide testimony for the U.S. House Select Committee on Energy Independence and Global Warming. In addition, Parmesan has been an active member of the Intergovernmental Panel on Climate Change (IPCC). As a lead author for the *Third Assessment Report*, she shared the Nobel Peace Prize awarded to the IPCC in 2007.

Photo courtesy of Camille Parmesan

versity and species endemism. For example, by estimating climate change velocity (the geographic rate of climate displacement) since the last glacial period, Sandel et al. (2011) reported that regions with high climate velocities are also characterized by absences of range-restricted amphibians, mammals, and birds. The association between endemism and climate change velocity was weakest in highly mobile birds and strongest in weakly dispersing amphibians. High velocity was also associated with low endemism at regional scales. These results emphasize that areas experiencing relatively low rates of climate change over geological periods are essential refuges for maintaining species over broad periods of time.

Wildlife managers may rightly question the extent to which knowledge of climate-driven geological events might extrapolate to the current management of wildlife populations. However, increasing evidence suggests that life history traits such as breeding phenology, clutch size, and body size are sensitive to long-term trends in climate (Sheldon 2010). Migratory behavior is an excellent example of a trait that has strong heritability

and might be particularly sensitive to climate change (Pulido et al. 1996). In a study of blackcaps (*Sylvia atricapilla*), Pulido and Berthold (2010) used a common garden and captive breeding experiment to demonstrate a genetic reduction in migratory activity and evolutionary change in phenotypic plasticity of migration onset. They reported that residency will evolve rapidly in completely migratory bird populations if selection for shorter migration distance persists. These findings suggest that, over time, climate change might favor genotypes that can winter closer to the breeding grounds.

Changes over Time: Phenological Responses

Here we examine the evidence for phenological changes in vertebrate populations resulting from climate warming: alterations of migration patterns, changes in the timing of breeding and reproductive success, and changes in hibernation patterns. Substantial evidence exists that seasonality of many temperate regions has been altered already (Hansen et al. 2010), leading to changes in phenology of plants and animals. Complicating the problem, the rate and the direction of the phenological change will vary across species or even within populations of the same species (Both 2010).

The best evidence that climate warming has already affected wildlife populations comes from long-term studies of butterflies (Parmesan 1996), birds (Thomas and Lennon 1999), and a few examples for mammals (Moyes et al. 2011).

Changes in Migration Patterns
BIRDS

How do we expect birds to respond? Birds may advance their arrival date to breeding sites (either by departing earlier from wintering grounds or by shortening the migration distance) or delay the onset of their fall migration (Lehikoinen and Sparks 2010). Short-distance migrants are more likely to adjust the timing of migration because they are thought to rely more on weather conditions, whereas long-distance migrants are cued by photoperiod, thereby being less likely to attune the timing of their migration to increasing temperatures. For this reason, long-distance migrants are expected to suffer higher population losses as they are unable to track the changing climate as closely as short-distance migrants or resident species.

There is mounting evidence that birds have advanced the time of arrival to breeding grounds (Sparks et al. 2005, Lehikoinen and Sparks 2010, DeLeon et al. 2011, Hurlbert and Liang 2012). Lehikoinen and Sparks (2010) compiled over 3,800 data series spanning 455 species and 19 countries. Their analysis revealed that 82% of trends in first arrival dates and 76% of trends in mean arrival dates were toward earliness. The recorded advancement was on average 2.8 and 1.8 days per decade for the first and mean dates of arrival, respectively. Hurlbert and Liang (2012) reported shifts in the arrival dates for 18 North American species of 0.8 days earlier for every 1°C of warming of spring temperature. While these changes are small, when combined with IPCC predictions for the coming decades, climate change will have a profound effect on migratory species.

From a management perspective, how much time a species spends in wintering habitat is important. Kobori et al. (2012) reported that six species wintering in Yokohama, Japan, delayed their arrival date by nine days and advanced their departure date by 21 days over a period of 23 years. As a result, the duration of their stay in wintering grounds shortened by approximately one month, and presumably (though this was not tested) the duration of their stay in the breeding grounds lengthened (Kobori et al. 2012).

Where species winter is also important to managers. An increasing number of European long-distance migrants, formerly wintering entirely in tropical and southern Africa, are now wintering in the Mediterranean region (Berthold 2001). Similarly, approximately 33% of short-distance or partial migrants breeding in Germany favor wintering at higher latitudes, while only 10% choose to winter in more southerly latitudes (Fiedler et al. 2004). Some migrants such as Canada goose (*Branta canadensis*) in North America and white stork (*Ciconia ciconia*) in Spain have ceased their migration and stay year-round in breeding or stopover sites (Ferrer et al. 2008).

MAMMALS

Can we expect migratory mammals to respond to climate change in the same way as birds? The evidence suggests that we can, although the data are scarce. As an example, snow depth triggers seasonal migration of white-tailed deer (*Odocoileus virginianus*) in northern latitudes, and in the absence of snow, deer remain resident on summer range throughout the year (Tierson et al. 1985). Moose (*Alces alces*) will start migrating to their winter ranges earlier and start leaving their winter grounds later if future snow conditions are affected by climate change (Ball et al. 1999). Sharma et al. (2009) expected that delays in or incomplete formation of ice in the Arctic will alter the timing of migration of migratory caribou (*Rangifer tarandus*). To ensure safe passage over large water bodies, migratory caribou will either have to advance the spring migration and delay the fall migration or shift their ranges farther north (Sharma et al. 2009).

Changes in Timing of Breeding and Reproductive Success

Changing the timing of breeding is yet another example of phenotypic plasticity resulting from climate change. The phenology of reproduction has been widely studied, and evidence for climate warming–induced changes in the timing of breeding is unequivocal (Root et al. 2003, Parmesan 2007). For example, a literature review by Root et al. (2003) revealed that more than 80% of species experienced significant shifts of the phenological events resulting from climate change. The consequences of phenological changes will include altered trophic interactions, because the extent of changes will likely differ between species.

What is the evidence for climate warming–mediated changes in breeding and reproductive success? A summary of change in laying dates of 68 species of birds from long-term studies reveals that 59% have significantly advanced their laying date, while 79% advanced their laying date in warmer years (Dunn and Winkler 2010). On average, laying dates for these species have advanced by 0.13 days/yr, which amounts to a shift of 2.4 days earlier for every 1°C in temperature increase. The timing of breeding affects other reproductive traits in birds, such as the number and size of clutches, incubation behavior, or recruitment, making them indirectly vulnerable to climate change (Dunn and Winkler 2010).

Studies report similar phenotypic shifts in wild mammals (Reale et al. 2003, Moyes et al. 2011). For example, populations of red squirrel (*Sciurus vulgaris*) in Canada have advanced their breeding by 18 days over a decade (Reale et al. 2003). This trend toward earlier breeding was partly explained by increased food abundance associated with increased temperatures, although selection favoring earlier breeders also could have played a role.

A recent study by Moyes et al. (2011) is one of the few to provide evidence for phenological advances in a population of an ungulate, red deer (*Cervus canadensis*). Six phenological traits (estrus date and parturition date in females and antler cast date, antler clean date, rut start date, and rut end date in males) advanced between five and 12 days over a 28-year study period, and local climate measures explained a significant portion of the variation in these traits.

Hibernation

Hibernation allows animals to conserve energy during winter, and there is evidence for changes in hibernation patterns corresponding to climate change. The timing of hibernation of Arctic ground squirrels (*Spermophilus parryii*) varied between populations living in slightly different environmental conditions in northern Alaska (Sheriff et al. 2011). Squirrels living in slightly warmer conditions (i.e., sites characterized by shorter periods with snow cover) emerged from hibernation nine days earlier and entered hibernation five days later than those living in the cooler region. Such differential response offers evidence that individuals have the ability to respond to variation in environmental conditions.

Yellow-bellied marmots (*Marmota flaviventris*) advanced their emergence from hibernation by 38 days over a period of 25 years (Inouye et al. 2000). The advancement was positively correlated with increasing spring air temperatures and could not be explained by the annual date of snow melt and plant flowering, because these variables did not change during the study period. Yet another study reported the termination of hibernation of dormouse (*Glis glis*) to have advanced eight days per decade (Adamik and Kral 2008) and to be related to rising spring temperatures.

Trophic Mismatch Hypothesis

We have shown that wildlife phenology (e.g., migration patterns, timing of breeding, timing and length of hibernation) will be altered by climate change, with some populations already responding to the warming. How do these adjustments play out at the ecosystem level? The responses of individual species reflect individual traits and are expected to differ across species because of high variability in life history traits and physiological tolerances (Parmesan and Yohe 2003). Species currently living in synchrony may find those relationships altered in the coming century; life cycles of predators and prey, herbivores and host plants, parasitoids and hosts, and symbiotic organisms may be disrupted through the differential impacts of climate change (Harrington et al. 1999, Parmesan and Yohe 2003, Visser and Both 2005).

Evidence for altered synchrony in wildlife populations is mounting. A review by Visser and Both (2005) revealed that in seven of 11 cases (nine predator–prey interactions and two insect host–plant interactions), responses of interacting species to climate change were different enough to put them out of synchrony with one another. In another study, pied flycatcher (*Ficedula hypoleuca*), a generalist long-distance migrant, declined significantly in seasonal environments characterized by short food peak but remained stable in less seasonal wetlands where the peak of food availability was longer (Both et al. 2009). In addition, resident and short-distance migratory birds did not decline in any of the habitats, suggesting that it was not the quality of the habitat causing the declines of the long-distance migrant but rather the trophic mismatch that these birds were facing. Sharper declines were also recorded in populations in western Europe, where spring temperatures increased more than in northern Europe (Both et al. 2009).

Post and Forchhammer (2008) provided evidence for a trophic mismatch between migratory caribou (*Rangifer tarandus*) and their foraging plants. Migratory herbivores are likely to develop trophic mismatches, because photoperiod cues the timing of their spring migration to the breeding grounds, while increasing spring temperatures cue the onset of plants. Caribou lagged behind climate warming–induced changes in vegetation, and they were not able to keep pace with advancement in plant-growing season on their breeding ranges. Consequences included a 4-fold decrease in offspring production and an increase in mortality of those young that were born.

Changes over Space: Distributional Responses

Temporal responses to climate change are often coupled with changes in spatial extents of vegetation and wildlife ranges. Species are likely to respond to changing climate space by moving their range in the direction of changing climatic niche. Let us look at the evidence for latitudinal and elevational shifts in the range boundaries of wildlife and vegetation communities.

Range Boundaries: Latitudinal and Elevational Shifts

Current hypotheses about the location of species distributional boundaries emphasize abiotic and biotic factors (MacArthur 1972, Mittelbach 2012). Thus far, evidence suggests that range boundaries are set, at least partially, by climatic conditions (Gaston 2003, Cunningham et al. 2009, Geber 2011, Baselga et al. 2012). Therefore we expect that species will move their distributions northward and up in elevation in response to recent climate warming. Range shifts may ul-

timately lead to changes in community composition, because each species will adjust its distribution in accord with its own ecological tolerances. Some species have broad ecological tolerances and will expand their distribution, while there will be range contraction for species that have narrow tolerances and are unable to effectively track climate change. Ultimately, extinctions will occur if there is no possibility for the species to move poleward or to higher elevations.

How far can we expect species to move? Predictably, the answer to this question will depend on the studied organism, because the responses will likely vary between different taxonomic groups. Recent studies drawing on Breeding Bird Atlas data reported that songbirds in New York State have shifted their ranges northward an average of 3.6 km between the early 1980s and 2000s (Zuckerberg et al. 2009). Parmesan and Yohe (2003) reported range shifts averaging 6.1 km per decade toward the poles across a range of three different taxa—birds (Aves), butterflies (Rhopalocera), and alpine herbs. Average northward shifts of 31–60 km and elevational shifts of 25 m over a 40-year period were reported for 16 different vertebrate and invertebrate taxonomic groups—dragonflies and damselflies (Odonata), grasshoppers and allies (Orthoptera), lacewings (Neuroptera), butterflies, spiders (Araneae), herptiles (Amphibia and Squamata), freshwater fish (Teleostei), mammals (Mammalia), woodlice (Isopoda), ground beetles (Carabidae), harvestmen (Opiliones), millipedes (Diplopoda), longhorn beetles (Cerambycidae), soldier beetles and allies (Cantharoidea and Buprestoidea), aquatic bugs (Heteroptera), and birds (Hickling et al. 2006). Of 329 species studied by Hickling et al. (2006), 84% showed a significant poleward shift.

Some of the most-convincing pieces of evidence for elevational range shifts resulting from climate change are studies of populations of small mountain mammals. For example, populations of American pika (*Ochotona princeps*) have been monitored since 1898 throughout the Great Basin ecoregion (Beever et al. 2011). Between 1998 and 2008, the rate of upslope range retraction accelerated 11-fold. The species is now experiencing an upward elevational shift at an average rate of 145 m per decade.

Even though distributional shifts in ranges of large charismatic fauna have rarely been reported, one study deserves mention here. Towns et al. (2010) reported poleward and eastward shifts of spatial distributions of polar bears (*Ursus maritimus*) in Hudson Bay between 1986 and 2004. The observed distributional changes were a result of three-week advancement of the sea ice breakup since the late 1970s.

Shifts in Vegetation Communities

Habitat suitability is a key determinant of the distribution and abundance of species. Vegetation lies at the heart of the habitat, often driving wildlife distributions at regional and local scales. Climate change is expected to trigger changes in the species composition of plant communities, leading to changes in overall habitat quality and, ultimately, composition of wildlife species. Plant communities might be altered as changes in temperature and moisture regimes impinge on physiological tolerances of existing plant species, changing the competitive edge to favor invasion by other species. However, much of the change in plant communities is likely to occur as warmer winters allow expansion of pests and pathogens. As one example, climate change models predict as much as a 700-km northward expansion of the gypsy moth (*Lymantria dispar*), and similar expansions are expected for many other insect pests (Vanhanen et al. 2007). Other studies also predict range expansion of the hemlock woolly adelgid (*Adelges tsugae*) resulting from increasing winter temperatures (Paradis et al. 2008, Fitzpatrick et al. 2012). This range expansion will result in further invasion of stands of eastern hemlock (*Tsuga canadensis*) and Carolina hemlock (*T. caroliniana*), lowering the quality of habitat to various wildlife species. Dukes et al. (2009) reported climate change is likely to affect the composition and structure of forests by benefiting several pest species such as hemlock woolly adelgid, forest tent caterpillar (*Malacosoma disstria*), and pathogens such as *Armillaria* spp., beech bark (*Cryptococcus fagisuga*), and invasive plant species such as glossy buckthorn (*Frangula alnus*) and oriental bittersweet (*Celastrus orbiculaturs*).

The shifts in competitive advantage among plant species will result in geographic redistribution of major plant biomes. Most evidence for latitudinal shifts has come from the studies in the boreal biome of the Northern Hemisphere. As an example, Suarez et al. (1999) reported an increase in tree encroachment into Alaskan tundra resulting in the shift of the forest–tundra biome by approximately 80–100 m over the past 200 years. Similarly, there has been pervasive advancement in the tree line of white spruce (*Picea glauca*) since 1800 along the forest–tundra border in Alaska (Lloyd and Fastie 2003, Payette 2007). Similar shifts are evident in the vegetation zones of the West African Sahel, where ranges are changing at an annual rate of approximately 500–600 m (Gonzalez 2001): arid species at the northern boundary of the biome are shifting farther north, while mesic species present at the southern boundary are retracting.

The evidence for elevational shifts in the range of biomes is even more compelling. For example, the location and species composition of the northern hardwood–boreal forest biome in Vermont changed significantly over a 40-year study period, experiencing a shift of approximately 90–120 m up in elevation (Beckage et al. 2008). Likewise, the evidence for an advance was found in Yukon, Canada, where white spruce tree lines on south-facing slopes moved up in elevation by 65–85 m over a period of 30 years (Danby and Hik 2007). Similar trends have been reported in Eurasia. For example, forest–tundra ecotone in the Polar Urals underwent an upward shift of 20–60 m during the 20th century. This shift was positively correlated with increases in average summer temperatures and precipitation (Devi et al. 2008). In the southern Swedish Scandes, the tree lines of three species—mountain birch (*Betula pubescens* spp. *Czerepanovii*), white spruce, and Scots pine (*Pinus sylestris*)—moved up in elevation by on average 70–90 m during the 20th century (Kullman and Oberg 2009). In the Montseny Mountains in Spain, Penuelas and Boada (2003) reported a replacement of cold-temperature ecosystems by ecosystems associated with warm Mediterranean climates between 1945 and present times.

Interestingly, climate change–driven changes in vegetation communities are not evident for the Southern Hemisphere. For example, no significant changes in biomes boundaries or in vegetation composition were found in savanna woodland ecotone in Australia (Sharp and Bowman 2004) or in silver beech (*Nothofagus menziesii*)–dominated tree lines in New Zealand (Cullen et al. 2001). The difference between Northern and Southern Hemispheres fits with the general hypothesis that climates in the Southern Hemisphere are changing more slowly because there is less land mass to reradiate energy into the atmosphere.

Population and Community Dynamics

As climate change alters species abundance and distributions, biotic interactions such as predation, parasitism, competition, and mutualism will be subject to disruption, ultimately leading to changes in community composition, biodiversity, and ecosystem functioning. Such indirect effects of climate change, mediated through impacts on species relationships, might in fact have far more severe consequences than direct effects of warming (Bretagnolle and Gillis 2010). In this section, we discuss potential consequences of climate change to predator–prey dynamics, parasite–host interactions, and patterns of interspecific competition.

Predator–prey Dynamics

What mechanisms can lead to alterations in predator–prey interactions? As we saw with trophic mismatches, both predator and prey may experience changes in their phenologies or spatial distributions, causing either a temporal or spatial mismatch between the species. As a result, changes in the trophic structure are difficult to predict.

Evidence for mismatches in the predator–prey trophic system is growing. As an example, Both et al. (2009) reported increasing temporal mismatch in a four-level trophic system of plants, caterpillars, four species of passerines that prey on caterpillars, and a raptor that preys on passerines. Climate change resulted in annual advances of 0.17 days in budburst, 0.75 days in caterpillars, and 0.36–0.5 days in passerines hatching dates. Raptors showed no trend toward earliness. Overall, the phenological response of the higher trophic levels (passerines and the raptor) was weaker than that of their prey, resulting in increasing asynchrony between the food demand and food availability.

A study by Vors and Boyce (2009) listed increased predation pressure among the repercussions of climate change for migratory caribou. Currently, predation is not a contributing factor to the caribou population regulation, because the geographic range of the potential predators does not generally overlap with the distribution of the caribou. However, if the frontier of the boreal forest shifts poleward, spatial overlap between wolf (*Canis lupus*) and caribou will increase, resulting in significant caribou mortality on its winter range.

Evidence is also growing for altered predator–prey dynamics resulting from changes in population cycles. For example, changes in the population dynamics of lemmings (Lemmus lemmus) are attributable, in part, to winter weather conditions. Increasingly warm winters cause less regular peaks in the lemming cycles, which in turn affect the dynamics of its bird predator (Kausrud et al. 2008).

Changes in rainfall patterns associated with climate warming are likely to produce dramatic effects on predator–prey dynamics. Wild turkeys (Meleagris gallopavo) provide a good example. In northern portions of wild turkey range, rainfall in May appears to affect nest predation. Predators using olfactory cues are more effective in finding nests when moisture conditions are high (Roberts and Porter 2001, Fleming and Porter 2007). In contrast, in southwestern portions of the range of wild turkeys, low rainfall reduces insect abundance, resulting in decreases in wild turkey populations (Pattee and Beasom 1979).

Host–parasite Dynamics

Host–parasite interactions are likely to be affected by climate change through similar mechanisms as predator–prey relationships. These mechanisms include (1) changes in phenology of parasite or host, (2) changes in spatial distribution of parasite or host, (3) changes in the prevalence and intensity of parasitism, and (4) changes in virulence or antiparasite defenses of hosts (Merino and Moller 2010).

There is compelling evidence for increasing temperatures affecting the timing of emergence of parasites and vectors (Mouritsen and Poulin 2002, Moller 2010). For example, the hippoboscid fly (Ornitholyia avicularia), a parasite of a barn swallow (Hirudo rustica) in Denmark, has advanced its phenology to emerge during egg laying and early incubation period, rather than during late stages of the swallow breeding cycle (Moller 2010). Consequences of such earlier emergence may potentially include lower reproductive success and recruitment of the host. On the other hand, brood parasitic cuckoo (Cuculus canorus) will likely be affected by climate warming–mediated changes in the phenology of its host, short-distance migratory birds (Saino et al. 2009). Short-distance migrants have advanced their arrival time to breeding grounds on average more than long-distance migratory birds such as cuckoo, and are now starting their breeding season prior to cuckoo's arrival. This mistiming might ultimately lead to cuckoo population declines because of a lack of opportunities for successful parasitism.

The prevalence of parasites is also expected to change with increasing temperatures. One of the primary vectors of zoonotic diseases, recently on the rise in Europe, is the castor bean tick (Ixodes ricinus), which transmits Borrelia burgdorferi, tick-borne encephalitis virus, and louping ill virus (Gilbert 2010). B. burgdorferi is a spirochete noted for causing Lyme disease and is carried by a tick of the Ixodes genus (I. dammini) in the United States. Gilbert (2010) tested whether increasing abundance of I. ricinus in Scotland is a result of recent climate change and reported it to be nega-

tively associated with elevation, which they used as proxy for climatic conditions. Given future climate change scenarios, I. ricinus is likely to become more abundant at higher altitudes. This elevational shift might result in higher prevalence of louping ill virus, because additional competent hosts—red grouse (Lagopus lagopus scotica) and mountain hares (Lepus timidus)—occur in higher numbers at higher elevations. Similar elevational shifts and increased densities of I. ricinus were reported over a 15-year study period in Sweden (Lindgren et al. 2000). Relatively mild climatic conditions during the study period in Sweden were thought to be one of the primary reasons for the observed increase of density and geographic range of I. ricinus. In the United States, winter ticks (Dermacentor albipictus) were reported to significantly influence moose population growth (Garner 1994). Mild winters favor higher tick survival, contributing to increased mortality in moose populations (Musante et al. 2010). Climate change is likely to benefit winter tick populations, indirectly contributing to increased mortality in wild ungulates.

Inter- and Intraspecific Competition

Climate change will likely disrupt competitive interactions via similar mechanisms as predator–prey or host–parasite interactions. Climate change may affect competition between migratory and resident birds by allowing short-distance migrants to return earlier to their shared breeding grounds (Forchhammer et al. 2002, Hubalek 2003), and possibly by enhancing overwinter survival of birds wintering in Europe (Lemoine and Bohning-Gaese 2003). The impact of climate change could therefore leave long-distance migrants at a competitive disadvantage. Furthermore, shifts in species geographic ranges might create a spatial overlap between species potentially competing for the same resource.

Ahola et al. (2007) studied interspecific nest-hole competition between resident great tit (Parus major) and migrant pied flycatchers (Ficedula hypoleuca) in Finland over five decades (1953–2005). Decreasing the interval of the laying date between the species increased the likelihood of pied flycatchers being killed by the great tit after an unsuccessful attempt to take over the tit's nest. Research shows that climate change will likely alter the interval of the interspecific laying date, potentially leading to altered competitive interactions between the species. An increasing overlap in the use of nesting boxes was also found for edible dormouse and some species of birds (Adamik and Kral 2008).

There are similar examples of altered competitive interactions among mammals, as well. Killengreen et al. (2007) reported that populations of Arctic fox (Vulpes lagopus) residing at the southern margin of the Arctic tundra have been declining as a result of invasion of red foxes (Vulpes vulpes). Red foxes started shifting their ranges poleward as a consequence of increasing temperatures. Another study predicted that geographic distributions of two herds of migratory caribou will change following increased temperatures and earlier sea ice breakup (Sharma et al. 2009). These spatial changes will cause an overlap between the herds, leading to increased competition on the calving grounds.

TOOLS FOR STUDYING WILDLIFE RESPONSES TO CLIMATE CHANGE

A wide range of methods exists to study climate change impacts on wildlife communities. The most popular methods include bioclimatic envelope models that allow for predicting species distributions under different climate change scenarios, and vulnerability assessments that quantify species exposure, sensitivity, and the ability to adapt to climate change. In this section, we discuss the applications, strengths, and limitations of both methods.

Bioclimatic Envelope Modeling

Bioclimatic envelope models (also known as ecological niche models, habitat suitability models, or species distribution models) attempt to define climatic conditions that best describe species range limits by correlating the current species distributions with selected climate variables. Projecting these current relationships to future climate change scenarios, then, can help forecast future species' ranges (Thuiller and Munkemuller 2010).

Different types of algorithms have been used in bioclimatic envelope modeling. The performance of these algorithms is dependent on the availability of the species distributional data; therefore it is difficult to judge which method is the best. Factors other than the availability and quality of species distributional data also should be considered while selecting the modeling approach. These factors include the availability of the environmental data, spatial and temporal scale of the study, and the goals of the study (Pearson and Dawson 2003).

Because bioclimatic models are simplified versions of the reality, they are necessarily based on assumptions (Pearson and Dawson 2003, Heikkinen et al. 2006, Pearson et al. 2007, Araújo and Peterson 2012). Bioclimatic envelope models have three basic assumptions: species' distributions are determined wholly or partly by climate, species inhabit all areas of suitable climate, and species' climatic envelopes remain constant over time (Araújo and Peterson 2012, Brotons et al. 2012).

Are species distributions determined solely by climate? The existing evidence suggests that species ranges are, at least partially, driven by climatic conditions. However, land cover, vegetation, and biotic interactions are important variables in determining habitat suitability. Some studies have incorporated land cover in models intended to explain species distributions, and the principal conclusion of these studies is that the value of land cover as a variable in the model is dependent on the geographic scale at which it is measured (Pearson et al. 2004, Luoto et al. 2007). Climatic conditions seem to dictate species distributions at coarse or large geographic scales (e.g., 40–80 km; Luoto et al. 2007). The predictive power of bioclimatic envelope models at large geographic extents is usually not improved by the addition of land-cover variables to the model (e.g., Pearson et al. 2004, Luoto et al. 2007). However, incorporating land-cover variables at finer scales (e.g., 10 km) can significantly improve predictions of geographic distributions for some species (Pearson et al. 2004, Luoto et al. 2007).

Bioclimatic envelope models are likely to produce some errors because the ecological systems they approximate are complex. Despite their limitations, bioclimatic envelope models are a valuable tool for determining species distributions and the effects of climate warming.

Vulnerability Assessments

In addition to bioclimatic envelope models, another strategy to quantify the effects of climate change is to assess species vulnerability. Vulnerability assessments usually comprise three elements: assessment of sensitivity, adaptive capacity, and potential exposure of individual species to climate change. While traits intrinsic to the species determine sensitivity and adaptive capacity, environmental variables and local microhabitat conditions govern exposure (Williams et al. 2008).

Sensitivity

Factors intrinsic to each species, such as physiological tolerance, behavioral traits, or genetic diversity, will determine sensitivity to climate warming. However, not all of these traits are easily characterized. As an example, information on reproductive success, the timing of breeding, or the timing of migration is readily available for many taxonomic groups, whereas data on physiological tolerances or genetic diversity are more scarce and difficult to obtain. Williams et al. (2008) suggested that where data are lacking for a particular species, it might be reasonable to use information available for a closely related species as a proxy for the responses of the species of interest. For example, thermal tolerances tend to be similar for closely related groups of organisms, and it might suffice to obtain the thermal tolerances only for a number of representatives of the taxonomic groups.

Resilience and Adaptive Capacity

Resilience is the ability of a species to survive or recover from an ecological perturbation. Traits such as high reproductive rates, fast life history, or short life span are thought to promote higher resilience and therefore lower risk of extinction (McKinney 1997, Williams et al. 2008). Dispersal abilities are also crucial in maintaining viable populations across the landscapes, especially in the context of climate change, where species will have to track their preferred climatic envelope (Williams et al. 2008).

Adaptive capacity of a species can be expressed either through phenotypic plasticity or genetic change. All organisms have some capacity to adapt to changing conditions. Phenotypic plasticity may include changes in the timing of migration or breeding, or changes in hibernation patterns. Both phenotypic plasticity and evolutionary adaptation have already occurred in a variety of species in response to climate change (Root et al. 2003, Bradshaw and Holzapfel 2006, Parmesan 2007), although a rapid evolutionary response is less likely than phenotypic change to occur for the majority of species (Williams et al. 2008).

Exposure

Two forces drive exposure: the degree of climate change across the geographic range of a species and the ability of local microhabitat to reduce exposure (Williams et al. 2008). For instance, the availability of thermal refugia or shelters (e.g., small pools, boulder fields, rocks, logs) can buffer a species from the full magnitude of regional climate change. As an example, brushtail possums (*Trichosurus vulpecula*) are buffered from extreme temperature by choosing to den in tree hollows that are $1.6°C$ cooler than other den locations (Isaac et al. 2008). American pikas are able to tolerate wider temperature ranges in locations with abundant rock-ice features in comparison with sites where those formations are less abundant or absent (Millar and Westfall 2010). Rock-ice features create local environments cooler than expected for mean summer temperatures at this elevation and warmer than expected for winter temperatures, reducing pikas' exposure to regional climate warming.

Together, sensitivity, adaptive capacity, and exposure determine the vulnerability of a species to climate change and allow prediction of potential impacts. Conservation organizations and management agencies have already started designing vulnerability assessment frameworks (Glick and Stein 2011, Rowland et al. 2011). For example, NatureServe has created a climate change vulnerability index (Young et al. 2009) that assesses exposure using projection of temperature and precipitation/water balance, sea level rise, and changes in land use. Information on dispersal capabilities, physical tolerances, biotic interactions, and genetics assesses sensitivity. The U.S. Department of Agriculture Forest Service has designed a vulnerability index (Finch et al. 2011, Glick and Stein 2011) that assesses sensitivity and exposure using variables related to four broad categories: habitat, physiology, phenology, and biotic interactions. Finally, the assessment tool developed by the U.S. Environmental Protection Agency evaluates vulnerability of species designated as threatened or endangered under the U.S. Endangered Species Act. The tool specifically determines how species conservation status might be altered by climate change (U.S. Environmental Protection Agency 2009, Glick and Stein 2011, Rowland et al. 2011).

CLIMATE CHANGE ADAPTATION

Climate change adaptation is a unique challenge for wildlife management (Mawdsley et al. 2009, Knutson and Heglund 2012). For many wildlife managers, incorporating climate change into planning and management represents a nuisance and distraction from the more obvious problems of habitat loss, point-source pollution, and invasive species management. As one wildlife manager recently stated when discussing climate change, "I'd rather not worry too much about cancer while I'm driving off a cliff." The assumption here is that the effects of climate change are a distant concern when com-

pared to the more pressing problems facing wildlife populations. This perspective, however, is both flawed and worrisome. The ecological consequences of climate change are undeniable (Parmesan 2006, Beever and Belant 2012), and the role of wildlife managers will only become increasingly important; the most dangerous action is inaction.

Climate Change Uncertainty

Although modern climate change is a relatively new environmental challenge, wildlife managers have a diverse toolbox for incorporating environmental variability in managing natural resources (Yoccoz et al. 2001, Williams et al. 2002). As a first step, Nichols et al. (2011) suggests that, rather than viewing climate change as an entirely new source of uncertainty affecting wildlife populations, we should view it as a phenomenon capable of exacerbating the sources of uncertainty already associated with management. Wildlife managers should avoid the broader conversation of identifying the effects of climate change per se, and rather focus on those existing stressors that may be more or less affected by climate change. As an example, species' climate sensitivity can be modified by local site conditions such as topography (Lawson et al. 2012) or predator control (Pearce-Higgins et al. 2010). Initial steps should identify and prioritize which species might be more or less sensitive to climate variability, and how climate change may interact with other nonclimatic stressors.

For many wildlife managers, the primary concern is that climate change represents a source of environmental uncertainty that is difficult to predict and incorporate into management. Managers are generally comfortable with environmental variation, and its influence on wildlife survival and reproduction, and they use a number of modeling approaches for predicting how these vital rates might change in the future (Williams et al. 2002). However, an assumption of many of these models is that the historic range of environmental variability will remain constant (i.e., stationary). Yet what we know now suggests that climate change and its associated changes in precipitation, snow cover, sea level rises, and extreme weather events will likely alter historic ranges of variability, and perhaps even produce future environmental states with no current analog (Williams et al. 2007b). As such, management and planning should include not only the biological variables of specific interest (e.g., population size, abundance), but also the climatic influences that are expected to change (Conroy et al. 2011, Nichols et al. 2011). Future modeling will require innovative thinking and a decreased ability to use historical data when developing new models to deal with system change.

Decision making in the face of climate change is complicated by the fact that ecological systems are changing, and that there is strong uncertainty associated with the magnitude and direction of that change. Two main sources of uncertainty associated with climate change are problems associated with downscaled regional- to local-scale predictions of future climate change, and uncertainty associated with predicting the biological responses to these climatic changes (McLennan 2012). Downscaled climate products are being developed for many regions, but reducing the uncertainty associated with biological responses to climate change will only result from further efforts in empirical research and monitoring. A lack of understanding of the biological impacts of climate change should not serve as a roadblock; dealing with imperfect knowledge is nothing new to wildlife management. Adaptive management (Williams et al. 2007*a*) and structured decision making (Lyons et al. 2008, Martin et al. 2009) are frameworks that have been explicitly designed for accounting for uncertainty (Chapter 5). As an example, McDonald-Madden et al. (2011) demonstrated the ability of dynamic optimization methods and alternative models of system changes to make informed decisions on the benefits and timing associated with relocating species faced with climate change.

The Importance of Monitoring When Facing Climate Change

Climate change has reinvigorated the need for long-term monitoring in wildlife management (Leptcz et al. 2009, Beever and Woodward 2011, Conroy et al. 2011). Monitoring has always been a central component of adaptive management, but climate change will likely test the flexibility of existing monitoring programs. As an example, monitoring efforts often establish survey areas based on historical patterns of where species are more likely to be found during certain times of the year (e.g., migratory corridors, breeding sites). Many species, however, are shifting their distributions and migratory patterns (Root et al. 2003, Walther et al. 2005). Some species of the Northern Hemisphere are exhibiting contractions along southern boundaries and expansions along northern boundaries (Thomas and Lennon 1999, Brommer 2004, Zuckerberg et al. 2009). These geographic shifts will force managers to reconsider their survey strata and boundaries. In addition to spatial considerations, phenological shifts are even more plastic than distributional changes (Penuelas and Filella 2001, Cleland et al. 2007, Visser et al. 2010), and monitoring programs will have to account for species that might be arriving to their breeding grounds earlier, singing or displaying at a different time, or using alternative food resources. These changes will add sources of variation in sampling (e.g., changes in detection probability) when collecting data for monitoring purposes.

Managers should not worry about identifying all sources of variation associated with monitoring but instead understand better how these sources may vary in response to climate change and plan accordingly. There should be a continuous re-evaluation of local, state, and regional efforts to develop a more comprehensive network of monitoring sites. Monitoring should be designed in such a way to reflect the seasonally and geographically complex nature of future climate change, capturing a full range of latitudinal, land-use, and seasonal variation. Methods such as occupancy and estimation (MacKenzie et al. 2006) provide enough flexibility to be used for multiple taxa and can account for heterogeneous detectability across species and areas. In addition, many wildlife agencies have a treasure trove of historical surveys and data sets that offer unique opportunities to resurvey sites and to document long-term shifts in species distributions (Tingley and Beissinger 2009, Tingley et al. 2009). These data can be particularly useful for identifying which species are showing the strongest response to climate change.

A NEED FOR REGIONAL COLLABORATION

Climate change is an environmental process that transcends political and jurisdictional boundaries, and as such, management and planning efforts should reflect this increasing complexity. Agency regulations that were developed to address straightforward environmental problems are limited when dealing with complex problems—such as climate change—that involve many stakeholders, jurisdictions, and ecosystems. As an example of regional collaboration, Landscape Conservation Cooperatives (LCCs; http://www.doi.gov/lcc/index.cfm) are public–private partnerships composed of states, tribes, federal and state agencies, nongovernmental organizations, and universities. The LCC network was established in 2009 as a response to landscape-scale stressors, including climate change. The cooperatives are intended to work interactively with Department of Interior Climate Science Centers to help coordinate regional adaptation efforts. LCCs identify common conservation goals, develop tools and strategies to inform landscape-scale planning and management, link science to management, and facilitate information exchange among partners. These collaborative efforts are also being implemented at the state level. In 2007, the Wisconsin Initiative for Climate Change Impacts (WICCI; www.wicci.wisc.edu) was established as a partnership between the University of Wisconsin, Wisconsin Department of Natural Resources, and other state agencies and institutions. WICCI assesses and anticipates climate change impacts on specific Wisconsin natural resources, ecosystems, and regions; evaluates potential effects on industry, agriculture, tourism, and other human activities; and develops and recommends adaptation strategies that can be implemented by businesses, farmers, resource managers, and other stakeholders. These types of partnerships involving multiple stakeholders and experts are critical for climate change adaptation to be effective.

SUMMARY

The evidence that climate change is occurring and that wildlife species and populations are beginning to respond is now overwhelming. Despite the political and social debate surrounding the causes of climate change, there is strong scientific consensus that environmental tipping points are being crossed, and many species are adapting (or failing to adapt) to novel climatic conditions.

We began the chapter with a question about the biological implications of climate change. We are already seeing these implications in the form of geographic redistribution of species and changes in migration and breeding patterns. Models suggest more changes are ahead, and with them a restructuring of ecosystems.

As wildlife managers, we face the question of how we should adapt. The answer is that we must enhance our abilities to anticipate changes and to be flexible in our responses. Managing wildlife populations under climate change has no single, optimal solution. Instead, both the process of climate change and its resultant ecological consequences are regionally dependent on many variables and difficult to predict. As with any environmental problem, there are multiple stakeholders with disparate values and objectives. Regional and collaborative decision making is more likely to lead to satisfactory management outcomes in the future. Climate change will require wildlife managers and stakeholders to make decisions despite long-range uncertainty in both the ecological and human systems.

Literature Cited

Adamik, P., and M. Kral. 2008. Climate- and resource-driven long-term changes in dormice populations negatively affect hole-nesting songbirds. Journal of Zoology 275:209–215.

Ahola, M. P., T. Laaksonen, T. Eeva, and E. Lehikoinen. 2007. Climate change can alter competitive relationships between resident and migratory birds. Journal of Animal Ecology 76:1045–1052.

Araújo, M. B., and A. T. Peterson. 2012. Uses and misuses of bioclimatic envelope modelling. Ecology 93:1527–1539.

Ball, J. P., G. Ericsson, and K. Wallin. 1999. Climate changes, moose and their human predators. Ecological Bulletins 47:178–187.

Barriopedro, D., E. M. Fischer, J. Luterbacher, R. Trigo, and R. Garcia-Herrera. 2011. The hot summer of 2010: redrawing the temperature record map of Europe. Science 332:220–224.

Baselga, A., J. M. Lobo, J. C. Svenning, and M. B. Araujo. 2012. Global patterns in the shape of species geographical ranges reveal range determinants. Journal of Biogeography 39:760–771.

Beckage, B., B. Osborne, D. G. Gavin, C. Pucko, T. Siccama, and T. Perkins. 2008. A rapid upward shift of a forest ecotone during 40 years of warming in the Green Mountains of Vermont. Proceedings of the National Academy of Sciences 105:4197–4202.

Beever, E. A., and J. L. Belant. 2012. Ecological consequences of climate change: mechanisms, conservation, and management. CRC Press, Boca Raton, Florida, USA.

Beever, E. A., C. Ray, J. L. Wilkening, P. F. Brussard, and P. W. Mote. 2011. Contemporary climate change alters the pace and drivers of extinction. Global Change Biology 17:2054–2070.

Beever, E. A., and A. Woodward. 2011. Design of ecoregional monitoring in conservation areas of high-latitude ecosystems under contemporary climate change. Biological Conservation 144:1258–1269.

Berthold, P. 2001. Bird migration: a general survey. Second edition. Oxford University Press, Oxford, United Kingdom.

Bindoff, N. L., J. Willebrand, V. Artale, A. Cazenave, J. Gregory, et al. 2007. Observations: oceanic climate change and sea level. Pages 385–432 in S. Solomon, D. Qin, M. Manning, Z. Chen, M. Marquis, K. B. Averyt, M. Tignor, and H. L. Miller, editors. Climate change 2007: the physical science basis. Contribution of Working Group I to the Fourth Assessment Report of the Intergovernmental Panel on Climate Change. Cambridge University Press, Cambridge, United Kingdom.

Both, C. 2010. Food availability, mistiming, and climatic change. Pages 129–148 in A. P. Moller, W. Fiedler, and P. Berthold, editors. Effects of climate change on birds. Oxford University Press, Oxford, United Kingdom.

Both, C., M. van Asch, R. G. Bijlsma, A. B. van den Burg, and M. E. Visser. 2009. Climate change and unequal phenological changes across four trophic levels: constraints or adaptations? Journal of Animal Ecology 78:73–83.

Bradshaw, W. E., and C. M. Holzapfel. 2006. Climate change—evolutionary response to rapid climate change. Science 312:1477–1478.

Bretagnolle, V., and H. Gillis. 2010. Predator–prey interactions and climate change. Pages 227–248 in A. P. Moller, W. Fiedler, and P. Berthold, editors. Effects of climate change on birds. Oxford University Press, Oxford, United Kingdom.

Brommer, J. E. 2004. The range margins of northern birds shift polewards. Annales Zoologici Fennici 41:391–397.

Brotons, L., M. De Caceres, A. Fall, and M. J. Fortin. 2012. Modeling bird species distribution change in fire prone Mediterranean landscapes: incorporating species dispersal and landscape dynamics. Ecography 35:458–467.

Cleland, E. E., I. Chuine, A. Menzel, H. A. Mooney, and M. D. Schwartz. 2007. Shifting plant phenology in response to global change. Trends in Ecology and Evolution 22:357–365.

Conroy, M. J., M. C. Runge, J. D. Nichols, K. W. Stodola, and R. J. Cooper. 2011. Conservation in the face of climate change: the roles of alternative models, monitoring, and adaptation in confronting and reducing uncertainty. Biological Conservation 144:1204–1213.

Cullen, L. E., G. H. Stewart, R. P. Duncan, and J. G. Palmer. 2001. Disturbance and climate warming influences on New Zealand Nothofagus tree-line population dynamics. Journal of Ecology 89:1061–1071.

Cunningham, H. R., L. J. Rissler, and J. J. Apodaca. 2009. Competition at the range boundary in the slimy salamander: using reciprocal transplants for studies on the role of biotic interactions in spatial distributions. Journal of Animal Ecology 78:52–62.

Danby, R. K., and D. S. Hik. 2007. Variability, contingency and rapid change in recent subarctic alpine tree line dynamics. Journal of Ecology 95:352–363.

DeLeon, R. L., E. E. DeLeon, and G. R. Rising. 2011. Influence of climate change on avian migrants' first arrival dates. Condor 113:915–923.

Devi, N., F. Hagedorn, P. Moiseev, H. Bugmann, S. Shiyatov, V. Mazepa, and A. Rigling. 2008. Expanding forests and changing growth forms of Siberian larch at the Polar Urals treeline during the 20th century. Global Change Biology 14:1581–1591.

Dong, B. W., J. M. Gregory, and R. T. Sutton. 2009. Understanding land-sea warming contrast in response to increasing greenhouse gases. Part I: transient adjustment. Journal of Climate 22:3079–3097.

Dukes, J. S., J. Pontius, D. Orwig, J. R. Garnas, V. L. Rodgers, et al. 2009. Responses of insect pests, pathogens, and invasive plant species to climate change in the forests of northeastern North America: what can we predict? Canadian Journal of Forest Research 39:231–248.

Dunn, P. O., and D. W. Winkler. 2010. Effects of climate change on timing of breeding and reproductive success in birds. Pages 113–128 in A. P. Moller, W. Fiedler, and P. Berthold, editors. Effects of climate change on birds. Oxford University Press, Oxford, United Kingdom.

Ferrer, M., I. Newton, and K. Bildstein. 2008. Climatic change and the conservation of migratory birds in Europe: identifing effects and conservation priorities. Convention on the conservation of European wildlife and natural habitats. 1 July 2008, Strasbourg, France.

Fiedler, W., F. Bairlein, and U. Koppen. 2004. Using large-scale data from ringed birds for the investigation of effects of climate change on migrating birds: pitfalls and prospects. Pages 49–67 in A. P. Moller, W. Fiedler, and P. Berthold, editors. Birds and climate change. Volume 35. Advances in ecological research. Academic Press, San Diego, California, USA.

Finch, D. M., M. Friggens, and K. E. Bagne. 2011. Case study 3: species vulnerability assessment for the Middle Rio Grande, New Mexico. Pages 96–103 in P. Glick and B. A. Stein, editors. Scanning the conservation horizon: a guide to climate change vulnerability assessment. National Wildlife Federation, Washington, D.C., USA.

Fitzpatrick, M. C., E. L. Preisser, A. Porter, J. Elkinton, and A. M. Ellison. 2012. Modeling range dynamics in heterogeneous landscapes: invasion of the hemlock woolly adelgid in eastern North America. Ecological Applications 22:472–486.

Fleming, K. K., and W. F. Porter. 2007. Synchrony in a wild turkey population and its relationship to spring weather. Journal of Wildlife Management 71:1192–1196.

Folland, C. K., T. R. Karl, and M. Jim Salinger. 2002. Observed climate variability and change. Weather 57:269–278.

Forchhammer, M. C., E. Post, and N. C. Stenseth. 2002. North Atlantic Oscillation timing of long- and short-distance migration. Journal of Animal Ecology 71:1002–1014.

Garner, D. L. 1994. Population ecology of moose in Algonquin Provincial Park, Ontario, Canada. Dissertation, State University of New York, Syracuse, USA.

Gaston, K. J. 2003. The structure and dynamics of geographic ranges. Oxford University Press, Oxford, United Kingdom.

Geber, M. A. 2011. Ecological and evolutionary limits to species geographic ranges. American Naturalist 178:S1–S5.

Gilbert, L. 2010. Altitudinal patterns of tick and host abundance: a potential role for climate change in regulating tick-borne diseases? Oecologia 162:217–225.

Glick, P., and B. A. Stein. 2011. Scanning the conservation horizon: a guide to climate change vulnerability assessment. National Wildlife Federation, Washington, D.C., USA.

Gonzalez, P. 2001. Desertification and a shift of forest species in the West African Sahel. Climate Research 17:217–228.

Hall, M. H. P., and D. B. Fagre. 2003. Modeled climate-induced glacier change in Glacier National Park, 1850–2100. Bioscience 53:131–140.

Hansen, J., R. Ruedy, M. Sato, and K. Lo. 2010. Global surface temperature change. Reviews of Geophysics 48:RG4004, doi:10.1029/2010RG000345.

Harrington, R., I. Woiwod, and T. Sparks. 1999. Climate change and trophic interactions. Trends in Ecology and Evolution 14:146–150.

Heikkinen, R. K., M. Luoto, M. B. Araujo, R. Virkkala, W. Thuiller, and M. T. Sykes. 2006. Methods and uncertainties in bioclimatic envelope modelling under climate change. Progress in Physical Geography 30:751–777.

Henson, R. 2011. The rough guide to climate change. Third edition. Rough Guides, London, United Kingdom.

Hewitt, G. M. 1996. Some genetic consequences of ice ages, and their role in divergence and speciation. Biological Journal of the Linnean Society 58:247–276.

Hickling, R., D. B. Roy, J. K. Hill, R. Fox, and C. D. Thomas. 2006. The distributions of a wide range of taxonomic groups are expanding polewards. Global Change Biology 12:450–455.

Hubalek, Z. 2003. Spring migration of birds in relation to North Atlantic Oscillation. Folia Zoologica 52:287–298.

Hurlbert, A. H., and Z. F. Liang. 2012. Spatiotemporal variation in avian migration phenology: citizen science reveals effects of climate change. PLoS ONE 7:e31662, doi:10.1371/journal.pone.0031662.

Inouye, D. W., B. Barr, K. B. Armitage, and B. D. Inouye. 2000. Climate change is affecting altitudinal migrants and hibernating species. Proceedings of the National Academy of Sciences 97:1630–1633.

IPCC. Intergovernmental Panel on Climate Change. 1995. Climate change 1995: impacts, adaptations and mitigation of climate change: scientific-technical analyses. Contribution of Working Group II to the Second Assessment Report of the Intergovernmental Panel on Climate Change. R. T. Watson, M. C. Zinyowera, and R. H. Moss, editors. Cambridge University Press, Cambridge, United Kingdom.

IPCC. Intergovernmental Panel on Climate Change. 2000. Summary for policy makers: emissions scenarios. A special report of IPCC Working Group III. N. Nakicenovic and R. Swart, editors. Cambridge University Press, Cambridge, United Kingdom.

IPCC. Intergovernmental Panel on Climate Change. 2007. Climate change 2007: the physical science basis. Contribution of Working Group I to the Fourth Assessment Report of the Intergovernmental Panel on Climate Change. S. Solomon, D. Qin, M. Manning, Z. Chen, M. Marquis, K. B. Averyt, M. Tignor, and H. L. Miller, editors. Cambridge University Press, Cambridge, United Kingdom.

Isaac, J. L., J. L. De Gabriel, and B. A. Goodman. 2008. Microclimate of daytime den sites in a tropical possum: implications for the conservation of tropical arboreal marsupials. Animal Conservation 11:281–287.

Kausrud, K. L., A. Mysterud, H. Steen, J. O. Vik, E. Ostbye, et al. 2008. Linking climate change to lemming cycles. Nature 456:93–97.

Killengreen, S. T., R. A. Ims, N. G. Yoccoz, K. A. Brathen, J. A. Henden, and T. Schott. 2007. Structural characteristics of a low Arctic tundra ecosystem and the retreat of the Arctic fox. Biological Conservation 135:459–472.

Klein, A. G., and J. L. Kincaid. 2006. Retreat of glaciers on Puncak Jaya, Irian Jaya, determined from 2000 and 2002 IKONOS satellite images. Journal of Glaciology 52:65–79.

Knutson, M. G., and P. J. Heglund. 2012. Resource managers rise to the challenge of climate change. Pages 261–284 in E. A. Beever and J. L. Belant, editors. Ecological consequences of climate change: mechanisms, conservation, and management. Taylor and Francis, Boca Raton, Florida, USA.

Kobori, H., T. Kamamoto, H. Nomura, K. Oka, and R. Primack. 2012. The effects of climate change on the phenology of winter birds in Yokohama, Japan. Ecological Research 27:173–180.

Kullman, L., and L. Oberg. 2009. Post–Little Ice Age tree line rise and climate warming in the Swedish Scandes: a landscape ecological perspective. Journal of Ecology 97:415–429.

Lawson, C. R., J. J. Bennie, C. D. Thomas, J. A. Hodgson, and R. J. Wilson. 2012. Local and landscape management of an expanding range margin under climate change. Journal of Applied Ecology 49:552–561.

Lehikoinen, E., and T. H. Sparks. 2010. Changes in migration. Pages 89–112 in A. P. Moller, W. Fiedler, and P. Berthold, editors. Effects of climate change on birds. Oxford University Press, Oxford, United Kingdom.

Lemke, P., J. Ren, R. B. Alley, I. Allison, J. Carrasco, et al. 2007. Observations: changes in snow, ice and frozen ground. Pages 338–383 in S. Solomon, D. Qin, M. Manning, Z. Chen, M. Marquis, K. B. Averyt, M. Tignor, and H. L. Miller, editors. Climate change 2007: the physical science basis. Contribution of Working Group I to the Fourth Assessment Report of the Intergovernmental Panel on Climate Change. Cambridge University Press, Cambridge, United Kingdom.

Lemoine, N., and K. Bohning-Gaese. 2003. Potential impact of global climate change on species richness of long-distance migrants. Conservation Biology 17:577–586.

Lepetz, V., M. Massot, D. S. Schmeller, and J. Clobert. 2009. Biodiversity monitoring: some proposals to adequately study species' responses to climate change. Biodiversity and Conservation 18:3185–3203.

Lindgren, E., L. Talleklint, and T. Polfeldt. 2000. Impact of climatic change on the northern latitude limit and population density of the disease-transmitting European tick *Ixodes ricinus*. Environmental Health Perspectives 108:119–123.

Lloyd, A. H., and C. L. Fastie. 2003. Recent changes in treeline forest distribution and structure in interior Alaska. Ecoscience 10:176–185.

Luoto, M., R. Virkkala, and R. K. Heikkinen. 2007. The role of land cover in bioclimatic models depends on spatial resolution. Global Ecology and Biogeography 16:34–42.

Lyons, J. E., M. C. Runge, H. P. Laskowski, and W. L. Kendall. 2008. Monitoring in the context of structured decision-making and adaptive management. Journal of Wildlife Management 72:1683–1692.

MacArthur, R. H. 1972. Geographical ecology: patterns in the distribution of species. Harper and Row, New York, New York, USA.

MacKenzie, D. I., J. D. Nichols, J. A. Royle, K. H. Pollock, L. L. Bailey, and J. E. Hines. 2006. Occupancy estimation and modeling: inferring patterns and dynamics of species occurence. Elsevier, Burlington, Maine, USA.

Martin, J., M. C. Runge, J. D. Nichols, B. C. Lubow, and W. L. Kendall. 2009. Structured decision making as a conceptual framework to identify thresholds for conservation and management. Ecological Applications 19:1079–1090.

Mawdsley, J. R., R. O'Malley, and D. S. Ojima. 2009. A review of climate-change adaptation strategies for wildlife management and biodiversity conservation. Conservation Biology 23:1080–1089.

McDonald-Madden, E., M. C. Runge, H. P. Possingham, and T. G. Martin. 2011. Optimal timing for managed relocation of species faced with climate change. Nature Climate Change 1:261–265.

McKinney, M. L. 1997. Extinction vulnerability and selectivity: combining ecological and paleontological views. Annual Review of Ecology and Systematics 28:495–516.

McLennan, D. 2012. Dealing with uncertainty: managing and monitoring Canada's northern national parks in a rapidly changing world. Pages 209–236 in E. A. Beever and J. L. Belant, editors. Ecological consequences of climate change: mechanisms, conservation, and management. Taylor and Francis, Boca Raton, Florida, USA.

Merino, S., and A. P. Moller. 2010. Host–parasite interactions and climate change. Pages 213–226 in A. P. Moller, W. Fiedler, and P. Berthold, editors. Effects of climate change on birds. Oxford University Press, Oxford, United Kingdom.

Millar, C. I., and R. D. Westfall. 2010. Distribution and climatic relationships of the American Pika (Ochotona princeps) in the Sierra Nevada and western Great Basin, USA: periglacial landforms as refugia in warming climates. Reply. Arctic, Antarctic, and Alpine Research 42:493–496.

Mittelbach, G. G. 2012. Community ecology. Sinauer, Sunderland, Massachusetts, USA.

Moller, A. P. 2010. Host–parasite interactions and vectors in the barn swallow in relation to climate change. Global Change Biology 16:1158–1170.

Montzka, S. A., E. J. Dlugokencky, and J. H. Butler. 2011. Non-CO$_2$ greenhouse gases and climate change. Nature 476:43–50.

Mosier, A. R., J. M. Duxbury, J. R. Freney, O. Heinemeyer, K. Minami, and D. E. Johnson. 1998. Mitigating agricultural emissions of methane. Climatic Change 40:39–80.

Mouritsen, K. N., and R. Poulin. 2002. Parasitism, climate oscillations and the structure of natural communities. Oikos 97:462–468.

Moyes, K., D. H. Nussey, M. N. Clements, F. E. Guinness, A. Morris, S. Morris, J. M. Pemberton, L. E. B. Kruuk, and T. H. Clutton-Brock. 2011. Advancing breeding phenology in response to environmental change in a wild red deer population. Global Change Biology 17:2455–2469.

Musante, A. R., P. J. Pekins, and D. L. Scarpitti. 2010. Characteristics and dynamics of a regional moose Alces alces population in the northeastern United States. Wildlife Biology 16:185–204.

National Aeronautics and Space Administration. 2010. 2009: second warmest year on record; end of warmest decade. http://www.giss.nasa.gov/research/news/20100121/. Accessed 25 October 2012.

National Oceanic and Atmospheric Administration. 2011. The NOAA annual greenhouse gas index. http://www.esrl.noaa.gov/gmd/aggi/. Accessed 25 October 2012.

National Oceanic and Atmospheric Administration. 2013. Trends in atmospheric carbon dioxide. http://www.esrl.noaa.gov/gmd/ccgg/trends/. Accessed 13 May 2013.

Nicholls, R. J., P. P. Wong, W. R. Burkett, J. O. Codignotto, J. E. Hay, R. F. McLean, S. Ragoonaden, and C. D. Woodroffe. 2007. Coastal systems and low-lying areas. Pages 315–356 in M. L. Parry, O. F. Canziani, J. P. Palutikof, P. J. van der Linden, and C. E. Hanson, editors. Climate change 2007: impacts, adaptation and vulnerability. Contribution of Working Group II to the Fourth Assessment Report of the Intergovernmental Panel on Climate Change. Cambridge University Press, Cambridge, United Kingdom.

Nichols, J. D., M. D. Koneff, P. J. Heglund, M. G. Knutson, M. E. Seamans, J. E. Lyons, J. M. Morton, M. T. Jones, G. S. Boomer, and B. K. Williams. 2011. Climate change, uncertainty, and natural resource management. Journal of Wildlife Management 75:6–18.

Oenema, O., N. Wrage, G. L. Velthof, J. W. van Groenigen, J. Dolfing, and P. J. Kuikman. 2005. Trends in global nitrous oxide emissions from animal production systems. Nutrient Cycling in Agroecosystems 72:51–65.

Paradis, A., J. Elkinton, K. Hayhoe, and J. Buonaccorsi. 2008. Role of winter temperature and climate change on the survival and future range expansion of the hemlock woolly adelgid (Adelges tsugae) in eastern North America. Mitigation and Adaptation Strategies for Global Change 13:541–554.

Parmesan, C. 1996. Climate and species' range. Nature 382:765–766.

Parmesan, C. 2006. Ecological and evolutionary responses to recent climate change. Annual Review of Ecology Evolution and Systematics 37:637–669.

Parmesan, C. 2007. Influences of species, latitudes and methodologies on estimates of phenological response to global warming. Global Change Biology 13:1860–1872.

Parmesan, C., and G. Yohe. 2003. A globally coherent fingerprint of climate change impacts across natural systems. Nature 421:37–42.

Pattee, O. H., and S. L. Beasom. 1979. Supplemental feeding to increase wild trukey productivity. Journal of Wildlife Management 43:512–516.

Payette, S. 2007. Contrasted dynamics of northern Labrador tree lines caused by climate change and migrational lag. Ecology 88:770–780.

Pearce-Higgins, J. W., P. Dennis, M. J. Whittingham, and D. W. Yalden. 2010. Impacts of climate on prey abundance account for fluctuations in a population of a northern wader at the southern edge of its range. Global Change Biology 16:12–23.

Pearson, R. G., and T. P. Dawson. 2003. Predicting the impacts of climate change on the distribution of species: are bioclimate envelope models useful? Global Ecology and Biogeography 12:361–371.

Pearson, R. G., T. P. Dawson, and C. Liu. 2004. Modelling species distributions in Britain: a hierarchical integration of climate and land-cover data. Ecography 27:285–298.

Pearson, R. G., C. J. Raxworthy, M. Nakamura, and A. T. Peterson. 2007. Predicting species distributions from small numbers of occurrence records: a test case using cryptic geckos in Madagascar. Journal of Biogeography 34:102–117.

Pederson, G. T., L. J. Graumlich, D. B. Fagre, T. Kipfer, and C. C. Muhl-feld. 2010. A century of climate and ecosystem change in western Montana: what do temperature trends portend? Climatic Change 98:133–154.

Penuelas, J., and M. Boada. 2003. A global change-induced biome shift in the Montseny mountains (NE Spain). Global Change Biology 9:131–140.

Penuelas, J., and I. Filella. 2001. Phenology—responses to a warming world. Science 294:793–795.

Post, E., and M. C. Forchhammer. 2008. Climate change reduces reproductive success of an Arctic herbivore through trophic mismatch. Philosophical Transactions of the Royal Society B 363:2369–2375.

Pulido, F., and P. Berthold 2010. Current selection for lower migratory activity will drive the evolution of residency in a migratory bird population. Proceedings of the National Academy of Sciences 107:7341–7346.

Pulido, F., P. Berthold, and A. J. vanNoordwijk. 1996. Frequency of migrants and migratory activity are genetically correlated in a bird population: evolutionary implications. Proceedings of the National Academy of Sciences 93:14,642–14,647.

Reale, D., A. G. McAdam, S. Boutin, and D. Berteaux. 2003. Genetic and plastic responses of a northern mammal to climate change. Proceedings of the Royal Society of London B 270:591–596.

Roberts, S. D., and W. F. Porter. 2001. Annual changes in May rainfall as an index to wild turkey harvest. National Wild Turkey Symposium 8:43–52.

Root, T. L., J. T. Price, K. R. Hall, S. H. Schneider, C. Rosenzweig, and J. A. Pounds. 2003. Fingerprints of global warming on wild animals and plants. Nature 421:57–60.

Rostom, R., and S. Hastenrath. 2007 Variations of Mount Kenya's glaciers 1993–2004. Erdkunde 61:277–283.

Rowland, E. L., J. E. Davison, and L. J. Graumlich. 2011. Approaches to evaluating climate change impacts on species: a guide to initiating the adaptation planning process. Environmental Management 47:322–337.

Saino, N., D. Rubolini, E. Lehikoinen, L. V. Sokolov, A. Bonisoli-Alquati, R. Ambrosini, G. Boncoraglio, and A. P. Moller. 2009. Climate change effects on migration phenology may mismatch brood parasitic cuckoos and their hosts. Biology Letters 5:539–541.

Sandel, B., L. Arge, B. Dalsgaard, R. G. Davies, K. J. Gaston, W. J. Sutherland, and J. C. Svenning. 2011. The influence of late Quaternary climate-change velocity on species endemism. Science 334:660–664.

Sharma, S., S. Couturier, and S. D. Cote. 2009. Impacts of climate change on the seasonal distribution of migratory caribou. Global Change Biology 15:2549–2562.

Sharp, B. R., and D. Bowman. 2004. Patterns of long-term woody vegetation change in a sandstone-plateau savanna woodland, Northern Territory, Australia. Journal of Tropical Ecology 20:259–270.

Sheldon, B. C. 2010. Genetic perspective on the evolutionary consequences of climate change in birds. Pages 149–168 in A. P. Møller, W. Fiedler, and P. Berthold, editors. Effects of climate change on birds. Oxford University Press, Oxford, United Kingdom.

Sheriff, M. J., G. J. Kenagy, M. Richter, T. Lee, O. Toien, F. Kohl, C. L. Buck, and B. M. Barnes. 2011. Phenological variation in annual timing of hibernation and breeding in nearby populations of Arctic ground squirrels. Proceedings of the Royal Society B 278:2369–2375.

Smith, K. A., and F. Conen. 2004. Impacts of land management on fluxes of trace greenhouse gases. Soil Use and Management 20:255–263.

Sparks, T. H., F. Bairlein, J. G. Bojarinova, O. Huppop, E. A. Lehikoinen, K. Rainio, L. V. Sokolov, and D. Walker. 2005. Examining the total arrival distribution of migratory birds. Global Change Biology 11:22–30.

Suarez, F., D. Binkley, M. W. Kaye, and R. Stottlemyer. 1999. Expansion of forest stands into tundra in the Noatak National Preserve, northwest Alaska. Ecoscience 6:465–470.

Sutton, R. T., B. W. Dong, and J. M. Gregory. 2007. Land/sea warming ratio in response to climate change: IPCC AR4 model results and comparison with observations. Geophysical Research Letters 34:L02701, doi:10.1029/2006GL028164.

Thomas, C. D., and J. J. Lennon. 1999. Birds extend their ranges northwards. Nature 399:213.

Thuiller, W., and T. Munkemuller. 2010. Habitat suitability modeling. Pages 77–85 in A. P. Moller, W. Fiedler, and P. Berthold, editors. Effects of climate change on birds. Oxford University Press, Oxford, United Kingdom.

Tierson, W. C., G. F. Mattfeld, R. W. Sage, and D. F. Behrend. 1985. Seasonal movements and home ranges of white-tailed deer in the Adirondacks. Journal of Wildlife Management 49:760–769.

Tingley, M. W., and S. R. Beissinger. 2009. Detecting range shifts from historical species occurrences: new perspectives on old data. Trends in Ecology and Evolution 24:625–633.

Tingley, M. W., W. B. Monahan, S. R. Beissinger, and C. Moritz. 2009. Colloquium papers: birds track their Grinnellian niche through a century of climate change. Proceedings of the National Academy of Sciences 106:19,637–19,643.

Towns, L., A. E. Derocher, I. Stirling, and N. J. Lunn. 2010. Changes in land distribution of polar bears in western Hudson Bay. Arctic 63:206–212.

U.S. Environmental Protection Agency. 2009. A framework for categorizing the relative vulnerability of threatened and endangered species to climate change. National Center for Environmental Assessment, Washington, D.C., USA.

Vanhanen, H., T. O. Veleli, S. Paivinen, S. Kellomaki, and P. Niemela. 2007. Climate change and range shifts in two insect defoliators: gypsy moth and nun moth—a model study. Silva Fennica 41:621–638.

Visser, M. E., and C. Both. 2005. Shifts in phenology due to global climate change: the need for a yardstick. Proceedings of the Royal Society B 272:2561–2569.

Visser, M. E., S. P. Caro, K. van Oers, S. V. Schaper, and B. Helm. 2010. Phenology, seasonal timing and circannual rhythms: towards a unified framework. Philosophical Transactions of the Royal Society B 365:3113–3127.

Vors, L. S., and M. S. Boyce. 2009. Global declines of caribou and reindeer. Global Change Biology 15:2626–2633.

Walther, G. R., S. Berger, and M. T. Sykes. 2005. An ecological "footprint" of climate change. Proceedings of the Royal Society B 272:1427–1432.

Walther, G. R., E. Post, P. Convey, A. Menzel, C. Parmesan, T. J. C. Beebee, J. M. Fromentin, O. Hoegh-Guldberg, and F. Bairlein. 2002. Ecological responses to recent climate change. Nature 416:389–395.

Williams, B. K., J. D. Nichols, and M. J. Conroy. 2002. Analysis and management of animal populations: modeling, estimation, and decision making. Academic Press, San Diego, California, USA.

Williams, B. K., R. C. Szaro, and C. D. Shapiro. 2007a. Adaptive management: the U.S. Department of the Interior technical guide. U.S. Department of the Interior, Washington, D.C., USA.

Williams, J. W., S. T. Jackson, and J. E. Kutzbacht. 2007b. Projected distributions of novel and disappearing climates by 2100 AD. Proceedings of the National Academy of Sciences 104:5738–5742.

Williams, S. E., L. P. Shoo, J. L. Isaac, A. A. Hoffmann, and G. Langham. 2008. Towards an integrated framework for assessing the vulnerability of species to climate change. PLoS Biology 6:2621–2626.

World Meteorological Organization. 2011. Weather extremes in a changing climate: hindsight on foresight. WMO-No. 1075. Geneva, Switzerland. http://www.wmo.int/pages/mediacentre/news/extreme
weathersequence_en.html. Accessed 25 October 2012.

Yoccoz, N. G., J. D. Nichols, and T. Boulinier. 2001. Monitoring of biological diversity in space and time. Trends in Ecology and Evolution
16:446–453.

Young, B., E. Byers, K. Gravuer, K. Hall, G. Hammerson, and A. Redder.
2009. Guidelines for using the NatureServe climate change vulnerability index, release 1.0. http://www.natureserve.org/prodServices
/climatechange/ClimateChange.jsp#v1point2. Accessed 25 October
2012.

Zemp, M., W. Haeberli, M. Hoelzle, and F. Paul. 2006. Alpine glaciers to
disappear within decades? Geophysical Research Letters 33:L13504,
doi:10.1029/2006GL026319.

Zuckerberg, B., A. Woods, and W. F. Porter. 2009. Poleward shifts in
breeding bird distributions in New York State. Global Change Biology 15:1866–1883.

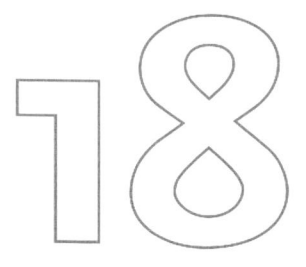

CONSERVATION PLANNING FOR WILDLIFE AND WILDLIFE HABITAT

SUSAN P. RUPP, ANNA M. MUÑOZ, AND ROEL R. LOPEZ

INTRODUCTION

Aldo Leopold illustrated the importance of the careful execution of habitat management, saying "I have read many definitions of what is a conservationist, and written not a few myself, but I suspect the best one is written not with a pen, but with an axe . . . A conservationist is one who is humbly aware that with each stroke he is writing his signature on the face of the land. Signatures of course differ, whether with axe or pen, and this is as it should be" (Leopold 1949:68). As land managers, such careful execution can enhance and protect natural resources.

This volume so far has reviewed basic principles in wildlife–habitat relationships (Chapter 15), approaches to measuring and evaluating wildlife habitat, and commonly applied habitat modification practices (Chapter 16). The integration of theory and practice is typically accomplished through conservation planning. This chapter describes the process of conservation planning for wildlife and wildlife habitat and its associated management. Conservation planning is a creative process that requires experience, analysis, intuition, and inspiration. Even though planning can sometimes be unstructured, management plans can add structure to the overall planning process and are an integral part of conservation (Lopez et al. 2005). Conservation planning is a progression of steps to determine what we have (e.g., animals, plants, physical resources), what we want (e.g., more or fewer white-tailed deer [*Odocoileus virginianus*]), and how to get there (e.g., increase or decrease cover or forage requirements). Although there is no single method or standard approach to developing a conservation plan, there are some general guidelines that can aid the practitioner in successfully engaging in the process and can increase their awareness regarding their signature on the face of the land. We introduce some of these basic concepts in the development and careful execution of habitat management plans.

WHAT IS PLANNING?

A good working definition of planning might include "the deliberate social or organizational activity of developing an optimal strategy for solving problems and achieving a desired set of objectives" (Yoe and Orth 1996:11). Although this definition oversimplifies an often-complex process, it does emphasize a few important elements of planning.

1. *Future control.* Planning focuses on the control of future consequences through present actions, suggesting that planning and action together are necessary to ensure and increase control in the future of the targeted lands. As such, uncertainties and how to best manage them must be part of the planning process.
2. *Problem solving.* Planning is a problem-solving approach that addresses a given natural resource issue. Generally, wildlife management problems can include increasing population numbers (e.g., endangered species), decreasing impacts from nuisance or invasive species, or maintaining a sustained yield for a game species. The need to address a problem or natural resource issue often drives the reason for conservation planning.
3. *Team effort.* Individuals often do planning in a team environment that considers the opinions and desires of a wide-ranging set of stakeholders (e.g., landowners, state and local governments). This, of course, depends on the spatial extent of the property and landowner (i.e., a plan for small, private property may be done by an individual). Planning should not be done without the consideration of various perspectives and approaches. Increasing the level of stakeholder input in conservation planning can ensure the careful review of all viable options.
4. *No single approach.* Planning is uniquely tailored to a specific situation and set of objectives and assumptions. Unfortunately, there is no standard approach in conservation planning. General rules and procedures can be applied, however, and the land manager should be willing to adapt and modify a management plan to the specific situation to improve its overall effectiveness.
5. *Adaptive framework.* Planning is adaptive and involves feedback loops. Continued monitoring, evaluation, and

adjustment are critical components to good management plans. As mentioned before, management plans are part of a process. This process typically has a predetermined endpoint (i.e., goal) that is re-evaluated at the end of the target timeline and restarts the management process.

6. *Intention to implement.* Planning is done with the intention to implement the strategies outlined in the plan, and is not done for planning's sake. Management plans require a substantial amount of time, effort, and funding. Land managers should develop management plans with the intention to follow them, and should not simply put them on a shelf.

MISSION AND VISION FORM THE FOUNDATION

Every agency and organization has a specific mission statement and vision, the heart and basic tenet of any conservation planning effort. A clearly stated mission and vision are what clarify the direction in any plan, eliminating circular arguments and allowing a clear development of the desired condition, for which goals and objectives become much easier to craft. In short, an agency's mission and vision are the foundation on which conservation planning develops (Fig. 18.1).

The authority for an agency or organization is usually (if not always) stated through a mission statement. A mission statement based on correct principles is like a personal constitution, the basis for making major, life-directing decisions in the midst of the circumstances and challenges that affect an agency or organization. It is the lens through which agencies and organizations see the world. For example, the mission statement for the U.S. Fish and Wildlife Service (USFWS) is "working with others to conserve, protect and enhance fish, wildlife, and plants and their habitats for the continuing benefit of the American people" (http://www.fws.gov/planning /Mission.html). This mission statement explains what the organization does, for whom, and the benefit of doing it, which guides decision making within the USFWS. It acts as a sieve through which everything is filtered. If a potential decision or action does not fall under the premises outlined in the mission statement, it is not considered to be a priority or, more importantly, not considered at all. You would not find the USFWS involved with developing a high-rise building in the middle of a major metropolitan area unless it is designed to assist their primary mission in some form or another (e.g., to house employees that assist with the USFWS mission).

Unlike a mission statement that concerns what an organization is all about, a vision statement is what the organization wants to become. Vision statements define the organization's purpose in terms of its values rather than bottom-line measures. It describes how the future will look if the organization achieves its mission. The USFWS vision is to "unite all service programs to lead or support ecosystem level conservation . . . by becoming a more technically capable and culturally diverse organization; through involving stakeholders; through scientific expertise; through land and water management; and, through appropriate regulation" (http://www.fws.gov /planning/Mission.html). Once the vision and mission of an agency or organization are established, it becomes much easier to design conservation plans that help to achieve those goals.

GOALS AND OBJECTIVES

Our working definition of planning includes a strategy for achieving a desired set of objectives. Management plans should be objective and goal oriented. The terms "goals" and "objectives" are often used interchangeably; however, there are some subtle differences between them when used in conservation planning. A goal can be defined as the final or desired end purpose. Goals are overarching statements that establish the overall direction for (and focus of) a conservation plan and define the scope of what the plan should achieve. Conversely, an objective is a quantifiable action used to achieve a goal. Objectives are specific, measurable, and have a defined completion date. They outline who will make what change, by how much, where, and by when in order to reach the goal. From a conservation planning perspective, both definitions convey the same basic intent: do the right thing for the resource. But the subtle differences in their definitions also establish a hierarchical structure that suggests we set goals first and then establish objectives to attain them. In other words, goals are overarching outcomes, and objectives are the specific actions—the building blocks—we use to achieve the goals. The best goals are consistent with the mission of the agency or organization. The objectives are then the stepping-stones to achieve the goals.

Below is an example that differentiates a goal from an objective from the perspective of two land managers. Both share the common goal of improving quail (*Colinus* spp.) habitat but differ in how they individually will strive to reach this goal.

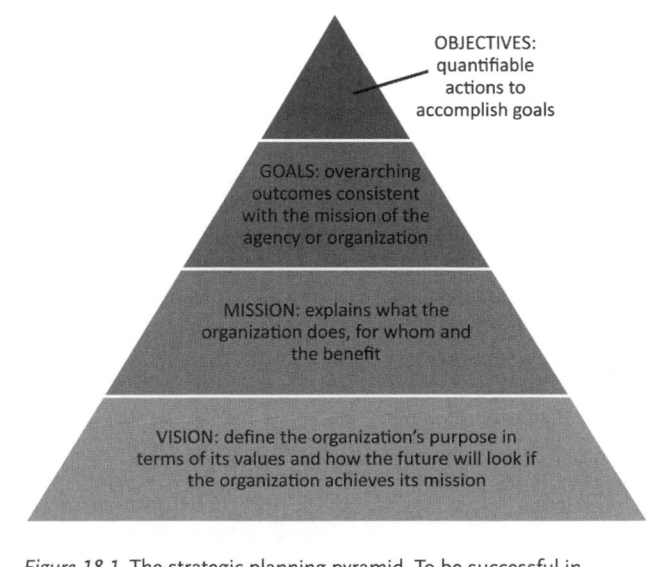

Figure 18.1. The strategic planning pyramid. To be successful in conservation planning, each project should be driven by the agency or organization's mission, vision, goals, and objectives.

Each land manager has the same goal (i.e., improve quail habitat), yet their objectives are different. In both cases, their individual objectives can allow them to reach their common goal because they are specific, measurable, and have a defined completion date. As Leopold (1949:68) so eloquently said, "signatures of course differ, whether with axe or pen, and this is as it should be."

Land Manager 1
- Goal: improve quail habitat
- Objective 1: apply 16.2-ha prescribed fire in April next year
- Objective 2: disc 4 ha annually on a rotational basis
- Objective 3: maintain brush piles in the northwest corner on annual basis

Land Manager 2
- Goal: improve quail habitat
- Objective 1: control predators through active trapping two weeks per year
- Objective 2: plant 2-ha food plots in spring of next year
- Objective 3: increase native grasses by 30% across property by end of five-year period

TYPES OF PLANNING

There are many different types of plans. Land-use plans maintain the most suitable types of land uses (e.g., developed versus natural areas) and specific activities relative to the management of those lands. Transportation plans may address public transportation strategies to support the projected growth of a region or county. Historic preservation plans might be used in a community to protect a historic area like a Civil War battlefield or the remnants of an old mining community. These are a few examples of the many types of plans used in conservation planning. A common challenge with these types of planning efforts is the lack of coordination and integration among these various processes within the same geographic area. The potential for conflicting approaches and recommendations between transportation and conservation plans, for example, requires a concerted effort by agencies and other resource managers to work collaboratively. Part of this challenge is in large part due to the varying missions of agencies and organizations responsible for the management of natural resources. Consider the multitude of agencies responsible for some form of conservation planning, each with their own varying mission and vision (Table 18.1). Organizations or agencies responsible for conservation planning can range from federal and state agencies to nonprofit organizations, land trusts, and other local or regional partnerships. This wide breadth of organizations and agencies involved in the management of natural resources can be challenging and even daunting at times. Questions regarding who is responsible for what and mission creep (i.e., the expansion of a project or mission beyond its original goals) can make the planning process difficult.

In addition to the multitude of agencies involved, the involvement of the private landowner in successful conservation planning is essential. In many areas of the country, the majority of lands are privately owned (Fig. 18.2). According to the Partners for Fish and Wildlife Act (PL 109-294) passed by Congress in October 2006, roughly 60% of fish and wildlife reside on private lands. In addition to the wildlife present there, private lands also provide many ecosystem services (e.g., clean water, open space). This makes working with private landowners a critical component of conservation planning. Often, private lands can serve as critical corridors for connecting existing public lands. However, because these lands cannot be acquired owing to private ownership or lack of funding to obtain them, private lands provide their own stewards (i.e., the landowner) responsible for managing those lands. Even if those lands could be obtained, federal and state budgets are stretched to the breaking point, and public agencies often struggle to manage the properties they already have. There is greater pressure for wildlife managers to work with private landowners to design effective management plans that can benefit the landowner through strategic planning and careful consideration of the social and economic values of existing lands to supplement essential habitat for wildlife. Despite their importance, however, private lands are often overlooked because agencies tend to think within their own boundaries and may get frustrated because they lack inventories and control of resources on private lands.

Planning also can vary in spatial scale or a particular focus. For example, conservation plans can range from multistate ecosystem-level plans (e.g., America's Longleaf Restoration Initiative) to local habitat conservation plans that focus on a single endangered species (Table 18.2). All levels of planning are important, but each individual plan is likely to be more effective when it connects with other planning efforts. Including information from a local habitat conservation plan in an ecoregional plan maintains consistency and increases coordination in conservation management activities. This was the case in the development of The Nature Conservancy's (TNC) Conservation by Design approach. TNC realized funding expenditures would be more effective if they had a long-range plan for land acquisition and restoration based on the best available science at larger regional or global scales. As a result, TNC initiated Conservation by Design, a science-based conservation approach at multiple scales that uses three complementary analytical methods: global habitat assessments, ecoregional assessments, and conservation action planning. Global habitat assessments provide a baseline against which TNC can measure progress toward their mission, identify conservation gaps, and establish priorities for allocating resources on a global scale. Within the global habitat assessments, TNC identifies which specific ecoregions require attention as well as threats to biodiversity and strategic opportunities that affect one or more major landscapes and demand immediate attention. In some cases, ecoregions span multiple states and countries, and they identify and prioritize conservation land on the current ecology and threat levels within that region. Finally, global and ecoregional priorities are translated into conservation strategies and actions through conservation action planning. This method is used to design and manage conservation projects that advance conservation at any scale, from efforts to con-

Table 18.1. Examples of agencies and organizations responsible for conservation planning

Federal agencies	State agencies	Municipal governments	Nonprofit organizations	Local land trusts	Coalitions and partnerships
U.S. Forest Service	Maine Bureau of Parks and Lands	Conservation commissions	Trust for Public Lands	Lower Kennebec Regional Land Trust	Mount Agamenticus to the Sea Conservation Initiative
U.S. Fish and Wildlife Service	Maine Forest Service	Town forest committees	The Nature Conservancy	Mahoosuc Land Trust	Quabbin to Cardigan Conservation Collaborative
Natural Resource Conservation Service	Maine Department of Inland Fisheries and Wildlife	Open space committees	The Conservation Fund	Maine Wilderness Watershed Trust	Great Bay Partnership
U.S. Army Corps of Engineers	Florida Fish and Wildlife Conservation Commission		Forest Society of Maine	Sebasticook River Watershed Association	Vermont Housing and Conservation Board
Bureau of Land Management	Florida Park Service		Maine Audubon Society	Five Rivers Land Trust	Prairie Coteau Habitat Partnership
Department of Defense	Florida Department of Environmental Conservation		American Farmland Trust	Lakes Region Conservation Trust	
National Park Service	Texas Parks and Wildlife Department		Society for the Protection of New Hampshire	Monadnock Conservancy	
	Texas Forest Service		Texas Land Trust	Piscataquog River Watershed Association	
	Texas Commission on Environmental Quality		Audubon Vermont	Jericho Underhill Land Trust	
	South Dakota Department of Game, Fish and Parks		Audubon Society of North Carolina	Lake Champlain Land Trust	
	South Dakota Department of Environment and Natural Resources			Middlebury Area Land Trust	
	California Conservation Corps				
	California Department of Boating and Waterways				
	California Department of Conservation				
	California Department of Fish and Game				
	California Department of Forestry and Fire Protection				
	California Department of Parks and Recreation				
	California Department of Water Resources				

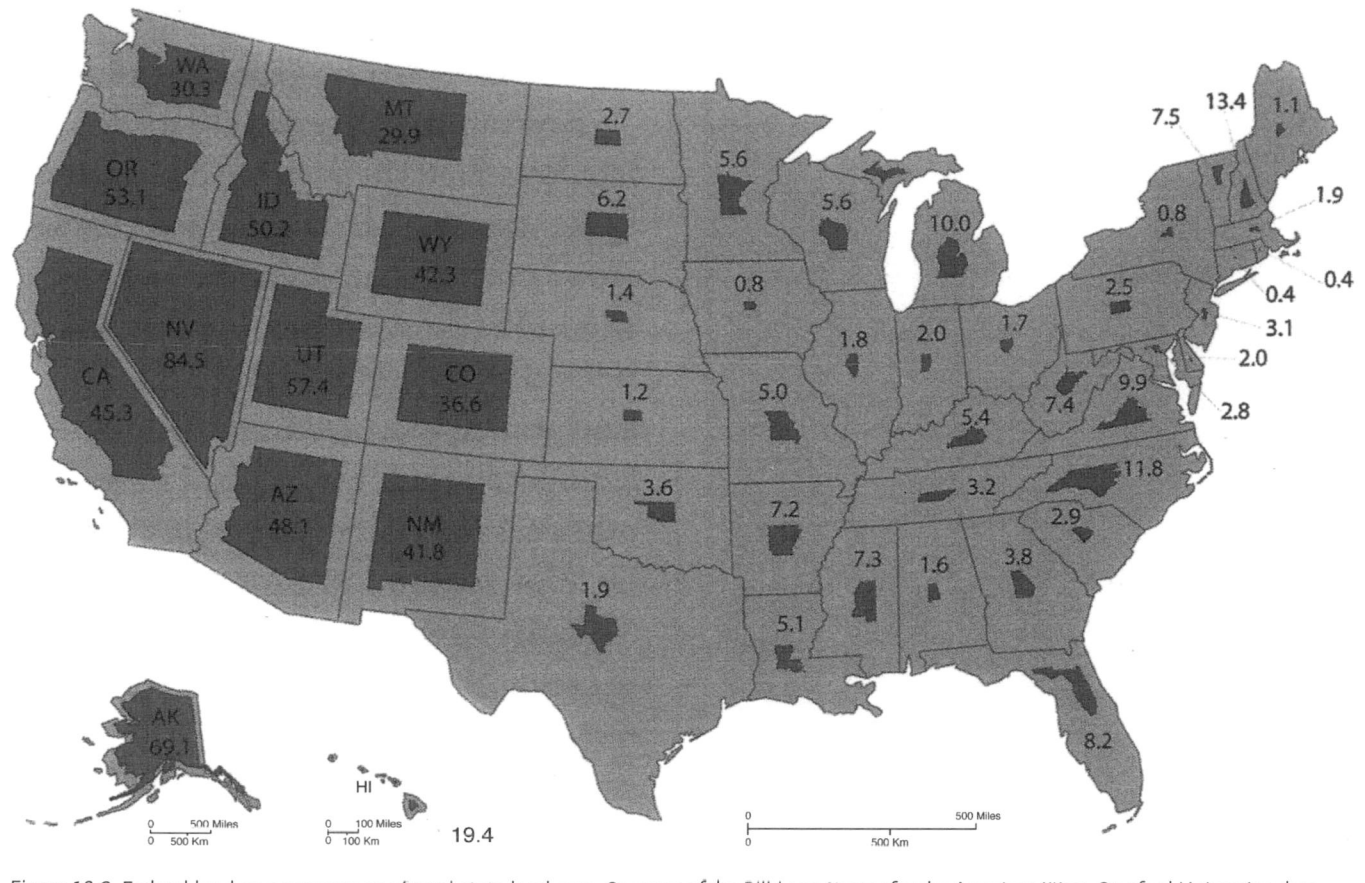

Figure 18.2. Federal land as a percentage of total state land area. Courtesy of the Bill Lane Center for the American West, Stanford University; data source is the U.S. General Services Administration, *Federal Real Property Profile 2004*, and excludes trust properties

Table 18.2. Examples of conservation planning efforts by focus or scale

Scale or focus	Example	Description	Source
International	North American Waterfowl Management Plan	An international plan to conserve waterfowl and migratory birds in North America. It was established in 1986 by Canada and the United States, and expanded to include Mexico in 1994.	http://www.fws.gov/birdhabitat/NAWMP/index.shtm
Landscape	America's Longleaf Restoration Initiative	A voluntary, collaborative effort of >20 organizations and agencies that seeks to define, catalyze, and support coordinated longleaf pine conservation efforts across a nine-state region.	http://www.americaslongleaf.org
Regional	State wildlife action plans	A proactive approach to habitat conservation and species preservation. These plans outline a strategy for protecting priority habitats and species that are at risk but not yet on the endangered species list in conjunction with state wildlife agencies from all 50 states, all U.S. territories, and the District of Columbia.	http://www.wildlifeactionplan.org
State	The Nature Conservancy's ecoregional assessments	Long-range plan for land acquisition and restoration based on natural ecoregions. Each ecoregional assessment identifies and prioritizes conservation lands based on ecology and threat levels.	http://www.nature.org
Local	Sonoran Desert Conservation Plan	In 1998, planners and ecologists came together to create a plan that protects critical environmental and cultural resources in Pima County, Arizona. The plan must be updated to remain current, and questions about habitat management on these newly protected lands remain.	http://www.pima.gov/cmo/sdcp/
Species specific	Endangered species recovery plans	A critical part of endangered species protection. They outline goals and actions to recover endangered species to a self-sustaining level for one or multiple species.	http://www.fws.gov/endangered/species/recovery-plans.html

serve species and ecosystems in a single watershed or landscape to efforts to reform regional or multinational policies.

Major Legislation and Policy Directing Wildlife Planning

We have discussed what planning is and have provided examples of several entities typically responsible for or involved in conservation planning. Another question to ask is, why plan? In addition to some of the aforementioned benefits of conservation planning, there are three basic frameworks that commonly drive conservation planning: agency directives or authorities (e.g., laws or policy that influence an organization's mission), environmental compliance (e.g., regulatory laws), and incentive programs (e.g., Farm Bill programs). We provide a brief description of each of these influences, but to understand why planning is necessary, it is important to realize that most conservation funding to support implementation of management activities typically requires a management plan that outlines the activities the funding will support. A summary of federal and state conservation funding programs illustrates the value and reason to develop management plans (Table 18.3). Familiarity with these programs is important for obtaining financial support for conservation or, if working with private landowners, encouraging certain behaviors that can benefit conservation of wildlife populations and their habitats. A review of these programs follows.

Directives and Agency Guidance

Federal and state agencies work under the authority of laws or policies in support of the organizations' missions. Nongovernmental organizations do not necessarily work under a given authority but may be influenced by some federal or state law (e.g., Endangered Species Act, or ESA) or may stand to benefit from authorities that provide financial or technical assistance (e.g., the Pittman-Robertson and Dingell-Johnson Acts; see Chapter 2). To receive Pittman-Robertson and Dingell-Johnson funds, a state must submit a comprehensive fish and wildlife resource management plan (e.g., state wildlife action plans). The plans must be for at least five years and must be based on long-range projections regarding the desires and needs of the public. In the expenditure of both Pittman-Robertson and Dingell-Johnson funding, management plans are critical components in state conservation agencies obtaining and supporting their wildlife programs through these funds. These federal acts direct agencies how to manage their natural resources.

Environmental Compliance

A second set of laws that influences conservation planning can be labeled under the environmental compliance umbrella. Two landmark pieces of legislation include the National Environmental Policy Act (NEPA) of 1969 and the ESA of 1973. The NEPA was passed during the environmental movement of the 1960s in response to several environmental concerns ranging from clean water and air, pesticide contamination, and declining wildlife species to protection of wilderness areas. The act declares a national policy to "encourage pro-

ductive and enjoyable harmony between man and his environment; to promote efforts which will prevent or eliminate damage to the environment and biosphere and stimulate the health and welfare of man; [and] to enrich the understanding of the ecological systems and natural resources important to the Nation" (42 U.S.C. § 4321). NEPA requires federal agencies to identify, analyze, describe (i.e., consider), and publically disclose the environmental impacts associated with federal actions through publication in the *Federal Register* and a series of public scoping meetings. The act also establishes the Council on Environmental Quality to review government policies and programs for conformity with NEPA. When any activities may have a significant impact to the environment, the development of an environmental impact statement (EIS) is required. An EIS includes an examination of environmental impacts and alternatives available to the proposed action. Prior to preparing an EIS, the agency should coordinate with other state and federal agencies having expertise on any environmental impact involved (e.g., USFWS and impacts to endangered species). Conservation planning conducted by federal agencies typically includes preparation of an EIS when developing management plans.

Another example of federal law that influences conservation planning is the ESA. The ESA provides for the conservation of plants and animals considered threatened with or in danger of extinction. The authority for implementing and executing the act is delegated to the USFWS for plants and animals or the National Marine Fisheries Service (NMFS) for marine life and anadromous fishes. However, the ESA declares that all federal departments and agencies must also seek to conserve endangered and threatened species and to use their authorities in furtherance of the purposes of the act (16 U.S.C. § 1531–1544; Sec. 2). The USFWS and NMFS are responsible for determining which species are listed as threatened or endangered, as well as for delineating critical habitats necessary for their conservation. The ESA comprises several sections that provide further guidelines to the aforementioned agencies. These sections offer direction for listing a species as threatened or endangered, designation of critical habitat, and recovery planning (section 4), prohibited actions for listed species (section 9), and penalties and enforcement procedures (section 11). Two sections of the ESA that are noteworthy with regard to conservation planning include sections 7 and 10. The USFWS has developed handbooks (USFWS and NMFS 1996, 1998) to guide applicants and planners through the section 7 and habitat conservation plan (HCP) processes.

Section 7 of the ESA outlines procedures for interagency cooperation to conserve federally listed species and designated critical habitats. Section 7 requires federal agencies to consult with USFWS to ensure that they are not undertaking, funding, permitting, or authorizing actions likely to jeopardize the continued existence of listed species or to destroy or adversely modify designated critical habitat. Section 7 outlines the processes for consultation to address potential effects to listed species and designated critical habitats, and conferences to address potential effects to species proposed for listing or proposed critical habitat. For projects that are likely to adversely affect a listed species or designated critical habitat, formal consultation is required to ensure the proposed federal actions will not jeopardize the continued existence of a species or destroy or modify habitat. In these cases, the action agency must develop and submit a biological assessment to USFWS or NMFS that outlines the proposed action and how it may affect listed species, including whether the action may result in the take of a species. Take includes the physical removal of an individual animal from the wild or killing thereof, removal of habitat so it can no longer be used by the species, and activities (e.g., recreation, machinery noise) that make an area unsuitable for habitation by the species. Take that is incidental to a proposed federal action may be permitted as long as the agency agrees to comply with the specific measures and conditions set forth by USFWS or NMFS when they issue their biological opinion.

Take can be permitted for scientific purposes and for non-federal projects (e.g., commercial or residential development on private lands) if the approved plans minimize and mitigate for the incidental take (usually of habitat) of the species. Under section 10 of the ESA, for example, incidental take is authorized through a variety of voluntary agreements to conserve or minimize and mitigate impacts upon fish and wildlife, including: candidate conservation agreements, safe harbor agreements, and HCPs with implementation agreements. A brief description of each of these types of agreements and how they influence or are a part of conservation planning is provided next.

Candidate Conservation Agreements with Assurances

The USFWS (and, to a lesser degree, NMFS) offers nonfederal landowners a policy option of entering into prelisting or candidate conservation agreements with assurances (CCAAs) that provide regulatory assurances to landowners who voluntarily agree to protect habitat for candidate fish and wildlife species before they are listed for protection under the ESA. Under the USFWS policy for CCAAs, successful applicants will receive an enhancement of survival permit if they agree to actions that will provide a conservation benefit to specified candidate species, so listing the species would be unnecessary if other landowners within the range of the species were to manage their land in the same fashion to remove known threats to the species. If species covered under CCAAs are eventually listed for protection, the enhancement of survival permit authorizes incidental take of those species by any action in accordance with the CCAA.

Safe Harbor Agreements

Safe harbor agreements are voluntary arrangements between the USFWS or NMFS and cooperating nonfederal landowners. Their purpose is to promote voluntary management for threatened and endangered species on nonfederal property while offering assurances to landowners regarding future regulatory restrictions in the management of these species on their land. The value of safe harbor agreements is that they allow landowners to provide a net conservation benefit and

contribute to recovery of a listed species in exchange for the assurance that a return to the baseline habitat conditions at the time of permit inception will not result in liability for unlawful take. For example, if a landowner has five breeding pairs of an endangered species on their property, she may sign a safe harbor agreement. If that landowner were to improve habitat to increase the population to ten breeding pairs, she would not be penalized if, at the end of the permit duration, this number was reduced to the baseline (i.e., five breeding pairs) because of an action (e.g., development of species habitat). As in CCAAs, parties to approved safe harbor agreements will receive an enhancement of survival permit that authorizes incidental take by actions consistent with the terms of the agreement.

Habitat Conservation Plans

The final kind of conservation plan under the ESA is an HCP. Section 10(a)(2)(B) of the ESA allows incidental take of listed species resulting from nonfederal actions through the development of an HCP, a planning document required as part the incidental take permit application and that can cover a single action or a number of similar activities that will occur over a broad area (e.g., expanding commercial and residential development zones or transmission line rights-of-way that cross multiple jurisdictions). HCPs describe the anticipated effects of the proposed taking, how those impacts will be minimized and mitigated, and how the plan is to be funded, as explained in section 10(a)(2)(A) of the ESA. Frequently, these plans require habitat protection, restoration, and enhancement in an area in exchange for some lost habitat in another, if it is determined that these actions are required to offset the impacts of the taking to the species. In addition to receiving an incidental take permit, an HCP applicant is also provided with long-term regulatory assurances that, if unforeseen circumstances arise, the USFWS will not require additional mitigation from the permittee beyond the level otherwise agreed to in the plan. Programmatic HCPs are developed when a number of similar activities will occur over a broad area, such as when city and county governments want to expand their zone of commercial and residential development, or when a transmission line (and associated right-of-way corridor) will cross numerous governmental jurisdictions. HCPs can be prepared for single species or multiple species and may include unlisted species. In this case, the nonlisted species is treated as if it were listed, including the required effects analysis and mitigation. In this way, the permittee can be assured that subsequent listing would not result in additional administrative or conservation requirements. Single-species HCPs are usually easier to develop and implement because they tend to be less complicated, but multispecies HCPs must be developed for areas containing one or more listed or candidate species, and for long-term projects such as phased developments where there is potential for additional listings in the future. Failure to address all listed species that are likely to be incidentally taken by the proposed activities would preclude the USFWS or NMFS from being lawfully able to issue the permit on the

Table 18.3. Summary of federal and state conservation funding programs

Acronym	Program	Agency[a]	Description
REPI	Readiness and Environmental Protection Initiative	DOD	Cost-sharing program for the acquisition of easements from willing sellers as a way to preserve high-value habitat and to limit incompatible development around military ranges and installations.
LWCF	Land and Water Conservation Fund Stateside Assistance Program	NPS	State matching grant program (at least 50%) where states request funds from the NPS for specific projects.
LWCF	LWCF federal land acquisition	NPS	Program to acquire new federal recreation lands in cooperation with the NPS.
UPARR	Urban Park and Recreation Recovery	NPS	Grants to urban communities for rehabilitation of facilities, parks planning, and innovative programs. Fiscal year 2002 was the last year that grants were awarded.
TAP	Transportation Alternatives Program	DOT	The TAP represents 10% of total Surface Transportation Program (STP) funds. STP is a formula apportionment (maximum 80%) to the states through the federal aid highway program. TAP projects include construction, but not maintenance, of various modes of transportation, including the rail trail program, which is funded by an excise tax from the Highway Trust Fund. Projects involve conversion of abandoned railroad corridors into multiuse trails available for recreation.
RTP	Recreational Trails Program	DOT	Apportionments (maximum 80%) to the states to benefit outdoor recreation, including hiking, biking, in-line skating, and equestrian use.
FLP	Forest Legacy Program	USFS	Competitive grant program and direct payments to support land acquisition (fee purchase and easement) to protect important scenic, cultural, fish, wildlife, and recreation resources as well as riparian areas.
FSP	Forest Stewardship Program	USFS	Provides technical and educational assistance to nonindustrial private forest owners to develop forest management plans.
FEP	Forestland Enhancement Program	USFS	Complements the FLP with additional technical, educational, and cost-share assistance to nonindustrial private forest owners to implement management plans. Only $40 million of the original $100 million budgeted was actually distributed, all by fiscal year 2006. The program was not renewed in the 2008 Farm Bill.
UCF	Urban and Community Forestry	USFS	Forest-related technical, financial, research, and educational services to local government, nonprofits, community groups, and educational institutions.
CFOSCP	Community Forest Open Space Conservation Program	USFS	Matching grants (50/50) for local governments, tribes, and nonprofit organizations for full-fee purchase of forestlands; differs from the FLP in its community focus and the requirement of fee purchase plus public access. The program is part of the 2008 Farm Bill; no appropriations yet.
CESCF	Cooperative Endangered Species Conservation Fund	USFWS	Grants to private landowners and groups to implement conservation projects for listed species and at-risk species. Funded activities include developing habitat conservation plans, land acquisition, habitat restoration, research, and wildlife management.
MBCF	Migratory Bird Conservation Fund	USFWS	Land and water acquisition or rental as recommended by the U.S. secretary of interior for the protection of migratory bird species. Funding comes from Federal Duck Stamp revenues, import duties on arms and ammunition, and refuge admission fees.
NCWCA	National Coastal Wetlands Conservation Grants	USFWS	Matching grants to states for acquisition, restoration, and enhancement of coastal wetlands. Funding comes from the Sport Fish Restoration and Boating Trust Fund, which is supported by excise taxes on fishing equipment, motorboat and small engine fuels, and import duties.
NAWCA	North American Wetland Conservation Act grants	USFWS	Matching grants to organizations and individuals to implement wetlands conservation projects in the United States, Canada, and Mexico. Funding comes from congressional appropriations as well as fines and penalties collected under the Migratory Bird Treaty Act of 1918, the Sport Fish Restoration and Boating Trust Fund, and interest accrued on the Wildlife Restoration Trust Fund.
SFR	Sport Fish Restoration Program	USFWS	Apportionments to the states for fishery projects, boating access, and aquatic education. This program is funded by the Sport Fish Restoration and Boating Trust Fund, which is supported by excise taxes on fishing equipment, motorboat and small engine fuels, and by import duties. The annual allocation to the Sport Fish Restoration Program from legislation, called the Dingell-Johnson Act or the Wallop-Breau Act, is equal to 57% of the trust fund's receipts (after annual deductions).
WRP	Wildlife Restoration Program	USFWS	Apportionments to the states to restore, conserve, and manage wild birds and mammals and their habitat. This program is funded by the Wildlife Restoration Trust Fund, which is supported by excise taxes, authorized by the Pittman-Robertson Act, on hunting equipment.
SWGP	State Wildlife Grant Program	USFWS	Grants to plan and implement programs that benefit wildlife and habitats, including species not hunted or fished. Funding comes through appropriations from the Land and Water Conservation Fund.

Table 18.3 continued

Acronym	Program	Agency[a]	Description
LIP	Landowner Incentive Program	USFWS	State grants to protect and restore habitats on private lands to benefit at-risk species (including federally listed, proposed, or candidate species). The program was discontinued after fiscal year 2007.
NFHAP	National Fish Habitat Action Plan	USFWS and others	Currently includes restoration activities only but, upon congressional approval of National Fish Habitat Conservation Act, it will include land acquisition.
NERRS	National Estuarine Research Reserves System	NOAA	Grants to coastal states to acquire lands and waters necessary to ensure long-term management of an area as a national estuarine reserve and for operations, construction, and education programs.
CELCP	Coastal and Estuarine Land Conservation Program	NOAA	Competitive state and local grants to acquire property or conservation easements from willing sellers within a state's coastal zone or coastal watershed boundary.
CRP	Conservation Reserve Program	NRCS	Pays farmers to take environmentally sensitive cropland out of production and to plant long-term resource-conserving covers (e.g., grasses and trees). The CRP offers ten- to 15-year contracts with annual rental payments, incentive payments for certain activities, and cost-share assistance to establish approved ground cover on eligible cropland in order to reduce erosion on sensitive lands, to improve soil and water, and to provide significant wildlife habitat.
GRP	Grassland Reserve Program	NRCS	Long-term rental agreements or easements on private lands to restore and protect grassland while maintaining areas for livestock grazing and hay production. The program was initially authorized at $254 million for fiscal year 2003 through fiscal year 2007; $30 million was reappropriated for the Wetlands Reserve Program and the CRP in 2003.
EQIP	Environmental Quality Incentives Program	NRCS	Technical assistance, cost-share payments, and incentive payments to assist crop and livestock producers with environmental and conservation improvements. The EQIP, established by the 1996 Farm Bill, offers financial, educational, and technical help to install or implement structural, vegetative, and management practices and is one of the several voluntary conservation programs which are part of the USDA "Conservation Toolbox."
FPP	Farmland Protection Program	NRCS	State and local grants to help purchase easements that would preclude nonfarm development of productive farmland.
WRP	Wetlands Reserve Program	NRCS	Acquires long-term or permanent easements and provides cost sharing to producers who agree to restore wetlands on agricultural land. The WRP is designed to restore and protect wetlands on private property in place of marginal agricultural land. The WRP also provides fish and wildlife habitat, improves water quality, protects biological diversity, and provides recreational opportunities.
WHIP	Wildlife Habitat Incentives Program	NRCS	Technical assistance and cost sharing in consultation with the local conservation district for development and improvement of wildlife habitat. WHIP is for landowners who want to develop and improve wildlife habitat on private lands.
CTA	Conservation Technical Assistance	NRCS	Provides technical assistance to help private landowners conserve, maintain, and improve their natural resources, including direct conservation planning, design, and implementation assistance. NRCS provides assistance to land users for developing and implementing conservation plans on their lands, as well as animal waste and nutrient management plans, and works with producers during planning to consider the overall impact on the land and the other plants, animals, and wildlife who live there, as well as economic factors and the sustained use and productivity of the resources.
HFRP	Healthy Forests Reserve Program	NRCS	Restores private forest ecosystems to: (1) promote the recovery of threatened species, (2) improve biodiversity, and (3) enhance carbon sequestration.
CSP	Conservation Security Program	NRCS	Financial and technical assistance to promote the conservation of soil, water, air, energy, plant, and animal life on private working lands. Replaced by the Conservation Stewardship Program in the 2008 Farm Bill.

[a]DOD, Department of Defense; NPS, National Park Service; DOT, Department of Transportation; USFS, U.S. Forest Service; USFWS, U.S. Fish and Wildlife Service; NOAA, National Oceanic and Atmospheric Administration; NRCS, Natural Resources Conservation Service.

grounds that the activities would be in violation of the ESA (e.g., unauthorized take).

Incentive Programs

The third set of laws that influences conservation planning is incentive programs. Conservation planning on private lands often involves incentivizing private landowners to manage their natural resources to their benefit and the benefit of the public at large. Incentive programs allow federal and state agencies to encourage sustainable use of these ecosystem services by providing financial and technical assistance to landowners.

As mentioned previously, the Partners for Fish and Wildlife Act (PL 109-294) states that roughly 60% of fish and wildlife occur on private lands. The mission of the Partners for Fish and Wildlife Program, which is administered by USFWS, is "to efficiently achieve voluntary habitat restoration on private lands, through financial and technical assistance, for the benefit of Federal Trust Species" (http://www.fws.gov/partners). The Partners Program (as it has come to be called) is active in all 50 states and U.S. territories. It provides technical assistance and delivers on-the-ground restoration projects, particularly to the nation's private landowners, farmers, ranchers, and corporations. In 2007, the USFWS finalized a five-year strategic plan for the Partners Program consisting of three parts: a vision document (USFWS 2010) that describes the program and its five major goals, regional strategic plans that highlight conservation priorities in each of eight geographic regions, and a national summary document that reflects a national overview of habitat priorities and targets based on the regional strategic plans. It is important to recognize that even though the Partners Program has its own mission, the program still operates under the overarching mission of the USFWS.

Several state agencies also have incentive programs to assist their landowners with conservation on private lands. Many of these programs are designed to meet specific habitat needs for wildlife species. For example, as part of their wetland and grassland habitat program, wildlifers in South Dakota work with landowners and other conservation partners (e.g., nonprofit organizations, USFWS, Natural Resources Conservation Service, or NRCS) to implement wetland and grassland conservation practices that will benefit breeding waterfowl and other wildlife species dependent upon these landscapes, as well as to meet the needs and management goals of landowners. Up to 100% cost share is provided to the landowner for wetland restoration projects involving removal of drainage tile or plugging drainage ditches. Other states such as Texas have state wildlife tax reductions that have been initiated by landowners to reduce taxes on their lands through the development of a formal wildlife plan. Furthermore, adjacent landowners can come together and form a "wildlife management property association" with one wildlife management plan for the group, but every landowner is required to sign it. Similarly, landowners in Florida can apply for a conservation exemption in which land dedicated in perpetuity and used exclusively for conservation purposes entitles landowners to a full property tax exemption. We encourage readers to become knowledgeable about the landowner programs available in their own state so they can make the most effective use of all the resources available to them.

A landmark piece of legislation supporting conservation on private lands is the Food, Conservation, and Energy Act of 2008, also known as the 2008 Farm Bill. It is a $288 billion, five-year agricultural policy bill. The 2008 Farm Bill is a continuation of the 2002 Farm Bill and follows a long history of agricultural subsidy in energy, conservation, nutrition, and rural development. One of the three major components of the Farm Bill is providing baseline funding for conservation and working lands programs ($4 billion in 2008). The majority of these programs are managed by the NRCS. Originally called the Soil Conservation Service, the NRCS was established by Congress in 1935 with the mission of conserving natural resources on private lands. The NRCS has since expanded to provide landowners with technical and financial assistance to improve soil, water, air, plants, and animals that result in productive lands and healthy ecosystems. The more common programs outlined below provide a brief description of each program and how funding and technical assistance are delivered (Fig. 18.3).

Conservation Reserve Program

The Conservation Reserve Program (CRP) pays farmers annual rental payments under ten- to 15-year contracts to set aside previously cropped land that is considered to be marginally productive or highly erodible. In return for establishing and maintaining conservation practices that address soil erosion, water quality, wetland and forest enhancement, and wildlife management, landowners receive annual rental payments, cost-share assistance (not to exceed 50% of the eligible costs), and under certain conditions incentives for enrolling land, undertaking particular practices, and performing certain maintenance practices. Land must meet eligibility requirements to qualify for the CRP. The CRP contains four programs: the general sign-up CRP, continuous CRP, Conservation Reserve Enhancement Program, and the Farmable Wetlands Program. The CRP and continuous CRP practices include establishing vegetation cover or trees on erodible cropland, planting native grasses, or placing buffer strips along stream banks to reduce pollution. According to the Farm Service Agency, over 12.5 million ha were enrolled in the CRP program nationwide in 2010.

Grassland Reserve Program

The Grassland Reserve Program is a voluntary program that enables landowners to restore or protect native grasslands on portions of their property through long-term or permanent easements. Maintaining and restoring native grasslands provides important wildlife habitat in addition to hunting and other recreational land uses. In 2010, approximately 1.2 million ha were enrolled in the Grassland Reserve Program nationwide.

Conservation Programs at a Glance

Program Name	Program Purpose	Contract Length	Eligibility (Applicant)	Eligibility (Land)	Sign-up Period	Evaluation of Applications	Cost Share Rate	Funding Limit	Special Notes
CRP Continuous	To protect soil, water, wildlife and to improve water quality	10 to 15 years	Individuals and/or groups who own eligible land for at least one year	Cropland (cropped 4 out of 6 years from 1996 to 2001) or marginal pastureland	Continuous, throughout the year	Noncompetitive. Based on eligibility	50% of predetermined average cost, plus additional practice installation incentive and sign-up bonus.	$50,000 per person per fiscal year	1, 2, 4, 6
CRP General	To apply high priority conservation practices that benefit large areas	10 to 15 years	Individuals and/or groups who have owned land for at least 1 year	Cropland (cropped 4 out of 6 years from 1996 to 2001) re-enrollments & extensions	Announced by Secretary of Agriculture	Competitive, environmental benefits/costs	50% of predetermined average cost	$50,000 per person per fiscal year	1, 2, 4
CREP	To improve water quality and wildlife habitat in selected watersheds in North Carolina	10 to 15 years, 30 years, or permanent easements by the State	Individuals and/or groups who own eligible land in Neuse, Tar-Pam, Chowan, or Jordan Lake watersheds, owned land for at least 1 year	Cropland (cropped 4 out of 6 years from 1996 to 2001) or marginal pasture grazed 2 years between 1996 and 2001.	Continuous, throughout the year	Priority based on environmental benefits	Up to 100% of predetermined average cost. 50% USDA & 50% State. Easement payments available.	$50,000 per person per fiscal year	1, 2, 5
CSP	To reward producers who undertake and maintain high levels of conservation stewardship	5 to 10 years	All who meet the quality criteria	Cropland, grazing lands and small incidental forest areas	Announced by Secretary of Agriculture	Based on meeting of program quality criteria	Calculated based on combination of: annual base payment, existing practice payment, one-time new practice payment, and enhancement component	Tier I - $20,000 Tier II- $35,000 Tier III- $45,000	2
EQIP	To address significant natural resource concerns on agricultural land	1 year after last practice is installed	Livestock or agricultural producers and landowners	All agricultural land uses	Continuous, with periodic designated ranking periods	Competitive, based on resource concerns, cost, & environmental benefits	Varies, from 50 to 75% of average installation cost. Up to 90% for limited resource and beginning farmers	$450,000 total per individual or entity from fiscal year 2002 to 2007	2
GRP	To restore and conserve grassland for agriculture and wildlife	Permanent 30 Year 20 Year 15 Year 10 Year	Private landowners with at least 10 acres of grassland	Permanent grassland used for grazing, hay or wildlife land	Continuous with periodic designated ranking periods	Competitive, environmental benefits and risk of conversion	100% land value for permanent easement; 30% for 30 year easement; or rental payments	Lump sum payment available. Value determined by appraisal	4
WHIP	Improving habitat for at-risk landscape species at landscape scale.	5-10 years	Individuals, clubs, companies, partnerships, NGO's, tribes, state and local governments	All land where important wildlife habitat can be provided.	Continuous with periodic designated ranking periods	Competitive, based on resource concerns, cost, & environmental benefits	75% of predetermined average cost	$100,000 per contract.	1, 2
WRP	Restore and protect wetlands and riparian areas for wildlife and improve water quality	Permanent 30 year 10 year	Individuals and/or groups who have owned land for at least 1 year	Prior converted wetland (PC), farmed wetland (FW), riparian areas	Continuous	Competitive	Up to 100% of restoration cost + 100% of appraised land value for permanent easement. Rates for 30 yr. and 10 yr. are lower	Lump sum payment available. $5,100 per acre cap for easement. Restoration cost determined by restoration plan components.	2
NC Ag Cost Share Program	Reduce agriculture nonpoint pollution and improve the water quality in NC	Typically 3 year contracts, some annual contracts	Landowner or tenant of an agricultural operation existing for more than 3 years	Cropland or pastureland	Continuous, throughout the year	Based on county priorities	75% of predetermined average cost. Up to 90% for Limited Resource & Beginning Farmers & participants in Enhanced Voluntary Ag. Districts.	$75,000 per individual per year. Some practices and SWCD's have specific caps.	3

Figure 18.3. U.S. Department of Agriculture Natural Resource Conservation Service 2011 conservation programs. Though North Carolina (NC) is highlighted, programs are similar across other states. Note that all plans require the development of a conservation plan. For further specifics regarding plan requirements and eligible practices, visit http://www.nrcs.usda.gov/. CRP = Conservation Reserve Program, CREP = Conservation Reserve Enhancement Project, CSP = Conservation Stewardship Program, EQIP = Environmental Quality Incentives Program, GRP = Grassland Reserve Program, WHIP = Wildlife Habitat Incentives Program, WRP = Wetland Reserve Program.

All Programs Require a Conservation Plan. See Your Local NRCS and SWCD Office for Specifics Regarding Plan Requirements and/or Eligible Practices, or visit http://www.nc.nrcs.usda.gov/programs

1) Contract length depends on practices selected, and option of applicant.

2) Maximum limits for each of the different programs are independent of each other. Federal programs can not be used to fund practices on the same acreage.

3) Federal funds can be used to supplement state cost-share, but total can not exceed 75% of actual cost of practice(s) implemented

4) Annual rental payments based on soils rental rates.

5) Annual rental payment based on soils rental rates plus incentive bonus for easements.

6) One time payment applied to certain practices.

Wetland Reserve Program

The Wetland Reserve Program is a voluntary program that provides farmers the opportunity to restore, maintain, and protect wetlands on their property. Farmers can sign up for a ten-year cost-share agreement, or 30-year or permanent conservation easements, with restoration cost-share funding. Most lands restored under the Wetland Reserve Program are marginal, high-risk, flood-prone lands. The Wetland Reserve Program enables landowners to take these lands out of production and restore them for wildlife populations. In 2010, approximately 809,000 ha were enrolled in the Wetland Reserve Program nationwide.

Wildlife Habitat Incentive Program

The Wildlife Habitat Incentive Program is a voluntary program that pays up to 75% of the cost to private landowners for enhancing wildlife habitat on their land. The program is not limited to agricultural lands but is open to any private landowners who would like to enhance wildlife habitat on a portion of their land. The NRCS works with the participant to develop a wildlife habitat development plan for their land to promote the restoration of native prairie grasses, to perform forest management practices, or to protect, restore, or improve aquatic areas for wildlife. In 2010, approximately $51 million was allocated to the states for Wildlife Habitat Incentive Program projects on about 405,000 ha of land.

Conservation Stewardship Program

The Conservation Stewardship Program differs from the programs discussed above by rewarding farmers and ranchers for undertaking additional conservation activities and improving and maintaining existing conservation systems. Through five-year contracts, the program offers payments to producers who maintain a high level of conservation on their land and who agree to adopt higher levels of stewardship. It provides two possible types of payments. An annual payment is available for installing new conservation activities (e.g., no-till farming) and maintaining existing practices, and a supplemental payment is available to participants who also adopt a resource-conserving crop rotation (e.g., planting soybeans every third year to add nitrogen back to the soil). Eligible lands include cropland, grassland, prairie land, improved pastureland, rangeland, nonindustrial private forestland, and agricultural land under the jurisdiction of an Indian tribe. The NRCS makes the Conservation Stewardship Program available on a nationwide basis through a continuous sign-up process. There was roughly 10.1 million ha of land actively enrolled in Conservation Stewardship Program contracts in fiscal year 2010.

Environmental Quality Incentives Program

Similar to the Conservation Stewardship Program, the Environmental Quality Incentives Program provides technical assistance, incentive payments, and cost sharing to farmers and ranchers to implement conservation practices on their lands. Contracts can be up to ten years in duration, and allowable practices are based on a set of national priorities that are adapted to each state. These priorities range from reduction of point and nonpoint source pollution to watersheds and groundwater to the improvement of wildlife habitat for at-risk species.

MAJOR TYPES OF PLANNING

In this section, we review examples of the types of conservation planning that are conducted by state and federal natural resource agencies (e.g., organizational background, the basic processes followed in the development of that organization's management plan, and funding to support management plan activities). Although the process for conservation planning is specific to each agency, there are some general approaches or aspects that each has in common (Tables 18.4 and 18.5). Natural resource students will likely work for one of these agencies in their careers; understanding how agencies operate is fundamental to professional success.

State Wildlife and Forest Action Plans
State Wildlife Action Plans

U.S. laws and policies place the primary responsibility for wildlife management with the appropriate wildlife or natural resource agency in each state (Freyfogle and Goble 2009). State fish and wildlife agencies have a long history of success in conserving game species through the support of hunter and angler license fees and federal excise taxes (e.g., the Pittman-Robertson and Dingell-Johnson Acts). How are species that are not hunted or fished (nearly 90%) managed on state and private lands? There is a gap in wildlife conservation funding and the management of nongame species. There are

Table 18.4. Overview of the formal planning process for state agencies

Step	State wildlife action plans	Statewide forest resource assessments
1	Distribution / abundance of wildlife (particularly low / declining populations)	Description of the priority landscape areas and issues
2	Habitats and community descriptions essential to species conservation	Glossary of terms and acronyms
3	Review of problems and factors adversely affecting species	Investing resources
4	Proposed conservation actions and priorities for identified species and their habitats	List of other plans consulted
5	Monitoring plans for species and their habitats	Listing and description of stakeholder involvement
6	Procedures for plan reassessment and evaluation	Monitoring and reporting
7	Coordination plans with other agencies	Protocol for translating strategies into actions
8	Public participation plans	Strategies to address the priority landscape areas and issues

Table 18.5. Overview of the formal planning process for federal agencies[a]

Step	BLM RMP	NPS GMP	DOD INRMP	USFWS CCP	USACE land-use plans	USFS LRMP	NRCS conservation plan
1	Introduction	Identify relevant laws	Description of the installation, its history, and its current mission	Background (introduction, purpose, need)	Specification of the water and related land resource problems and opportunities (relevant to the planning setting) associated with the federal objective and specific state and local concerns	Introduction	Collection and analysis a. Identify problems and analysis b. Determine objectives c. Inventory resources d. Analyze resource data
2	Purpose statement	Identify issues and concerns (scoping)	Management goals and associated timeframes	Refuge overview (location and size, physical resources, biological resources, socioeconomic environment)	Inventory, forecast, and analysis of water and related land resource conditions within the planning area relevant to the identified problems and opportunities.	Desired future conditions, goals, and objectives (forest wide)	Decision support a. Formulate alternatives b. Evaluate alternatives c. Make decisions
3	Authority	Collect data	Projects to be implemented and estimated costs	Plan development (public involvement, planning process, review, and revision)	Formulation of alternative plans	Standards and guidelines	Application and evaluation a. Implement the plan b. Evaluate the plan
4	Organization and scope of an RMP document	Identify alternatives	Review of military mission and training requirements supported	Management direction (vision, goals, objectives, strategies)	Evaluation of the effects of the alternative plans		
5	Project history	Prepare draft plan	Legal requirements and biological needs	Plan implementation (proposed projects, funding and personnel, monitoring and adaptive management)	Comparison of alternative plans	Monitoring, evaluation, research, and implementation	
6	Location/setting	Revise and consult	Role of the installation's natural resources in the context of the surrounding ecosystem	Environmental assessment (NEPA)	Selection of a recommended plan based upon the comparison of alternative plans	Appendix and glossary	
7	Overview of public involvement efforts	Approve final plan	Input from the USFWS, state fish and wildlife agency, and the general public	Appendices			
8	Overview of consultation efforts	Implement the plan					
9	Management framework						
10	Planning process						
11	Opportunities and constraints						
12	Issues and issue categories						
13	Existing resource inventory						
14	Goals and objectives						
15	Desired future conditions						
16	Management action/direction						
17	Implementation procedures (monitoring, standards and guides, and plan revision or amendment)						

[a]BLM RMP, Bureau of Land Management Resource Management Plan; DOD INRMP, Department of Defense Integrated Natural Resources Management Plan; NEPA, National Environmental Policy Act; NPS GMP, National Parks Service General Management Plan; NRCS, Natural Resources Conservation Service; USACE, U.S. Army Corps of Engineers; USFS LRMP, U.S. Forest Service Land and Resource Management Plan; USFWS CCP, U.S. Fish and Wildlife Service Comprehensive Conservation Plan.

two funding programs used to address this gap in funding for the management of nongame and nonlisted species: the State Wildlife Grants Program and Landowner Incentive Program. The State Wildlife Grants Program (created by Congress in 2001) provides federal funding to states to support projects that prevent nongame wildlife from declining to the point of being endangered. Some examples include the restoration of degraded habitat, translocations of native wildlife populations, development of conservation partnerships with private landowners, and data collection on declining species or species of concern. Congress charged each state and territory with developing a statewide wildlife action plan to make the best use of program funding. These plans, technically referred to as comprehensive wildlife conservation strategies, assess the health of each state's wildlife and associated habitat, identify threats, and outline the actions needed to conserve them over the long term. Like the Pittman-Robertson funds, funds annually appropriated by Congress under the State Wildlife Grants Program are based on a predetermined formula that evaluates the state's size and population. State wildlife grants require a nonfederal match, which is also determined by Congress based on annual appropriations, to assure local ownership and leverage state and private funds to support conservation in each state. A specific format for state wildlife action plans, which comprise eight sections, is required for funding support.

A complement to the State Wildlife Grants Program is the Landowner Incentive Program, which was designed to benefit at-risk wildlife species and the habitats critical to their survival. The USFWS defines at-risk species as species of greatest conservation need (high priority) in a state's wildlife action plan. Landowner Incentive Program funds are allocated by Congress to USFWS for distribution to state fish and wildlife agencies. Funding for the Landowner Incentive Program is collected from revenues of the Outer Continental Shelf Oil and Gas Leasing Program royalties deposited into the Land and Water Conservation Fund Act of 1965. Funds can go toward development and administration of a dedicated private lands habitat program that provides professional, technical, and financial assistance to private landowners. The Landowner Incentive Program includes two funding tiers. Tier 1 is noncompetitive, and tier 2 is competitive nationally. Under tier 1, each state may receive funding for eligible projects up to $200,000 annually. When funding is available, the program will rank tier 2 grants and award grants through a national competition.

Statewide Forest Resource Assessment and Strategy

The 2008 Farm Bill resulted in the amendment of the Cooperative Forestry Assistance Act (CFAA) of 1973. The purpose of the act is to authorize the U.S. secretary of agriculture to assist in establishing a cooperative federal, state, and local forest stewardship program for management of nonfederal forestlands. Examples of funding programs include the Forestry Incentives Program, Urban and Community Forestry Assistance, Rural Fire Prevention and Control, and direct assistance to state forestry agencies. The State and Private Forestry Organization of the U.S. Department of Agriculture (USDA) Forest Service manages the majority of these programs. States are required to complete a statewide forest resource assessment and strategy plan to receive funds under CFAA. Like state wildlife action plans, statewide forest assessments provide an analysis of forest conditions and trends in the state and delineate priority rural and urban forest landscape areas. The statewide assessment outlines long-term plans for investing state, federal, and other resources in the most efficient manner. Each statewide forest resource assessment and strategy comprise eight sections that can be used to obtain funding support.

Federal Planning: An Overview

There are several federal agencies responsible for the management of natural resources directly or indirectly in the United States. For each of these agencies, conservation planning occurs as mandated by agency authorities or directives in meeting their mission. Though each agency uses a different term to describe its management plan, all federal agencies are subject to general requirements as dictated by federal laws (e.g., NEPA) that outline a general process in the development and implementation of an agency's management plan. In general, a federal agency begins its planning process by publishing a notice of intent in the *Federal Register* and in local newspapers to prepare or revise their version of the management plan. Such notices invite the public to identify issues and to submit comments to the agency for consideration during the planning process. This process of public review is called the "scoping" process. Based upon the information gathered (e.g., species data, public input), the agency then prepares a reasonable set of alternatives for managing the proposed public resources within the planning area. Management plan alternatives are designed to address issues identified by the agency during scoping and to comply with applicable laws and agency policy guidance. One alternative typically identified in the management plan is the no-action alternative, which maintains the current management direction. Approaches to the identification of the preferred management alternative may differ between federal agencies, but at some point a given set of alternatives, including the preferred option, are presented to the public as a draft management plan and draft EIS. Public meetings, mass mailings, and other informal discussions are used to solicit input into the prepared draft management plan and EIS. A public review period (e.g., 60–90 days) is usually conducted to receive final public input prior to accepting the final version of the management plan and issuance of the final EIS.

U.S. Fish and Wildlife Service Comprehensive Conservation Plan

The USFWS mission is to work with others to conserve, protect, and enhance fish, wildlife, plants, and their habitats for the continuing benefit of the American people. Major agency responsibilities include the management of migratory birds, endangered species, freshwater and anadromous fishes, and wetlands. One major aspect of the agency is the management of the National Wildlife Refuge System. Established in 1903 when President Theodore Roosevelt designated Pelican Island as the first national wildlife refuge (NWR), the system has grown to include more than 60.7 million ha comprising 553 national wildlife refuges and 38 wetland management districts. Each refuge requires a management plan to direct activities on USFWS-owned properties. The National Wildlife Refuge System Improvement Act of 1997 (also known as the Refuge Improvement Act) requires USFWS to prepare a comprehensive conservation plan for each NWR by the year 2012. The comprehensive conservation plan is a 15-year refuge management plan that describes the purposes of each refuge; the distribution, migration patterns, and abundance of fish, wildlife, plant populations, and related habitats within the planning unit; significant problems that may adversely affect those populations and the actions necessary to correct or mitigate such problems; the archaeological and cultural values of the planning unit; and opportunities for compatible wildlife-dependent recreational uses (PL 105-57; 9 October 1997). The development of a refuge comprehensive conservation plan generally takes about a year, which does not include the time required for completing the NEPA process. There are five basic steps to the comprehensive conservation plan process.

Natural Resources Conservation Service Conservation Plans

As mentioned previously, NRCS was established with the mission of conserving natural resources on private lands. Because 70% of land is privately owned in the United States (Fig. 18.2), providing support to private landowners is an important aspect of the NRCS mission. The NRCS field agents provide conservation planning expertise and manage incentive programs under the 2008 Farm Bill. Management plans by NRCS are outlined in the *National Biology Manual*, which contains policies and procedures for biological resource activities, and the *National Planning Procedures Handbook*, which provides specific policies and procedures to management plan development. The planning process used by NRCS is a three-phase, nine-step process (Fig. 18.4; Table 18.5). The planning process is dynamic and not necessarily conducted in a linear or chronological order.

U.S. Army Corps of Engineers Land-use Plans

The U.S. Army Corps of Engineers (USACE) is a federal agency and U.S. Army command comprising civilian and military personnel involved in a wide range of public works support to the nation and the Department of Defense (DOD) throughout the world. The USACE mission is to provide public engineering services and strengthen the nation's security, energize the economy, and reduce risks from disasters. Some activities of the USACE include planning, designing, building, and operating locks and dams, design and construction of flood protection systems, design and construction management of military facilities for the Army and Air Force, and environmental regulation and ecosystem restoration. The USACE is the leading provider of outdoor recreation (e.g., boating, fishing, hunting, camping, swimming) and hydropower capacity in the United

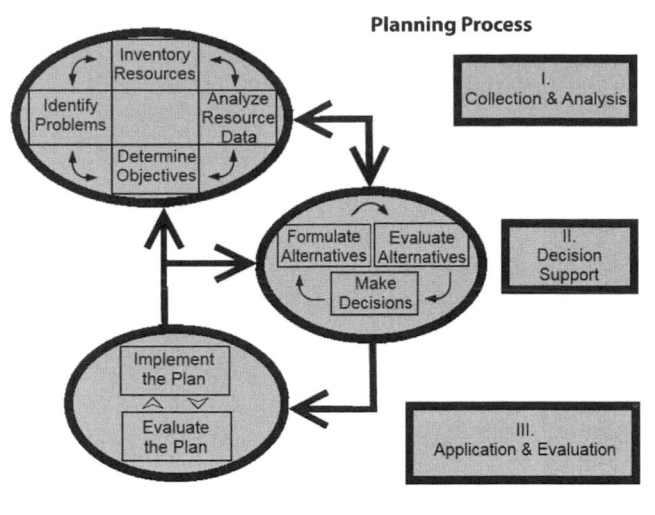

Figure 18.4. Example of the conservation planning process used by the Natural Resource Conservation Service.

States, managing 4.9 million ha of public lands and waters at more than 400 lake and river projects in 43 states (USACE 2011). Through its Civil Works program, the USACE carries out a wide array of projects that provide coastal protection, flood protection, hydropower, navigable waters and ports, recreational opportunities, and water supply. Water conservation planning for USACE initially evolved from the Flood Control Acts (1928, 1936) and Rivers and Harbors Acts (1925) to the more modern "308 reports," which are comprehensive studies of U.S. river basins. USACE (1983) articulates water conservation planning and involves a six-step planning process.

Bureau of Land Management Resource Management Plans

The U.S. Bureau of Land Management (BLM) was created in 1946 when the Grazing Service was merged with the General Land Office to form the agency. The agency's mission is to sustain the health, diversity, and productivity of the nation's public lands for the use and enjoyment of present and future generations (U.S. BLM 2011). The BLM manages over 99 million ha of public lands (most of these lands are in the West) and approximately 283 million ha of the subsurface minerals estate of public and private lands. When the BLM was initially created, there were more than 2,000 authorities for managing these public lands. As a result, Congress enacted the Federal Land Policy and Management Act of 1976 (FLPMA). In it, Congress recognized the value of the remaining public lands by declaring BLM lands would remain in public ownership, and they gave the agency the mandate of multiple use management. Under the FLPMA, the BLM also was required to develop comprehensive land-use plans called resource management plans. Resource management plans are developed for all BLM-owned lands to include resource areas, national monuments, and national conservation areas. Like other federal agencies, the BLM resource management plans offer opportunities for public input through the NEPA process. There are 17 components of a resource management plan.

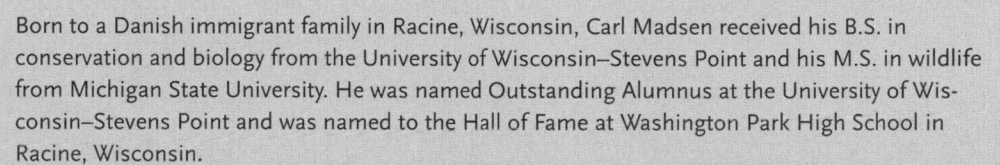

CARL MADSEN (b. 1937)

Born to a Danish immigrant family in Racine, Wisconsin, Carl Madsen received his B.S. in conservation and biology from the University of Wisconsin–Stevens Point and his M.S. in wildlife from Michigan State University. He was named Outstanding Alumnus at the University of Wisconsin–Stevens Point and was named to the Hall of Fame at Washington Park High School in Racine, Wisconsin.

He served as a wildlife biologist in Wisconsin, Minnesota, and the Dakotas for the U.S. Fish and Wildlife Service (USFWS) from 1967 to 2004. Madsen served in various positions in their migratory bird and wetland habitat protection programs, and was a leader for the Midcontinent Waterfowl Management Project, where he field-tested management techniques and helped develop agricultural programs to benefit wildlife. He served as the habitat coordinator for the North American Waterfowl Management Plan during its early years. He was a visionary and early pioneer of the USFWS's successful Partners for Fish and Wildlife Program, setting the standard for program accomplishments at the field level in South Dakota.

Madsen has received many awards for leadership in innovative developments in cooperative wildlife management, including both the Meritorious Service Award and the Distinguished Service Award, the highest recognition given by the U.S. Department of the Interior. He received national awards from Ducks Unlimited and the National Association of Conservation Districts as well as many state and local level awards, including the Professional Award from the Minnesota and South Dakota chapters of The Wildlife Society, where he served as president of both chapters.

He and his wife of 50 years, Aileen, raised three children and enjoy their family at their rural home in Brookings, South Dakota. He owns and operates a tree nursery, is active in wildlife and community issues, and enjoys gardening, hunting, and fishing.

Photo by Madsen Graphics & Photography

National Park Service General Management Planning

The National Park Service (NPS) is responsible for the management of national parks, many national monuments, and other conservation and historical properties. Established through the authority of the National Park Service Organic Act (1916), the NPS oversees 393 units (58 designated as national parks) comprising approximately 34 million ha. The agency's mission is to "conserve the scenery and the natural and historic objects and the wild life therein and to provide for the enjoyment of the same in such manner and by such means as will leave them unimpaired for the enjoyment of future generations" (16 U.S.C. § 1.1). The planning process for NPS stems from a base document called a statement for management that outlines procedures with the general management planning (GMP) document. NPS units rely on GMP to direct management and development for 10–15 years before re-evaluation. Each general management plan is a collection of eight action plans that focus on the following key resources: wilderness, wildlife, history, archeology, paleontology, geology, recreation, and access. The NPS prepares for GMP using an eight-step process.

U.S. Forest Service Land and Resource Management Plans

The U.S. Forest Service (USFS) is an agency of the USDA that oversees 155 national forests and 20 national grasslands that collectively encompass approximately 78.1 million ha (USFS

2011). Major divisions of the agency include the National Forest System, State and Private Forestry, and Research and Development branches. The Forest Reserve Act (1891) originally authorized withdrawing land from the public domain as forest reserves managed by the Department of Interior. Later, the Transfer Act (1905) transferred the management of forest reserves from the Department of Interior to the Bureau of Forestry, which later became the USFS under USDA.

The National Forest Management Act (1976, 1990) reorganized and expanded the management of national forestlands. This act requires the U.S. secretary of agriculture to assess forestlands and to develop a resource management plan based on multiple-use, sustained-yield principles for each USFS unit. Land and resource management plans, also known as forest plans, are the product of a comprehensive notice and comment process established under the National Forest Management Act. Land and resource management plans provide direction for all future decisions in the planning area. The secretary must revise and update the management plans at least once every 15 years. The general outline for land and resource management plans comprises six sections.

Department of Defense Integrated Natural Resources Management Plans

The DOD is responsible for coordinating and supervising all agencies and functions related to national security and the U.S. Armed Forces. The DOD comprises the Office of the Secretary of Defense, the Army, Navy, Air Force, and other support agencies. The DOD manages over 12 million ha of various ecosystems that represent the major land and climate types in which military personnel may be expected to fight wars. The primary function of DOD lands is to support the test and training mission for the U.S. Armed Forces; however, DOD also is responsible for conserving and protecting biological resources under the authority of the Sikes Act (1960) with assistance from USFWS. In 1997, the Sikes Act was amended to require military installations to develop and implement mutually agreed-upon integrated natural resource management plans through voluntary cooperative agreements between the DOD installation, the USFWS, and the respective state fish and wildlife agencies. Integrated natural resource management plans are planning documents that allow DOD installations to implement landscape-level management of their natural resources while coordinating with various stakeholders. The plans are reviewed and updated annually, and reapproved (to include the tripartite signatures of the DOD, the USFWS, and the appropriate state agency) every five years. The integrated natural resource management plan process also takes into account military mission requirements, installation master planning, environmental planning, and outdoor recreation. The basic elements of an integrated natural resource management plan include seven steps.

GENERAL CHARACTERISTICS OF PLANNING

From our review of formal plans for state and federal agencies (Tables 18.4 and 18.5), we exposed some common elements or themes to management plans. In practice, conservation planning is somewhat analogous to the scientific method, which lays out a systematic process for increasing our knowledge and understanding of the world around us. First, you observe a condition and form a hypothesis. You test your hypothesis in an experiment and compare the results to your hypothesis. You either confirm your hypothesis or repeat the process with a revised hypothesis.

Conservation planning is simply the scientific method dressed up, modified, and recycled! Take, for example, the formal conservation planning process used by TNC (Fig. 18.5). TNC's Conservation Action Planning process involves four steps: identify and define the problem, develop strategies and measures, implement or test those measures, and learn and improve from the results you observed. By redefining the problem in their first step with more specific goals and objectives, we propose modifying their four-step approach into five, more distinct steps we call the conservation planning process: state goals and objectives, identify and assign tasks, conduct the tasks, evaluate the results, and modify the plan as needed (Fig. 18.6).

Conservation Planning Process
The five steps of the conservation planning process frame the major components of any management plan, regardless of whom you may work for as a wildlife manager. By en-

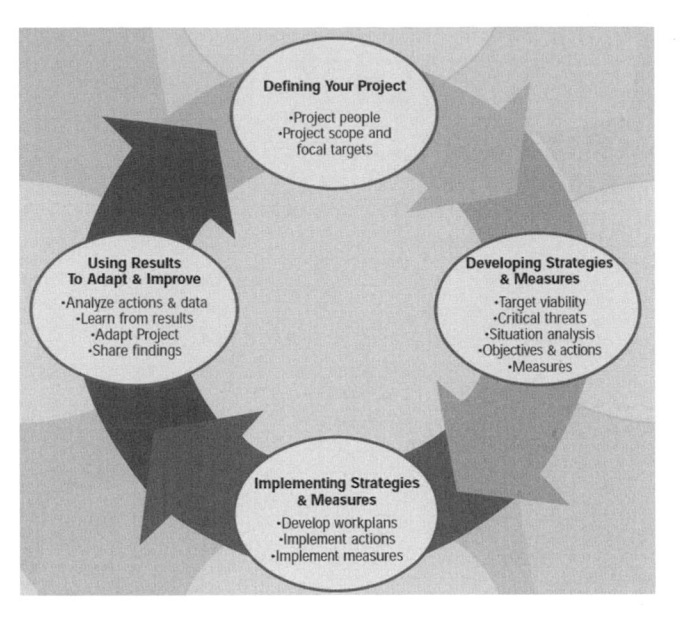

Figure 18.5. Example of the conservation planning process used by The Nature Conservancy.

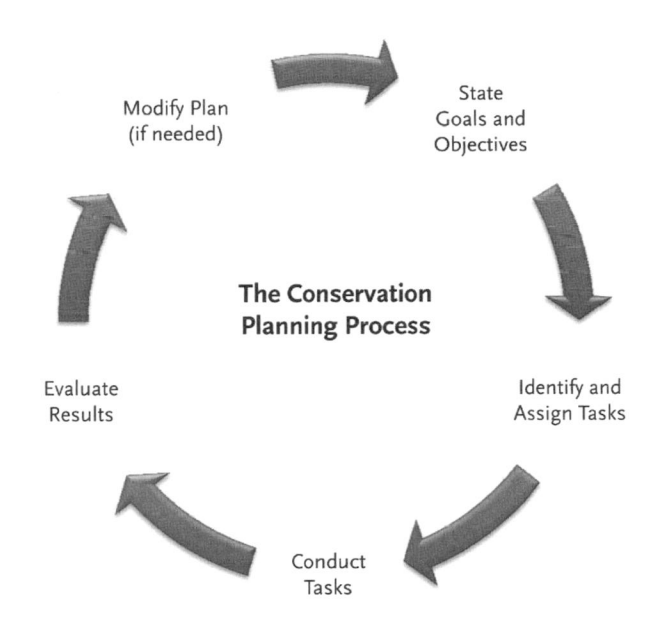

Figure 18.6. The five basic steps of the conservation planning process.

compassing these five steps into five distinct sections, we can generalize and define a generic approach in the development of a conservation plan: (1) the introduction and purpose section encompasses goals and objectives, (2) the recommendations section identifies and assigns tasks, (3) the implementation section discusses how tasks will be conducted, (4) the monitoring section discusses how the plan will be evaluated and modified based on its implementation, and (5) the supporting documents section provides supplemental materials (Table 18.6). These major sections describe most management plan formats you are likely to encounter throughout your career. We provide a brief description of each section below.

Table 18.6. Example outline used in developing a wildlife management plan for a landowner

Major sections	Plan elements	Description
Introduction and purpose	Purpose	Introduces the reader to the purpose of the plan. Answers the question, why are we doing this plan?
	Owner information	Provides information about property owner or client. Describes the purpose of the property from the landowner's perspective (e.g., recreational use, consumptive).
	Property description	Provides information about property location, including general location and vegetation cover maps, and legal description, area size, historic and current land uses, and other relevant biological descriptions (e.g., soils, topography, cover type, flora, fauna).
	Target species	Identifies species targeted for management and provides relevant life history information. Normally, target species are stated within the landowner's objective(s).
	Goals and objectives	Describes the wildlife management goals and objectives for property. A landowner *goal* may include increasing the number of blue birds. A set of *objectives* used to address this goal may include (1) identifying one or more viable management practices and (2) recommending practice(s) to the landowner that would best achieve their goals.
Recommendations	Feasibility assessment	Reviews species requirements in relation to environmental constraints that will serve to frame or "bound" management recommendations. This section describes feasible options for landowner consideration. A management constraints section also may be included to identify and discuss any limitations to proposed management activities.
	Alternative recommendations	Outlines proposed management alternatives or options based on the feasibility assessment (i.e., species requirements, constraints, etc.) to support recommendations. A "no management" option is a viable alternative.
	Proposed target areas	Identifies target areas where the proposed management activities are to occur. A map is often included to aid in identifying those areas.
Implementation	Final recommendations	Outlines final management recommendations to the landowner. Identifies how recommendation(s) will support and achieve management goal(s).
	Timeline	Outlines a schedule of events of all proposed activities or necessary steps needed to implement the plan. Typically includes a table or some other calendar that is organized by practice / activity (e.g., installation of nest box, maintenance, monitoring) and when that activity should be conducted.
	Budget	Provides a summary of costs for implementing the management plan. Typically includes a table with proposed number of acres treated, cost per treatment organized by management practice. "Matching" or "cost share" is often reflected here, as well.
Monitoring	Plan evaluation and modification	Outlines a monitoring protocol based on results that will be observed following implementation of the plan, which is essential in the feedback loop process. Monitoring is essential to determine success and clearly defines the metrics that will be monitored to realize success (e.g., successfully fledging quail).
Supporting documents	Appendices, glossary, and literature cited	Includes supporting information relevant to the wildlife management plan, including popular literature, information from websites, or scientific publications. Some examples include technical assistance sources and contacts, material / equipment sources, blueprints for construction of nest boxes / platforms, or census and monitoring forms.

Management Plan Overview

INTRODUCTION AND PURPOSE

The first section of a management plan introduces the reader to the purpose of the plan. It answers the question, why are we doing this plan? It also describes the current land uses and specific ecological site descriptions for the planning area in question. Ecological sites, defined through the establishment of the *Rangeland Interagency Ecological Site Manual* by the NRCS, USFS, and BLM, are distinctive lands with specific soil and physical characteristics that produce certain kinds and amounts of vegetation that are able to respond similarly to management actions and natural disturbances (Bestelmeyer and Brown 2010). Unlike vegetation classification, ecological site classification uses climate, soil, geomorphology, hydrology, and vegetation information to describe the ecological potential of land areas. Ecological site descriptions are reports that describe the biophysical properties of ecological sites, vegetation and surface soil properties of reference conditions (e.g., pre-European vegetation and historical range of variation or proper functioning condition or potential natural vegetation), state-and-transition model graphics and text, and ecosystem services provided by the ecological site and other interpretations. Plant-soil inventory data, long-term experimental or management studies, historical reconstructions, and local knowledge go into developing the ecological site descriptions.

In describing current land uses, use of remote sensing (e.g., aerial photography) or other geographic information system data can be used to physically describe the planning area to include general vegetation cover, area of management units, soil types, property boundaries, hydrology, topography, transportation access (e.g., roads, trails), and other relevant spatial features. This review of the planning area serves to identify any site constraints prior to developing plan alternatives. Maps in this section of the management plan often help tell the story and describe the current state of the area in question. The first part of the management plan also introduces to the reader descriptions of the property owners (i.e., federal agency, private landowner) as well as their goals and objectives. A brief to extensive review of life history requirements for target wildlife species sometimes is included as background information. This biological background is important, as it will later serve to justify management recommendations based on the biology of the target species. Lastly, the overall goal(s) of the plan, and the more specific objectives that will lead to attaining the goal(s), are specified. In short, the first section of our generic management plan outline describes the current situation for further consideration by the wildlife practitioner.

RECOMMENDATIONS

The second major section of our generic management plan describes the process in developing and identifying feasible alternatives for further development and consideration. This section is an important part of any management plan. The wildlife practitioner, through this process, is made keenly aware of their "signature on the face of the land" (Leopold 1949:68). Two primary factors likely frame the development of management plan alternatives and their feasibility: the biological needs of the species (reference to the life history requirements of the first section) and the physical or site constraints of the planning area (both abiotic and biotic). The development of sound alternatives will have a profound effect on the quality of the management plan's final decision (Lichfield et al. 1973). A good plan cannot be selected from a poor set of alternatives in conservation planning, which is why the recommendations section is so important. During the initial stages of developing reasonable alternatives, a laundry list of possible actions is developed. Alternatives come from people with varied backgrounds and experiences; however, not all alternatives are feasible. A second review of alternatives is typically done on the basis of known constraints and available resources. Feasibility criteria are any factors that may prevent the implementation of the plan (e.g., regulatory constraints, budget or labor constraints) and can be based on cost–benefit analyses, anticipated social or political support, or other predetermined criteria. In addition to the review of feasible alternatives, the recommendations section of the management plan also may identify target areas for proposed management activities.

IMPLEMENTATION

The implementation section of our generic management plan identifies final recommendations for the landowner or manager to consider and apply. In addition to the review of specific management recommendations and supporting management actions or prescriptions, the management plan outlines the proposed timeline of events or management plan milestones, costs or budget requirements for implementing the plan, and management guidelines (i.e., best management practices).

MONITORING

The monitoring section addresses what follows implementation of the plan, plan evaluation, and modification. This section is critical because it provides the details needed to successfully evaluate the management plan and whether desired goals and objectives are achieved. Unfortunately, monitoring is often forgotten in planning activities. Management plans are similar to roadmaps. You need to know where you are going, how you will get there (i.e., reach your goals) based on evaluation criteria that are measurable and related to the goals and objectives of the management plan, and whether you reached your destination. If you failed to reach your destination, you must re-create the roadmap (i.e., plan) using alternatives scenarios to ensure you get there.

SUPPORTING DOCUMENTS

The last section in our generic management plan consists of supporting materials or documents related to the main body of the plan. Though supporting documents can seem unimportant, they are necessary to the fulfill understanding and implementation of the management plan. Appendices, glossary of terms, and literature cited bookend your management plan. In more formal management plans for federal agencies, a copy of an EIS or biological opinion may be included in this section. In the case of a management plan for a private landowner, popular literature, how-to publications, equipment lists and suppliers, building plans (e.g., nest boxes), data forms, and other relevant planning documents such as application forms for incentive programs are typically included (Table 18.6).

SUMMARY

Conservation planning is an essential component of wildlife management, but it is a process that requires experience, analysis, intuition, and even inspiration. Conservation planning guides management actions in a way that ensures careful evaluation and execution of management plans, which are the basis for conservation planning. We reviewed some formal management plan outlines currently used by state and federal agencies, and the authorities by which these agencies implement conservation planning on private and public land. We also emphasized a few of these common elements and proposed a generic approach that can help develop a management plan. We hope that this process has provided a reasonable roadmap to determine what you have, what you want, and how you can get there during the conservation planning process so you can be "humbly aware that with each stroke [you are] writing [your] signature on the face of the land" (Leopold 1949:68).

Conservation planning begins with the notion that we are dissatisfied with the status quo (i.e., conservation planning is problem driven). A declining endangered population, for example, may provide the reason for plans or actions on the part of the wildlife manager. Because environments are always changing (naturally, human induced, both), one must plan ahead even if currently satisfied. In either case, addressing a management problem or keeping things the way they are into the future require some sort of image of a desired state (i.e., goals) and how to get there (i.e., objectives). The conservation planning process begins with a situational diagnosis framed by the clear articulation of goals and objectives. Based on our evaluation of goals and objectives, we then begin to formulate predictions of likely outcomes from the suite of alternatives identified. Next we conduct a feasibility analysis, where we determine what we ideally desire, any limiting factors that may inhibit implementation of the plan, and reasonable options for further consideration. We evaluate all possible options and, based on our findings that include reality constraints (e.g., budget) and consultation with stakeholders, we select and implement a given set of alternatives in our plan. We monitor and repeat (i.e., conservation planning is an iterative process), because we realize that plans need to be adjusted and revisited from time to time. In essence, this summarizes the conservation planning process.

Management plans are an integral component in managing wildlife habitat and their populations, and developing them is a skill that wildlife practitioners should obtain and refine in the practice and application of wildlife management. The basic elements common to all conservation planning efforts can be encompassed in five sections of a management plan: introduction and purpose, recommendations, implementation, monitoring, and supporting documents (Table 18.6). Even though planning can sometimes be unstructured, management plans add structure to the overall planning process and are an integral part of the conservation planning process.

Literature Cited

Bestelmeyer, B. T., and J. R. Brown. 2010. An introduction to the special issue on ecological sites. Rangelands 32:3–4.

Freyfogle, E. T., and D. D. Goble. 2009. Wildlife law: a primer. Island Press, Washington, D.C., USA.

Leopold, A. 1949. A Sand County almanac: and sketches here and there. Oxford University Press, New York, New York, USA.

Lichfield, N., P. Kettle, and M. Whitebread. 1973. Evaluation in the planning process. Oxford University Press, Oxford, United Kingdom.

Lopez, R. R., B. Hayes, M. W. Wagner, S. L. Locke, R. A. McCleery, and N. J. Silvy. 2005. Integrating land conservation planning in the classroom. Wildlife Society Bulletin. 34:223–228.

USACE. U.S. Army Corps of Engineers. 1983. Economic and environmental principles and guidelines for water and related land resources implementation studies. http://www.usace.army.mil/CECW/Documents/pgr/pg_1983.pdf. Accessed 23 December 2011.

USACE. U.S. Army Corps of Engineers. 2011. USACE website. http://www.usace.army.mil/Missions/Environmental.aspx. Accessed 20 November 2012.

U.S. BLM. U.S. Bureau of Land Management. 2011. Who we are, what we do. http://www.blm.gov/wo/st/en/info/About_BLM.html. Accessed 26 October 2012.

USFS. U.S. Forest Service. 2011. About us. http://www.fs.fed.us/about us/. Accessed 26 October 2012.

USFWS. U.S. Fish and Wildlife Service. 2010. Strategic plan. The Partners for Fish and Wildlife Program: stewardship of fish and wildlife through voluntary conservation. Washington, D.C., USA. http://www.fws.gov/partners/docs/783.pdf. Accessed 20 November 2012.

USFWS and NMFS. U.S. Fish and Wildlife Service and U.S. National Marine Fisheries Service. 1996. Habitat conservation planning and incidental take permit processing handbook. Washington, D.C., USA. http://www.nmfs.noaa.gov/pr/pdfs/laws/hcp_handbook.pdf. Accessed 26 October 2012.

USFWS and NMFS. U.S. Fish and Wildlife Service and U.S. National Marine Fisheries Service. 1998. Consultation handbook: procedures for conducting consultation and conference activities under section 7 of the Endangered Species Act. Washington, D.C., USA. http://www.fws.gov/endangered/esa-library/pdf/esa_section7_handbook.pdf. Accessed 26 October 2012.

Yoe, C. E., and K. D. Orth. 1996. Planning manual. Report 96-R-21. Institute for Water Resources, Alexandria, Virginia, USA.

19

MANAGING POPULATIONS

WILLIAM P. KUVLESKY JR., LEONARD A. BRENNAN,
BART M. BALLARD, TYLER A. CAMPBELL, DAVID G. HEWITT,
CHARLES A. DEYOUNG, SCOTT E. HENKE, FIDEL HERNANDEZ,
AND FRED C. BRYANT

INTRODUCTION

Managing wildlife populations is not a novel concept. The idea that free-ranging populations could be manipulated arose in the 1930s, when Leopold (1933) introduced the modern philosophy of wildlife management. Leopold noted that the concept of conserving wildlife emerged when humans were hunters and gathers. He and others suggested that social groups or tribes survived because they regulated their harvest of wild animals in a manner that conserved the resources on which their survival depended.

For early Native American cultures, there were no laws to regulate and conserve wildlife populations. As long 10,000 years ago, the Native Americans of the Sierra Nevada Mountains of California were sustained by the natural resources of the ecosystem they inhabited, so they intensively managed fish, game, vegetation, and building materials in a manner that had important ecological and evolutionary consequences (Anderson and Moratto 1996). Many of the Native American cultures that inhabited North America centuries prior to Euro-American exploration and colonization viewed themselves as part of the lands they inhabited (McHugh 1972, Jorgenson 1995, McCorquondale 1997). Because their survival depended on fish and wildlife, many cultures considered these species sacred, and even as brothers and sisters (Jorgensen 1995). For example, the Tlingit people of southeastern Alaska (Jorgensen 1995) and the Native Americans of the Great Plains developed and maintained a spiritual link to sea otters (*Enhydra lutris*) and bison (*Bison bison*), respectively, and recognized that to maintain sustainable populations of both species, careful management was necessary.

Leopold (1933) also indicated that the ancient Greeks and Romans had game laws, but with different motivations. The Greeks enacted hunting laws because they regarded hunting as good preparation for war, while the Romans instituted them to protect landowner rights, rather than to conserve wildlife. Perhaps the first attempts to manage wildlife for conservation occurred during the 8th century, when Charlemagne enacted elaborate game regulations that included bag limits and habitat preservation in the forests of his realm in Europe

(Caughley 1985). Another pioneer of wildlife population management was Genghis Khan, who decreed in the 11th century that Mongols of his empire could only hunt during the four months of winter to ensure population sustainability (Caughley 1985). The Mongol Empire evidently managed wildlife populations, because Marco Polo wrote that in the 14th century Kublai Khan decreed that hunting certain wildlife species was prohibited between March and October in an effort to ensure that these species increased and multiplied (Leopold 1933).

About a century after Marco Polo's observations about Kublai Khan, wildlife population management became better defined in feudal England, where wildlife was managed for the benefit of the aristocracy (Leopold 1933). During the 14th century, seasonal as well as age and sex restrictions were established for red deer (*Cervus elaphus*). Hunting was viewed as an elite sport, and hunting customs were not conservation measures; they were used as a political tool to exclude the commoner (Wolfe 1995) or at least to restrict what species, sex, or age class commoners could harvest (MacKenzie 1988). Leopold (1933) indicated that formal written laws were established later, during the 16th century and the reign of Henry VIII, who implemented seasonal waterfowl hunting restrictions. Following the lead of Henry VIII, King James I decreed during the late 16th and early 17th centuries that pheasants and partridges would receive seasonal protection. In addition to implementing hunting seasons and bag limits, supplementing wildlife populations via artificial propagation or restocking appeared in England during the reign of Henry VIII in the 16th century; mallard (*Anas platyrhynchos*) propagation was evident in 17th-century England (Maxwell 1913).

Efforts to conserve and maintain wildlife populations through instituting seasons and bag limits were not the only population management actions introduced in Europe during the early 16th century. Henry VIII placed bounties on several bird species determined to be pests during his reign, and Queen Elizabeth I retained these bounties and extended them to include additional birds and mammals in the middle to late

E. CHARLES MESLOW (b. 1937)

Dr. E. Charles "Chuck" Meslow was born in Waukegan, Illinois. His love of nature and wildlife started during the time he spent on a Wisconsin farm during his early teenage years. After spending three years in the Navy, he received B.S. and M.S. degrees in wildlife management at the University of Minnesota and then a Ph.D. in wildlife ecology in 1970 from the University of Wisconsin, where he studied snowshoe hare populations with Dr. L. B. Keith. Meslow began his academic career at North Dakota State University, but moved to Oregon State University, where he served as assistant unit leader (1971–1975) and then unit leader of the Oregon Cooperative Wildlife Research Unit. In addition, he held a position as professor of wildlife ecology until he retired in 1994. After retirement, he served as the northwest regional representative for the Wildlife Management Institute until 1999.

Meslow has made a significant contribution to the field of wildlife population management through his research on numerous wildlife species inhabiting the forests of the Pacific Northwest. Undoubtedly, his leadership during the contentious years of spotted owl conservation and management was his most important contribution to wildlife conservation. Spotted owls might not have received threatened status under the Endangered Species Act in the absence of Meslow's quiet professionalism, enthusiasm, and optimism. Because of his work on spotted owls, he became a key member of the Forest and Ecosystem Management Assessment Team, which wrote the Northwest Forest Plan that has resulted in the conservation of old forest in the Pacific Northwest. He has authored and coauthored over 80 peer-reviewed publications and has advised and mentored more than 50 M.S. and Ph.D. students. He has been involved in The Wildlife Society for almost 50 years, serving on numerous committees and as an officer at the state and national levels. The Wildlife Society has bestowed upon him numerous honors during his distinguished career, most notably the organization's most prestigious award, the Aldo Leopold Memorial Award, in 2005.

Meslow may be retired, but his influence on the field of wildlife population management and other professionals continues today. His advice and counsel continue to be sought not only by his former students and professional associates, many of whom are leaders in the wildlife profession, but also by executives in government and in the timber industry who respect his professionalism and honesty.

Photo courtesy of Bob Anthony

16th century (Leopold 1933). Furthermore, formal bounties to control wild birds and mammals deemed threats to humans or their livelihoods were apparently first implemented in Finland during the mid-17th century, and this organized persecution was considered an important component of game management (Pohja-Mykra et al. 2005). Caughley (1985) indicated that reducing predator populations in particular was a way of decreasing nonhunting mortality on game animals. These bounty systems were vigorously pursued and were so effective that many large predator populations were significantly reduced or extirpated. Wolves (Canis lupus) and brown bears (Ursus arctos arctos) were soon extirpated from England (Caughley 1985) and within a short time were either extirpated or reduced to small, remnant populations in a number of Western European countries.

With the exception of the cultural mores that Native Americans used to manage wildlife populations in North America, wildlife populations were exploited for subsistence or to secure safe environments when Europeans began to colonize the continent (Lund 1980). However, when population management was initiated via hunting, Euro-Americans followed the traditions established in Europe during the 17th century, except that recreational hunting was no longer restricted to

the privileged and could be enjoyed by anyone regardless of social standing (Lund 1980). Wildlife was basically available to anyone who wanted to hunt and, because game laws were virtually unknown in the United States until the second half of the 19th century (Caughley 1985), wildlife populations were decimated. For example, unregulated market hunting was a major factor in the extinctions of the Labrador duck (Camptorhynchus labradorius) and great auk (Pinguinus impennis; Mahoney 2009). The general public's view—that wildlife was an inexhaustible resource—continued to prevail as American settlers moved west. As European descendants transformed wilderness into farms and settlements, populations of many native bird and mammal species declined significantly, and populations of some species never recovered from unregulated hunting. For example, market hunting and egg collecting have been attributed to the extinction of the passenger pigeon (Ectopistes migratorius), which was the most abundant bird species on earth during the 18th century; extirpation of this species required only a few decades during the mid-1800s (Askins 2000). The unregulated trapping of beavers (Castor canadensis) and market hunting of American bison almost resulted in the extirpation of these two species during the middle to late 1800s (Mahoney 2009). Similarly, federal campaigns to eradi-

NOVA J. SILVY (b. 1941)

Dr. Nova J. Silvy was born on a dairy farm in Kansas, where he received an undergraduate degree in zoology at Kansas State University. He received graduate degrees at Kansas State University and Southern Illinois University, where he worked on greater prairie chickens and Florida Key deer, respectively. Silvy has spent almost his entire professional career as a professor of wildlife management in the Department of Wildlife and Fisheries Sciences at Texas A&M University. As a population ecologist, he has worked on more than 100 species over his 30-year career. Research conducted by Silvy and his many graduate students has provided greater insight into the population ecology and management of many wildlife species, including mourning and white-winged doves, wild turkeys, bobwhite quail, lesser prairie chickens, wood ducks, bobcats, and several small mammal species. Considered a world authority on the ecology of the endangered Attwater's prairie chicken and Florida Key deer, Silvy has authored or coauthored 240 refereed publications that have made significant contributions to wildlife population management.

Although a gifted researcher, Silvy considers himself first and foremost a teacher and mentor. He has advised hundreds of graduate and undergraduate students during his tenure at Texas A&M University, and he has taught five undergraduate and graduate courses. Silvy encourages his students to be active professionals and leads by example. He has served on numerous professional committees for The Wildlife Society, has served as an officer at the state and national level, and in 2003 was recognized for his dedication to the wildlife profession by being named the recipient of the Aldo Leopold Award. Because of his dedication to his students, Silvy has received numerous teaching awards, culminating in recognition as Regents Professor in 2001, the highest teaching award the Texas A&M University Board of Regents bestows on its faculty. Silvy's influence on his students has been enormous; many have gone on to become exceptional researchers, teachers, and administrators, and numerous former students have played influential roles on endangered species recovery teams. Clearly, his legacy to the wildlife profession and to wildlife population management is realized by the mentoring, guidance, and compassion he has demonstrated toward the many students who represent the next generation of wildlife professionals.

Photo courtesy of Roel Lopez

cate predators during the late 19th and early 20th centuries resulted in the extinction of the plains grizzly (*U. horribilis horribilis*; Brown 1996) and the extirpation of the Mexican gray wolf (*C. l. baileyi*; Brown and Parsons 2001).

Although game laws were rare until the second half of the 19th century, a few individuals realized that wildlife required protection, and subsequently established laws that protected certain species. For example, the Ohio legislature determined that better management of furbearer populations was required and passed laws in 1829 and 1833 that closed seasons on muskrats (*Ondatra zibethicus*), beavers, mink (*Neovison vison*), and otters (*Lontra canadensis*; Dambach 1948).

In addition, New York lawmakers began managing white-tailed deer (*Odocoileus virginianus*) populations in 1865 by requiring that hunters purchase hunting licenses. Iowan personnel began managing greater prairie chickens (*Tympanuchus cupido*) in 1878 by instituting a bag limit of 25 birds per day (Leopold 1933). However, concerted efforts to better manage wildlife populations did not begin until the late 19th century, when Theodore Roosevelt, an avid hunter, became concerned with the plight of many game species. As a consequence, Roosevelt and others started the Boone and Crockett Club. Many of the initial members of the club exerted political influence on federal and state agencies to institute hunting seasons

and bag limits, and to establish protected areas that resulted in population recoveries of many big game species in North America.

Theodore Roosevelt and other influential figures developed a conservation ethic that laid the foundations upon which Leopold developed his philosophy of wildlife management during the 1920s and 30s. Leopold's ideas stimulated the creation of many of the basic wildlife population management principles that were adopted almost universally by many state game and fish agencies in the United States during the 1940s and 50s. Effectively managing wildlife populations required a scientific approach, which eventually yielded more accurate and precise surveys and census procedures, population monitoring techniques (i.e., radiotelemetry, satellite telemetry, computer modeling, modern genetic procedures) that rapidly advanced the science of wildlife population management.

Wildlife population management has come a long way since the early efforts by Charlemagne and Genghis Khan to manage wildlife populations. However, such historical efforts indicate that managing wildlife populations is nothing new and has been practiced for centuries. In fact, this abbreviated summary of the history of wildlife population management (see Chapter 2 for a detailed account of the history of wildlife management in North America) reveals that three basic themes of wildlife population management have been practiced for the past 800 years: managing to maintain populations, managing to increase populations, and managing to reduce populations. Caughley (1985) recognized these three themes and one more—monitoring populations without applying management—which will not be discussed in this chapter because it does not involve active measures to manipulate a population, a central characteristic of the other three themes. Managing to maintain and increase wildlife populations was the intent of Genghis Khan, Kublai Khan, and the rulers of feudal Europe. Maintaining wildlife populations is the most common form of wildlife population management that state game and fish agencies try to accomplish on an annual basis today. Closing and establishing hunting seasons and instituting bag limits represented attempts by managers in the 1800s to increase wildlife populations, and today state and federal natural resource agencies also attempt to increase endangered species populations via strict protection by suspending harvest, or otherwise significantly limiting take, and establishing refuges to protect critical habitats. Finally, the vermin control programs instituted by the rulers of 16th-century England and 17th-century Finland, and the predator eradication programs established by federal agencies in the United States in the late 19th and early 20th centuries, were similar to what present-day state and federal agencies do to reduce populations of pest species such as feral hogs.

This chapter provides an overview of basic wildlife population management practiced by wildlife professionals by selecting some common examples of how they manage populations of prominent wildlife species. The basic themes of wildlife population management that have been practiced for centuries—management to increase populations, management to sustain populations, and management to reduce populations—provide a structure for our discussion. Within each theme we use case studies to illustrate examples of how specific species or classes of wildlife are managed. The case studies for each of the three themes have three sections that include background information, including a brief description of the natural history of the animal or class of animals for each case study, a brief overview of techniques used to manage populations of select species, and an overview of the ecological and cultural importance of the subject animal.

INCREASING WILDLIFE POPULATIONS

We selected the gray wolf, red-cockaded woodpecker (RCW; *Picoides borealis*), and piping plover (*Charadrius melodus*) as case studies because each represents a species whose numbers had diminished to the point that focused efforts were needed to achieve population increases. Moreover, each was determined to be an important component of their respective ecosystems and important manifestations of national or regional cultures. Consequently, considerable research and management have been devoted to increasing populations of gray wolves, RCWs, and piping plovers. Each case study provides a short

discussion outlining the reasons, methodology, and ecological and cultural justification for increasing populations.

Gray Wolf
Background
Wolves and people have shared an intricate association through the millennia. It is a relationship in which wolves have been viewed as both friend and foe. The gray wolf, for example, is the ancestral species from which the domestic dog originated. Domestic dogs historically provided people with protection, food, fur, and labor. Ironically, the gray wolf also has been the target of human persecution. Gray wolves are carnivores that prey largely on wild ungulates and sometimes domestic livestock. Consequently, poisoning and shooting through government control programs killed thousands of wolves during the turn of the 19th century, which led to their extermination in most of the 48 contiguous United States.

Gray wolves reappeared in the United States during the 1990s naturally and through translocations. The translocations of gray wolves are one of the most important acts of wildlife conservation in the 20th century (Smith et al. 2003). However, they are laden with controversy. It is a case study that involves a multifaceted decision-making process—decisions based on scientific, ecological, political, social, cultural, and economic considerations. The gray wolf translocation issue extends beyond basic questions of ecology and penetrates deep into the core of human values (Wilson 1997). Gray wolf translocations embody the complex and often-sociopolitical nature of wildlife conservation and management.

Although gray wolf translocation occurred in various regions of the United States such as the northern Rocky Mountains (U.S. Fish and Wildlife Service 1994), upper Midwest (U.S. Fish and Wildlife Service 1992), and Southwest (U.S. Fish and Wildlife Service 1996), no translocation has captured more public and scientific interest than the gray wolf translocation into the greater Yellowstone ecosystem (GYE). Gray wolf translocations into Yellowstone National Park captured headlines in written and televised media and in scientific journals worldwide. We use this case study to illustrate the complexity of restoring an endangered species in a modern world.

Management Approach
The gray wolf is the largest wild member of the family Canidae (Nowak 1999). Wolves also had the largest natural geographic distribution of living large terrestrial mammals (besides humans) in the world, until active measures during the past few centuries eradicated them (Boitani 1995). Historically, gray wolves lived throughout the Northern Hemisphere (north of 20°N latitude) and in most habitats excluding deserts (Mech 1974). The historic range of the species in North America extended from the Arctic tundra to central Mexico and from the Pacific coast to the eastern seaboard. Human persecution, however, eliminated the species from much of its historic range. Gray wolf populations declined in the lower 48 contiguous states from about two million to a few hundred animals by the 1950s (McIntyre 1995).

Yellowstone National Park was established in 1872 (Haines 1977). It was the nation's first national park, and intense control of wolves occurred there almost since its establishment. Wolves were extirpated from Yellowstone National Park during the mid-1920s, and for the next 70 years wolves were absent from Yellowstone. However, with the passing of the Endangered Species Act in 1973, gray wolves subsequently received federal protection and, in 1978, the gray wolf became listed as an endangered or threatened species in the lower 48 states.

This federal listing set in motion the development of the Rocky Mountain Wolf Recovery Plan, which was completed in 1987. Its primary objective was the establishment of wolf populations into appropriate habitats in the western United States (U.S. Fish and Wildlife Service 1987). For the northern Rocky Mountain region, which included Montana, Idaho, Wyoming, and the GYE, the U.S. Fish and Wildlife Service established a recovery goal of ten or more breeding pairs per area. However, Yellowstone National Park was designated as a nonessential experimental population, which represented a compromise to address local concerns and opposition to wolf recovery (U.S. Fish and Wildlife Service 1987). Designating the Yellowstone National Park wolf population as a nonessential experimental population permitted flexibility in the management of gray wolf. For example, landowners were permitted to kill wolves on their private property if wolves were observed depredating livestock (U.S. Fish and Wildlife Service 1987). In addition, the 10j Rule (Subsection J, Section 10 of the Endangered Species Act), which defines the treatment of endangered species by agency personnel or private citizens, provides agencies with the legal flexibility for the killing of wolves that threaten livestock or big game. However, when wolves naturally dispersed from Canada into Idaho and Montana, legal complications arose, because the wolf populations were no longer strictly experimental.

This conservation and bureaucratic process eventually resulted in the release of 31 gray wolves from southwestern Canada into Yellowstone National Park during 1996–1997 (Bangs et al. 1998). The wolf population responded favorably to translocation and experienced rapid growth (Fig. 19.1). Within seven years after their release, wolves recolonized much of Yellowstone National Park and several adjacent areas (Smith et al. 2003).

In 2003, gray wolves were downgraded from endangered to threatened status in the eastern and western United States but remained listed as endangered in the Southwest. The population increases in GYE raised the issue of whether the number of breeding pairs outside of GYE was sufficient to grant gray wolves the status of recovered. Population counts in 2009 suggested there were 115 breeding pairs in Idaho, Wyoming, and Montana, an estimate that far exceeded recovery goals. Environmental groups challenged these estimates in court in the ensuing years. However, in early 2011, Congress delisted the species, which removed the issue from the court system where it had been stalled for years, and gray wolves in the Rocky Mountain region were removed from the endangered species list.

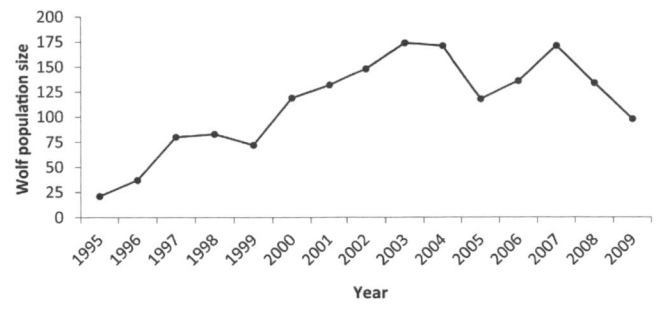

Figure 19.1. Estimated population size of the gray wolf in Yellowstone National Park, 1995–2009. Data were obtained from the Annual Wolf Project Reports, Yellowstone National Park.

Management of the species currently rests with the states in this region, and sustainable hunting will soon resume in Idaho, Montana, and parts of Utah, Oregon, and Washington. The recovery of the gray wolf in the United States is an ongoing conservation success story.

Ecological and Cultural Context

The controversy surrounding the return of gray wolves into the United States involved ecological and cultural components. Ecologically, the controversy centered on whether wolf translocations would restore ecosystem function via a trophic cascade. Culturally, the heart of the gray wolf translocation controversy lay in the nature of the relationship between humans and the environment. A brief treatment of the gray wolf controversy from an ecological and cultural perspective gives richer understanding of the complexity of managing a declining population in a society with diverse cultural values.

Community structure and its causes (i.e., "top-down" or "bottom-up" processes) have been a source of ecological debate for a long time. A top-down influence of community structuring suggests that top predators regulate primary consumers, which in turn allows for a diverse plant community. In contrast, a bottom-up influence proposes that the productivity of plants influences community structure by determining the number of primary consumers, which in turn influences the number of predators. In Yellowstone National Park, the ecological controversy involved the role that reintroduced gray wolves would play in the recovery of declining deciduous trees in the park.

Aspen (*Populus tremuloides*), cottonwood (*Populus* spp.), and willow (*Salix* spp.) experienced dramatic declines during much of the 20th century, and intense elk (*Cervus canadensis*) browsing played a crucial role in the decline of aspen (Ripple and Beschta 2003, Kauffman et al. 2010). Because this decline is coincident with the extirpation of gray wolves in Yellowstone National Park, ecologists proposed that high levels of herbivory (due to unchecked elk populations) could have been responsible for the low recruitment of aspen in Yellowstone National Park (Wagner et al. 1995). However, alternative explanations exist for the decline, such as fire suppression, climate change, and natural plant-community dynamics. The restoration of ecosystem function following gray wolf reintroduction (i.e., if aspen recruitment would increase following gray wolf translocations because of decreased elk browsing) captures the essence of the ecological debate.

Preliminary findings suggest that that aspen, cottonwood, and willow populations are recovering in Yellowstone National Park (Ripple and Beschta 2003, 2005). However, whether wolves are responsible for these trends and whether these trends are still occurring are still being debated (Kauffman et al. 2010). In addition, the ecological interactions appear to be more complex and to involve other factors (e.g., fire) beyond mere predation and herbivory (Mao et al. 2005).

Aspen stem density increases dramatically following fire because of resprouting. Intense herbivory, however, can decrease stem densities to prefire levels within a relatively short time (Bartos 1994). Because predators can influence plants by altering herbivore densities or foraging behavior, herbivory, fire, and predation risk appear to play an interactive role in aspen recruitment. The working hypothesis is that the coupling of fire and predation risk may create a positive feedback mechanism that results in improved aspen recruitment (Halofsky et al. 2008). Fire creates open areas of increased predation risk (Kauffman et al. 2010), and fire also stimulates aspen resprouting. Elk avoid burned areas because of increased predation risk, which in turn permits the regrowth of aspen following fire and the formation of dense thickets (Halofsky et al. 2008). These dense thickets further increase predator risk because of decreased elk maneuverability, resulting in stronger avoidance. This coupling of fire and predation risk is suggested to be an important link in the apparent recovery of aspen in Yellowstone National Park (Halofsky et al. 2008). However, some investigators have challenged the existence of this behaviorally mediated trophic cascade, and recent research suggests that aspen recruitment in Yellowstone National Park might not be occurring as suggested or related to this "landscape of fear" created by gray wolves (Kauffman et al. 2010).

The cultural context of gray wolf translocation is just as intricate as the ecological context. Sociologists have proposed that the controversy surrounding the return of the gray wolf is sociopolitical and not biological. Sociologists propose that the gray wolf is merely a symbol representing a conflict of two opposing social movements: environmentalism and wise use (Wilson 1997, Nie 2001). Environmentalists, who are mostly urban, middle-class citizens (Brick 1995), view the GYE as one of the last remaining wilderness areas that provide critical habitat for a large diversity of organisms. Conversely, the wise use movement, a group comprising people from rural America, perceives the economic viability of the GYE region as inextricably linked to such extractive industries as grazing, forestry, and mining (Power 1991). The core, underlying issues of gray wolf translocation are a divergence of social values between the two social movements relative to social power, private property rights, and nature (Wilson 1997), with the gray wolf merely a surrogate for broader cultural issues (e.g., preservation versus resource use, recreation versus extraction

economies, rural versus urban values; Primm and Clark 1996, Nie 2001).

From a conservation perspective, the translocation of gray wolves into Yellowstone National Park may be hailed as a conservation achievement; an extirpated species has been restored into part of its former range. Ecologically and culturally, however, the implications and consequences of such translocations are not as clearly delineated and likely will be debated for years.

Red-cockaded Woodpecker
Background
The RCW is a cardinal-sized woodpecker that was distributed throughout the coastal plains of the southeastern United States. Today, the RCW is an endangered species whose population abundance is likely less than 3% of what it was before European settlement of the United States.

RCWs are endangered because of widespread loss of habitat. It is closely linked with the longleaf pine (*Pinus palustris*) forests of the southeastern United States, which is one of the most endangered ecosystems in the world; less than 1% of the original longleaf pine forests remain (Simberloff 1993).

Efforts to sustain, elevate, and recover RCW populations rank among the most important wildlife management efforts in the world. Dozens of leading wildlife science professionals from a variety of agencies and organizations have collaborated to develop a plan to recover RCWs on federal, state, and private lands throughout the southeastern coastal plain (U.S. Fish and Wildlife Service 2003).

The RCW was endangered and on track for extinction in 1970, three years before passage of the Federal Endangered Species Act. Populations continued to decline through the 1970s and 80s. During the 1990s, RCW recovery efforts such as aggressive habitat management, creation of roosting and nesting cavities, and translocations began to have a positive effect, and populations started to increase. For example, in 1993, there were 4,700 known clusters of breeding RCWs; the population increased to 6,100 by 2006, and continues to increase today.

Management Techniques
The devastation of the longleaf pine forest and the decline of RCW populations are linked. The RCW habitat requires mature, old-growth trees (i.e., longleaf pine trees >100 years old) in an open, park-like configuration that is maintained by prescribed fire every two to three years.

Beginning in the late 1980s, biologists began to realize the critical importance of the remaining longleaf pine forests and the need to replant longleaf for restoration. For more than a century, longleaf pine was largely clear-cut and replaced with loblolly pine (*Pinus taeda*), which was preferred by foresters because it grew faster initially, producing commercial returns more rapidly than longleaf pine. However, the high density of the closed-canopy loblolly plantations, and exclusion of fire, resulted in a habitat structure and configuration that did not benefit RCWs. As loblolly pine dominated the southeastern states, RCW populations declined.

Red-cockaded woodpeckers need mature longleaf pine forests because they make their nests in living pine trees. It may take a RCW one year or more to excavate a nest cavity in a living pine tree. The birds typically select older pine trees infected with red heart fungus (*Phellinus pini*), presumably because this makes excavation of the heartwood at the core of the stem somewhat easier once they make it through the xylem or outer sapwood zone (phloem) of the tree.

Living pine trees are used by RCWs for nesting and roosting cavities because they maintain an active series of scars or resin wells around the entrance of the cavity, and sometimes around the entire tree. These resin wells allow the pinesap to exude and form a dense, sticky layer around the cavity entrance. This layer of sap makes it nearly impossible for one of the RCWs primary nest predators, rat snakes (*Elaphe* spp.), to access their nests.

Cavity Provisioning

During the 1980s and 90s, the development of artificial cavities "revolutionized the management of red-cockaded woodpeckers" (U.S. Fish and Wildlife Service 2003:81), because loss of natural cavities for RCW nesting and roosting limited the recovery of many RCW populations.

Researchers developed two different approaches to artificial cavity provisioning: drilling and inserts. The drilling technique (Copeyon 1990, Taylor and Hooper 1991) uses a configuration of two different paths to excavate a nest cavity that will be attractive to the birds (Fig. 19.2). The insert technique (Allen 1991) uses prefabricated cavity boxes that are inserted into a section of the tree that is cut out with a chainsaw (Fig. 19.2). Both techniques have been highly successful at providing cavities that attract RCWs for roosting and nesting.

Translocations

The RCW Recovery Plan (U.S. Fish and Wildlife Service 2003) lists four applications of translocations for RCW recovery: augmentation of a population in immediate danger of extirpation, development of a better spatial arrangement of groups to reduce isolation of groups or subpopulations, translocation of birds to suitable habitat within their historic range, and management of genetic resources. Individual RCWs are not trapped and translocated to a new area unless there is a specific reason or objective related to doing so. Since 2000, translocation of RCWs has been a successful tool for supplementing small and isolated populations that were declining. The combination of provisioning artificial cavities in "recipient" areas has been an essential component of this success.

Finally, for artificial cavities and translocation to be successful tactics for RCW recovery, they must have suitable habitat for nesting and foraging. Without suitable habitat, which is nearly always mature longleaf pine forest maintained by frequent prescribed fire, no amount of artificial cavity construction or translocation will be successful for RCW recovery.

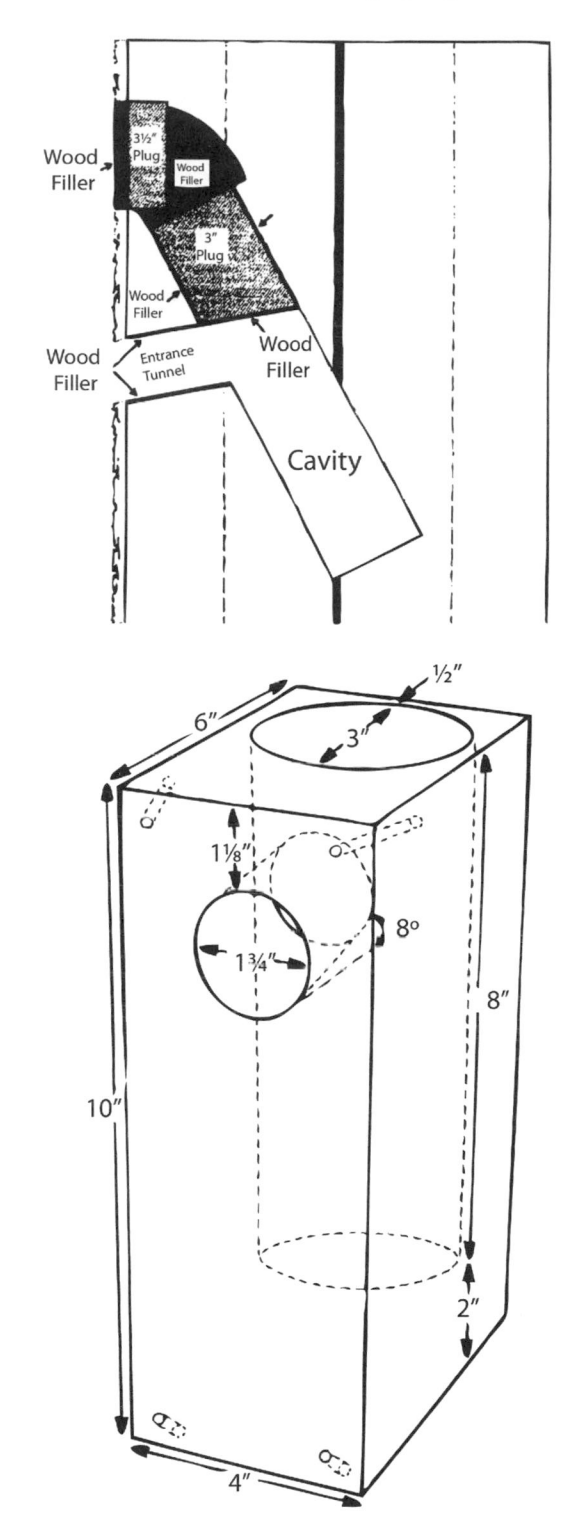

Figure 19.2. *Top*: schematic cross section of a tree with a drilled artificial nest cavity for a red-cockaded woodpecker, developed by Copeyon (1990) and modified by Taylor and Hooper (1991). Note that the hole was constructed from drilling in two directions, and then the top hole was plugged. *Bottom*: schematic and dimensions for an artificial nest cavity insert for the red-cockaded woodpecker. This nest cavity is installed into a rectangular cavity carved into a pine tree with a chainsaw. See Allen (1991) for further details.

Ecological and Cultural Context

The last decade of the 20th century and the first decade of the 21st century represent a breakthrough in RCW conservation and recovery. The first and most critical cultural aspect related to this breakthrough was that people realized longleaf pine forests were an endangered ecosystem that needed urgent conservation attention to steward what little remained and to restore it wherever possible.

Artificial cavity construction and translocation were largely technical, but they also had a cultural context. The idea of providing artificial cavities for RCWs had been discussed for many years but was not implemented because biologists thought it either would not work, would be too expensive, or artificial cavities would become occupied by other animals. However, the devastation from Hurricane Hugo in 1989 prompted biologists to install artificial cavities because so many natural RCW cavities were lost to the storm in the Francis Marion National Forest, South Carolina. To everyone's surprise and delight, RCWs readily took to the artificial cavities.

Researchers for many years had also discussed translocation as a RCW population recovery technique. The safe harbor approach to endangered species management (Chapter 18) on private lands was, in our view, a critical component that helped push translocation from a hypothetical tool to a practical one for RCW conservation. Despite recent successes in sustaining and elevating RCW populations, full recovery and removal of the species from the endangered species list is not complete. The RCW Recovery Plan (U.S. Fish and Wildlife Service 2003) suggests that it will take until 2050 or later for RCWs to be sufficiently abundant to be able to be removed from the endangered species list.

Piping Plover

Background

The piping plover is a small shorebird inhabiting open beaches, sand, algal, and alkali flats (Drake et al. 2001, Elliott-Smith and Haig 2004). Piping plovers breed in three geographically distinct populations: along the Atlantic coast from New Brunswick to North Carolina, beaches along the Great Lakes, and widely throughout the northern Great Plains of the United States and Canada. The plovers breeding along the Atlantic coast are considered a separate subspecies (*C. m. melodus*) because of their genetic differentiation and geographic separation from individuals breeding throughout the northern Great Plains and Great Lakes areas (*C. m. circumcinctus*; U.S. Fish and Wildlife Service 2009). The population breeding in the United States and Canadian Great Lakes is federally listed as endangered, and the populations breeding in the northern Great Plains and Atlantic coast are considered threatened. Piping plover abundance appears to be stable or slightly increasing; however, extinction risk is sensitive to any declines in adult or juvenile survival rates (Larson et al. 2002). Many areas on which piping plovers rely are highly susceptible to anthropogenic impacts, although piping plovers tend to avoid developed areas or areas with frequent human disturbance (Drake et al. 2001). Consequently, habitat loss and degradation, disturbance by humans, and predation appear to be the greatest threats to piping plovers (U.S. Fish and Wildlife Service 2009).

Management Techniques

Piping plovers breed and winter across a wide geographic area, including many habitats where access can be difficult. The substantial effort to survey the entire breeding and wintering areas precludes conducting annual surveys. Consequently, a range-wide survey is conducted every five years to monitor changes in population size. The breeding survey occurs during a two-week period in spring when biologists attempt to survey all known and potential piping plover breeding habitat. Winter surveys are conducted during the same years in late January to early February (U.S. Fish and Wildlife Service 2009). The five-year gaps between surveys limit the ability to understand annual changes in population sizes.

Most management for piping plovers has been focused on breeding areas. Designation of critical habitat through the Endangered Species Act has been important for acquiring and protecting essential piping plover habitat. Considerable amounts of habitat have gained protection in breeding and wintering areas because of their critical habitat designation.

Breeding habitat creation projects mechanically place dredged sand in strategic locations to create new sandbars for the large number of plovers in the northern Great Plains that nest along rivers. Piping plover breeding habitat along rivers is related to the amount of river flow, with high flows resulting in less habitat and low flows exposing more sandbars, equating to more breeding habitat. Biologists are currently working with U.S. Army Corps of Engineers to develop management plans for river flows that favor more consistently available nesting habitat. Biologists also practice vegetation control on existing breeding habitats, but with varying success (U.S. Fish and Wildlife Service 2009). Because piping plovers nest in open habitats, vegetation encroachment can result in suboptimal breeding conditions. Exotic species such as beach vitex (*Vitex rotundifolia*) are potential problems on migration, and wintering areas and control measures have been implemented to manage these exotic plants (U.S. Fish and Wildlife Service 2009).

Piping plover biologists are also concerned about predation during the breeding season, so they often employ predator control to reduce this risk in breeding habitats. Predator control at breeding sites has included placing exclosures around nests to reduce nest predation. Nest exclosures can improve nest success of piping plovers, but they may increase predation risk to adults (Murphy et al. 2003). Predator removal and relocation also increase nest, chick, and adult survival during breeding. High predation rates are often related to limited, high-quality nesting habitat (Murphy et al. 2003). Appropriate habitat management objectives should be coupled with predator control.

Closing portions of beaches to the public to protect piping plover nests during the nesting season is practiced on public lands where human access can be regulated. Ropes cordon off sections of beaches, and signs explain the need for reduced disturbance in areas where piping plovers nest. Piping plover use of private lands is less understood, particularly on wintering areas (U.S. Fish and Wildlife Service 2009). Increased communication with landowners, including education about piping plovers, is an important aspect to the recovery of piping plovers, particularly as optimal habitat declines.

Another technique being used to recover piping plovers is a captive-rearing program initiated in 1998 to help increase the Great Lakes piping plover population. Abandoned eggs found at monitored nests are collected, incubated, and reared to fledging stage before releasing young birds back to the population. There are indications that the program has produced breeding adults (U.S. Fish and Wildlife Service 2009).

Ecological and Cultural Context

Piping plovers use habitats that are attractive to humans for development and recreation. But many practices used to enhance shorelines for anthropogenic purposes negatively impact piping plover habitat. Incidentally, habitat loss and degradation throughout breeding, migration, and wintering areas are considered the greatest threat to this species (U.S. Fish and Wildlife Service 2009). Beach stabilization and nourishment practices, inlet dredging, and construction of jetties and groins modify sediment deposition patterns and impact piping plover wintering habitat. Additionally, beach cleaning practices remove debris that provides important roosting and foraging habitat for piping plovers and other shorebirds (Drake 1999, U.S. Fish and Wildlife Service 2009). Also, the mechanical process of beach cleanup has a tendency to destabilize the beach by reducing its natural ability to trap sand. Most of these practices affect the ability of a shoreline to retain the dynamic processes of coastal formation (U.S. Fish and Wildlife Service 2009).

Another high priority in most recovery plans for piping plovers is to implement recreation management techniques that reduce disturbances to plovers from pets and off-road vehicles on wintering areas. The use of off-road vehicles allows human access to areas that otherwise would not be disturbed. Piping plovers are sensitive to human disturbances and change their distribution and behaviors in response (Zonick 2000, Drake et al. 2001, Cohen et al. 2008). Disturbance will become a larger issue as human use of coastal areas increases.

Management for piping plovers involves habitat management, predator control, human recreation management, and public education. Continued habitat degradation and disturbance by humans will likely provide increasing challenges for managers and biologists tasked with this species' recovery.

SUSTAINING WILDLIFE POPULATIONS

Endangered and threatened species populations require human intervention and active management to achieve population increases and recovery. However, the goal for managing many wildlife populations is simply to sustain them at specific levels to achieve management goals. Moose (*Alces alces*) are large mammals that are an important component of the ecosystems they inhabit, and sustainable populations of moose represent an important resource to recreational and subsistence hunters. Waterfowl populations also fulfill important ecological and recreational functions, and managing harvests—a complex annual activity—is integral to sustaining North American waterfowl populations. Prairie dogs (*Cynomys* spp.) were once treated as vermin, and concerted efforts were aimed to reduce populations. Now, however, they are recognized as a keystone species, and researchers are attempting to manage prairie dog populations because they are critical to the sustainability of prairie ecosystems in parts of North America.

Moose
Background
Moose are the largest member of the deer family (Cervidae) and are popular with consumptive and nonconsumptive wildlife enthusiasts. Hunter harvest of moose is important in many areas of the species' range because local hunters, including Native Americans, rely on moose meat for sustenance. Recreational hunters enjoy the thrill of pursuing such a majestic animal, and local economies benefit from money spent by hunters, photographers, and others interested in moose. For these reasons, wildlife managers seek to sustain productive moose populations through census, habitat and predator management, and hunting.

To fully understand moose management, one must understand moose ecology. Moose are distributed in boreal and mixed coniferous-deciduous forests of Canada, Alaska, the northern United States, and the Rocky Mountains. Large predators, particularly wolves and bears, are found throughout most of moose range in North America. Moose have many adaptations that enable them to live in areas that are extremely cold and that accumulate deep snow during winter. Moose's large size and well-insulated winter coat enable them to endure cold temperatures. In fact, moose are so well adapted to cold temperatures that some research suggests they begin to experience heat stress at temperatures above $-5°C$ during winter and $14°C$ during summer (Renecker and Hudson 1986). Large body size and long legs enable moose to move through deep snow and to live in areas inaccessible to other ungulates during winter. Moose eat aquatic vegetation and leaves, buds, and twigs of woody plants. Buds and twigs can be particularly important, because they are available during winter. Reliance on browse means that moose are dependent on early succession vegetation communities, because forage biomass and quality are high in such communities.

Management Techniques
State or provincial wildlife agencies oversee moose management. Knowing the size and composition of a moose population, in conjunction with habitat and nutrition information, facilitates agency management decisions (Boertje et al. 2007). Most moose populations are surveyed from aircraft along transects or in quadrats, because large areas can be surveyed quickly and the expansive areas of moose range often have poor road access. Timing of surveys varies from November

to March when snow cover increases visibility, but in some areas it must be completed before snow accumulation forces moose to use forested habitats (Timmermann and Buss 1998). Because not all moose within a survey area are seen, sightability models may correct for unseen animals (Anderson and Lindzey 1996). Moose are classified as calves, cows, and bulls. The number of calves per 100 females is a measure of calf production. The number of adult males per 100 females can index the potential harvest intensity of males relative to females and could suggest when composition is sufficiently skewed to influence breeding. Some agencies use the sex ratio as a management objective and adjust the harvest accordingly. Because of their expense, surveys may only be conducted in a given area every two to five years.

Sustaining a moose population requires knowledge of the population size relative to the carrying capacity of vegetation. Population size relative to forage resources can be assessed through a browse survey that measures the proportion of high-quality browse consumed each year (Seaton et al. 2011). Another approach is to use reproductive data, particularly twinning rates, body mass of young moose, and body condition to assess the ability of vegetation to support moose populations (Schwartz and Renecker 1998, Boertje et al. 2007).

Because moose diets comprise primarily browse, and because forest succession reduces the amount and quality of browse over time, forest manipulation is often necessary to sustain moose populations (Thompson and Stewart 1998). Fire was the primary means of creating moose habitat before large-scale forestry operations, and in many areas of the species' range, prescribed and wild fire remain important processes to create early successional habitat. Timber harvest can also convert mature forests to an early seral stage useful to moose, with the greatest benefit from cut areas interspersed with mature forest. If cuts are too large, moose will not use the interior until regeneration provides sufficient cover. In areas without marketable timber, mechanical treatment of woody vegetation with some type of drum roller and cutting blades can stimulate regrowth.

Moose habitat also has attributes that require protection as opposed to creation. Overstory cover, usually including conifer trees, is important during winter to reduce snow depth and to provide thermal and hiding cover; at least some blocks of mature forest should be left undisturbed. Moose use wetlands for feeding and thermoregulation during spring and summer; such wetlands should be protected, and sufficient cover should remain around wetlands to ensure moose are comfortable using them. Finally, moose have a strong sodium appetite during spring and summer to meet deficiencies acquired during winter and from consumption of large amounts of succulent vegetation during spring. This sodium appetite is met at salt licks and through consumption of aquatic vegetation (Jordan 1987); sites providing minerals should be maintained.

Predator control is another moose management technique. In areas where wolf and bear populations are near carrying capacity, moose populations are often lower than predicted by the quality and amount of moose habitat (Gasaway et al. 1992,

Hayes et al. 2003). Reducing densities of wolves and bears in such areas may increase the moose harvest several fold as the moose population increases from low density caused by predation (Boertje et al. 2009). Predator control is not necessary to sustain moose populations but can increase the sustainable harvest.

Harvest can be structured to provide subsistence and recreational opportunity, to ensure population persistence, and sometimes to promote a change in the moose population's size. Because of long life spans, often-low reproductive and recruitment rates, and low population density, moose are susceptible to overharvest. Moose hunting opportunities in most areas are allocated through a drawing process, which determines when and where a hunter may hunt, and the number, sex, and age of moose to be harvested (Timmermann and Buss 1998). Female moose harvest can change population size; a larger female harvest causes a decline in population size, and a smaller one can increase populations. In Ontario, harvest is designated for calf moose in some units with declining moose populations, because calf moose are less likely to survive the winter and will be less likely than an adult female to produce offspring the following year (Ontario Ministry of Natural Resources 2011).

Providing opportunities to harvest large male moose is a management objective in some areas. Antler criteria are one means by which to reduce harvest of young and middle-aged moose, as antler size increases with age. In some management areas of Alaska, male moose must have antlers with a spread of over 127 cm or a specific number of brow tines (antler tines arising from the brow palm). In many areas of Alaska, small male moose with spike or fork antlers are legal to kill for subsistence hunting (Alaska Department of Fish and Game 2011).

Ecological and Cultural Context

The geographic range of moose across large expanses of relatively undeveloped areas of North America raises intriguing and often-controversial management issues. First, the presence of large predators, specifically wolves and bears, adds a dimension to ungulate management missing in many areas of North America. While control of predators appears to increase moose density and hunter harvest, attitudes about wolves vary widely (Kellert et al. 1996), such that some people and groups oppose shooting or trapping predators for this purpose (Musiani and Paquet 2004, Decker et al. 2006, Boertje et al. 2010).

Traditional and subsistence use of moose is part of the management milieu in many parts of moose range. For example, the Alaska Department of Fish and Game has a directive to accommodate subsistence needs of Alaska residents before recreational harvest of wildlife. Moose and caribou (*Rangifer tarandus*) meat is an important part of the diet of many rural Alaskans, with residents of some communities consuming 100 kg per person annually (Titus et al. 2009). Many jurisdictions have separate regulations and seasons for traditional and subsistence hunting by Native Americans, some of which date back to treaty obligations in the 1800s and early 1900s (Timmermann and Buss 1998). Incorporating subsistence hunting can be difficult and contentious, as demonstrated by the political and judicial activity at the state and federal level as Alaska's policy developed (Crichton et al. 1998).

Moose management has economic ramifications. Money spent by moose hunters is difficult to quantify, but economic estimates of more than $400 million (Canadian) annually as a result of moose hunting in North America during the 1980s and 90s suggest the value may be substantial, especially in rural economies (Timmermann and Buss 1998). Funds generated from license sales for moose hunting help support state and provincial wildlife agencies and thereby management of many wildlife species. Tourism related to moose also generates funds for local communities. Guidelines and regulations on timber harvest to accommodate habitat management objectives for moose also have a cost, albeit rarely quantified.

Mortality along the southern periphery of the species' range is an emerging issue potentially challenging management for sustainable moose populations. For example, moose populations in northwestern Minnesota declined from over 3,500 in the 1980s to less than 100 in the 2000s (Lenarz 2007), apparently because interactions among nutritional deficiencies, parasites, and disease reduced adult survival (Murray et al. 2006, Lenarz et al. 2010). A moose population in New Hampshire appears stable despite favorable habitat and conservative harvest. Heavy infestations of winter ticks cause low calf survival and poor reproduction by yearling females, limiting this population's growth rate (Musante et al. 2010). A population in Wyoming had negative population growth rates, most likely due to high adult female mortality and poor calf production (Becker 2008).

Although human-induced habitat changes could help explain some declining or stable moose populations, even in the absence of harvest, mortality is positively correlated to ambient temperature, suggesting climate change may affect moose along the southern edge of their range (Murray et al. 2006, Lenarz et al. 2009, Musante et al. 2010). The large size and excellent insulation of moose are adaptations to the long, cold winters of northern North America, but these traits could make moose susceptible to warming temperatures. Evidence for direct mortality caused by heat stress is lacking (Murray et al. 2006, Lowe et al. 2010), but complex interactions among diet quality, foraging behavior, competition with white-tailed deer, parasites, and disease could be influenced by a warming climate. Whatever the cause, poor survival or reproduction of moose on the southern edge of their range could result in a shift of moose range to the north and represent a challenge to sustainable management of the world's largest deer species (Lenarz et al. 2010).

Waterfowl
Background
Migratory waterfowl are an important natural resource in North America providing recreational and economic opportunities across international boundaries. The migratory nature of waterfowl makes them a shared resource across North America. The United States, Canada, and Mexico developed treaties that acknowledge their shared accountability toward the responsible management and sustainability of North America's migratory birds. Sport harvest regulations have been developed to ensure the long-term sustainability of populations, while at the same time providing an opportunity for hunters to harvest migratory birds. The treaties define the general guidelines for sport hunting regulations of waterfowl, including which species may be harvested and when the hunting seasons can occur. The process of regulating harvest of waterfowl is conducted annually through extensive monitoring of waterfowl populations, breeding habitat conditions, and hunter success (Blohm et al. 2006).

Management Techniques
The breeding waterfowl survey in North America is the oldest and most extensive wildlife survey in the world. It covers over 5,400,000 km² of important duck breeding habitat in the northern United States and Canada. The survey estimates the size of breeding populations of waterfowl and evaluates their breeding habitat conditions. The survey was initiated in an experimental phase in 1947 and became operational in 1955. The U.S. Fish and Wildlife Service, Canadian Wildlife Service, many state and provincial agencies, and tribal organizations cooperate to conduct this intensive survey, which has been taken each year since its inception (U.S. Fish and Wildlife Service 2010).

The survey is conducted in May via fixed-wing aircraft flying at low altitude. All species of ducks observed on over 89,000 km of transects are recorded. Not all species are readily observed on aerial surveys; biologists on the ground (or using helicopters in some areas) conduct intensive ground-truthing surveys in portions of the survey area to deal with this disparity in detection probabilities. The differences between aerial and ground counts are used to develop visibility correction factors for each species. After the aerial counts are adjusted by the visibility correction factors, they are extrapolated across the survey area to derive estimates of breeding populations for all species of waterfowl observed on the survey. In July, a production survey is conducted on a subset of transects to count the number of broods produced. Breeding habitat conditions (i.e., number and types of ponds) are also surveyed along the transects to predict productivity which, in combination with population estimates, are used in models to estimate the fall flight and help set hunting regulations (i.e., season length and species' daily bag limits; U.S. Fish and Wildlife Service 2010).

Waterfowl biologists across North America also mark individual ducks with aluminum leg bands containing unique numeric codes. More than 200,000 ducks are banded annually in North America (Blohm et al. 2006). Based on the number of individuals banded, number of bands recovered, and band reporting rates, biologists are able to estimate survival and harvest rates for individual species.

Another assessment of harvest is through surveying hunters. The U.S. Fish and Wildlife Service selects a sample of about 60,000 of the roughly 3.5 million migratory bird hunters each year who purchased a hunting license in the United States (Padding et al. 2006). The participants are asked to keep records of the date, location, and number of ducks and geese they harvest each day afield. The sample is stratified by state and by the hunter's success during the previous hunting season, and responses are used to estimate the average number of ducks and geese harvested per hunter within each stratum. Researchers then estimate overall harvests of ducks and geese by multiplying the average harvests per hunter by the number of active hunters in each stratum and adding stratum estimates. Because hunters vary in their ability to identify waterfowl they harvest, another sample of hunters is surveyed in the Waterfowl Parts Collection Survey. These hunters are asked to mail a wing from each duck and a tail fan from each goose they harvest that hunting season to the U.S. Fish and Wildlife Service, resulting in approximately 90,000 duck wings and 20,000 goose tails annually (Padding et al. 2006). Waterfowl biologists are then able to determine the species, age, and gender on the basis of characteristics of each duck wing, and species and age from goose tail fans. The two harvest surveys provide species-specific harvest estimates and age and sex ratios of harvested species.

In the United States, adaptive harvest management helps the decision-making process on waterfowl hunting regulations each year. Using this approach, decisions about waterfowl bag limits and hunting season length are made on the basis of results from statistical models used to select from a fixed set of regulatory alternatives. Each year, information from the annual monitoring programs discussed above informs the models in an iterative fashion. Subsequent monitoring then assesses the impacts of the regulatory decisions on populations. The strength of the decision process is its adaptive quality, which is designed to reduce uncertainty on how the monitoring results relate to harvest (Blohm et al. 2006, Nichols et al. 2007).

Ecological and Cultural Context

Waterfowl hunting regulations are set independently within each country based on the results of the monitoring programs described above. Subsistence harvest is important to the culture of some native tribes in northern Canada and Alaska, but it is not regulated to the degree of sport harvest. Subsistence harvest of birds and eggs is legal and allowed outside of the regulatory framework dates established by the treaties between countries that define the regulations.

Several aspects of waterfowl ecology provide a unique array of management challenges. For instance, there is a large diversity of waterfowl species, each one varying with regard to their abundance, population trend, distribution, and use of habitats. Harvest regulations that take into account these differences can become complicated. Also, the migratory nature of waterfowl allows them to cross state, provincial, and international boundaries that have varying harvest regulations, and often-different management interests. Additionally, the large geographic distribution of waterfowl in North America necessitates a large monitoring effort and can relegate different

populations of a species to varying pressures such as habitat alterations or hunting intensity. To address these challenges, the regulation of the harvest of waterfowl in North America likely represents the most complicated and extensive wildlife management effort in the world (Blohm et al. 2006).

Prairie Dogs

Prairie dogs make an interesting case study because they have received a gambit of wildlife management practices, including population reduction, maintenance, and increase. Historically, five species inhabited the Great Plains and Rocky Mountain regions of North America: the black-tailed prairie dog (*Cynomys ludovicianus*), Gunnison's prairie dog (*C. gunnisoni*), the Mexican prairie dog (*C. mexicanus*), the white-tailed prairie dog (*C. leucurus*), and the Utah prairie dog (*C. parvidans*; Pizzimenti 1975). Of these, black-tailed prairie dogs had the most extensive range. Their distribution extended from southern Saskatchewan, Canada, to western Texas and north-central Mexico (Hoogland 2003). Today, all species of prairie dogs are rare, occurring in small, remnant colonies dotted throughout their former range of 150 years ago. In fact, populations have become so reduced that the black-tailed prairie dog now occupies less than 2% of its former range (Miller et al. 2000). In 2012, Mexican prairie dogs received endangered status, Utah prairie dogs received threatened status, and Gunnison's prairie dogs were warranted for protection (not yet listed), but black-tailed and white-tailed prairie dogs have no federal listing status (U.S. Fish and Wildlife Service 2012).

Background

Prairie dogs are colonial and fossorial rodents that belong to the family Sciuridae (Hafner 1984). They are a large rodent species with an average total length of 360 mm and average weight of 400–1,350 g (Pizzimenti 1975). Although there are five species of prairie dog, the black-tailed prairie dog is the most commonly referred to, likely because it was the most numerous and had the largest distribution.

Black-tailed prairie dogs live in a territorial, polygynous harem society with family groups called coteries (King 1955). A coterie typically consists of seven individuals—one breeding male, two to three adult females, and three to four offspring (Hoogland 1995)—maintaining a territory of 0.05–1.0 ha (King 1955, Hoogland 1995). The number of burrow entrances per coterie ranges from five to 214 (Hoogland 1995). Black-tailed prairie dogs are selectively herbivorous, eating a variety of plants and underground roots (King 1955, Fagerstone et al. 1981), but they also will eat insects (Costello 1970, O'Meilia et al. 1982) and frequently cannibalize conspecific offspring (Hoogland 1995). In addition, black-tailed prairie dogs cut down plants more than 20 cm within their territory (Hoogland 1995). This mowing behavior is believed to facilitate predator detection.

Historically, prairie dogs were an important food source for certain Native American tribes and early European settlers, presumably because black-tailed prairie dogs do not hibernate and were abundant throughout the year (Scheffer 1945).

As the North American continent was settled, large tracts of prairie dog habitat were converted to farmland. From 1900 to 1960, an estimated 40,000,000 ha of prairie dog habitat had been reduced to less than 600,000 ha (Miller et al. 2000). In addition, ranchers replaced bison on the Great Plains with domestic livestock. Believing that prairie dogs competed with cattle for forage and that prairie dog burrows would result in injuries (i.e., broken legs) to cattle and horses, ranchers began an intensive eradication campaign. Methods to rid the Great Plains of prairie dogs included shooting, drowning, poisoning, and habitat destruction (Swenk 1915, Randall 1976a,b). In addition, prairie dogs are highly susceptible to bubonic plague (*Yersinia pestis*), a bacteria transmitted most commonly by fleas (Cully et al. 1997). Prairie dogs have little to no immunity to bubonic plague, so once introduced into a colony, it decimates the residents within weeks. Although considered an accidental introduction into the United States about 100 years ago via rodent-infested cargo from European ships (Olsen 1981), bubonic plague has been intentionally introduced into prairie dog colonies as an eradication method. The combination of all eradication methods has been very successful, resulting in the small remnant prairie dog populations observed today.

Management Techniques

Public education is the key to prairie dog survival. Ranchers must be re-educated to understand the critical role that prairie dogs play within grassland ecosystems. In addition, several myths about prairie dogs need to be dispelled. Eradication of prairie dogs was advocated because some people believed that prairie dogs compete with livestock for food, destroy grazing habitat, injure livestock with their burrows, and spread disease to livestock and humans.

Prairie dogs consume some plant species that are also consumed by livestock. However, the majority of the diet of livestock is avoided by prairie dogs and vice versa (O'Meilia et al. 1982). Prairie dogs are selectively herbivorous and improve the quality of certain plants; therefore livestock prefer to forage within prairie dog colonies (King 1955, Koford 1958). Second, because prairie dogs clip tall vegetation within their territory, ranchers have claimed that prairie dogs destroy grazing habitat. However, black-tailed prairie dogs prefer to colonize areas of existing low vegetation (Koford 1958). In addition, clipping tall vegetation keeps plants in a young growing state, which can aid digestibility of plants for livestock. Third, prairie dogs do create holes in the ground; the number of burrow entrances can exceed 100/ha (Hoogland 1995). However, even in areas with high numbers of burrow entrances, there have been few documented cases where livestock have fractured their legs from stepping into burrows of prairie dogs (Hoogland 1995). It appears that this complaint has been much exaggerated. Lastly, prairie dog colonies harbor disease, in particular, bubonic plague. Unfortunately, plague has been used as an eradication tool, which has enhanced its spread across the North American plains. However, vector control programs that treat prairie dog burrows with insecticide can reduce flea populations, which in turn can reduce the probability of bu-

bonic plague outbreaks (Barnes et al. 1972). Many complaints against prairie dogs can be dismissed. However, private landowners must buy into these facts about prairie dogs and cease eradication efforts for repopulation attempts to have any chance of success.

Other management strategies to alleviate declines of prairie dog populations include habitat restoration, developing corridors for dispersal between locations of suitable habitat, translocations of young male prairie dogs from their natal range, and supplemental feeding programs during times when body condition becomes poor. Habitat restoration can be easy for prairie dogs, because they colonize areas that have been overgrazed by livestock (Koford 1958, Costello 1970). Natal dispersal is male biased (King 1955), and mammalian and avian predators of prairie dogs are numerous (Hoogland 2003). Therefore translocation of juvenile males from their natal range to nearby territories can help reduce predation risks. Overwinter survivorship varies with body mass and can be a hardship on prairie dogs (Hoogland 1995).

Ecological and Cultural Context

Unfortunately, prairie dog eradication was conducted without adequate knowledge of ecosystem function of grasslands. Early biologists and wildlife managers could have learned from Aldo Leopold's message, "to keep every cog and wheel is the first precaution to intelligent tinkering" (Leopold 1993:145–146). Prairie dogs, especially black-tailed prairie dogs, are now considered an essential component of grassland ecosystems.

Prairie dogs meet the definition of a keystone species. Keystone species have a disproportionate effect relative to their abundance, which results in significant impacts on ecosystem structure, function, and composition (Paine 1980). Removing prairie dogs from grasslands has resulted in cascading effects, some of which are still being identified. Prairie dogs dig tunnels to protect themselves from weather and predators. Burrow systems per coterie have been as long as 33 m and 5 m deep (King 1955), and about 225 kg of soil per burrow system can be removed (Whicker and Detling 1993). This activity increases soil turnover, soil aeration, and soil macroporosity (Munn 1993). In addition, prairie dogs cache food and often defecate in their burrows, which changes soil chemistry, increases soil organic and nitrogen content, and promotes the abundance of nematodes and other invertebrates within soils (Ingham and Detling 1984, Munn 1993, Outwater 1996). Such soil improvements result in an improved vegetation community by increasing plant nutritional content, plant digestibility, and the ratio of live to dead plants within the community (Whicker and Detling 1993). This activity coupled with prairie dog behavior of cutting down tall vegetation promotes greater plant diversity. Prairie dogs also retard the spread of mesquite (*Prosopis* spp.; List 1997, Weltzin et al. 1997). Enhanced vegetational diversity results in greater species richness, density, and diversity of small mammals, which then cascades into a greater diversity and abundance of mammalian and avian predators (Manzano 1996, Ceballos et al. 1999). Prairie dogs produce significant bottom-up effects on soils, vegetation structure, plant productivity, nutrient cycling, and ecosystem functions where they occur.

To date, researchers have identified about 170 species that potentially rely on prairie dogs for some aspect of their survival (Miller et al. 2000). Most notably are black-footed ferrets (*Mustela nigripes*), whose diet is almost exclusively prairie dogs (Clark et al. 1982), mountain plovers (*Charadrius montana*), which selectively use prairie dog habitat (Knowles et al. 1982), and burrowing owls (*Athene cunicularia*), which rely on prairie dog burrows for homes (Plumpton and Lutz 1993).

The general public must be made aware of the keystone role that prairie dogs play within grassland ecosystems, and the public and policymakers alike should understand the concept that Leopold espoused concerning intelligent tinkering.

REDUCING POPULATIONS

Just as wildlife management practices increase some populations while sustaining others, some management practices reduce the populations of wildlife species determined to interfere with human interests or to threaten desirable wildlife species or ecosystems. Feral swine (*Sus scrofa*) have invaded almost every U.S. state, where they have become threats to rural economies and native ecosystems. Active population management is ongoing to reduce feral hog numbers. Like feral swine, urban and suburban white-tailed deer populations threaten human interests, so populations require reduction in many cities and towns. Brown-headed cowbirds (*Molothrus ater*) pose an ethical dilemma to wildlife managers. Native to North American grasslands but brood parasites, cowbirds pose a significant threat to populations of many native nongame bird species. To protect many of these avian populations, it is necessary to reduce cowbird populations.

Feral Swine
Background
Expanding populations of vertebrate invasive species—including feral swine, feral and free-ranging cats, and feral horses and burros—pose one of the greatest emerging threats to wildlife conservation and management in North America (Pimentel 2011). We focus on feral swine because of their transcontinental distribution, burgeoning populations, game status in some states and provinces, and rich history in North America.

Feral swine are a highly adaptable and mobile species capable of exploiting resources and causing ecological and agricultural damage under a wide range of environmental conditions (Campbell and Long 2009, Campbell et al. 2010b). Several biological, ecological, and cultural features of feral swine contribute to their invasiveness. First, feral swine are fecund, reaching sexual maturity at a young age, having large litter sizes, and reproducing up to two times per year (Taylor et al. 1998, Delgado-Acevedo et al. 2010). Second, feral swine have no natural predators in North America, and many large predators that do occur are declining or are at population levels too low to be an effective regulating factor. Third, because feral swine are prized game animals that are hunted year-round in many states and provinces, they are often illegally translocated to new locations to augment existing populations (Sweeney et al. 2003).

Management Techniques
An integrated and comprehensive management program should control or eradicate feral swine populations (Campbell and Long 2009). Techniques to implement such a plan may include lethal and nonlethal approaches. Lethal methods include live trapping followed by euthanasia (Wyckoff et al. 2006, Williams et al. 2011), snaring (Barrett and Birmingham 1994), shooting from the ground by hunters (Engeman et al. 2007), shooting at night with the aid of night vision equipment and noise suppression devices (Adams et al. 2006), shooting from the air (Campbell et al. 2010b), and shooting with the aid of trained dogs (Katahira et al. 1993). Nonlethal methods include primarily exclusion fencing. Researchers reported that two strands of electrified fencing reduced feral swine damage to a row crop (Reidy et al. 2008a), and 86-cm-high paneling contained feral swine during simulated depopulations (Lavelle et al. 2011) and excluded feral swine from a localized resource (Rattan et al. 2010). Each technique has distinct advantages and disadvantages (Campbell and Long 2009). These nonlethal management approaches, plus social and legal considerations, should be known and anticipated prior to their use. Several techniques aimed at controlling feral swine populations are being developed, including fertility control agents (Campbell et al. 2010a, Sanders et al. 2011) and toxicants (Cowled et al. 2008). For these developing methods to be registered for use, a feral swine–specific oral delivery system must also be identified (Campbell et al. 2006, 2011b; Campbell and Long 2007; Long et al. 2010).

Ecological and Cultural Context
Feral swine are one of only 14 mammals noted in a recent list of 100 of the world's worst invasive alien species because of the damage they cause to natural resources, agriculture, livestock health and production, and human health and safety (Lowe et al. 2000). Feral swine are opportunistic omnivores and forage at or below the soil surface in a destructive manner called rooting. Natural resource managers are beginning to recognize the impacts that rooting and other feral swine behaviors have on sensitive species, habitats, and ecosystems (Kaller and Kelso 2006, Campbell and Long 2009, Jolley et al. 2010). Among agricultural producers, feral swine rooting causes damage to crops (Adams et al. 2005). Feral swine also pose a threat to livestock producers because they can transmit diseases to livestock (Campbell et al. 2008, 2011a). In southern Texas, feral swine regularly come into contact with domestic swine at facilities with low biosecurity where disease transmission may occur (Wyckoff et al. 2009). Feral swine threats to human health and safety are direct (e.g., swine–vehicle collisions or defensive attacks) and indirect (e.g., zoonotic disease transmission). Feral swine harbor zoonotic diseases like type A influenza virus (Hall et al. 2008) and *Escherichia coli* O157:H7 (Jay et al. 2007), among others.

Several factors complicate feral swine population management. First, similar to other exotic invasive organisms, early detection of their presence is paramount to successful management. Unfortunately, feral swine are primarily nocturnal (Campbell and Long 2010) and signs of their presence (i.e., tracks, scats, rooting, wallows, rubs) often go unnoticed, allowing for populations to become well established before they are detected (Campbell and Long 2009). Second, feral swine populations are difficult to survey using traditional techniques (e.g., spotlight counts, aerial surveys), which hinders population monitoring. However, the recent development of biomarkers may allow for feral swine mark-recapture population analyses relative to control activities (Wiles and Campbell 2006, Reidy et al. 2008b, 2011). Third, hunting regulations and feral swine regulations in general vary among states and provinces. In some states such as California, feral swine are managed by the wildlife agency as a game animal; in other states such as Kansas and Nebraska, it is illegal to hunt feral swine; and in still other states, there are no regulations for feral swine. Efforts to unify regulations between adjacent states and provinces with feral swine are needed. Lastly, control or eradication of feral swine populations might not be acceptable to individuals who place high value on feral swine recreational opportunities, who depend upon feral swine as a source of protein, or who disagree with population control measures (Lowe et al. 2000).

Populations of feral swine are often managed in a reactionary and piecemeal manner with no cohesive objectives and minimal planning (Campbell and Long 2009). The value of federal and state agencies forging partnerships among agricultural, conservation, industry, public health, and landowner groups cannot be overstated (Hartin et al. 2007). Additional regulatory efforts and enforcement of existing laws are needed to curb the illegal translocation and expansion of feral swine populations. Decisive actions to fulfill realistic and obtainable objectives, and stable funding to implement management and to conduct research, are also necessary.

Deer in Human-dominated Landscapes
Background
Deer populations occupy urban (more than 250 homes per square kilometer), suburban (25–249 homes per square kilometer), and exurban (6–24 homes per square kilometer) landscapes (Bowman 2011). The area of these landscapes increased as much as five times from 1950 to 2000 to over one million square kilometers (Brown et al. 2005). Deer populations in human-dominated areas create a variety of conflicts with human residents, requiring management actions to reduce deer populations to a tolerable level. Whereas all species and subspecies of North American deer live to some extent in human-dominated landscapes, we emphasize the impact of white-tailed deer.

White-tailed deer have a huge range, north to south from Canada to Peru, and east to west from Atlantic to Pacific coasts (Baker 1984). They are an ancient species, having successfully persisted for over two million years. Through this long history, they have adapted—even thrived—through ice ages and subsequent warming periods. They endured the large predator suite of the Pleistocene (Haynes 1983) and seemingly benefited from the reduced competition brought on by the extinction of large herbivores in the late Pleistocene (Owen-Smith 1987).

Over their large distribution, white-tailed deer have different life-history characteristics in different environmental settings (DeYoung 2011). Some populations migrate to winter deeryards to survive winter weather in the northern portions of their range (Hoskinson and Mech 1976). Whitetails occur in semiarid and tropical habitats at low to moderate densities (DeYoung et al. 2008). Whitetails are best known from the large body of research conducted in the midwestern lake states, as well as the eastern forests of the United States (Russell et al. 2001). In the United States they respond to secondary forest succession, exhibit rapid population growth, and are often overpopulated.

White-tailed deer were severely reduced or eliminated from much of their North American range in the late 1800s and early 1900s because of market and subsistence hunting and habitat modification (McCabe and McCabe 1984). In the 1940s, state wildlife departments in the United States began successfully transplanting deer to unoccupied habitat. Their efforts, along with male-only hunting seasons, led to the increase of white-tailed deer populations.

Many intrinsic and extrinsic factors affect white-tailed deer population dynamics, but fundamental to all habitats is their relationship to forage plants (DeYoung 2011). Often considered a browser of woody plants, white-tailed deer actually prefer forbs and fruits in habitats where they are available. Urban, suburban, and exurban populations benefit from the protection afforded by these human-modified habitats, which often contain enhanced artificial or natural food sources. White-tailed deer in these habitats have a high reproductive potential that may include breeding by female fawns and occasional triplet litters by mature females (Ozoga 1987).

In urban, suburban, and exurban landscapes, human–deer conflicts can be numerous, including damage to garden and landscape plantings, vehicle collisions, and spread of disease (DeNicola et al. 2000). As deer density increases in human-dominated habitats, moderate to severe damage to residential trees and herbaceous vegetation commonly occurs (Butfiloski et al. 1997). This damage can worsen when residents provide supplemental feed to attract deer. White-tailed deer also spread the seeds of exotic and potentially invasive plants in their feces (Williams et al. 2008).

Deer–vehicle collisions are a serious problem in human-dominated landscapes (DeNicola et al. 2000) and worsen as deer densities increase. Collisions result in significant property damage, human injury, and even death. The average costs of deer–vehicle collision exceed $3,000 (Bissonette et al. 2008), and annual vehicle repair costs in the United States exceed $1 billion (Conover et al. 1995). As of 1995, it was estimated that deer–vehicle collisions in the United States annually resulted in 29,000 human injuries and over 200 deaths (Conover et al. 1995). Deer mortality from collisions can also be substantial (e.g., >74% of all white-tailed deer mortality in the Florida Keys; Lopez et al. 2003; Chapter 10).

Overlapping landscape use by deer and humans also leads to disease concerns. Lyme disease, which is transmitted by the deer tick (*Ixodes scapularis*), is a common concern (Magnarelli et al. 1995). Lyme disease is widespread but occurs commonly in the United States in the northeastern, mid-Atlantic, and upper midwestern states (DeNicola et al. 2000). Incidence of Lyme disease in humans has increased substantially in recent decades (Dennis 1998). Another disease, bovine tuberculosis, which can be contracted by humans, also occurs in deer (Schmitt et al. 1997; Chapter 8).

Management Techniques

There are several techniques used to reduce human–deer conflicts that do not require killing deer, including fencing, repellents, translocation, and fertility control.

Fencing can be an effective barrier to deer, thus reducing damage, but much depends on the motivation of deer to enter an area (VerCauteren et al. 2006). Many types of fencing can control deer movements, but most fences are woven mesh and electrified wire. Permanent, woven wire fencing that is 2.4–3.0 m in height forms an effective deer barrier for many applications (VerCauteren et al. 2006). However, because of its high cost, which can exceed $10/m, the use of woven wire fencing is usually restricted to highways, airports, game preserves, and orchards (Craven and Hygnstrom 1994). Electric fencing is generally cheaper but not as effective. High-tensile wire powered by a high-voltage energizer has been widely used in New Zealand and can form an effective deer barrier (Palmer et al. 1985).

Odor or taste repellents can often reduce deer damage to high-value plants (Bowman 2011). Repeated application to foliage is usually necessary, as rainfall and plant growth weaken repellents. No repellent is 100% effective, and much depends on deer motivation (Ward and Williams 2010).

Trapping individual deer that are causing conflicts with humans and translocating them to a rural area makes sense to the general public (Beringer et al. 2002), because translocation can be a nonlethal way to reduce deer abundance. However, depending on a variety of factors, mortality of translocated deer can exceed 50% (Jones and Witham 1990). Factors affecting survival include capture stress, condition of deer, and mortality sources at the release site. It is often difficult to find a release site that is not already occupied by a dense deer population. Cost per translocated deer can range from $500 to $1,000 or more (Beringer et al. 2002), putting the technique beyond the budget of many organizations.

Extensive research has been done on fertility control as a way to reduce deer populations in human-dominated landscapes (Fagerstone et al. 2010). The objective of fertility control is to lower the reproductive success of a deer population so that it is less than or equal to the mortality rate (Bowman 2011). Methods researched have included surgical sterilization, synthetic steroid hormones, immunocontraceptive vaccines, and abortion induction (Fagerstone et al. 2010, Bowman 2011). Any type of drug or chemical used in the United States outside the research arena for fertility control in deer must be approved or labeled by regulatory agencies. Currently, the only approved product for deer is GonaCon, an immunocontraceptive vaccine developed by the U.S. Department of Agriculture and registered by the U.S. Environmental Protection Agency in 2009. It is registered as a pesticide and can be legally used only by certified pesticide applicators. A single injection of GonaCon suppresses reproduction in both sexes (Miller et al. 2004). However, treatment of males may result in dropped antlers or antlers that remain in velvet. GonaCon can suppress reproduction for two or more years, and it is safe for humans to consume injected animals (Fagerstone et al. 2010). Fertility control may be an effective option for small, isolated populations, but it would likely be prohibitively expensive for large deer populations (Bowman 2011).

Lethal methods for reducing deer populations in human-dominated landscapes include sport hunting and sharpshooters. A deer population of 15/km^2 has been suggested as a goal to limit damage while considering the desires of the public to view deer (Hansen and Beringer 1997).

If effective, sport hunting is the least expensive method of reducing and controlling deer populations where deer–human conflicts are significant (Ishmael and Rongstad 1984). However, there are several factors that can inhibit or prevent implementation of hunting. Hunting in human-dominated areas often raises human safety concerns, leading to restrictions on weapon type to shotguns, muzzle-loaders, or bows (Hansen and Beringer 1997). The size of undeveloped parcels of deer habitat interspersed across the landscape also affects hunting effectiveness. Hunting may be most appropriate in exurban areas with large undeveloped tracts (Bowman 2011), but tightly controlled hunts, where feasible, have been employed in some instances (Ebersole et al. 2007). These typically involve control over hunt timing, and length, number, and location of hunters (Bowman 2011).

Agency personnel or contractors perform sharpshooting as a means of population control. Personnel are generally highly trained, resulting in a high degree of human safety and rapid removal of deer (Bowman 2011). A comprehensive sharpshooting program can attain significant reductions in deer–vehicle collisions (DeNicola and Williams 2008). However, sharpshooting efforts can be expensive (Bowman 2011), with efficiency varying with deer density and other factors. Sharpshooting may initially lower a deer population, which then can be maintained at lower density with nonlethal techniques. Alternatively, lowering a deer population may produce a density-dependent response of higher productivity, requiring regular sharpshooting through time.

Recently, some researchers suggested regulated commercial harvesting of overabundant white-tailed deer herds in urban areas (VerCauteren et al. 2011). This proposal is controversial but has the potential to be successful in some situations.

Ecological and Cultural Context

Urban, suburban, and exurban development converts deer habitat into roads, yards, and building sites with concomitant effects on ecological processes (Hansen et al. 2005). Fragmentation occurs as blocks of natural habitat become smaller and smaller as human density increases. Other effects on ecological processes result from exclusion from fire, alteration of flood regimes, and changes in nutrient cycles (Hansen et al. 2005). Predators may become more common in human-dominated landscapes because of access to pet food. Finally, just the presence of humans and their pets, especially dogs, may deter deer use of portions of the urban, suburban, or exurban landscape. For example, deer in exurban areas may shift activity periods to become more nocturnal during weekends, when human disturbance is greatest (Bowman 2011).

Deer management in human-dominated landscapes requires more effort than traditional state agency management in rural areas. Hunters are the main constituents of many state wildlife agencies. In human-dominated landscapes, however, homeowners and landowners have a strong voice. Residents generally enjoy the presence of deer (Stout et al. 1993) and have an ethical bias against lethal methods used to reduce deer populations, and so conflicts with humans occur (Stout et al. 1997). Suburban residents typically favor nonlethal techniques (e.g., translocation, fertility control), which are often prohibitively expensive. Acceptance of lethal techniques is greater for people who have experienced a vehicle collision with a deer or other serious conflict (Lischka et al. 2008). Acceptance may also increase as people become more knowledgeable about deer ecology (Bowman 2011).

Deer managers of human-dominated landscapes often deal with some type of homeowner's association. They also commonly form task forces or committees to study human–deer conflicts and to recommend solutions (Butfiloski et al. 1997). Task forces must comprise certain individuals if they are to be effective (Bowman 2011). They often contain a wildlife biologist either as a committee member, advisor, or facilitator. Task force membership should be balanced with diverse points of view so that the ultimate management recommendations have a greater chance of acceptance. Unanimous support for deer management plans is usually not possible, but obtaining the widest possible support is key (DeNicola et al. 2000; Chapter 4).

Reducing deer populations in human-dominated landscapes has emerged as a common task of wildlife biologists during the past 50 years. It is likely to continue to be a priority in many areas as development in urban, suburban, and exurban areas continues to grow. Deer management in these landscapes requires a specialized set of skills on the part of a wildlife biologist. People management is important in all wildlife management, and such skills are even more important in human-dominated landscapes.

Brown-headed Cowbirds
Background
The brown-headed cowbird uses brood parasitism as a reproductive strategy by forgoing construction of nest and by laying its eggs in the nests of other bird species (the host). This behavior evolved in response to the close association cowbirds historically had with nomadic herbivores (e.g., bison; Friedmann 1929). Because of the nomadic nature of theses herbivores, cowbirds could not stay in a location long enough to raise young, so they evolved an efficient reproductive strategy where they lay their eggs in host species' nests as they followed the nomadic herds. Cowbirds feed in open habitats and have a maximum daily traveling distance of 15 km (Curson et al. 2000). Historically, large contiguous tracts of forested habitat protected many forest-nesting species from parasitism by cowbirds because interior portions were too far from cowbird foraging sites. Cowbird parasitism decreases with increasing forest patch size and increases as forest edge density increases (Robinson et al. 1995, Thompson et al. 2000). Considerable habitat fragmentation since civilization of North America has permitted forest-nesting species to become more available as hosts to cowbird parasitism.

With a brood parasitic reproductive strategy, the host incubates and raises the cowbird young as if it is one of its own. Smaller-bodied hosts and hosts with incubation periods longer than that of a cowbird are typically impacted most, because in these situations cowbird nestlings have a physiological advantage when food is delivered to the nest. The number of host offspring successfully fledged is typically reduced in a brood parasitized by cowbirds (Marvil and Cruz 1989, Peer and Bollinger 1997).

Cowbirds parasitize a large number of bird species and have been implicated in the decline of several songbird species that breed in North America. In particular, the numbers of Kirtland's warbler (*Dendroica kirtlandii*), black-capped vireo (*Vireo atricapilla*), golden-cheeked warbler (*D. chrysoparia*), least Bell's vireo (*V. bellii*), and southwestern willow flycatcher (*Empidonax traillii extimus*) have declined because of parasitism by brown-headed cowbirds. Active cowbird management programs attempt to reduce the impacts of cowbird parasitism in areas where these species breed.

Management Techniques
Cowbird management strategies include habitat improvement for host species, reducing human influences that often attract cowbirds, and direct control measures (Siegle and Ahlers 2004). Increasing breeding habitat of host species is often a management need when cowbirds significantly impact a host species. There is commonly a goal to manage for large, contiguous tracts of habitat, but that goal can be unrealistic because of differences in land ownership across large areas. Promoting vegetation characteristics that help conceal the hosts' nests is also a management strategy. Dense vegetation around the nest that provides concealment results in lower rates of cowbird parasitism (Whitfield and Sogge 1999). Additionally, prescribed fire can be effective for species like the Kirtland's warbler and back-capped vireo, which require fire to produce optimal breeding habitat (Clotfelter et al. 1999).

Anthropogenic influences on habitat such as livestock, agriculture, and recreational activities often create situations that attract cowbirds. These land uses fragment the habitat and provide feeding areas for cowbirds. Rotating grazing to relocate livestock during a host's breeding season can effectively reduce foraging opportunities for cowbirds (Siegle and Ahlers 2004). Educating people who feed birds during the breeding season to use feed that is not attractive to cowbirds may limit cowbird densities in some areas. Reducing the amount of mowing (e.g., roadsides, parks, campgrounds) across the landscape may reduce feeding areas. Although these methods will not eliminate cowbirds from an area, they may help reduce their impacts on host species.

Direct control typically involves trapping and killing cowbirds in areas to reduce rates of parasitism on host populations. Cowbirds are highly gregarious and are relatively easy to capture with decoy traps. Trapping is primarily conducted at feeding sites in an effort to attract cowbirds to areas distant from preferred nesting habitats of host species. Trapping in areas associated with cattle increases capture success of brown-headed cowbirds. Live cowbirds can be placed in the traps as decoys to attract other individuals. When traps are active, they should be checked frequently to release nontarget species.

Removal or addling of cowbird eggs are techniques that can be effective for small populations of hosts. Addling eggs is often a preferred method when there is potential for nest abandonment by host if eggs are removed (Siegle and Ahlers 2004). The accessibility of host nests must also be considered for egg removal or addling methods. For instance, host species that nest high in trees or use delicate vegetation to support their nests are likely inappropriate candidates for these techniques. On the other hand, egg removal and addling may limit effects to nontarget species.

Ecological and Cultural Context

Control of cowbirds through lethal means has been controversial since its first recommendation in the Kirtland's warbler recovery plan (U.S. Fish and Wildlife Service 1976). Within the scientific community there is controversy over cowbird control. Many researchers studying cowbird ecology have struggled to find empirical evidence that cowbirds have actually caused declines in songbird populations (Morrison et al. 1999), and they have questioned the efficacy of cowbird control. In most cases, changes in habitat are likely the main cause for the population decline of the host species (Franzreb 1990, Probst and Weinrich 1993, Hatten and Paradzick 2003), with brood parasitism by brown-headed cowbirds being secondary. Alternatively, managers trying to increase abundance of threatened host populations favor intensive control of cowbirds, primarily through trapping. Because of these issues, cowbird management brings about challenging ethical and ecological decisions. The lethal control of one species to help perpetuate another always brings ethical issues to the forefront. Management becomes complicated for species like the brown-headed cowbird, a native species that has not exhibited increases in population abundance (Rothstein and Peer 2005). Also, in an ecological context, the question arises as to how actions aimed at significantly reducing parasitism in certain areas will affect the evolution of host–parasite relationships.

This impact may be especially pertinent for species that have evolved with cowbirds in grassland ecosystems and that seem to have strategies to deal effectively with parasitism (Peer et al. 2000). Large-scale cowbird control programs could negatively impact these species if selective pressures from cowbirds are reduced for long periods of time (Ortega et al. 2005).

Cowbird control will likely remain a common management technique for threatened songbird populations that experience frequent brood parasitism by brown-headed cowbirds. However, the controversial nature of this technique, particularly in regard to the poor understanding of how suppressing cowbird parasitism impacts host populations, may make this technique less viable in some situations.

LOOKING TO THE FUTURE

Most wildlife scientists believe that climate change, or global warming, is beginning to have an impact on wildlife populations, especially in the northern and southern regions of the globe. Examples are legion—shrinking polar ice caps and glaciers, shifts in phenology of plants that break bud and flower earlier than ever, migratory birds that now arrive on their breeding grounds weeks before their usual timing, and so on.

The management context for species of wildlife that are likely to be negatively impacted by climate change is one of the greatest challenges faced by wildlife scientists and managers today. Will governmental policies and incentives result in a reduction of atmospheric carbon and a reversal of the warming trends we are seeing? Probably not. It is more likely that the emerging economies in such countries as India and China will continue to mature and develop at a breakneck pace, and that the atmospheric carbon issue will worsen before it improves.

From the standpoint of wildlife, it is probably safe to assume that some species will be winners and that some will be losers if global warming trends continue. Species such a polar bears (*U. maritimus*) and penguins (Spheniscidae) are clearly in potential peril from global warming, while other species may simply experience more northward or southward shifts in their geographic distributions. Understanding and coping with these changes is something that all people, not just wildlife professionals, must do as we move forward into the middle of the 21st century (Chapter 17).

SUMMARY

There are three important emergent themes from each of the population management case studies outlined in this chapter. The first is that an understanding of life history attributes of a species is essential for implementing a successful management program, regardless of whether that program is aimed at increasing, sustaining, or reducing populations.

In addition, effective management techniques always use some aspect of an animal's life history to identify a limiting factor, and then manipulate that factor to increase, sustain, or reduce populations. The RCW is a classic example of managers identifying that nesting and roosting cavities were limiting, and then using this concept by provisioning artificial cavities. Red-cockaded woodpecker recovery efforts could then progress at a far greater rate than thought possible.

Finally, for any wildlife population management action to be effective, it must be conducted in a cultural and ecological context. In the North American Model of Wildlife Conservation (Chapter 3), the public owns the wildlife resources in the context of a public trust. All citizens are therefore stakeholders when it comes to issues related to wildlife population management. Attempting to manage wildlife populations in the absence of considering interactions with people is folly.

Human activity is essentially the reason that a need exists to increase, sustain, or reduce wildlife populations. Recovering populations of gray wolves and RCWs would not be necessary had humans not almost extirpated wolves from North America or removed virtually the entire habitat required to sustain RCW populations. Similarly, the recreational and subsistence needs of humans are largely the reasons that moose and waterfowl populations are managed at sustainable levels. Furthermore, feral swine and urban deer populations require reduction because humans introduced feral swine to North America and also created exceptional habitats for deer in areas virtually free of predators and where hunting is prohibited. People have a vested interested in managing wildlife populations whether they realize it or not, so encouraging public input and making an effort to engage citizens should be an integral part of any wildlife population management plan.

Literature Cited

Adams, C. E., B. J. Higginbotham, D. Rollins, R. B. Taylor, R. Skiles, M. Mapston, and S. Turman. 2005. Regional perspectives and opportunities for feral hog management in Texas. Wildlife Society Bulletin 33:1312–1320.

Adams, C. E., K. J. Lindsey, and S. J. Ash. 2006. Urban wildlife management. Taylor and Francis, Boca Raton, Florida, USA.

Alaska Department of Fish and Game. 2011. Identifying a legal moose in antler restricted hunts. http://www.adfg.alaska.gov/static/regulations/wildliferegulations/pdfs/mooseid.pdf. Accessed 30 October 2012.

Allen, D. H. 1991. Constructing artificial red-cockaded woodpecker cavities. General Technical Report SE-73. U.S. Department of Agriculture Forest Service, Asheville, North Carolina, USA.

Anderson, C. R., Jr., and F. G. Lindzey. 1996. Moose sightability model developed from helicopter surveys. Wildlife Society Bulletin 24:247–259.

Anderson, M. K., and M. J. Moratto. 1996. Native American land-use practices and ecological impacts. Pages 187–206 in Sierra Nevada Ecosystem Project: final report to Congress. Volume II. Assessments and scientific basis for management options. Center for Water and Wildland Resources, University of California, Davis, USA.

Askins, R. A. 2000. Restoring North American birds. Yale University Press, New Haven, Connecticut, USA.

Baker, R. H., 1984. Origin, classification and distribution. Pages 1–18 in L. K. Halls, editor. White-tailed deer: ecology and management. Stackpole Books, Harrisburg, Pennsylvania, USA.

Bangs, E. E., S. H. Fritts, J. A. Fontaine, D. W. Smith, K. M. Murphy, C. M. Mack, and C. C. Niemeyer. 1998. Status of gray wolf restoration in Montana, Idaho, and Wyoming. Wildlife Society Bulletin 26:785–798.

Barnes, A. M., L. J. Ogden, and E. G. Campos. 1972. Control of the plague vector, Opisocrostis hirsutis, by treatment of prairie dog (Cynomys ludovicianus) burrows with 2% carbaryl dust. Journal of Medical Entomology 9:330–333.

Barrett, R. H., and G. H. Birmingham. 1994. Wild pigs. Pages 65–70 in S. Hyngstrom, R. Timm, and G. Larsen, editors. Prevention and control of wildlife damage. Cooperative Extension Service, University of Nebraska, Lincoln, USA.

Bartos, D. L. 1994. Twelve year biomass response in aspen communities following fire. Journal of Range Management 47:79–83.

Becker, S. A. 2008. Habitat selection, condition, and survival of Shiras moose in northwest Wyoming. Thesis, University of Wyoming, Laramie, USA.

Beringer, J., L. P. Hansen, J. A. Demand, J. Sartwell, M. Wallendorf, and R. Mange. 2002. Efficacy of translocation to control urban deer in Missouri: costs, efficiency, and outcome. Wildlife Society Bulletin 30:767–774.

Bissonette, J. A., C. A. Kasser, and L. J. Cook. 2008. Assessment of costs associated with deer–vehicle collisions: human death and injury, vehicle damage, and deer loss. Human–Wildlife Conflicts 21:17–27.

Blohm, R. J., D. E. Sharp, P. I. Padding, R. W. Kokel, and K. D. Richkus. 2006. Integrated waterfowl management in North America. Pages 199–203 in G. C. Boerne, C. A. Galbraith, and D. A. Stroud, editors. Waterbirds around the world. Stationary Office, Edinburgh, United Kingdom.

Boertje, R. D., M. A. Keech, and T. F. Paragi. 2010. Science and values influencing predator control for Alaska moose management. Journal of Wildlife Management 74:917–928.

Boertje, R. D., M. A. Keech, D. D. Young, K. A. Kellie, and C. T. Seaton 2009. Managing for elevated yield of moose in interior Alaska. Journal of Wildlife Management 73:314–327.

Boertje, R. D., K. A. Kellie, C. T. Seaton, M. A. Keech, D. D. Young, B. W. Dale, L. G. Adams, and A. R. Aderman. 2007. Ranking Alaska moose nutrition: signals to begin liberal antlerless harvests. Journal of Wildlife Management 71:1494–1506.

Boitani, L. 1995. Ecological and cultural diversities in the evolution of wolf–human relationships. Pages 3–11 in S. H. Fritts and D. R. Selp, editors. Ecology and conservation of wolves in a changing world. Proceedings of the second North American symposium on wolves. Canadian Circumpolar Institute Press, Edmonton, Alberta. Canada.

Bowman, J. L. 2011. Managing white-tailed deer: exurban, suburban, and urban environments. Pages 599–622 in D. G. Hewitt, editor. Biology and management of white-tailed deer. CRC Press, Boca Raton, Florida, USA.

Brick, P. 1995. Taking back the rural West. Pages 61–65 in J. A. Baden and R. B. Eby, editors. Let the people judge: wise use and the private property rights movement. Island Press, Washington, D.C., USA.

Brown, D. E. 1996. The grizzly of the Southwest: documentary of extinction. University of Oklahoma Press, Norman, USA.

Brown, D. K., M. Johnson, T. R. Loveland, and D. M. Theobald. 2005. Rural land-use trends in the conterminous United States, 1950–2000. Journal of Wildlife Management 15:1851–1863.

Brown, W. M., and D. R. Parsons. 2001. Restoring the Mexican gray wolf to the mountains of the Southwest. Pages 169–186 in D. S. Maehr, R. F. Noss, and J. L. Larkin, editors. Large mammal restoration: ecological and sociological challenges in the 21st century. Island Press, Washington, D.C., USA.

Butfiloski, J. W., D. I. Hall, D. M. Hoffman, and D. L. Forster. 1997. White-tailed deer management in a coastal Georgia residential community. Wildlife Society Bulletin 25:491–495.

Campbell, T. A., R. W. DeYoung, E. M. Wehland, L. I. Grassman, D. B. Long, and J. Delgado-Acevedo. 2008. Feral swine exposure to selected viral and bacterial pathogens in southern Texas. Journal of Swine Health and Production 16:312–315.

Campbell, T. A., M. R. Garcia, L. A. Miller, M. A. Ramirez, D. B. Long, J. Marchand, and F. Hill. 2010a. Immunocontraception of male feral swine with a recombinant GnRH vaccine. Journal of Swine Health and Production 18:118–124.

Campbell, T. A., S. J. Lapidge, and D. B. Long. 2006. Using baits to deliver pharmaceuticals to feral swine in southern Texas. Wildlife Society Bulletin 34:1184–1189.

Campbell, T. A., and D. B. Long. 2007. Species-specific visitation and removal of baits for delivery of pharmaceuticals to feral swine. Journal of Wildlife Diseases 43:485–491.

Campbell, T. A., and D. B. Long. 2009. Feral swine damage and damage management in forested ecosystems. Forest Ecology and Management 257:2319–2326.

Campbell, T. A., and D. B. Long. 2010. Activity patterns of wild boar in southern Texas. Southwestern Naturalist 55:564–567.

Campbell, T. A., D. B. Long, L. R. Bazan, B. V. Thomsen, S. Robbe-Austerman, R. B. Davey, L. A. Soliz, S. R. Swafford, and K. C. VerCauteren. 2011a. Absence of Mycobacterium bovis in feral swine (Sus scrofa) from the southern Texas border region. Journal of Wildlife Diseases 47:974–978.

Campbell, T. A., D. B. Long, and B. R. Leland. 2010b. Feral swine behavior relative to aerial gunning is southern Texas. Journal of Wildlife Management 74:337–341.

Campbell, T. A., D. B. Long, and G. Massei. 2011b. Efficacy of the Boar-Operated-System to deliver baits to feral swine. Preventive Veterinary Medicine 98:243–249.

Caughley, G. 1985. Harvesting of wildlife: past, present and future. Pages 2–14 in S. L. Beasom and S. L. Roberson, editors. Game harvest management. Caesar Kleberg Wildlife Research Institute, Kingsville, Texas, USA.

Ceballos, G., J. Pacheco, and R. List. 1999. Influence of prairie dogs (Cynomys ludovicianus) on habitat heterogeneity and mammalian diversity in Mexico. Journal of Arid Environments 41:161–172.

Clark, T. W., T. M. Campbell, D. G. Socha, and D. E. Casey. 1982. Prairie dog colony attributes and associated vertebrate species. Great Basin Naturalist 42:572–582.

Clotfelter, E. D., K. Yasukawa, and R. D. Newsome. 1999. The effects of prescribed burning and habitat edges on brown-headed cowbird parasitism of red-winged blackbirds. Studies in Avian Biology 18:275–281.

Cohen, J. B., S. M. Karpanty, D. H. Catlin, J. D. Fraser, and R. A. Fischer. 2008. Winter ecology of piping plovers at Oregon Inlet, North Carolina. Waterbirds 31:472–479.

Conover, M. R., W. C. Pitt, K. K. Kessler, T. J. DuBow, and W. A. Sanborn. 1995. Review of human injuries, illnesses, and economic losses caused by wildlife in the United States. Wildlife Society Bulletin 23:407–414.

Copeyon, C. K. 1990. A technique for constructing cavities for the red-cockaded woodpecker. Wildlife Society Bulletin 18:303–311.

Costello, D. F. 1970. The world of the prairie dog. Lippincott, Philadelphia, Pennsylvania, USA.

Cowled, B. D., P. Elsworth, and S. J. Lapidge. 2008. Additional toxins for feral pig (Sus scrofa) control: identifying and testing Achilles' heels. Wildlife Research 35:651–662.

Craven, S. R., and S. E. Hygnstrom. 1994. Deer. Pages 25–40 in S. E. Hygnstrom, R. M. Timm, and G. E. Larson, editors. Prevention and control of wildlife damage. University of Nebraska Cooperative Extension, Lincoln, USA.

Crichton, V. F. J., W. E. Regelin, A. W. Franzmann, and C. C. Schwartz. 1998. The future of moose management and research. Pages 665–663 in A. W. Franzmann and C. C. Schwartz, editors. Ecology and management of the North American moose. Smithsonian Institution Press, Washington, D.C., USA.

Cully, J. F., A. M. Barnes, T. J. Quan, and G. Maupin. 1997. Dynamics of plague in a Gunnison's prairie dog complex from New Mexico. Journal of Wildlife Diseases 33:706–719.

Curson, D. R., C. B. Goguen, and N. E. Mathews. 2000. Long-distance commuting by brown-headed cowbirds in New Mexico. Auk 117:795–799.

Dambach, C. A. 1948. The relative importance of hunting restrictions and land use in maintaining wildlife populations in Ohio. Ohio Journal of Science 6:209–229.

Decker, D. J., C. A. Jacobson, and T. L. Brown. 2006. Situation-specific "impact dependency" as a determinant of management acceptability: insights from wolf and grizzly bear management in Alaska. Wildlife Society Bulletin 34:426–432.

Delgado-Acevedo, J., A. Zamorano, R. W. DeYoung, T. A. Campbell, D. G. Hewitt, and D. B. Long. 2010. Promiscuous mating in feral pigs (Sus scrofa) from Texas, USA. Wildlife Research 37:539–546.

DeNicola, A. J., K. C. VerCauteren, P. D. Curtis, and S. E. Hygnstrom. 2000. Managing white-tailed deer in suburban environments: a technical guide. Cornell Cooperative Extension, Cornell University, Ithaca, New York, USA.

DeNicola, A. J., and S. C. Williams. 2008. Sharpshooting suburban white-tailed deer reduces deer–vehicle collisions. Human–Wildlife Conflicts 2:28–33.

Dennis, D. T. 1998. Epidemiology, ecology, and prevention of Lyme. Pages 7–43 in J. Evans, editor. Lyme disease. College of Physicians, Philadelphia, Pennsylvania, USA.

DeYoung, C. A. 2011. Population dynamics. Pages 147–180 in D. G. Hewitt, editor. Biology and management of white-tailed deer. CRC Press, Boca Raton, Florida, USA.

DeYoung, C. A., D. L. Drawe, T. E. Fulbright, D. G. Hewitt, S. W. Stedman, D. R. Synatzske, and J. G. Teer. 2008. Density dependence in deer populations: relevance for management in variable environments. Pages 203–222 in T. E. Fulbright and D. G. Hewitt, editors. Wildlife science: linking ecological theory and management applications. CRC Press, Boca Raton, Florida, USA.

Drake, K. R. 1999. Time allocation and roosting habitat of sympatrically wintering piping plovers (Charadrius melodus) and snowy plovers (C. alexandrinus). Thesis, Texas A&M University, Kingsville, USA.

Drake, K. R., J. E. Thompson, K. L. Drake, and C. Zonick. 2001. Movements, habitat use and survival of non-breeding piping plovers. Condor 103:259–267.

Ebersole, R., J. L. Bowman, and B. Eyler. 2007. Efficacy of an exurban controlled hunt. Proceedings of the Annual Conference of Southeastern Association of Fish and Wildlife Agencies 61:68–75.

Elliott-Smith, E., and S. M. Haig. 2004. Piping plover (Charadrius melodus). In A. Poole, editor. The birds of North America online. Cornell Lab of Ornithology, Ithaca, New York. http://bna.birds.cornell.edu/bna/species/002/articles/introduction. Accessed 10 December 2011.

Engeman, R. M., A. Stevens, J. Allen, J. Dunlap, M. Daniel, D. Teague, and B. Constantin. 2007. Feral swine management for conservation of an imperiled wetland habitat: Florida's vanishing seepage slopes. Biological Conservation 134:440–446.

Fagerstone, K. A., L. A. Miller, G. Killian, and C. A. Yoder. 2010. Review of issues concerning the use of reproductive inhibitors, with particular emphasis on resolving human–wildlife conflicts in North America. Integrative Zoology 1:15–30.

Fagerstone, K. A., H. P. Tietjen, and O. Williams. 1981. Seasonal variation in the diet of black-tailed prairie dogs. Journal of Mammalogy 62:820–824.

Franzreb, K. E. 1990. An analysis of options for reintroducing a migratory, native passerine, the endangered least Bell's vireo *Vireo bellii pusillus* in the Central Valley, California. Biological Conservation 53:105–123.

Friedmann, H. 1929. The cowbirds: a study in the biology of social parasitism. Charles C. Thomas, Springfield, Illinois, USA.

Gasaway, W. C., R. D. Boertje, D. V. Grangaard, D. G. Kelleyhouse, R. O. Stephenson, and D. G. Larsen. 1992. The role of predation in limiting moose at low densities in Alaska and Yukon and implications for conservation. Wildlife Monographs 120:1–59.

Hafner, D. J. 1984. Evolutionary relationships of the Nearctic Sciuridae. Pages 3–23 *in* J. O. Murie and G. R. Michener, editors. The biology of ground dwelling squirrels. University of Nebraska Press, Lincoln, USA.

Haines, A. L. 1977. The Yellowstone story. Colorado Associated University Press, Boulder, USA.

Hall, J. S., R. B. Minnis, T. A. Campbell, S. Barras, R. W. DeYoung, K. Palilonia, M. L. Avery, H. Sullivan, L. Clark, and R. G. McLean. 2008. Influenza exposure in feral swine from the United States. Journal of Wildlife Diseases 44:362–368.

Halofsky, J. S., W. J. Ripple, and R. L. Beschta. 2008. Recoupling fire and aspen recruitment after wolf reintroduction in Yellowstone National Park, USA. Forest Ecology and Management 256:1004–1008.

Hansen, A. J., R. L. Knight, J. M. Marzluff, S. Powell, K. Brown, P. H. Gude, and K. Jones. 2005. Effects of exurban development on biodiversity: patterns, mechanisms, and research needs. Ecological Applications 15:1893–1905.

Hansen, L., and J. Beringer. 1997. Managed hunts to control white-tailed deer populations on urban public areas in Missouri. Wildlife Society Bulletin 25:484–487.

Hartin, R. E., M. R. Ryan, and T. A. Campbell. 2007. Distribution and disease prevalence of feral hogs in Missouri. Human–Wildlife Conflicts 1:186–191.

Hatten, J. R., and C. E. Paradzick. 2003. A multi-scaled model of southwestern willow flycatcher breeding habitat. Journal of Wildlife Management 67:774–788.

Hayes, R. D., R. Farnell, R. M. P. Ward, J. Carey, M. Dehn, G. W. Kuzyk, A. M. Baer, C. L. Gardner, and M. O'Donoghue. 2003. Experimental reduction of wolves in the Yukon: ungulate responses and management implications. Wildlife Monographs 152:1–35.

Haynes, G. 1983. A guide for differentiating mammalian carnivore taxa responsible for gnaw damage to herbivore limb bones. Paleobiology 9:164–172.

Hoogland, J. L. 1995. The black-tailed prairie dog: social life of a burrowing mammal. University of Chicago Press, Chicago, Illinois, USA.

Hoogland, J. L. 2003. Black-tailed prairie dog. Pages 232–247 *in* G. A. Feldhamer, B. C. Thompson, and J. A. Chapman, editors. Wild mammals of North America: biology, management, and conservation. Johns Hopkins University Press, Baltimore, Maryland, USA.

Hoskinson, R. L., and D. L. Mech. 1976. White-tailed deer migration and its role in wolf predation. Journal of Wildlife Management 40:429–441.

Ingham, R. E., and J. K. Detling. 1984. Plant herbivore interactions in a North American mixed-grass prairie III. Soil nematode populations and root biomass on *Cynomys ludovicianus* colonies and adjacent uncolonized areas. Oecologia 63:307–313.

Ishmael, W. E., and O. J. Rongstad. 1984. Economics of an urban deer-removal program. Wildlife Society Bulletin 12:394–398.

Jay, M. T., M. Cooley, D. Carychao, G. W. Wiscomb, R. A. Sweitzer, et al. 2007. *Escherichia coli* O157:H7 in feral swine near spinach fields

and cattle, central California coast. Emerging Infectious Diseases 13:1908–1911.

Jolley, D. B., S. S. Ditchkoff, B. D. Sparklin, L. B. Hanson, M. S. Mitchell, and J. B. Grand. 2010. Estimate of herpetofauna depredation by a population of wild pigs. Journal of Mammalogy 91:519–524.

Jones, J. M., and J. H. Witham. 1990. Post-translocation survival and movements of metropolitan white-tailed deer. Wildlife Society Bulletin 18:434–441.

Jordan, P. A. 1987. Aquatic foraging and the sodium ecology of moose: a review. Swedish Wildlife Research (Supplement) 1:119–137.

Jorgenson, C. J. 1995. How Native Americans as an indigenous culture consciously maintained a balance between themselves and their natural resources. Pages 31–33 *in* J. A. Bissonette and P. R. Krausman, editors. Integrating people and wildlife for a sustainable future. The Wildlife Society, Bethesda, Maryland. USA.

Kaller, M. D., and W. E. Kelso. 2006. Swine activity alters invertebrate and microbial communities in a coastal plain watershed. American Midland Naturalist 156:163–177.

Katahira, L. K., P. Finnegan, and C. P. Stone. 1993. Eradicating feral pigs in montane mesic habitat at Hawaii Volcanoes National Park. Wildlife Society Bulletin 21:269–274.

Kauffman, M. J., J. F. Brodie, and E. S. Jules. 2010. Are wolves saving Yellowstone's aspen? A landscape-level test of a behaviorally mediated trophic cascade. Ecology 91:2742–2755.

Kellert, S. R., M. Black, C. R. Rush, and A. J. Bath. 1996. Human culture and large carnivore conservation in North America. Conservation Biology 10:977–990.

King, J. A. 1955. Social behavior, social organization, and population dynamics in a black-tailed prairie dog town in the Black Hills of South Dakota. Contribution No. 67. Laboratory of Vertebrate Biology, University of Michigan, Ann Arbor, USA.

Knowles, C. J., C. J. Stoner, and S. P. Gieb. 1982. Selective use of black-tailed prairie dog towns by mountain plovers. Condor 84:71–74.

Koford, C. B. 1958. Prairie dogs, whitefaces, and blue grama. Wildlife Monographs 3:1–78.

Larson, M. A., M. R. Ryan, and R. K. Murphy. 2002. Population viability of piping plovers: effects of predator exclusion. Journal of Wildlife Management 66:361–371.

Lavelle, M. J., K. C. VerCauteren, J. W. Fischer, G. E. Phillips, T. Hefley, S. E. Hygnstrom, S. R. Swafford, D. B. Long, and T. A. Campbell. 2011. Evaluation of fences for containing motivated feral pigs during depopulations. Journal of Wildlife Management 75:1200–1208.

Lenarz, M. S. 2007. 2007 Aerial moose survey. http://files.dnr.state .mn.us/recreation/hunting/moose/moose_survey_2007.pdf. Accessed 30 October 2012.

Lenarz, M. S., J. Fieberg, M. W. Schrage, and A. J. Edwards. 2010. Living on the edge: viability of moose in northeastern Minnesota. Journal of Wildlife Management 74:1013–1023.

Lenarz, M. S., M. E. Nelson, M. W. Schrage, and A. J. Edwards. 2009. Temperature mediated moose survival in northeastern Minnesota. Journal of Wildlife Management 73:503–510.

Leopold, A. 1933. Game management. Charles Scribner's Sons, New York, New York, USA.

Lischka, S. A., S. J. Riley, and B. A. Rudolph. 2008. Effects of impact perception on acceptance capacity for white-tailed deer. Journal of Wildlife Management 72:502–509.

List, R. 1997. Ecology of kit fox (*Vulpes macrotis*) and coyote (*Canis latrans*) and the conservation of the prairie dog ecosystem in northern Mexico. Dissertation, University of Oxford, Oxford, United Kingdom.

Long, D. B., T. A. Campbell, and G. Massei. 2010. Evaluation of feral swine-specific feeder systems. Rangelands 32:8–13.

Lopez, R. R., M. E. P. Viera, N. J. Silvy, P. A. Frank, B. D. Whisenant, and D. A. Jones. 2003. Survival, mortality, and life expectancy of Florida Key deer. Journal of Wildlife Management 67:34–45.

Lowe, S., M. Browne, S. Boudjelas, and M. De Poorter. 2000. 100 of the world's worst invasive alien species: a selection from the global invasive species database. Invasive Species Specialist Group of the Species Survival Commission of the World Conservation Union, Auckland, New Zealand.

Lowe, S., B. R. Patterson, and J. A. Schaefer. 2010. Lack of behavioral responses of moose (Alces alces) to high ambient temperatures near the southern periphery of their range. Canadian Journal of Zoology 88:1032–1041.

Lund, T. A. 1980. American wildlife law. University of California Press, Berkley, USA.

MacKenzie, J. M. 1988. The empire of nature. Manchester University Press, New York, New York, USA.

Magnarelli, L. A., A. D. Denicola, K. C. Stafford III, and J. F. Anderson. 1995. Borrelia burgdorferi in an urban environment: white-tailed deer with infected ticks and antibodies. Journal of Clinical Microbiology 33:541–544.

Mahoney, S. P. 2009. Recreational hunting and sustainable wildlife use in North America. Pages 266–281 in B. Dickson, J. Hutton, and W. Adams, editors. Recreational hunting, conservation and rural livelihoods: science and practice. Wiley-Blackwell, Hoboken, New Jersey, USA.

Manzano, P. 1996. Avian communities associated with prairie dog towns in northwestern Mexico. Thesis, University of Oxford, Oxford, United Kingdom.

Mao, J. S., M. S. Boyce, D. S. Smith, F. J. Singer, D. J. Vales, J. M. Vore, and E. H. Merrill. 2005. Habitat selection by elk before and after wolf reintroduction in Yellowstone National Park. Journal of Wildlife Management 69:1691–1707.

Marvil, R. E., and A. Cruz. 1989. Impact of brown-headed cowbird parasitism on the reproductive success of the solitary vireo. Auk 106:476–480.

Maxwell, A. E. 1913. Pheasants and covert shooting. Adams and Charles Black, London, United Kingdom.

McCabe, R. E., and T. R. McCabe. 1984. Of slings and arrows: an historical perspective. Pages 19–72 in L. K. Halls, editor. White-tailed deer: ecology and management. Stackpole Books, Harrisburg, Pennsylvania, USA.

McCorquondale, S. M. 1997. Cultural contexts of recreational hunting and native subsistence and ceremonial hunting: their significance for wildlife management. Wildlife Society Bulletin 25:568–573.

McHugh, T. 1972. The time of the buffalo. University of Nebraska Press, Lincoln, USA.

McIntyre, R. 1995. War against the wolf: America's campaign to exterminate the wolf. Voyageur Press, Stillwater, Minnesota, USA.

Mech, D. 1974. Canis lupus. Mammalian Species 37:1–6.

Miller, B., R. Reading, J. Hoogland, T. Clark, G. Ceballos, R. List, S. Forrest, L. Hanebury, P. Manzano, J. Pacheco, and D. Uresk. 2000. The role of prairie dogs as a keystone species: response to Stapp. Conservation Biology 14:318–321.

Miller, L. A., J. Rhyan, and G. Killean. 2004. GonaCon™: A versatile GnRH contraceptive for a large variety of pest animal problems. Proceedings of Vertebrate Pest Conference 21:269–273.

Morrison, M. L., L. S. Hall, S. K. Robinson, S. I. Rothstein, D. C. Hahn, and T. D. Rich. 1999. Introduction. Studies in Avian Biology 18:2–3.

Munn, L. C. 1993. Effects of prairie dogs on physical and chemical properties of soil. Pages 11–17 in J. L. Oldmeyer, D. E. Biggins, and B. J. Miller, editors. Management of prairie dog complexes for the reintroduction of the black-footed ferret. U.S. Department of the Interior, Washington, D.C., USA.

Murphy, R. K., I. M. G. Michaud, D. R. C. Prescott, J. S. Ivan, B. J. Anderson, and M. L. French-Pombier. 2003. Predation on adult piping plovers at predator exclosure cages. Waterbirds 26:150–155.

Murray, D. L., E. W. Cox, W. B. Ballard, H. A. Whitlaw, M. S. Lenarz, T. W. Custer, T. Barnett, and T. K. Fuller. 2006. Pathogens, nutritional deficiency, and climate influences on a declining moose population. Wildlife Monographs 166:1–30.

Musante, A. R., P. J. Pekins, and D. L. Scarpitti. 2010. Characteristics and dynamics of a regional moose Alces alces population in the northeastern United States. Wildlife Biology 16:185–204.

Musiani, M., and P. C. Paquet. 2004. The practices of wolf persecution, protection, and restoration in Canada and the United States. BioScience 54:50–60.

Nichols, J. D., M. C. Runge, F. A. Johnson, and B. K. Williams. 2007. Adaptive harvest management of North American waterfowl populations: a brief history and future prospects. Journal of Ornithology 148:343–349.

Nie, M. A. 2001. The sociopolitical dimensions of wolf management and restoration in the United States. Research in Human Ecology 8:1–12.

Nowak, R. M. 1999. Walker's mammals of the world. Sixth edition. Volume 1. Johns Hopkins University Press, Baltimore, Maryland, USA.

Olsen, P. F. 1981. Sylvatic plague. Pages 232–243 in J. W. Davis, L. H. Karstadt, and D. O. Trainer, editors. Infectious diseases of wild animals. Iowa State University Press, Ames, USA.

O'Meilia, M. E., F. L. Knopf, and J. C. Lewis. 1982. Some consequences of competition between prairie dogs (Cynomys ludovicianus) and beef cattle. Journal of Range Management 35:580–585.

Ontario Ministry of Natural Resources. 2011. Moose regulations. http://www.mnr.gov.on.ca/stdprodconsume/groups/lr/@mnr/@fw/documents/document/239848.pdf. Accessed 30 October 2012.

Ortega, C. P., A. Cruz, and M. E. Mermoz. 2005. Issues and controversies of cowbird (Molothrus spp.) management. Ornithological Monographs 57:6–15.

Outwater, A. 1996. Water: a natural history. Basic Books, New York, New York, USA.

Owen-Smith, N. 1987. Pleistocene extinctions: the pivotal role of megaherbivores. Paleobiology 13:351–362.

Ozoga, J. J. 1987. Maximum fecundity in supplementally fed northern Michigan white-tailed deer. Journal of Mammalogy 68:878–879.

Padding, P. I., J. F. Gobeil, and C. Wentworth. 2006. Estimating waterfowl harvest in North America. Pages 849–852 in G. Boere, C. Galbraith, and D. Stroudt, editors. Waterbirds around the world. Stationery Office, Edinburgh, United Kingdom.

Paine, R. T. 1980. Food webs: linkage, interaction strength and community infrastructure. Journal of Animal Ecology 49:667–685.

Palmer, W. L., J. M. Payne, R. G. Wingard, and J. L. George. 1985. A practical fence to reduce deer damage. Wildlife Society Bulletin 13:240–245.

Peer, B. D., and E. K. Bollinger. 1997. Explanations for the infrequent cowbird parasitism on common grackles. Condor 99:151–161.

Peer, B. D., S. K. Robinson, and J. R. Herkert. 2000. Egg rejection by cowbird hosts in grasslands. Auk 117:892–901.

Pimentel, D. 2011. Biological invasions. Second edition. CRC Press, Boca Raton, Florida, USA.

Pizzimenti, J. J. 1975. Evolution of the prairie dog genus Cynomys. Occasional Papers of the Museum of Natural History, University of Kansas 39:1–73.

Plumpton, D. L., and R. S. Lutz. 1993. Nesting habitat use by burrowing owls in Colorado. Journal of Raptor Research 27:175–179.

Pohja-Mykra, M., T. Vuorisalo, and S. Mykra. 2005. Hunting bounties as a key measure of historical wildlife management and game conservation. Oryx 39:284–291.

Power, T. M. 1991. Ecosystem preservation and the economy of the Greater Yellowstone Area. Conservation Biology 5:395–404.

Primm, S. A., and T. W. Clark. 1996. Making sense of the policy process for carnivore conservation. Conservation Biology 10:1036–1045.

Probst, J. R., and J. Weinrich. 1993. Relating Kirtland's warbler population to changing landscape composition and structure. Landscape Ecology 8:257–271.

Randall, D. 1976a. Poison the damn prairie dogs. Defenders 51:381–383.

Randall, D. 1976b. Shoot the damn prairie dogs. Defenders 51:378–381.

Rattan, J. M., B. J. Higginbotham, D. B. Long, and T. A. Campbell. 2010. Exclusion fencing for feral hogs at white-tailed deer feeders. Texas Journal of Agriculture and Natural Resources 23:83–89.

Reidy, M. M., T. A. Campbell, and D. G. Hewitt. 2008a. Evaluation of electric fencing to inhibit feral pig movements. Journal of Wildlife Management 72:1012–1018.

Reidy, M. M., T. A. Campbell, and D. G. Hewitt. 2008b. Tetracycline as an ingestible biological marker for feral pigs. Proceedings of the Vertebrate Pest Conference 23:210–212.

Reidy, M. M., T. A. Campbell, and D. G. Hewitt. 2011. A mark-recapture technique for monitoring feral swine populations. Rangeland Ecology and Management 64:316–318.

Renecker, L. A., and R. J. Hudson. 1986. Seasonal energy expenditure and thermoregulatory response of moose. Canadian Journal of Zoology 64:322–327.

Ripple, W. J., and R. L. Beschta. 2003. Wolf reintroduction, predation risk, and cottonwood recovery in Yellowstone National Park. Forest Ecology and Management 184:299–313.

Ripple, W. J., and R. L. Beschta. 2005. Willow thickets protect young aspen from elk browsing after wolf reintroduction. Western North American Naturalist 65:118–122.

Robinson, S. K., S. I. Rothstein, M. C. Brittingham, L. J. Petit, and J. A. Grzybowski. 1995. Ecology and behavior of cowbirds and their impact on host populations. Pages 428–460 in T. E. Martin and D. M. Finch, editors. Ecology and management of neotropical migratory birds: a synthesis and review of critical issues. Oxford University Press, New York, New York, USA.

Rothstein, S. I., and B. D. Peer. 2005. Conservation solutions for threatened and endangered cowbird (Molothrus spp.) hosts: separating fact from fiction. Ornithological Monographs 57:98–114.

Russell, F. L., D. B. Zippin, and N. L. Fowler. 2001. Effects of white-tailed deer (Odocoileus virginianus) on plants, plant populations and communities: a review. American Midland Naturalist 146:1–26.

Sanders, D. L., F. Xie, R. E. Mauldin, L. A. Miller, M. R. Garcia, R. W. De-Young, D. B. Long, and T. A. Campbell. 2011. Efficacy of ERL-4221 as an ovotoxin for feral swine. Wildlife Research 38:168–172.

Scheffer, T. H. 1945. Historical encounter and accounts of the plains prairie dog. Kansas History Quarterly 13:527–537.

Schmitt, S. M., S. D. Fitzgerald, T. M. Cooley, C. S. Bruning-Fann, L. Sullivan, D. Berry, T. Carlson, R. B. Minnis, J. B. Paycur, and J. Sikarskie. 1997. Bovine tuberculosis in free-ranging white-tailed deer from Michigan. Journal of Wildlife Diseases 33:749–758.

Schwartz, C. C., and L. A. Renecker. 1998. Nutrition and energetics. Pages 441–478 in A. W. Franzmann and C. C. Schwartz, editors. Ecology and management of the North American moose. Smithsonian Institution Press, Washington, D.C., USA.

Seaton, C. T., T. F. Paragi, R. D. Boertje, K. Kielland, S. DuBois, and C. L. Fleener. 2011. Browse biomass removal and nutritional condition of Alaska moose Alces alces. Wildlife Biology 17:1–12.

Siegle, R., and D. Ahlers. 2004. Brown-headed cowbird management techniques manual. Bureau of Reclamation, U.S. Department of Interior, Denver, Colorado, USA.

Simberloff, D. 1993. Species-area and fragmentation effects on old growth forests: prospects for longleaf pine communities. Pages 1–13 in S. M. Hermann, editor. The longleaf pine ecosystem: ecology, restoration, and management. Proceedings of the Tall Timbers Fire Ecology Conference. Tall Timbers Research Station, Tallahassee, Florida, USA.

Smith, D. W., R. O. Peterson, and D. B. Houston. 2003. Yellowstone after wolves. Bioscience 53:330–340.

Stout, R. J., B. A. Knuth, and P. D. Curtis. 1997. Preferences of suburban landowners for deer management techniques: a step towards better communication. Wildlife Society Bulletin 25:348–359.

Stout, R. J., R. C. Stedman, D. J. Decker, and B. A. Knuth. 1993. Perceptions of risk from deer-related vehicle accidents: implications for public preferences for deer herd size. Wildlife Society Bulletin 21:237–249.

Sweeney, J. R., J. M. Sweeney, and S. W. Sweeney. 2003. Feral hog. Pages 1164–1179 in G. A. Feldhamer, B. C. Thompson, and J. A. Chapman, editors. Wild mammals of North America. Johns Hopkins University Press, Baltimore, Maryland, USA.

Swenk, M. H. 1915. The prairie dog and its control. Nebraska Agricultural Experiment Station Bulletin 154:3–38.

Taylor, R. B., E. C. Hellgren, T. M. Gabor, and L. M. Ilse. 1998. Reproduction of feral pigs in southern Texas. Journal of Mammalogy 79:1325–1331.

Taylor, W. E., and G. Hooper. 1991. A modification of Copeyon's drilling technique for making artificial red-cockaded woodpecker cavities. Technical Report SE-72. Southeastern Forest Experiment Station, U.S. Department of Agriculture Forest Service, Asheville, North Carolina. USA.

Thompson, F. R., III, S. K. Robinson, T. M. Donovan, J. R. Faaborg, D. R. Whitehead, and D. R. Larson. 2000. Biogeographic, landscape, and local factors affecting cowbird abundance and host parasitism levels. Pages 271–279 in J. N. M. Smith, T. L. Cook, S. I. Rothstein, S. K. Robinson, and S. G. Sealy, editors. Ecology and management of cowbirds and their hosts. University of Texas Press, Austin, USA.

Thompson, I. D., and R. W. Stewart. 1998. Management of moose habitat. Pages 377–401 in A. W. Franzmann and C. C. Schwartz, editors. Ecology and management of the North American moose. Smithsonian Institution Press, Washington, D.C., USA.

Timmermann, H. R., and M. E. Buss. 1998. Population and harvest management. Pages 559–615 in A. W. Franzmann and C. C. Schwartz, editors. Ecology and management of the North American moose. Smithsonian Institution Press, Washington, D.C., USA.

Titus, K., T. L. Haynes, and T. F. Paragi. 2009. The importance of moose, caribou, deer, and small game in the diet of Alaskans. Pages 137–143 in R. T. Watson, M. Fuller, M. Pokras, and W. G. Hunt, editors. Ingestion of lead from spent ammunition: implications for wildlife and humans. Peregrine Fund, Boise, Idaho, USA.

U.S. Fish and Wildlife Service. 1976. Kirtland's warbler recovery plan. Twin Cities, Minnesota, USA.

U.S. Fish and Wildlife Service. 1987. Northern Rocky Mountain wolf recovery plan. Denver, Colorado, USA.

U.S. Fish and Wildlife Service. 1992. Recovery plan for the eastern timber wolf. Twin Cities, Minnesota, USA.

U.S. Fish and Wildlife Service. 1994. The reintroduction of gray wolves to Yellowstone National Park and central Idaho: final environmental impact statement. Helena, Montana, USA.

U.S. Fish and Wildlife Service. 1996. Reintroduction of the Mexican wolf within its historic range in the southwestern United States: final environmental impact statement. U.S. Department of the Interior, Albuquerque, New Mexico, USA.

U.S. Fish and Wildlife Service. 2003. Recovery plan for the red-cockaded woodpecker (*Picoides borealis*). Second revision. Atlanta, Georgia, USA.

U.S. Fish and Wildlife Service. 2009. Piping plover (*Charadrius melodus*) 5-year review: summary and evaluation. Hadley, Massachusetts, and East Lansing, Michigan, USA.

U.S. Fish and Wildlife Service. 2010. Waterfowl population status, 2010. U.S. Department of the Interior, Washington, D.C., USA.

U.S. Fish and Wildlife Service. 2012. Endangered Species Program. http://www.fws.gov/endangered/. Acccssed 29 October 2012.

VerCauteren, K. C., C. W. Anderson, T. R. Van Deelen, D. Drakew, D. Walter, S. M. Vantassel, and S. E. Hygnstrom. 2011. Regulated commercial harvest to manage overabundant white-tailed deer: an idea to consider? Wildlife Society Bulletin 35:185–194.

VerCauteren, K. C., M. J. Lavelle, and S. Hygnstrom. 2006. Fences and deer-damage management: a review of designs and efficacy. Wildlife Society Bulletin 34:191–200.

Wagner, F. W., R. Forester, R. B. Gill, D. R. McCullough, M. R. Pelton, W. F. Porter, and H. Salwasser. 1995. Wildlife Policies in U.S. National Parks. Island Press, Washington, D.C., USA.

Ward, J. S., and S. C. Williams. 2010. Effectiveness of deer repellents in Connecticut. Human–Wildlife Interactions 41:56–66.

Weltzin, J. F., S. Archer, and R. K. Heitshmidt. 1997. Small mammal regulation of vegetation structure in a temperate savanna. Ecology 78:751–785.

Whicker, A., and J. K. Detling. 1993. Control of grassland ecosystem processes by prairie dogs. Pages 18–27 *in* J. L. Oldmeyer, D. E. Biggins, and B. J. Miller, editors. Management of prairie dog complexes for the reintroduction of the black-footed ferret. U.S. Department of the Interior, Washington, D.C., USA.

Whitfield, M. J., and M. K. Sogge. 1999. Range-wide impacts of brown-headed cowbird parasitism on the southwestern willow flycatcher (*Empidonax traillii extimus*). Studies in Avian Biology 18:182–190.

Wiles, M. C., and T. A. Campbell. 2006. Liquid chromatography-electrospray ionization mass spectrometry for direct identification of iophenoxic acid in serum. Journal of Chromatography B 832:144–157.

Williams, B. L., R. W. Holtfreter, S. S. Ditchkoff, and J. B. Grand. 2011. Trap style influences wild pig behavior and trapping success. Journal of Wildlife Management 75:432–436.

Williams, S. C., J. S. Ward, and U. Ramakrishnan. 2008. Endozoochory by white-tailed deer (*Odocoileus virginianus*) across a suburban/woodland interface. Forest Ecology and Management 255:940–947.

Wilson, M. A. 1997. The wolf in Yellowstone: science, symbol, or politics? Deconstructing the conflict between environmentalism and wise use. Society and Natural Resources 10:453–468.

Wolfe, M. L. 1995. Legal structures for managing wildlife. Pages 327–331 *in* J. A. Bissonette and P. R. Krausman, editors. Integrating people and wildlife for a sustainable future. The Wildlife Society, Bethesda, Maryland, USA.

Wyckoff, A. C., S. E. Henke, T. A. Campbell, D. G. Hewitt, and K. C. VerCauteren. 2009. Feral swine contact with domestic swine: a serologic survey and assessment of potential for disease transmission. Journal of Wildlife Diseases 45:422–429.

Zonick, C. A. 2000. The winter ecology of piping plovers (*Charadrius melodus*) along the Texas Gulf Coast. Dissertation, University of Missouri, Columbia, USA.

INDEX

Page numbers followed by *t* indicate tables.